Synthesis of Results from Scientific Drilling in the Indian Ocean

Geophysical Monograph Series

Including

IUGG Volumes

Maurice Ewing Volumes
Mineral Physics Volumes

GEOPHYSICAL MONOGRAPH SERIES

Geophysical Monograph 70

T.M

Synthesis of Results from Scientific Drilling in the Indian Ocean

Robert A. Duncan
David K. Rea
Robert B. Kidd
Ulrich von Rad
Jeffrey K. Weissel

Editors

American Geophysical Union

Published under the aegis of the AGU Books Board.

Library of Congress Cataloging-in-Publication Data

Synthesis of Results from Scientific Drilling in the Indian Ocean
/ Robert A. Duncan . . . [et al.], editors
    p.   cm. — (Geophysical monograph : 70)
Includes bibliographical references.
ISBN 0-87590-822-5
1. Geology—Indian Ocean.   2. Paleoceanography—Indian Ocean.
I. Duncan, Robert A.   II. Series.
QE350.5.I52   1992   92-34422
551.46'08'095——dc20   CIP

ISSN: 0065-8448
ISBN: 0-87590-822-5

CONTENTS

CONTENTS

..ism, and (5) geochemical characterization of a complete crustal section (sediments plus igneous rocks) about to be subducted in the Java Trench, for comparison with Indonesian volcanoes and estimates of elemental fluxes between ocean, crust and mantle.

Unexpected discoveries were: (1) a detailed tephrochronologic record of the volcanic activity of the Indonesian arc, and (2) a forty million-year record of volcanic activity near Broken Ridge. Lithosphere objectives not achieved were to determine the age and composition of basement rocks underlying the northern Kerguelen Plateau and of the Australian marginal plateaus.

Inevitably, not all highly ranked proposals could be integrated into drilling legs. For instance, investigations of new ocean basin formation and metallogenesis of hot brines in the Red Sea and of the tectonics of the sedimentary wedge caught between collision of the Arabian peninsula and Pakistan (Makran) were top priorities but could not be drilled because of safety and political factors. Future drilling opportunities in the Indian Ocean identified by the workshop participants and the authors of this volume include: (1) a return to the Atlantis-II Fracture Zone, where rapid penetration and excellent recovery of gabbroic rocks demonstrated the feasibility of offset drilling in tectonic windows to sample sections of the deep crust, (2) the Somali Basin, where seismic imaging shows promise of continuous stratigraphic sections for studies of Mesozoic climate and evolutionary processes, (3) the continental margin and continent-ocean transition south of Australia, to investigate the history of Australia-Antarctica rifting and separation, and (4) the Southeast Indian Ridge, where crustal sampling arrays along and perpendicular to the spreading ridge might determine the geometry of upper mantle flow away from the Kerguelen hotspot.

It is certain that the Indian Ocean will continue to be a rich treasury of information about the earth system.

R. A. Duncan, D. K. Rea, R. B. Kidd, U. Von Rad, and J. K. Weissel.
(Corvallis, Oregon, June 18, 1992)

REFERENCES

Heirtzler, J. R., H. M. Bolli, T. A. Davies, J. B. Saunders, and J. G. Sclater (Eds.), *Indian Ocean Geology and Biostratigraphy*, 616 pp., American Geophysical Union, Washington, D.C., 1977.

JOIDES, *Report of the Conference on Scientific Ocean Drilling (COSOD)*, 110 pp., Joint Oceanographic Institutions, Washington, D.C., 1981.

JOIDES, *Report of the Second Conference on Scientific Ocean Drilling (COSOD II)*, 142 pp., European Science Foundation, Strasbourg, 1987.

Nairn, A. E. M., and F. G. Stehli, *The Ocean Basin and Margins: The Indian Ocean*, 6, 794 pp., Plenum, New York, 1982.

von der Borch, C. C. (Ed.), Synthesis of Deep-Sea Drilling Results in the Indian Ocean, *Marine Geology*, 26, 1-175, 1978.

ACKNOWLEDGMENTS

The Ocean Drilling Program (ODP) is an international scientific effort that is supported financially by the governments of 20 nations. The planning, logistics, analysis of data, and dissemination of results requires a significant effort from a large number of people, and this program is a preeminent example of what can be accomplished by wide-ranging international scientific cooperation. The planning effort involves years of scientific preparation, including site-survey cruises, and is critical to the success of each Leg of the Program.

The drillship the *JOIDES Resolution (SEDCO/BP 471)* is at the heart of the drilling program; without it nothing else is possible. The men and women who serve on that vessel as the ship's crew, the drilling crew and the technical staff are of fundamental importance to ODP. They work long hours under often difficult conditions with interest and enthusiasm and are directly responsible for the many accomplishments of the overall program such as those described in this volume. The dedication to the project of Lamar Hayes, Operations Superintendent, who died on the Kerguelen Plateau during Leg 120 typified the spirit that makes ODP a success.

This effort to synthesize the important results of scientific drilling in the Indian Ocean was supported by the U.S. Science Advisory Committee of the Joint Oceanographic Institutions, the National Environmental Research Council of the United Kingdom, and the University of Wales, College of Cardiff, all of which contributed to the workshop effort. The editors would also like to thank Ms. Sue Pullen of Oregon State University, College of Oceanography, who was responsible for putting all manuscripts into camera-ready copy, assuring constancy of format and many miscellaneous editorial tasks. Dave Reinert redrafted many of the illustrations and fitted all of the graphics to appropriate size and format. The National Geophysical Data Center of the National Oceanic and Atmospheric Administration provided the color bathymetric map of the Indian Ocean region fronticepiece. The quality of the manuscripts published here is a direct result of rigorous reviews conducted to the standard of scientific societies such as our publisher, the American Geophysical Union. By way of thanking all the reviewers for their hours spent improving the manuscripts we list them below.

Reviewers for this Volume

Peter Barrett
Wolfgang Berger
Gilbert Boillot
David Boote
Marita Bradshaw
Rainer Brandner
Jean-Pierre Caulet
J.A. Crux
William Curry
Pierre de Graciansky
Robert Detrick
Robert Dunbar
Kay Emeis
Elisabetta Erba
Horacio Ferriz
Martin Fisk
Donald Forsyth
Brian Funnell
Jeffrey Gee
David Graham
Alice Gripp
Terri Hagelberg
David A. Hodell
Peter Hooper
Roy Hyndman

Hugh Jenkyns
David Johnson
Gerta Keller
Keith Kvenvolden
Keith Louden
Jean Marcoux
John Mahoney
Georges Mascle
Brian McGowran
Theodore Moore, Jr.
Daniel Moos
Richard Muhe
Carolyn Zehnder Mutter
Richard Naslund
Katharina Perch-Nielsen
Nicklas Pisias
Isabella Premoli Silva
Mark Richards
William Ruddiman
Bernd Simoneit
Rudiger Stein
Colin Summerhayes
Jorn Thiede
Roy Wilkins
Frank Wind

The Plutonic Foundation of a Slow-Spreading Ridge

HENRY J.B. DICK

Department of Geology and Geophysics,Woods Hole Oceanographic Institution, Woods Hole, MA 02543,
USA

PAUL T. ROBINSON

Centre For Marine Geology, Dalhousie University, Nova Scotia, Canada, B3H 3J5

PETER S. MEYER

Department of Geology and Geophysics, Woods Hole Oceanographic Institution, Woods Hole, MA 02543,
USA

Hole 735B drilled 500 m of gabbroic layer 3 at the SW Indian Ridge. The section consists of small intrusions with no evidence for a large, steady-state magma chamber. The complex stratigraphy represents multiple phases of magmatism, alteration, and ongoing deformation in the zone of lithospheric necking and crustal accretion beneath a slow-spreading ocean ridge. Magma evolution was by fractional crystallization of intercumulus melt in crystal mush, by melt-rock reaction and wall rock assimilation as batches of melt migrated upward through the crust. The major form of igneous layering, undeformed olivine gabbro cut by numerous layers of sheared ferrogabbro, formed by synkinematic differentiation in deforming, partially molten gabbro. This process drove late intercumulus melt into shear zones, where reaction, crystallization, oxide precipitation and melt trapping transformed the rock to ferrogabbro.

The section underwent extensive brittle-ductile deformation, with shear zones forming while the section was still partially molten, under anhydrous granulite conditions, and at higher water-rock ratios in the amphibolite facies. The shear zone is extensively controlled both magmatic and subsolidus fluid flow and alteration. Alteration and circulation of seawater was tectonically enhanced, with extension and lithospheric necking superimposed on the dilational thermal stress available for cracking, resulting in high permeabilities.

Alteration decreased abruptly in mid-amphibolite facies with the end of brittle-ductile deformation as the section was transferred into the rift valley wall and the zone of block uplift. Alteration conditions then closely resembled those of the statically cooled Skaergaard intrusion, with diopside replacing amphibole as the principal mafic hydrothermal vein mineral. This is attributed to low permeability and more reacted fluid with cracking driven only by thermal dilation during static cooling.

Late trondhjemite veins formed by fractional crystallization and by wall rock anatexis of amphibolites during reintrusion of the section. This provides direct evidence that the section underwent multiple alteration and magmatic events as magmatism waxed and waned.

INTRODUCTION

The composition, internal stratigraphy, and petrogenesis of the plutonic foundation of the ocean crust is one of the fundamental questions of earth science. After an initial debate in the late 50's and early 60's, a consensus was achieved for a fairly straightforward "layer cake" ocean crust stratigraphy of gabbro, sheeted dikes and pillow lavas overlying the residual mantle. This model was based on the match of P- and S-wave velocities and densities of gabbros, diabase and basalt dredged from the seafloor and found in ophiolites (on-land sections of fossil ocean crust) to the physical properties of the various ocean crust seismic layers [e.g., Raitt, 1963; Christensen, 1972, 1978]. Detailed geologic models were then constructed by extrapolating the simple observed layered seismic structure to the inferred geology of tectonically disrupted ophiolites [e.g., Coleman, 1977].

Synthesis of Results from Scientific
Drilling in the Indian Ocean
Geophysical Monograph 70

Based on this thinking the lowermost crustal seismic layer 3 was believed to be a uniform layer of magmatic cumulates deposited on the floor and walls of a large semi-continuous magma chamber, overlain by more evolved isotropic gabbros crystallized down from its roof [e.g., Fox et al., 1973; Rosendahl, 1976; Fox and Stroup, 1981]. Conceptually, this internal stratigraphy was, in reality, more dependent on that of the layered intrusions found on the continents, than it was on the internal stratigraphy of gabbros in ophiolites. Layered intrusions have dominated the thinking of the petrologic community, with their systematic progression in composition from bottom to top, and their historic role in establishing the key role of fractional crystallization in the evolution of magmas [e.g., Wager and Brown, 1967]. It has been a natural impulse to impose their stratigraphy on that of lower crust in both ophiolites and the present day ocean basins. Thus evolved the extremely attractive model of the "infinite onion": a large continuous magma chamber underlying the global ocean ridge system, disrupted only by the largest of ocean fracture zones, from which layers of ocean crust continuously grew at top, sides, and bottom to form a uniform layer of coarse gabbroic rocks comprising two-thirds of the ocean crust [Cann, 1970].

Two decades of testing, observations of the ocean crust, and detailed study of ophiolites has thrown this model into question. The provenance of ophiolites has become ambiguous, with an atypical supra-subduction zone ocean crust environment now favored for most [e.g., Miyashiro, 1973; Robinson et al., 1983]. The ocean crust has become increasingly viewed as three-dimensional [e.g., Whitehead et al., 1984], highly dependent on magma supply and spreading rates, and without large steady-state magma chambers [e.g., Sinton and Detrick, 1991]. Observed seismic structure has become increasingly complex, and layer 3 has been found to thin at large and zero-offset fracture zones [Mutter et al., 1985]. Compilations of dredge results [Dick, 1989; Cannat, 1992] and gravity lows centered over ridge segments [Kuo and Forsyth, 1988; Lin et al., 1990] also suggest that a continuous gabbroic layer does not exist at slow spreading ridges, and that its internal stratigraphy is governed by dynamic processes of alteration and tectonism as much as by igneous processes.

The recovery of an intact 500 m section of gabbros at ODP Site 735 at the Southwest Indian Ridge, unroofed and uplifted on the transverse ridge flanking the Atlantis II Fracture Zone, by the *JOIDES Resolution* is playing a historic role: producing a new paradigm for the lower ocean crust. The complex internal structure and stratigraphy of the section provides a first look *in situ* at the processes of crustal accretion and on-going tectonism, alteration, and ephemeral magmatism at a slow-spreading ocean ridge. The section was not formed from a large steady-state magma chamber, but by numerous small, crosscutting, rapidly crystallized intrusions. There is little evidence of the process of magmatic sedimentation, believed important in layered intrusions. Instead, magmas were intruded into, and initially supercooled by, a lower ocean crust consisting of already crystalline rock and relatively cool semi-solidified crystal/melt aggregate. This led to rapid undercooling and initial crystallization of new magma to form a highly viscous crystal mush. This effectively prevented the formation of magmatic sediments, and instead there was a long, and petrologically important, period of intercumulus melt evolution.

We infer from the Hole 735B section, and its similarity to dredged gabbros, that long-lived magma chambers or melt lenses were virtually absent beneath the Southwest Indian Ridge (SWIR). The high viscosity and rigidity of the intrusions throughout most of their crystallization explains the near absence of erupted highly evolved lavas such as ferrobasalt along the SWIR [Dick, 1989], as opposed to their common occurrence along fast spreading ridges, where a long-lived melt lens is believed to underlie the ridge axis [e.g., Sinton and Detrick, 1992].

Wall rock assimilation, occurring while small batches of melt worked their way up through the partially solidified lower crust, left a significant physical imprint on the section and therefore may have an important role in the chemical evolution of mid-ocean ridge basalt. This process has largely not been evaluated for basalt petrogenesis to date, and throws into question simple models for the formation of MORB drawn from experimental studies which assume equilibrium crystallization and melting processes throughout MORB genesis.

An unanticipated feature of the 735B section is evidence of deformation and ductile faulting of still partially molten gabbro. This deformation was petrologically significant over a narrow window, late in the cooling history of the gabbros, when they became rigid enough to support a shear stress, probably at about 60% solidification [e.g., Sinton and Detrick, 1992]. This produced numerous small and large shear zones, creating zones of enhanced permeability into which late iron-rich intercumulus melt moved. This migration of intercumulus melt into and intrusion up shear zones transformed the gabbro there into oxide-rich ferrogabbro by variable degrees of melt/rock reaction and impregnation due to crystallization of the melt. This synkinematic igneous differentiation produced a complex igneous stratigraphy where undeformed, oxide-free olivine gabbro and microgabbro is cut by bands of sheared ferrogabbro. This form of igneous differentiation and layering is probably ubiquitous in the lower crust at slow-spreading ocean ridges, and should be observed in ophiolites from similar tectonic regimes.

At Site 735, ductile deformation and shearing continued into the sub-solidus regime, causing local recrystallization of the primary igneous assemblage under granulite facies conditions, and the formation of amphibole-rich shear zones [Stakes et al., 1991; Dick et al., 1991a; Cannat et al., 1991]. Here again, formation of ductile shear zones localized late fluid flow, with the most intense alteration occurring in the ductile faults [Dick et al., 1991a]. Undeformed sections of gabbro also underwent enhanced alteration at this time, principally by replacement of pyroxene and olivine by amphibole.

A consequence of simultaneous extension and alteration of the gabbro in the zone of lithospheric necking beneath SWIR rift valley, has been far more extensive high temperature alteration than found in statically intruded and cooled layered intrusions. An abrupt change in alteration conditions of the 735B gabbros, however, occurred in the middle amphibolite facies with the cessation of shearing and ductile deformation. Mineral vein assemblages changed from amphibole-rich to diopside-rich, reflecting different fluid compositions. Continued alteration and cooling occurred under static conditions similar to those found for layered intrusions. These changes likely occurred due to an inward jump of the master faults defining the rift valley walls, thus transferring the gabbros out of the extension zone into a zone of simple block uplift in the adjoining rift mountains. Ongoing hydrothermal circulation, no longer enhanced by stresses related to extension, was greatly reduced, and driven only by thermal dilation cracking as the section cooled to ambient seafloor temperature.

The complex 735B gabbro section formed beneath the very-slow spreading SW Indian Ridge (0.8 cm/yr half-rate),

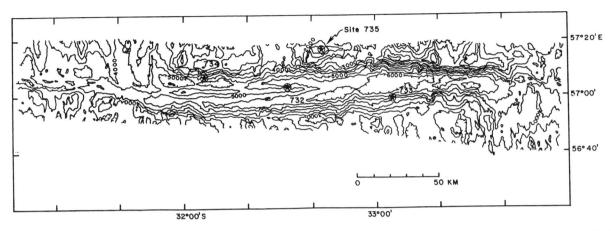

Fig. 1. Bathymetric map of the Atlantis II Fracture Zone showing the locations of ODP Leg 118 drill sites. Hand contoured from original Seabeam data [Dick et al., 1991b]. Contour interval is 500 m. Northern rift valley is at 32°53'S and southern rift valley is at 33°40'S.

far from any hotspot. It thus represents the end of the ocean ridge tectonic spectrum with the lowest rates of magma supply and the heaviest influence of deformation and alteration on crustal stratigraphy. Judged from the results of Hole 735B, the critical brittle-ductile transition migrated from the base of the sheeted dikes up and down throughout the lower crust with the waxing and waning of magmatism. In contrast, at the opposite end of the spreading rate spectrum where the majority of the seafloor has formed (7-9 cm/yr), we anticipate that the higher rate of magma supply would produce a very different igneous stratigraphy [e.g., Sinton and Detrick, 1991], with a relatively stable brittle-ductile transition lying close to the base of the sheeted dikes during crustal accretion. Thus, faulting, brittle-ductile deformation and hydrothermal alteration would rarely extend into the lower crust beneath the spreading axis, and consequently would play only a minor role in its evolution.

The conclusions drawn from Hole 735B are based on only a small part of what may be a 2- to 4-km thick plutonic section and thus may represent only part of a more complex overall stratigraphy. Seismic layer 3 may also not be a uniform layer of gabbroic rock, but consist, at least in part, of residual peridotite cross-cut by numerous small gabbro intrusions. This is the case close to the SWIR transforms, where 43 dredge hauls from the walls of 10 fracture zones recovered 65.2% peridotite, 8.3% gabbro, and 26.5% diabase and basalt. Although gabbro may be significantly harder to dredge than peridotite and basalt, these data cannot be reconciled with a uniform gabbroic layer beneath the SWIR [Dick, 1989]. Moreover, while the measured density and P and S-wave velocities of Hole 735B gabbros are appropriate for layer 3, seismic attenuation measurements on the core [Goldberg et al., 1991] and attenuation calculated from a vertical seismic experiment [Swift and Stephen, 1992] are too high, indicating that the gabbros are atypical for layer 3. The latter suggest, alternatively, that upper layer 3 is metadolerite, and the lower portion interbedded gabbro and ultramafics.

The conclusions of Swift and Stephen [1991], however, are valid only insofar as the 735B gabbros are representative of gabbros in oceanic seismic layer 3. While their igneous petrogenesis is identical to gabbros dredged elsewhere along the SWIR, they were emplaced to the seafloor at the ridge-transform intersection, rather than undergoing whatever alteration and maturation occurs beneath a full crustal section during spreading beneath the rift mountains. In effect, the 735B gabbros represent only a snapshot of crustal formation up until the middle amphibolite facies, not the complete story.

TECTONIC SETTING
Southwest Indian Ridge (SWIR)

The SWIR has existed since the breakup of Gondwanaland in the Mesozoic [e.g., Norton and Sclater, 1979]. Shortly before 80 Ma, plate readjustment in the Indian Ocean connected the newly formed Central Indian Ridge to the SWIR and the Southeast Indian Ridge, forming the Indian Ocean Triple Junction [Fisher and Sclater, 1983]. Steady migration of the triple junction to the northeast created a succession of new ridge segments and fracture zones including the Atlantis II. Thus, the Atlantis II F.Z. and the adjacent ocean crust is entirely oceanic in origin, free from any potential complications associated with continental breakup [e.g., Bonatti 1971].

Over the last 34 m.y., the SWIR half spreading rate has been relatively constant near 0.8 cm/yr [Fisher and Sclater, 1983]. All the characteristic features of slow-spreading ridges, including rough topography, deep rift valleys, and abundant exposures of plutonic and mantle rocks, are present. Significantly, two-thirds of the rocks dredged from the fracture zones are altered mantle peridotite, while the remainder are largely pillow basalt. This abundance of peridotite and lack of gabbro, compared to similar dredge collections from the North Atlantic, suggests an unusually thin crustal section and that magma chambers were generally small or absent near SWIR transforms [Dick, 1989].

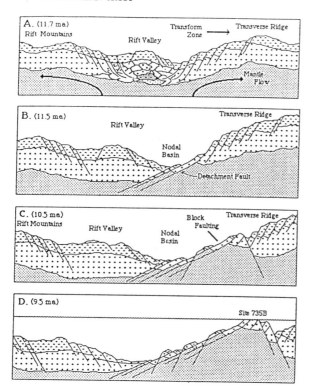

Fig. 2. Temporal cross-sections across the SWIR rift valley drawn parallel to the spreading direction (not across the fracture zone) showing the evolution of the transverse ridge and Hole 735B from Dick et al. [1991a]. The sections are drawn 18 km from the transform fault. Crust spreading to the right is uplifted into the transverse ridge parallel to the transform valley. Crust spreading to the left is uplifted into the rift mountains of the SWIR parallel to the inactive extension of the Atlantis II Fracture Zone. (A) Initial symmetric spreading, possibly at the end of a magmatic pulse: late-magmatic brittle-ductile faulting occurring due to lithospheric necking above and in the vicinity of whatever passes for a magma chamber at these spreading rates, hydrothermal alteration at high temperatures accompanies necking and ductile flow in subsolidus regions. (B) At some point, the shallow crust is welded to the old cold lithosphere to which the ridge axis abuts, causing formation of a detachment structure: initiation of low-angle faulting, continued brittle-ductile faulting and amphibolite facies alteration of the Hole 735B section. (C) Initiation of block uplift terminates extension and drastically reduces permeability and alteration of the Hole 735B rocks, effectively terminating most circulation of seawater and alteration in the ductily deformed rocks. (D) At peak of uplift, erosion at inside-corner high creates wave-cut platform, which will later subside to its present depth at Site 735B.

Thin crust adjacent to transform offsets may reflect segmented magmatism producing rapid along-strike changes in the structure and stratigraphy beneath spreading centers [e.g., Whitehead et al. 1984; Francheteau and Ballard, 1982; Crane, 1985; Schouten et al., 1985; Macdonald, 1987]. Such a model views the SWIR as comprised of a series of regularly spaced, long-lived shield volcanoes and underlying magmatic centers undergoing continuous extension to form ribbons of crust rather than

large constructional volcanic edifices [Dick, 1989]. The anticipated length of these segments is 35 km for the SWIR [Schouten et al., 1985], and thus, if the transform boundary marks one end, the Hole 735B gabbros may have formed near the mid-point of a magmatic cell.

GEOLOGY OF THE ATLANTIS II FRACTURE ZONE AND SITE 735

The 220 km offset Atlantis II transform was surveyed in detail by Dick et al. [1991b]. It is typical of many large offset SWIR fracture zones [e.g., Engel and Fisher, 1975; Sclater et al., 1978; Fisher et al., 1986], with a 6 km deep transform valley flanked by steep walls and high transverse ridges exposing abundant plutonic rocks (Figure 1). The paleo-fracture zone valley was surveyed in detail only to the north, where it is comprised of a generally steep, but discontinuous western paleo-transform wall, and by more gently dipping and irregular eastern non-transform wall. The latter contrasts sharply with the transform wall, and is comprised of east-west volcanic ridges and fault scarps sloping gently down and curving into the floor of the fracture zone valley. Due to asymmetric spreading of the SWIR to the east and west (6 mm/yr away from the transform and 10 mm/yr towards the transform), the transform is growing at a rate of 4 mm/yr.

Hole 735B is located on a shallow wave-cut platform, informally named Atlantis Bank, on the crest of the 5-km-high transverse ridge forming the eastern wall of the Atlantis II transform valley (Plate 1, Figure 1). The bank consists of a north-south 9 km long, 4 km wide platform, and is the shoalest of a series of uplifted blocks and connecting saddles forming the transverse ridge. The platform is flat, with little more than 100 m relief over about 20 km^2. A 200 x 200 m video survey in the vicinity of the hole showed a smooth, wave-cut surface exposing foliated and massive jointed gabbro locally covered by sediment drifts [Shipboard Scientific Party, 1989]. A best guess, based on the video survey, is that the foliation is oriented E-W. The platform probably formed by erosion of an island similar to St. Paul's Rocks in the central Atlantic, before subsiding to its present depth [Dick et al., 1991b].

The site is located between magnetic anomalies 5 and 5a, 93 km south of the present axis of the SWIR and 18.4 km from the inferred axis of transform faulting on the floor of the Atlantis II F.Z. [Dick et al., 1991b]. A single 11.3 Ma Pb-zircon age [Stakes et al., 1991] for a trondhjemite near the top of the hole confirms the plate reconstruction.

The complementary paleo-position on the conjugate lithospheric flow-line is 56 km due north of the SWIR rift valley axis. At this position, there is no geomorphologic expression of the transform to the west. Instead, the seafloor is comprised of an intact volcanic terrain with well preserved small volcanic seamounts, and east-west oriented volcanic and tectonic ridges parallel to the spreading center to the south. The volcanic and fault lineations are consistent with an east-west basement foliation at the top of the 735B gabbros on the transverse ridge 149 km to the south. This geomorphology is typical of rift valley floors and walls along the SWIR, and indicates that the 735B

Plate 1. Shaded relief image of gridded Seabeam bathymetry of the Atlantis II Bank [Dick et al., 1991b] showing the location of Site 735B, prepared with the NECOR Seabeam group at the University of Rhode Island. This bank consists of a wave-cut platform lying along the transverse ridge on the east flank of the Atlantis II transform valley. The bank was created by block uplift of a gabbro massif from beneath the paleo-rift valley floor of the SWIR at 11.5 Ma into the rift mountains flanking the transform. The water depth at the top of the bank where Site 735B is located is 700 m, while the average depth of the transform valley floor is 6 km.

gabbros crystallized beneath a typical rift valley segment unaffected by transform tectonics up until the unroofing process.

The foliation seen in the video survey at Site 735 was found in the upper 100 m of the hole. A similar foliation exposed subaerially in an identical tectonic setting is found in the peridotite mylonites on St. Paul's Rocks. This foliation is also orthogonal to the adjacent St. Paul's transform [Melson et al., 1970], which suggests that such foliations could be typical of the internal structure of transverse ridges.

The Hole 735B gabbros are nearly unique for ODP and DSDP crustal holes in having a single uniform magnetic in lination throughout the section. This demonstrates that there has been no late tectonic disruption of the section since cooling below the Curie point, although the relatively steep inclination suggests block rotation of about 18° [Pariso et al., 1991]. Thus, unlike the tectonically disrupted rocks dredged from fracture-zone walls, the Hole 735B gabbros preserve an intact metamorphic and tectonic stratigraphy recording brittle-ductile deformation and alteration at high temperatures beneath the rift valley.

Assuming a normal thickness of sheeted dikes and pillow lavas (2 km) and a rift valley depth similar to today's (4 km), the gabbros have been uplifted from beneath the rift valley a minimum of 6 km to sea level to form the wave-cut platform. The present 700 m water depth is close to that anticipated for simple subsidence of the platform along the age-depth curve over 11 my, suggesting that uplift to sea level occurred at the inside-corner high near the ridge-transform intersection. A similar wave-cut platform is presently situated on such an inside-corner high on the SWIR at 207 m depth at the southern end of the transverse ridge flanking the DuToit F.Z. [Fisher et al., 1986].

EMPLACEMENT MECHANISM OF THE 735B GABBROS

There are two possibly related phenomena occurring during the formation of transverse ridges. The first is the unroofing and emplacement to seafloor of deep crustal rocks and mantle which we discuss in detail below. The second, is the enormous total uplift of the lower crust and shallow mantle. No entirely satisfactory mechanism has been proposed to explain this. The high gravity of these ridges preclude that they are serpentinite diapirs intruded along the flank of fracture zones [Louden and Forsyth, 1982], while plate reconstructions show that many formed during periods of transtensional motion across the transform, and thus they cannot be an expression of compressional tectonics [Dick et al., 1991b]. Bercovici et al. [1992] have alternatively proposed formation by visco-elastic rebound accompanying corner flow at the ridge-transform plate boundary, but this is based only on 2-D modeling.

The entire transverse ridge flanking the Atlantis II F.Z. exposes largely plutonic rocks [Dick et al., 1991b], while the crustal section spreading in the opposite direction at the ridge transform intersection preserves an intact rift valley volcanic terrain. This striking contrast is matched by a physiographic contrast, where the volcanic terrain exposed in the rift mountains to the north on the non-transform wall slopes gently down to the floor of the fracture zone, while deep crustal and plutonic rocks are uplifted in a series of 4-10 km-wide blocks at the inside corner high to form the transverse ridge to the south. (This is true at both intersections - e.g., Plate 2). Site 735B is the shoalest of these blocks, but is otherwise typical. This profound asymmetry is interpreted as the result of a crustal weld which periodically forms between the shallow levels of the ocean crust and the old, cold lithospheric plate at the ridge-transform intersection. This weld causes the ocean crust to be dismembered, with the shallow levels spreading with the older plate away from the active transform, unroofing the deep ocean crust on low-angle normal faults as it is uplifted to form the transverse ridge flanking the transform (Figure 2).

This asymmetric spreading of the crustal section appears to be a general phenomenon at ridge-transform intersections on slow spreading ridges. On the rift valley wall of the inside-corner high at the Kane F.Z. a gently dipping fault surface exposing gabbro was observed by submersible which is believed to be the surface expression of a detachment or low-angle normal fault [Dick et al., 1981; Mevel et al., 1991]. Detachment faults have also been suggested to form periodically within rift valleys by fault capture during amagmatic periods [Harper, 1985, Karson et al., 1991]. Thus, the structures and fabrics seen in the Hole 735B core are likely to be representative of the kinds of fabrics generally found in lower crustal sections formed at slow spreading ridges. It is true, however, that due to the proximity to the transform, the extent of the ductile shear may be greater than elsewhere beneath the rift valley. On the other hand, Hole 735B is located within a crustal block, and its internal structures likely relate more to processes occurring prior to unroofing than to the formation of the block and late uplift.

IGNEOUS STRATIGRAPHY

Over 500 discrete lithologic intervals were identified in the 435m of Hole 735B core. These are grouped into six major lithologic units [Figure 3, Dick et al., 1991a]. Three principal classes of rocks are present in the section: (1) primary medium- to very-coarse-grained gabbro, (2) late intrusive fine- to medium-grained microgabbro, and (3) synkinematic oxide-bearing gabbro. The first two types consist chiefly of isotropic, equigranular olivine gabbro with minor gabbronorite, gabbro, and troctolitic gabbro. These mostly oxide-free lithologies underwent local late-magmatic deformation which channeled iron-rich intercumulus melt through brittle-ductile shear zones. This resulted in the formation of evolved oxide-bearing gabbros due to precipitation of iron oxides and iron-rich silicates and chemical exchange between the migrating melt and the sheared gabbro. The first order crystal-chemical control on melt composition, however, appears to have been

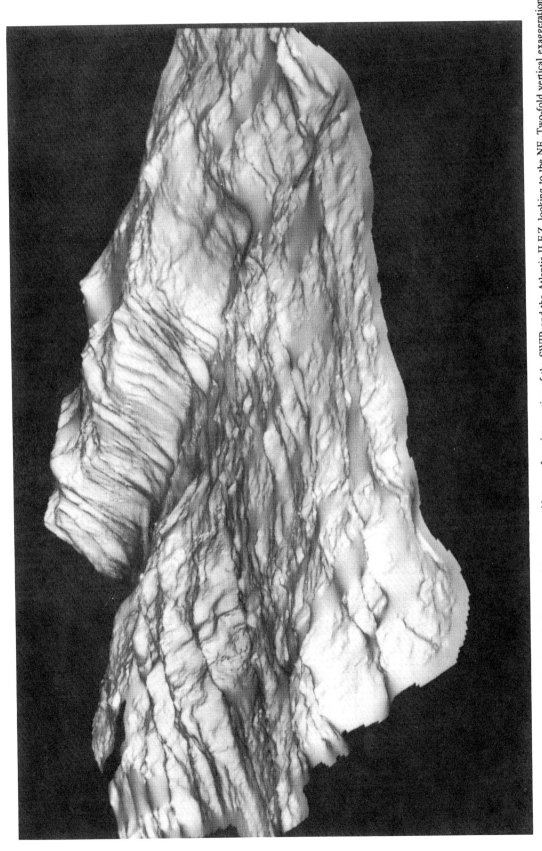

Plate 2. Shaded relief image of gridded seabeam bathymetry of the western ridge-transform intersection of the SWIR and the Atlantis II F.Z. looking to the NE. Two-fold vertical exaggeration. Image prepared at the NECOR Seabeam Facility. Deep L-shaped valley marks the transform valley (trending to upper left) and rift valley (trending to lower left). Note the subdued bathymetry of crust spreading south of the neovolcanic zone in contrast to the rugged steeply uplifted topography north of the neovolcanic zone where the crust is spreading parallel to the transform. This is accompanied by exposure of an essentially intact volcanic terrain of pillow basalts to the south, and plutonic rocks and diabase to the north.

Table I
Hole 735B Average and Bulk Compositions

Lithology	# Int.	Total Thick.	Ave. Thick.	Std. Dev.	Vol. % Recovery	Ave. Density	St.Dev.	N	Wt.% Recovery	SiO2	TiO2	Al2O3	Fe2O3	FeO	FeO*	MnO	MgO	CaO	Na2O	K2O	P2O5	H2O	CO2	LOI	Mg#	Ca#	Ca/Al
Primary Gabbros																											
Gabbronorite	20	26.86	1.34	1.76	5.364	2.924	0.031	6	5.256	52.8	0.63	16.29	1.98	5.81	7.59	0.16	6.93	10.66	3.79	0.14	0.05	1.02	0.08	0.85	68.0	60.9	0.59
Gabbro	4	0.40	0.1	0.02	0.080				0.078	50.5	0.34	16.95	0.93	4.39	5.19	0.11	10.07	13.33	2.49	0.04	0.01	0.82	0.06	3.79	80.3	74.8	0.73
Olivine Gabbro	176	303.5	1.72	3.73	60.617	2.953	0.052	31	59.975	50.5	0.34	16.95	0.93	4.39	5.19	0.11	10.07	13.33	2.49	0.04	0.01	0.82	0.06	3.79	80.3	74.8	0.73
Patchy Olivine Gabbro	14	8.07	0.58	0.57	1.612	2.919	0.030	5	1.577	49.7	0.28	16.52	0.94	4.45	4.73	0.09	11.64	14.37	1.95	0.02	0.01	0.46	0.08	0.60	82.3	80.3	0.80
Diopsidic Olivine Gabbro	4	4.39	1.1	0.86	0.877			0	0.859	50.2	0.30	15.85	0.59	3.16	3.69	0.09	10.62	16.34	2.01	0.02	0.01	0.80	0.00	0.68	85.7	81.8	0.94
Late Intrusives					0.000																						
Oxide Microgabbronorite	4	1.52	0.38	0.23	0.304	3.021	0.047	2	0.307	50.3	1.95	14.66		12.23	12.23	0.22	6.60	9.95	3.69	0.08	0.10				49.0	59.8	0.62
Microgabbro	5	1.10	0.22	0.11	0.220			0	0.215	51.7	0.48	15.78	1.41	6.28	7.55	0.14	8.47	10.84	3.43	0.03	0.00	1.48	0.00	1.57	70.6	63.6	0.62
Olivine Microgabbro	15	5.68	0.38	0.37	1.134	2.924	0.025	6	1.111	50.1	0.26	16.96	1.19	4.83	5.90	0.11	10.72	11.48	2.92	0.10	0.00	1.33	0.06	1.23	78.6	68.5	0.62
Troctolite	9	5.61	0.62	0.67	1.120	2.993	0.182	2	1.124	45.7	0.17	16.90	0.90	6.33	6.67	0.11	17.82	9.81	1.74	0.03	0.01	1.89	0.00	1.41	83.4	75.7	0.53
Troctoliitic Gabbro	9	5.47	0.61	0.61	1.092	2.906		1	1.064	48.7	0.20	20.15	0.61	4.32	4.68	0.09	11.60	12.74	2.03	0.02	0.01	0.65	0.00	0.59	82.7	77.6	0.57
Troctoliitic Microgabbro	9	1.42	0.16	0.1	0.284	2.911	0.073	2	0.277	44.9	0.28	12.67	5.45	11.88	14.33	0.24	16.36	8.60	1.56	0.07	0.01	0.74	0.00	0.11	71.1	75.3	0.62
Trondhjemite	1									70.8	0.22	17.28	1.24	0.00	1.12	0.02	0.17	1.93	6.95	1.48	0.01	n.d.	n.d	0.40	13.3		
Diabase	2	1.53	0.77		0.306	2.955		1	0.303	49.8	1.73	14.91	2.38	7.18	9.32	0.19	7.76	11.26	3.01	0.08	0.36	1.28	0.00	1.21	65.8	67.4	0.69
Synkinematic Gabbros					0.000																						
Oxide Gabbronorite	11	2.77	0.25	0.2	0.553	3.103		1	0.575	43.5	4.50	11.29	8.72	13.49	21.33	0.33	7.66	6.98	2.74	0.06	0.07	0.74	0.00	0.46	49.6	58.5	0.56
Oxide Gabbro	36	8.08	0.22	0.24	1.614	3.331	0.466	5	1.802	48.3	3.30	14.37	1.96	10.41	11.59	0.18	7.25	10.47	3.41	0.10	0.02	0.93	0.13	0.47	55.6	62.9	0.66
Disseminated Oxide Olivine Gabbro	36	41.40	1.15	2.75	8.268	2.945	0.036	15	8.160	51.9	0.61	15.58	1.63	6.87	8.34	0.17	8.30	10.95	3.29	0.05	0.01	0.55	0.10	1.29	68.3	64.8	0.65
Oxide Olivine Gabbro	121	78.43	0.65	1.63	15.664	3.114	0.116	22	16.344	42.9	6.10	10.96	6.10	16.47	20.73	0.30	6.07	9.07	2.74	0.06	0.15	0.67	0.00	0.39	40.8	64.7	0.79
Patchy Oxide Olivine Gabbro	3	2.37	0.79	0.49	0.473			0	0.528	54.0	0.49	16.56	0.92	5.95	6.78	0.16	7.02	10.19	3.90	0.06	0.06	0.62	0.00	1.22	67.8	59.1	0.56
Oxide Olivine Microgabbro	9	1.65	0.18	0.23	0.330	3.134	0.124	2	0.346	45.6	4.18	12.56	4.77	13.98	16.84	0.24	7.03	9.60	2.97	0.04	0.00	0.64	0.07	0.16	47.3	64.1	0.69
Oxide Pyroxenite	3	0.24	0.08	0.02	0.048			0	0.054																		
Oxide-Rich Vein	3	0.20	0.07	0.07	0.040			0	0.045																		
Bulk Hole 735B		500.70				2.984				49.3	1.41	15.72	1.97	6.92	8.44	0.15	9.15	12.12	2.67	0.05	0.04	0.80	0.05	2.57	71.4	82.9	0.72
Bulk Hole with Trondhjemite										49.4	1.40	15.73	1.97	6.88	8.40	0.15	9.10	12.07	2.69	0.06	0.04						

Notes: # Int. is the number of discrete lithologic intervals of each lithology. N is the number of separate observations of the density. Bulk hole composition computed first without trondhjemite.

Lithology	V	Cr	Ni	Cu	Zn	Rb	Sr	Y	Zr	Nb
Primary Gabbros										
Gabbronorite	190	13	45	34	53	2	183	22	186	1.0
Gabbro	130	383	129	45	28	1	161	11	22	1.0
Olivine Gabbro	130	383	129	45	28	1	161	11	22	1.0
Patchy Olivine Gabbro	108	335	137	108	29	1	161	6	8	<0.5
Diopsidic Olivine Gabbro	162	2857	190	19	18	<0.8	124	9	11	<0.5
Late Intrusives										
Oxide Microgabbronorite										
Microgabbro	224	49	47	9	30	<0.8	168	13	13	<0.5
Olivine Microgabbro	104	241	161	41	27	<0.8	168	9	10	0.7
Troctolite	43	1314	506	52	41	1	120	4	7	0.6
Troctolitic Gabbro	75	986	260	63	31		146	7	22	0.6
Troctolitic Microgabbro	102	96	318	93	128	2	47	11	18	2.8
Trondhjemite										
Diabase	242	190	91	61	68	1	157	43	134	3.7
Synkinematic Gabbros										
Oxide Gabbronorite	880	41	47	122	153	<0.8	141	17	42	3.0
Oxide Gabbro	615	52	107	95	44	<0.8	145	32	32	2.2
Disseminated Oxide Olivine Gabbro	220	37	48	49	44	1	179	14	19	0.7
Oxide Olivine Gabbro	619	7	20.9	72.3	117	0.9	162	28	54	4.2
Patchy Oxide Olivine Gabbro	173	10	24	29	43	<0.8	203	10	14	<0.5
Oxide Olivine Microgabbro	650	65	35	72	85	0.9	167	16	20	1.7
Oxide Pyroxenite										
Oxide-Rich Vein										
Bulk Hole 735B	234	295	105	51	47		162	15	36	

fractional crystallization [Hebert et al., 1991; Ozawa et al., 1991; Bloomer et al., 1991; Natland et al. 1991; Dick et al., 1991a], although there is textural evidence for local assimilation of wall rocks at the contacts between late intrusives and earlier crystallized gabbros.

Shown in Table I are the proportions of different rock types, average interval thickness, average compositions, and a calculated bulk hole composition based on shipboard XRF and unpublished analyses [Meyer, personal communication]. The latter was calculated with the lithologic log and densities measured on board ship. Reasonable values were substituted for minor unanalyzed lithologies. General descriptions of the principal lithologies are given below. We note that molecular Ca/(Ca+Na) is recalculated from Dick et al. [1991], where due to a programming error the wrong atomic weight for sodium was used.

PRIMARY GABBROS (GABBRONORITE AND OLIVINE GABBRO)
Olivine Gabbro

Medium to coarse-grained olivine gabbro comprises 61% of the core, and is composed of calcic plagioclase and augite with from 1 to 20% olivine (7% ave.) and less than 0.1% oxide. Overall, the olivine gabbros have isotropic texture, similar to the upper gabbros of some ophiolites. The typical olivine gabbro is equigranular with interlocking anhedral plagioclase, pyroxene, and olivine grains of roughly equal size. Some varieties have anhedral to subhedral plagioclase laths subophitically enclosed in pyroxene, with subhedral to intergranular olivine intergrown with the plagioclase and pyroxene. A few very coarse-grained pegmatitic gabbros contain spectacular augite oikocrysts up to 20 cm long enclosing mats of plagioclase laths. Minor interstitial brown hornblende rims some olivine grains.

Mineral compositions are homogeneous, with standard deviations of only about 5% for key element ratios. Plagioclase ranges from An_{48} to An_{66}, while olivine and clinopyroxene are fairly magnesian (Fo_{60-80}, Mg_{77-86}). The olivine gabbros lie toward the primitive end of the compositional spectrum, and are noticeably richer in chromium and nickel than the gabbronorites and microgabbros, but are less magnesian than the troctolitic rocks (Figure 4) and contain significantly more calcic anorthite.

The overall coarse grain size and the frequent occurrence of coarse pegmatoidal patches is a striking feature [Bloomer et al., 1991], which is not typical of layered intrusions, which are finer-grained on average. Such pegmatoidal textures in continental intrusives are commonly associated with rapid crystal growth and devolatilization of magma bodies rather than prolonged cooling. Together with the isotropic texture of the olivine gabbro, this may be evidence that the olivine gabbro underwent rapid initial crystallization to form a crystal mush: likely the result of emplacement and rapid undercooling in previously solidified or semi-solidified gabbros.

Patchy olivine gabbro (18.1 meters in 14 intervals) is a textural subvariety with irregular patches of pegmatoidal gabbro and microgabbro. These could represent some other

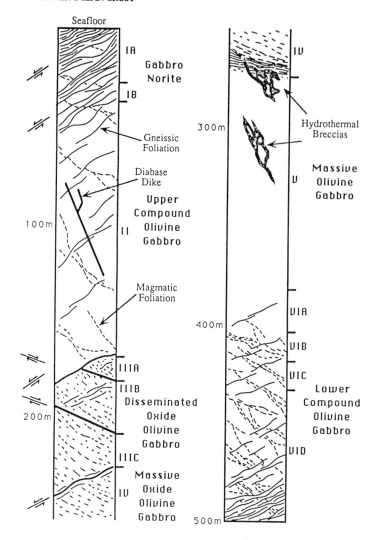

Fig. 3. A. Summary lithologic column for Hole 735B. Short dashed lines in all units indicate approximate orientation of magmatic foliation in oxide-bearing gabbros, while long wavy lines indicate the locus of sub-solidus amphibolite facies ductile deformation. The strike and dip-direction of the subsolidus and late magmatic ductile deformations are assumed constant, though this might not be the case. Approximate location of several hydrothermal breccias shown for reference - note crosscutting relationship at the base of Unit V. Approximate locations of major subunit boundaries deduced from microfaults in the core are shown for Units III and VI. Contacts within Unit I and II are presumed to lie in the foliation plane.

form of heterogeneous nucleation, for example, intrusion of new melt into the semi-solidified olivine gabbro when it was a crystal mush sufficiently viscous to inhibit rapid mixing of new and old melts, but insufficiently rigid for intrusive contacts to form. A lack of flow fabrics suggest that these patches are not the product of slumping in a magma chamber.

Some olivine gabbros have modal and grain size layering superficially similar to that in layered intrusions. Olivine, plagioclase, and clinopyroxene, however, tend to be uniform in size, coarsening or fining together as the grain size of the rock varies [Bloomer et al., 1991], while in magmatic sediments different minerals typically have different grain sizes in a layer, reflecting their respective densities and settling velocities. No unambiguous

magmatic sedimentary features were identified in the olivine gabbro, and most of the textural types can be explained by local variations in the proportions and growth rates of phases nucleating in a crystal mush.

Gabbronorite

Twenty gabbronorite intervals were identified, making up a little more than 5% of the total section. In most cases the gabbronorite is deformed and recrystallized to augen gneiss and mylonite. The primary lithology, where preserved, is medium- to coarse-grained gabbro with an intergranular to subophitic texture having interlocking high- and low-calcium pyroxene crystallized with and around a mat of plagioclase. Plagioclase is sodic, typically around An$_{42}$, and clinopyroxene is iron-rich with a Mg/(Mg+Fe) ratio of

Legend: Hole 735B Igneous Lithostratigraphy

Unit I = Gabbronorite
  Subunit IA = Massive gabbronorite
  Subunit IB = Olivine gabbro and gabbronorite
Unit II = Upper compound olivine gabbro
  Minor: intrusive microgabbro and olivine micro-gabbro synkinematic oxide and oxide-olivine gabbro
Unit III = Disseminated oxide-olivine gabbro
  Subunit IIIA = Disseminated-oxide-olivine and olivine gabbro
  Subunit IIIB = Massive disseminated-oxide-olivine gabbro
  Subunit IIIC = Disseminated-oxide-olivine and olivine gabbro
Unit IV = Massive oxide olivine gabbro
Unit V = Massive olivine gabbro
Unit VI = Lower compound olivine gabbro
  Subunit VIA = Compound olivine gabbro
  Minor: Intrusive oxide microgabbronorite and olivine microgabbro synkinematic oxide and ox-ide-olivine gabbro
  Subunit VIB = Compound olivine, oxide-olivine, and disseminated oxide-olivine gabbro
  Minor: Intrusive olivine and oxide-olivine micro-gabbro
  Subunit VIC = Compound troctolitic and olivine gabbro
  Minor = Synkinematic oxide-olivine gabbro and oxide-olivine microgabbro intrusive troctolite and troctolitic microgabbro
  Subunit VID = Compound olivine and oxide-olivine gabbro
  Minor = Intrusive troctolite, diopsidic-olivine and troctolitic gabbro synkinematic disseminated-ox-ide-olivine and oxide gabbro

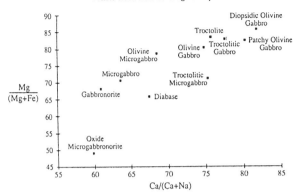

Fig. 4. Whole-rock XRF analyses for oxide-free gabbros, and microgabbros, averaged from Table 1. Molecular ratios calculated excluding ferric iron.

about 70. The gabbronorites are the most evolved of the primary and late intrusive gabbros, with moderate silica (52.8% ave.) and titanium (0.63%) and low Mg/(Mg+Fe) and Ca/(Ca+Na) ratio (Figure 4). The contact relationships between the gabbronorite and the olivine gabbro with which it is intercalated are obscured by deformation.

Late Intrusives

Three major groups of late intrusive rocks are recognized: (1) various kinds of related troctolites, gabbros and microgabbros, (2) trondhjemites, and (3) diabase dikes. The late intrusive gabbros and microgabbros make up about 4% of the core, whereas trondhjemites constitute less than 1% and diabase 0.3%.

TROCTOLITES, GABBROS AND MICROGABBROS

The late gabbro, microgabbro and troctolitic intrusives include a wide variety of rock types, such as olivine gabbro and microgabbro, oxide microgabbro, troctolite, and troctolitic microgabbro. They are generally small, averaging 0.37 m thick when measured vertically down the core. They include the troctolite and troctolitic microgabbro intrusives found in the lower portion of the hole where they become systematically coarser grained with depth. These are believed to be precursors of a second major intrusive body just at, or just below, the bottom of the hole [Dick et al., 1991a; Bloomer et al., 1991]. The gabbros and microgabbros intruded over a long period and cross-cut the olivine gabbros and gabbronorites, and locally one may cut a synkinematic oxide-gabbro near the base of the hole.

These late intrusives have fine- to medium-grained granular textures, and highly irregular contacts with the coarse-grained olivine gabbros (Plate 3). Contacts generally mark a sharp change in grain size, but individual mineral grains commonly interlock across the contact, producing a sutured igneous contact. Undeformed contacts are typically irregular, and upper and lower contacts may dip in opposite directions. Locally, the gabbros and

microgabbros have sharp sub-parallel, sub horizontal contacts giving an appearance of magmatic layering. Close examination of these layers, however, shows that they are generally deformed, with what were originally sutured, irregular intrusive contacts transposed into the plane of deformation.

The primary contact relationships of the late gabbro and microgabbro intrusives are similar to those seen in alpine-type peridotites where podiform dunites crosscut tectonized harzburgite. There, magmas migrating through the mantle have reacted with and partially assimilated mantle peridotite, at the same time precipitating olivine and spinel [e.g., Quick, 1981]. Similarly, the Hole 735B microgabbros we believe represent cumulates left by batches of melt of diverse composition migrating upward through and reacting with the solidifying olivine gabbro.

The late gabbro and microgabbro intrusives have a wide range in mineral and rock composition (Table I, Figure 4) with a systematic variation in Ca/(Ca+Na) versus Mg/(Mg+Fe) which suggests that they can be related through fractional crystallization of similar parent magmas [Hebert et al., 1991; Natland et al., 1991]. In general, incompatible element concentrations are too low at a given Ca/(Ca+Na) ratio for the large majority of these rocks to represent liquid compositions (Figure 5a) and therefore they are cumulates from which residual liquid was removed [Dick et al., 1991a; Bloomer et al., 1991]. One troctolite interval contains a spinel-rich layer which could be attributed to magmatic sedimentation.

The gabbro and microgabbro varieties are granular fine to medium grained with variable olivine contents. Pyroxene tends to be stubby, and is reddish brown to yellow-light brown and pale green. Locally, fine-grained intergranular chrome diopside is rimmed by reddish pyroxene, and some of the finer-grained titanaugite might have been produced by reaction between late liquids and chrome diopside.

The oxide-microgabronorites are fine- to medium-grained, equigranular, and contain abundant augite, hypersthene, and plagioclase enclosed by subophitic to ophitic pyroxene. Unlike the synkinematic oxide-bearing gabbros and related rock types described below [Bloomer et al., 1991; Dick et al., 1991a; Natland et al. 1991], there is no evidence of a deformation related origin. Contacts are clearly intrusive, with small partially digested clasts of olivine gabbro occurring near the olivine gabbro wall rock (Plate 3). The bulk composition determined by XRF (Table 1) is close to ferrobasalt magma, and the abundant iron oxide appears to be an early cumulus phase, with subhedral crystal morphologies. The oxide microgabbronorites are believed to have formed by fractionation of a primary MORB magma which was intruded into the section where it crystallized without later modification by late magmatic liquids.

The troctolite, troctolitic gabbro, and troctolitic microgabbro (troctolites) are fine- to coarse-grained, equigranular rocks, consisting of about 60% plagioclase, 20-40% olivine, and 0-20% clinopyroxene. They range from pyroxene-poor, xenomorphic granular, olivine-

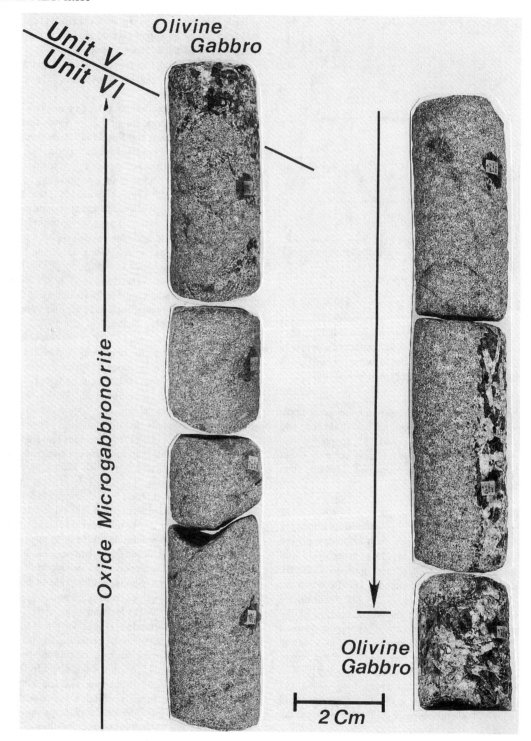

Plate 3. Oxide microgabbronorite intruding olivine gabbro at the top of Unit VI showing irregular upper and lower contacts and lack of a chilled margin around the microgabbro intrusion.

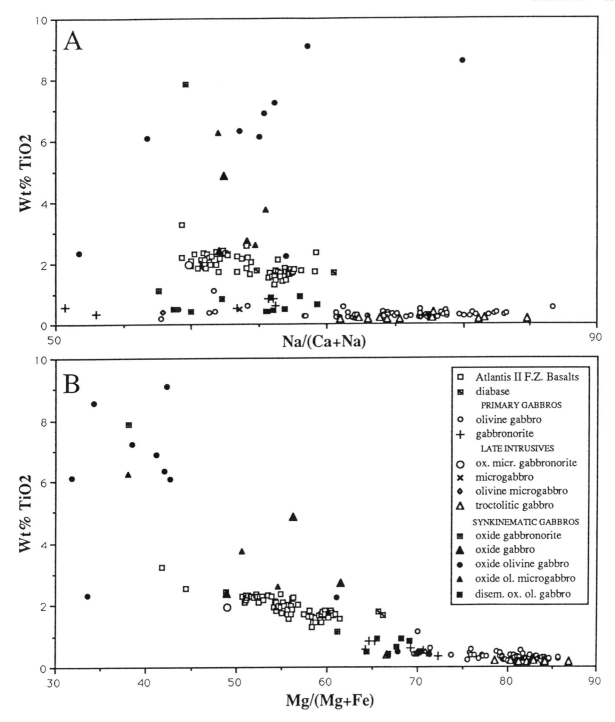

Fig. 5. A. Plot of TiO2 versus CaO/(CaO+Na2O) and B. MgO/(MgO+FeO) for Hole 735B gabbros. Pillow basalts dredged from the SWIR in the vicinity of the Atlantis II F.Z. are shown by open squares; ferrogabbros are shown by solid symbols, while oxide-free undeformed gabbros by open symbols and crosses. Abbreviations are for lithologies in Table 1.

plagioclase adcumulates and mesocumulates to pyroxene-rich, heteroadcumulates. Texturally, olivine is subhedral to anhedral, locally occurring as granular to wormy blebs. It may be intergranular to subophitic to plagioclase. Subophitic to oikocrystic, intergranular, red clinopyroxene (titaniferrous augite) is abundant, while

medium-grained green chrome diopside intergrown locally with coarser stubby plagioclase laths may also be present. The intergranular to subophitic chrome diopside stands out due to its bright green color and comparatively coarse grain size. Chromian spinel is common, but is partly to completely oxidized to a titaniferrous ferritchromite. In one 2 m troctolitic layer, a single 10 cm thick spinel-rich zone occurs (83-7 pc.4d-4c) containing about 2% spinel.

The troctolites (Figure 4) are the most primitive rocks from Hole 735B, with whole-rock $Mg/(Mg+Fe)$ ranging from 78.6 to 86.9 and $Ca/(Ca+Na)$ ranging from 83 to 91.8. Olivine is magnesian (Fo_{80-85}) and plagioclase is calcic (An_{66-76}), and spinel ($Cr/(Cr+Al)$ = 49 to 61) is chrome-rich, extending to the upper limit for spinel in abyssal basalts [Dick and Bullen, 1984]. The troctolites are noticeably poor in silica and alumina and rich in iron and magnesium compared to basaltic liquids.

Troctolite contacts are sometimes gradational to olivine gabbro, and the latter may be unusually olivine-rich near troctolite and troctolitic intrusives with patchy texture and large variations in grain size and phase proportions. In some cases the olivine gabbro also contains coarse patches of kelly green diopside in addition to the typical brown or reddish brown augite. We suggest that these diopsidic olivine gabbros (Table 1) are hybrid rocks where semi-rigid olivine gabbro mush was infiltrated by the same primitive magmas from which the troctolite intrusives crystallized.

Trondhjemite

Trondhjemite, though volumetrically minor (0.43%), is present at many levels throughout the core. Typical mineral assemblages are 75% sodic plagioclase, 20% quartz, minor clinopyroxene, brown amphibole or biotite, magnetite, trace amounts of zircon, and apatite, all of which are compatible with an igneous origin. Euhedral zircon grains are locally abundant and acicular hornblende crystals are present in some hand samples. Most trondhjemites have granular aplitic igneous textures with interstitial quartz and blocky subhedral plagioclase crystals strongly zoned at their rims [Shipboard Scientific Party, 1989]. Clinopyroxene is largely replaced by actinolite and biotite poikilitically encloses plagioclase crystals. The one sample analyzed has a granitic melt composition with a very high albite content and 70.8% SiO_2. The trondhjemites usually occur in net vein complexes or form intrusion breccias enclosing clasts of gabbro (Plate 4). Individual veins are typically 1-2 cm wide. Contacts are sharp, formed by brittle fracture during intrusion. Locally the veins enclose partly dissolved xenoliths of wall rock [Dick et al., 1991a].

Trondhjemite might form by liquid immiscibility in the highly evolved magmas that impregnated the oxide gabbros [Natland et al., 1991], consistent with the composition gap between them and other Hole 735B gabbros and their often intimate association with oxide gabbro. New trace element data, however, show that they contain rare earth element enriched pyroxene [P.S. Meyer, unpublished data] consistent with the high levels of Zr and P implied by the presence of zircon and apatite. These elements should partition into the iron-rich liquid, not the silica-rich one, during liquid immiscibility, and thus we favor an origin by fractional crystallization of basalt. On the other hand, homogenization temperatures of quartz fluid inclusions from some trondhjemitic veins range from 227°

to 358°C and those for inclusions in plagioclase cluster around 270°C [Vanko and Stakes, 1991]. These temperatures indicate formation from either hydrothermal or magmatic fluids that had undergone significant cooling, so some of the trondhjemites, could have a hydrothermal origin. The trondhjemites are late, crosscutting all of the gabbros including the synkinematic oxide-gabbros. Most trondhjemites bodies are deformed with the amphibolite gneisses, but some in the upper amphibolite gneiss occur in ptygmatic folds or lenses (Plate 4). Unlike the enclosing amphibolite, these trondhjemites, however, do not appear extensively recrystallized. The folds, and the trondhjemites they contain, contrast sharply to the style of deformation and the appearance of trondhjemite throughout the rest of the core. Dick et al. [1991a] suggest that these trondhjemites appear to have formed by local anatexis of the amphibolites due to a nearby late-stage intrusion.

Diabase Dikes

At about 25 mbsf, a 0.5-m-thick interval of partially recrystallized and weakly foliated diabase is intercalated with highly deformed gneisses. The contacts have been obscured by deformation, but apparently lie in the gneissic foliation plane. Minor recrystallization and rotation of feldspar in the matrix, as well as stretching and grain boundary recrystallization of the coarser plagioclase phenocrysts, produce a weak foliation and gneissic appearance. At 105 mbsf, olivine gabbro is crosscut by an undeformed diabase dike dipping 50° to 60°, with sharp subparallel contacts and well-developed chilled margins. The second diabase is similar in composition to the deformed one in Unit I, except for a higher water content due to secondary hydrothermal phases. The average diabase composition (Table I) is a moderately primitive MORB with $Mg/(Mg+Fe)$ of 0.66 and relatively high TiO_2 (1.73 wt. %).

Synkinematic Oxide-Bearing Gabbros

These rocks are strongly associated with penetrative high-temperature deformation, and contain ubiquitous iron-titanium oxides. They occur in brittle-ductile shear zones in the olivine gabbro and gabbronorite, which range from a few millimeters to over 100 m thick. They commonly have well developed feldspar lineations, coarse equant pyroxene porphyroclasts, and occasional boudins and phacoids of more competent undeformed oxide-free gabbro (Plate 5). Interstitial magmatic oxide filling extension cracks in stretched and broken pyroxene augen demonstrate syntectonic precipitation of titanomagnetite. In many cases, deformation is not evident in thin section, and the rock has a medium- to coarse-grained, equigranular igneous texture. Macroscopically, the same rock can clearly be seen to have undergone deformation and flow, and it appears that past-deformation precipitation of feldspar, pyroxene and titanomagnetite have produced an igneous overprint on the earlier tectonite fabric.

All of the synkinematic gabbros are evolved (Figure 5b) and contain relatively sodic plagioclase (An_{30-45}), iron-

Plate 4. A. Trondhjemite intrusion breccia crosscutting foliated oxide olivine gabbro in Unit IV (Core 53-4, pcs 5 and 6). B. Ptygmatically-folded, zircon-bearing trondhjemite in Unit I amphibole gneiss (Core 7, sec. 1, pc. 2).

rich olivine (Fo_{30-60}) and iron-rich pyroxene. Their titanium content is highly variable and is directly related to the amount of oxide present. Mineralogically, the rocks range from oxide gabbro and gabbronorite to oxide-olivine gabbro. There are varieties of synkinematic gabbros texturally corresponding to a sheared or slightly deformed version of each of the oxide-free gabbros and gabbronorite, though we only describe the major divisions below. These are believed to have formed largely by impregnation of and reaction with the latter by intercumulus melt migrating into and along brittle-ductile shear zones which extended into the lower crust when it was sufficiently solidified to support a shear stress [Dick et al., 1991a; Natland et al., 1991; Bloomer et al., 1991].

Oxide-Olivine Gabbro

Oxide olivine gabbro is the most abundant synkinematic gabbro, with 78.5 m occurring in 121 separate intervals in all but the uppermost lithologic unit. Most of these rocks have a very-weak to strong penetrative foliation, with equigranular interlocking, pyroxene, plagioclase and olivine crystals and some pigeonite. In addition, this gabbro contains 0.2-50% (ave. 8%) intergranular iron-titanium oxide, mostly ilmenite. The oxide is anhedral and locally may fill cracks were coarse pyroxene appears to have been pulled apart [Natland et al., 1991; Dick et al., 1991a]. Highly anhedral, the intergranular oxide often defines a foliation in the gabbro and may vary radically in abundance on a centimeter scale. The oxide content overlaps with the disseminated oxide olivine gabbro below only because of local textural and mineralogic heterogeneity in these coarse grained often pegmatitic gabbros which generally contain greater than 5% oxide in hand specimen scale. Compositionally the oxide-olivine gabbros represent some of the most evolved rocks, short of trondhjemite, in Hole 735B, representing the end-member of the synkinematic gabbro spectrum.

A problem arises in the interpretation of these rocks in that the end-member compositions are very close to pure ferrobasalt cumulates. This could occur by in-situ crystallization of a ferrobasalt liquid, or where high melt-rock ratios in the shear zones result in complete reaction of a earlier olivine gabbro protolith with the migrating liquid such that it is transformed completely. Thus, the only criterion for discriminating these different origins for a particular rock is textural. Most, but not all, do exhibit textural evidence of an earlier deformed protolith, for which there is at times no chemical evidence. Moreover, on the thin section scale it can be very hard or impossible to see such evidence, which is often observed only macroscopically in larger sections of core. This is because deformation appears to have stopped before complete solidification of the invasive liquid in some examples, resulting in a tectonite with undeformed igneous grain boundary relations.

A similar problem was encountered in the Rhum Layered Intrusion where Bedard et al.[1988] have shown that were layered troctolites are intruded by peridotite sills crystallized from picritic melts, the troctolites are locally transformed to gabbro. This occurred by chemical exchange and in-situ replacement of plagioclase and olivine by pyroxene by reaction with liquid expelled from the intrusive into the troctolite. Original layering in the troctolites is pseudomorphed by the gabbros, and can be followed texturally across gabbro/troctolite contacts. This evidently occurred when the troctolite was still partially molten, and thus permeable to the intruding picritic melt. Plagioclase in these replacement gabbros shows no zoning. Thus chemical exchange between an invaded protolith (the troctolite at Rhum - the olivine gabbro in Hole 735B) can be quite rapid and complete. The critical difference here, however, is that flow of the invasive magma at Rhum was controlled by pre-existing permeability in a partially molten crystal mush, while that in the 735B was controlled by permeability created by faulting and penetrative deformation. Some evidence of similar metasomatic exchange has been seen in thin section in Hole 735B where chemical mapping shows where late melt was locally channeled through undeformed olivine gabbro Meyer et al. [1992].

Disseminated Oxide-Olivie Gabbro

Disseminated oxide olivine gabbro comprises a total of 41.4 m in 36 intervals. These contain less than 2% oxide, are characterized by the intermittent appearance of pigeonite, and generally occur in fairly thick intervals (1.2 m average) compared to much of the oxide-olivine gabbros (0.65 m average). These rocks are also generally finer grained. This gabbro has a characteristic penetrative foliation and lineation of feldspar laths that can be mistaken for magmatic layering [Dick et al., 1991a; Cannat, 1991]. The lamination persists, albeit weakly, even where the texture looks most igneous. These gabbros are similar in composition to the gabbronorites at the top of the hole and intermediate between oxide gabbros and olivine gabbros.

Oxide Gabbro

Oxide gabbro is generally equigranular, and largely comprised of plagioclase, augite, ilmenite, occasional pigeonite and rarely orthopyroxene. It tends to be intermediate in composition between oxide-olivine gabbro and gabbronorite, and in most cases it has a weak foliation due to deformation. It occurs in Unit II, and locally in Unit VI as deformed and undeformed crosscutting veins crosscutting. In all there is 8.1 m of oxide gabbro with the average interval only 0.2 m thick, occurring largely as thin seams in olivine gabbro. The absence of olivine in oxide gabbro in shear zones cutting undeformed olivine gabbro is likely due to reaction between the infiltrating late magmatic liquid and pre-existing olivine to form pyroxene. This indicates that the infiltrating liquid which produced the oxide gabbro was of different composition than that which infiltrated other shear zones to produce the oxide-olivine gabbro.

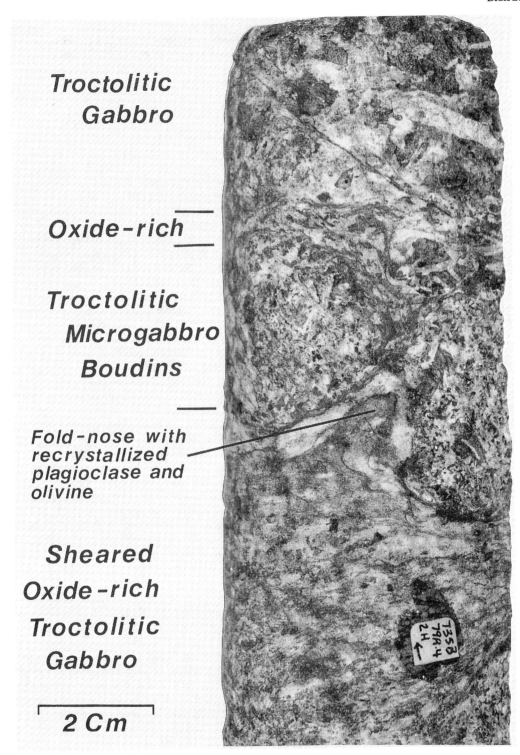

Troctolitic
Gabbro

Oxide-rich ___

Troctolitic
Microgabbro
Boudins

___

Fold-nose with
recrystallized
plagioclase and
olivine

Sheared
Oxide-rich
Troctolitic
Gabbro

2 Cm

Plate 5. Similar folding of ductily-deformed olivine and plagioclase between resistant olivine gabbro boudins in Core 79-4 pc. 2f. Plagioclase in fold is lighter mineral, olivine is darker mineral, while black mineral may be either pyroxene or iron-titanium oxide. Note that oxide is concentrated locally along the edge of the shear zone a contact with boudins. Upper shear zone, about 0. 5 cm thick, separates coarse oxide free troctolitic gabbro from troctolitic microgabbro boudins.

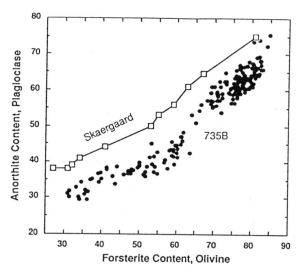

Fig. 6. Olivine and plagioclase compositions in Site 735B gabbros. Trend for the Skaergaard Intrusion [Wager and Brown, 1967] is shown for comparison. The systematic vertical offset of the trends is consistent with offsets in liquidus trends of spatially associated basalt glasses and is believed to reflect different primary melt compositions and higher degrees of mantle melting [Meyer et al., 1989].

Oxide-rich veins and dikelets

Narrow undeformed oxide-rich veins, seams and lenses locally crosscut olivine gabbro at several levels in the core. The pyroxene in these veins has a bleached appearance and a pale green-brown color and is often quite coarse-grained. Olivine is only present in some, and there may be several types. Where the veins crosscut undeformed olivine gabbro, they are sometimes oriented subperpendicular to the foliation in adjacent deformed oxide-olivine gabbros. Core 36-1 (pc3), for example, contains an undeformed 1.5- to 2.0-cm-thick vertical dike of almost pure oxide and pyroxene in undeformed olivine gabbro. This dikelet tapers out upward in the olivine gabbro and ends downward in a sheared contact with a deformed sub-horizontal oxide-olivine gabbro. These veins can reasonably be attributed to a late magmatic, iron-titanium rich melt migrating along an active shear zone in-filling and crystallizing within extension cracks in resistant shear polyhedra and phacoids and in the undeformed wall rocks [Dick et al., 1991a].

Petrogenetically, these veins are important as many may not have a hybrid origin, and permit a direct look at cumulate minerals crystallized from the late magmatic liquid infiltrating the shear zones. The pyroxene in at least some of these veins is close to equilibrium with the mineralogy of the trondhjemites [P.S. Meyer, unpublished data]. Once ilmenite begins to precipitate, it takes relatively little fractional crystallization to form trondhjemite from ferrobasalt, which suggests that the late magmatic liquid infiltrating the shear zones was indeed a highly evolved ferrobasalt produced by fractional crystallization in the intercumulus pore space in the adjacent olivine gabbro or in similar rocks below the hole.

This is consistent with the occurrence of undeformed or very weakly deformed trondhjemitic net vein complexes in the massive oxide-olivine gabbros, which would logically form as a consequence of the infiltration and crystallization of intercumulus melts in the shear zones.

IGNEOUS CHEMISTRY

Nearly all the 735B gabbros have cumulate compositions; that is to say, few could represent a true igneous liquid. This means that some portion of liquid was removed prior to complete solidification - not that they are magmatic sediments, which are a sub-class of igneous cumulate rocks. In this light, a striking feature of the 735B gabbros is their diverse chemical compositions, which span the range of known oceanic gabbros. Similar large ranges in oceanic gabbro composition have been seen elsewhere, and have been attributed to fractional crystallization in a large magma chamber [e.g., Miyashiro and Shido, 1980]. In part, this is based on comparison with layered intrusions, which have a similar range in chemistry. Plotted with the 735B data in Figure 6 are compositional data for the Skaergaard Complex, a layered intrusion in Greenland. As can be seen, the two groups of rocks show very similar patterns and ranges of chemical variation. It is notable, however, that while the Skaergaard rocks represent a 3.5 km thick section of a single intrusion, the 735B section is only 0.5 km thick.

In this light, the strong bimodality of both whole rock and mineral compositions down-hole in the 735B section (Figure 7) [Ozawa et al., 1991; Dick et al., 1991a] is important. There is no systematic distribution of iron-rich and iron poor sequences, rather the iron-rich layers are strongly associated with petrographic evidence of hyper-solidus deformation at irregular intervals down the core. Calculation of liquids in equilibrium with the 735B gabbros, gives two populations (Figure 8), the more primitive of which is in equilibrium with the olivine gabbros and troctolites and corresponds to the compositions of basalts dredged from the SWIR near the Atlantis II Fracture Zone[P.S. Meyer, personal communication, 1989; Dick et al., 1992]. These liquids, then, are believed to be analogous to the parental liquids for the 735B section.

The second set of calculated liquids correspond to highly evolved ferrobasalt, and are more evolved than any basalts dredged from the SWIR in this region. These liquids would be in equilibrium with the 735B ferrogabbros, and it is therefore suggested that their source is the late magmatic intercumulus liquid at the time the section underwent hypersolidus deformation. This condition occurred near the rigidus of a crystallizing olivine gabbro, commonly believed to lie between 60 and 80% crystallization [Sinton and Detrick, 1992]. At such percentages, the composition of the liquid for closed system fractional crystallization would be ferrobasalt [e.g., Clague and Bunch, 1976]. Where such late liquids are dispersed throughout a crystalline mush, as the would be in small ephemeral rapidly crystallized intrusions, they would be virtually

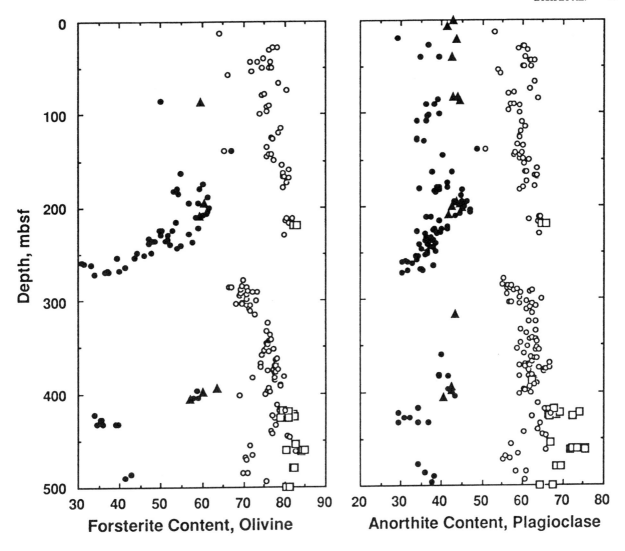

Fig. 7. Downhole plot of olivine and plagioclase chemistry for Hole 735B gabbros using data from Ozawa et al. [1991], Bloomer et al. [1991], Dick and Angeloni [unpublished data] and Meyer et al. [unpublished data], sorted by rock type. The strong bimodality of oxide-bearing synkinematic gabbros and oxide-free gabbros is a characteristic feature of most downhole chemistry plots for Hole 735B.

uneruptable [see review by Sinton and Detrick, 1991]. The absence of ferrobasalts along most of the SWIR is generally attributed to the absence of long-lived melt lenses or magma chambers as thought to exist beneath the EPR [Dick, 1989; Sinton and Detrick, 1992].

The chemical diversity of the Skaergaard Intrusion and the 735B gabbros both reflect fractional crystallization. But the physical petrogenesis is entirely different: static crystallization in a large magma chamber versus intrusion and reintrusion of small rapidly cooled migrating melt bodies in a tectonically active semi-solidified crust. The

strong bimodality of mineral and rock compositions downhole due to synkinematic differentiation is a unique characteristic of this oceanic plutonic sequence, attributed to the dynamic environment [Dick et al., 1991a; Natland et al., 1991; Bloomer et al., 1991].

The bulk composition of the 735B section (Table 1) is very similar to that of the diabase dikes, with the exception of the most incompatible trace elements: Y, Zr, and P. The latter, normally concentrated in late silicic liquids like trondhjemite, are a factor 3 to 9 lower in the gabbro than the diabase. The proportions of plagioclase, pyroxene and

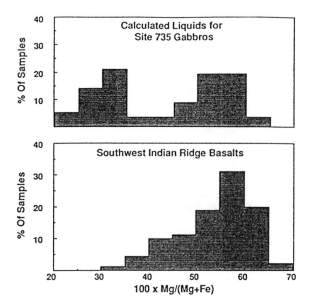

Fig. 8. Comparison of trapped and erupted liquids: Mg/(Mg+Fe) of dredged SWIR pillow basalt glasses and ratios calculated from clinopyroxene compositions for liquids coexisting with Hole 735B gabbros.

olivine in a three phase gabbro cumulate is such that its major element composition is constrained to be close to that of basalt. This is why, for example, the basalt silica contents vary little over a broad range of fractional crystallization. As a consequence, low incompatible elements concentrations, which are not incorporated in the major silicates, are critical for distinguishing a cumulate (a rock from which liquid has been removed during crystallization) from a non-cumulate (an igneous rock representing a solidified magma).

Insofar as the concentration of TiO_2 in the bulk hole is similar to that of the diabase, it is unlikely that any significant volume of late liquid was expelled from the section (e.g., >10%), particularly given the relatively primitive melt composition which would be in equilibrium with the olivine gabbro cumulates. The very low concentrations of highly incompatible elements, Zr, Y, P, on the other hand, do require that some amount of highly evolved liquid is missing, perhaps 1 or 2% trondhjemite, which could easily have been missed during drilling. In summary, the chemistry of the section is consistent with *in situ* crystallization intruded melts in this section of the lower crust.

DEFORMATION

Brittle-Ductile Deformation

Low-angle brittle-ductile shear zones and related veins formed both above and below the gabbro solidus are the principal deformation features of the 735B core [Dick et al., 1991a, Cannat, 1991]. Macroscopically, much of this deformation can be treated as pure ductile shear. Microscopically, however, there is abundant evidence of cataclasis, cracking and stretching of grains, the formation of pressure shadows, and local microfaulting which show that deformation was not purely ductile, and is locally heterogeneous. The shear zones are often associated with extension cracks in the adjacent undeformed gabbro. Locally, amphibole veins and crack networks may crosscut, or be cut by, the shear zones, indicating that shearing and crack and vein formation occurred contemporaneously. As a result, the terms ductile and brittle-ductile are sometimes used interchangeably here. It is important, however, to distinguish the two when discussing permeability. Pure ductile deformation cannot create permeability, whereas brittle-ductile deformation in the Hole 735B gabbros controlled and channeled flow of late intercumulus melt and later hydrothermal fluids.

To determine extent and location of deformation, each core fragment was classified in hand specimen on a six-point scale ranging from undeformed to ultramylonite (Figure 9). Figure 10 presents downhole plots of a 0.5-m running average of the deformation intensity down the hole, and the dip of the foliation wherever it is sufficiently developed to be measured. The dip and character of foliation and deformation vary considerably with depth. The large majority of Hole 735B gabbros are undeformed to slightly deformed (71%), lacking foliation and essentially retaining their primary igneous texture. In thin section, many gabbros show very slight grain boundary recrystallization, mortar texture, and kinking of mineral grains so that the cutoff point between undeformed and slightly deformed is arbitrary. At the opposite end of the scale, less than 2% are mylonites or ultramylonite, where recrystallization is so intense that grains are no longer visible in hand specimen. The degree of deformation varies greatly, often erratically, even on a scale of centimeters. and a running average is somewhat misleading. Such heterogeneity may be characteristic of deformed ocean ridge gabbros, with mylonite bands and undeformed igneous textures commonly found together on a hand sample scale [e.g., Helmstaedt and Allen, 1977; Malcolm, 1980; Mevel, 1988].

The hypersolidus and subsolidus shear zones are candidate seismic reflectors as a consequence of the high densities of the late magmatic shear zones, and the strong preferred mineral orientations in the subsolidus shear zones. The razor sharp contacts associated with the late magmatic shear zones make them particularly interesting. While the density difference between the oxide-olivine gabbro and the olivine gabbro is only 5.5%, the velocity contrast may be enhanced by the preferred orientation of the oxide-olivine gabbro. The shallow dips of both hyper- and subsolidus shear zones are consistent with the shallow dipping basement reflectors in the north Atlantic [e.g., NAT Study Group, 1985].

Hypersolidus Deformation

During hypersolidus deformation, faulting apparently extended into partially molten rocks resulting in migration of late intercumulus melt up the shear zones [Dick et al.,

Deformation Intensity in Hole 735B

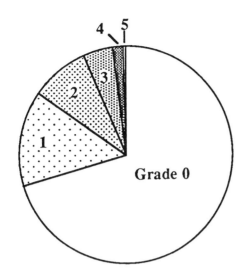

71.2% 0. Undeformed to
            slightly Deformed

14.5% 1. Weakly Foliated

8.7% 2. Strongly Foliated

4.6% 3. Porphyroclastic
            Mylonite

1.6% 4. Mylonite

0.2% 5. Ultramylonite

Deformation Zones in Hole 735B

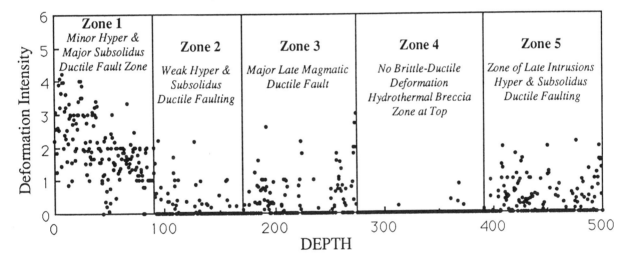

Fig. 9. A. Relative proportion of deformed and undeformed gabbros in Hole 735B, estimated by visual inspection of the core at the ODP Core Repository [Dick et al., 1991b]. B. Half-meter averages of deformation intensity plotted against depth below seafloor.

1991a; Natland et al., 1991; Bloomer et al., 1991]. These liquids then underwent crystallization and reaction with the deforming protolith to form ferrogabbro. An alternative interpretation is that the ferrogabbros were intruded first, and then acted as zones of preferential weakness for hypersolidus and subsolidus deformation [Cannat, 1991]. This late-magmatic deformation occurred locally throughout the core, but is most extensive in Units III, IV and VI. Generally, late-magmatic ductile deformation is found along small intermittent shear bands, ranging from a few millimeters to tens of centimeters wide. Morphologically these appear to be braided shear zones,

similar to the lower-temperature amphibolite facies deformation zones. Locally, the late magmatic shear zones are much thicker (53 m in Unit 3 and 50.5 m in Unit 4) with dramatic effects on fabric and rock chemistry. The high-temperature deformation zones are locally overprinted and texturally obliterated by late subsolidus brittle ductile deformation and alteration, particularly in the upper 100 m of the core, where there are highly deformed oxide-bearing intervals compositionally identical to the synkinematic ferrogabbros. Texturally, hypersolidus ductilly-deformed gabbros are significantly different from subsolidus ductily deformed gabbros [e.g., Benn and Allard, 1989]. Although

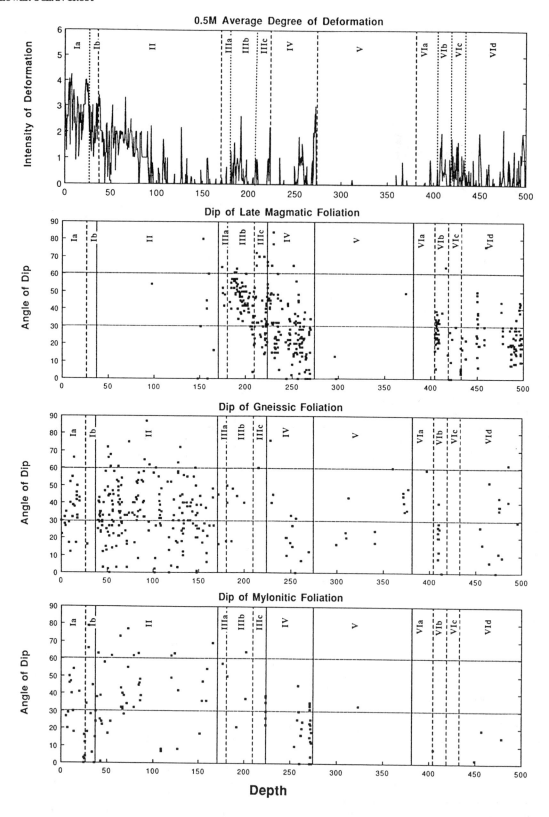

Fig. 10. Downhole structural logs (meters below seafloor) showing the average 0.5 m deformation intensity, angle of dip of the principal foliations and the divisions between the major structural zones.

coarse plagioclase and pyroxene grains in both hypersolidus and subsolidus deformed gabbros exhibit local fracture and pull-apart texture, in the former case the cracks between grain fragments may be filled by magmatic oxides, whereas in the latter they are often filled by amphibole [Natland et al., 1991; Dick et al., 1991a]. The oxide-bearing gabbros deformed under hypersolidus conditions may also exhibit a uniform penetrative deformation over long intervals, whereas the subsolidus fabrics are extremely variable. While both lamination and strong lineation are seen in the hypersolidus deformed gabbros, direct evidence for accompanying recrystallization is sometimes scarce. Instead some oxide-bearing gabbros may have an igneous texture, characterized by unstrained grains, interlocking grain boundaries, and sometimes subhedral mineral morphologies. This strongly suggests deformation while the gabbro was still partially molten [Cannat, 1991; Dick et al., 1991a], with the igneous textures reflecting post-tectonic crystallization of trapped intercumulus melt. Although hypersolidus and subsolidus deformed gabbros are often texturally distinct, a complete gradation exists. With increasing deformation intensity (probably reflecting increasing strain rate) hypersolidus and subsolidus mylonites can be virtually indistinguishable in hand specimen.

Subsolidus Deformation

Subsolidus brittle-ductile deformation was extensive at many levels, and is reported to have begun under granulite facies conditions (700-1000°C) with recrystallization and re-equilibration of the primary igneous phases [Stakes et al., 1991; Cannat, 1991]. Brittle-ductile deformation continued with the onset of hydrothermal alteration at about 720°C to mid-amphibolite grade (450°-500°C) [Vanko and Stakes, 1991]. Accompanying subsolidus brittle-ductile deformation was pervasive amphibolite facies hydrothermal alteration. This tectonically enhanced alteration affected the entire core to some degree, both as dynamic alteration in the shear zones, and as replacement of unrecrystallized primary igneous minerals in the undeformed sections. The overall degree of alteration greatly exceeds that generally reported in statically cooled layered intrusions like the Skaergaard Intrusion [e.g., Bird et al., 1986, 1988]. This is attributed to the imposed tensile stress during brittle-ductile deformation enhancing crack formation and seawater circulation throughout the section during hydrothermal alteration [Dick et al., 1991a]. Amphibolitization is most extensive in sheared and recrystallized gabbro, while many undeformed gabbros appear virtually unaltered, suggesting that shear zones strongly controlled hydrothermal circulation and alteration.

Amphibolite facies brittle-ductile deformation produced a major 30-m-thick section of amphibolite gneiss at the top of the hole, and a much smaller zone near the base of the hole. Smaller amphibolitized brittle-ductile shear zones are common, except for Units V and VIa where for 130 m the core is nearly devoid of brittle-ductile deformation. Where

the sense of shear could be determined from rolled porphyroclasts with tails (fish and snails), microfaults and cross foliations, it was normal in all but one case [Dick et al., 1991a]. These gabbros are often heavily recrystallized, and are characterized by deformed mineral grains, undulose extinction, grain boundary recrystallization, kinking of mineral grains, cataclastic textures, and extensive polygonalization and recrystallization fabrics.

Contacts between deformed and undeformed rock may be razor sharp, with undeformed igneous textures abutting ultramylonite, or they may be gradational from undeformed gabbro through weakly deformed rock to gneissic and mylonitic textures over a scale of 50 cm. Gneissic zones tend to be thicker than mylonitic zones, and as the overall degree of deformation drops in a section, deformation becomes less penetrative, with the thickness and number of the deformed zones decreasing sharply. At the same time, however, the intensity of deformation within a zone increases, with a higher percentage of mylonite.

At several levels in the core narrow 1-3 cm mylonite bands with opposing dip crosscut late-magmatic foliations. The bands have normal shear sense and are similar to mylonites in the amphibolite gneisses. They might represent: (1) relatively late faults antithetic to the principal plane of late-magmatic and subsolidus brittle-ductile deformation, (2) late outward-facing faults complementary to the inward-facing faults seen on rift valley walls - produced during block uplift at the inside corner high, (3) the result of a small ridge jump, after late magmatic deformation, reversing the spreading direction and dip of faulting.

Cannat and Paraiso [1991] attempted to reorient the foliations in the core using the stable magnetic declination which was assumed to be constant consistent with the uniform dip of this vector downhole. Different sets of structures downhole gave consistent dip orientations and suggest that while the early magmatic and granulite facies foliations have similar orientation, the amphibolite facies foliations are oblique to these. This suggests either a rotation of the principal strain axes during evolution of the section [Cannat and Paraiso, 1991], or a rotation of the earlier foliation due to faulting prior to formation of the latter. The precise orientation of the foliations cannot be determined, however, unless the azimuth of the stable paleomagnetic vector is known.

Brittle Deformation

Crosscutting amphibole veins, trondhjemitic intrusion breccias, and late hydrothermal breccias and veins are the most obvious manifestations of brittle deformation. Brittle deformation accompanying ductile deformation produced innumerable amphibole-filled cracks, crack networks, and veins which both crosscut and are crosscut by amphibolitized shear zones. Amphibole veins have a highly variable dip, averaging about 60° (Figure 11), at right angle and complementary to that of the predominant foliation in the amphibolites - consistent with their penecontemporaneous formation. Thus, the brittle-ductile

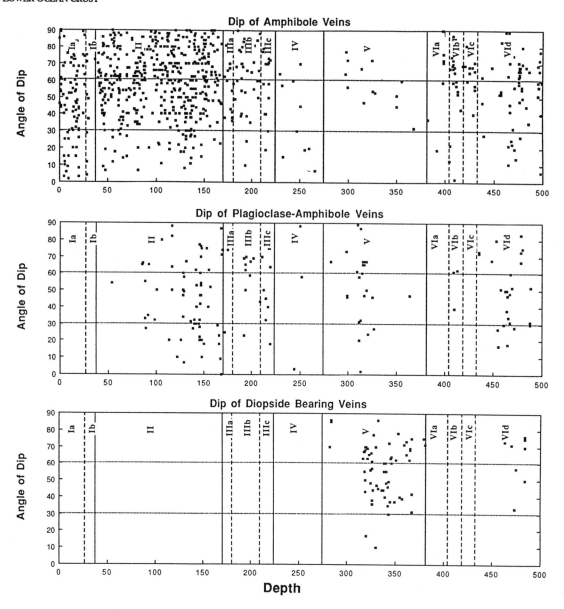

Fig. 11. (and opposite page) Downhole structural logs showing the angle of dip (inclination from horizontal) for Hole 735B hydrothermal veins. Horizontal lines show the divisions between the major lithostratigraphic units.

shear zones, while they are "ductile faults" [e.g., Ramsay and Graham, 1970] do not represent pure ductile flow, but rather complex heterogeneous deformation in the brittle-ductile transition zone. Cracking and vein formation occurred in lesser amounts in unsheared gabbros during this period, but continued there below middle amphibolite facies while it did not in the recrystallized rocks. Only the trondhjemites at the base of Unit II and III, have a significantly different dip from the other hydrothermal veins, perhaps because they were emplaced by igneous intrusion, rather than hydrofracturing under an imposed extensional stress. The trondhjemites and related felsic veins crosscut the late-magmatic foliation in the oxide-olivine gabbros in Units III and IV, but in some cases were also very slightly deformed during the late-magmatic deformation. The trondhjemites, however, are always deformed with the gabbro protolith where subsolidus amphibolite facies brittle-ductile deformation occurs.

Evidence of late brittle deformation is largely limited to breccias and crack networks associated with felsic veins in the middle amphibolite to greenschist facies. After cessation of brittle-ductile deformation in the middle amphibolite facies, vein formation virtually stopped in recrystallized gabbro, whereas it continued at progressively lower temperatures with the formation of veins in the relatively undeformed gabbros. The lack of

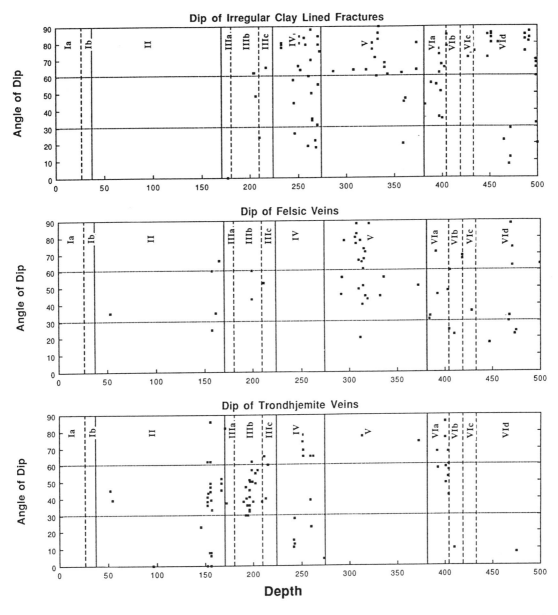

Dip of Irregular Clay Lined Fractures

Dip of Felsic Veins

Dip of Trondhjemite Veins

Depth

lower amphibolite and greenschist alteration and filling of late cracks in the amphibolites is attributed to their finer grain size (resulting in short microcrack length and lower stress concentration at crack tips to drive fracture growth), and low stored elastic thermal strain due to the annealing effect of recrystallization during deformation [Dick et al., 1991a]. At the end of brittle-ductile deformation, in the coarser-grained undeformed gabbro larger microcracks and thermal elastic strain, unrelieved by recrystallization, continued to drive cracking as cooling continued. In the recrystallized sections, however, dilational stress buildup for thermal cracking was insufficient over the remaining cooling interval for much failure and crack growth to occur.

Relatively pure amphibole veins are common, particularly between 100 to 210 m and locally at 471 m near the lowest zone of amphibolites (Figure 11). Some sheared amphibole

veins reopened and are lined with smectite, and others have free vuggy spaces lined with black, acicular lineated hornblende crystals lying in the plane of the vein. These can be used to determine slip-direction and the shear-sense using surface roughness (shingle structure). The shear sense is mostly normal and the slip-direction averages $48.5°\pm14.7°$ from the dip direction on the slip plane (Figure 12).

Morphologically, Site 735B is bounded east and west by steep slopes which suggest conjugate normal faults and simple vertical block uplift. This block is unlikely, therefore, to have been back-tilted away from the transform the minimum of 49° required to explain the average non dip-slip component of the amphibole lineations. Non dip-slip lineations, not created by later block rotations, occur on faults when both strike-slip and dip-slip components of

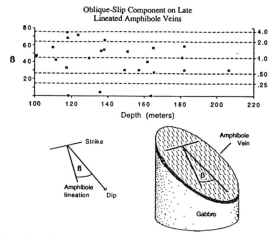

Fig. 12. Plot of oblique-slip on sheared amphibole veins crosscutting Hole 735B gabbros and gneisses showing the difference between the dip direction and the amphibole lineation β and depth in the hole. Dashed lines and scale on right give the ratio of strike-slip to dip-slip motion which is equal to the tangent of β. Cartoon shows a sheared vein with amphibole lineated at angle β oblique to the dip direction.

motion occur. This could arise where the median valley wall is at an angle to the rift valley trend and fault surfaces are oblique to the spreading direction. The ratio of strike-slip to dip-slip motion, however, is generally slightly greater than one and locally up to four. Thus, these late veins likely reflect a stress field different from that for simple extensional faulting and flow beneath the rift valley floor - one where principal stress directions have rotated.

Non dip-slip motion can also occur near ridge-transform intersections where fault orientation rotates with the stress field as the principal motion changes from normal to strike-slip [e.g., the oblique faults at the Kane ridge-transform intersection - Karson and Dick, 1983]. However, the amount of shear on these veins is small, and oblique to the principal foliation in the amphibolites. We suggest, then, that the veins were produce at the end of brittle-ductile deformation by small components of oblique shear and mineralization on joints during block uplift into the rift mountains at the ridge-transform intersection.

While striking in appearance, the amount hydrothermal breccia was overestimated in the Leg 118 Scientific Results, where it was based on preliminary shipboard core descriptions (437 m of rock described in 16 days by 12 different scientists). Relogging the breccias at the core repository, strictly defining them as fragmented rocks containing rotated clasts, showed that they are only a very minor component of the core. Felsic net vein complexes, without rotated clasts, of similar composition to the breccia matrix are more common, but still minor. The late hydrothermal breccias are generally undeformed. and we believe these late breccias, and associated vein networks represent local hydrothermal up-flow zones. Seventy open fractures greater than 0.5 cm width were seen in the hole with the borehole televiewer with an average strike subparallel to the transform and dipping steeply west-

southwest [Goldberg et al., 1991]. The hole is situated near the western wall of the wave-cut platform, and these fractures can be attributed to joints and faults associated with block uplift on this wall at the inside-corner high.

Thin late-stage smectite-lined subvertical cracks are present at many intervals in the core [Dick et al., 1991a], and often extend for a meter or more down the core. They differ from earlier veins as they are irregular features that break around mineral grains rather than through them, except where they have formed during reopening of earlier veins. They have a mean dip of about 70°C, and probably have a true subvertical mean-dip when drilling bias is accounted for. Generally absent in the highly recrystallized gneissic intervals, they are most common in the coarsely crystalline undeformed rocks near the base of the hole. The formation of the irregular cracks and the reopening of the sheared amphibolite veins is likely to have happened at the same time, and probably reflect late low-temperature release of stored elastic thermal strain during uplift, unloading and erosion of the wave cut platform.

TECTONIC STRATIGRAPHY

There are five major deformation zones based on intensity and character of deformation. Within these are a major and a minor interval of subsolidus brittle-ductile deformation, and two minor and one major intervals of late-magmatic brittle-ductile deformation (Figure 9b).

Zone 1

Zone 1, from 0 to 90 mbsf, consists of the largest subsolidus shear zone in Hole 735B. The gneissic and mylonitic foliations in Zone 1 dip about 35°, but show tremendous variability. The igneous stratigraphy shows a lithologic gradation through the zone, and the stratigraphy is thinned without any major discontinuity demonstrating that it is a ductile fault. The total amount of displacement across the zone cannot be determined; however, the rocks at the top of the zone are petrologically related to those at the base, suggesting a small displacement (<500 m) given how small we believe these intrusions are, and the position of the fault within a major fault block on the transverse ridge. The range in recrystallization temperatures from granulite to middle amphibolite facies discussed earlier, indicate that this was a long-lived fault, possibly the root of a major listric normal fault bounding the paleo rift valley beneath the SWIR.The numerous oxide-bearing horizons, as discussed earlier, demonstrate that there was earlier late magmatic ductile deformation in the zone overprinted by subsolidus deformation and amphibolitization.

Zone 2

At 90 mbsf the intensity of deformation drops abruptly and is only intermittent for the next 80 meters to the top of Unit IIIA (170 mbsf). This zone is chiefly comprised of weakly to undeformed olivine gabbro. The gneissic foliation in the intermittent deformed intervals varies widely, but has an average dip of about 35°. Amphibole-filled veins and cracks are predominantly oriented orthogonal to the foliation. The orientations and variability of the foliations and veins in Zone 2 are indistinguishable from those in Zone 1 despite the drop in intensity of deformation in this zone (Figure 10).

Zone 3

From 170 to 274 mbsf is a region of intense penetrative late-magmatic brittle-ductile deformation and transformation of an olivine gabbro protolith to ferrogabbro. Although large intervals in this zone are logged as undeformed, the entire section exhibits a weak penetrative deformation, lamination, and lineation. Local zones of intense deformation, locally mylonite at the bottom of the zone, may also reflect only late-magmatic deformation, but need more study [Cannat et al., 1991; Dick et al., 1991]. As the deformation log was based on stretching and recrystallization of mineral grains, suppressed during hypersolidus deformation, it does not fully reflect the amount of ductile deformation here, which appears to have been greatest near the base. The uniform development of the late magmatic foliation, and its systematic flattening or fanning downhole, is striking. This structure may be incomplete as the lower portion of the zone is obscured by alteration, brecciation, and poor recovery.

There is a remarkable correlation between rock composition and dip in Zone 3 (compare Figure 7 and 10). Where the foliation is flattest, and deformation most intense, the gabbro is the most iron-rich [Dick et al., 1991]. This can be explained if the foliation initially formed at an angle to the maximum resolved shear stress in a hypersolidus ductile fault, and, with on-going shear, was rotated into the principal plane of displacement at the center of the deformation zone. Ramsay and Graham [1970] described shear zones in metagabbros and granites with this geometry. This would mean that formation of ferrogabbro by late intercumulus melt flowing into the olivine gabbro in the shear zone would be most complete where the intensity of deformation and therefore the flow of melt were greatest.

Locally foliations in the ferrogabbros in Zone 3 with different orientations are juxtaposed and appear to crosscut. These are not igneous crossbeds, but reflect discrete episodes of deformation and melt-impregnation, most likely where various shear systems in a braided shear zone were operative at different times during a deformation cycle, leading to local crosscutting relationships. This supports an interpretation that this is a major long-lived late magmatic fault zone. Moreover, the relatively flat dips (0-30°) in the zone of most intense deformation, as compared to the higher dips in the gneissic intervals (Figure 10), is consistent with listric normal fault flattening downward as temperature and ductility increase.

Zone 4

Deformation Zone 4 extends approximately 116 m from the base of Unit 4 to the middle of Unit 6a, and is

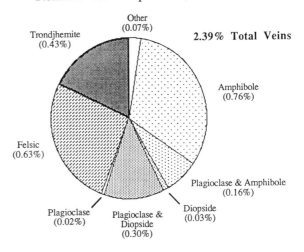

Fig. 13. Relative vein proportions, Hole 735B, visually estimated on a fragment by fragment basis for the entire core at the ODP Core Repository and then summed [Dick et al., 1991a].

characterized by the near absence of brittle-ductile deformation, either late or post-magmatic, and an absence of crosscutting late intrusives. Prominent in this horizon, however, is a thick hydrothermal breccia, locally with clasts in excess of 10 cm and a matrix of felsic hydrothermal cement consisting mostly of plagioclase, with some diopside, and an assortment of lower-temperature later hydrothermal phases [Stakes et al., 1991]. Numerous hydrothermal veins are also present including a series of apparently related compound plagioclase-diopside and diopside veins, and less abundant greenschist to zeolite facies assemblages associated with the higher-temperature breccias and veins.

Zone 5

From 390 to 50.4 mbsf is a very complex zone exhibiting both subsolidus and late magmatic phases of ductile deformation, one frequently superimposed on the other. Numerous late microgabbro and troctolitic intrusions also occur, coarsening downward, and their emplacement may overlap the late magmatic deformation. The amphibolite facies deformation becomes sufficiently intense in the lower portion of Zone 5 to constitute a secondary ductile fault, which must significantly thin the local igneous stratigraphy. The dip of the subsolidus gneissic and mylonitic foliation in Zone 5 is highly variable and similar to that in Zones 1 and 2, averaging about 30°.

ALTERATION

Five major styles of alteration or metamorphism are present in the rocks of Hole 735B: 1) high temperature metamorphism associated with brittle-ductile deformation; 2) in-situ replacement during which olivine and some orthopyroxene were partly to completely replaced by Fe-

Mg amphiboles, talc and secondary magnetite in the unrecrystallized gabbro; 3) infilling of hydrothermal veins formed during brittle deformation; 4) deposition of green smectite in subvertical cracks; 5) late stage replacement of olivine and some pyroxene grains by reddish iron oxides, carbonate and clay minerals.

High Temperature Metamorphism

Subsolidus high temperature deformation and associated alteration is most pronounced in Unit 1. It is also present in Unit 2 and prominent in local thin shear zones which occur throughout the core except for an interval extending from the top of Unit 4 (223.53 m) to the base of Unit 6B (419.28 m). Two major mineral assemblages have been recognized, one formed at low water/rock ratios, the other at high water/rock ratios, although there are gradations between them [Stakes et al., 1991; Vanko and Stakes, 1991].

The assemblage formed at low water/rock ratios consists chiefly of igneous minerals that have been partially to completely recrystallized and re-equilibrated to somewhat lower P-T conditions. These rocks consist chiefly of plagioclase, clinopyroxene, orthopyroxene with rare olivine, brown amphibole and iron oxides. Variations in mineral assemblages largely reflect the compositions of the protoliths (chiefly gabbronorite in Unit 1 and olivine gabbro in Unit 2). Round to ovoid porphyroclasts of igneous minerals are surrounded by neoblasts of plagioclase, pyroxene, olivine and brown amphibole. Most neoblasts are compositionally similar to adjacent porphyroclasts although plagioclase may be slightly more sodic and clinopyroxene slightly more diopidic. Orthopyroxene typically occurs as small crushed crystals around porphyroclasts rather than as neoblasts. Small polygonal neoblasts of red-brown hornblende are interpreted as being primary igneous phases [Stakes et al., 1991].

With increasing water/rock ratio, pyroxene is progressively replaced by amphibole, forming an amphibolite gneiss. Typical mineral assemblages in these rocks include plagioclase, iron oxide, green and brown amphibole and some brown mica. Some samples contain relict porphyrocasts of pyroxene, but in most cases these are also replaced by amphibole. Many of these foliated rocks also contain amphibole veins which cross-cut or are cut by the foliation. The vein amphibole is similar in composition to that defining the foliation and both are aluminous magnesio-hornblende. Some of the aluminous magnesio-hornblendes contain significant amounts of chlorine [a few tens of a percent up to 1.5 wt.%; Vanko and Stakes, 1991] suggesting formation by seawater/rock interaction [Batiza and Vanko, 1985; Vanko, 1986; Mevel, 1988]. A few amphibolite gneisses also contain small amounts of actinolitic hornblende and intermediate to sodic plagioclase. The actinolitic hornblende commonly replaces porphyroclasts or rims previously formed magnesio-hornblende. Stakes et al. [1991] interpret this

assemblage as marking the terminal stage of hydrous ductile deformation.

The earliest stage of ductile deformation and anhydrous metamorphism, marked by recrystallization of orthopyroxene, may have occurred within the granulite facies. Augite porphyroclasts in these rocks still have magmatic oxygen isotopes with $\delta^{18}O$ values between 5.08 and 5.4 per mil [Vanko and Stakes, 1991] indicating little or no reaction with seawater: consistent with a low water/rock ratio. However, as water penetrated into the system water/rock ratios increased and the temperatures of metamorphism decreased to amphibolite grade. Actinolitic hornblende and actinolite in these rocks have $\delta^{18}O$ values as low as 1.6 per mil and much of the plagioclase is also isotopically depleted, indicating considerable exchange with seawater [Vanko and Stakes, 1991]. Temperature estimates of the amphibolite grade metamorphism based on mineral compositional data range from about 590°C to 720°C and cluster round 640°C [Vanko and Stakes, 1991]. Hornblende-plagioclase veins in these rocks yield temperatures of about 640°C, whereas later-stage alteration associated with the brecciated horizons took place at temperatures less than 500°C [Vanko and Stakes, 1991].

High-temperature dynamic hydrothermal alteration and pseudomorphing of *undeformed* gabbro mineralogy also occurred throughout the core. Partial to complete replacement of olivine and rarely orthopyroxene is the most widespread and obvious form of this alteration. Much, and possibly all of this is not true "static-alteration", as the local amphibolitized shear zones provide evidence of substantial externally imposed stress on the entire section during alteration. Even though there may be little evidence of deformation in thin section, an externally imposed stress may greatly enhance cracking, permeability, and water-rock ratio over that occurring as a result of simple thermal dilation during static cooling.

Olivine grains commonly exhibit at least some marginal alteration, chiefly to colorless amphibole, talc and secondary magnetite. In other cases, the primary minerals are completely pseudomorphed by secondary minerals. Tremolite is the most common amphibole in the pseudomorphs although cummingtonite and anthophyllite are also present. The amphiboles commonly form networks of randomly oriented, tabular grains or bands of grains oriented either parallel or perpendicular to the original boundary of the primary crystal. In the latter case, there is frequently zonation in the olivine pseudomorphs, with mixtures of tremolite and talc in the cores and bands of tremolite, cummingtonite or anthophyllite on the rim. The secondary magnetite occurs in minute grains, ordinarily arranged in concentric bands parallel to the outline of the original crystal.

Neither origin or timing of alteration in unrecrystallized gabbro, however, is entirely clear. Because it largely affects olivine (and to a lesser extent orthopyroxene), its occurrence is largely related to variations in primary mineralogy. In general, the distribution of this type of alteration does not reflect vein density or the proximity to

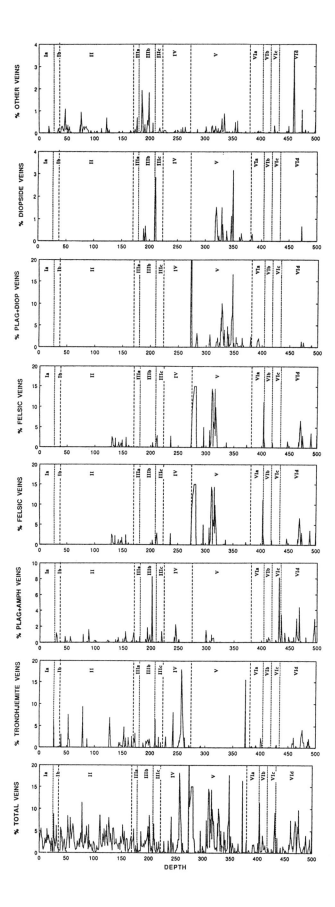

Fig. 14. Downhole log of Hole 735B showing 1 meter average density for Hole 735B vein assemblages, and the 1 meter running-average for deformation intensity. Horizontal lines show the divisions between the major lithostratigraphic units [Dick et al., 1991a]. "Other" refers to low-temperature veins including carbonate, clay, and zeolites.

specific veins. Thus, it may be, in part, deuteric, i.e., olivine and orthopyroxene reacting with late-stage magmatic fluids. Alternatively, it may mark seawater influx into cooling rock along minute cracks and grain boundaries. Where pyroxene porphyroclasts have been pseudomorphed by actinolite or actinolitic hornblende in the ductilely deformed zones, the amphibole has $\delta^{18}O$ values as low as 1.6 per mil, indicating extensive exchange with seawater [Vanko and Stakes, 1991] and affirming the strong control on seawater circulation by shear zones.

The replacement of the olivine in the unrecrystallized gabbro may have been penecontemporaneous with or slightly later than the brittle-ductile deformation and amphibolite facies metamorphism. It is best preserved in the undeformed zones but some olivine pseudomorphs are present in the foliated rocks. Where these occur, the pseudomorphs are elongated parallel to the foliation but the individual secondary minerals are not. This suggests that the pseudomorphs may have formed by true static alteration of deformed olivine grains after deformation ceased. If earlier static alteration occurred it has apparently been overprinted by the amphibole-facies metamorphism.

Other types of replacement alteration of unrecrystallized gabbro are locally present but volumetrically insignificant. Small amounts of pargasitic amphibole and phlogopite occur along some olivine-plagioclase grain boundaries. Minute amounts of clinozoisite locally replace feldspar grains, and in some amphibole gneisses, plagioclase is rimmed by actinolite and minor chlorite and clinopyroxene porphyroclasts may be partly replaced by actinolite or actinolitic hornblende [Stakes et al., 1991]. Based on the dynamic analysis of the 735B section, alteration below middle amphibolite grade took place under true static conditions after ductile deformation ceased during uplift of the gabbro massif into the rift mountains.

Filling of Hydrothermal Veins

Both hydrothermal and trondhjemite veins are ubiquitous in Hole 735B (Figure 13), making up 2.4% of the core [Dick et al., 1991a]. These veins have a mean dip of about 59° (Figure 11) and exhibit highly variable mineral assemblages. These hydrothermal veins occur throughout the core except in the highly brecciated zone near the base of Unit 4. Overall, veins occur rather uniformly down the hole, however, each variety has a unique stratigraphic distribution (Figure 14). The tabulation in Figure 13 is somewhat misleading because some veins exhibit sequential assemblages reflecting precipitation of successively lower temperature minerals as the hydrothermal fluids cooled down. This is particularly true of the plagioclase-diopside and felsic veins, many of which are strongly zoned and some of which contain narrow bands of zeolite, carbonate or clay in their centers.

Amphibole veins

Almost a third of the veins in Hole 735B are monomineralic amphibole veins. These are concentrated in or near brittle-ductile deformation zones, and are most abundant in the upper 150 m of the core. The amphibole veins take several forms, and are both cut by and cut the gneissic and mylonitic foliation. Amphibole veins generally do not cut the foliation in the intensely deformed gneisses in the upper 40 m. Any early cross-cutting veins apparently were rotated into the plane of foliation. The reverse, however, is true in the smaller gneissic zones in Unit VI. In less deformed rocks, amphibole veins typically filled 0.5-3 mm wide long cracks oriented sub-perpendicular to foliation. Amphibole-filled microcracks also occur in porphyroblasts and in more continuous cracks that extend across porphyroblast and into the enclosing recrystallized matrix. The monomineralic amphibole veins are typically filled with green or greenish-brown magnesio-hornblende. Some relatively large euhedral crystals are compositionally zoned [Stakes et al, 1991; Robinson, unpublished data]. As pointed out earlier, however, there is little difference in composition between vein hornblende and that replacing primary phases.

The close association of hornblende veins and brittle-ductile deformation indicates that the veins are closely related to the amphibolite facies metamorphism. Since the veins both cross-cut and are cut by the foliation in these rocks, vein formation must have overlapped with and likely extended somewhat beyond the period of deformation. This interpretation is supported by the overlap in estimated temperatures of formation for metamorphic amphibole (590° to 720°C) and vein hornblende (about 640°C) [Vanko and Stakes, 1991]. In addition to pure amphibole veins, numerous polymineralic veins contain some amphibole in the lower parts of the core. In these, amphibole is relatively minor and typically rims other minerals such as diopside.

Trondhjemite Veins

Presumably igneous trondhjemite veins and intrusion breccias constitute slightly more than a third of all veins. In the uppermost amphibolite gneiss zone, where the trondhjemites are transposed into the plane of foliation, however, they are difficult to distinguish from other felsic layers and their total abundance is not known. A few ptygmatic trondhjemite veins there are morphologically distinct, and likely formed by anatexis during late intrusion and reheating of amphibolite. This suggests formation of the section involved multiple cycles of intrusion.

Felsic and Related Veins

Of more problematic origin is suite of felsic veins with variable mineral assemblages. These range from what appear to be highly altered trondhjemites to clearly hydrothermal monomineralic plagioclase veins. Their mineral assemblages include sodic-plagioclase (oligoclase), diopside, amphibole, epidote, sphene, magnetite, and zircon. Some of these are essentially monomineralic plagioclase veins, but most are compound veins with obvious disequilibrium assemblages. They grade into plagioclase-diopside, plagioclase-amphibole and even small monomineralic diopside veins. While Dick et al.

[1991a] distinguished these different varieties, they are lumped simply to emphasize their overlapping or related paragenesis. A complete gradation exists, likely due to ongoing late hydrothermal alteration and overprinting of earlier veins. The typical felsic vein occurs in brecciated or highly fractured zones of the core and consists largely of sodic plagioclase and accessory minerals. These veins and net-vein complexes generally are overprinted by lower temperature greenschist and zeolite facies assemblages.

Diopside-bearing veins are often compound, often with coarse green diopside crystals along their margins, followed by tabular crystals of epidote, and then by aggregates of small anhedral feldspar grains at the center. Elsewhere, spectacular centimeter thick veins have plagioclase lining their walls and coarse green diopside at their center. Some diopside crystals are euhedral and show strong zoning of iron and magnesium; others are barrel-shaped or anhedral and relatively uniform in composition. Many diopside crystals are rimmed with green actinolite and actinolite may occur on the vein walls where they cut primary pyroxene grains in the host rocks. Scattered crystals of sphene, ilmenite, pyrite, allanite and biotite are commonly present in the central parts of the veins. Pale-green, brown or colorless chlorite is also locally present and appears to either replace plagioclase or fill irregular cavities in the veins. In many of the veins diopside and amphibole appear to have formed chiefly by epitaxial growth inward from the walls of fractures on augite. Replacement of primary wall-rock minerals adjacent to the vein sometimes also occurs, particularly augite by hornblende.

Fluid inclusions in vein diopsides homogenize at temperatures ranging from 321° to 420°C, but most are between about 380°C [Vanko and Stakes, 1991]. Inclusions in vein plagioclase from the same specimens are slightly less saline than those in diopside and homogenize at distinctly lower temperatures, generally between 207° and 285°C [Vanko and Stakes, 1991]. Many of these veins also contain very low temperature secondary minerals such as zeolites, clay minerals or carbonates. In some cases, these occur as irregular patches apparently replacing pre-existing minerals, particularly feldspar; in others, low temperature minerals fill cracks in reopened early veins.

Stakes et al. [1991] and Dick et al. [1991a] argue that some of these veins may represent magmatic veins extensively modified by hydrothermal fluids to produce such minerals as epidote and chlorite. However, Vanko and Stakes [1991] found that the oxygen isotope characteristics of most felsic veins suggest a hydrothermal origin. Only zircon-bearing veins have near magmatic isotopic compositions whereas others have delta $\delta^{18}O$ values that suggest formation from seawater.

Distinguishing felsic veins from the trondhjemites is often very difficult or impossible, and it is reasonable to suspect that many of the felsic vein assemblages formed by alteration and overgrowth of the former. We suggest, however, that only those felsic veins containing quartz and/or zircon are likely to have formed from late magmatic

fluids. Most, if not all, of the diopside-bearing veins have textures, mineralogies and isotopic compositions that suggest a hydrothermal origin. Such an interpretation is supported by the temperature estimate of 505°-480°C for formation of the vein diopside [Vanko and Stakes, 1991].

Lower middle amphibolite facies veins, including late felsic, plagioclase-diopside, diopside, and veins with secondary overprinted greenschist and zeolite facies assemblages are not ductilely deformed and clearly postdate ductile deformation. We believe these and the undeformed hydrothermal breccias, and associated vein networks largely represent minor local hydrothermal up-flow zones. These veins appear to have formed under true static conditions during uplift of the massif into the rift mountains of the SWIR. Early felsic veins and some amphibole-plagioclase veins may have been involved in the ductile deformation, and their formation could overlap the high-temperature ductile deformation.

The precipitation of veins, and matrix alteration drop exponentially with the end of brittle-ductile deformation and the initiation of static conditions and block uplift in the middle amphibolite facies. From this point, the extent and nature of alteration closely resembles that in the statically cooled Skaergaard Intrusion described by Bird et al. [1986, 1988]. The appearance of diopside as the principal mafic vein phase also suggests a shift to Skaergaard-like conditions. The precipitation of diopside rather than amphibole veins at Skaergaard is attributed to the low permeability of the section and consequently highly reacted hydrothermal fluids. Similarly, alteration of 735B gabbros in the middle amphibolite facies involved formation of talc, amphibole and sodic plagioclase - reactions releasing calcium and consuming magnesium raising fluid Ca/Mg. Thus, with a shift to static conditions, lower permeability, and greater extents of fluid reaction in the 735B gabbros, diopside (Ca/Mg = 0.9) eventually replaced amphibole (Ca/Mg = 0.3-0.5) as the stable phase precipitating in hydrothermal upflow zones.

Smectite-Lined Fractures

The late irregular subvertical fractures are clay lined, containing only green smectite, suggesting crack formation and infilling at temperatures less than about 50°C. In addition, many sheared amphibole veins with amphibole oriented sub-parallel to the vein walls re-opened and are also lined with smectite along the fracture [Robinson et al., 1991].

Late Stage Oxidation

The last recognizable stage of alteration involves local oxidative alteration of olivine and orthopyroxene. In hand specimen this is easily recognizable by the presence of red to reddish-brown clots in an otherwise light gray to light brown rock. The altered zones occur at definite intervals in the core and their distribution presumably reflects access of oxygenated seawater along cracks or fractures. The major zones of alteration occur in cores 1-8, 9-10, 37-39, 57, 63-67 and 70-73. However, the degree of alteration varies

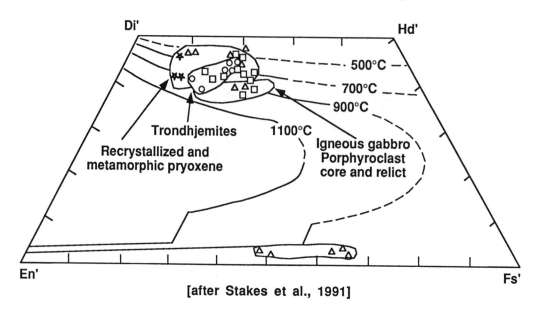

A

735B Relict Igneous and Metamorphic Pyroxenes

[after Stakes et al., 1991]

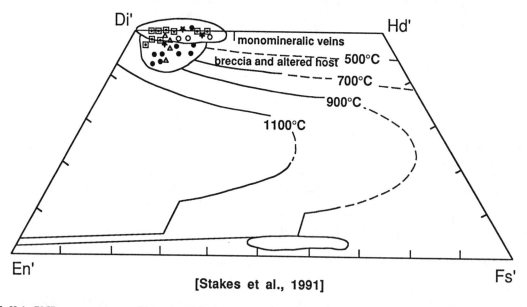

B

735B Hydrothermal Veins and Breccia Pyroxenes

[Stakes et al., 1991]

Fig. 15. Hole 735B pyroxene compositions plotted in the pyroxene quadrilateral, modified from Stakes et al. [1991]. A. Relict igneous pyroxene, trondhjemite pyroxenes, and recrystallized igneous and metamorphic pyroxene. B. Hydrothermal veins and breccia. Temperature contours after [Bird et al., 1986] show approximate temperature relationships.

considerably within these intervals and the most intense alteration is typically concentrated along relatively narrow bands.

In thin section, altered pyroxene grains are reddened and partly replaced by oxides and clay minerals, but the original mineral is still readily identifiable. Olivine, on the other hand, is often completely pseudomorphed by secondary minerals, resulting in opaque clots. Where the olivine was partly replaced during the earlier alteration, the opaque clots are rimmed by coronas of talc, tremolite and secondary magnetite. In a few cases, the olivine is replaced by green-brown clay minerals (smectite) or calcite as well as ferric oxides. Calcite veinlets are common in many of the oxidized zones and these presumably acted as the channelways along which seawater penetrated into the rock. Most of this alteration probably took place at temperatures close to ambient seawater.

Chemical Effects of Alteration

Although there have been no systematic studies of the chemical effects of alteration in Hole 735B, some conclusions can be drawn from the shipboard geochemical data [Robinson, von Herzen et al., 1989, Robinson et al., 1991]. A significant increase in H_2O is apparent in the upper parts of the core in lithologic Units 1 and 2. Although variable, water contents of these rocks range up to 2.4 wt.%, significantly above the average background value of about 0.5 wt.%. Lower in the core, where deformation is limited, the rocks are significantly less hydrated. The highest recorded water content is about 1.4 wt.%, and these values occur only in narrow shear zones which acted as
conduits for water flow. Some hydration is recorded in the veins which contain such minerals as amphibole, epidote, chlorite, zeolite and clay minerals but the veins make up only about 2.5% of the entire core. Thus, their influence on bulk rock compositions is limited. The hydration reflects an influx of water during brittle-ductile deformation and amphibolite facies metamorphism. Relatively high Cl contents in some amphiboles indicates formation from seawater-derived fluids. Many of the undeformed gabbros appear pristine in thin section at first glance, entirely lacking amphibole and greenschist facies alteration. Water contents in these rocks is rarely less than 0.3%, however, which suggests more extensive low-temperature alteration than previously supposed - probably finely dispersed clay minerals along grain boundaries and within plagioclase.

Carbon dioxide and K_2O contents are also considerably elevated above background levels in the upper 150 m of core. Background levels of CO_2 are less than 0.1 wt.% and maximum values range up to 0.35 wt.%. K_2O shows a small but steady decrease downhole from an average of about 0.08 wt.% at the top to about 0.04 wt.% at the base. Higher values, up to 0.36 wt.% occur in the upper, hydrated section and sporadically in the lower parts of the core. The relatively high CO_2 and K_2O contents in the upper parts of the core reflect seawater-rock interaction.

Ferric/ferrous iron ratios range from a low of about 0.05 to a high of about 0.46. The highest values occur in Unit 4, the sequence of Fe-Ti oxide gabbro. Since the rocks of this interval are not hydrated and only sporadically enriched in CO_2 or K_2O, the high ferric/ferrous ratios certainly reflect original magmatic conditions and the abundance of primary titanomagnetite and ilmenite. In some cases, the elevated ratios in the upper part of the core, and sporadically elsewhere could reflect late-stage oxidative alteration where oxygenated seawater reacted to form olivine and orthopyroxene pseudomorphs of ferric oxides, clay minerals and carbonates.

It is interesting to note that although the rocks of lithologic Unit 1 are enriched in H_2O, CO_2, K_2O and ferric oxides, individual samples show wide variations in these alteration indices. As previously noted for sea floor basalts [e.g., Robinson, et al., 1977], the alteration-sensitive oxides do not necessarily vary sympathetically but rather follow quite different trends. Rocks may be very hydrated but show little if any enrichment in CO_2, K_2O or ferric oxide. Conversely, enrichment in CO_2, K_2O and ferric oxide can occur without significant hydration. This confirms that alteration in these rocks is not pervasive and that many micro domains are present, each characterized by fluids of specific compositions.

The systematic and progressive change in alteration conditions from high to low temperature are well illustrated by the progressive change in the composition of clinopyroxene towards increasing wollastonite content with decreasing temperature [Stakes et al., 1991; Figure 15]. Porphyroclast augite retains an essentially igneous composition only slightly modified by thermal re-equilibration. The anhydrous recrystallized pyroxene has a range of composition, extending from that of the porphyroclasts down to highly calcic compositions which overlap the lower temperature igneous diopsides found in trondhjemites and in hydrothermal felsic veins and breccias. Nearly pure diopside in monomineralic veins is essentially identical in composition to late low-temperature hydrothermal veins in the Skaergaard Intrusion [e.g., Bird et al., 1986, 1988].

Alteration of plagioclase to more sodic compositions is widespread. Although often referred to as albitization, due to the cloudy white appearance of the more sodic plagioclase the composition is generally greater than An_{15}, however, consistent with largely amphibolite facies alteration [Stakes et al., 1991]. As with clinopyroxene, the composition of the plagioclase reflects a range of alteration temperatures, and extends from igneous compositions where the plagioclase is recrystallized under anhydrous (possibly granulite) conditions, to nearly pure albite where plagioclase has been replaced under greenschist facies conditions in a mylonite. Relatively sodic plagioclase is also found as a common vein mineral, sometimes with a halo of albitized igneous anorthite adjacent to the vein.

Conclusions

The scientific results of Hole 735B bring a new perspective to the understanding of the composition, structure and formation of the lower ocean crust. While numerous past proposals for drilling a representative section of the ocean crust emphasized the importance of a tectonically undisrupted section of the ocean crust (a section preserving intact igneous stratigraphy, free of internal disruption and foreshortening due to faulting and extension), such a section is not likely to exist. The ocean ridges are not only the zone of crustal accretion, but of continuous brittle-ductile deformation and lithospheric necking. While deformation, extension, and hydrothermal alteration are omnipresent with magmatism at fast-spreading ridges, where magmatism is ephemeral as at slow-spreading ridges, they are probably the only continuous processes. What is seen in the Hole 735B section is entirely consistent with this, and demonstrates that tectonic processes and alteration are not merely superimposed on an igneous stratigraphy formed at an ocean ridge, but are an integral part of its development.

Igneous Petrogenesis

The 735B gabbros formed as a series of ephemeral, short-lived, intrusions. Their size and shape are poorly known from a single vertical drill hole, but individual cumulate bodies range from less than 0.5 m pods of crosscutting microgabbro, to the large 400 m olivine gabbro unit. The longevity of these bodies is not well constrained, other than by the length of time it takes for crust to form and spread out of the rift valley. The present day inner rift valleys, consisting of axial highs and flanking deeps are 6.7 and 4.3 km wide east and west of the Atlantis II F.Z. At a spreading rate of 8 mm/yr, it would take about 400,000 years for crust to form and spread from the midpoint of the valley to the flanking walls, where it would be exposed by a detachment fault on the inside-corner high of the paleoridge-transform intersection. This period would encompass all igneous intrusion, metamorphism, and deformation activity up to the end of ductile deformation (Figure18). Subsequent uplift to the crest of the inside-corner high, which presently is situated 6 to 8 km from the present day neovolcanic zones, would take an additional 400 to 600 thousand years, by which time the section would have likely cooled to near ambient sea floor temperatures.

The overall section consists of a single 400 m olivine gabbro body which apparently intruded a gabbronorite of unknown size, and was intruded from below by a troctolite, again of unknown size. In addition, the body was reintruded by numerous small batches of magma which worked their way up through the crystallizing pile. The pegmatoidal textures and lack of magmatic sediments, and isotropic textures of the olivine gabbro all suggest fairly rapid initial crystallization of the different intrusives to form a crystal mush. It is likely, given the complex igneous stratigraphy, and extensive re-intrusion while the olivine gabbro was still partially molten, that the initial rapid crystallization was followed by a longer period of cooling to the solidus.

Synkinematic differentiation of the intrusions into deformed oxide olivine gabbro and undeformed oxide-free olivine gabbro occurred once the mush was sufficiently rigid to sustain a yield stress and the master faults bounding the rift valley propagated downward into the semi-solid plutonic section. The source of the late magmatic liquid which intruded and impregnated the shear zones at all scales is believed to be intercumulus liquid which moved into the ductile faults due to enhanced permeability created during

shearing. The latter phenomenon, and the subsequent movement of fluid into and along shear zones is well known and documented for hydrothermal fluids in the literature on ore deposits, but is poorly understood mechanically. Individual bands of ferrogabbro in the 735B section may have been impregnated by liquid derived from the adjacent wall rocks, or by intercumulus liquid tapped by the shear zone deeper in the section. In some instances, a good argument can be made for the latter case for the Unit III and IV gabbros, where chemical composition and extent of shearing are directly related.

Reintrusion of the olivine gabbro by small batches of melt was an ongoing process (Figure 16). These "microgabbros" are cumulates, and have a wide range of composition, from troctolite to oxide gabbronorite. Their morphology and contact relationships with an absence of chill zones and sutured grain boundaries across the contacts all suggest intrusion while the olivine gabbro was still partially molten. These micro-intrusions grade into patchy pegmatoidal gabbros which are believed to represent reintrusion of the crystal mush before it was sufficiently solidified for intrusive contacts to form, but at a sufficient viscosity to inhibit rapid mixing [Figure 17, Dick et al., 1991a].

Local anatexis and the formation of trondhjemite in the amphibolites suggest that there may have been more than one cycle of intrusion and alteration, providing additional evidence for the waxing and waning of magmatic activity even over the short time period allowable for crustal formation at an ocean ridge. Thus it is likely that the brittle-ductile transition migrated both up and down during formation of this section. It would therefore seem unreasonable to postulate even a steady-state crystal mush zone at the level of the 735B section beneath the rift valley floor.

The irregular contacts and the evidence for partially digested xenoliths in the microgabbros all indicate that the melts from which they crystallized underwent extensive chemical interaction by assimilation and reprecipitation as they moved upward through the crust. This may explain the broad range of microgabbro compositions, and be a petrologically significant igneous process for MORB. Models for formation of MORB based on fractional crystallization and basalt phase equilibria alone, then, may lead to erroneous conclusions, with respect to depth of origin, depth of fractionation and the crystallization sequence.

Lower crust melt-wall rock interaction, may help to explain why MORB glasses from individual ridge segments plot near a single liquidus trend in many composition projections [e.g., Bryan and Dick, 1982; Dick et al., 1984; Klein and Langmuir, 1987]. These liquidus trends are not fixed, but shift systematically along ridges with respect to mantle hotspots and other geophysical anomalies, reflecting variations in degree of mantle melting, initial temperature, initial composition, and depth of melt segregation [Dick et al., 1984; Klein and Langmuir, 1987; Dick, 1989; Klein and Langmuir, 1989]. At the same time,

Evolution of the Layer 3 Section at Site 735B

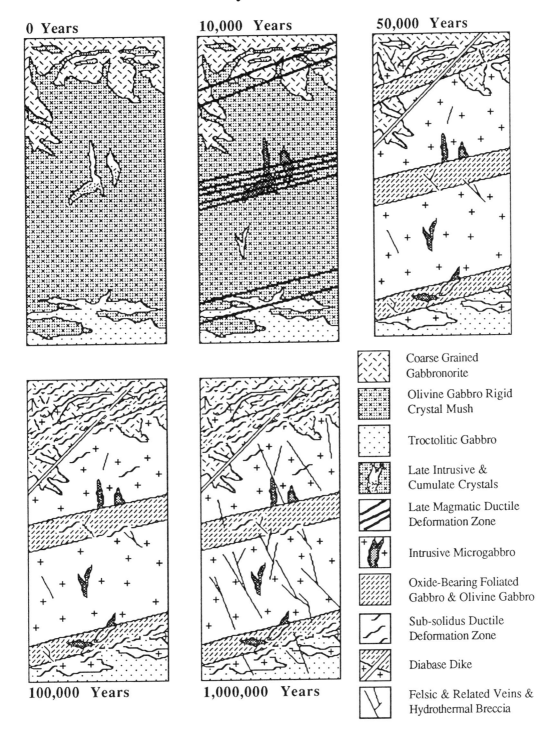

Fig. 16. Sequential cross-sections showing the geologic evolution of Hole 735B [Dick et al., 1991a]. Time scale is a very rough estimate, broadly constrained by spreading rate.

Intrusive Relationships In Hole 735B

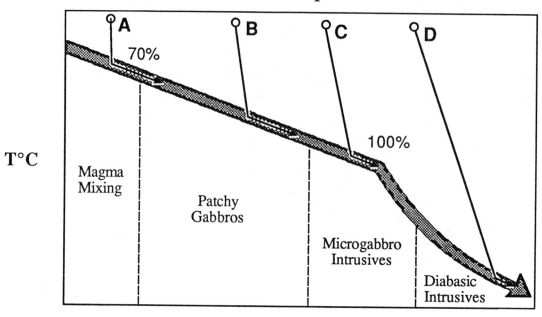

Time

Fig. 17. Cartoon showing possible textural relationships in the Hole 735B gabbros due to the periodic intrusion of new melt into a solidifying gabbro body [Dick et al., 1991a]. Shown are four batches of melt, each intruded at roughly the same initial temperature into a slowly crystallizing pluton. Each melt batch undergoes a period of rapid cooling and nucleation of new crystals until it reaches the temperature of the crystallizing pluton, at which point it solidifies with the pluton. Shaded arrow shows the cooling curve for the pluton, with an inflection at the point at which it is solidified and hydrofracting and the circulation of seawater enhances the rate of cooling (the point at which the cooling curve is dominated by advective rather than conductive heat transfer). The boundary between different textural fields is only generally indicated, with the boundary between microgabbros and diabase intrusive taken to be the point at which undercooling is sufficient to form intrusive chilled margins (presumably at a temperature close to the solidus of the pluton).

isotopic and incompatible trace element ratios indicate an underlying heterogeneous mantle source and melt input into the crust [e.g., le Roex et al., 1984]. Where melts migrating through the crust assimilate and reprecipitate gabbroic cumulates, their major element composition may effectively be buffered close to a single low pressure plagioclase-olivine-pyroxene pseudocotectic reflecting the overall bulk composition of the section.

In summary, the picture of the lower ocean crust beneath a slow spreading ridge drawn from Hole 735B is one of ephemeral magma chambers and longer-lived, semi-rigid crystal mush zones, with intervening periods where the crust may be entirely solidified. Overall, deformation and shearing are integral to the development of the igneous stratigraphy, unlike continental intrusions emplaced in static environments. A major consideration for future work on the genesis of MORB arising from this work should be a more concerted effort to understand the extent and implications of assimilation processes on models for ridge basalt formation.

Subsolidus Deformation and Alteration

As the rocks of Hole 735B crystallized and cooled they underwent subsolidus high-temperature, brittle-ductile deformation. This produced strongly foliated rock with porphyroclastic and mylonitic textures. Where temperatures were high (650°-750°C) and water-rock ratios low, the primary igneous minerals were extensively recrystallized to neoblasts of approximately the same composition. Where water-rock ratios were significantly higher, and temperatures probably somewhat lower (600°-650°C), the original minerals were partly to completely replaced by brown and green hornblende to form amphibolite gneisses. Although this deformation and metamorphism is concentrate in the upper 150 m of the core, narrow shear zones also occur at several intervals deeper in the section. These shear zones strongly controlled circulation of seawater and localized alteration above the middle amphibolite facies.

Because the brittle-ductile deformation and associated amphibolite facies metamorphism took place at high temperatures it must have occurred shortly after the rocks crystallized beneath the ridge axis. We believe that the high temperature deformation reflects extension of listric normal faulting along the rift valley walls down into the brittle-ductile transition. Supporting evidence for this is provided by Pariso et al. [1991] who found the magnetic inclination of the core too steep for this latitude, and cannot be explained by secular wandering of the geomagnetic pole. This suggests rotation of the section by up to 18°, consistent with listric faulting. Moreover, the uniform inclination of the magnetic vector downhole indicates block uplift with no tectonic disruption below the Curie point, suggesting that Hole 735B recovered a largely intact metamorphic stratigraphy.

The exact nature, development, relation to high angle listric normal faulting, and geometry of the detachment structures on which the gabbroic rocks are unroofed and uplifted to the seafloor at ridge-transform intersections are not known [compare the various sections of Dick et al., 1981, 1991b; Karson et al., 1990; Mevel et al., 1991; Tucholke and Lin, in preparation]. The ductile fault in deformation zone could represent part of such a detachment structure, rather than extension of a simple listric fault into the brittle-ductile transition. Lying within a major tectonic block, however, and given the cut off of alteration and deformation in the middle amphibolite facies, this fault cannot have been active at a shallow level. Thus, motion on this section of the fault must have been displaced by an inward imbrication if it is part of a detachment structure before it reached shallow level. This is consistent with the uniform magnetic inclination within and below the fault zone which suggests no significant fault motion below the Curie point.

The ductile deformation and accompanying alteration are believed to have occurred while the 735B section was beneath the inner rift valley floor. Currently, the rift valley on this portion of the SWIR is about 8 km wide [Dick et al., 1991b] and the half-spreading rate is about 8 mm/yr [Sclater et al., 1978]. If the same conditions prevailed at 11-12 Ma, it would take a segment of newly formed crust about 400,000 years to migrate out of the inner rift into the flanking mountains (Figure 2). Given the loose constraints of a simplified plate reconstruction, Dick et al [1991b] suggest that it might have taken on the order of 50,000 to 100,000 years for the gabbro section to form and cool into the upper amphibolite facies, where subsolidus deformation would dominate. These shear zones allowed deep penetration of seawater, formation of abundant amphibole and accelerated cooling. Cooling and crustal thinning continued as the section moved across the rift valley floor. Early amphibole veins probably formed at this time since some are cut by the foliation produced by brittle-ductile deformation.

At a spreading rate of 10 mm/y, by about 400,000 years, the section would lie beneath the edge of the inner rift valley wall, transferred out of the zone of lithospheric necking and undergoing uplift into to the rift mountains. This would mark the end of brittle-ductile deformation and is reflected in the formation of additional hydrothermal veins only in the unrecrystallized portions of the section. This and the shift to diopside replacing amphibole as the primary hydrothermal vein mineral was a response to a drop in rock permeability where cracking and hydrothermal circulation were no longer enhanced by extensional stresses in the zone of lithospheric necking [Dick et al., 1991a].

As the crust continued to cool, many late hydrothermal veins re-opened and low temperature assemblages of calcite, zeolite and clay minerals were deposited. The oxidative alteration of olivine is commonly associated with calcite veining and probably took place at about the same time. The subvertical, clay-lined fractures probably formed late in this period at the time of formation of the wave-cut platform on the crest of the inside corner high due to unloading of the section by erosion. By about 2 Ma, the section would have been uplifted close to sealevel to form a wavecut platform (Figure 2).

The extent of alteration of the 735B section is much greater than that found in statically cooled layered intrusions. This reflects hydrothermal alteration in the dynamic environment of a ridge axis. Previous studies of oceanic gabbros have also shown this extensive alteration, primarily in the amphibolite facies [e.g., Ito and Anderson, 1983; Malcolm, 1980; Helmstaedt and Allen, 1977; Mevel, 1987,1988; Stakes and Vanko, 1986; Vanko and Batiza, 1982], and it can be concluded that this is a general phenomenon beneath slow-spreading ocean ridges. Newly recognized is the distinction between true static and dynamic alteration in this environment and the sharp shift in permeability, solution chemistry, and alteration accompanying transfer of the plutonic section out of the zone of lithospheric necking. This is different from previous descriptions of oceanic gabbros where all matrix alteration of unrecrystallized gabbro was attributed to static alteration, despite the dynamic extensional environment and enhanced state of stress and cracking in which much of it undoubtedly occurred.

Acknowledgments. Preparation of this paper and additional research on the alteration of gabbros from Hole 735B was funded by the Naitonal Science Foundation (NSF/OCE 91-01340). Assistance in the preparation of figures and data reduction was provided by Linda M. Angeloni. We thank M.R. Fisk for proof-reading. The paper is based on a presentation at the JOI/USSAC Workshop on Indian Ocean Drilling at Cardiff, Wales in July, 1991.

REFERENCES

Alt, J. C., and Anderson, T. F., Mineralogy and isotopic composition of sulfur in layer 3 gabbros from the Indian Ocean, Hole 735B, in *Proc. ODP, Sci. Results, 118,* Von Herzen, R. P., and Robinson, P. T. et al., p. 113-126, 1991.

Angeloni, L. M., and Dick, H. J. B., Troctolitic gabbros from Hole 735B, Atlantis II Fracture Zone, Southwest Indian Ridge. *EOS, Trans. Am. Geophys. Union, 71,* 704, 1990.

Batiza, R. and Vanko, D. A., Petrologic evolution of large failed rifts in the Eastern Pacific: Petrology of volcanic and plutonic rocks from the Mathematician Ridge area and the Guadeloupe Trough, *J. Petrol., 26,* 564-602, 1985.

Bedard, J. H. Sparks, R. S. J., Cheadle, M. J., and Hallworth, M.A., Peridotite sills and metasomatic gabbros in the Eastern Layered Series of the Rhum complex, *J. Geol. Soc., 145,* 207-224, 1988.

Bird, D. K., Rogers, R. D. and Manning, C. E., Mineralized fracture systems of the Skaergaard intrusion, East Greenland, Meddelelser om Gronland, *Geoscience, 16,* 1-68, 1986.

Bird, D. K., Manning, S. and Rose, N., Hydrothermal alteration of Tertiary layered gabbros. *Am. J. Sci., 288,* 405-457, 1988.

Bonatti, E., Ancient continental mantle beneath oceanic ridges, *J. Geophys. Res. 76,* 3825-3831, 1971.

Bloomer, S. H., Meyer, P. S., Dick, H. J. B., Ozawa, K., and Natland, J. H., Textural and mineralogical variations in gabbroic rocks from Hole 735B, in *Proc. ODP, Sci. Results, 118,* edited by R. P. von Herzen, and P. T. Robinson et al., p. 21-40, 1991.

Bryan, W. B., and Dick H. J. B., Contrasting liquidus trends for abyssal basalts, evidence for mantle heterogeneity, *Earth Planet. Sci. Lett., 58,* p. 15-26, 1982.

Cann, J. R., New model for the structure of the ocean crust, *Nature, 226,* p. 928-930, 1970.

Cannat, M., Mevel, C. and Stakes, D., Normal ductile shear zones at an oceanic spreading ridge: tectonic evolution of Site 735 gabbros (southwest Indian Ocean, in *Proc. ODP, Sci. Results, 118,* edited by R. P. von Herzen, and P. T. Robinson et al., p. 415-431, 1991.

Christensen, N., The abundance of serpentinites in the ocean crust, *J. Geol., 80,* 709-719, 1972.

Christensen, N. I., Ophiolites, seismic velocities and oceanic crustal structure, *Tectonophysics, 47,* 131-157, 1978.

Clague, D. A., and Bunch, T. E., Formation of ferrobasalt at East Pacific midocean spreading centers, *J. Geophys. Res., 81,* 4247-4256, 1976.

Coleman, R. G., *Ophiolites,* 229 pp., Springer Verlag, New York, 1977.

Dick, H. J. B., Bryan, W. B., and Thompson, G., Low-angle faulting and steady-state emplacement of plutonic rocks at ridge-transform intersections, *EOS, Trans. Am. Geophys. Union, 62,* 406, 1981.

Dick, H. J. B., Abyssal Peridotites, Very-slow spreading ridges and ocean ridge magmatism, in *Magmatism in the Ocean Basins, Geol. Soc. London Spec. Publ. No. 42,* edited by A. D. Saunders, and M. J. Norry, p. 71-105, 1989.

Dick, H. J. B., Meyer, P. S., Bloomer, S., Kirby, S., Stakes, D. and Mawer, C., Lithostratigraphic evolution of an in-situ section of oceanic layer 3, in *Proc. ODP ,Sci. Results, 118,* edited by R. P. von Herzen, and P. T. Robinson et al., p. 439-538, 1991.

Dick, H. J. B., Schouten, H., Meyer, P. S., Gallo, D. G., Bergh, H., Tyce, R. Patriat, P., Johnson, K. T. M., Snow, J, and Fisher, A., Tectonic evolution of the Atlantis II Fracture Zone, in *Proc. ODP, Sci. Results, 118,* edited by R. P. von Herzen, and P. T. Robinson et al., p. 359-398, 1991.

Engel, C. G., Fisher, R. L., Granitic to ultramafic rock complexes of the Indian Ocean ridge system, western Indian Ocean, *Geol. Soc. Am. Bull., 86,* 1553-1578, 1975.

Fisher, R. L, Dick, H. J. B., Natland, J. H., and Meyer, P. S., Mafic/ultramafic suites of the slowly spreading Southwest Indian Ridge: PROTEA Exploration of the Antarctic plate boundary, 24°-47°E, *Ophioliti, 11,* 147-178, 1986.

Fox, P. J., and Stroup, J. B., The plutonic foundation of the oceanic crust, in *The Sea, 7,* edited by C. Emiliani, pp. 119-218, Wiley, New York, 1981.

Fox, P. J., Schreiber, E., and Peterson, J. J., The geology of the ocean crust: compressional wave velocities of oceanic rocks, *J. Geophys. Res. 78,* 5155-5172, 1973.

Goldberg, D., Broglia, C., and Becker, K., Fracturing, alteration, and permeability: in-situ properties in Hole 735B, Atlanis II Fracture Zone, southwest Indian Ocean, in *Proc. ODP, Sci. Results, 118,* edited by R. P. von Herzen and P. T. Robinson et al., p. 261-270, 1991.

Goldberg, D., Baldri, M., and Wepfer, W., Ultrasonic attenuation measurements in gabbros from Hole 735B, in *Proc. ODP, Sci. Results, 118,* edited by R. P. von Herzen, and P. T. Robinson et al., p. 253-259, 1991.

Harper, G. D., Tectonics of slow spreading mid-ocean ridges and consequences of a variable depth to the brittle/ductile transition, *Tectonics, 4,* 395-409, 1985.

Hebert, R., Constantin, M., and Robinson, P. T., Primary mineralogy of Leg 118 gabbroic rocks and their place in the spectrum of oceanic mafic igneous rocks, 1991.

Helmstaedt, H., and Allen, J. M., Metagabbronorites from DSDP hole 334, an example of high temperature deformation and recrystallization near the Mid-Atlantic Ridge, *Can. J. Earth Sci. 14,* 886-898, 1977

Hodges, F. N., and Papike, J. J. Magmatic cumulates from oceanic layer 3, *J. Geophys. Res. 8,* 4135-4151, 1976.

Ito, E. and Anderson, A. T., Jr., Submarine metamorphism of gabbros from the Mid-Cayman Rise: Petrographic and mineralogic constraints on hydrothermal processes at slow-spreading ridges, 371-388, 1983.

Karson, J., Seafloor spreading on the Mid-Atlantic Ridge: Implications for the structure of ophiolites and oceanic lithosphere produced in slow-spreading environments, in *Ophiolites: Oceanic Crustal Analogues,* edited by J. Malpas, E. M. Moores, A. Panyiotou, and C. Xenophontas, p. 547-555, Geol. Survey Dept., Nicosia, Cyprus, 1990.

Karson, J. A., Thompson, G., Humphris, S. E., Edmond, J. N., Bryan, W. B., Brown, J. R., Winters, A. T., Pockalny, R. A., Casey, J. F., Campbell, A. C., Klinkhammer, G., Palmer, M. R., Kinsler, R. J, and Sulanowska, M. M., Along-axis variations in seafloor spreading in the MARK area, *Nature, 328,* 681-685, 1987.

Kempton, P. D., Hawkesworth, C. J., and Fowler, M., Geochemistry and isotopic composition of gabbros from layer 3 of the Indian Ocean crust, Leg 118, Hole 735B, in *Proc. ODP Sci. Results, 118,* edited by R. P. von Herzen, and P. T. Robinson et al., p. 127-144, 1991.

Klein, E., and Langmuir, C. H., Global correlations of ocean ridge basalt chemistry with axial depth and crustal thickness, *J. Geophys. Res., 92,* 8089-8115, 1987.

Kuo B.-Y., Forsyth D. W. Gravity anomalies of the ridge-transform system in the South Atlantic between 31 and 34.5°S: Upwelling centers and variations in crustal thickness, *Mar. Geophys. Res., 10,* 205-232, 1988.

le Roex, A. P., Dick, H. J. B., Reid, A. M., Erlank, A. J., Ferrobasalts from the Speiss Ridge segment of the Southwest Indian Ridge, *Earth Planet. Sci. Lett., 60,* p. 437-451, 1982.

le Roex, A. P., Dick, H. J. B., Erlank, A. J., Reid, A. M., Frey, F. A., and Hart, S. R., Geochemistry, mineralogy and petrogenesis of lavas erupted along the Southwest Indian Ridge between the Bouvet Triple Junction and 11°, *J. Petrol., 24,* p. 267-318, 1983.

Lin J., Purdy G. M., Schouten H., Sempere J.-C., Zervas C., Gravitational evidence of focused magmatic accretion along the Mid-Atlantic Ridge, *Nature, 344,* 627-632, 1990.

Lister, C. R. B., On the penetration of water into hot rock. *Geophys. J. R. Astr. Soc., 39,* 465-509, 1974.

Louden, K. E. and Forsyth, D. W., Crustal structure and isostatic compensation near the Kane Fracture Zone from topography and gravity measurements, I, Spectral analysis approach, *Geophys J. R. Astron. Soc. 68,* 725-750, 1982.

Malcolm, F. L., Microstructures of the Cayman Trough gabbros, *J. Geol. 89,* 675-688, 1981

Meyer, P. S., Dick, H. J. B., and Thompson, G., Cumulate gabbros from the Southwest Indian Ridge, 54°S-7°16'E: implications for magmatic processes at a slow spreading ridge, Contrib. Min. Petr. 103, 44-63, 1989.

Mevel, C., Evolution of oceanic gabbros from DSDP Leg 82: influence of the fluid phase on metamorphic crystallizations, *Earth Planet. Sci. Lett., 83,* 67-79, 1987.

Mevel, C., Metamorphism in oceanic layer 3, Gorringe Bank, Eastern Atlantic, Contrib. Mineral. Petrol., 100, 496-509, 1988.

Mevel, C., Cannat, M., Gente, P., Marion, E., Auzende, J. M., and Karson, J. A., Emplacement of deep crustal and mantle rocks on the west wall of the MARK area (Mid-Atlantic Ridge, 23°N). *Tectonophysics, 190*, 31-53, 1991.

Miyashiro, A., The Troodos ophiolitic complex was probably formed in an island arc, *Earth Planet. Sci. Lett. 19*, 218-224, 1973.

Miyashiro, A., and Shido, F., Differentiation of gabbros in the Mid-Atlantic Ridge near 24!N. Geochem. J., 14, 145-154, 1980.

Moores, E. M., and Vine, F. J., The Troodos Massif, Cyprus and other ophiolites as ocean crust, *Phil. Trans. R. Soc. London, A268*, 443-466, 1971.

Mutter, J. C., and North Atlantic Transect Study Group, Multichannel seismic images of the oceanic crust's internal structure: evidence for a magma chamber beneath the mesozoic Mid-Atlantic Ridge, *Geology, 13*, 629-632, 1985.

NAT Study Group, North Atlantic Transect: A wide aperture, two-ship Multichannel seismic investigation of the oceanic crust, *J. Geophys. Res. 90*, 10,321-10,341, 1985.

Natland, J. H., Meyer, P. S., Dick, H. J. B., and Bloomer, S. H., Magmatic oxides and sulfides in gabbroic rocks from Hole 735B and the later development of the liquid line of descent., in *Proc. ODP, Sci. Results, 118*, edited by R. P. von Herzen, and P. T. Robinson et al., p. 75-112, 1991.

Norton, I. O., and Sclater, J. G., A model for the evolution of the Indian Ocean and the breakup of Gondwanaland, *J. Geophys. Res. 84*, 6803-6830, 1979.

Ozawa, K., Meyer, P. S., and Bloomer, S. H., Mineralogy and textures of iron-titanium oxide gabbros and associated olivine gabbros from Hole 735B, in *Proc. ODP, Sci. Results, 118*, edited by R. P. von Herzen, and P. T. Robinson et al., p. 41-74, 1991.

Pariso, J. E., Scott, J. H., Kikawa, E., and Johnson, H. P., A magnetic logging study of Hole 735B gabbros at the Southwest India Ridge, in *Proc. ODP, Sci. Results, 118*, edited by R. P. von Herzen, and P. T. Robinson et al., p. 309-322, 1991.

Quick J. E., The origin and significance of large, tabular dunite bodies in the Trinity peridotite, northern California, *Contrib. Min. Petrol. 78*, 413-422, 1981.

Raitt, R. W., The crustal rocks, in *The Sea, 3*, edited by M.N. Hill, pp. 85-102, Wiley, New York, 1963.

Ramsay, J. G., and Graham, R. H., Strain variation in shear belts, *Can. J. Earth Sci. 7*, 786-813, 1970.

Robinson, P. T., Flower, M. F. J., Schmincke, H.-U., and Ohnmacht, W., Low temperature alteration of oceanic basalts, DSDP Leg 37, in *Init. Repts. DSDP, 37*, edited by F. Aumento,

W. G. Melson et al., 775-793, U. S. Government Printing Office, Washington, D.C., 1977.

Robinson, P. T., Melson, W. G., O'Hearn, T., and Schminke, H.-U., Volcanic glass compositions of the Troodos ophiolite, Cyprus, *Geology 1*, 400-404, 1983.

Robinson, P. T., von Herzen, R. et al., *Proc. ODP, Init. Repts., Part A, 118*, Ocean Drilling Program, College Station, TX, 1989.

Rosendahl, B. R., Evolution of oceanic crust 2, Constraints, implications, and inferences, *J. Geophys. Res. 81*, 5305-5314, 1976.

Schouten, H., Klitgord, K. D., and Whitehead, J. A. Segmentation of mid-ocean ridges, *Nature, 317*, 225-229, 1985.

Slater, J. G., Dick, H. J. B., Norton, I, and Woodroffe, D., Tectonic structure and petrology of the Antarctic Plate Boundary near the Bouvet Triple Junction, *Earth Planet. Sci. Lett., 36*, 393-400, 1978.

Shipboard Scientific Party, Site 735B, in *Proc. ODP, Init. Repts., 118*, edited by P. T. Robinson, R. von Herzen, p. 117-119, Ocean Drilling Program, College Station, TX, 1989.

Sinton, J., and Detrick, R. S., Ocean ridge magma chambers, *J. Geophys. Res. 97*, 197-216, 1992.

Stakes, D., Mevel, C., Cannat, M., and Chapu, T., Metamorphic stratigraphy of Hole 735B, in *Proc. ODP Sci. Results, 118*, edited by R. P. von Herzen, and P. T. Robinson et al., pp. 153-180, 1991.

Stakes, D., and Vanko, D. A., Multistage hydrothermal alteration of gabbroic rocks from the failed Mathematician Ridge, *Earth Planet. Sci. Lett., 79*, 75-92., 1986.

Swift, S. A., and Stephen, R. A., How much gabbro is in ocean seismic layer 3?, *Geophys. Res. Lett.*, in press, 1992.

Tucholke, B. E., and Lin, J., A geologic model for the structure of slow spreading oceanic crust, *J. Geophys. Res.*, in press, 1992.

Vanko, D. A., and Stakes, D. S., Fluids in oceanic layer 3: evidence from veined rocks, Hole 735B, Southwest Indian Ridge, in *Proc. ODP, Sci. Results, 118*, edited by R. P. von Herzen, and P. T. Robinson et al., p. 181-218, 1991.

Vanko, D. A., High-chlorine amphiboles from oceanic rocks: product of highly saline hydrothermal fluids?, *Am. Mineral., 71*, 51-59, 1988.

Vanko, D., and Batiza, R., Gabbroic rocks from the Mathematician Ridge failed rift, *Nature, 300*, 742-744, 1982.

Wager, L. R. and Brown, G. M., *Layered Igneous Rocks*, 588 p., W.H. Freeman and Co., San Francisco, 1967.

Physical Properties and Logging of the Lower Oceanic Crust: Hole 735B

R. P. VON HERZEN

Woods Hole Oceanographic Institution, Woods Hole, MA, 02543, USA

D. GOLDBERG

Lamont-Doherty Geological Observatory, Palisades, NY, 10964, USA

M. MANGHNANI

Dept. of Geophysics, University of Hawaii, Honolulu, HI, 96822, USA

Results from downhole logging instrumentation and physical properties measurements on samples recovered from a 500-m-thick section of gabbros at Site 735 on the SW Indian Ridge are compared. Here we emphasize particularly the seismic, electrical, and nuclear logging measurements to deduce the physical state and evolution of this crustal section over the 11-12 m.y. since its formation. Various seismic methods give compressional velocities ranging from 6.5 to >7 km/s, typical of lower oceanic crustal velocities determined from marine refraction measurements. Except for the unusually low intrinsic electrical resistivity (<10 ohm-m) of some Fe-Ti-oxide gabbros, the relatively high range of resistivities (~3x10^2 - 2x10^4 ohm-m) for most of the section deeper than 150 meters below seafloor is consistent with low porosities (few percent) derived from the neutron log. The decrease with depth of thin, relatively high porosity (20-25%) zones, low temperature (sea water) rock alteration, and fluid permeability suggests that overburden stress is an important factor maintaining closed fractures in young ocean crust.

INTRODUCTION

The Ocean Drilling Program (ODP) provides one of the few opportunities to obtain samples and investigate the in-situ properties of ocean crust deep beneath the seafloor. Although both the ocean crust and the underlying mantle have been (and continue to be) subjects of intensive study by marine geophysical methods, these necessarily involve remote observations with a corresponding loss of resolution, compared to in-situ measurements, primarily depending on the distance between the measuring instrumentation and the subject of investigation. Of course, marine geological and geophysical research of the last few decades has provided much of the impetus for ODP, to verify deductions and to investigate hypotheses in detail brought from these broader-scale studies.

The drilling results described in this volume derive from a scientific need for better understanding of the oceanographic and tectonic evolution of the Indian Ocean. Many of the tectonic objectives of this drilling campaign required sampling and observations of the igneous ocean crust of the region. However, only one drilling leg of this program had objectives focussed primarily on deep penetration of ocean crust: Leg 118 in the Atlantis II transform fault on the southwest Indian Ridge (Figure 1). Many of the observations and results of ODP Leg 118

Synthesis of Results from Scientific
Drilling in the Indian Ocean
Geophysical Monograph 70

drilling have been recently published in the Initial Reports [Robinson, Von Herzen et al., 1989] and the Scientific Results [Von Herzen, Robinson et al., 1991]. The section drilled at Site 735 on the elevated eastern wall of the transform formed at the spreading ridge axis 11-12 Ma (based on magnetic anomalies), and was uplifted several km as the crust migrated southward from the ridge. The recovered gabbros are of varied compositions, and were subject to high-temperature ductile deformation as well as some minor amount of low-temperature sea water alteration mostly in the upper 100-150 m below seafloor (mbsf). The gabbro section was exposed at the seafloor probably by a combination of crustal thinning by tectonism (listric faulting?) and wave erosion.

The achievement of a 500-m-deep penetration in gabbros at Site 735 has given us a unique view of rock believed to be representative of the lower oceanic crust. In addition, the unusually high recovery (87%) of core allowed unprecedented detailed description and measurements of oceanic crustal stratigraphy compared to any previous crustal drilling legs. However, the crustal depths represented by the drilled section is unknown. The physical properties measurements and logging in Hole 735B were summarised briefly by Von Herzen et al. [1991]. The present paper is an attempt to provide a more quantitative, in-depth comparison of these measurements, incorporating additional shore-based laboratory data on samples from Hole 735B that have recently become available [Manghnani et al., unpublished manuscript, 1992]. In

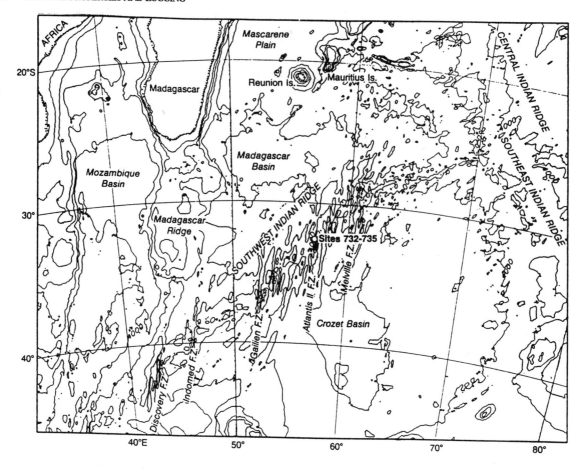

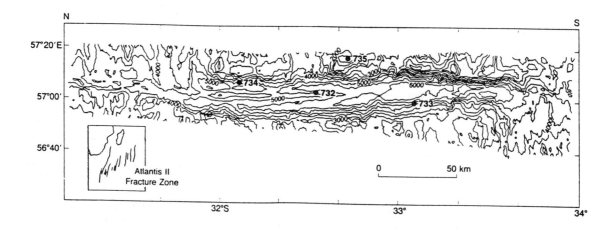

Fig. 1. Upper panel: Regional geographic and topographic setting of ODP Leg 118 drill sites in the SW Indian Ocean. Bathymetric contours at 1 km intervals. Lower panel: Detailed location of Leg 118 drill sites in the Atlantis II transform fault on the SW Indian Ridge. Bathymetric contour interval 500 m. After Shipboard Scientific Party [1989b, Figures 1 and 3].

addition, we contrast these measurements in gabbros with similar investigations of basalts in other deep-penetration crustal holes with extensive logging measurements, i.e., Holes 504B (E. Pacific), and 395A and 418 (N. Atlantic).

SEISMIC MEASUREMENTS

Measurements of seismic velocities are one of the most powerful techniques to characterize oceanic crustal rocks. Indeed, even the definition and subdivision of ocean crust are based on the systematic variations of velocity with depth [e.g., Spudich and Orcutt, 1980]. Especially for the lower oceanic crust, seismic measurements are almost always by necessity remote from the rock material under investigation. Hence only the average or gross seismic properties, not the detailed variations, of the lower crust are known with any certainty. The gabbros exposed at shallow depths (~700 m) beneath sea level made Site 735 a particularly attractive drilling location to characterize the details of lower oceanic crustal variability. The success of drilling Hole 735B, with a high recovery of relatively unaltered gabbros, established a unique and useful site to investigate detailed seismic properties of the lower oceanic crust.

Seismic measurements at Site 735 consisted of laboratory measurements on samples, and downhole measurements following drilling. The sample measurements were carried out aboard ship during drilling [Shipboard Scientific Party, 1989a], and subsequently at shorebased laboratories [Iturrino et al., 1991; Goldberg et al., 1991a; Manghnani et al., unpublished manuscript, 1992]. Except for the investigation by Goldberg et al., [1991a] which focussed on seismic attenuation and is not included in this review, these measurements emphasized seismic velocities. The in-situ downhole seismic measurements were made with a multi-channel sonic logging tool [Shipboard Scientific Party, 1989a; Goldberg et al., 1991b], and with a borehole seismometer recording an airgun sound source operated near the sea surface [Shipboard Scientific Party, 1989a; Swift et al., 1991]. In this paper we compare the sample measurements made in the various laboratories , and relate those to the in-situ seismic measurements, with resulting implications for the physical state of the gabbros that may constitute the lower oceanic crust.

Somewhat different selection, treatment, and measurement of samples occurred in the different investigations. The shipboard measurements utilized 1) cylindrical mini-cores with axes oriented perpendicular to the main core axis (horizontal) and parallel to foliation (if it existed), 2) cubes with faces normal and parallel to lineation and foliation, and 3) half-round sections from the main core with vertical axes [Shipboard Scientific Party, 1989b]. The orientations of the mini-cores and cubes were selected to determine effects of rock fabrics on the velocity anisotropy, which was found to be significant. All shipboard samples were measured in the saturated state (i.e., before dehydration of the pore water in samples), as were the measurements of mini-cores made by Manghnani et al. [unpublished manuscript, 1992]. However, most of the measurements on

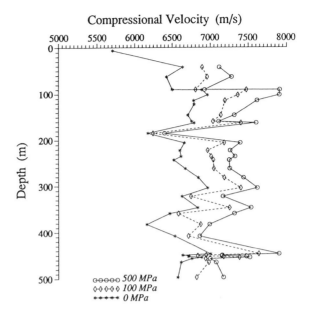

Fig. 2. Compressional velocities vs. depth measured at hydrostatic pressures of 0, 100, and 500 MPa in the laboratory on the same samples of cores from Hole 735B [Manghnani et al., unpublished manuscript, 1992]. The in-situ hydrostatic pressure averages about 10 MPa over Hole 735B.

mini-cores by Iturrino et al. [1991] were carried out in the dry (dehydrated) state, using a rubber jacket covering the samples to prevent saturation under hydrostatic pressure. These different methodologies allow some comparisons of the effects of porosity on the seismic velocities. Only compressional velocities with a pulse source were measured on the shipboard samples, whereas both compressional and shear velocities were determined in the shorebased investigations.

Samples at ambient laboratory conditions usually exhibit lower seismic velocities than when they are measured at in-situ conditions, primarily a result of the reduced ambient pressure on rock velocities. Temperature changes usually cause smaller effects on velocity than those of pressure, except for some unusual environmental situations. Also, the processes of drilling and sampling may cause significant effects similar to those of pressure (i.e., reduced velocities after sampling), although in the case of Hole 735B the relatively unaltered condition of most of the rock drilled would seem to preclude any significant physical effects of sampling. The effects of pressure can be clearly seen (Figure 2) in the measurements carried out on the same samples in the laboratory at different hydrostatic pressure conditions. The compressional velocity (V_p) increase between 0 and 100 MPa is 6-7% (Table 1), and the well-known non-linear effects are evident (Figure 3), with larger changes in velocity occuring at lower pressures, a result of closing of micro-cracks in addition to the effects of pressure on the constituent mineral velocities [Birch, 1960; and the following discussion].

TABLE 1. Statistics of Seismic Velocity Measurements, Hole 735B.

Measurement[1] Description	Mean SD[4] (m/s)	N	Reference
A. Sample compressional velocities (horizontal)			
1. S, 0 MPa	6799 342	151	Shipboard Scientific Party [1989a, Table 8]
2. D, 20 MPa	6602 292	67	Iturrino et al. [1991]
3. D, 100 MPa	6822 299	67	Iturrino et al. [1991]
4. S, 0 MPa	6604 320	59	Manghnani et al., unpublished manuscript [1992]
5. S, 100 MPa	7044 286	30	Manghnani et al., unpublished manuscript [1992]
6. S, 500 MPa	7306 320	30	Manghnani et al., unpublished manuscript [1992]
B. Sample shear velocities (horizontal)[2]			
7. D, 20 MPa	3701 129	34	Iturrino et al. [1991]
8. S, 0 MPa	3627 252	59	Manghnani et al., unpublished manuscript [1992]
C. Logs (MCS) and vertical seismic profile (VSP)			
9. MCS (V_p)	6816 490	298	This paper
10. MCS (V_s)	3701 129	310	This paper
11. VSP	6506 74[3]	16	Swift et al. [1991]

[1] D=dry, S=saturated, MPa=megaPascals, V_p=compressional velocity, V_s=shear velocity,
    MCS=multichannel sonic tool, VSP=vertical seismic profile.
[2] For Iturrino et al. [1991], shear wave propagating horizontally and vibrating parallel to
    foliation.
[3] Standard deviation of all VSP data about best fit line.
[4] SD = standard deviation.

Although different samples were measured by the respective shorebased laboratories, the effects of pore saturation may be estimated from statistical comparisons of samples measured in a wet (saturated) condition [Manghnani et al., unpublished manuscript, 1992] with those measured in a dry condition [Iturrino et al., 1991]. First, we note that the mean horizontal V_p of dry samples measured at 100 MPa in the laboratory by Iturrino et al. [1991] is nearly the same as that for the saturated samples measured at ambient laboratory conditions on board ship [Table 1; Shipboard Scientific Party, 1989a], about 6.8 km/s. Only velocities measured in the horizontal direction were used because of the small but significant velocity anisotropy caused by fabrics in these samples [Shipboard Scientific Party, 1989a, Table 10]. Furthermore, except for a few low values, this comparison seems remarkably valid over the entire depth range of Hole 735B (Figure 4). This suggests that the effect on compressional velocities of saturating the pore space of these low-porosity samples [generally <2%; Shipboard Scientific Party, 1989a, Table 9] is comparable to imposing ~100 MPa of hydrostatic pressure on the dry samples. This relationship in V_p between saturated and dry samples is seen in more detail in the measurements of the Shipboard Scientific Party [1989a, Table 8] compared to those of Iturrino et al. [1991, Table 4].

This apparent contrast in V_p between dry and saturated samples is also seen in the comparison of the means between the shorebased lab results at 100 MPa (Table 1). The dry sample mean is 6822 m/s with a standard error (SE = SD/$\sqrt{N}$) of 37 m/s, and the saturated mean is 7044 m/s with SE = 52 m/s; the means are not the same statistically at a 95% confidence level (±2 SE's). From the depth plot of the samples measured in these different laboratories (Figure 5), it may be seen that these differences apply throughout most of the section, and are not the result of large aberrations of only a few samples. A caveat on such statistical deductions is the observation that the mean V_p for shipboard samples (6799 m/s, SE = 28) and the saturated samples measured by Manghnani et al. [unpublished manuscript, 1992] at 0 MPa (6604 m/s, SE = 42) also appear significantly different (Table 1); within the data variability, these means do not overlap at the 95% confidence level. It is not clear whether such differences are a result of not using the same samples, or whether they might be caused by some small systematic differences in

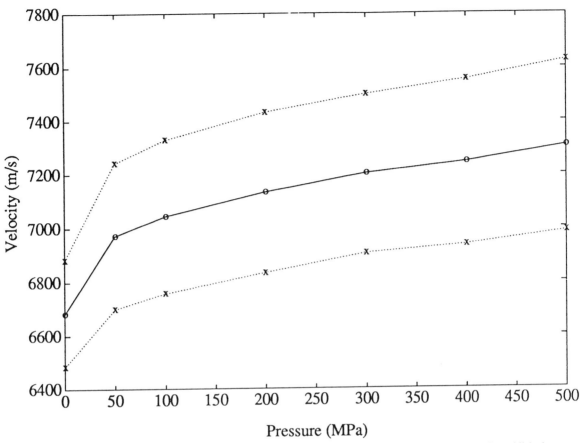

Fig. 3. Mean compressional velocities vs. pressure for all Hole 735B samples measured by Manghnani et al. [unpublished manuscript, 1992, Table 1]. Mean measured values are given by the open circles connected by a solid line, including ± 1 standard deviation by the x symbols connected by dotted lines.

methodologies.

A significant relationship between V_p and porosity for the shipboard samples was documented by the Shipboard Scientific Party [1989a], reproduced here as Figure 6. The relatively large decrease in mean velocity with increasing porosity, about 15% velocity decrease from 0 to 2% porosity, may be explained by low aspect ratios (the ratio between minimum and maximum diameters) of the cracks or voids that constitute the porosity. The mean effective aspect ratios may be inferred to be in the range of .02-.05 based on calculations by Wilkens et al. [1991, Figure 8], as derived from a model of Kuster and Toksoz [1974]. Lower aspect ratios amplify the effects of porosity on seismic velocity. The decrease of seismic velocity with increasing porosity for the gabbros at Site 735 is comparable to, or even larger than, that at Sites 504 and 418 for basalts [Wilkens et al., 1991, Figure 5].

Velocities measured on samples, which may be disturbed by their acquisition and processing, or as a result of measurement in an environment different than in-situ, are useful to compare with logging results. The acoustic velocity was logged in-situ in Hole 735B with a multi-channel seismic tool [Shipboard Scientific Party, 1989a]. Data were acquired at about 0.3 m vertical intervals with this tool, and the various types of waves were identified and

arrival times processed later by semblance analysis. We averaged the compressional and shear velocity logs over vertical intervals of about 1.5 m to produce a data set at intervals comparable with the sample measurements.

The mean logged compressional velocity for the entire hole is nearly the same as the mean calculated from all shipboard samples (Table 1). The closeness of this agreement is probably coincidental, as indicated by plotting the log data together with the sample values over depth (Figure 7). The log shows velocity increasing somewhat with depth over most of the hole, punctuated with sharp dips to lower velocities. The latter could be simply the result of signal dropouts (i.e., missed first arrivals) during data acquisition, which are probably caused for the most part by real features of the section intersected by the hole, producing low semblance in the processed log. The evidence of a relatively good correlation of these minima in the acoustic log with maxima in the neutron porosity log (see below) indicates that they are caused primarily by the in-situ stratigraphy, some elements of which are probably fractures or voids intersecting an otherwise dense rock section. That the log gives velocities significantly less than those of the samples in the uppermost ~150 m of the hole probably indicates the greater frequency of fractures, and/or a higher degree of

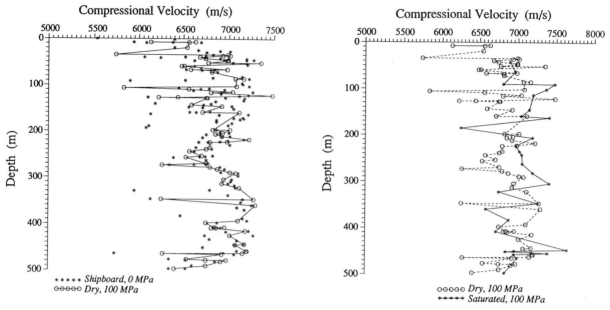

Fig. 4. Horizontal compressional velocities vs. depth measured on Hole 735B core samples in a saturated condition at 0 MPa on board ship [Shipboard Scientific Party, 1989a], compared to those measured in a shorebased laboratory in a dry condition at 100 MPa hydrostatic pressure [Iturrino et al., 1991]. The solid line connects the dry sample measurements.

Fig. 5. Comparison of compressional velocities vs. depth measured on Hole 735B dry samples [open circles connected by dashed line, Iturrino et al., 1991] and saturated samples [star symbols connected by solid line, Manghnani et al., unpublished manuscript, 1992], at 100 MPa hydrostatic pressure.

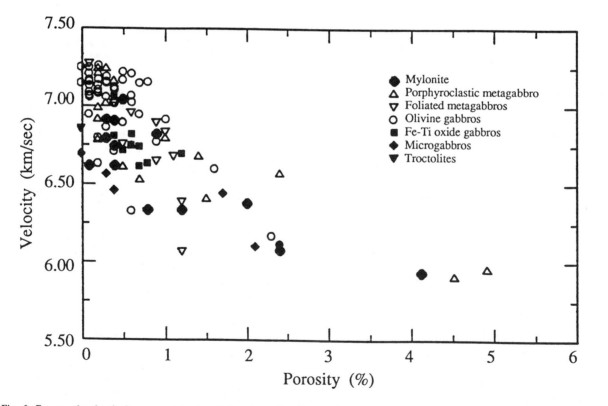

Fig. 6. Compressional velocity vs. porosity for shipboard samples of cores from Hole 735B. Symbols indicate different gabbro types as shown. From Shipboard Scientific Party [1989a, Figure 63].

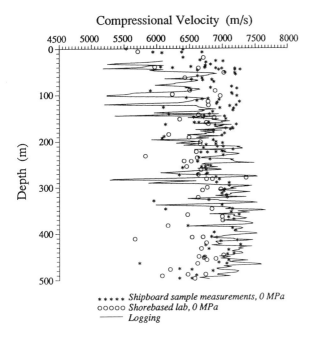

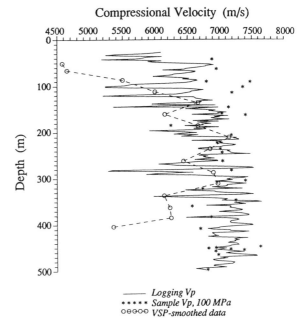

Fig. 7. Comparison of Hole 735B acoustic log compressional velocities (solid line) with those measured on core samples at 0 MPa hydrostatic pressure, both aboard ship [star symbols, Shipboard Scientific Party, 1989a] and at a shorebased laboratory [open circles, Manghnani et al., unpublished manuscript, 1992]. The log data have been averaged over ~1.5 m depth intervals.

Fig. 8. Logged compressional velocities (solid line) at Hole 735B, compared with smoothed VSP velocities [open circles connected by dashed line; Swift et al., 1991] and the horizontal velocities measured on core samples at 100 MPa hydrostatic pressure [star symbols; Manghnani et al., unpublished manuscript, 1992]. The log data are averaged over ~1.5 m depth intervals, and the VSP data represent running means over ~115 m.

alteration as shown by the petrography and other logs [Goldberg et al., 1991b] over this interval. Conversely, the relatively high velocities of the log compared with sample measurements for the lowermost ~130 m of the hole suggests that the sample velocities, particularly those measured later in the shore-based labs, may be different from their values at in-situ conditions. For at least 2 reasons, we would expect the logged velocities to be lower than those measured on samples: (1) the log data give the vertical component of in-situ velocity, whereas the sample measurements represent the horizontal component, which for the measured sample anisotropy [Shipboard Scientific Party, 1989a, Table 10] should decrease the former somewhat compared to the latter; (2) the log data include effects of in-situ fractures or voids not present in the samples. The latter could also cause large-scale seismic anisotropy, although no data are available to determine whether it exists at Site 735.

There is better agreement between the saturated sample compressional velocities measured at 100 MPa hydrostatic pressure with the log data below about 150 m in the hole (Figure 8). The increased pressure on the samples, although greater than the in-situ hydrostatic pressure (~10 MPa), may provide a better comparison by closing any microfractures in the samples that were opened by raising them to the surface, which would tend to lower seismic velocities. On the other hand, there is quite good agreement (Table 1, Figure 9) between the shear velocity log and the

shear velocities measured on samples at 0 MPa [Manghnani et al., unpublished manuscript, 1992]. There are no measurements of shear velocities on saturated samples at elevated hydrostatic pressures, and, as previously noted [Shipboard Scientific Party, 1989a], the logged shear velocities may have artificially low variability from the effects of guided waves in high velocity rock.

Some deductions about the mechanisms of velocity changes in rock with pressure may be made by comparing the data in detail. The two primary mechanisms of velocity changes in rocks with pressure are 1) the closure of cracks, and 2) the effects of pressure on the constituent minerals. Crack closure should affect the highest porosity (lowest velocity) rock the most, whereas the magnitude of the pressure effect on most mineral velocities is generally monotonic with the unpressured velocity value [Press, 1966]. Plots of the normalized sample velocity variation with confining pressure (Figure 10) show that the velocity increase over the range 0-50 MPa is about the same as that over 50-500 MPa, probably caused by the greater effect on velocities of closing microcracks at the lower pressure range. The normalized velocity difference is defined as $(V_2 - V_1)/V_1$, where V_2 and V_1 are the velocities measured at the high and low confining pressures, respectively. Note the considerable scatter in the data, without significant correlation of the relative velocity differences in samples with their intrinsic velocity for either pressure range, suggesting that more than a single mechanism is

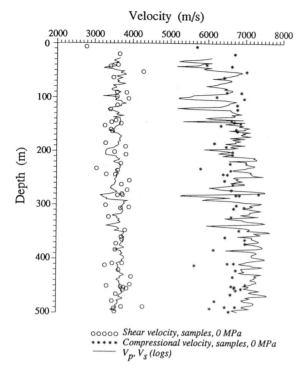

Velocity (m/s)

○○○○○ *Shear velocity, samples, 0 MPa*
***** *Compressional velocity, samples, 0 MPa*
——— V_p, V_s *(logs)*

Fig. 9. Logged compressional and shear velocities plotted with measurements on saturated samples at 0 MPa hydrostatic pressure [Manghnani et al., unpublished manuscript, 1992]. Logged data are averaged over ~1.5 m depth intervals.

responsible for the change in velocity with confining pressure.

Comparisons with the results of the vertical seismic profile (VSP) conducted in the hole [Swift et al., 1991] are somewhat enigmatic (Figure 8). Although the mean VSP velocities found over the mid-depth range (150-350 mbsf) of the hole are comparable to those found from logging and from measurements on samples, they appear significantly lower over the shallowest and deepest parts of the hole. However, as a direct result of the relatively high velocities at Site 735, the usual timing uncertainties (several ms) give relatively large interval velocity uncertainties. It now appears that the lower interval velocities initially deduced for the upper and lower sections of the hole are not significant [S. Swift, personal communication, 1992].

A more robust estimate is obtained by using the VSP travel time vs. depth slope over the entire hole from 44 to 479 mbsf [Shipboard Scientific Party, 1989a], which gives an overall mean V_p of 6.51 ± .10 km/s (95% confidence). This mean compressional velocity appears significantly lower than that obtained with the sonic log (MCS, Table 1), a result which does not have obvious explanation. One possibility is lateral variability around Site 735, since the VSP samples a much larger volume of rock around the hole than the MCS log. However, the relatively smooth overall variation of velocity with depth for the MCS log suggests that lateral variability at similar scales is also minimal. Another possibility is that velocity dispersion exists,

which has been modeled to explain the lower apparent velocities deduced from VSPs vs. logs in commercial boreholes [e.g., Stewart et al., 1984]. At Site 735 the frequency contrast of the MCS log and the VSP is ~20 kHz vs. 20-40 Hz, respectively, which could explain the apparent difference in velocities obtained with these techniques for reasonable values of attenuation [Q~40; Goldberg et al., 1991a].

ELECTRICAL AND MAGNETIC MEASUREMENTS

The electrical and magnetic properties of oceanic basement rocks also contain important information on the structure, composition, and evolution of the crust. The intrinsic electrical resistivity of the most common crustal rock minerals at low temperatures (<200°C) is rather high (10^4-10^7 ohm-m), and lower values found in-situ and in samples usually reflect the effects of porosity and alteration to hydrous phases [e.g., Kirkpatrick, 1979]. The intensity of magnetization and magnetic susceptibility of crustal rocks are primarily dependent on mineral composition and grain size, as well as the extent of rock alteration. In particular, magnetic minerals such as magnetite and ilmenite, even in trace amounts, usually dominate the magnetic behavior of rock. Since these minerals are subject to alteration by sea water, the aging of oceanic crust is frequently accompanied by significant changes, particularly in magnetic properties.

Some of the deductions that may be made from measurements of electrical conduction and magnetic properties at Site 735 were briefly summarized in Von Herzen et al. [1991]. Electrical conductivity, natural remanent magnetization (NRM), and magnetic susceptibility all varied by several orders of magnitude at this site. Good correlation was found among these properties measured on samples in the laboratory, as well as downhole in-situ measurements, with significantly higher values for each related to the presence of Fe-Ti oxide gabbros. The porosity, which was quite low (<5%) throughout most of the section (see below), had a smaller effect on conductivity.

At the time of publication of the scientific results from Leg 118 [Von Herzen, Robinson et al., 1991], electrical resistivity of samples had been investigated only in the study by Pezard et al. [1991]. Recently, the measurements by Manghnani et al. [unpublished manuscript, 1992] on other samples from Hole 735B have become available for comparison. Overall, both sets of measurements show the same spatial variability downhole as the logging results (Figure 11), although there is considerable small-scale variability that is obviously not correlated. Detailed correlation would not necessarily be expected, because the samples give the resistivity over only a small volume (few cm³), whereas the logging tool (deep laterolog) averages over a much greater volume (few m³) of rock extending away from the hole axis. Furthermore, the log data shown in Figure 11 have been averaged over ~1.5 m vertical intervals. This averaging of the log data may partially explain why the sample values cover a greater range than

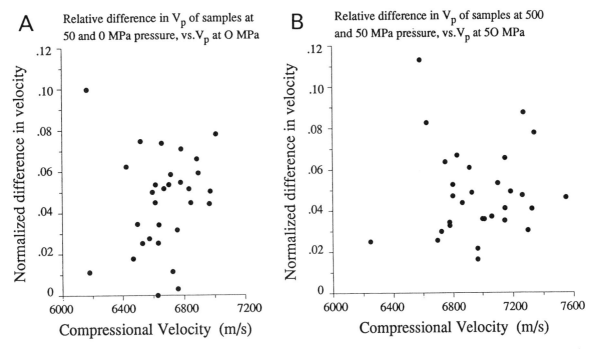

Fig. 10. A. Normalized difference in compressional velocity (V_p) measured on Hole 735B samples at 50 and 0 MPa, vs. V_p measured at 0 MPa. Mean = .0453 ± .023 (SD). B. Same as for A, for pressures of 500 and 50 MPa. Mean = .048 ± .021 [Manghnani et al., unpublished manuscript, 1992, Table 1].

the logging results.

A comparison of the sample resistivity values with depth appears to show systematic differences between the measurements of Pezard et al. [1991] and those of Manghnani et al. [unpublished manuscript, 1992]. Over most of the section, the latter study gives resistivities about 1 order of magnitude higher than the former. Assuming that resistivity values have a log normal distribution (Figure 12), the mean of the Pezard et al. [1991] values is 1.99 ± 0.51 (SD, n=29), and that of Manghnani et al. [unpublished manuscript, 1992] is 2.71 ± 0.83 (n=32), with values given in log_{10} of resistivity (ohm-m). These compare to the deep laterolog mean of 2.92 ± 0.82 (n=298). As a result of the large variability in each data set, none of these means differ from each other at the 95% confidence level, and this is true even when the data contributed by the unusually low resistivity Fe-Ti oxide gabbro layer (~225-280 mbsf) are removed. Even though most of the values of Manghnani et al. [unpublished manuscript, 1992] are higher than those obtained by Pezard et al. [1991], a few are even lower and may represent some of the samples with unusually high Fe-Ti content. These sample values are comparable with the lowest resistivities measured by the induction log, which is probably more accurate than the laterolog for lower resistivities. Otherwise it seems that there may be some systematic differences between the shorebased measurements of these laboratories, perhaps a result of somewhat different sample preparation and handling affecting the porosity or state of saturation. We note that the samples of Pezard et al. [1991] were apparently dried before measurements to determine density and porosity,

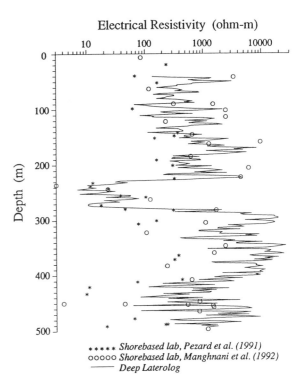

Fig. 11. Electrical resistivity measured on samples in shorebased laboratories [star symbols, Pezard et al., 1991; open circles, Manghnani et al., unpublished manuscript, 1992] plotted with the deep laterolog (LLD, solid line) data vs. depth in Hole 735B. Logging data are averaged over ~1.5 m depth intervals. Note logarithmic scale for resistivity.

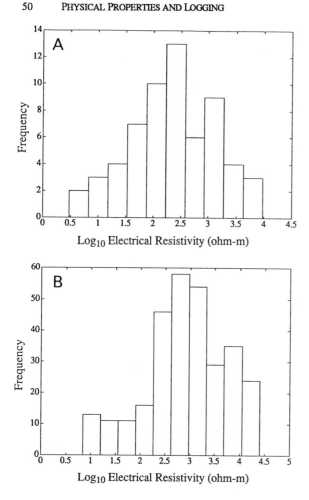

Fig. 12. Logarithmic (base 10) distributions of Hole 735B electrical resistivity values obtained from (A) samples and (B) logging (LLD) data (Figure 11). Note that both of these distributions appear approximately log-normal.

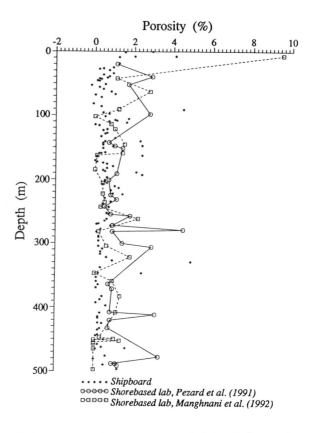

Fig. 13. Porosity measured on Hole 735B core samples by different investigators, using the technique of wet/dry weighing, plotted vs. depth below the seafloor. Small star symbols are shipboard values [Shipboard Scientific Party, 1989, Table 9], open circles connected by a solid line are those of Pezard et al. [1991], and open squares connected by a dashed line are from Manghnani et al. [unpublished manuscript, 1992].

then re-saturated, whereas those of Manghnani et al. [unpublished manuscript, 1992] were kept under continuous sea water saturation before measurement. It might be expected that the difficulties of re-saturating the former low-porosity samples would cause them to have higher resistivity than the latter, although mostly the opposite relationship was apparently observed.

Even though they may constitute only a small fraction of lower oceanic crust, the Fe-Ti-oxide gabbros might be expected to contribute significantly to its electrical conduction since their resistivity is about 2 orders of magnitude lower than normal gabbros. The evidence from active source electromagnetic soundings on the seafloor [e.g., Young and Cox, 1981] suggests that the lower crust is rather resistive, ~10^3 to 10^4 ohm-m. Although thin conductive layers at lower crustal depths are not resolved with such experiments, if present they would contribute to the mean conduction. However, they would not be important if they are laterally discontinuous over small (< few km) scales.

The magnetic properties of the rock samples from Hole 735B documented by Kikawa and Pariso [1991], and the magnetic logs and comparisons with the sample measurements given by Pariso et al. [1991], were briefly summarized by Von Herzen et al. [1991]. Here we describe only some of the highlights of these magnetic studies. Although the natural remanent magnetization (NRM) inclinations of 264 samples were about equally divided between normal and reversed directions, almost all showed stable reversed magnetization after they were subjected to alternating field demagnetization in the laboratory. This is interpreted to indicate that the entire 500-m thick drilled section cooled during a period when the earth's field was reversely magnetized (probably Chron 5a, based on the site survey data [Dick et al., 1991] , over a time interval of 0.5 m.y. or less. This seems to be a reasonable time period for crust that has been cooled by hydrothermal circulation (see below). The unstable NRM removed by the alternating field cleaning in the laboratory was probably induced by the drilling and coring procedures.

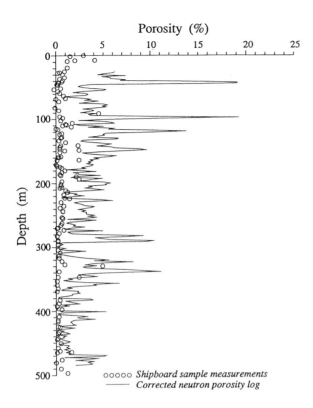

Porosity (%)

Fig. 14. Porosity vs. depth determined from shipboard measurements on Hole 735B core samples (open circles) and derived from in-situ measurements using the neutron log (solid line). The log data have been averaged over ~1.5 m depth intervals, and have been corrected for effects of rock alteration by sea water [Broglia and Ellis, 1990].

The mean stable NRM of the samples is about 1.6 A/m, a relatively high value, as pointed out by Kikawa and Pariso [1991]. If distributed uniformly over a 4-5 km thick lower crust, this level of magnetization would explain much of the magnetic anomaly amplitudes measured in the deep sea (± 100 nT), without any additional contributions from the upper crustal rocks. However, the latter may be needed to explain the anomaly amplitudes up to several hundred nT measured during the site survey around Hole 735B [Dick et al., 1991]. Kikawa and Pariso [1991] noted that the NRM intensities of the samples are apparently not affected by rock alteration, consistent with the minimal seawater alteration for most of Hole 735B samples.

The magnetic logging measurements in Hole 735B [Pariso et al., 1991] mostly confirmed the deductions from the sample measurements. The magnitude of the total magnetic field measured with the downhole magnetometer is closely correlated with the NRM of samples. Except for the portion of the section dominated by Fe-Ti-oxide gabbros (~225-280 mbsf), the relationship between these parameters appears linear. The variability in magnetic susceptibility is also well correlated between the core sample measurements and the downhole susceptibility log. However, the susceptibilities measured by downhole logging are systematically greater in magnitude by a factor

of ~3 than those of the samples at the same depths; the sample values were confirmed by repeat measurements using different instrumentation, and the logging tool calibration was checked after Leg 118 (J. Scott, personal communication, 1992), so that the discrepancy remains unresolved. Finally, the mean stable inclination of the magnetization over the section drilled at Site 735, corrected for a small (3-6 degrees) non-verticality of the borehole, is about 70 degrees, compared to an inclination of 52 degrees expected from a geocentric axial dipole field. Neither the southward movement of Site 735 since formation of the crust ~11.5 Ma, nor the secular variation of the earth's magnetic field, seem likely to explain the discrepancy; tectonic rotation of the section about a horizontal axis is suggested by Pariso et al. [1991].

POROSITY AND PERMEABILITY

These parameters are important to characterize the physical state, geological evolution, and dynamic processes that occur in oceanic crust. Since the first in-situ logging measurements that showed the relatively high porosity of the uppermost portion of oceanic crust [Kirkpatrick, 1979], the generally high porosity of pillow basalts has been confirmed. Seismic measurements [Purdy, 1987] suggest that the porosity of upper crustal rock may decrease with time, perhaps a result of alteration ('weathering') by circulating sea water. However, the thermal neutron logging tool commonly used to measure porosity in boreholes determines the concentration of hydrogen, which is also present in the hydrated alteration products, such that the true porosity may not be obtained when alteration is present. Only recently have attempts been made to correct such logs for the effects of alteration in oceanic crustal rock [Broglia and Ellis, 1990].

The magnitude and distribution of fluid permeability in oceanic crust are critical to evaluate the possibility and intensity of dynamic processes such as hydrothermal circulation [Lowell, 1991]. In the common marine sediments, fluid permeability has a general direct relationship to porosity [Bryant et al., 1974], and similar relationships could be expected for crustal rock, but few data are available to establish this even for frequently drilled oceanic basalt. Crustal rock permeability is best determined from in-situ pressure vs. time measurements [Becker, 1991], using borehole packers developed for hydrofracturing [e.g., Healy and Zoback, 1988]. Obviously, the thermal history of oceanic crust, and the related potential for rock alteration, are both affected substantially where the permeability is sufficient for fluid circulation. Unfortunately, the fluid permeability of oceanic crust is perhaps its least well established parameter, and may vary over many orders of magnitude with location and depth. The samples and logging measurements at Site 735 have given us the first opportunity for correlating measurements with different techniques in the gabbros which may represent the bulk of the lower oceanic crust.

The porosity of core samples is routinely determined aboard the drilling vessel as part of a standard suite of

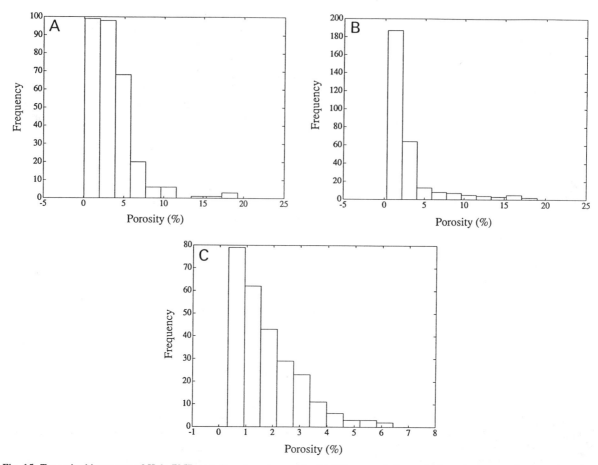

Fig. 15. Ten-point histograms of Hole 735B porosity values measured with different techniques: A. from the corrected neutron log (Figure 14), B. computed from the laterolog (LLD) data (Figure 11), C same as B, except data from low-resistivity, high Fe-Ti-oxide section (225-280 mbsf) omitted.

physical properties measurements; a total of 130 values were obtained during the drilling of Hole 735B [Shipboard Scientific Party, 1991a, Table 9]. Most of the values are very low, less than 1%, and never exceed 5% (Figure 13); the mean is 0.703 ± 0.822% (SD). A few values are even slightly negative, apparently a result of inaccurate weighing of wet and dried samples. The porosities measured on different samples in shorebased laboratories are similarly low: a mean of 1.466 ± 0.991% (SD, n=29) was obtained by Pezard et al. [1991], and 1.028 ± 1.676% (SD, n=32) by Manghnani et al. [unpublished manuscript, 1992] [0.755 ± 0.713% (SD, n=31) if one value differing from the mean by more than 2 standard deviations is omitted]. These low mean values are near the limits of resolution with the standard techniques of wet/dry weighing on small samples, and do not differ significantly from each other.

Obviously, the sample porosity does not necessarily represent in-situ values; in most cases, the former is significantly less than the latter, as a result of fracturing and other larger voids not present in core samples. This relationship between sample and in-situ porosity also appears valid for Hole 735B based on the porosity derived from the corrected neutron log (Figure 14). Peak values up to nearly 20% appear in the log data, and even higher

values in the data that have not been averaged over depth, particularly in the upper 100 m of the hole where other logging data indicate that significant fracture porosity exists. The neutron log data, corrected for the effects of alteration following Broglia and Ellis [1990], give a mean of 3.477 ± 2.860% (SD, n=302). This mean in-situ porosity is much less than that determined for the upper oceanic crust, either measured in-situ or deduced from seismic data [Salisbury et al., 1985; Moos, 1990; Carlson and Herrick, 1990, Table 3], but is comparable to values of a few percent inferred for the lower crust. However, even when the peaks that may represent fracture porosity or alteration in the uppermost 150 mbsf of Hole 735B are excluded, comparison with other data (see below) indicates that the mean porosity derived from the corrected neutron log may be somewhat higher than the sample values.

Another measure of porosity may be derived from the resistivity log (LLD). The common minerals in most rocks have relatively high resistivity, such that lower values measured in-situ are a result of conduction through the pore waters, in this case sea water. A simple way to estimate porosity (ϕ) from in-situ resistivity (ρ) is to use empirical equations, such as Archie's law, relating these parameters:

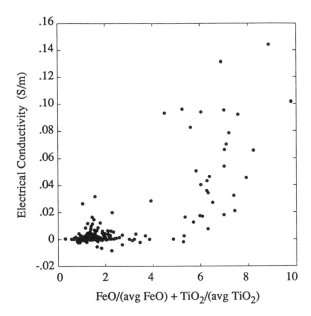

Fig. 16. Electrical conductivity corrected for neutron log porosity (Figure 12), vs. normalized FeO and TiO2 content of gabbros as measured by the geochemical log [GLT, Pelling et al., 1991, Figure 5]. Data points are derived from log values averaged over ~1.5 m depth intervals. Note that a few of these calculated matrix conductivities are slightly negative.

$$\rho = \rho_{sw} \cdot \phi^{-n} \qquad (1)$$

where ρ_{sw} is the resistivity of the pore water, and the exponent n can be determined by log-linear regression if ρ_{sw} is known, and ρ and ϕ have been measured. Although Archie's law originally was derived from resistivity data in sedimentary rocks, subsequent investigations [Hyndman and Drury, 1976; Kirkpatrick, 1979; Becker et al., 1982; Cann and Von Herzen, 1983; Pezard et al., 1991] show that it also appears to fit the relationships between these parameters for upper crustal rocks. The exponent n commonly has a value close to 2 for sedimentary rocks, but a wide range, from about 1.0 to 2.5, has been determined by these studies of oceanic rocks. From modeling, low values of n may be indicative of a high aspect ratio of the cracks that contribute to the electrical conduction. For Hole 735B, Pezard et al. [1991] obtained the very low value of 1.08, although the number of samples were relatively few (22) and the possible effects of Fe-Ti-oxide samples apparently were excluded using only preliminary petrological descriptions.

We calculated porosities from the resistivity log using Archie's law with exponent n=2. The resistivity of the pore water is assumed to be 0.25 ohm-m, approximately that of sea water at ~10°C [Von Herzen and Scott, 1991]. The resulting porosities have a mean of 2.800 ± 3.407% (SD, n=298), a value that compares reasonably well with the statistics of the porosity data derived from the neutron log, described above. If the data variability followed a Gaussian

distribution, which it apparently does not (see below), the porosity means from the neutron and resisitivity logs would not differ from one another at the 95% confidence level. It is also important to consider that the low resistivity in the Fe-Ti-oxide gabbros is caused primarily by conduction in the solid rock, not in the pore fluid. Error is introduced by using Archie's law to calculate porosities in such gabbros. To estimate this bias for Hole 735B, we eliminated the main Fe-Ti-oxide layer (~225-280 mbsf) from the porosity calculation. This gave a mean porosity of 1.734 ± 1.198% (SD, n=261), significantly lower than the mean given above with the layer included. Apparently all porosity values greater than about 6.5% derived from the resistivity log using Archie's law come from this Fe-Ti-oxide layer, and are therefore probably spurious (Figure 15). Some bias may remain from Fe-Ti-oxide gabbros that occur outside of this layer, however.

The main difference between the distributions of porosities derived from the corrected resistivity log (i.e., without the main Fe-Ti-oxide layer) and the neutron log (cf. Figure 15A and C) is the increased frequency of calculated porosity values for the neutron log in the 2 to 6% range, which gives the neutron log porosity a higher mean than that calculated from Archie's law. For these very low porosities, there may be at least 2 possible explanations for the discrepancy: 1) the corrections made to the neutron log for sea water alteration [Broglia and Ellis, 1990] may be biased for the low-porosity Site 735 gabbros, and 2) Archie's law may have a different exponent, or it does not hold for such porosities. For low porosities, it may be that a greater proportion of the porosity is not connected, which would exclude it from contributing to the electrical conduction, the implicit assumption in Archie's law. We did not attempt to optimize the numerical value of the exponential factor in Archie's law for Site 735 data, because of the additional uncertainties introduced by the highly conducting Fe-Ti-oxide matrix of some gabbros in the section (Figure 16). There is clearly a general correlation of matrix electrical conductivity with Fe-Te-oxide content, but there is considerable scatter, and even some negative values of calculated conductivity apparently because the neutron porosity is too high for the measured in-situ conductivity. Additional careful measurements and comparisons will be required to confirm these possibilities. Magnetic data may be useful to identify and eliminate Fe-Ti-oxide rocks at Site 735 in a careful test of the relationship between electrical resistivity and porosity in these gabbros.

The bulk fluid permeability of ocean crust is an important parameter controlling the possibility for fluid flow, i.e., hydrothermal circulation. It is also one of the most difficult parameters to measure directly, and drill holes that penetrate the ocean crust provide opportunities for such measurements, including the depth distribution of permeability. The measurements carried out in Hole 735B, described by Becker [1991], show that permeability generally decreases with depth in these gabbros over several orders of magnitude, from about 10^{-13} to 10^{-16} m^2, in the same range as values measured for basalts in other oceanic crustal holes. The measurements in oceanic crustal rocks summarized by Becker [1991, Figure 8] are reproduced here as Figure 17. We note that some of the basalts drilled in other holes, generally at greater depth below the seafloor, have even lower permeabilities than the gabbros drilled at Hole 735B.

The somewhat limited data on in-situ permeability summarized in Figure 17, measured on both basalts and gabbros, shows a systematic trend with depth. For the lowest measured values (~10^{-17} m^2) below a few hundred m depth, hydrothermal circulation is probably very sluggish or non-existent. This conclusion seems consistent with the geothermal data from Hole 735B [Von Herzen and Scott, 1991], indicating that the temperature gradient in the upper part of the hole primarily shows the effects of the surrounding ocean temperatures but increasingly reflects the conducted geothermal flux from below at greater depths. It also agrees with the modeling for Site 504 [Fisher et al., 1990], indicating that only a relatively thin layer of high permeability in the uppermost crust beneath the sediments can explain most of the observations.

All 3 holes in oceanic crust having permeability measurements are located in relatively young (<10-12 Ma) seafloor, although they include a range of different seafloor environments. Hole 395A is drilled in a relatively thin (~90 m) sediment-covered topographic depression on the mid-Atlantic ridge surrounded by high mountains, Hole 504B is on a relative high in modest topography but has relatively thick (~275 m) overlying sediment cover, and Hole 735B is on an uplifted ridge without sediments bounding a deep transform fault. It seems reasonable to expect that the low porosity gabbros comprising the lower oceanic crust would have lower permeability than upper level basalts with higher porosity, but Figure 17 indicates that this may not be the case. The unusual tectonic uplift of gabbros at Site 735 [Dick et al., 1991] may have increased their permeability by increasing the aperture and number of interconnected fractures. However, it seems unlikely that this process would leave the rocks at Site 735 with a permeability similar to that of oceanic basalts elsewhere. We tentatively hypothesize that depth below the seafloor, directly related to overburden confining stress, may be the controlling factor for permeability. Clearly, more data are needed to substantiate this or other hypotheses for permeability distributions, particularly for older seafloor, since time may also exert an important control on the effective permeability in crustal rock through the mechanism of sea water alteration [e.g., Thompson, 1983].

Although crustal rocks appear to follow a similar porosity vs. electrical resistivity relationship (Archie's law) as do sediments, their respective relationships of porosity to permeability are markedly different (Figure 18). For a given porosity, at least young (<15 Ma) oceanic crust is much more permeable than sediments. This difference is most likely a result of the contrasting porosity structure in these media. For the sediments, small particles are surrounded by the pore fluid of equal or greater volume, with pathways for

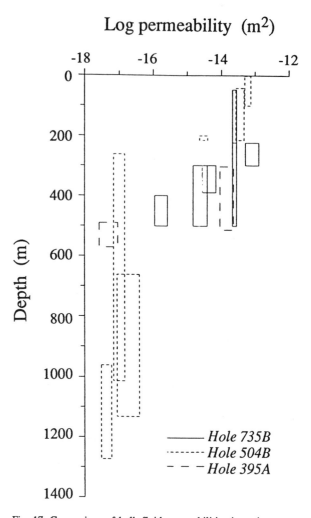

Fig. 17. Comparison of bulk fluid permeabilities in various ocean crustal drill holes vs. depth below the upper surface of the crust [from Becker, 1991, Figure 8]. Widths of rectangles indicate uncertainties in permeability values, and depth ranges over which measurements or calculations of permeability were made.

fluid flow that have dimensions comparable to that of the solid particles. For the crustal rocks, the fractures that make up much of the connected porosity have much larger dimensions, giving the medium a substantially higher permeability for a given porosity since the resistance to fluid flow decreases with the inverse square of the dimension confining the flow. Uncovering the systematic variation of crustal permeability with aging and other parameters is likely to be a focus of marine geophysical research for at least the next decade or so.

CONCLUSIONS

We have collated and summarized results from physical properties and downhole measurements at Hole 735B and other deep holes in ocean crust relevant to the in-situ physical state of crustal rocks. We focussed on measurements by seismic, electrical, and nuclear methods, both in the laboratory and in-situ downhole. Principal derived crustal properties are seismic velocities, electrical conductivity, and porosity (including its relationship to fluid permeability). The main results are:

1) Seismic velocities measured both on core samples and in-situ are comparable to those measured previously on gabbros and in refraction measurements of the lower oceanic crust. Even for relatively low (few percent) porosities, sample velocities are lowered significantly by drying; subjecting the dehydrated samples to ~100 MPa hydrostatic pressure gives approximately the same velocities as the wet samples. The increasing velocity with confining pressure up to 500 MPa is probably a result of crack closure as well as pressure effects on constituent minerals.

2) The distribution of electrical resistivity values, both from samples and logging (LLD), appears approximately log normal, with a high mean (at least several hundred ohm-m) as a result of low porosity. Unusual Fe-Ti-oxide gabbros have low intrinsic resistivity (<10 ohm-m). The stable remanent magnetization of samples is almost uniformly reversed from that of the present field, with a high mean amplitude ~1.6 A/m; the high magnetization suggests that the lower crust may contribute significantly to marine magnetic anomalies. Oversteepening of the mean magnetic inclination (~70°) at Site 735, compared to that expected from a geocentric dipole field (~52°), is probably a result of tectonic rotation.

3) Porosity is generally low (few percent) for both core samples and logs, except for a few thin zones with higher values measured in-situ in the upper 150 m of Hole 735B that probably represent fracture porosity. The corrected neutron porosity log systematically gives slightly higher porosities than those calculated from the resistivity log (LLD, excluding the Fe-Ti-oxide gabbro layer), using Archie's law. The magnitude and depth dependence of the fluid permeability of the gabbro section at Site 735 are comparable to those of basalt sections at other oceanic drill sites, suggesting that the effects of overburden stress on fracture aperture is the main parameter controlling permeability for most of ocean crust.

Acknowledgements. This study obviously would not have taken place without the skills and efforts of the many persons with different talents who make ODP deep ocean drilling so successful, supported internationally by dedicated administrators and citizens of the ODP member countries. Many of the scientific party from Leg 118, as well as our professional colleagues, provided ideas and encouragement. The Indian Ocean Drilling Results workshop held in Cardiff, U.K., in July 1991, was a particular stimulus. We are grateful to R. Duncan as editor, and R. Hyndman, S. Swift, and two anonymous reviewers for their reviews and suggestions to improve the manuscript. Contribution 8042 of the Woods Hole Oceanographic Institution.

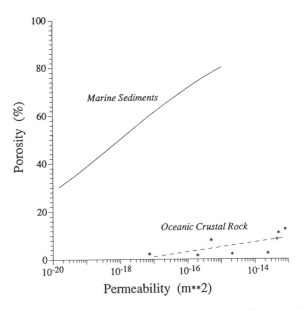

Fig. 18. Bulk permeability vs. porosity for marine sediments and for oceanic crustal rocks measured in-situ in drill holes. Solid line is the average relationship for marine sediments given by Bryant et al. [1974]. Dashed line is a linear fit to the data for crustal rocks shown as star symbols, taken from Anderson et al. [1985], Hickman et al. [1984], and Becker [1989, 1991].

REFERENCES

Anderson, R. N., Zoback, M. D., Hickman, S. H., and Newmark, R. L., Permeability versus depth in the upper oceanic crust: in-situ measurements in DSDP Hole 504B, eastern equatorial Pacific, *J. Geophys. Res., 90*, 3659-3669, 1985.

Becker, K., et al., In-situ electrical resistivity and bulk porosity of the oceanic crust, Costa Rica rift, *Nature, 300*, 594 -598, 1982.

Becker, K., Measurements of the permeability of the sheeted dikes in Hole 504B, ODP Leg 111, in *Proc. ODP, Sci. Results, 111*, edited by K. Becker, H. Sakai et al., 317-325, Ocean Drilling Program, College Station, TX, 1989.

Becker, K., In-situ bulk permeability of oceanic gabbros in Hole 735B, ODP Leg 118, in *Proc. ODP, Sci. Results, 118*, edited by R. P. Von Herzen, P. T. Robinson et al., 333-347, Ocean Drilling Program, College Station, TX, 1991.

Birch, F., The velocity of compressional waves in rocks to 10 kilobars, 1, *J. Geophys. Res., 65*, 1083-1102, 1960.

Broglia, C., and Ellis, D., Effect of alteration, formation absorption, and standoff on the response of the thermal neutron porosity log in gabbros and basalts: examples from Deep Sea Drilling Project - Ocean Drilling Program sites, *J. Geophys. Res., 95*, 9171-9188, 1990.

Bryant, W. R., DeFlache, A. P., and Trabant, P. K., Consolidation of marine clays and carbonates, in *Deep-sea Sediments, Physical and Mechanical Properties*, edited by A. L. Inderbitzen, *Marine Science, 2*, Plenum Press, 209-244, 1974.

Cann, J. R., and Von Herzen, R. P., Downhole logging at DSDP Sites 501, 504, and 505, near the Costa Rica rift, In *Init. Repts. DSDP, 69*, edited by J. R. Cann et al., 281-299, 1983.

Carlson, R. L., and Herrick, C. N., Densities and porosities in the oceanic crust and their variations with depth and age, *J. Geophys. Res., 95*, 9153-9170, 1990.

Dick, H. J. B. et al., Tectonic evolution of the Atlantis II fracture zone, in *Proc. ODP, Sci. Results, 118*, edited by R. P. Von Herzen, P. T. Robinson et al., 359-398, Ocean Drilling Program, College Station, TX, 1991.

Fisher, A. T., Becker, K., Narasimhan, T. N., Langseth, M. G., and Mottl, M. J., Passive, off-axis convection through the southern flank of the Costa Rica rift, *J. Geophys. Res., 95*, 9343-9370, 1990.

Goldberg, D., Badri, M, and Wepfer, W., Ultrasonic attenuation measurements in gabbros from Hole 735B, in *Proc. ODP, Sci. Results, 118*, edited by R. P. Von Herzen, P. T. Robinson et al., 253-259, Ocean Drilling Program, College Station, TX, 1991a.

Goldberg, D., Broglia, C., and Becker, K., Fracturing, alteration, and permeability: in-situ properties in Hole 735B, in *Proc. ODP, Sci. Results, 118*, edited by R. P. Von Herzen, P. T. Robinson et al., 261-269, Ocean Drilling Program, College Station, TX, 1991b.

Healy, J. H., and Zoback, M. D., Hydraulic fracturing in situ stress measurements to 2.1 km depth at Cajon Pass, California, *Geophys. Res. Lett., 15*, 1005-1008, 1988.

Hickman, S. H., Langseth, M. G., and Svitek, J. F., In-situ permeability and pore-pressure measurements near the mid-Atlantic ridge, DSDP Hole 395A, in *Init. Repts. DSDP, 78B*, edited by R. D. Hyndman, M. H. Salisbury et al., 699-708, U.S. Government Printing Office, Washington, D.C., 1984.

Hyndman, R. D., and Drury, M., The physical properties of oceanic basement rocks from deep-drilling on the mid-Atlantic ridge, *J. Geophys. Res., 81*, 4042-4052, 1976.

Kikawa, E., and Pariso, J. E., Magnetic properties of gabbros from Hole 735B, southwest Indian ridge, in *Proc. ODP, Sci. Results, 118*, edited by R. P. Von Herzen, P. T. Robinson et al., 285-307, Ocean Drilling Program, College Station, TX, 1991.

Kirkpatrick, R. J., The physical state of the oceanic crust: results of downhole geophysical logging in the mid-Atlantic ridge at 23°N, *J. Geophys. Res., 84*, 178-188, 1979.

Kuster, G. T., and Toksoz, M. N., Velocity and attenuation of seismic waves in two phase media, 1, Theoretical formulations, *Geophysics, 39*, 587-606, 1974.

Lowell, R. P., Modeling continental and submarine hydrothermal systems, *Rev. Geophys., 29*, 457-476, 1991.

Moos, D., Petrophysical results from logging in DSDP Hole 395A, ODP Leg 109, in *Proc. ODP, Sci. Results, 106/109*, edited by R. Detrick, J. Honnorez, W. B. Bryan, T. Juteau et al., 237-253, Ocean Drilling Program, College Station, TX, 1990.

Pariso, J. E., Scott, J. H., Kikawa, E., and Johnson, H. P., A magnetic logging study of Hole 735B gabbros at the southwest Indian ridge, in *Proc. ODP, Sci. Results, 118*, edited by R. P. Von Herzen, P. T. Robinson et al., 309-321, Ocean Drilling Program, College Station, TX, 1991.

Pelling, R., Harvey, P. K., Lovell, M. A., and Goldberg, D., Statistical analysis of geochemical logging tool data from Hole 735B, Atlantis fracture zone, SW Indian Ocean, in *Proc. ODP, Sci. Results, 118*, edited by R. P. Von Herzen, P. T. Robinson et al., 271-283, Ocean Drilling Program, College Station, TX, 1991.

Pezard, P. A., Howard, J. J., and Goldberg, D., Electrical conduction in oceanic gabbros, Hole 735B, Southwest Indian Ridge, in *Proc. ODP, Sci. Results, 118*, edited by R. P. Von Herzen, P. T. Robinson et al., 323-331, Ocean Drilling Program, College Station, TX, 1991.

Press, F., Seismic velocities, in Handbook of Physical Constants, edited by S. P. Clark Jr., 195-218, *Geol. Soc. Amer. Memoir 97*, 1966.

Purdy, G. M., New observations of the shallow seismic structure of young oceanic crust, *J. Geophys. Res., 92*, 9351-9362, 1987.

Robinson, P. T., Von Herzen, R. P. et al., *Proc. ODP, Init. Repts., 118*, Ocean Drilling Program, College Station, TX, 1989.

Salisbury, N. H., Christensen, N. I., Becker, K., and D. Moos, The velocity structure of layer 2 at DSDP Site 504 from logging and laboratory measurements, in *Init. Repts., 83*, edited by R. N. Anderson, J. Honnorez, and K. Becker et al., U.S. Government Printing Office, Washington, D.C., 1984.

Shipboard Scientific Party, Site 735, in *Proc. ODP, Init. Reports, 118*, edited by P. T. Robinson, R. Von Herzen et al., 89-222, Ocean Drilling Program, College Station, TX, 1989a.

Shipboard Scientific Party, Introduction and explanatory notes, in *Proc. ODP, Init. Reports, 118*, edited by P. T. Robinson, R. Von Herzen et al., 3-23, Ocean Drilling Program, College Station, TX, 1989b.

Spudich, P., and Orcutt, J., A new look at the seismic velocity structure of the oceanic crust, *Rev. Geophys., 18*, 627-645, 1980.

Stewart, R. R., Huddleston, P. D., and Kan, T. K., Seismic versus sonic velocities: a vertical seismic profiling study, *Geophysics, 49*, 1153-1168, 1984.

Swift, S. A., Hoskins, H., and Stephen, R. A., Seismic stratigraphy in a transverse ridge, Atlantis II Fracture Zone, in *Proc. ODP, Sci. Results, 118*, edited by R. P. Von Herzen, P. T. Robinson et al., 219-226, Ocean Drilling Program, College Station, TX, 1991.

Thompson, G., Basalt-seawater interaction, in Hydrothermal Processes at Seafloor Spreading Centers, edited by P. A. Rona et al., 225-278, *Marine Sciences, 12*, Plenum Press, 1983.

Von Herzen, R. P., Dick, H. J. B., and Robinson, P. T., Downhole measurements and physical properties, Hole 735B: Summary and tectonic relationships, in *Proc. ODP, Sci. Results, 118*, edited by R. P. Von Herzen, P. T. Robinson et al., 553-556, Ocean Drilling Program, College Station, TX, 1991.

Von Herzen, R. P., Robinson, P.T. et al., *Proc. ODP, Sci. Results, 118*, Ocean Drilling Program, College Station, TX, 1991.

Von Herzen, R. P., and Scott, J. H., Thermal modeling for Hole 735B, in *Proc. ODP, Sci. Results, 118*, edited by R. P. Von Herzen, P. T. Robinson et al., 349-356, Ocean Drilling Program, College Station, TX, 1991.

Wilkens, R. H., Fryer, G. J., and Karsten, J., Evolution of porosity and seismic structure of upper oceanic crust: importance of aspect ratios, *J. Geophys. Res., 96*, 17981-17995, 1991.

Young, P. D., and Cox, C. S., Electromagnetic active source sounding near the East Pacific Rise, *Geophys. Res. Lett., 8*, 1043-1046, 1981.

The Influence of Mantle Plumes in Generation of Indian Oceanic Crust

DOMINIQUE WEIS

Laboratoires Associés Géologie-Pétrologie-Géochronologie, Earth and Environment Sciences Department, Brussels Free University, CP 160/02, Avenue F. D. Roosevelt, 50, B-1050 Brussels, Belgium

WILLIAM M. WHITE

Department of Geological Sciences, Cornell University, Ithaca, NY, 14853, USA

FREDERICK A. FREY

Department of Earth, Atmospheric and Planetary Sciences, Massachusetts Institute of Technology, Cambridge, MA, 02139, USA

ROBERT A. DUNCAN AND MARTIN R. FISK

College of Oceanography, Oregon State University, Corvallis, OR, 97331, USA

JOHN DEHN

GEOMAR, Wischhofstrasse 1-3, D-2300 Kiel 14, Germany

JOHN LUDDEN

Université de Montréal, Faculté des Arts et des Sciences, Département de Géologie, C.P. 6128 succursale A, Montréal (Québec) H3C 3J7, Canada

ANDREW SAUNDERS AND MICHAEL STOREY

Department of Geology, University of Leicester, Leicester LE1 7RH, UK

The physical and chemical behavior of mantle plumes and hotspots, and their role in the generation of oceanic crust, has been investigated through deep ocean drilling in the Indian Ocean. High density sampling of crustal rocks has established the continuity of volcanic activity along the two major hotspot tracks in this basin - those formed over the Réunion and Kerguelen plumes. The volcanic lineaments link present hotspots to huge flood basalt provinces which are the initial occurrence of mantle plumes. Drilling in the Indian Ocean has provided the first significant sampling of oceanic plateaus, such as large portions of the Kerguelen Plateau, showing that these are entirely volcanic in origin and are probably oceanic equivalents of continental flood basalt provinces. These are significant components in intraplate ocean crust formation and are derived from melting of plume and asthenosphere mantle mixtures.

Major and trace element, and isotopic compositions of erupted products changed with time along these two hotspot tracks. The compositional variation in Réunion hotspot products is consistent with early entrainment of asthenospheric mantle within the plume, and subsequent gradual increase in the proportion of plume material in the mantle melted. A similar evolution in Kerguelen hotspot magmas is observed, but compositional variations are more complex and require three mantle components (asthenosphere and two distinct plume compositions). Both hotspots are now located well away from plate boundaries but earlier lay near or at spreading ridges. Some of the compositional changes along hotspot tracks could have resulted from varying proportions of asthenosphere and plume mantle mixing, correlated with the plate tectonic setting.

INTRODUCTION

The Indian Ocean basin offers excellent opportunities to investigate both the chemical and physical evolution of mantle plumes, their associated hotspot tracks, and their interaction with the surrounding mantle. In addition, the temporal evolution of Indian Ocean crust formed at

Synthesis of Results from Scientific
Drilling in the Indian Ocean
Geophysical Monograph 70
Copyright 1992 American Geophysical Union

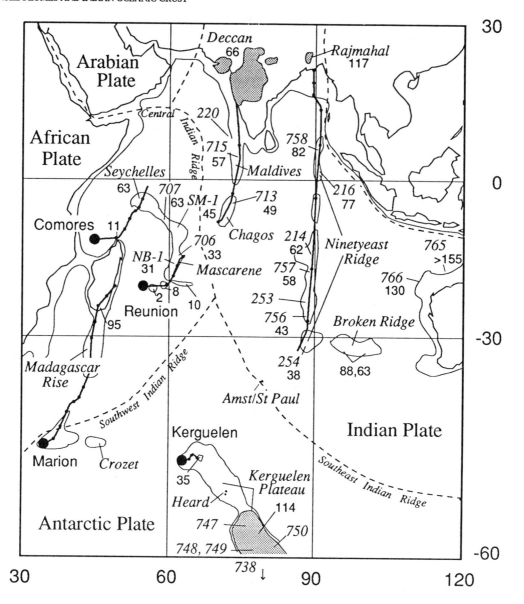

Fig. 1. Bathymetric and plate tectonic features, including modeled hotspot tracks, of the Indian Ocean basin and the Ocean Drilling Program sites (and selected Deep Sea Drilling Project sites) that reached oceanic crust (large italics). Ages shown are radiometric age determinations (Ma) for basalts, mainly from the two prominent volcanic lineaments: the Réunion and the Kerguelen hotspot tracks. Flood basalt provinces (stippled) mark the initiation of each of these hotspots. Computer-modeled tracks (dots are 10 m.y. increments of volcanic activity) assume fixed hotspots [see Duncan and Storey, this volume]. SM-1 and NB-1 are industry wells on the Mascarene Plateau.

spreading ridge axes can be inferred by comparing the compositions of recent axial lavas with the 155 Ma old basalts recovered during ODP Leg 123. Prominent hotspot tracks, such as the Chagos-Maldives and Ninetyeast ridges, and the associated basalt outpourings of the Deccan Traps, Rajmahal Traps and Kerguelen Plateau, record substantial parts of the life history of major, individual plumes, presently located beneath the volcanic islands of Réunion and Kerguelen. Rapid, northward migration of the Indian Plate over these hotspots during Cretaceous and early

Cenozoic times produced parallel volcanic traces that allow not only high-resolution testing of the concept of hotspot fixity and comparison of the hotspot and paleomagnetic reference frames, but also investigation of the compositional evolution of mantle sources. Furthermore, in the Indian Ocean, plume-related volcanism has occurred in a large variety of plate tectonic settings ranging from continental, where magmas have been erupted through thick, old Gondwanan lithosphere, to ocean spreading ridge, where the high extension factors have allowed the plume mantle to undergo substantial decompression and melt generation. Subsequently, both Réunion and Kerguelen plumes have been overridden by oceanic lithosphere of different characteristics, and are now erupting lavas in the intraplate setting. We thus have an opportunity to test models of plume evolution and plume-lithosphere interactions, ranging from those models which advocate a sudden onset of voluminous magmatism associated with the arrival at the base of the lithosphere of a large 'start-up' plume head [Richards et al., 1989; Griffiths and Campbell, 1990], to those models which propose the build-up of a plume head by a more steady-state plume flux, and melt release via lithosphere extension [e.g., Kent, 1991; White and McKenzie, 1989].

Ocean drilling, sampling basalt that hitherto had been hidden beneath a thick carapace of sedimentary rocks, therefore complements on-land studies. We have now assembled a substantial chemical database with which to evaluate the evolution of these two prominent plume systems. Questions we particularly wish to address are:

(1) What is the nature of the interaction between the plume and the overlying lithosphere; what physical influence does this interaction have on the volume and composition of mantle melts? For example, if a plume impinges on the base of thick lithosphere, it may not be able to decompress sufficiently for the peridotite to begin to melt. Alternatively, if the plume interacts with a spreading axis, the melting may be extensive. Thus, for a constant plume flux rate, the magma output should vary as a function of lithosphere thickness. Furthermore, the mineralogy of the melting assemblage may vary with pressure (e.g. garnet-present or garnet-absent), resulting in variations in the composition of melt produced.

(2) What is the chemical influence of plume-lithosphere interactions? In particular, does the plume mobilize low-melting point fractions within the lithospheric mantle? This is particularly important because both Réunion and Kerguelen plumes have at some stage underlain old, possibly isotopically distinct continental lithosphere.

(3) Does the plume mantle show compositional and physical evolution with time? Are there fluctuations of plume flux with time? For example, large igneous provinces such as the Kerguelen and Mascarene plateaus may be the oceanic equivalent of continental flood basalts, forming when large thermal anomalies in the mantle decompress and undergo rapid melting.

(4) How does the plume mantle interact with the surrounding asthenospheric mantle? Have the plumes been responsible for contaminating large volumes of the upper mantle beneath the Indian Ocean, or are the effects more local, and restricted to adjacent ridge segments?

SITES DRILLED

Understanding the relative importance of isolated reservoirs and convective mixing in the mantle requires abundant data on the temporal and spatial scale of ocean crust compositional heterogeneities. With this objective, portions of five legs of the Ocean Drilling Program in the Indian Ocean were focused on sampling the oldest ocean crust in the basin (Leg 123) and a large age range of products from two vigorous, long-lived hotspots (Legs 115, 119, 120, and 121). A major aspect of this program was to determine the longevity of geochemical signatures of hotspots that reflect the degree of convective isolation of mantle reservoirs. Both Réunion and Kerguelen hotspots have been overridden by spreading ridges, so the interaction between asthenosphere and hotspot mantle material can be evaluated by comparing volcanic products erupted in on-ridge and intraplate settings. Important topographic components of the ocean basins that appear to record very significant mantle convection events are large volcanic plateaus that may be oceanic equivalents of continental flood basalt provinces, forming as the initial products of mantle plume activity. Two of these plateaus (Mascarene Plateau and Kerguelen Plateau) were investigated during three legs (115, 119 and 120) of the Indian Ocean drilling program.

Figure 1 illustrates the bathymetric and plate tectonic features of the Indian Ocean basin and the Ocean Drilling Program and some comparative Deep Sea Drilling Project sites that recovered oceanic crust.

Leg 115 drilling recovered basaltic rocks at four sites along the Réunion hotspot track [Backman, Duncan et al., 1988]; two of these sites confirmed the volcanic character of the Mascarene Plateau. These new sites complement studies made of the geology of the Deccan flood basalts of western India, the islands of Réunion, Mauritius, and Rodrigues, and volcanic rocks supplied by industry wells from the Nazareth and Saya de Malha banks (Mascarene Plateau), making a temporal series of ten sites with which to examine the history of this hotspot.

Legs 119 and 120 recovered basalts from five sites on the central and southern Kerguelen Plateau [Barron, Larsen et al., 1989; Schlich, Wise et al., 1989], confirming the oceanic origin of this huge feature and its relation to the Kerguelen hotspot. Leg 121 sampled three sites along the Ninetyeast Ridge portion of the Kerguelen hotspot track [Peirce, Weissel et al., 1989]. Previous DSDP drilling at four sites (Legs 22 and 26), dredged samples from Broken Ridge, and subaerial samples from the Rajmahal flood basalts of eastern India and the Kerguelen Archipelago provide a sampling density for the Kerguelen hotspot system comparable to that for the Réunion system.

An important characteristic of several Ninetyeast Ridge drill sites is the abundance of ash that immediately overlies the lavas. At Site 756, only trace amounts of volcaniclastic debris occur in the sediments, but at Site 253, there is 405 m of mid-Eocene [approximately 46 Ma; all ages assigned from the timescale of Harland et al., 1990] vitric volcanic ash and lapilli and at Site 214, 100 meters of interbedded lignite and volcaniclastic material are present. Drilling at Site 757 penetrated 369 meters of sediments before encountering basaltic basement and the lowermost sedimentary unit comprises 157 meters of basaltic volcaniclastic material with upper Paleocene (approximately 55-59 Ma) microfossils. At Site 758, 67

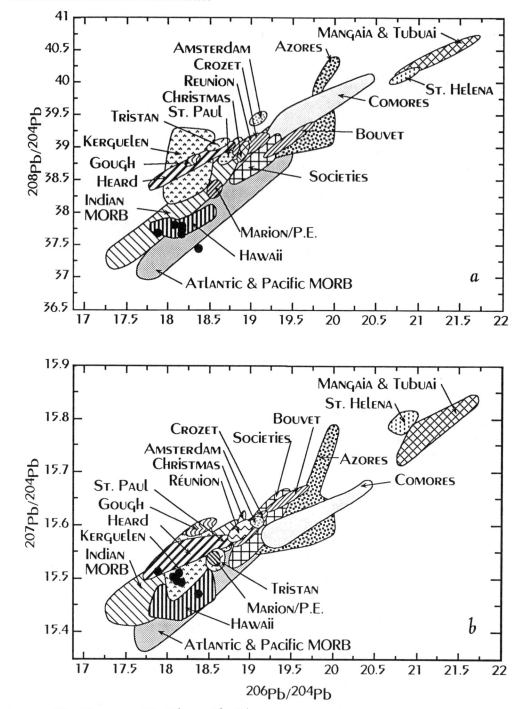

Fig. 2. Plot of (a) $^{208}Pb/^{204}Pb$ and (b) $^{207}Pb/^{204}Pb$ vs. $^{206}Pb/^{204}Pb$ for MORB and various ocean islands. Note the higher $^{207}Pb/^{204}Pb$ and $^{208}Pb/^{204}Pb$ of Indian MORB and Indian Ocean islands (labelled) relative to Atlantic and Pacific MORB and islands. Filled circles are data for Leg 123 basalts [Ludden and Dionne, 1992].

meters of volcanic ash overlie the basement lavas while the oldest sediments recovered are Campanian (73-83 Ma) tuffs with minor interbeds of ashy chalk.

Deep crustal drilling in the Argo Abyssal Plain during Leg 123 has allowed definition of the composition of magmas associated with the final stages of rifting at a continental margin and formation of the earliest Indian Ocean crust. In addition, the total geochemical budget of sediments and crust at this site provides a reference of material entering the subduction zone just to the north (Java Trench) for use in global mass-balance, lithosphere recycling models [e.g., Hofmann and White, 1982].

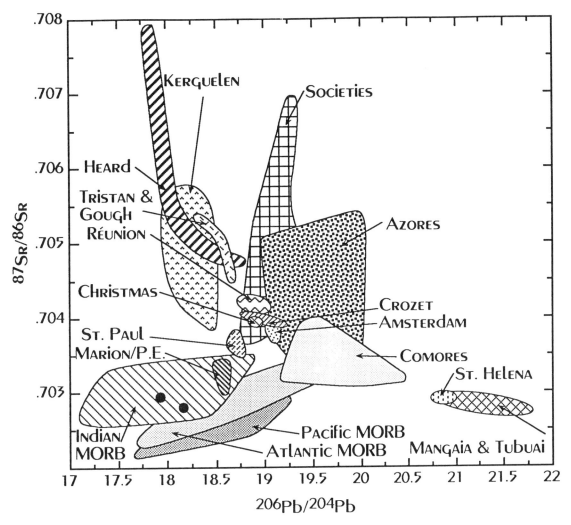

Fig. 3. Plot of $^{87}Sr/^{86}Sr$ vs. $^{206}Pb/^{204}Pb$ for MORB and various ocean islands. Indian Ocean MORB and islands tend to have high $^{87}Sr/^{86}Sr$ and low $^{206}Pb/^{204}Pb$ relative to Atlantic and Pacific MORB and islands.

Detailed site descriptions, sample recovery, and preliminary analyses can be found in the initial reports for specific legs.

INDIAN OCEAN CRUST

Basaltic ocean crust forms principally at plate boundary spreading ridges by passive asthenospheric upwelling but can be locally augmented by mantle plumes. The crustal composition reflects the mix of underlying mantle compositions contributing to melting at that location and time. Mantle geochemical heterogeneity of a variety of length scales has been documented along presently active spreading ridges where the volcanic samples are accessible with dredge and submersible [e.g., Dupré and Allègre, 1983; Hart, 1984; Schilling et al., 1983; Hanan et al., 1986; Zindler et al., 1984; Langmuir et al., 1986].

Young oceanic crust forming at spreading centers consists of tholeiitic basalt characterized by low concentrations of incompatible elements such as K, Rb, Cs, Ba, and the light

rare earth elements (REE). This material is referred to as normal mid-ocean ridge basalt or N-MORB. Because of this low abundance of incompatible elements, the term "depleted" is applied to most MORB and to the mantle that melts to produce MORB. Isotopic ratios of Sr, Nd, and Hf in N-MORB further demonstrate that the incompatible-element-depleted nature is inherited from a portion of the mantle, presumably the asthenosphere, that acquired its depleted character long ago [e.g., Gast, 1968; DePaolo and Wasserburg, 1976; Patchett and Tatsumoto, 1980]. This depletion is conventionally thought to have resulted from previous melting episodes, with the complementary incompatible-element inventory now resident in the continental crust. Other so-called E- or P-MORB are relatively enriched (less depleted) and are generally, though not universally, associated with mantle plumes [Schilling et al., 1983; le Roex et al., 1985].

Most basalts erupting at modern Indian Ocean spreading centers have low abundances of incompatible elements similar to Atlantic and Pacific MORB, although there may be subtle differences in incompatible elements between Indian MORB and Atlantic and Pacific MORB which have not yet been documented in detail. Indian MORB, however, are distinct from Atlantic and Pacific MORB in their Sr, Nd, and Pb isotopic composition. This distinctive nature of Indian MORB was first noticed by Subbarao and Hedge [1973], but because of the limited data available and questions about possible seawater alteration, it was not until the work of Dupré and Allègre [1983] that the distinctiveness of Indian Ocean MORB was firmly established. According to the compilation of Ito et al. [1987], Indian MORB has, on average, higher $^{87}Sr/^{86}Sr$ (0.7029) and lower ε_{Nd} (8) and $^{206}Pb/^{204}Pb$ (17.97) than Atlantic and Pacific MORB (0.7026, 10, and 18.40 respectively). In addition, Indian MORB has higher $^{208}Pb/^{204}Pb$ and $^{207}Pb/^{204}Pb$ for a given $^{206}Pb/^{204}Pb$ than Pacific or Atlantic MORB. These features can be seen in Figures 2 and 3.

One of the more remarkable features of the Indian Ocean is that at least some of the unique isotopic characteristics of Indian MORB are shared by Indian Ocean island basalts. Although the large range in $^{87}Sr/^{86}Sr$ and ε_{Nd} of oceanic island basalts precludes ready comparisons, it is apparent that many Indian Ocean island basalts share the Pb isotopic features of Indian MORB, namely high $^{208}Pb/^{204}Pb$ and $^{207}Pb/^{204}Pb$ for a given $^{206}Pb/^{204}Pb$ (Figure 2). Furthermore, the low $^{206}Pb/^{204}Pb$ is generally associated with high $^{87}Sr/^{86}Sr$ (Figure 3). Hart [1984] called the area in which these geochemical features dominate the "DUPAL anomaly", and suggested that they occurred in a globe-encircling belt centered at 30°S. The South Atlantic islands, Tristan da Cunha and Gough, clearly share these characteristics, as do Atlantic MORB from south of about 20-30°S [Hanan et al., 1986], so it seems fairly clear that this distinctive mantle signature extends into the South Atlantic. These geochemical characteristics are, however, rare among Pacific oceanic islands, and are completely absent from South Pacific MORB sampled so far [e.g., MacDougall and Lugmair, 1986; White et al., 1987, Klein et al., 1988]. DUPAL-like isotopic features are also largely absent in basalts from other localities [e.g., Northeast Pacific seamounts, Desonie and Duncan, 1990; the Azores, Dupré, 1983; the Samoan and Cook islands, Palacz and Saunders, 1986; Wright and White, 1987], but they are especially common in the Indian Ocean and South Atlantic, and only in the Indian Ocean is the DUPAL character so pervasive in MORB.

The southeastern boundary of the DUPAL geochemical province is well defined and occurs at the Australian-Antarctic Discordance zone of the Southeast Indian Ridge [Klein et al., 1988; Pyle et al., 1992]. The western boundary consists of a more gradual transition in the South Atlantic [Hanan et al., 1986]. The transition to normal Atlantic-Pacific isotope geochemistry that appears to occur on the Southwest Indian Ridge at about 26°E [Mahoney et al., 1992], may represent the southern boundary of the DUPAL geochemical province. Red Sea MORB have no DUPAL-like characteristics despite being an extension of the Indian ridge crest [Eissen et al., 1989].

The nature of this large-scale geochemical feature and its implications for mantle evolution and dynamics are of considerable interest. Important questions include: What is the origin of this unique isotopic signature? What communication is there between the asthenosphere, presumably the source of MORB, and the deep mantle, presumably the source of plumes producing oceanic island volcanism? How long has this feature been in existence? In answer to the first question, Dupré and Allègre [1983] suggested that recycling of oceanic crust and sediments might be the ultimate cause of the isotopic signature associated with the DUPAL anomaly. Ben Othman et al. [1989] found that this possibility is consistent with isotope and parent-daughter ratios in modern marine sediments, but further work will be necessary to establish this hypothesis unequivocally. This and other questions have been addressed by other workers and in part by ODP drilling in the Indian Ocean.

Storey et al. [1989] suggested that the unique isotope geochemistry of Indian MORB resulted from 'contamination' of Indian Ocean asthenosphere by the Kerguelen mantle plume. However, the $^{206}Pb/^{204}Pb$ ratio in recent Kerguelen plume products is not as low as in many Indian MORB. Also the growth of Indian Ocean asthenosphere corresponds to the waning activity of the Kerguelen plume flux [Duncan and Storey, this volume], making it difficult to see how a DUPAL flavor could be maintained. An alternative hypothesis is that the low $^{206}Pb/^{204}Pb$ is derived from the sub-Gondwanan lithosphere [e.g., Mahoney et al., 1989; Mahoney et al., 1992; Storey et al., 1989, 1992]. It is, however, difficult to understand how continental lithosphere could remain fixed beneath a spreading center for 200 m.y. given the tendency for asthenospheric motion to drive it away. Neither of these hypotheses explain why the DUPAL signature is present in both Indian MORB and Indian oceanic island basalts (OIB). Yet another alternative hypothesis is that the mass flux from several Indian Ocean plumes, rather than just Kerguelen, gives the Indian Ocean asthenosphere its DUPAL flavor [Mahoney et al., 1992]. This might be more viable; it at least explains why the DUPAL signature is present in both Indian OIB and MORB.

Determining the nature of the earliest Indian Ocean crust was one of the objectives of Leg 123. The time of the first appearance of the DUPAL signature in the Indian Ocean would be an important constraint on the origin of this anomaly. The results, however, are ambiguous. Figures 2 and 3 show that 155 Ma basalts from Sites 765 and 766 in the Argo Abyssal Plain have Pb and Sr isotope ratios which plot predominantly in the area of overlap between Atlantic-Pacific MORB and Indian MORB [Ludden and Dionne, 1992]. Thus it remains unclear whether the DUPAL signature was present in the earliest Indian MORB.

Drilling on Leg 123 also provided insight into magmatic processes associated with continental rifting and opening of ocean basins. Basalts at Sites 765 and 766 are

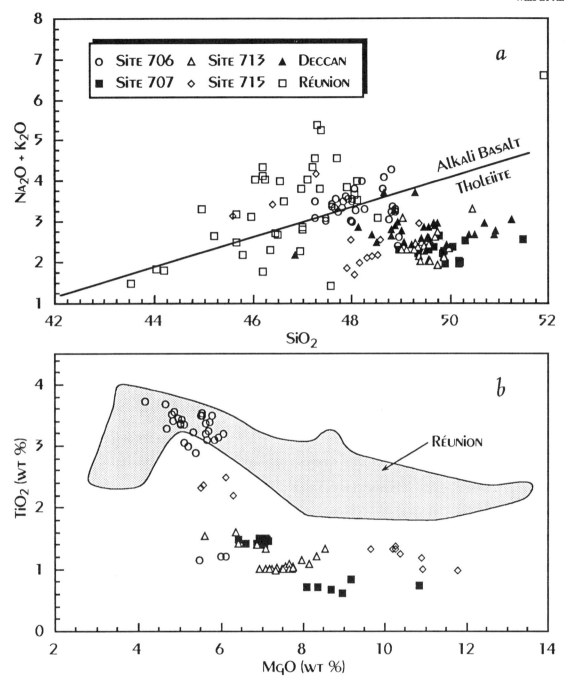

Fig. 4. (a) Alkali-silica diagram comparing Leg 115 basalts with those of Réunion Island and the Deccan flood basalts. Line divides Hawaii tholeiites and alkali basalts [Macdonald and Katsura, 1964]. Deccan basalts fall mostly in the tholeiite field, Réunion mostly in the alkali basalt field. Leg 115 basalts fall mainly in the tholeiite field. Data from Backman, Duncan et al. [1988], Fisk et al. [1988], Cox and Hawkesworth, [1985]. (b) TiO_2 in Leg 115 basalts as a function of MgO. The plot illustrates the intrasite variations that occur. Data from Backman, Duncan et al. [1988]. Shaded area is field for Réunion [data from Fisk et al., 1988].

characterized by higher CaO, FeO and lower Na_2O, Al_2O_3, $(La/Sm)_N$ (normalized to chondritic abundances) and incompatible element concentrations than basalts erupted at the present Indian Ocean spreading centers. In this respect, they share many geochemical similarities with tholeiitic basalts of the Red Sea rift [Eissen et al., 1989].

Both Site 765 and the Red Sea basalts were formed at the onset of rifting of the Indian Ocean at 155 Ma and 3-5 Ma, respectively. Ludden and Dionne [1992] interpreted these geochemical characteristics as reflecting high degrees of melting of asthenosphere resulting from the enhanced thermal regime. In the case of the Red Sea, the thermal anomaly is reflected by the presence of the Afar hotspot; for the northwestern Australian margin, excess melting is indicated by extensive plateaus of volcanic rocks that were formed at the time of continental break-up (160-130 Ma). Thus the chemical characteristics of the Red Sea, Site 765 and Site 766 may be applicable generally to oceanic crust formed at the onset of rifting. Basalts from Site 707 of Leg 115 were erupted in a similar environment (as the Seychelles Bank was being rifted from India) and have similar chemical characteristics.

PLUME-RELATED MAGMATISM AND HOTSPOT TRACKS

The Indian Ocean basin contains numerous ocean islands which may be linked to mantle plumes. In several instances, the islands are contiguous with, or form age-progressive lineaments with submarine volcanic ridges and large outpourings of continental or oceanic flood basalts. Paramount among these are the Réunion/Mauritius Islands-Mascarene Plateau and Chagos/Maldives Ridge to Deccan Traps trend, the Kerguelen Islands-Ninetyeast Ridge and Broken Ridge-Rajmahal Traps and Kerguelen Plateau trend; and the Marion/Prince Edward Island-Madagascar Rise-Madagascar Basalt Province trend (Figure 1). Other islands probably associated with mantle plumes include the Crozet and Comores Islands groups and St Paul/Amsterdam Islands. The total volume of crust generated by these plume systems is large: for example, the combined crustal volume of Kerguelen Plateau, Broken Ridge and Ninetyeast Ridge may exceed 80×10^6 km^3 [Schubert and Sandwell, 1989]. Therefore any model which attempts an integrated study of plume-mantle evolution must consider the processes associated with the generation and extraction of large volumes of melt from the mantle.

LEG 115: THE RÉUNION HOTSPOT TRACK

Among the principal objectives of drilling on Leg 115 were: (a) determination of the nature and origin of three prominent bathymetric features in the western Indian Ocean: the Mascarene Plateau, the Saya de Malha Bank and the Chagos-Maldives Ridge; (b) examination of the possible relationship between these features and the activity of the mantle plume presently located beneath Réunion Island; and (c) examination of the possible relationship between the Réunion mantle plume and Deccan flood basalt volcanism 65 million years ago. Accordingly, basement drilling was undertaken at four locations (Figure 1). Site 706, located on the northern margin of the Nazareth Bank, penetrated 77 m of pillow basalt. Site 707 penetrated 63 m of subaerially erupted basalts from the saddle between Saya de Malha and Seychelles banks. Site 713, on the northern margin of the Chagos Bank, penetrated 85 m of basalts. These were intercalated with sediment, indicating submarine eruption, though their vesicular nature suggests eruption at comparatively shallow (<1000 m) depth. Site 715, on the northeastern edge of the Maldives Ridge, penetrated 77 m of subaerially erupted basalt [Backman, Duncan et al., 1988; Shipboard Scientific Party, 1989]. In addition, samples from two

industry drill sites, SM-1, which penetrated 830 m of basalt on the Saya de Malha Bank, and NB-1, which penetrated 160 m of basalt and trachyte on the Nazareth Bank, were also studied. Thin ash horizons were distributed throughout the lower sedimentary sequence of Sites 706 and 713. They become rare upsection and are probably derived from the nearby volcanic centers. The absence of thick ash deposits similar to those seen in the sites along the Ninetyeast Ridge is probably because Sites 706 and 713 were at the margin of the volcanic edifice but the Ninetyeast Ridge sites were drilled close to the volcanic centers.

Geochronology and Age Relationships. Duncan and Hargraves [1990] determined crystallization ages for basaltic rocks recovered from the Réunion hotspot track by ^{40}Ar-^{39}Ar incremental heating experiments. The ages agree closely with biostratigraphic ages of sediments immediately overlying the volcanic rocks, and no variation in age could be detected within each of the sites. The age distribution of the volcanism matches that predicted from a stationary hotspot and plate reconstructions based on seafloor spreading data [Duncan, 1990; Duncan and Richards, 1991; Royer et al., 1991].

From a comparison of the age of volcanism at each of the drilling sites and the age of the seafloor on which the volcanoes were constructed, inferred from the magnetic anomaly pattern of the western Indian Ocean basin [Schlich, 1982], it is probable that the hotspot was close to the ancestral central Indian spreading ridge between 63 and 33 Ma. Site 707 basalts were erupted at 63.2±1.3 Ma, about the time seafloor spreading along the nascent Carlsberg Ridge was rifting the Seychelles from western India, some 2 million years after the main pulse of Deccan volcanism. Site 715 basalts were erupted on an oceanic island at 56.8±1.8 Ma, perhaps no more than a few hundred kilometers northeast of the ancestral Carlsberg–Central Indian Ridge. Site 713 and industry well SM-1 (Saya de Malha Bank) are nearly contemporaneous at 49 and 45 Ma, respectively, and they may have formed a large volcanic plateau similar in size to Iceland before spreading of the Central Indian Ridge separated them [McKenzie and Sclater, 1971]. At this time the hotspot lay approximately beneath the spreading ridge. Site 706 basalts were erupted at 33.4±0.5 Ma in a deep submarine environment, perhaps also near the Central Indian Ridge, which began the present phase of northeast-southwest spreading at that time. Lavas at Site NB-1 erupted a few million years later, at 31.5±0.2 Ma.

General Compositional Features and Intrasite Variability. Most basalts recovered during Leg 115 are hypersthene-normative tholeiites [Baxter, 1990] and fall in the tholeiite field of an alkali-silica diagram (Figure 4a). Although some Site 706 and 715 samples scatter into the alkali basalt field, they are petrologically and chemically more closely related to the Deccan tholeiites than to the mildly alkalic basalts of Réunion. Plagioclase is the most abundant phenocryst in most lithologic units from 3 of the 4 sites

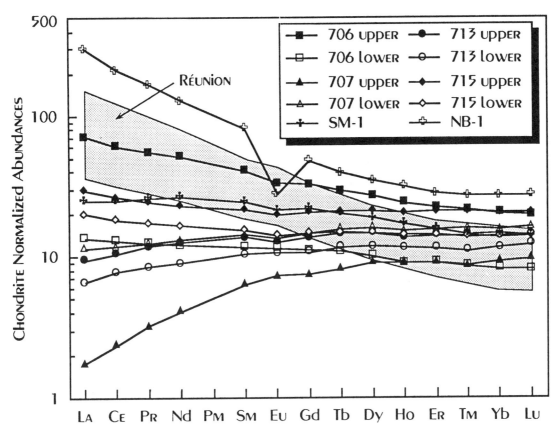

Fig. 5. Average chondrite-normalized rare earth abundances for the identified chemical groups in Leg 115 basalts (W. M. White, unpublished ICP-MS analyses). Shaded field shows range of rare earth patterns in Réunion lavas [Fisk et al., 1988].

(706, 707, and 713) and in this sense most Leg 115 basalts differ from those of Site 715, Deccan, Mauritius, and Réunion, where olivine is the dominant phenocryst. The mineralogy of lavas at Sites 706, 707, and 713 suggests they are volcano-capping flows similar to the last eruptive phase of Piton des Neiges, Réunion Island [Upton and Wadsworth, 1972]. On Réunion the isotopic and trace element ratios of these volcano-capping flows, however, are representative of the main phase of volcanism [Fisk et al., 1988] and presumably this is also true at Sites 706, 707, and 713.

The alteration of Leg 115 basalts ranges from slight to moderate with loss-on-ignition typically less than 1 wt. %. Feldspars are rarely albitized and fresh glass is present in some samples. Leg 115 lavas appear to be much fresher than those recovered from the Ninetyeast Ridge. The difference may be related to distance from center of volcanic activity to the sample location. All samples from Leg 115 were drilled from the margins of the hotspot track because the thick carbonate cap prevented drilling near the axis of the track. Ninetyeast basalts were drilled near the center of the hotspot track and therefore could have experienced more significant hydrothermal alteration at slightly elevated temperatures relative to Leg 115 basalts.

Significant intrasite compositional variability occurs at all these sites [Baxter, 1990; Fisk and Howard, 1990]. This

is evident in both variation diagrams such as TiO_2 vs. MgO (Figure 4b), and in the rare earth element patterns (Figure 5). At Site 706, the flow units are divided into 2 groups. The upper units, which have Mg# ranging from 31 to 42, are more incompatible-element-rich, and show slightly greater light-rare-earth-element enrichment than the lower units, which have Mg# ranging from 49 to 54. Though the upper units have clearly experienced more fractional crystallization than the lower ones, the differences in incompatible elements reflect mainly differences in the composition of the primary magmas. This conclusion derives from the differences in $^{87}Sr/^{86}Sr$ and ϵ_{Nd} of the upper and lower units [White et al., 1990], and the failure of fractional crystallization models to derive the upper unit from the lower unit. The isotope data indicate that the incompatible element differences in primary magmas is at least partly due to differences in mantle sources, but differences in degree of melting and extent of fractional crystallization may also be important.

The situation at Site 707 is similar, although here it is the lower group of lavas that is more fractionated and more incompatible-element-enriched (Figures 4 and 5). Again, fractional crystallization alone cannot adequately account for the incompatible element differences between upper and lower lavas [Baxter, 1990] and the two groups have significantly different ϵ_{Nd} values [White et al., 1990;

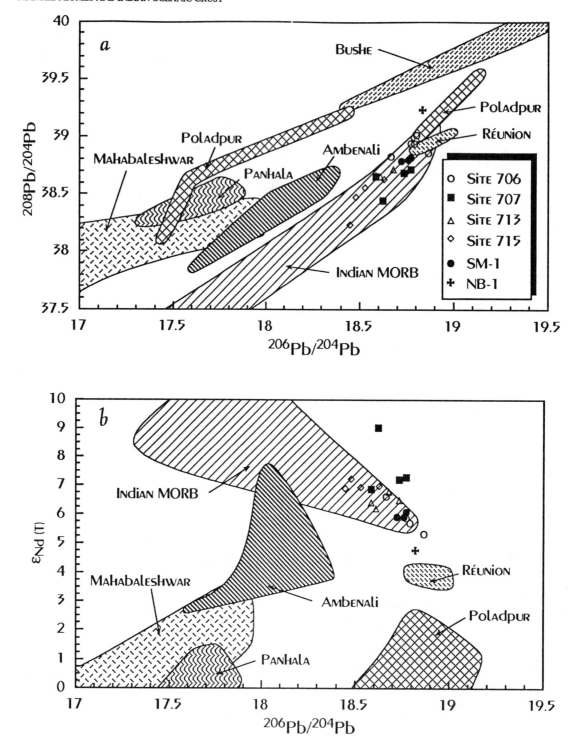

Fig. 6. (a) $^{208}Pb/^{204}Pb$ vs. $^{206}Pb/^{204}Pb$ and (b) ε_{Nd} (T, age-corrected) vs. $^{206}Pb/^{204}Pb$ for Leg 115 basalts with fields for Réunion, Indian Ocean MORB and Deccan (Ambenali, Mahabaleshwar, Poldapur, Panhala, and Bushe formations) basalts. Leg 115 data are from White et al. [1990 and unpublished data, 1992]; Deccan data from Lightfoot and Hawkesworth [1988] and Lightfoot et al. [1990]; Réunion data from Fisk et al. [1988] and W. M. White [unpublished data, 1992]; Indian MORB data compiled from the literature. Much of the variation in the Deccan reflects crustal assimilation. Note the trend of the Leg 115 data from Réunion toward Indian MORB.

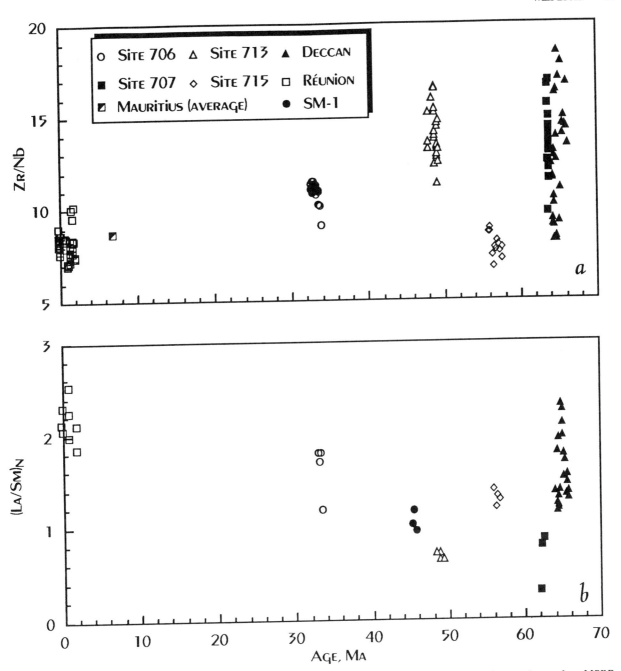

Fig. 7. Variation of (a) Zr/Nb and (b) (La/Sm)$_N$ as a function of eruption age for Leg 115 basalts. Overall this is a trend away from MORB toward OIB values (low Zr/Nb, high La/Sm) of these ratios through time. Leg 115 Zr/Nb data are from Backman, Duncan et al. [1988]; Deccan data from Cox and Hawkesworth [1985]; Réunion data from Fisk et al. [1988]; Mauritius data from Baxter [1990]. Leg 115 (La/Sm)$_N$ is unpublished ICP-MS data of W.M. White; Réunion data are from Fisk et al. [1988], and Deccan data are from Lightfoot and Hawkesworth [1988] and Lightfoot et al. [1990].

Tatsumi and Nohda, 1990]; this suggests that the upper and lower groups were derived from different mantle sources.

Three chemical groups can be recognized at Site 713, the uppermost group being the most distinctive and having incompatible element concentrations 1.5-2 times greater than the lower group. Though there are differences in absolute abundances, the REE patterns of these three groups are essentially parallel. Isotopic variations are

small and are not systematically related to other chemical parameters. Simple fractional crystallization models fail to account for the incompatible element differences between groups, but it seems likely that open system fractionation could account for these differences [Baxter, 1990].

Two compositional groups also can be recognized in basalts recovered at Site 715, a low-MgO group (<6.5 wt %) and a high-MgO group (>9.5 wt %). A third group, the uppermost flows, can be recognized petrographically, but were too altered for analysis. Incompatible element abundances in the low-MgO group are higher than in the high-MgO group, and these differences can be accounted for in fractional crystallization models. The ε_{Nd} is nearly constant in all analyzed samples [White et al., 1990; Tatsumi and Nohda, 1990], and rare earth element patterns are parallel, observations consistent with the notion that fractional crystallization is primarily responsible for compositional variability at Site 715.

Relationship to Deccan Volcanism. The geochronological results of Leg 115 and dated subaerial sites clearly link Réunion and Deccan volcanism in time and space, and provide convincing confirmation of the widely held view that continental flood basalt volcanism is ultimately related to mantle plume activity. An outstanding problem in continental flood basalt genesis is the degree to which the lavas have assimilated continental lithosphere. For the Deccan basalts, crustal assimilation can be unambiguously established for some units [e.g., the Bushe Formation, Cox and Hawkesworth, 1985], but it has been argued that some magmas (e.g., the Ambenali Formation) represent essentially uncontaminated mantle melts [e.g., Mahoney, 1988; Lightfoot and Hawkesworth, 1988], although there is some uncertainty about the relative contribution from plume and subcontinental lithospheric mantle. Drilling at Site 707 provided an opportunity to examine the question of the composition of Deccan magmas in the absence of contamination by continental crust and lithospheric mantle, because the basalts were probably erupted on new oceanic crust produced at the beginning of rifting of the Seychelles Bank from India a few million years after the main Deccan magmatic event.

Figure 6 compares Pb and Nd isotopic compositions of Leg 115 basalts with those of modern Indian MORB, Deccan, and Réunion basalts. The Ambenali formation has Nd isotope ratios overlapping those of Site 707 basalts. However, there is no overlap in Pb isotopic compositions, and a continental 'flavor', characterized by high $^{208}Pb/^{204}Pb$ and low ε_{Nd}, is apparent in all Deccan lavas. Those Ambenali samples having the highest ε_{Nd} (~ +7) overlap the Indian MORB $^{206}Pb/^{204}Pb$-ε_{Nd} array, but not the field of Site 707 basalts. It is possible that the composition of the plume shifted between production of the Ambenali and Site 707. Perhaps a more plausible explanation for the difference in Pb isotopic compositions is that even the Ambenali basalts with highest ε_{Nd} have assimilated some continental crust. The continental crust is much more enriched in Pb than Nd relative to that mantle; assimilation of crust would therefore affect Pb isotope ratios of mantle-derived magmas much more readily than Nd isotope ratios.

Intersite and Temporal Variations. Figure 7 shows the variation of Zr/Nb and (La/Sm)$_N$ as a function of age for Leg 115, Réunion, and Deccan basalts. With the exception of Site 715, there is a clear trend through time from relatively

high and variable Zr/Nb ratios (especially in the Deccan basalts), and low (La/Sm)$_N$ to the low Zr/Nb and high (La/Sm)$_N$ ratios characteristic of Réunion. Similar variations were also observed in Ba/Ti and Nb/Y [Fisk et al., 1989]. Figure 8 shows the variation of initial ε_{Nd} as a function of eruption age. As with the (La/Sm)$_N$ and Zr/Nb ratios, there is a systematic and regular decrease in ε_{Nd} from values of around +7 characteristic of Site 707 and the least contaminated Ambenali lavas at 65 Ma to values of +4 that characterize the recent Réunion lavas. Data from Mauritius, as well as DSDP Site 220 (eastern Arabian Sea) are consistent with this trend. The Deccan lavas scatter to much lower ε_{Nd} (<-16), reflecting assimilation, as discussed above, so it is only those basalts with highest ε_{Nd} that are of interest. Lavas from the upper units at Site 707 also fall off the general trend, having ε_{Nd} around +9. These high, essentially MORB-like ε_{Nd} values and strong light REE depletion suggest an asthenospheric source essentially free of any admixture of plume material.

The compositional changes accompany a decrease in eruption rate from approximately 5 km^3/yr at Deccan time (65 Ma) to about 0.04 km^3/yr now on the island of Réunion [Duncan and Storey, this volume]. Much of this decrease apparently occurred immediately at the close of the Deccan flood event. However, a consideration of the change in morphology of the hotspot track, from an almost continuous ridge for the early activity to discrete volcanoes over the last 10 million years suggests that eruption rate subsequently decreased further.

The data in Figures 7 and 8 raise several questions: why has the composition of magmas associated with the Réunion mantle plume varied with time; to what degree are the changes in composition and eruption rate related; and why is the variation in ε_{Nd} so much more regular than the variation in (La/Sm)$_N$ or Zr/Nb? We discuss these questions and compare the compositional variations of the Réunion and Kerguelen mantle plumes in a subsequent section of this paper.

LEGS 119, 120 AND 121: KERGUELEN PLATEAU, BROKEN RIDGE AND NINETYEAST RIDGE: THE KERGUELEN HOTSPOT TRACK

The Kerguelen Plateau (including its conjugate Broken Ridge) and the highly linear 5000-km-long Ninetyeast Ridge are two of the most remarkable features of the eastern Indian Ocean floor (Figure 1). On the basis of paleomagnetic, geochronological and plate reconstruction studies [Luyendyk and Rennick, 1977; Duncan, 1978, 1981; Peirce, 1978; Morgan, 1972] and earlier geochemical studies on dredge and DSDP samples [e.g., Mahoney et al., 1983; Davies et al., 1989; Storey et al., 1989; Weis et al., 1989a], the most popular hypothesis for the origin of both of these structures is that they represent the early products of the Kerguelen hotspot. The Rajmahal flood basalts, which lie at the northern terminus of the Ninetyeast Ridge, were probably joined with the southern Kerguelen Plateau as part of the initial volcanic efflorescence of the Kerguelen hotspot (see Duncan and

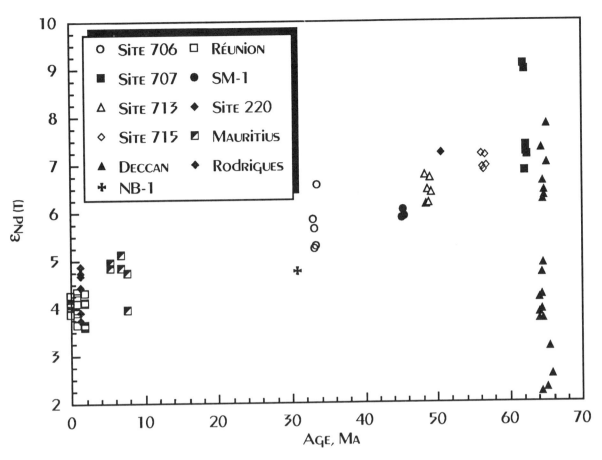

Fig. 8. Variation of ε_{Nd} as a function of age for Leg 115 [White et al., 1990 and unpublished data 1992; Tatsumi and Nohda, 1990], Réunion [Fisk et al., 1988], Mauritius [Mahoney et al., 1989], Rodrigues [Baxter et al., 1985], and Deccan [Mahoney et al., 1982; Cox and Hawkesworth, 1985]. DSDP Site 220 is also shown (data from Mahoney et al., 1989). A systematic shift away from MORB toward OIB isotopic signatures is apparent. Deccan scatters to much lower ε_{Nd} due to crustal assimilation.

Storey, this volume). Subsequent activity related to this plume produced the Tertiary to Quaternary volcanism of the Kerguelen archipelago [e.g., Watkins et al., 1974; Giret and Lameyre, 1983; Storey et al., 1988; Gautier et al., 1990]. Although it has been previously proposed that Heard Island, some 400 km south of the Kerguelen archipelago, represents the present-day location of the plume, Duncan and Richards [1991] have shown from modeling volcanic traces with stationary hotspots that the present location of the Kerguelen hotspot more likely lies a short distance to the west of Kerguelen Island. It appears then that the Kerguelen plume has been active for nearly 120 m.y., and thus represents one of the longest lived and most vigorous of the Indian Ocean hotspots.

One of the major achievements of ODP Legs 119 and 120 was to obtain the first drill-core of the basement of the Kerguelen Plateau. Specifically, Leg 119 penetrated 38.2 m of basement lavas and breccias at Site 738 at the southernmost tip of the plateau, while Leg 120 drilled basement at Sites 747 (53.9 m penetration), 749 (47.5 m) and 750 (34.2 m) on the central part of the plateau (Figure 1). In addition, a highly altered alkali basalt was

encountered at Site 748 some 150-200 m above the predicted basement depth.

Kerguelen Plateau — Tectonic Setting and Age. Kerguelen Plateau and Broken Ridge have an estimated combined crustal volume of as much as 50×10^6 km^3 [Schubert and Sandwell, 1989] and, if this estimation is correct, are comparable in size to the Ontong Java Plateau in the Pacific. The Kerguelen Plateau has been divided into northern and southern sectors by Schlich [1975] and Houtz et al. [1977]. The northern part generally lies in water depths of less than 1000 m and contains the Eocene to Quaternary volcanic islands of Kerguelen, McDonald and Heard. The boundary between the northern and southern sectors of the plateau lies immediately south of Heard Island. The transition zone exhibits a complex morphology with a large west-trending spur, the Elan Bank, extending west from the main plateau over a distance of some 600 km. The southern portion of the plateau is generally deeper, with water depths between 1000 and 3000 m, and has a more subdued topography. It consists of a broad anticlinal arch affected by multiple stages of normal faulting, resulting in horst and graben development [e.g.,

Coffin et al., 1986]. The axis of this arch is marked by the 77° Graben.

Seismic refraction studies on and in the vicinity of the Kerguelen Archipelago by Recq et al. [1983], Recq and Charvis [1986] and Recq et al. [1990] suggest a crustal thickness of between 15-23 km for this part of the plateau. These authors note that the seismic velocity-versus-depth profile is typical of oceanic islands. On the basis of gravity data, Houtz et al. [1977] inferred a crustal thickness of 20-23 km for the southern Kerguelen Plateau. These estimates imply a total crustal volume of about 25 x 10^6 km^3, or half of the Schubert and Sandwell [1989] estimate.

The northwest part of the Kerguelen Plateau is bordered by magnetic anomaly 34 [Schlich, 1982], indicating a minimum age of 84 Ma for the adjacent ocean crust. A recent revision of the magnetic profiles in the Enderby Basin on the southwestern margin of the plateau [Nogi et al., 1991] has indicated the presence of anomalies M0 to M8 (c. 118 to 129 Ma).

Radiometric dating of the Kerguelen Plateau basement has so far been restricted to a few samples from the central-southern parts of the plateau [Leclaire et al., 1987; Whitechurch et al., 1992]. Whole-rock dating on the least altered Leg 120 basalts gives ^{40}Ar/^{39}Ar 'plateau ages' of 109.2±0.7 Ma for Site 749 located on the Banzare Bank and 118.2±5 Ma for Site 750 in the Raggatt Basin [Whitechurch et al., 1992]. The Leg 120 results show a close correspondence with a plagioclase K-Ar age of 114±1 Ma from a dredged tholeiite (dredge station 5) on the 77° Graben [Leclaire et al., 1987], and with ^{40}Ar/^{39}Ar ages for the Rajmahal basalts of northeast India [117±1 Ma; Baksi, 1986]. However, all three dated Kerguelen Plateau samples are within 400 km of each other (Figure 1). Two dredge samples from Broken Ridge gave ages of 83-88 and 62-63 Ma [Duncan, 1991]. Duncan considered the older ages to represent part of the plateau construction event, while the younger ages correspond to volcanism associated with early rifting between Broken Ridge and Kerguelen Plateau [Mutter and Cande, 1983]. These dates indicate that the construction of Kerguelen Plateau-Broken Ridge took place over a period of ~ 50 m.y.

Kerguelen Plateau — Petrology and Geochemistry of Basalts. The basement lavas recovered by ODP Legs 119 and 120 consist mainly of olivine-normative tholeiites. They range from moderately fresh to highly altered [zeolite facies metamorphism; Sevigny et al., 1992]. The presence of oxidized flow tops at Site 747 and overlying non-marine, middle-Albian sediments at Site 750 has been interpreted as evidence that some parts of the Kerguelen Plateau were constructed subaerially. Full descriptions of the Leg 119 and 120 cores are given by Barron, Larsen et al. [1989] and Schlich, Wise et al. [1989], respectively.

Sr, Nd, and Pb isotopic data for Kerguelen Plateau tholeiites show a remarkably large variation in composition; for example ε_{Nd} ranges from +5.2 to - 8.5 [Alibert, 1991; Davies et al., 1989; Weis et al., 1989a; Salters et al., 1992; Storey et al., 1992]. In terms of

individual locations, Site 747 basalts have high ^{87}Sr/^{86}Sr ratios (0.7056 to 0.7058) while ε_{Nd} ranges from +2.7 to -4, overlapping with the most enriched[1] basalts from Kerguelen Island (Figure 9). By comparison, ε_{Nd} is higher and more variable for basalts from Sites 749 (ε_{Nd} = +1.9 to +4.5) and 750 (ε_{Nd} = +1.4 to +5.2). Site 750 tholeiites differ from those of Site 749 in having higher ^{87}Sr/^{86}Sr ratios; acid-leaching experiments suggest this is only partly due to alteration. In marked contrast to the tholeiites from the central and northern part of the Kerguelen Plateau, which have Sr and Nd isotopic compositions within the oceanic mantle array, isotopic analysis of a tholeiite from Site 738 [Alibert, 1991], at the southern tip of the Kerguelen Plateau, gave extreme ^{87}Sr/^{86}Sr and ε_{Nd} values of 0.7090 and -8.5, respectively, well outside the range of oceanic (hotspot-related) basalts. Although the Pb isotopic compositions of basalts from Sites 748 and 749 overlap with data for Kerguelen Island, the tholeiites from Sites 738, 747 and 750 are characterized by lower ^{206}Pb/^{204}Pb and more grouped (17.47 to 17.74), despite major differences in their Nd and Sr isotopic compositions. The single analysis of the Site 738 tholeiite [Alibert, 1991] also has a very high ^{207}Pb/^{204}Pb (15.71).

In summary, Kerguelen Plateau basalts have Sr and Nd isotopic compositions lying outside the field for Indian Ocean MORB but, with the exception of Site 738, the Sr and Nd isotopic data are within the range of hotspot-related oceanic basalts; specifically, those from Sites 747 and 748 overlap with the field for the most recent basalts from Kerguelen Island (Figure 9). While the Sr and Nd isotopic data on the drilled and dredged samples from the Kerguelen Plateau generally support the Kerguelen hotspot hypothesis [e.g., Davies et al., 1989; Storey et al., 1989; Weis et al., 1989a; Salters et al., 1992; Storey et al., 1992], it is notable that basalts from Kerguelen Island are characterized by higher ^{206}Pb/^{204}Pb ratios than plateau samples from Sites 738, 747, and 750. Low ^{206}Pb/^{204}Pb and sometimes high ^{207}Pb/^{204}Pb ratios are a feature of old continental mantle lithosphere [e.g., Nelson et al., 1986; Hawkesworth et al., 1990]. Storey et al. [1992] suggest that the low ^{206}Pb/^{204}Pb values of some Kerguelen Plateau tholeiites represent a contribution from Gondwanan continental mantle lithosphere underlying the plateau at the time of its formation. Although plate reconstructions for Gondwana indicate that the Kerguelen Plateau must be essentially oceanic [e.g., de Wit et al., 1988; Powell et al., 1988], the extreme isotopic compositions reported by Alibert [1991] for a Site 738 tholeiite [and the very high La/Ta ratios [21-30] for Site 738 given by Mehl et al., 1991] may indicate that the southernmost tip of the plateau is underlain by continental crust, possibly in a setting analogous to the North Atlantic Rockall Plateau [Morton and Taylor, 1987; Merriman et al., 1988] and Vøring Plateau sequences [Viereck et al., 1989].

---

[1]Enriched and depleted are used here relative to bulk earth for Sr and Nd isotopes and relative to the Northern Hemisphere reference line [NHRL; Hart, 1984] for Pb isotopes. See also Weis and Frey [1991].

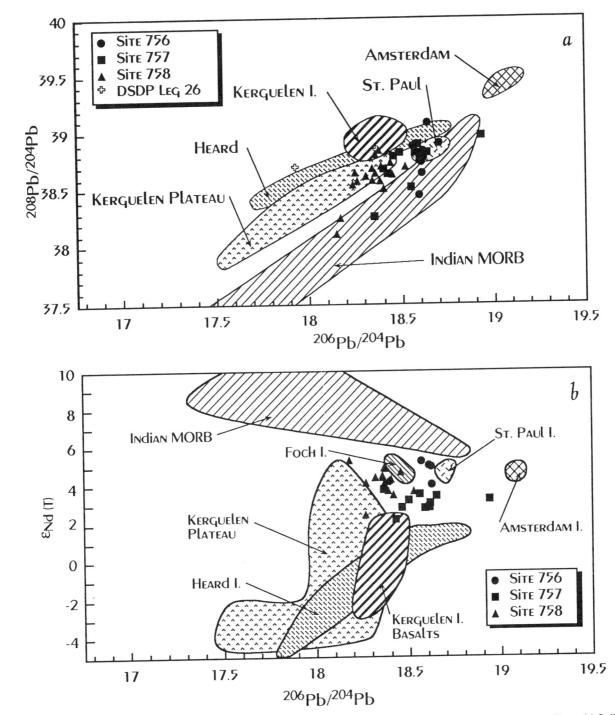

Fig. 9. (a) $^{208}Pb/^{204}Pb$ vs. $^{206}Pb/^{204}Pb$ and (b) ε_{Nd} vs. $^{206}Pb/^{204}Pb$ for Leg 121 Ninetyeast Ridge basalts. Data sources as in Figure 14. Indian MORB data compiled from the literature.

The low abundances of incompatible elements such as Th, Nb, Ta, Zr and the light REE in some of the Kerguelen Plateau basalts (e.g., Sites 749 and Site 750) are similar to basalts from other large-volume, hotspot related oceanic magmatic provinces, such as the Caribbean Plateau, the Ontong Java Plateau and Nauru Basin [e.g., Floyd, 1989].

This observation is consistent with a near- or on-ridge setting for the Kerguelen Plateau (perhaps analogous to Iceland), an environment which would favor extensive melting of an upwelling mantle plume [McKenzie and Bickle, 1988; Storey et al., 1991].

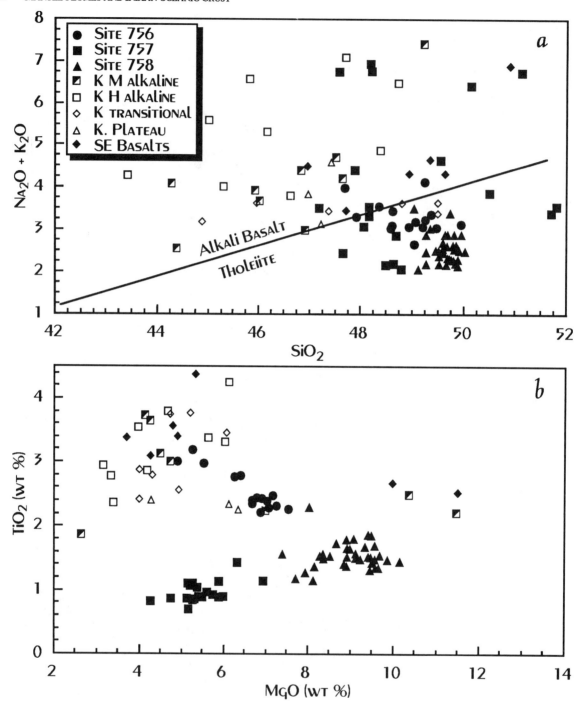

Fig. 10. (a) Alkali-silica diagram comparing Leg 121 basalts, Ninetyeast Ridge, with those of Kerguelen Archipelago and Plateau. Kerguelen basalts, including the young (22-8 Ma) SE Province samples, fall mostly in the alkali basalt field while Ninetyeast Ridge basalts, except a few altered samples, fall in the tholeiite field. (b) TiO_2 as a function of MgO in Leg 121 and Kerguelen Archipelago basalts. The plot shows that strong intrasite variations occur and illustrates the less evolved character of Site 758 basalts. Data from Davies et al. [1989], Frey et al. [1991], Gautier et al. [1990] and Weis et al. [1989b and 1992].

Ninetyeast Ridge: Objectives and Results of ODP Drilling.
The principal objectives of drilling basement during Leg 121 were the following: (a) determine the composition of the basement of the Ninetyeast Ridge, an aseismic ridge that extends for ~5000 km from the Bay of Bengal to about

30°S in the central Indian Ocean; (b) assess any temporal variation in chemical and isotopic values along the Ninetyeast Ridge; (c) examine the possible relationships between the basalts of the Ninetyeast Ridge, Kerguelen Plateau, and Rajmahal Traps to the putative mantle plumes presently associated with Kerguelen Island and Amsterdam/St. Paul Islands. Three sites were drilled along Ninetyeast Ridge during ODP Leg 121: Site 756 at 27°21'S, Site 757 at 17°01'S and Site 758 at 5°23'N (Figure 1). At Site 756, located near the crest at the southern end of the Ninetyeast Ridge, 32.9 m of basement was recovered and a total of 16 flow units were identified. At Site 757, 25.0 m with 20 flow units of plagioclase-phyric basalt were recovered, and at Site 758, the northernmost drill site on the Ninetyeast Ridge, 118.5 m, including a total of 29 basaltic flows and 8 interbedded tuff units. Ages of 43.2±0.5 Ma for Site 756 basalts, around 58 Ma for Site 757 basalts and of 81.8±2.6 Ma for Site 758 basalts were determined by ^{40}Ar-^{39}Ar incremental heating experiments [Duncan, 1991]. Complementing the four previous basement holes drilled during DSDP Legs 22 and 26 along the Ninetyeast Ridge (Figure 1), the ODP drilling shows that the ridge is composed of basalt ranging in age from ~38 Ma in the south to ~81-83 Ma in the north in an almost linear progression [Duncan and Storey, this volume]. Interpretation of the seafloor magnetic anomaly pattern of the eastern Indian Ocean suggests that along much of its length the Ninetyeast Ridge is contemporaneous with the adjacent seafloor to the west [Sclater and Fisher, 1974]. The crust to the east (the Wharton Basin) ages southwards and is separated from the Ninetyeast Ridge by the Ninetyeast Fault. A progressive south to north increase in age [Duncan and Storey, this volume] along the Ninetyeast Ridge is consistent with its formation as a hotspot trace [Morgan, 1972; Frey et al., 1977; Peirce, 1978] resulting from the northward migration of the Indian Plate over a mantle plume. Prior to ODP drilling [Weis et al., 1991], it was uncertain whether the plume responsible for the Ninetyeast Ridge was the Kerguelen plume [e.g., Duncan, 1978], a plume under Amsterdam/St. Paul Islands [Frey et al., 1977] or a combination of the two [Luyendyk and Rennick, 1977]. Evidence supporting a dominant Kerguelen plume is the latitude of magnetization (about 50° S) for the Ninetyeast Ridge lavas [Peirce, 1978], and for the ashes recovered above Sites 757 and Site 758 basalts [Klootwijk et al., 1991].

Ninetyeast Ridge: General Compositional Features and Intrasite Variability. For detailed lithological and petrological descriptions, the reader is referred to Volume 121 of the ODP Initial Reports [Peirce, Weissel et al., 1989] and Volumes 22 and 26 of the DSDP Initial Reports [Von der Borch, Sclater et al., 1974; Davies, Luyendyk et al., 1974].

The lavas recovered from the Ninetyeast Ridge are tholeiitic basalts (Figure 10a). They range from aphyric olivine tholeiites at Site 756 to strongly plagioclase-phyric at Site 757, while at Site 758 the basalts are sparsely to strongly plagioclase-phyric. At Site 758, the basalts were clearly erupted in a submarine environment (presence of pillow basalts). All Ninetyeast Ridge basalts have been extensively altered (H_2O+ typically >1%) in a

low-temperature environment under both reducing and oxidizing conditions. This alteration, particularly late-stage albitization of the abundant feldspars in these lavas, has shifted a few of the Site 757 lavas into the alkali basalt field of Figure 9. Only one of the seven drill sites (DSDP Site 214) recovered evolved lavas (andesites). An important observation is that the Ninetyeast Ridge basalts are, in terms of major element compositions, distinctly more tholeiitic than the three main magmatic series (transitional, mildly alkaline and highly alkaline; Gautier et al., 1990), forming the Kerguelen Archipelago (Figure 10).

Although the effects of post-magmatic alteration preclude detailed petrogenetic inferences based on major element compositions, significant (magmatic or pre-alteration) intrasite variability occurs. This is evident in both variation diagrams such as TiO_2 vs. MgO (Figure 10b) and in rare earth element patterns (Figure 11). Clearly, all lavas from Site 758 have higher MgO contents than lavas from Sites 756 and 757. Also, the lower units of Site 757 are significantly different from the upper units, requiring two compositionally distinct parental magmas, which is also reflected in all the isotope systems [Weis and Frey, 1991]. Most of the lavas at Site 757 are plagioclase-rich cumulates and there is compelling evidence for plagioclase settling within individual flow units [Frey et al., 1991; Saunders et al., 1991]. Settling of a low density phase like plagioclase is unusual, and requires a low density melt such as the high-Al_2O_3, low-iron parental melt inferred for the Site 757 lavas. Such melt compositions are characteristic of melts segregated from a peridotite source at pressures less than 10 kb [e.g., Falloon and Green, 1988]. Consistent with this inference, Ninetyeast Ridge basalts do not have the trace element signature of equilibrium with residual garnet. Although derived from a mantle with a plume component (heat and/or mass input), the primary magmas that led to the formation of the Ninetyeast Ridge were segregated at low pressures similar to those inferred for MORB [e.g., Klein and Langmuir, 1987].

For major elements, Site 756 basalts define the most coherent trends, reflecting their relative freshness and absence of phenocrysts, which qualitatively fit a simple fractional crystallization model. Nevertheless, some REE complexities (crossing REE patterns, for instance) together with slight isotope differences, imply some source control. The homogeneity of the basalts at Site 758 is striking, but one unit, F4, is compositionally and, to some extent, isotopically different.

In summary, both crystal fractionation and different extents of melting can explain most of the intrasite compositional variability at Ninetyeast Ridge drilling sites. However, isotopic data (Figure 12) indicate that some intrasite variability is source controlled or results from mixing of distinct mantle source reservoirs.

Ninetyeast Ridge: Intersite and Temporal Variations. In addition to the intrasite compositional variation, there are also important temporal trends defined by lavas from the

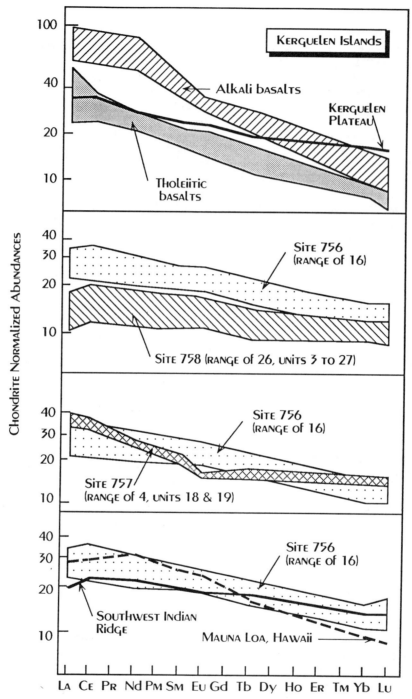

Fig. 11. Average chondrite normalized rare earth abundances for the different Leg 121 Ninetyeast Sites [Frey et al., 1991] and a few representative samples of Kerguelen different magmatic series for comparison [Gautier et al., 1990].

Ninetyeast Ridge and other lavas presumed to have been derived from the activity of the Kerguelen mantle plume. There are trends through time from high Zr/Nb and low $(La/Sm)_N$ in lavas from the Rajmahal volcanics, the Kerguelen Plateau and the oldest sites of Ninetyeast Ridge, to the low Zr/Nb and high $(La/Sm)_N$ in lavas characteristic of the Kerguelen Archipelago (Figure 13). In these and

other ratios involving alteration-resistant, incompatible elements the Ninetyeast Ridge lavas are intermediate between MORB and Kerguelen Archipelago lavas.

Initially, these results led to models of two-component mixing but in isotopic diagrams such as the variation of initial ε_{Nd} values with time (Figure 14), the trend is more complex. Another example is that at Site 756 $^{206}Pb/^{204}Pb$

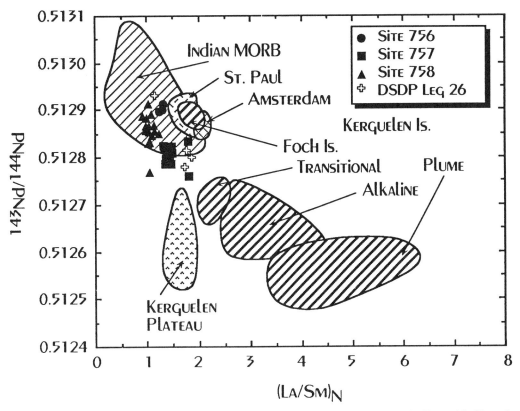

Fig. 12. ε_{Nd} plotted against chondrite-normalized La/Sm ratios for Leg 121 basalts (data sources as in Figure 14), Kerguelen lavas, the Southeast Indian Ridge and St. Paul Island [data from Dosso et al., 1988; Frey et al., 1991; Gautier et al., 1990 and Weis et al., 1992]. The Kerguelen plume field is defined by the Upper Miocene series of the SE Province.

ratios peak at values which exceed the ratios in lavas from the Kerguelen Archipelago (Figure 9); a component with high $^{206}Pb/^{204}Pb$ is required [Saunders et al., 1991; Weis et al., 1991]. In the Nd isotopic system, Site 756 lavas have ratios comparable to those of the Foch Island tholeiites of Kerguelen Archipelago [White and Hofmann, 1982]. Lavas from the older sites along the Ninetyeast Ridge exhibit a larger range in isotopic compositions and extend to lower ε_{Nd} than Site 756 lavas. Although they require a source component with a depleted composition based on their incompatible element compositions, these 757 and 758 lavas have ε_{Nd} values pointing towards those of the Kerguelen plume, as sampled by the upper Miocene series of the SE Province of the archipelago [Gautier et al., 1989; Weis et al., 1992]. It is then necessary to invoke mixing between at least three source components: a depleted, MORB-type component such as the one erupted today on the Southeast Indian Ridge; a very enriched, high-$^{87}Sr/^{86}Sr$, low-ε_{Nd}, OIB-type component with moderate $^{206}Pb/^{204}Pb$ (the Kerguelen plume); and a higher $^{206}Pb/^{204}Pb$ OIB-type component comparable to that sampled by the St. Paul and Amsterdam Island lavas [Saunders et al., 1991; Weis and Frey, 1991]. A difficulty pointed out by Saunders et al. [1991] is that precise definition of the isotopic composition of the Kerguelen plume through time is not possible, thus frustrating any attempt at binary mixing modeling. Barling and Goldstein [1990] have shown that

some lavas from Heard Island have much higher $^{206}Pb/^{204}Pb$ ratios than the Kerguelen Island lavas. Thus, in the unlikely event that the samples recovered from Heard Island are derived from the same plume, and have not been contaminated during their ascent through the thickened, Cretaceous lithosphere of the Kerguelen Plateau, they would extend the array of isotopic compositions from the Kerguelen plume considerably. Essentially, the three components required by the trace element and isotopic data are potentially present within the plume and the surrounding asthenosphere; the plume carries two (or perhaps all three) isotopic components. An important point, and one developed in a later section, is that the Kerguelen plume - as represented by its most recent products - is compositionally far more diverse than the Réunion plume.

Successful petrogenetic models must also be able to explain the considerable intrasite chemical heterogeneity that is superimposed on the long-term temporal trends (Figures 13 and 14). Most of the Ninetyeast Ridge drill sites with significant (>25 m) basement penetration require at least two distinct parental magmas to explain differences in radiogenic isotope ratios and crossing chondrite-normalized REE patterns. As observed in Réunion hotspot products, some decoupling between temporal variations in isotope ratios and incompatible element ratios may reflect differences in the conditions of

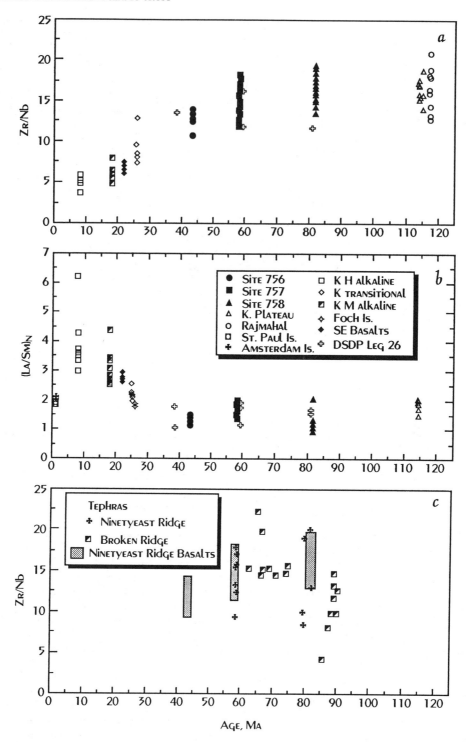

Fig. 13. Variation of (a) Zr/Nb and (b) (La/Sm)$_N$ as a function of eruption age for Leg 121 basalts. There is a trend through time from the intermediate ratios, between MORB and OIB, for the Kerguelen hotspot-related oldest Ninetyeast Ridge lavas, Rajmahal traps and Kerguelen Plateau toward the Kerguelen plume values. Leg 121 data are from Frey et al. [1991]; DSDP Ninetyeast Ridge data are from Frey et al. [1977]; Rajmahal from Mahoney et al. [1983]; Kerguelen Plateau from Weis et al. [1989a] and Leg 119 from Alibert [1991]; Kerguelen Islands Gautier et al. [1990] and Weis et al. [1992]. St. Paul and Amsterdam data from White and Dupré [unpublished]. (c) Zr/Nb variations in the ashes sampled along Broken Ridge and Ninetyeast Ridge.

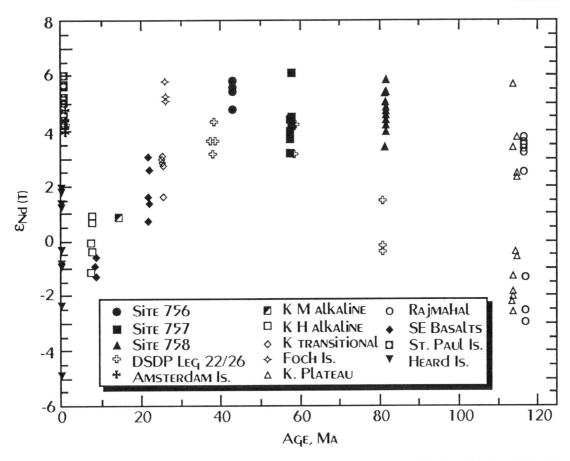

Fig. 14. Variation of ε_{Nd} as a function of age for Rajmahal Traps, Kerguelen Plateau, Ninetyeast Ridge Leg 121, St. Paul, New Amsterdam, and Kerguelen Archipelago lavas. Ninetyeast Ridge data from Weis and Frey (1991). Previous Ninetyeast Ridge data [Dupré and Allègre, 1983; Hart, 1988] and new data [D. Weis, unpublished data, 1992]. Other data sources as follows: St. Paul and Amsterdam Islands [Dupré and Allègre, 1983; Hamelin et al., 1985/1986; Michard et al., 1986; Dosso et al., 1988; White and Dupré, unpublished data], Kerguelen Archipelago [Foch Island, White and Hofmann, 1982; transitional, mildly alkaline and highly alkaline basalts (Kerguelen Basalts), Weis et al., 1989b, Gautier et al., 1990], Heard [Storey et al., 1988; Barling and Goldstein, 1990], Kerguelen Plateau [Weis et al., 1989a] and previous Ninetyeast Ridge data [Mahoney et al., 1983; Hart, 1988] and new data [D. Weis, unpublished data, 1992].

melting, and to a lesser degree, subsequent fractional crystallization. These variations are superimposed on those reflecting different proportions of mixing of the three mantle components which may be determined by the relative position of the Kerguelen plume and the spreading ridge.

Relationship between the Kerguelen Hotspot and the Rajmahal Flood Basalts. Among the continental volcanic provinces attributed to the initial activity of the Kerguelen hotspot and which accompanied the breakup of eastern Gondwana and the formation the Kerguelen Plateau, the Rajmahal basalts are the most significant in terms of their volume. The eruption of both the plateau and the flood basalts was probably contemporaneous (115-117 Ma), and the Rajmahal Traps also probably represent the northern end of the Ninetyeast Ridge [e.g., Mahoney et al., 1983].

The Rajmahal flood basalts themselves are exposed near the continental margin of northeast India, cropping out over an area of about 4,300 km² in the Rajmahal Hills,

Bihar. Seismic reflection profiles and borehole data for the Bengal Basin [Biswas, 1963; Sengupta, 1988] also show the presence of a seaward-dipping reflector sequence on the continental margin, indicating a much greater southward extent for this province.

Geochemical studies on the volcanic rocks of the Rajmahal Traps and the probably once-contiguous Sylhet Traps [e.g., Mahoney et al., 1983; Baksi et al., 1987; Storey et al., 1992] have shown that they consist predominantly of quartz-normative tholeiites and basaltic andesites with a phenocryst assemblage of plagioclase, augite, Fe-Ti oxides and, more rarely, olivine. Analysis of a few samples from the Bengal Basin has also revealed the presence of alkali basalts and olivine-normative tholeiites [Baksi et al., 1987]. The low $^{206}Pb/^{204}Pb$ (17.30 to 17.69) and high $^{207}Pb/^{204}Pb$ (15.59 to 15.62) ratios of the Rajmahal basalts suggest the assimilation of continental material [Storey et al., 1992], which complicates the assessment of their relationship to the Kerguelen mantle plume.

Tephra from Broken Ridge and Ninetyeast Ridge.
Abundant basaltic tephra layers were recovered at Sites 752 through 755 on the Broken Ridge and at Sites 757 and 758 on the Ninetyeast Ridge. All of the pre-Eocene tephras are highly weathered basaltic glass intermixed with biogenic sediments. The primary weathering products of the glasses are clays and zeolites. Plagioclase crystals are present in many of the samples, particularly those from the Ninetyeast Ridge. Augite crystals were present only in the massive tuff layers at Site 757.

The ash layers of Broken Ridge range from the Turonian (ca. 90 Ma) to the upper Paleocene (ca. 58 Ma). They occur as (1) discrete fallout or distal turbidite deposits or (2) larger proximal turbidite layers rich in volcaniclastics. The fallout layers are difficult to differentiate from the distal turbidites since their emplacement mechanisms are very similar. They both exhibit sharp basal contacts, often slightly bioturbated, and grade upwards in grain size and content of volcanogenic material. These layers range from a few centimeters to a maximum of 50 cm thickness. The proximal turbidite layers are easily identified since they display cross bedding, a winnowing of fine grained particles, and in some cases, scoured basal contacts. These layers, only a few centimeters thick, are often grouped into larger deposits (e.g., Site 755).

The tephras of Ninetyeast Ridge occur as large-scale turbidite deposits (e.g., Site 758), near vent tuffs deposited in or near shallow water (e.g., Site 757), or as discrete fallout layers. The older (Campanian, ca. 80 Ma) volcaniclastics of Site 758 are mostly turbiditic, and display similar characteristics to the turbidites of Site 755 on Broken Ridge. The massive tuff layer recovered just above basement at Site 757 (ca. 60 Ma) shows clear signs of sea-level volcanic activity. Shattered shell fragments, angular basalt fragments, and blocky glass shards indicate explosive, phreatic eruptions capable of creating small eruption columns and dispersing ash over larger distances than most basaltic eruptions. More detailed descriptions of the tephra layers can be found in Peirce, Weissel et al. [1989].

The occurence of these tephra layers raises the following questions: Were some or all of the tephra layers on the Broken Ridge formed by the volcanic activity along the Ninetyeast Ridge, and if yes, how can they be correlated? Is there a chemical or physical method to differentiate the fallout tephras from the fine reworked distal turbidite layers?

Most of the volcaniclastic layers could be dated by their associated microfossils and by paleomagnetic means, particularly because the iron-rich volcanic sediments held a strong magnetic signal [Peirce, Weissel et al., 1989]. The massive tephra layers however, presented a problem in conventional dating methods due to the absence of microfossils. Unfortunately, there are no radiometric dates yet available for these layers. Approximate dates can be calculated by interpolating the biogenic sedimentation rate and allowing for a steady input of volcanic material. This method supplies only a rough estimate of the ages for the ash layers since the frequency of each eruption and the volume of dispersed tephra can vary drastically. Indeed, early estimates in the ash flux change on Broken Ridge show that the input of volcanic material can vary significantly in only a few million years [Rea et al., 1990]. Therefore a conservative estimate of the amount of volcanism will yield only the youngest possible ages for the tephra layers.

The inherent difficulties in separating the strongly weathered volcanogenic material from the host sediment limit the accuracy of whole rock XRF analyses. In that light, trace elements that are not present in the biogenic sediments provide the best geochemical information about the tephras. Single shard microprobe analyses also provide useful information on the major element geochemistry. All of the tephras analyzed are tholeiitic basalts. The Zr/Nb ratio is similar to that of the underlying basalts (Figure 13c), though a few samples from Broken Ridge (at Sites 752 and 755) differ markedly because of their very low Nb concentrations (detection limit 2 ppm). These samples do not however differ in any of the related major oxides (FeO, MgO and CaO all conform to the other samples). Shape analysis of the glass shards shows that the shards in the anomalous layers have undergone mechanical weathering not observed in the other discrete tephras. Assuming that eruption style and fragmentation process were similar for all the layers (as a result of their similar chemistry and tectonic setting) this excess weathering indicates secondary transport, probably as distal turbidites. This transport would also further sort the tephra by minute changes in the density of the glass shards, and may have winnowed out the denser shards containing Nb-bearing Fe- and Ti- oxide microphenocrysts.

The correlation of the tephras from Broken Ridge and Ninetyeast Ridge is supported by their similar chemistries, and by their coinciding time frames. The paleotectonic reconstruction of the Indian Ocean at 56 Ma [Royer and Sandwell, 1989] shows at least 500 km between the then active Ninetyeast Ridge and the drilled Broken Ridge sites. Calculation of the transport distances for these tephras shows that explosive phreatic eruptions can transport particles ca. 200 km, until they settle on the ocean surface. Settling through the water column (a matter of days to weeks) could also transport the particles up to 300 km further. It is therefore possible that the fallout layers of the Broken Ridge resulted from the explosive volcanism along the Ninetyeast Ridge, although local volcanic activity at 83-88 Ma and 62-63 Ma has been documented from dredged samples [Duncan, 1991]. No active volcanism from the northern Kerguelen Plateau is reported during this time [Bitschene et al., this volume], though Kerguelen Plateau could be the source of epiclastic volcanic sediments which form the discrete distal turbidite layers.

PLUME–RIDGE–ASTHENOSPHERE INTERACTIONS: IMPLICATIONS FOR VOLUME AND COMPOSITION

The lavas forming both the Ninetyeast Ridge and the Chagos-Maldives-Mascarene Ridge were erupted close to a

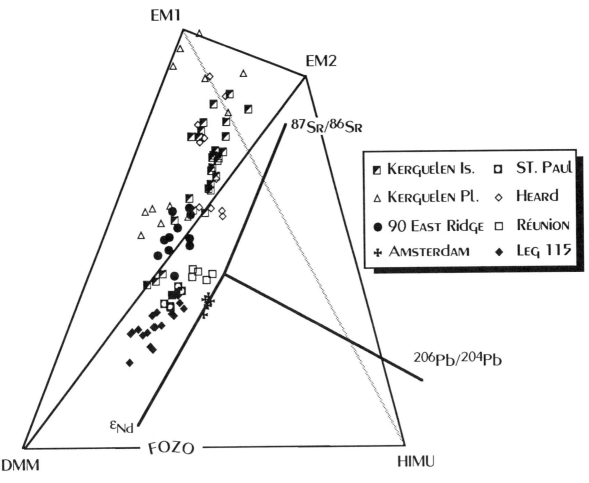

Fig. 15. Three-dimensional plot of ^{87}Sr/^{86}Sr, ε_{Nd}, and ^{206}Pb/^{204}Pb for basalts of the Réunion and Kerguelen plumes. Axes are tilted about 60° toward the viewer. The tetrahedron defines the volume in isotope space that can be produced by mixtures of the 4 components of Zindler and Hart [1986]: DMM, EM1, EM2, and HIMU [compare Figures 1 and 2 of Hart et al., 1992]. The Réunion plume products show a clear trend toward depleted mantle (DMM); no such clear trend exists for Kerguelen plume products. FOZO is the "Focus Zone" of Hart et al. [1992]. Data sources as in Figures 8 and 13.

spreading axis, consistent with plate tectonic reconstructions of the eastern and western Indian Ocean basins. Furthermore, both the Réunion and Kerguelen plumes have been crossed by at least one spreading axis. This could affect compositions in two ways: mixing of the plume with hot asthenosphere could shift isotope and trace element compositions toward more depleted values, and the higher asthenospheric temperatures in the vicinity of the ridge could allow higher degrees of melting, also producing more depleted trace element signatures. In the case of the Kerguelen plume, a third component, comparable to the one exhibited in lavas erupted at St. Paul and Amsterdam Islands, was involved in the genesis of the youngest lavas (Site 756) along the Ninetyeast Ridge.

Proximity to a spreading ridge may also affect melt volumes. The theoretical relationship between the volume

of melt production, the temperature of the mantle, and the thickness of the lithosphere as postulated by the models of McKenzie and Bickle [1988] may be tested by the two hotspot systems within the Indian Ocean. Two alternatives have been hypothesized: (1) the plume mantle flux is not constant, with an initial powerful flux related to a thermal diapir ('plume head') rising from the deep mantle and impinging on the base of the lithosphere. This plume head may entrain mantle as it ascends through the lower mantle. Subsequently, magmatism diminishes in relation to the decreasing flux of the 'plume tail' [Richards et al., 1989; Griffiths and Campbell, 1990]. (2) The plume flux is essentially steady state, and the volume of melt generated is primarily a function of the thickness of the lithosphere and the reservoir of warmer, plume-derived mantle trapped beneath the lithosphere [White and McKenzie, 1989]. In this model, a plume head may build beneath thick, stationary lithosphere with little or no magmatism, catastrophically releasing melt during initial extension. Variations in melt volume will be accompanied by changes in magma chemistry; for example, deeper, small-volume melts may exhibit the effects of garnet fractionation, with enhanced La/Yb ratios.

Isotopic Effects of Plume-Asthenosphere Mixing

Fisk et al. [1989] and White et al. [1990] argued that the temporal variation in incompatible element and isotopic ratios and the decrease in eruption rate reflected decreasing entrainment of asthenosphere in the plume. White et al. [1990] noted that the isotopic compositions of Leg 115 basalts are qualitatively consistent with mixing between a Réunion-like component (the mantle plume) and a MORB-like one (entrained asthenosphere), as is apparent in Figure 6. They cited the work of Griffiths [1986] showing that the plume head can entrain significant amounts of the surrounding mantle as it rises. This process increases the volume of the head of the plume, which already represents a much larger flux than the following tail.

A recent study by Hart et al. [1992] led us to re-examine the question of whether the Réunion plume is mixing with asthenosphere or some more exotic component. Many oceanic islands define sublinear Sr-Nd-Pb isotopic arrays that converge toward depleted compositions, implying mixing between mantle plumes and some depleted component. However, Hart et al. [1992], examining the data in multi-dimensional isotopic space, find that the depleted component in these mixing arrays is not the depleted upper mantle source of MORB, but a composition with distinctly higher $^{206}Pb/^{204}Pb$, which they call FOZO. This, however, is not the case for the Réunion plume. Figure 15 shows the Réunion-Leg 115 data in the three-dimensional plot of Hart et al. [1992]. The data define an array that points directly at DMM (Depleted MORB Mantle). Thus the isotopic evidence for Leg 115 basalts being derived from some sort of mixture of asthenosphere and plume material with the isotopic composition of Réunion basalts, and for the fraction of the former to decrease over time, is fairly firm.

Though the evidence for mixing between the Réunion plume and asthenosphere is apparent the exact mechanism by which this has occurred remains uncertain. There are several problems with the plume-head-entrainment hypothesis proposed by White et al. [1990]. First, plume flux and asthenosphere entrainment should drop off rapidly after the plume head dissipates, which should lead to a rapid decrease in eruption rate and a marked change in composition after the initial pulse of magmatism. Eruption rate apparently did drop markedly after the Deccan event, but as may be seen in Figure 8, compositional change was gradual. Second, if the plume rose from the core/mantle boundary, much of the entrainment, theoretically, should have occurred in the lower mantle [Griffiths and Campbell, 1990], and this entrained lower mantle material may not have had MORB-like isotopic compositions.

The Leg 115 data suggest that the proximity of the Réunion plume to a spreading center has, at best, only a secondary effect on the isotopic composition of the basalts produced by the plume. The plume was probably located within a few hundred km of the ancestral Central Indian spreading center between about 63 and 33 Ma. The ridge must have crossed the spreading center sometime during the interval between 50 and 40 Ma, and was perhaps ridge-centered for an extended period around this time. Subsequent to about 30 Ma, the distance between the plume and the hotspot has progressively increased. However, over the entire 65 million year period for which we have isotopic data, the isotopic signature of the basalts has become progressively and steadily less depleted, and there is no apparent relationship between isotopic composition and distance from a spreading center. There is no obvious change in slope of the ε_{Nd} versus age curve in Figure 8 corresponding to the time when the hotspot left the ridge at 30-35 Ma, nor is there any obvious discontinuity corresponding to the plume being ridge-centered at 40-50 Ma.

For the Kerguelen track, identifying plume-asthenosphere mixing is more difficult for several reasons. First, the intrasite variability is comparable to the intersite (temporal) variability. For example, the isotopic compositions of the Kerguelen Plateau basalts encompasses much of those of the remaining history of the plume (Figures 14 and 15). Second, the choice of the endmember Kerguelen plume composition is difficult because of the possibilities that the plume is either heterogeneous, has varied in composition with time, and that a third plume component (St. Paul—Amsterdam) may be involved. In contrast to the steady, nearly linear, change in isotopic composition of Réunion plume basalts, Kerguelen plume ε_{Nd} values appear to have passed through a maximum at 45-50 Ma (Sites 757 and 756). Storey et al. [1988], Weis et al. [1989b], and Gautier et al. [1990] conclude that isotopic signatures of the basalts reflect a decreasing contribution of depleted asthenosphere as the Southeast Indian Ridge migrated away from the Kerguelen plume after that time. A three-dimensional view of the data shows that Ninetyeast Ridge compositions are indeed shifted toward more depleted compositions, but there is no coherent trend toward DMM as there is for the Leg 115 data (Figure 15). The Foch Island tholeiites (>26 Ma)[White and Hofmann, 1982; White and Dupré, unpublished data, 1992] do show a shift toward DMM, indicating a stronger asthenospheric contribution to plume-derived magmas.

Partial Melting Effects on Incompatible Elements in Réunion Plume Basalts

The difference between temporal variations of isotope ratios and incompatible element ratios in Réunion mantle plume products could reflect differences in the degree and depth of the melting process and, to a lesser degree, subsequent fractional crystallization. Isotopic ratios are not affected by melting or crystallization, whereas

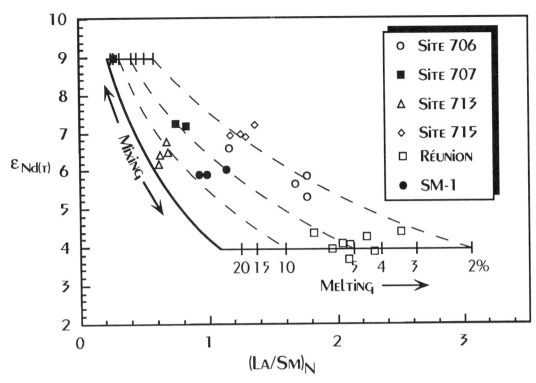

Fig. 16. ε_{Nd} plotted against chondrite-normalized La/Sm ratios for Leg 115 and Réunion basalts [data from Fisk et al., 1988; White et al., 1990; and W.M. White, unpublished data, 1992]. Results of calculations of the combined effects of mixing and variable degree of melting on $(La/Sm)_N$ and ε_{Nd} are superimposed on the data. Choice of end-member mantle source compositions is discussed in the text. Melting calculations assume equilibrium non-modal (eutectic) melting of spinel peridotite.

incompatible element ratios may be affected, most strongly by the former. Figure 16 explores the combined effects of mixing and variation in degree of melting on $(La/Sm)_N$ ratios. ε_{Nd} values for hypothetical end-members were chosen to match the compositions of Réunion and the most MORB-like Site 707 basalts. $(La/Sm)_N$ values for these end-members were chosen by considering (1) concentrations necessary to generate the basalts with less than 25% melting (assuming all magmas have experienced 25% fractional crystallization) and (2) the $(La/Sm)_N$ of the mantle source must be less than or equal to that of the basalts. The mixing curve shown is the one with the maximum curvature (i.e., closest fit to the data) still meeting these requirements. $(La/Sm)_N$ ratios for hypothetical 2 to 25% equilibrium melts of mixtures of these end-member mantle sources were then calculated. For simplicity, only melting in the spinel peridotite stability field was considered. Site NB-1 was not considered in this model because our only analyzed sample from this site is a trachyte that has clearly experienced extensive fractional crystallization.

The calculations show that the $(La/Sm)_N$ variations in Leg 115 basalts can be accommodated in a simple binary mixing model if degree of melting varies from about 20% (Site 713) to less than 5% (Sites 706, 715). Though this range is large, it is not implausible. Since the choice of 'source' compositions is in part arbitrary, it is the relative, rather than absolute, range in degree of melting that is

significant. The range in melt fraction could be reduced if (a) the end-members are not perfectly homogeneous, (b) the effects of fractional crystallization are variable, (c) melting sometimes occurs in the garnet stability field, or (d) more complex dynamic melting models are considered. All of these are plausible, and perhaps likely, meaning the range in degree of melting may be less than the order of magnitude found in the calculations. A further complication arises in that only the top of the volcanic pile may have been sampled [Fisk and Howard, 1990]. Nevertheless, the incomplete correlation between isotopic ratios and incompatible element ratios and these melting/mixing calculations point out the pitfalls in inferring compositional variability of sources solely from incompatible element ratios. Figure 16 also illustrates that most of the intrasite variability can be explained in terms of varying degrees of melting, with little change in source characteristics. Important exceptions are Site 706 and 707, where distinct differences between upper and lower lavas occur. In these cases, both degree of melting and the relative mix of plume and depleted mantle appear to have changed.

The calculations shown in Figure 16 suggest that the degree of melting along the Chagos-Maldives Ridge did not change systematically with time. There may be, however, some relation between degree of melting and proximity of the plume to a spreading center. Maximum degree of melting apparently occurs at Sites 707, 713 and SM-1. Site

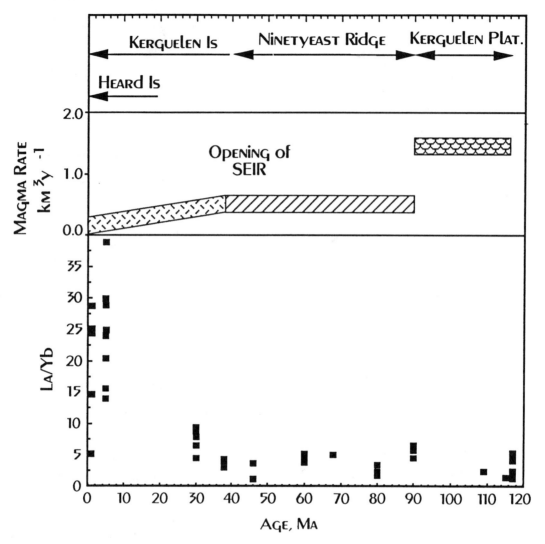

Fig. 17. Top: Volume flux of melt produced during the formation of the Kerguelen Plateau-Ninetyeast Ridge and Kerguelen Islands. Bottom: La/Yb vs. time [Saunders et al., 1991].

713 basalts were erupted to the east of the spreading center at 49 Ma, while SM-1 basalts were erupted only a few million years later to the west of the spreading center. As noted above, this suggests the plume was nearly ridge-centered at this time, which may explain the apparent relatively high degree of melting.

Because of the uncertainties in the composition of the Kerguelen plume mentioned above, applying this simple model to the Kerguelen plume is more problematic. Nevertheless, there is qualitative support for the idea that proximity to a spreading center affects degree of melting. The younger, more alkaline Kerguelen archipelago products are shifted the furthest from any reasonable source mixing curve (such as that shown in Figure 16), while the Ninetyeast Ridge basalts, erupted when the plume was near a spreading center, plot closest to a mixing curve, implying larger degrees of melting.

Volume and Compositional Variations in Kerguelen Plume Products

The volume of melt produced during the formation of Kerguelen Plateau-Ninetyeast Ridge-Kerguelen Island has clearly varied with time (Figure 17). Saunders et al. [unpublished manuscript, 1992] have suggested that the net magma production rates (in excess of those that would be produced at a normal ridge segment) range from a conservative estimate of 1.5 km³/yr during the formation of the Kerguelen Plateau, to about 0.5 km³/yr during the formation of Ninetyeast Ridge, and dropping rapidly to less than 0.1 km³/yr during the formation of the Kerguelen Islands. These figures correspond to location of the Kerguelen plume near or on a ridge segment, allowing extensive decompression melting. Hawaii, for example, has an output rate of about 0.15 km³/yr, despite being considered the most vigorous present-day plume [Sleep, 1990]. The high magma output of the Kerguelen Plateau and

Ninetyeast Ridge therefore implies shallow melt generation at a ridge setting, which is corroborated by the absence of any indication of final melt separation from a mantle with residual garnet and the high Al_2O_3, low FeO contents of inferred primary magmas.

The initial burst of hotspot activity during the formation of the Kerguelen Plateau does not appear to have been repeated during final separation of Broken Ridge from the Kerguelen Plateau at about 40 Ma, i.e., the initiation of the Kerguelen Archipelago. Although voluminous tholeiitic activity characterized the early stages of Kerguelen Island magmatism - and proximity of the plume to the ancestral Southeast Indian Ridge has been invoked for this - the volumes are substantially less than those of the Cretaceous Kerguelen plateau-building magmatism. Thus there appears to have been a sudden release of stored thermal energy at 115 to 120 Ma for the Kerguelen plume system, akin to the formation of the Deccan Traps some 50 m.y. later for the Réunion plume system.

Kent [1991] and Saunders et al. [unpublished manuscript, 1992] have suggested that this storage occurred at the base of the continental lithosphere over a protracted period before separation of India from Antarctica. Evidence for this comes from drainage patterns within eastern India and uplift ages for the Prince Charles Mountains, Antarctica, which indicate that this region of Gondwana was undergoing substantial uplift in Permian and possibly even Carboniferous times. This may reflect incubation of a (or more probably several) large plume head over tens (if not hundreds) of millions of years before the fragmentation of Gondwana allowing mantle upwelling, melting and release of the thermal energy. Such a long term accumulation of plume material is not consistent, however, with paleomagnetic evidence for plate motion across the mantle during this period [McElhinny, 1973].

An alternative explanation is that the Kerguelen Plateau initiated with the arrival approximately 120-115 m.y. ago of a large thermal or start-up plume [Richards et al., 1989; Griffiths and Campbell, 1990] at the base of the lithosphere. Interestingly, this coincides, within the range of uncertainty in basement ages, with the formation of the Ontong Java Plateau in the western Pacific [Tarduno et al., 1991], so the two episodes may ultimately be linked to a common deep mantle convective event [Duncan and Richards, 1990]. However, older ages for the currently poorly documented Naturaliste Plateau and Bunbury basalts of the southwest margin of Australia [Storey et al., 1992], if linked to the Kerguelen plume system, would prohibit application of the plume head model for plume initiation.

COMPARATIVE PLUMOLOGY: THE HAWAIIAN, LOUISVILLE, RÉUNION, AND KERGUELEN PLUMES

With the completion of ODP drilling in the Indian Ocean, there are now 4 plumes for which we have extensive compositional histories: Hawaii [e.g., Lanphere et al., 1980; Clague and Dalrymple, 1987], Louisville [Hawkins et al., 1987; Cheng et al., 1987, Mahoney et al.,

unpublished manuscript, 1992], Réunion, and Kerguelen. There are important differences between each and it is now clear that generalizations about plume life cycles should be made with caution.

(1) At Hawaiian volcanoes the isotopic signature of the plume is strongest in the initial shield-building tholeiites at each volcano [e.g., West et al., 1987] but within the Kerguelen Archipelago the isotopic signature of the plume is strongest in the young, highly alkaline lavas [basanites and their differentiation products, Weis et al., 1992]. In addition, the Kerguelen plume is characterized by greater than bulk earth $^{87}Sr/^{86}Sr$, less than bulk earth ε_{Nd}, and high ratios of $^{207}Pb/^{204}Pb$ and $^{208}Pb/^{204}Pb$ at a given $^{206}Pb/^{204}Pb$ ratio (Figures 2, 3 and 9). These characteristics differ substantially from the Hawaiian plume which apparently has $\varepsilon_{Nd} >0$ and $^{87}Sr/^{86}Sr$ less than bulk earth estimates [e.g., West et al., 1987].

On Réunion, Nd and Sr isotopic ratios are uniform at ε_{Nd} of +3.6 to +4.4 and $^{87}Sr/^{86}Sr$ of 0.7040 to 0.7044 [Fisk et al., 1988]. These values fall within the range for both Hawaii and for Kerguelen but are among the lowest Nd and highest Sr isotopic ratios for Hawaii and are among the highest Nd and lowest Sr isotopic ratios for Kerguelen. Lead isotopic ratios of Réunion Island are similar to Kerguelen in their high $^{207}Pb/^{204}Pb$ and $^{208}Pb/^{204}Pb$ at a given $^{206}Pb/^{204}Pb$ ratio but Réunion $^{206}Pb/^{204}Pb$ is higher than at Kerguelen. Unlike Hawaii and Kerguelen, however, there is no significant difference in isotopic signature of the old shield-building lavas and the late stage differentiated lavas of Réunion.

Though volcanic evolution of Mauritius, the second youngest product of the Réunion plume, seems more complex than the Hawaiian model, with 3 distinct phases of volcanism, each separated by 1 to 2 m.y. hiatuses [McDougall and Chamalaun, 1969; Duncan, unpublished data], the limited available isotopic data do suggest a pattern that is somewhat similar to the Hawaiian one. The Younger and Intermediate Series lavas of Mauritius have slightly more depleted isotopic character than those of the shield-building Older Series [Mahoney et al., 1989].

(2) The tholeiitic basalts forming the Hawaiian shields equilibrated with residual garnet [Frey and Roden, 1987], but the compositions of the tholeiitic basalts forming the Ninetyeast Ridge provide no evidence of residual garnet. Apparently, the pressure of melt segregation for Ninetyeast Ridge lavas was lower than the spinel to garnet transition (60 to 70 km). Basalts from the Réunion hotspot also appear to show no strong evidence of residual garnet in the source. All magmas can be derived from the spinel peridotite field.

(3) Individual Hawaiian volcanoes evolve from tholeiitic shields to post-shield alkalic volcanism. This change in lava composition is accompanied by a trend toward MORB isotopic ratios. These short-term compositional and isotopic changes are interpreted as resulting from a decrease in extent of melting and magma supply as the volcano moved away from the hotspot and a correspondingly greater influence of contamination from the

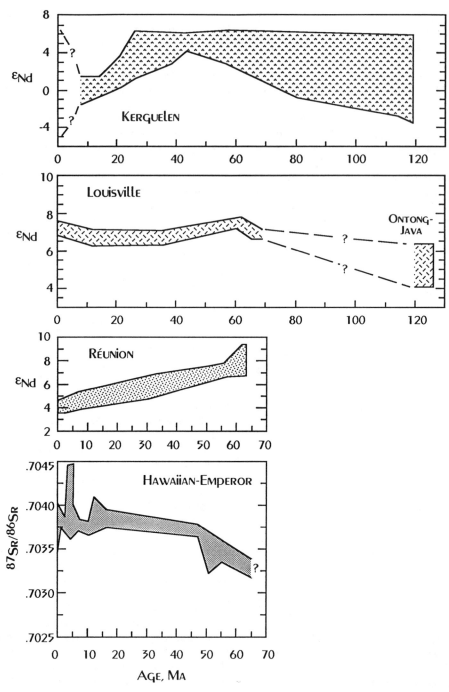

Fig. 18. Comparison of time-varying isotopic changes in the volcanic products of the Kerguelen, Louisville, Reunion and Hawaiian plumes. Sr data are compared for the Hawaiian plume because of limited Nd data. We see no common pattern in the evolution of basalt compositions.

lithosphere through which the latest melts rose [Chen and Frey, 1985]. The Ninetyeast Ridge drill sites do not show similar short-term temporal trends; i.e., alkalic lavas are absent and the youngest lavas do not trend toward MORB isotopic ratios. Late stage magmas from Réunion Island are slightly more evolved than the bulk of the island shield due to the effect of fractionation in crustal magma chambers,

but there is no change in isotopic composition with time in this volcanic system. Petrological evolution on Mauritius, though complex, is more similar to the Hawaiian evolution [Baxter, 1976]. The shield-building lavas of the Older Series are transitional between tholeiitic and alkali basalt. Eruption of the Intermediate Series, which vary from alkali basalt to nephelinite, followed a 2 m.y. hiatus, a pattern very reminiscent of the Hawaiian one. This was followed, after a 1.6 m.y. hiatus by the eruption of the Younger Series, which are only mildly alkalic. Eruption of the Intermediate Series followed the eruption of the oldest dated lava on Mauritius by 4.3 m.y. The oldest lavas on Réunion are less than 2 Ma [McDougall, 1971], so it is possible that a similar pattern will occur on Réunion.

Figure 18 compares the isotopic variations through time for the Kerguelen, Louisville, Réunion, and Hawaiian plumes. Clearly, the composition of the products of each plume has varied independently through time, and no overall pattern emerges. The systematic change toward more enriched isotopic signatures of the Réunion plume is unique, though there is a hint of similar variation in the early history of the Hawaiian plume and in the late history of the Kerguelen plume. The Louisville plume was apparently responsible for creation of the Ontong Java Plateau [Richards et al., 1989]. If so, magmas of the initial phase of this plume are less depleted than subsequent ones [Mahoney et al., unpublished manuscript, 1992]. Over the past 60 million years, Louisville Ridge plume products have been isotopically homogeneous [Cheng et al., 1987], in marked contrast to those of the other plumes.

The geochemical characteristics of the Hawaiian, Louisville, Kerguelen, and Réunion mantle plumes are very different. Some of the differences may be explained by the changes in proximity of the Indian Ocean plumes to spreading ridge axes through time. All of the studied Hawaiian-Emperor volcanoes formed in an intraplate setting; i.e., ranging from construction on 40 m.y. old oceanic crust for Suiko Seamount to 80-90 m.y. old oceanic crust for recent Hawaiian volcanoes [Clague and Dalrymple, 1987]. In contrast, these Indian Ocean hotspot traces formed on oceanic crust < 30 Ma, with the exception of Réunion and Mauritius islands (see Royer et al., this volume). Moreover, in this tectonic environment extensive mixing of MORB (tholeiitic basalt) and plume-related components can explain the absence of alkaline volcanism along the hotspot trace and the evidence for melt segregation at low pressures, similar to those postulated for MORB.

ACCOMPLISHMENTS AND INSIGHTS PROVIDED BY INDIAN OCEAN DRILLING

Deep ocean drilling in the Indian Ocean has revealed several new aspects of the physical and chemical behavior of mantle plumes and hotspots, and their role in the generation of oceanic crust. First, drilling has established the continuity of volcanic activity along the two major hotspot tracks in the Indian Ocean basin — those formed over the Réunion and Kerguelen plumes. The systematic progression in ages allows us to trace the volcanic lineaments back to huge flood basalt provinces which are the initial phase of activity of these hotspots. This firm connection between hotspots and flood basalt events has spawned a new model of mantle plumes as thermal diapirs that may account for large, episodic transfers of heat from

the core to the lithosphere [Courtillot and Besse, 1987; Richards et al., 1989; Larson, 1991]. There appears to be a similar relationship between several other plumes and flood basalt provinces as well (e.g., Tristan da Cunha—Parana, Galapagos—Caribbean, Iceland—North Atlantic Basalt Province, Louisville—Ontong Java), but other plumes (e.g., Azores, Comores) apparently did not have an initial flood basalt phase. Why not? Perhaps they are simply too small, though a relationship between present-day plume flux and the presence or absence of an initial flood basalt phase has not been demonstrated. Do these "second-class" plumes come from a shallower thermal boundary layer, perhaps near 670 km depth, and have no chance to develop a large-volume "head"?

The first significant sampling of oceanic plateaus, such as large portions of the Kerguelen Plateau, has shown that these are entirely volcanic in origin, and are probably oceanic equivalents of continental flood basalt provinces. These represent huge volumes (up to 50×10^6 km^3) of mantle melting and are significant components in intraplate ocean crust formation. Compositional data establish that these melts are derived from plume and asthenosphere mixtures. We do not yet know what the earliest melts are like, although there are indications from other flood basalt provinces that picritic compositions may predominate [Storey et al., 1991], as predicted by Campbell and Griffiths [1990].

We can now document in some detail the variation in compositions along these two hotspot tracks. Early products from the Réunion hotspot have more "depleted" isotopic signatures than present Réunion magmas. With time, these isotopic signatures have become progressively more plume-like. The isotopic variations in the Réunion plume products are broadly consistent with extensive entrainment of asthenosphere within the plume during its initial stages, with a subsequent gradual decline in the amount of material entrained. A similar interpretation might be applied to the Kerguelen mantle plume as well, as products of that plume also have become more "enriched" over time. However, the progression was much less simple than in the Réunion case, and three components are apparently involved. Thus these studies have not led to a universally applicable model for the evolution of plume volcanism. They do demonstrate, however, that the composition of plume products can vary systematically over time.

In both the Réunion and Kerguelen plumes, trace element abundance ratios also varied with time. The variation in the Kerguelen case is more systematic than in the Réunion case, which is the opposite of what was observed for isotope ratios. Trace element abundance ratios reflect variations in both source composition and the conditions of magma generation. Combining trace element and isotopic ratios allows the latter to be assessed. In the case of the Réunion plume, this approach suggests that the degree of melting was higher when the plume was close to a spreading center (Sites 707, 713 and SM-1) than at other times. Though the situation is more complex, there seems to be a similar relationship in the Kerguelen data: degree of melting was apparently higher when the plume was ridge-centered than it has been since the spreading center drifted off the plume 30-40 m.y. ago.

Acknowledgments. Samples from two industry sites SM-1 (Saya de Mahla Bank) and NB-1 (Nazareth Bank) have kindly been made available to Leg 115 investigators by Texaco. Weis was supported by the Belgian National Fund for Scientific Research. White, Frey, Duncan and Fisk were supported by funds from the National Science Foundation. WMW acknowledges the analytical assistance of M. M. Cheatham. The manuscript benefited very substantially from careful reviews by John Mahoney and David Graham.

REFERENCES

Allègre, C. J. and Turcotte, D. L., Implications of a two-component marble-cake mantle, *Nature, 323*, 123-127, 1986.

Alibert, C., Mineralogy and geochemistry of a basalt from Site 738: implications for the tectonic history of the southernmost part of the Kerguelen Plateau, in *Proc. ODP, Sci. Results, 119*, edited by J. Barron, B. Larsen et al., 293-298, Ocean Drilling Program, College Station, TX, 1991.

Backman, J., Duncan, R. A. et al., *Proc. ODP, Init. Repts., 115*, Ocean Drilling Program, College Station, TX, 1988.

Baksi, A. K., $^{40}Ar/^{39}Ar$ incremental heating study of whole-rock samples from the Rajmahal and Bengal Traps, eastern India, *Terra Cognita, 6*, 161, 1986.

Baksi, A. K., Barman, R. T., Paul, D. K. and Farrar, E., Widespread Early Cretaceous flood basalt volcanism in eastern India: geochemical data from the Rajmahal-Bengal-Sylhet traps, *Chem. Geol., 63*, 133-141, 1987.

Barling, J. and Goldstein, S. L. Extreme isotopic variations in Heard Island lavas and the nature of mantle reservoirs, *Nature, 348*, 59-62, 1990.

Barron, J., Larsen, B. et al., *Proc. ODP, Init. Repts., 119*, Ocean Drilling Program, College Station, TX, 1989.

Baxter, A. N., Geochemistry and petrogenesis of primitive alkali basalts from Mauritius, Indian Ocean, *Geol. Soc. Am. Bull., 87*, 1028-1034, 1976.

Baxter, A. N., Major and trace element variations in basalts from Leg 115, in *Proc. ODP, Sci. Results, 115*, edited by R. A. Duncan, J. Backman, L. C. Peterson et al., 11-21, Ocean Drilling Program, College Station, TX, 1990.

Baxter, A. N., Upton, B. J. G. and White, W. M., Petrology and geochemistry of Rodrigues Island, Indian Ocean, *Contrib. Mineral. Petrol., 89*, 90-101, 1985.

Ben Othman, D., White, W. M. and Patchett, J., The geochemistry of marine sediments, island arc magma genesis, and crust-mantle recycling, *Earth Planet. Sci. Lett., 94*, 1-21, 1989.

Biswas, B., Results of exploration for petroleum in the western part of the Bengal Basin, India, *ECAFE Miner. Resour. Devl. Ser., 18*, 241-250, 1963.

Campbell, I. H. and Griffiths, R. W., Implications of mantle plume structure for the evolution of flood basalts, *Earth Planet. Sci. Lett., 99*, 79-93, 1990.

Chen, C. Y. and Frey, F. A., Trace element and isotopic geochemistry of lavas from Haleakala volcano, east Mauri, Hawaii: implications for the origin of Hawaiian basalts, *J. Geophys. Res., 90*, 8743-8768, 1985.

Cheng, Q., Macdougall, J. D., Lugmair, G. W. amd Natland, J., Temporal variation in isotopic composition at Tahiti, Society Islands, *EOS Trans. Am. Geophys. Union, 69*, 1521, 1987.

Clague, D. A. and Dalrymple, G. B., The Hawaiian-Emperor Volcanic Chain, Part 1: geologic evolution, *Geol. Surv. Prof. Pap. U.S., 1350*, 5-54, 1987.

Coffin, M. F., Davies, H. L. and Haxby, W. F., Structure of the Kerguelen Plateau province from Seasat altimetry and seismic reflection data, *Nature, 324*, 134-136, 1986.

Cox, K. G. and Hawkesworth, C. J., Geochemical stratigraphy of the Deccan Traps at Mahabaleshwar, Western Ghats, India, with implications for open system magmatic processes, *J. Petrol., 26*, 355-377, 1985.

Courtillot, V. and Besse, J., Magnetic field reversals, polar wander, and core-mantle coupling, *Science, 237*, 1140-1147, 1987.

Davies, H., Sun, S. S., Frey, F. A., Gautier, I., McCulloch, M. T., Price, R. C., Bassias, Y., Klootwijk, C. T. and Leclaire, L., Basalt basement from the Kerguelen Plateau and the trail of a Dupal plume, *Contrib. Mineral. Petrol., 103*, 457-469, 1989.

Davies, T. A., Luyendyk, B. P. et al., *Init. Repts. DSDP, 26*, 295-325, U.S. Government Printing Office, Washington, D.C., 1974.

DePaolo, D. and Wasserburg, G.J., Inferences about magma sources and mantle structure from variations of $^{143}Nd/^{144}Nd$, *Geophys. Res. Lett., 3*, 743-746, 1976.

Desonie, D. L. and Duncan, R. A., The Cobb-Eichelberg seamount chain: hotspot with MORB affinity, *J. Geophys. Res., 95*, 12,697-12,711, 1990.

de Wit, M., Jeffrey, M., Bergh, H. and Nicolaysen, L., Geological map of sectors of Gondwana, reconstructed to their disposition at 150 Ma, AAPG, Tulsa, OK, 1988.

Dosso, L., Bougault, H., Beuzart, P., Calvez, J.-Y. and Joron, J.-L., The geochemical structure of the South-East Indian Ridge, *Earth Planet. Sci. Lett., 88*, 47-59, 1988.

Duncan, R. A., Geochronology of basalts from the Ninetyeast Ridge and continental dispersion in the eastern Indian Ocean, *J. Volcanol. Geotherm. Res., 4*, 283-305, 1978.

Duncan, R. A., Hotspots in the southern oceans - an absolute frame of reference for motion of the Gonwana continents, *Tectonophysics, 74*, 29-42, 1981.

Duncan, R.A., The volcanic record of the Réunion hotspot, in *Proc. ODP, Sci. Results, 115*, edited by R. A. Duncan, J. Backman, L. C. Peterson et al., 3-10, Ocean Drilling Program, College Station, TX, 1990.

Duncan, R.A., The age distribution of volcanism along aseismic ridges in the eastern Indian Ocean, in *Proc. ODP, Sci. Results, 121*, edited by J. Weissel, J. Peirce, E. Taylor, J. Alt et al., 507-517, Ocean Drilling Program, College Station, TX, 1991.

Duncan, R. A. and Hargraves, R. B., 1990. $^{40}Ar/^{39}Ar$ geochronology of basement rocks from the Mascarene Plateau, the Chagos Bank, and the Maldives Ridge, in *Proc. ODP, Sci. Results, 115*, edited by R. A. Duncan, J. Backman, L. C. Peterson et al., 43-51, Ocean Drilling Program, College Station, TX, 1990.

Duncan, R.A. and Richards, M.A., Early Cretaceous global flood basalt volcanism, *EOS, Trans. Am. Geophys. Union, 71*, 1668, 1990.

Duncan, R.A. and Richards, M.A., Hotspots, mantle plumes, flood basalts, and true polar wander, Rev. Geophys., 29, 31-50, 1991.

Dupré, B. and Allègre, C.J., Pb-Sr isotope variation in Indian Ocean basalts and mixing phenomena, *Nature, 303*, 346-349, 1983.

Dupré, B., Structure et évolution du manteau terrestre étudiées à l'aide des traceurs isotopiques couplés: Sr-Pb, PhD Thesis, 211 p., Univ. Paris, 1983.

Eissen, J-P., Juteau, T., Joron, J-L., Dupré, B., Humler, E. and Al'Mukhamedov, A., Petrology and Geochemistry of Basalts from the Red Sea Axial Rift at 18° North, *J. Petrol., 30*, 791-839, 1989.

Falloon, T. J. and Green, D. H., Anhydrous partial melting of peridotite from 8 to 35 kb and the petrogenesis of MORB, *J. Petrol. Spec. Lithosphere Iss.*, 379-414, 1988.

Fisk, M. R., Upton, B. G. J., Ford, C. E. and White, W. M., Chemistry of Réunion Island volcanic rocks: Major element, trace element, isotope and mineral chemistry, and experimental phase relations, *J. Geophys. Res., 93*, 4933-4950, 1988.

Fisk, M. R., Duncan, R. A., Baxter, A. N., Greenough, J. D., Hargraves, R. B. and Tatsumi, Y., Réunion hotspot magma chemistry over the past 65 m.y.: results from Leg 115 of the Ocean Drilling Program, *Geology, 17*, 934-937, 1989.

Fisk, M. R. and Howard, K. J., 1990. Primary mineralogy of Leg 115 basalts, in *Proc. ODP, Sci. Results, 115*, edited by R. A. Duncan, J. Backman, L. C. Peterson et al., 113-42, Ocean Drilling Program, College Station, TX, 1990.

Floyd, P. A., Geochemical features of intraplate oceanic plateau basalts, in *Magmatism in The Ocean Basins, Geol. Soc. London Spec. Publ. 42*, edited by A. D. Saunders, and M. J. Norry, 215-230, 1989.

Frey, F. A., Dickey, J. S. Jr., Thompson, G. and Bryan, W. B., Eastern Indian Ocean DSDP sites: correlations between petrography, geochemistry and tectonic setting, in *A Synthesis of Deep Sea Drilling in the Indian Ocean*, edited by J. R. Heirtzler, and J. G. Sclater, 189-257, U.S. Government Printing Office, Washington, D.C., 1977.

Frey, F. A. and Roden, M. F., The mantle source for the Hawaiian islands: Constraints from the lavas and the ultramafic inclusions, in *Mantle Metasomatism*, edited by C. J. Hawkesworth, and M. A. Menzies, 423-463, Academic Press, London, 1987.

Frey, F. A., Jones, W. B., Davies, H. and Weis, D., Geochemical and petrologic data for basalts from Sites 756, 757, and 758: implications for the origin and evolution of Ninetyeast Ridge, in *Proc. ODP, Sci. Results, 121*, edited by J. Weissel, J. Peirce, E. Taylor, J. Alt et al., 611-659, Ocean Drilling Program, College Station, TX, 1991.

Gast, P. W., Trace element fractionation and the origin of tholeiitic and alkaline magma types, *Geochim. Cosmochim. Acta, 32*, 1057-1086, 1968.

Gautier, I., Frey, F. A. and Weis, D., The differentiated lavas of the Southeast Province of Kerguelen Archipelago (South Indian Ocean), *EOS, Trans. Am. Geophys. Union, 70*, 1384, 1989.

Gautier, I., Weis, D., Mennessier, J.-P., Vidal, P., Giret, A. and Loubet, M., Petrology and geochemistry of the Kerguelen Archipelago basalts: evolution of the mantle sources from ridge to intraplate position, *Earth Planet. Sci. Lett., 100*, 59-76, 1990.

Giret, A., and Lameyre, J., A study of Kerguelen plutonism: petrology, geochronology and geological implications, in *Antarctic Earth Science*, edited by R. L. Oliver, P. R. James, and J. B. Jago, 646-651, Cambridge University Press, Cambridge, 1983.

Griffiths, R. W., The differing effects of compositional and thermal buoyancies on the evolution of mantle diapirs, *Phys. Earth Planet. Int., 43*, 261-273, 1986.

Griffiths, R. W. and Campbell, I. H., Stirring and structure in mantle plumes, *Earth Planet. Sci. Lett., 99*, 66-78, 1990.

Hanan, B. B., Kingsley, R. H. and Schilling, J.-G., Pb isotopic evidence in the South Atlantic for migrating ridge-hotspot interaction, *Nature, 322*, 137-144, 1986.

Hamelin, B., Dupré, B. and Allègre, C.-J., Pb-Sr-Nd isotopic data of Indian Ocean ridges: new evidence for large-scale mapping of mantle heterogeneities, *Earth Planet. Sci. Lett., 76*, 288-298, 1985/1986.

Harland, W.B., Armstrong, R.L., Cox, A.V., Craig, L.E., Smith, A.G. and Smith, D.G., *A Geologic Timescale 1989*, Cambridge University Press, Cambridge, U.K., 1990.

Hart, S. R., A large-scale isotope anomaly in the Southern Hemishpere mantle, *Nature, 309*, 753-757, 1984.

Hart, S. R., Heterogeneous mantle domains: signatures, genesis and mixing chronologies, *Earth Planet. Sci. Lett., 90*, 273-296, 1988.

Hart, S. R., Hauri, E. H., Oschmann, L. A. and Whitehead, J. A., Mantle plumes and entrainment: isotopic evidence, *Science, 256*, 517-520, 1992.

Hawkesworth, C. J., Kempton, P. D., Rogers, N. W., Ellam, R. W. and van Calsteren, P. W., Continental mantle lithosphere, and

shallow level enrichment processes in the Earth's mantle, *Earth Planet. Sci. Lett., 96*, 256-268, 1990.

Hawkins, J. W., Lonsdale, P. F. and Batiza, R., Petrologic evolution of the Louisville Seamount Chain, in *Seamounts, Islands, and Atolls, Geophys. Mon. Series, Vol. 43*, edited by B. H. Keating, P. Fryer, R. Batiza and G. W. Boehlert, pp. 235-254, AGU, Washington, D.C., 1987.

Hofmann, A. W. and White, W. M., Mantle plumes from ancient oceanic crust, *Earth Planet. Sci. Lett., 57*, 421-436, 1982.

Houtz, R. E., Hayes, D. E. and Markl, R. G., Kerguelen Plateau bathymetry, sediment distribution and crustal structure, *Mar. Geology, 25*, 95-130, 1977.

Ito, E., White, W. M. and Goepel, C., The O, Sr, Nd and Pb isotope geochemistry of MORB, *Chem. Geol., 62*, 157-176, 1987.

Kent, R. W., Lithospheric uplift in eastern Gondwana: evidence for a long-lived mantle plume system? *Geology, 19*, 19-23, 1991.

Klein, E. M. and Langmuir, C. H., Ocean ridge basalt chemistry, axial depth, crustal thickness and temperature variations in the mantle, *J. Geophys. Res., 92*, 8089-8115, 1987.

Klein, E. M., Langmuir, C. H., Zindler, A., Staudigel, H. and Hamelin, B., Isotope evidence of a mantle convection boundary at the Australian-Antarctic discordance, *Nature, 333*, 623-629, 1988.

Klootwijk, C. T., Gee, J. S., Smith, G. M. and Peirce, J. W., Constraints on the India-Asia convergence: paleomagnetic results from the Ninetyeast Ridge, ODP Leg 121, in *Proc. ODP, Sci. Results, 121*, edited by J. Weissel, J. Peirce, E. Taylor, J. Alt et al., 777-882, Ocean Drilling Program, College Station, TX, 1991.

Langmuir, C. H., Bender, J. F. and Batiza, R., Petrological and tectonic segmentation of the East Pacific Rise, 5°30'-14°30'N, *Nature, 332*, 422-429, 1986.

Lanphere, M. A., Dalrymple, G. B. and Clague, D. A., Rb-Sr systematics of basalts from the Hawaiian-Emperor volcanic chain, in *Init. Repts. DSDP, 55*, edited by E. D. Jackson, I. Koisumi et al., 695-706, U.S. Government Printing Office, Washington, D.C., 1980.

Larson, R. L., Latest pulse of earth: evidence for a mid-Cretaceous super plume, *Geology, 19*, 547-550, 1991.

Leclaire, L., Denis-Clocchiatti, M., Davies, H., Gautier, I., Gensous, B., Giannesini, P. J., Morand, F., Patriat, P., Segoufin, J., Tesson, M. and Wanneson. J., Nature et age do plateau de Kerguelen-Heard, secteur-sud. Resultats preliminaires de la campagne N.A.S.K.A.-MD 48, *C.R. Acad. Sci. Paris, 304*, 23-28, 1987.

Le Roex, A. P., Dick. H. J. B., Reid, A. M., Frey , F. A., Erlank, A. J. and Hart, S. R., Petrology and geochemistry of basalts from the American-Antarctic Ridge, Southern Indian Ocean: Implications for the westward influence of the Bouvet mantle plume, *Contrib. Mineral. Petrol., 90*, 367-380, 1985.

Lightfoot, P. and Hawkesworth, C., Origin of Deccan Trap lavas: evidence from combined trace element and Sr-, Nd- and Pb-studies, *Earth Planet. Sci. Lett., 93*, 89-104, 1988.

Lightfoot, P., Hawkesworth, C., Devey, C. W., Rogers, N. W. and van Calsteren, P. W. C., Source and differentiation of Deccan Trap lavas: implications of geochemical and mineral chemical variations, *J. Petrol., 31*, 1165-1200, 1990.

Ludden, J. N., and Dionne, B., The geochemistry of oceanic crust at the onset of rifting in the Indian Ocean, in *Proc. ODP, Sci. Res., 123*, edited by F. M. Gradstein, J. N. Ludden et al., Ocean Drilling Program, College Station, TX, 1992.

Luyendyk, B. P., and Rennick, W., Tectonic history of aseismic ridges in the eastern Indian Ocean, *Geol. Soc. Am. Bull., 88*, 1347-1356, 1977.

Macdonald, G. A. and Katsura, T., Chemical composition of Hawaiian lavas, *J. Petrol., 5*, 82-133, 1964.

Macdougall, J. D. and Lugmair, G. W., Sr and Nd isotopes in basalts from the East Pacific Rise: significance for mantle heterogeneity, *Earth Planet. Sci. Lett., 77*, 273-284, 1986.

Mahoney, J. J., Deccan Traps, in *Continental Flood Basalts*, edited by J. D. Macdougall, 151-194, D. Riedel and Co., Amsterdam, 1988.

Mahoney, J. J., Macdougall, J. D., Lugmair, G. W., Murali, A. V., Sankar Das, M. and Gopalan, K., Origin of the Deccan Trap flows at Mahabaleshwar inferred from Nd and Sr isotopic and chemical evidence, *Earth. Planet. Sci. Lett., 60,* 47-60, 1982.

Mahoney, J. J., Macdougall, J. D., Lugmair, G. W. and Gopalan, K., Kerguelen hotspot source for the Ninetyeast Ridge? *Nature, 303,* 385-389, 1983.

Mahoney, J. J., Natland, J. H., White, W. M., Poreda, R., Fisher, R. L. and Baxter, A. N., Isotopic and geochemical provinces of the western Indian Ocean Spreading Centers, *J. Geophys. Res., 94,* 4033-4052, 1989.

Mahoney, J. J., le Roex, A. P., Peng, Z., Fisher, R. L. and Natland, J. H., Western limits of Indian MORB mantle and the origin of low $^{206}Pb/^{204}Pb$ MORB: isotope systematics of the central Southwest Indian Ridge (17-50°E), *J. Geophys. Res., 97,* 1992.

McDougall, I., The geochronology and evolution of the young oceanic island of Réunion, Indian Ocean, *Geochim. Cosmochim. Acta, 35,* 261-270, 1971.

McDougall, I. and Chamalaun, F. G., Isotopic dating and geomagnetic polarity studies on volcanic rocks from Mauritius, Indian Ocean, *Geol. Soc. Am. Bull., 80,* 1419-1431, 1969.

McElhinny, M. W., *Palaeomagnetism and Plate Tectonics,* 358 p, Cambridge University Press, Cambridge, 1973.

McKenzie, D. P. and Sclater, J. G., The evolution of the Indian Ocean since the Late Cretaceous, *Geophys. J. R. Soc. Astron., 25,* 437-528, 1971.

McKenzie, D. and Bickle, M. J., The volume and composition of melt generated by extension of the lithosphere, *J. Petrol., 29,* 625-679, 1988.

Mehl, K. W., Bitschene, P. R., Schmincke, H.-U. and Hertogen, J., Composition, alteration, and origin of the basement lavas and volcaniclastic rocks at Site 738, southern Kerguelen Plateau, in *Proc. ODP, Sci. Results, 119,* edited by J. Barron, B. Larsen et al., 299-321, Ocean Drilling Program, College Station, TX, 1991.

Merriman, R. J., Taylor, P. N. and Morton, A. C., Petrochemistry and isotope geochemistry of early Palaeogene basalts forming the dipping reflector sequence SW of Rockall Plateau, NE Atlantic, in *Early Tertiary Volcanism and the Opening of the NE Atlantic ,* edited by A. C. Morton and L. M. Parson, 123-134, *Geol. Soc. London Spec. Publ. 39,* 1988.

Michard, A., Montigny, R. and Schlich, R., Geochemistry of the mantle beneath the Rodriguez Triple Junction and the South-East Indian Ridge, *Earth Planet. Sci. Lett., 78,* 104-114, 1986.

Morgan, W. J., Deep mantle convection plumes and plate motions, *Am. Assoc. Pet. Geol. Bull., 56,* 203-213, 1972.

Morton, A. C. and Taylor, P. N., Lead isotope evidence for the structure of the Rockall dipping-reflector passive margin, *Nature, 326,* 381-383, 1987.

Mutter, J. C. and Cande, S. C., The early opening between Broken Ridge and Kerguelen Plateau, *Earth Planet. Sci. Lett., 65,* 369-376, 1983.

Nelson, D. R., McCulloch, M. T. and Sun, S. S., The origins of ultrapotassic rocks as inferred from Sr, Nd and Pb isotopes, *Geochim. Cosmochim. Acta, 50,* 231-311, 1986.

Nogi, Y., Seama, N., Iaozaki, N., Hayashi, T., Funaki, M. and Kaminuma, K., 1991. Geomagnetic anomaly lineations and fracture zones in the Basin West of the Kerguelen Plateau, *EOS, Trans. Am. Geophys. Union, 72,* 445, 1991.

Ogg, J., Kodama, K. and Wallick, B., Lower Cretaceous magnetostratigraphy and paleolatitudes off Northwestern Australia, ODP Site 765 and DSDP Site 261, Argo Abyssal Plain, and ODP Site 766, Gascoyne Abyssal Plain, in *Proc. ODP, Sci. Res., 123,* edited by F. M. Gradstein, J. N. Ludden et al., Ocean Drilling Program, College Station, TX, 1992.

Palacz, Z. A. and Saunders, A. D., Coupled trace element and isotope enrichment in the Cook-Austral-Samoa islands, southwest Pacific, *Earth Planet. Sci. Lett., 79,* 270-280, 1986.

Patchett, P.J. and Tatsumoto, M., Hafnium isotope variations in oceanic basalts, *Geophys. Res. Lett., 7,* 1077-1080, 1980.

Peirce, J. W., The northward motion of India since the Late Cretaceous, *Geophys. J. R. Astr. Soc., 52,* 277-311, 1978.

Peirce, J. W., Weissel, J. K. et al., *Proc. ODP, Init. Repts., 121,* pp. 1000, Ocean Drilling Program, College Station, TX, 1989.

Powell, C. McA., Root, S. R. and Veevers, J. J., Pre-breakup continental extension in East Gondwanaland and the early opening of the eastern Indian Ocean, *Tectonophysics, 155,* 261-283, 1988.

Pyle, D. G., Christie, D. M. and Mahoney, J. J., Resolving an isotopic boundary within the Australian-Antarctic discordance, *Earth Planet. Sci. Lett., 112,* 161-178, 1992.

Rea, D. K., Dehn, J., Driscoll, N. W., Farrell, J. W., Janacek, T. R., Owen, R. M., Posphical, J. J., Resiwati, P. and ODP Leg 121 Shipboard Scientific Party, Paleoceanography of the eastern Indian Ocean from ODP Leg 121 drilling on Broken Ridge, *Geol. Soc. Am. Bull., 102,* 679-690, 1990.

Recq, M., Charvis, P. and Hirn, A., Preliminary results on the deep structure of the Kerguelen Ridge, from seismic refraction experiments, *C. R. Acad. Sci. Paris, Ser. 2, 297,* 903-908, 1983.

Recq, M. and Charvis, P., A seismic refraction survey in the Kerguelen Isles, southern Indian Ocean, *Geophys. J. R. Astron. Soc., 84,* 529-559, 1986.

Recq, M., Brefort, D., Malod, J. and Veinante, J.-L., The Kerguelen Isles (southern Indian Ocean): new results on deep structure from refraction profiles, *Tectonophysics, 182,* 227-248, 1990.

Richards, M. A., Duncan, R. A. and Courtillot, V., Flood basalts and hotspot tracks: plume heads and tails, *Science, 246,* 103-107, 1989.

Royer, J.-Y. and Sandwell, D. T., Evolution of the eastern Indian Ocean since the late Cretaceous: constraints from GEOSAT altimetry, *J. Geophys. Res., 94,* 13,755-13,782, 1989.

Royer, J.-Y., Peirce, J. W. and Weissel, J. K, 1991. Tectonic constraints on hotspot formation of the Ninetyeast Ridge, in *Proc. ODP, Sci. Results, 121,* edited by J. Weissel, J. Peirce, E. Taylor, J. Alt et al., 763-776, Ocean Drilling Program, College Station, TX, 1991.

Salters, V. J. M., Storey, M., Sevigny, J. H. and Whitechurch, H., Trace element and isotopic characteristics of Kerguelen-Heard Plateau basalts, in *Proc. ODP, Sci. Res., 120,* edited by S. W. Wise, Jr., R. Schlich et al., 55-62, Ocean Drilling Program, College Station, TX, 1992.

Saunders, A. D., Storey, Gibson, I. L., Leat, P., Hergt, J. and Thompson, R. N., Chemical and isotopic constraints on the origin of the basalts from the Ninetyeast Ridge, Indian Ocean: results from Deep Sea Drilling Project Legs 22 and 26, and Ocean Drilling Program Leg 121, in *Proc. ODP, Sci. Res., 121,* edited by J. Peirce, J. Weissel, E. Taylor et al., 559-590, Ocean Drilling Program, College Station, TX, 1991.

Schilling, J.-G., Zajac, M., Evans, R., Johnson, T., White, W. M., Devince, J. D. and Kingsley, R., Petrologic and geochemical variations along the Mid-Atlantic Ridge from 29°N to 73°N, *Am. J. Sci., 283,* 510-586, 1983.

Schlich, R., Structure et âge de l'Océan Indien occidental, *Mem. Hors Sér. Soc. Géol. France, 6,* 1-103, 1975.

Schlich, R., The Indian Ocean: aseismic ridges, spreading centers, and oceanic basins, in *The Ocean Basins and Margins, The Indian Ocean, 6,* edited by A. E. M. Nairn, and F. G. Stehli, 1-148, Plenum, New York, 1982.

Schlich, R., Wise, S. W. Jr. et al., *Proc. ODP, Init. Repts., 120,* Ocean Drilling Program, College Station, TX, 1989.

Schubert, G. and Sandwell, D., Crustal volumes of the continents and of oceanic and continental submarine plateaus, *Earth Planet. Sci. Lett., 92,* 234-246, 1989.

Sclater, J. G., and Fisher, R. L., The evolution of the east central Indian Ocean, with emphasis on the tectonic setting of the Ninetyeast Ridge, *Bull. Geol. Soc. Am., 85,* 683-702, 1974.

Sengupta, S., Upper Gondwana stratigraphy and paleobotany of Rajmahal Hills, Bihar (India), *Palaeontologica India. Geol. Soc. India Monogr., N°48 (new series),* 180 p., 1988.

Sevigny, J. H., Whitechurch, H., Storey, M. and Salters, V. J. M., 1992. Zeolite-facies metamorphism of central Kerguelen Plateau basalts, in *Proc. ODP, Sci. Res. 120*, edited by S. W. Wise, Jr., R. Schlich et al., 63-70, Ocean Drilling Program, College Station, TX, 1992.

Shipboard Scientific Party, Réunion hotspot activity through Tertiary time: initial results from the Ocean Drilling Program, Leg 115, *J. Volcan. Geotherm. Res., 36*, 193-198, 1989.

Sleep, N. H., Hotspots and mantle plumes: some phenomenology, *J. Geophys. Res., 95*, 6715-6736, 1990.

Storey, M., Saunders, A. D., Tarney, J., Leat, P., Thirlwall, M. F., Thompson, R. N., Menzies, M. A. and Marriner, G. F., Geochemical evidence for plume-mantle interactions beneath Kerguelen and Heard Islands, Indian Ocean, *Nature, 336*, 371-374, 1988.

Storey, M., Saunders, A. D., Tarney, J., Gibson, I. L., Norry, M. J., Thirlwall, M. F., Leat, P., Thompson, R. N. and Menzies, M. A., Contamination of Indian Ocean asthenosphere by the Kerguelen- Heard mantle plume, *Nature, 338*, 574-576, 1989.

Storey, M., Kent, R. W., Saunders, A. D., Salters, V. J., Hergt, J., Whitechurch, H., Sevigny, J. H., Thirlwall, M. F., Leat, P., Ghose, N. C. and Gifford, M., Lower Cretaceous volcanic rocks on continental margins and their relationship to the Kerguelen Plateau, in *Proc. ODP, Sci. Res. 120*, edited by S. W. Wise, R. Schlich et al., 33-54, Ocean Drilling Program, College Station, TX, 1992.

Storey, M., Mahoney, J. J., Kroenke, L. W. and Saunders, A. D., Are oceanic plateaus sites of komatiite formation?, *Geology, 19*, 376-379, 1991.

Subbarao, K. V. and Hedge, C. E., K, Rb, Sr and $^{87}Sr/^{86}Sr$ in rocks from the Mid-Indian Ocean ridge, *Earth Planet. Sci. Lett., 18*, 223-234, 1973.

Tarduno, J. A., Sliter, W. V., Kroenke, L., Leckie, M., Mayer, H., Mahoney, J. J., Musgrave, R., Storey, M. and Winterer, E. L., Rapid formation of the Ontong Java Plateau by Aptian mantle plume volcanism, *Science, 254*, 399-403, 1991.

Tatsumi, Y. and Nohda, S., Geochemical stratification in the upper mantle: evidence from Leg 115 basalts in the Indian Ocean, in *Proc. ODP, Sci. Results, 115*, edited by R. A. Duncan, J. Backman, L. C. Peterson et al., 63-69, Ocean Drilling Program, College Station, TX, 1990.

Viereck, L. G., Hertogen, J., Parson, L. M., Morton, A. C., Love, D. and Gibson, I. L., Chemical stratigraphy and petrology of the Vøring plateau tholeiitic lava and interlayered volcaniclastic sediments at ODP Hole 624E, *Proc. ODP, Sci. Results, 104*, 367-396, Ocean Drilling Program, College Station, TX, 1989.

von der Borch, C. C., Sclater, J. G. et al., Site 215, in *Init. Repts DSDP, 22*, 193-212, U.S. Government Printing Office, Washington, D.C., 1974.

Upton, B. G. J., and Wadsworth, W. J., Aspects of magmatic evolution on Réunion Island, *Phil. Trans. R. Soc. Lond. A, 217*, 105-130, 1972.

Watkins, N. D., Gunn, B. M., Nougier, J. and Baksi, A K, Kerguelen: continental fragment or oceanic island, *Geol. Soc. Am. Bull., 85*, 201-212, 1974.

Weis, D., Bassias, Y., Gautier, I. and Mennessier, J.-P., Dupal anomaly in existence 115 Ma ago: evidence from isotopic study of the Kerguelen Plateau (South Indian Ocean), *Geochim. Cosmochim. Acta, 53*, 2125-2131, 1989a.

Weis, D., Beaux, J.-F., Gautier, I., Giret, A. and Vidal, P., Kerguelen Archipelago: geochemical evidence for recycled material, in *Crust/Mantle Recycling at Convergence Zones: NATO ASI series*, edited by S. R. Hart, and L. Gülen, 59-63, Kluwer Academic Press, Amsterdam, 1989b.

Weis, D. and Frey, F.A., Isotope geochemistry of Ninetyeast Ridge basalts: Sr, Nd, and Pb evidence for the involvement of the Kerguelen hotspot, in *Proc. ODP, Sci. Results, 121*, edited by J. Weissel, J. Peirce, E. Taylor, J. Alt et al., 591-610, Ocean Drilling Program, College Station, TX, 1991.

Weis, D., Frey, F. A., Saunders, A. D., Gibson, I. and Leg 121 Scientific Shipboard Party, The Ninetyeast Ridge (Indian Ocean): a 5000 km record of a DUPAL mantle plume, *Geology, 19*, 99-102, 1991.

Weis, D., Frey, F. A., Leyrit, H. and Gautier, I., Kerguelen Archipelago revisited: geochemical and isotopic study of the SE Province lavas, *Earth. Planet. Sci. Lett.*, in press, 1992.

West, H. B., Gerlach, D. C., Leeman, W. P. and Garcia, M. O., Isotopic constraints on the origin of Hawaiian lavas from the Maui volcanic complex, Hawaii, *Nature, 330*, 216-220, 1987.

White, R. S. and McKenzie, D. P., Magmatism at rift zones: the generation of volcanic continental margins and flood basalts, *J. Geophys. Res., 94*, 7685-7730, 1989.

White, W. M. and Hofmann, A. W., Sr and Nd isotope geochemistry of oceanic basalts and mantle evolution, *Nature, 296*, 821-825, 1982.

White, W. M., Cheatham, M. C. and Duncan, R. A., Isotope geochemistry of Leg 115 basalts and inferences on the history of the Réunion mantle plume, in *Proc. ODP, Sci. Results, 115*, edited by R. A. Duncan, J. Backman, L. C. Peterson et al., 53-61, Ocean Drilling Program, College Station, TX, 1990.

White, W. M., Hofmann, A. W. and Puchelt, H., Isotope geochemistry of Pacific mid-ocean ridge basalts, *J. Geophys. Res., 92B*, 4881-4893, 1987.

Whitechurch, H., Montigny, R., Sevigny, J. H., Storey, M. and Salters, V. J. M., K-Ar and $^{40}Ar/^{39}Ar$ ages of central Kerguelen Plateau basalts, in *Proc. ODP, Sci. Res., 120*, edited by S. W. Wise, Jr., R. Schlich et al., 71-78, Ocean Drilling Program, College Station, TX, 1992.

Wright, E. and White, W. M., The origin of Samoa: new evidence from Sr, Nd, and Pb isotopes, *Earth Planet. Sci. Lett., 81*, 151-162, 1986.

Zindler, A., Staudigel, H. and Batiza, R., Isotopic and trace element geochemistry of young Pacific seamounts: implications for the scale of upper mantle heterogeneity, *Earth Planet. Sci. Lett., 70*, 175-195, 1984.

The Life Cycle of Indian Ocean Hotspots

ROBERT A. DUNCAN

College of Oceanography, Oregon State University, Corvallis, OR , 97331-55503, USA

MICHAEL STOREY

Department of Geology, University of Leicester, LE1 7RH, UK

The volcanic record of Indian Ocean hotspots offers the opportunity to study the complete dynamic, kinematic, chemical and thermal histories of individual mantle plumes. Plate motions over the Réunion and Kerguelen hotspots, since 65 Ma and 117 Ma, respectively, have produced linear, age-progressive chains of volcanic ridges and islands. Ocean drilling has documented the continuity of these provinces which provide a direct and simple frame of reference for plate reconstructions. Hotspots have remained stationary throughout their long history, and reflect a stable, mantle-wide pattern of convective upwelling. Comparison of hotspot and paleomagnetic reference frames reveals a small component of true polar wander during the early Tertiary. The Réunion and Kerguelen hotspots began with construction of massive flood basalt platforms, both on continental and oceanic lithosphere, produced at rates at least an order of magnitude faster than present hotspot activity.

INTRODUCTION

Linear, age-progressive volcanic activity in the ocean basins and on continents results from focused, relatively fixed upper mantle melting anomalies, termed hotspots. Their cosmopolitan occurrence argues that hotspots do not arise directly from plate tectonic processes. It is apparent that hotspots are very long-lived thermal features that produce magmas chemically distinct from those erupted at spreading ridges. The longevity, stability, and compositional identity of hotspots indicate a connection with upwardly convecting plumes from deeper regions of the mantle. However, the ultimate composition and source region for mantle plumes, the mechanism and time scale for convective instabilities that initiate plumes, and the size, lifespan, and motion of hotspots are all subjects of considerable speculation.

The long-term circulation and time-integrated structure of the mantle is reflected in the dynamic behavior of plumes/hotspots. How do hotspots start? What is the lifetime of plumes? Are hotspots stationary relative to each other? Quantitative information about the size, distribution, compositions, ages and origins of mantle reservoirs, and the efficiency of convective stirring can be learned from the volcanic record of hotspots. How do the flux and composition of hotspot products vary with time? How do hotspots and spreading ridges interact? How important are plumes to global elemental fluxes and long-term heat transport from earth's interior?

These questions are directed at global processes, but answers can best be learned through study of the volcanic record of well-sampled, long-lived hotspots, most clearly preserved in the ocean basins. While the products and effects of current hotspot activity can be observed in numerous young island chains, long-term aspects of hotspot/plume behavior can be studied only by sampling the older, submerged volcanic trails of hotspots. Deep ocean drilling is the only method available to penetrate often thick sediment cover to obtain representative volcanic samples. A major objective of the Ocean Drilling Program's (ODP) nine-leg Indian Ocean initiative was to investigate the life cycle of plume volcanism by targeting the well-preserved Réunion and Kerguelen hotspot tracks. Basaltic rocks were recovered at four sites along the Réunion hotspot track during Ocean Drilling Program (ODP) Leg 115. Along the Kerguelen hotspot track samples come from five locations on the Kerguelen Plateau during Legs 119 and 120, and from three new sites along the Ninetyeast Ridge during Leg 121, adding to four Deep Sea Drilling Program (DSDP) sites.

This paper reviews the contributions of ODP and DSDP drilling to our understanding of the dynamic, kinematic, and thermal histories of mantle plumes, through the volcanic record of these two long-lived hotspots. Specifically, we examine the evidence for the continuity of volcanism along the proposed hotspot tracks, the case for fixed hotspots as a mantle reference frame for plate motions, and the association between hotspots, flood basalts and continental rifting. Related papers in this volume discuss compositional variability and interaction of plume and asthenospheric mantle reservoirs [Weis et al.], and the structure and tectonic history of Kerguelen Plateau, a major oceanic flood basalt province [Coffin].

INDIAN OCEAN HOTSPOT TRACKS

The Indian Ocean basin is a particularly fruitful region to study the behavior of hotspots and mantle plumes. Rapid plate velocities have produced clear linear volcanic ridges, which strongly constrain estimates of inter-hotspot motion. Plate boundary shifts have caused spreading ridges

Synthesis of Results from Scientific Drilling in the Indian Ocean
Geophysical Monograph 70
Copyright 1992 American Geophysical Union

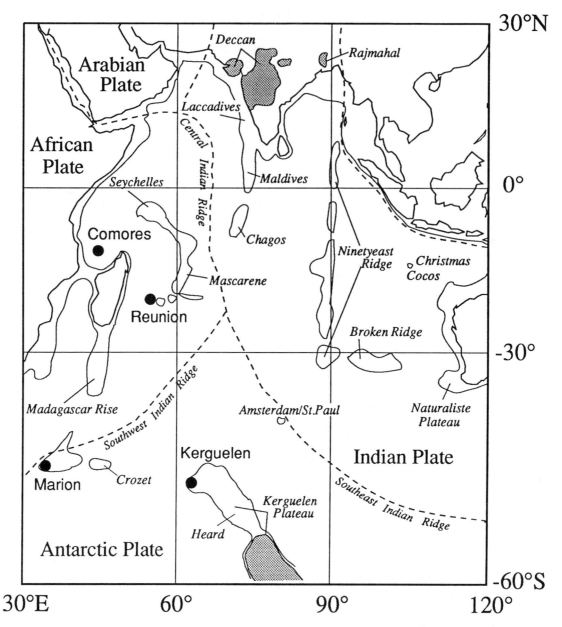

Fig. 1. Physiographic elements of the Indian Ocean basin including elevated volcanic structures associated with hotspot activity (filled circles). Plate boundaries are shown schematically as dashed lines. Continental and oceanic flood basalt provinces are shaded.

to migrate across hotspots, providing opportunities to examine variable mixing of possible mantle reservoir compositions. And, in the case of two prominent volcanic lineaments, the beginnings of hotspot activity can be traced to vast accumulations of continental and oceanic plateau flood basalts.

Hotspots are a common feature of the Indian Ocean basin and account for most of the elevated, intraplate bathymetric features. Persistent, sub-lithospheric thermal anomalies beneath Réunion, Kerguelen, Grand Comore, Prince Edward/ Marion and Balleny Islands have left clear, age-progressive volcanic trails on overlying plates and, thus,

qualify as members of the hotspot constellation. Other sites of young volcanism, at St. Paul/Amsterdam Islands on the Southeast Indian spreading ridge, Heard Island on the central Kerguelen Plateau, the Crozet Islands between Marion and Kerguelen hotspots, and Christmas and Cocos Islands in the northeastern Indian basin, do not appear to be connected to any long-term mantle upwelling. Their origin(s) are not understood; some (Heard, Cocos, Christmas) may be due to activation of older tectonic structures during lithospheric extension.

The two major hotspot traces in the Indian Ocean basin are those left by the Réunion and the Kerguelen hotspots

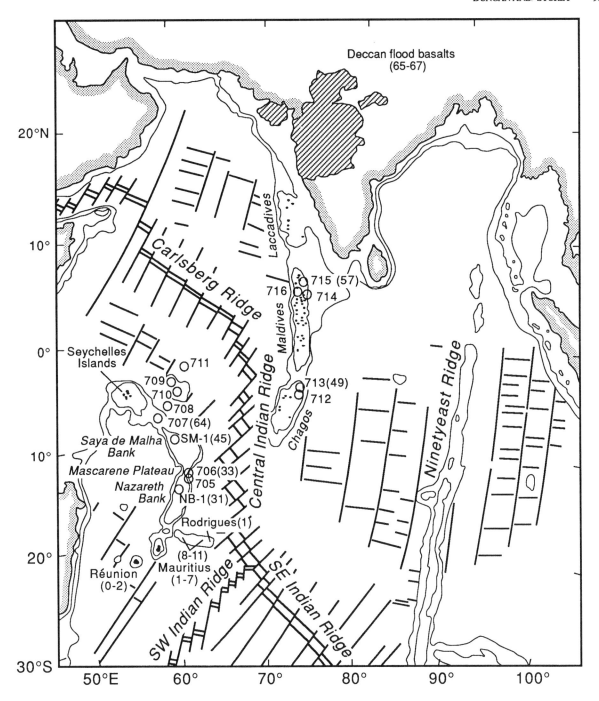

Fig. 2. Réunion hotspot track. ODP Leg 115 sites are shown, as are two industry wells (SM-1, NB-1). Radiometric ages (parentheses) were determined by ^{40}Ar-^{39}Ar dating of volcanic samples recovered by drilling, and subaerial exposures. The present extent of Deccan flood basalts in India and equivalent rocks in the Seychelles region are shown by diagonal lines.

(Figure 1). The western (Réunion) province includes the Mascarene Islands, the Mascarene Plateau, the Chagos-Maldives-Laccadives Ridges, and the Deccan flood basalts, which crop out across western and central India; the eastern (Kerguelen) province includes the Kerguelen archipelago, the Kerguelen Plateau, the Broken and Ninetyeast Ridges and the Rajmahal flood basalts of eastern India [Morgan,

1981; Duncan, 1978, 1981]. These volcanic trails preserve an unusually complete record of the activity of mantle plumes, from their birth in catastrophic flood basalt events [Richards et al., 1989], through maturity, feeding continuous volcanic ridges, into decline as discrete islands marking intermittent hotspot activity.

The Réunion Hotspot

Morgan [1972] hypothesized that the trace of the Réunion hotspot trends northeastward from the islands of Réunion and Mauritius, along the eastern limb of the Mascarene Plateau formed by Cargados Carajos and Nazareth Banks (Figure 2). The Saya de Malha Bank had been considered a submerged, southeastward extension of the late Precambrian Seychelles microcontinent [McKenzie and Sclater, 1971] or a volcanic platform formed either as part of the Deccan flood basalts [Courtillot et al., 1986] or during a later episode of the hotspot's activity [Morgan, 1981; Duncan, 1981]. An older section of the hotspot track, trending north from the Chagos Bank, was separated from the younger Mascarene Plateau by post-Eocene spreading at the Central Indian Ridge [McKenzie and Sclater, 1971; Schlich, 1982; Patriat et al., 1982]. From the Chagos Bank the hotspot trace follows the Maldives and Laccadives Ridges and finally terminates at the Deccan flood basalt province in western India (Figure 2).

In addition to the subaerially erupted basalts at the northern and southern ends of this volcanic lineament, deep drilling at industry wells on the Nazareth and Saya de Malha banks and on the Maldives Ridge have penetrated great thicknesses of basaltic rocks [Meyerhoff and Kamen-Kaye, 1981]. Volcanic rocks were dredged from the Rodrigues Ridge during the site survey cruise prior to Leg 115 (*RRS Charles Darwin* cruise 21/87). Submarine drilling during Leg 115 successfully sampled at four new locations the volcanic rocks underlying the carbonate platform, which covers the Mascarene Plateau and the Chagos-Maldives-Laccadives ridge system (Figure 2). Thus, there are samples from ten geographically and temporally distinct sites from which to examine the evolution of the Réunion hotspot.

The present Réunion hotspot lies beneath Paleocene-age oceanic lithosphere of the African plate [Schlich, 1982], on which the island of Réunion is being built. This island, rising to 7000 m from the ocean floor, is constructed from two coalesced volcanoes: Piton des Neiges, which is inactive and forms the northwest two-thirds of the island, and the active Piton de la Fournaise, which regularly erupts magma and forms the southeastern third of the island. A large unsampled seamount 160 km west of Réunion may be the newest volcanic product of the hotspot. Volcanic activity at Réunion is at least as old as 2 Ma [McDougall, 1971]. Estimates of average eruption rates over the hotspot are 0.01-0.04 km^3/yr [Richards et al., 1989; Gillot and Nativel, 1989]. Compositional aspects of the volcanic products of Réunion hotspot activity have been reported by Fisk et al. [1989] and Baxter [1990], and are compared with Kerguelen hotspot activity in Weis et al. (this volume).

The eroded volcanic island of Mauritius, some 220 km northeast of Réunion, has had an extended eruptive history, starting with a shield-building phase at 7-8 Ma, with later stages at 2-3.5 and 0.1-1.0 Ma [McDougall, 1971, Duncan, unpublished results, 1992]. The cause of these rejuvenescent phases of volcanism may seem perplexing in terms of a simple point-source hotspot model, for if the age

of the initial shield-building stage of island formation marks the time of location over the hotspot, then the later activity must derive from melts erupted when the island was several hundred kilometers downstream from the hotspot. For all contemporaneous eruptions to be directly related to the hotspot, the melting zone would have to be enormous (radius ~200 km), yet we see no evidence of volcanic activity between Réunion and Mauritius. It seems more probable that the rejuvenescent phases of volcanism at Mauritius occurred as a result of regional extension due to loading of the plate at Réunion, analogous to the mechanism proposed to explain Hawaiian post-erosional lavas [ten Brink and Brocher, 1987].

Rodrigues Island lies at the eastern end of the Rodrigues Ridge, a 450-km, east-west lineament that perpendicularly intersects the main hotspot track just north of Mauritius. It is now known from extensive dredging and surveying (*RRS Charles Darwin* cruise 21/87) that the Rodrigues Ridge is composed of basaltic to trachytic rocks, overlain by patchy development of carbonate reefs. Rodrigues Island is built of olivine basalts that erupted about 1.5 Ma [McDougall, 1971]. Dredged rocks from six sites distributed along the ridge yield radiometric ages in the range from 8 to 11 Ma (^{40}Ar-^{39}Ar analyses; Duncan, unpublished results, 1992) with no systematic age progression. Apparently, the entire Rodrigues Ridge was built simultaneously, just before the earliest volcanism at Mauritius; the younger Rodrigues Island activity must then be considered a rejuvenescent event. Morgan [1978] has proposed that the Rodrigues Ridge, and similar features such as the Wolf-Darwin lineament north of the Galapagos Islands, result from "channelized" asthenospheric flow from the hotspot to the nearby spreading ridge. The trend of the Rodrigues Ridge does not have any apparent relationship to transform faults or old spreading segments formed at the Central Indian Ridge. Morgan's [1978] suggestion that the orientation may be the vector sum of the African plate motion plus the relative motion of the spreading ridge, due to progressive construction at the eastern end of the lineation, is not supported by the age distribution of the volcanism.

The Mascarene Plateau and the Chagos-Maldives-Laccadives ridge system are the proposed northward continuation of this province. Submarine eruptions rapidly built up volcanoes, often into subaerial islands [Backman, Duncan et al., 1988; Meyerhoff and Kamen-Kaye, 1981], that eroded to sea level and subsided as the oceanic lithosphere cooled, accumulated sediments, and adjusted to the weight of the new load [Simmons, 1990]. Carbonate bank and reef deposits that cover the central portions of these plateaus are up to 2 km thick [Meyerhoff and Kamen-Kaye, 1981]. Duncan and Hargraves [1990] reported ^{40}Ar-^{39}Ar radiometric age determinations from volcanic rocks recovered from the six drill sites (Figure 2). These dates show an unequivocal south-to-north progression in the age of volcanic activity along the system of ridges, connecting young oceanic island volcanism at Réunion Island with flood basalt volcanism in India. This is strong confirmation of the hotspot model for formation of this volcanic lineament. Industry well site NB-1 on the Nazareth Bank is 31 Ma, while volcanism at Site 706 (33 Ma) was contemporaneous with nascent spreading at the Central Indian Ridge, which began during Chron 13N [Backman, Duncan et al., 1988]. Sites 713 (Chagos Bank) and industry well SM-1 (Saya de Malha Bank) are nearly contemporaneous at 49 and 45 Ma, respectively, and, on reconstruction of the Central Indian Ridge, merge to form a large volcanic plateau rather similar in size to Iceland. It appears that the hotspot lay beneath a spreading ridge segment for some substantial time during this period of the volcanic track. This time also coincided with the "hard" collision of India with Asia and a dramatic decrease in plate velocity over the hotspot [Klootwijk and Peirce, 1979]. Site 715 basalts were erupted at an oceanic island at about 57 Ma. Indications from geochemical data are that this volcanism occurred well away from a spreading ridge. The tholeiitic basalts at Site 707, well to the west of the main hotspot track, were erupted at about 64 Ma, just at the end of Deccan flood basalt activity. Seafloor spreading between India and the Seychelles Bank, a small continental fragment, began with a northward ridge jump at Chron 27N [63 Ma; Schlich, 1982] when Site 707 was adjacent to the western margin of India.

The earliest manifestation of the Réunion hotspot is the Deccan flood basalt province. Just prior to 65 Ma, the Réunion hotspot burst to life with massive eruptions of predominantly tholeiitic lavas in western India. This volcanic province is now a thick sequence of flat-lying basalt flows covering nearly 500,000 km^2. At its greatest thickness the section is over 2,000 m. Including correlative basalts identified in the Seychelles [Devey and Stephens, 1991], offshore (Arabian Sea) and lavas probably eroded, the original volume of the province may have exceeded 1.5×10^6 km^3 [Courtillot et al., 1986].

Several lines of evidence indicate that most of this enormous quantity of basalt was erupted rapidly. Interflow sedimentary beds, lateritic weathering, and erosional unconformities are scarce [Mahoney, 1988]. The flows are predominantly reversely magnetized, and only two polarity reversals (N-R-N) were identified in the entire sequence [Courtillot et al., 1986]. Radiometric ages (^{40}Ar-^{39}Ar incremental heating method) show that the Deccan volcanism occurred sometime between 65 and 67 Ma [Duncan and Pyle, 1988; Courtillot et al., 1988; Baksi and Kunk, 1988]. Combining the magnetostratigraphic (which indicate a very brief duration of volcanism) and absolute age data, it follows that the Deccan basalts may have accumulated in as little as 0.5 m.y. at the Cretaceous/Tertiary boundary [Jaeger et al., 1989]. The average eruption rate during Deccan volcanism was then close to 3 km^3/yr, or 2 orders of magnitude greater than at the hotspot today.

The Kerguelen Hotspot

Kerguelen is one of the longest lived and most vigorous of the Indian Ocean hotspots. It has been active for more

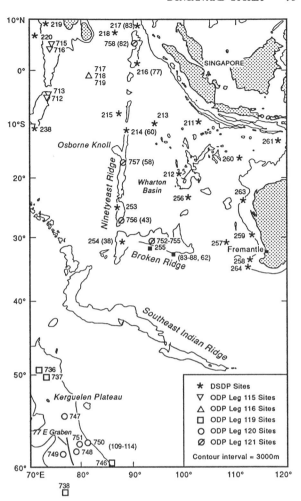

Fig. 3. Kerguelen hotspot track. ODP and DSDP sites are shown, as are dredge locations from the Broken Ridge (solid squares). The age distribution of basement samples is indicated in parentheses (Ma).

than 117 m.y. and has formed a significant proportion of the elevated eastern Indian Ocean lithosphere. The Ninetyeast Ridge, Broken Ridge, the Kerguelen Plateau, and the Rajmahal flood basalts (eastern India) (Figure 3) are all volcanic products of this long-lived hotspot presently located near the Kerguelen archipelago [Luyendyk and Rennick, 1977; Duncan, 1978, 1981; Morgan, 1981]. The rationale for connecting these volcanic elements to a common origin involves (1) a temporal pattern of increasing ages to the north along the Ninetyeast Ridge [Sclater et al., 1974; Davies, Luyendyk et al., 1974; Duncan, 1978; Peirce, Weissel et al., 1989] to the Rajmahal province [Baksi, 1986], (2) constant paleolatitudes of Ninetyeast Ridge basalts [Peirce, 1978], (3) compositional similarities among the pieces [Frey et al., 1977; Whitford and Duncan, 1978; Mahoney et al., 1983], and (4) plausible plate tectonic histories that could bring each volcanic element over the Kerguelen hotspot at appropriate times during the last 120 m.y. [Morgan, 1981; Davies et al., 1989].

The Kerguelen hotspot has been located beneath the Antarctic plate since the beginning of seafloor spreading at the present southeast Indian Ridge approximately 40 m.y. ago [McKenzie and Sclater, 1971; Schlich, 1982]. Since then extremely slow, clockwise motion of the Antarctic plate has resulted in the protracted construction of the northern Kerguelen Plateau and archipelago by hotspot volcanism since 35 to 40 Ma [Giret and Lameyre, 1983]. The islands are the remains of several coalesced shield volcanoes of prominently tholeiitic to transitional basalts [Watkins et al., 1974; Storey et al., 1988; Gautier et al., 1990]. These are overlain by more localized alkali basalt flows and their differentiates that erupted up to late Quaternary time. Plate reconstructions based on the hotspot reference frame put the current location of the hotspot a short distance to the west of the archipelago [Duncan and Richards, 1991], where a group of large seamounts with prominent geoid anomalies exist [Haxby, 1987]. Heard and McDonald Islands are young, isolated volcanoes 500 km southeast of the Kerguelen islands. It is difficult to relate these to hotspot activity; instead, we speculate that they and the Gaussberg intraplate volcano further southeast on the Antarctic continental margin may be forming in response to extension and graben formation over this interior part of the Antarctic plate.

The Kerguelen Plateau, one of the oceans' largest physiographic features, rises more than 3 km above the surrounding abyssal depths over an area of about 1×10^6 km^2 (Figure 3). Seismic refraction and gravity studies over the plateau [Recq et al., 1990; Houtz et al., 1977] suggest crustal thicknesses of between 15 and 23 km, implying a total volume (in excess of normal ocean crust) of 8 to 16 $\times$ 10^6 km^3. The plateau has been divided into a shallower, rougher northern and a deeper, smoother southern province [Schlich, 1975]. The southern province lies just south of Heard Island and consists of a broad anticlinal arch, broken by several stages of N-S normal faulting and graben formation [e.g., Coffin et al., 1986], including the central 77° E Graben. The southern plateau was constructed on young seafloor formed during the Early Cretaceous opening of the eastern Indian Ocean as indicated on the west flank by the presence of magnetic anomalies M0-M8 in the Enderby Basin [Nogi et al., 1991].

ODP Legs 119 and 120 recovered moderately fresh to highly altered basaltic rocks at Sites 738, 747, 748, 749, and 750 from the southern Kerguelen Plateau. Basaltic samples were also dredged from the 77° E Graben (central southern plateau) during *Marion Dufresne* cruise MD48 [Bassias et al., 1987]. Plagioclase separates from a dredged tholeiitic basalt gave a K-Ar age of 114±1 Ma [Leclaire et al., 1987], whereas two ^{40}Ar-^{39}Ar incremental heating ages for Leg 120 basalts range from 109 to 118 Ma [Whitechurch et al., 1992]. There is evidence at several of the sites (738, 747 and 750) for subaerial eruptions, followed by a depositional hiatus. Biostratigraphic data from overlying sediments provide minimum age estimates

from early Eocene (>55 Ma) to Turonian (>90 Ma).

The Broken Ridge is a volcanic fragment that rifted from the northeastern margin of the southern province of the Kerguelen Plateau in mid-Eocene time. The ridge has an asymmetric profile, sloping gently to the north but with a steep, fault-bounded southern escarpment. During site surveys for Leg 121 drilling (*Robert Conrad* cruise RC27-08) basaltic rocks were dredged from two locations along the southern scarp. Crystallization ages fall into two groups: 83-88 and 62-63 Ma [Duncan, 1991]. Given the Turonian age (~87 Ma) of oldest sediments at DSDP Site 255 on the crest of Broken Ridge, it seems likely that the older ages mark the volcanic construction age while the younger ages correspond to volcanism associated with early rifting between Broken Ridge and Kerguelen Plateau [Mutter and Cande, 1983]. Lithospheric extension culminated in the separation of Broken Ridge (Australian plate) and the Kerguelen Plateau (Antarctic plate) at 45-42 Ma [Munschy et al., 1992].

At over 4500 km long and some 50 to 100 km wide, the Ninetyeast Ridge is one of the most remarkable linear features on earth. Ocean drilling has now sampled volcanic rocks from seven widely-spaced sites along the Ninetyeast Ridge (Figure 3). ^{40}Ar-^{39}Ar radiometric age determinations [Duncan, 1978, 1991] have confirmed the biostratigraphic evidence for a clear north-to-south age progression in the construction of this lineament. The oldest site (ODP 758) was dated at 80-84 Ma, compatible with the earliest Campanian/latest Santonian age of basal sediments. Drilling at DSDP Site 217, some 400 km to the north, did not reach basement but oldest sediments were Campanian (83 Ma). Age determinations for basalts from DSDP Site 216 range between 72 and 81 Ma, compatible with the Maastrichtian age of basal sediments and the age of seafloor adjacent to the west [Sclater and Fisher, 1974]. We plot the mean of this age range (77 Ma) in Figure 3.

DSDP Site 214 basalts produced very reliable age determinations close to 60 Ma, compatible with overlying Paleocene sediments. ODP Site 757 basalts were dated at 58 Ma which matches the late Paleocene biostratigraphic age. DSDP Site 253 recovered very low-K (0.07%), altered basalt that did not produce reliable radiometric age determinations. The basal sediments here were identified as middle Eocene (44-51 Ma). Basalts at ODP Site 756 were dated at 43 Ma compatible with the late Eocene age of lowermost sediments. Finally, the basalts from DSDP Site 254, located just south of the intersection with Broken Ridge, yield crystallization ages of about 38 Ma. Hence, this southernmost site was contemporaneous with the oldest ages measured for the Kerguelen islands; the two locations were adjacent and rifted apart as the Southeast Indian Ridge migrated northward across the Kerguelen hotspot in concert with the Central Indian Ridge moving away from the Réunion hotspot.

The Rajmahal basalts are exposed over an area of about 4300 km^2 on the continental margin of northeast India (Figure 1), and reach a maximum thickness of 200-300m. These are subaerially-erupted, quartz tholeiite flows,

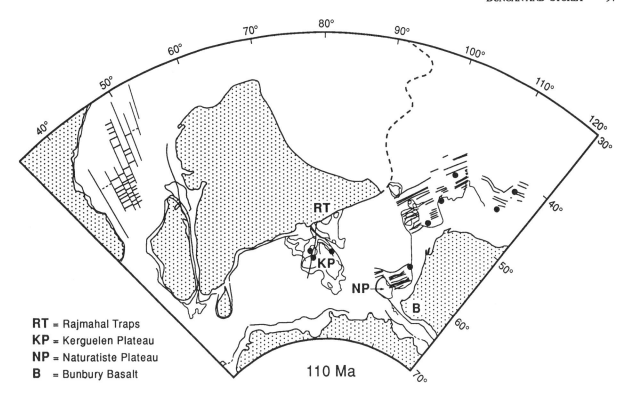

Fig. 4. Plate reconstruction for ~110 Ma for the eastern Indian Ocean region [from Royer and Coffin, 1992], showing the oceanic nature of most of the Kerguelen flood basalt province.

interbedded with thin volcaniclastic or sedimentary beds of Early Cretaceous age. ^{40}Ar-^{39}Ar incremental heating ages for basalts are 117±1 Ma [Baksi, 1986], contemporaneous with the southern Kerguelen Plateau. Seismic reflection prospecting in the Bengal Basin has identified a seaward-dipping reflector sequence corresponding to these basalts [Sengupta, 1966]. Paleomagnetic studies [Klootwijk, 1971] reveal that most of the flows were erupted during a normal magnetic interval, probably the beginning of the Cretaceous long normal period, at a latitude of ~48°S. Their age and paleolatitude argue that the Rajmahal basalts and Kerguelen Plateau were contiguous portions of a large flood basalt province that formed about 110 to 118 Ma. Figure 4 shows a plate reconstruction of the eastern Indian Ocean region at the time of this huge volcanic event.

Davies et al. [1989] and Storey et al. [1992] include the Naturaliste Plateau, a 2500 m rise off the southwestern corner of Australia (Figures 1,3,4) (and the probable on-land equivalent, the Bunbury basalt) among the earliest products of the Kerguelen hotspot. DSDP drilling at two sites (258 and 264) failed to reach basement but penetrated mid-Albian volcaniclastic sediments. Dredged basalts from the Naturaliste Plateau are quartz tholeiites and show some significant differences from Kerguelen Plateau and Rajmahal basalts [Storey et al., 1992] that may be due to continental lithospheric contamination of the parental melts. Radiometric ages for the basalts are not yet available, but will be necessary to test this proposed

connection with the Kerguelen Plateau.

The age distribution of volcanic rocks sampled by deep ocean drilling from along the submerged ridges and plateaus of the Indian Ocean has established the continuity of Réunion and Kerguelen hotspot activity. Each began with rapid, massive eruptions of tholeiitic basalts - the first in the young eastern Indian Ocean basin at ~118 Ma, and the second through cratonic India in the western Indian Ocean around the time of the Cretaceous/Tertiary boundary (65 Ma). Hence the lifespans of vigorous hotspots must be of order 100 m.y. or longer. From these detailed histories we can describe some fundamental aspects of mantle plume behavior.

MANTLE PLUME DYNAMICS

Richards et al. [1989], following Griffiths [1986], have proposed a general model for hotspot initiation with flood basalt volcanism. In this scheme plumes develop as density instabilities deep in the mantle (e.g., the core/mantle boundary). Initially, a plume forms a large "head," which establishes a conduit as it rises diapirically to the surface [Whitehead and Luther, 1975]. When the plume head reaches the base of the lithosphere, rapid, voluminous decompression melting occurs [an excess plume temperature of 250°C-300°C could generate 10%-20% melt fraction, McKenzie and Bickle, 1988] and lithospheric thinning and extension results. This event produces flood basalt volcanism within continents or

TABLE1. Production Rates for Indian Ocean Hotspots and Hawaii

Location	Volume* (10^6km^3)	Duration (m.y.)	Production Rate (km^3/yr)
Reunion hotspot			
Deccan flood basalts	2.5	0.5	5.0
Chagos-Laccadive R.	15.6	25	0.6
Mascarene Plateau	9.7	35	0.3
Reunion Island	0.1	2	<0.1
Kerguelen hotspot			
S. Kerguelen Plateau-Rajmahal flood basalts	45.9	8	5.7
Ninetyeast-Broken R.	36.1	75	0.5
N. Kerguelen Plateau	14.9	40	0.4
Hawaiian hotspot			
Flood basalt province?	?	?	?
Emperor Seamounts	13.4	35	0.4
Hawaiian Ridge	9.7	43	0.2
Hawaii	0.1	1	0.1

* from isostatic compensation model of Schubert and Sandwell (1989).

ocean basins, or presages rifting at plate margins. Once melting from the plume head is exhausted, a continuing but much lower flux of deep mantle material from the plume "tail" (conduit) maintains the hotspot, and produces the more familiar and far less catastrophic linear series of volcanic centers (i.e., the hotspot track).

This model requires that all hotspots begin with flood basalt volcanism, and many currently active hotspots can be traced back to such provinces (e.g., Tristan da Cunha through the Rio Grande Rise to the Parana basalts, Marion through Madagascar and back southwest to the Karoo basalts, and Iceland to the North Atlantic Tertiary basalts, [Morgan, 1981; Richards et al., 1989]. One of the major contributions of Indian Ocean drilling was to tie the Deccan and Kerguelen Plateau-Rajmahal flood basalts unequivocally to the birth of the Réunion and Kerguelen hotspots.

It is now clear that flood basalt volcanism can occur in the ocean basins and build large plateaus such as the Kerguelen and Ontong Java Plateaus [Tarduno et al., 1991; Mahoney et al., 1992], without the requirement of rifting thick continental lithosphere [White and McKenzie, 1989; Richards et al., 1989]. The material and heat budgets of these irregular events are enormous, and their importance to core-mantle-lithosphere coupling and, possibly, the biosphere is just starting to be appreciated [Courtillot and Besse, 1987; Larson, 1992; Larson and Olson, 1992].

The material flux through hotspots can be estimated by calculating volcanic eruption rates at locations along these tracks. We use crustal volumes for the ridges and plateaus of the Indian Ocean from Schubert and Sandwell [1989], based on a local isostatic compensation model. We consider the southern two-thirds of the Kerguelen Plateau, plus the Naturaliste Ridge and the Rajmahal basalts, to have formed in the initial phase of Kerguelen hotspot activity at 110-118 Ma. The combined volume divided by the duration gives an average eruption rate of about 6

km^3/yr (Table 1). For the Ninetyeast and Broken Ridges (40 to 115 Ma), however, the rate dropped to about 0.5 km^3/yr; the Kerguelen Islands and northern Kerguelen Plateau were built at an average rate of less than 0.4 km^3/yr. Through these rough estimates we observe an immediate decrease in plume supply following the flood basalt event, and a subsequent gradual tapering toward the present. A part of this apparent decrease in plume flux may be due to increasing lithosphere thickness with the change from on-ridge to intraplate hotspot location. A similar analysis of the Réunion track shows a decrease from 5 km^3/yr (Deccan flood basalts) to 0.6 km^3/yr (Chagos-Laccadives) to 0.3 km^3/yr (Mascarene Plateau) to <0.1 km^3/yr (Réunion) in plume melt supply. These flux changes appear to support the plume head and tail model [Whitehead and Luther, 1975; Griffiths, 1986; Richards et al., 1989].

Hotspot Kinematics

Because Indian plate velocities were so high in the Late Cretaceous and early Tertiary [e.g., Klootwijk and Peirce, 1979], the dramatically linear Chagos-Maldives-Laccadives Ridge and parallel Ninetyeast Ridge sections of the traces provide a particularly good estimate of inter-hotspot motion. We can examine the proposition that the Réunion and Kerguelen hotspots have been stationary with respect to other prominent hotspots by means of the simple test prescribed by Morgan [1981] and Duncan [1981]. To do this, we first recognize the motion of the African plate over hotspots in the South Atlantic from well-documented volcanic lineaments, such as the Tristan da Cunha hotspot and its trace, the Walvis Ridge [O'Connor and Duncan, 1990], consistent with the geometrical limits imposed by other traces such as the St. Helena-Cameroon, the Bouvet-Agulhas Plateau, and Marion-Madagascar Rise lines (Figure 5). Next we add the relative motion of the Indian and Antarctic plates away from the African plate for the last 120 m.y., based on seafloor-spreading data [e.g.,

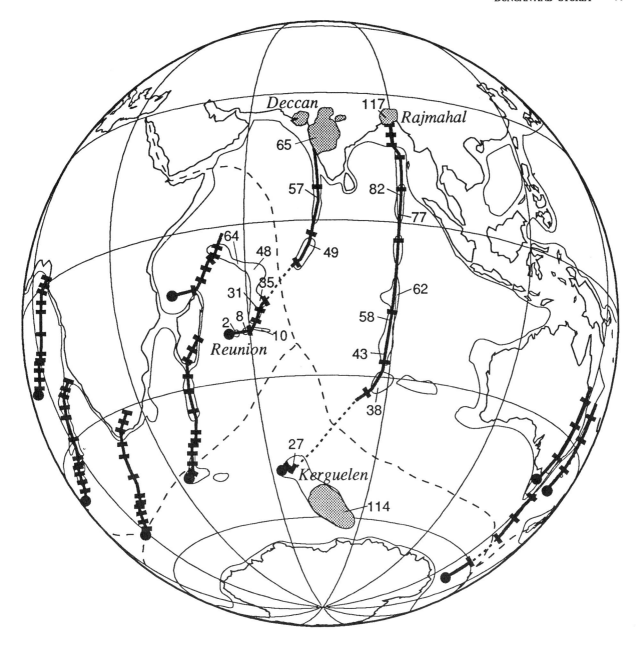

Fig. 5. Computer-modeled hotspot tracks are calculated assuming hotspots are stationary. The predicted trails (heavy dashes are 10-m.y. increments in volcano age) are determined from African plate motion over South Atlantic hotspots (principally from dated sites on the Walvis Ridge), and relative motion between the Indian and African plates. A complete set of rotation poles is given in Duncan and Richards [1991]. These compare well with the actual lineaments and documented ages (Ma) of volcanic activity along the Réunion and Kerguelen hotspot tracks. Hence, Atlantic and Indian Ocean hotspots have remained fixed relative to one another for at least the last 100 m.y. Flood basalts at the northern ends of the Réunion and Kerguelen tracks resulted from hotspot initiation.

Molnar et al., 1987; Besse and Courtillot, 1988].

The result is the predicted motion of the Indian plate with respect to the starting set of hotspots underlying the South Atlantic region. If the Indian region hotspots have not moved much with respect to their Atlantic cousins, the predicted plate motion should closely follow the direction and age of volcanism determined from the two hotspot tracks. From Figure 5, it is clear that the geometry and new

age estimates from sites along both hotspot tracks are in excellent agreement with those required by the stationary hotspot model.   Uncertainties in plate motions can be calculated with the Molnar and Stock [1985] approach. Molnar and Stock [1987] have estimated uncertainties in plate positions stemming from Africa-India relative motion data to be on the order of 1° (i.e., about 100 km) for the early Tertiary. Hence, the maximum allowed inter-hotspot motion between the Atlantic and Indian hotspots is less than 2 mm/yr, which is effectively motionless.

A series of reconstructions of Indian Ocean plate positions and boundaries for times of prominent magnetic anomalies, and tied to the hotspot reference frame, are presented as an appendix to this volume and described in Royer et al. [1992], Royer and Sandwell [1989], and Royer et al. [1988]. These are used to interpret geochemical variations along the hotspot track in terms of mantle mixing (Weis et al., this volume). Hotspots in this region thus constitute a mantle-fixed reference frame that very conveniently records Indian Ocean basin evolution for Late Cretaceous through Tertiary time.

Hotspot vs. Paleomagnetic Reference Frames: True Polar Wander?

From the volcanic records of many well-sampled, long-lived hotspots it appears that globally distributed mantle plumes are stationary relative to one another, to within 2 mm/yr, over periods as long as 100 m.y. This requires that the source region for plumes lie below that portion of the mantle that is convecting with large horizontal velocities, and the most likely origin is the lower mantle, perhaps the core/mantle boundary (seismic layer D"). Hotspot tracks then record the histories of plate motions over the lower (slowly convecting) mantle. This mantle reference frame is independent of the paleomagnetic reference frame and has certain advantages: it does not depend on the assumption of the geocentric axial dipole field, and it resolves east-west motion that is not detected in paleomagnetic studies. Differential motion between the mantle and the spin axis can, in principle, be determined by comparing plate motions recorded by the hotspot and paleomagnetic reference frames. Such motion has been termed true polar wander (distinguished from apparent polar wander of the geomagnetic axis inferred from time sequences of paleomagnetic field directions for given plates). True polar wander could occur through mass redistribution (e.g., mantle convection and plate motions) sufficient to shift the entire body relative to its spin axis [Goldreich and Toomre, 1969].

In the absence of true polar wander, mantle plumes would not move with respect to the geomagnetic (= spin) axis. Hence, every volcano generated along a given hotspot track would record the magnetic inclination, expressed as paleolatitude, of the present hotspot activity site. If such measured paleolatitudes are not constant, it follows that the mantle reference frame moves relative to the spin axis, which is true polar wander. Vandamme and Courtillot [1990] and Schneider and Kent [1990] measured paleolatitude data from basalts and sediments recovered at Leg 115 drilling sites. Results from three sites (706, 707, and 715) appear to be reliable and, together with magnetic studies of the Deccan basalts [Courtillot et al., 1986], indicate a possible, slow (8 mm/yr) northward motion of the Réunion hotspot since its birth at about 66 Ma (Figure 6).

The calculated true polar wander is just barely significant

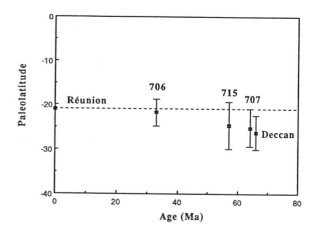

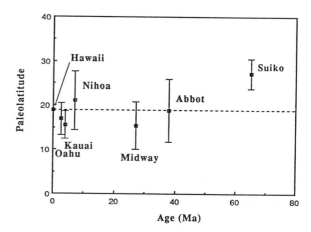

Fig. 6. Paleolatitudes of sample sites from the Réunion hotspot track, with α_{95} confidence limits, as a function of time. Data are from Schneider and Kent [1990] and Vandamme and Courtillot [1990] and indicate a slow northward motion of the Réunion hotspot relative to the geomagnetic axis of 5°-6° since 66 Ma. For comparison, sample sites from the Hawaiian-Emperor chain (Pacific Ocean) show a southward motion of the Hawaiian hotspot of ~8° for the same period (Kono, 1980). The results are consistent with a small component of true polar wander (i.e., motion of the entire earth with respect to the spin axis).

at the α_{95} confidence level of the paleomagnetic data. Considering the uncertainties in size and precise location of the hotspot during Deccan volcanism, the paleolatitudes may not be significantly different. However, the direction and magnitude (5-6°) of inferred mantle motion is consistent with other comparisons of the hotspot and paleomagnetic reference frames [see Courtillot and Besse, 1987 for a summary]. Of particular interest are magnetic studies of basalts recovered by DSDP drilling at Suiko Seamount (Site 433) in the north Pacific Ocean that yielded a 65-Ma paleolatitude for the Hawaiian hotspot of 27.1°N ± 3.4° [Kono, 1980], requiring some 8° southward motion of the hotspot through Tertiary time. Paleolatitudes of younger sites along the Hawaiian-Emperor chain are not

Fig. 7. True polar wander rotation for early Tertiary (8° clockwise about 0°N, 100°E).

significantly different from the present hotspot position (19°N). These data (Figure 6) are consistent with slow northward motion of hotspots in the Indian Ocean and southward motion of hotspots in the Pacific, which would be the case if the whole mantle rotated 8° clockwise about a pole located on the equator near 90°E [Courtillot and Besse, 1987 and Figure 7].

SUMMARY

The volcanic records of the Réunion and Kerguelen hotspots offer several new insights on the life cycle and behavior of mantle plumes. Drilling on the Mascarene Plateau, Chagos-Maldives Ridge, Kerguelen Plateau and Ninetyeast Ridge has demonstrated the volcanic continuity of the proposed trace of the Réunion hotspot on the Indian and African plates, and the Kerguelen hotspot on the Indian and Antarctic plates. A most important connection between flood basalt volcanism in western India (the Deccan basalts) and in the eastern Indian Ocean basin (Kerguelen Plateau-Rajmahal basalts), and the initiation of Réunion and Kerguelen hotspot activity, respectively, has now been firmly established and used in a generalized model of plume initiation from deep mantle density instabilities [Richards et al., 1989]. Volumetric and compositional variations along the hotspot track support a mantle diapir model, involving initial entrainment and stirring of upper mantle within a large plume head [Griffiths and Campbell, 1990] that rapidly melted over a broad region to produce the flood basalt event. Smaller volumes of plume material continued to rise within the conduit established by the initial diapir but opportunities

for mixing with the upper mantle were much more limited and the hotspot volcanic products reflect a progressively more plume-like composition with time [Weis et al., this volume].

The relatively high density of radiometrically dated sample sites and clear, linear volcanic traces produced by rapid plate velocities in early Tertiary time allow a rigorous evaluation of the magnitude of inter-hotspot motion. It appears that hotspots in the Atlantic and Indian Ocean basins have been virtually stationary (less than 2mm/yr relative motion) for periods as long as 100 m.y. It follows that the underlying plumes are not much affected by horizontal upper mantle flow; this is surprising given that active spreading ridges have crossed over both the Réunion and Kerguelen hotspots. A barely significant component of true polar wander (5-6° since 66 Ma) can be distinguished by comparing hotspot and paleomagnetic reference frames. This is, however, consistent with other perceived motions of hotspots with respect to the geomagnetic axis.

Basement drilling during the ODP Indian Ocean campaign has shown the value of detailed sampling of prominent hotspot tracks. In the future similar coverage should be achieved on the older portions (i.e. Cretaceous and early Tertiary) of several oceanic lineaments, such as the Emperor Seamounts, Louisville Ridge, and the Walvis Ridge. Ocean drilling will also focus on the major oceanic plateaus (e.g., Ontong Java, Caribbean) and volcanic rifted margins (e.g., North Atlantic) to examine the composition, timing, structure and tectonic history of these oceanic equivalents of the continental flood basalt events, signaling the beginning of hotspot activity.

Acknowledgments. We thank the crew, drillers and technical support staff of the JOIDES Resolution for their professionalism and enthusiasm for the scientific objectives of the Indian Ocean drilling program. In particular we are grateful for the experience and drilling supervision of the late Lamar Hayes. This research was supported by the National Science Foundation (U.S.A.) and the Natural Environment Research Council (U.K.).

REFERENCES
Backman, J., Duncan, R. A., et al., *Proc. ODP, Init. Repts., 115,* Ocean Drilling Program, College Station, TX , 1988.
Baksi, A. K., 40Ar-39Ar incremental heating study of whole-rock samples from the Rajmahal and Bengal Traps, eastern India, *Terra Cognita, 6,* 161 (Abstract), 1986.
Baksi, A. K., and Kunk, M. J., The age of initial volcanism in the Deccan Traps, India: preliminary 40Ar-39Ar age spectrum dating results, *Eos, 69,* 1487 (Abstract), 1988.
Bassias, Y., Davies, H. L., Leclaire, L., and Weis, D., Basaltic basement and sedimentary rocks from the southern sector of the Kerguele-Heard Plateau: new data and their Meso-Cenozoic paleogeographic and geodynamic implications, *Bull. Mus. Hist. Nat., 4/9,* 367-403, 1987.
Baxter, A. N., Major and trace element variations in basalts from Leg 115, *Proc. ODP., Sci. Results, 115,* 11-22, 1990.
Besse, J., and Courtillot, V., Paleogeographic maps of the continents bordering the Indian Ocean since the early Jurassic, *J. Geophys. Res., 93,* 11,791-11,808, 1988.
Coffin, M.F., Davies, H.L., and Haxby, W.F., Structure of the Kerguelen Plateau province from SEASAT altimetry and seismic reflection data, *Nature, 324,* 134-136, 1986.
Courtillot, V., and Besse, J., Magnetic field reversals, polar

wander, and core-mantle coupling, *Science*, *237*, 1140-1147, 1987.

Courtillot, V., Besse, J., Vandamme, D., Montigny, R., Jaeger, J. J., and Capetta, H., Deccan flood basalts at the Cretaceous-Tertiary boundary?, *Earth Planet. Sci. Lett.*, *80*, 361-374, 1986.

Courtillot, V., Feraud, G., Maluski, H., Vandamme, D., Moreau, M. G., and Besse, J., The Deccan flood basalts and the Cretaceous/Tertiary boundary, *Nature, 333*, 843-846, 1988.

Davies, H. L., Sun, S. S., Frey, F. A., Gautier, I., McCulloch, M. T., Price, R. C., Bassias, Y., Klootwijk, C. T., and Leclaire, L., Basalt basement from the Kerguelen Plateau and the trail of a Dupal plume, *Contrib. Mineral. Petrol.*, *103*, 457-469, 1989.

Davies, T., Luyendyk, B. et al., *Init. Repts. DSDP, 26*, U.S. Government Printing Office, Washington, D.C., 1974.

Devey, C. W., and Stephens, W. E., Tholeiitic dykes in the Seychelles and the original spatial extent of the Deccan, *J. Geol. Soc. Lond.*, *148*, 979-983, 1991.

Duncan, R. A., Geochronology of basalts from the Ninetyeast Ridge and continental dispersion in the eastern Indian Ocean, *J. Volcanol. Geotherm. Res.*, *4*, 283-305, 1978.

Duncan, R. A., Hotspots in the southern oceans - an absolute frame of reference for motion of the Gondwana continents, *Tectonophysics*, *74*, 29-42, 1981.

Duncan, R. A., Age distribution of volcanism along aseismic ridges in the eastern Indian Ocean, *Proc. ODP, Sci. Results*, *121*, 507-517, 1991.

Duncan, R. A., and Hargraves, R. B., $^{40}Ar/^{39}Ar$ geochronology of basement rocks from the Mascarene Plateau, Chagos Bank, and the Maldives Ridge, *Proc. ODP, Sci. Results, 115*, 43-51, 1990.

Duncan, R. A., and Pyle, D. G., Rapid eruption of the Deccan flood basalts at the Cretaceous/Tertiary boundary, *Nature, 333*, 841-843, 1988.

Duncan, R. A., and Richards, M. A., Hotspots, mantle plumes, flood basalts, and true polar wander, *Rev. Geophys.*, *29*, 31-50, 1991.

Fisk, M. R., Duncan, R. A., Baxter, A. N., Greenough, J. D., Hargraves, R. B., Tatsumi, Y. and Shipboard Scientific Party, Reunion hotspot magma chemistry over the past 65 m.y.: Results from Leg 115 of the Ocean Drilling Program, *Geology*, *17*, 934-937, 1989.

Frey, F. A., Dickey, J. S., Jr., Thompson, G., and Bryan, W. B., Eastern Indian Ocean DSDP sites: correlations between petrology, geochemistry and tectonic setting, in *Indian Ocean Geology and Biostratigraphy*, edited by J. R. Heirtzler, H. M. Bolli, T. A. Davies, J. B. Saunders, and J. G. Sclater, pp. 189-257, Am. Geophys. Union, 1991.

Gautier, I., Weis, D., Mennessier, J.-P., Vidal, P., Giret, A., and Loubet, M., Petrology and geochemistry of the Kerguelen Archipelago basalts (South Indian Ocean): Evolution of the mantle sources from ridge to intraplate position, *Earth Planet. Sci. Lett.*, *100*, 59-76, 1990.

Gillot, P.-Y., and Nativel, P., Eruptive history of the Piton de la Fournaise volcano, Réunion Island, Indian Ocean, *J. Volcanol. Geotherm. Res.*, *36*, 53-65, 1989.

Giret, A., and Lameyre, J., A study of Kerguelen plutonism: petrology, geochronology and geological implications, in *Antarctic Earth Science*, edited by R. L. Oliver, P. R. James, and J. B. Jago, pp. 646-651, Cambridge University Press, 1983.

Goldreich, P., and Toomre, A., Some remarks on polar wandering, *J. Geophys. Res.*, *74*, 2555-2567, 1969.

Griffiths, R. W., The differing effects of compositional and thermal buoyancies on the evolution of mantle diapirs, *Phys. Earth Planet. Int.*, *43*, 261-273, 1986.

Griffiths, R. W. and Campbell, I. H., Stirring and structure in mantle starting plumes, *Earth Planet. Sci. Lett.*, *99*, 66-78, 1990.

Haxby, W. F., *Gravity Field of World's Oceans* (color map), Lamont-Doherty Geological Observatory, Palisades, NY, 1987.

Houtz, R. E., Hayes, D. E., and Markl, R. G., Kerguelen Plateau bathymetry, sediment distribution, and crustal structure, *Mar. Geology*, *25*, 95-130, 1977.

Jaeger, J. J., Courtillot, V., and Tapponier, P., A paleontological view on the ages of the Deccan traps, of the Cretaceous-Tertiary boundary and of the India-Asia collision, *Geology*, *17*, 316-319, 1989.

Klootwijk, C. T., Paleomagnetism of the Upper Gondwana Rajmahal Traps, northeast India, *Tectonophysics*, *12*, 449-467, 1971.

Klootwijk, C. T., and Peirce, J. W., India's and Australia's pole path since the late Mesozoic and the India-Asia collision, *Nature*, *282*, 605-607, 1979.

Kono, M., Paleomagnetism of DSDP Leg 55 basalts and implications for the tectonics of the Pacific plate, in *Init. Repts. DSDP, 55*, edited by E. D. Jackson, I. Koisumi et al., pp. 737-752, U.S. Government Printing Office, Washington, D.C., 1980.

Larson, R. L., Latest pulse of earth: evidence for a mid-Cretaceous superplume, *Geology, 19*, 547-550, 1991.

Larson, R. L., and Olson, P., Mantle plumes control magnetic reversal frequency, *Earth Planet. Sci. Lett.*, *108*, 1992.

Leclaire, L., Bassias, Y., Denis-Clocchiatti, M., Davies, H. L., Gautier, I., Gensous, B., Giannesini, P.-J., Morand, F., Patriat, P., Segoufin, J., Tesson, M., and Wannesson, J., Lower Cretaceous basalt and sediments from the Kerguelen Plateau, *Geo-Marine Lett.*, *7*, 169-176, 1987.

Luyendyk, B. P., and Rennick, W., Tectonic history of aseismic ridges in the eastern Indian Ocean, *Geol. Soc. Amer. Bull.*, *88*, 1347-1356, 1977.

McDougall, I., The geochronology and evolution of the young oceanic island of Réunion, Indian Ocean, *Geochim. Cosmochim. Acta, 35*, 261-270, 1971.

McKenzie, D. P., and Bickle, M. J., The volume and composition of melt generated by extension of the lithosphere, *J. Petrol.*, *29*, 625-679, 1988.

McKenzie, D. P., and Sclater, J. G., The evolution of the Indian Ocean since the Late Cretaceous, *Geophys. J. R. Astron. Soc.*, *25*, 437-528, 1971.

Mahoney, J. J., Deccan traps, in *Continental Flood Basalts*, edited by J. D. Macdougall, pp. 151-194, Kluwer Academic, Amsterdam, 1988.

Mahoney, J. J., Macdougall, J. D., Lugmair, G. W., and Gopalan, K., Kerguelen hotspot source for the Ninetyeast Ridge?, *Nature, 303*, 385-389, 1983.

Mahoney, J. J., Storey, M., Duncan, R. A., Spencer, K. J., and Pringle, M., Geochemistry and geochronology of Leg 130 basement lavas: Nature and origin of the Ontong Java Plateau, *Proc. ODP, Sci. Results, 130* , in press, 1992.

Meyerhoff, A. A., and Kamen-Kaye, M., Petroleum prospects of Saya de Malha and Nazareth Banks, Indian Ocean, *Am. Assoc. Pet. Geol. Bull.*, *65*, 1344-1347, 1981.

Molnar, P., Pardo-Casas, F., and Stock, J., The Cenozoic and Late Cretaceous evolution of the Indian Ocean basin: uncertainties in the reconstructed positions of the Indian, African and Antarctic plates, *Basin Res.*, *1*, 23-40, 1987.

Molnar, P., and Stock, J., A method for bounding uncertainties in combined plate reconstructions, *J. Geophys. Res.*, *90*, 12,537-12,544, 1985.

Molnar, P., and Stock, J., Relative motion of hotspots in the Pacific, Atlantic and Indian Oceans since Late Cretaceous time, *Nature, 327*, 587-591, 1987.

Morgan, W. J., Deep mantle convection plumes and plate motions, *Am. Assoc. Pet. Geol. Bull.*, *56*, 203-213, 1972.

Morgan, W. J., Rodrigues, Darwin, Amsterdam..., a second type of hotspot island, *J. Geophys. Res.*, *83*, 5355-5360, 1978.

Morgan, W. J., Hotspot tracks and the opening of the Atlantic and Indian Oceans, in *The Sea, Vol. 7*, edited by Emiliani, C., pp. 443-487, Wiley, New York, 1981.

Munschy, M., Dymont, J., Boulanger, M., Boulanger, J. D., Tissot, J. D., Schlich, R., Rotstein, Y., and Coffin, M. F., Breakup and seafloor spreading between the Kerguelen Plateau-

Labuan Basin and the Broken Ridge-Diamantina Zone, *Proc. ODP, Sci. Results, 120*, 931-944, 1992.

Mutter, J. C., and Cande, S. C., The early opening between Broken Ridge and Kerguelen Plateau, *Earth Planet. Sci. Lett., 65*, 369-376, 1983.

Nogi, Y., Seama, N., Isezaki, N., Hayashi, T., Funaki, M., and Kaminuma, K., Geomagnetic anomaly lineations and fracture zones in the basin west of the Kerguelen Plateau, *EOS, Trans. Am. Geophys. Union, 72*, 445, 1991.

O'Connor, J. M., and Duncan, R. A., 1990. Evolution of the Walvis Ridge and Rio Grande Rise hotspot system: implications for African and South American plate motions over hotspots, *J. Geophys. Res., 95*, 17,475-17,502, 1990.

Patriat, P., Ségoufin, J., Schlich, R., Goslin, J., Auzende, J. M., Beuzart, P., Bonnin, J., and Olivet, J. L., Les mouvements relatifs de l'Inde, de l'Afrique et de l'Eurasie, *Bull. Geol. Soc. France, 24*, 363-373, 1982.

Peirce, J. W., The northward motion of India since the Late Cretaceous, *Geophys. J. R. Astron. Soc., 52*, 277-311, 1978.

Peirce, J., Weissel, J. et al., *Proc. ODP, Init. Repts., 121*, Ocean Drilling Program, College Station, TX, 1989.

Recq, M., Brefort, D., Malod, J., and Veinante, J.-L., The Kerguelen Isles (Southern Indian Ocean): new results on deep structure from refraction profiles, *Tectonophysics, 182*, 227-248, 1990.

Richards, M. A., Duncan, R. A., and Courtillot, V. E., Flood basalts and hotspot tracks: plume heads and tails, *Science, 246*, 103-107, 1989.

Royer, J.-Y., and Coffin, M. F., Jurassic to Eocene plate tectonic reconstructions in the Kerguelen Plateau region, *Proc. ODP, Sci. Results, 120*, 917-928, 1992.

Royer, J.-Y., Peirce, J. W., and Weissel, J. K., Tectonic constraints on the hotspot formation of Ninetyeast Ridge, *Proc. ODP, Sci. Results, 121*, 763-776, 1991.

Royer, J.-Y., and Sandwell, D. T., Evolution of the Eastern Indian Ocean since the Late Cretaceous: constraints from Geosat altimetry, *J. Geophys. Res., 94*, 13,755-13,782, 1989.

Royer, J.-Y., Patriat, P., Bergh, H. W., and Scotese, C. R., Evolution of the Southwest Indian Ridge from the Late Cretaceous (anomaly 34) to the middle Eocene (anomaly 20), *Tectonophysics, 155*, 235-260, 1988.

Schlich, R., Structure et âge de l'océan Indian occidental, *Mem. Hors Sér. Soc. Géol. France, 6*, 1-103, 1975.

Schlich, R., The Indian Ocean: aseismic ridges, spreading centers and oceanic ridges, in *The Ocean Basins and Margins: The Indian Ocean, Vol. 6*, edited by A. E. M. Nairn, and F. G. Stehli, pp. 51-147, Plenum Press, New York, 1982.

Schneider, D. A., and Kent, D. V., Paleomagnetism of Leg 115 sediments: implications for Neogene magnetostratigraphy and paleolatitude of the Réunion hotspot, *Proc. ODP, Sci. Results, 115*, 717-736, 1990.

Schubert, G., and Sandwell, D., Crustal volumes of the continents and of oceanic and continental submarine plateaus, *Earth Planet. Sci. Lett., 92*, 234-246, 1989.

Sclater, J. G., von der Borch, C., Veevers, J. J., Hekinian, R.,

Thompson, R. W., Pimm, A. C., McGowran, B., Gartner, S., Jr., and Johnson, D. A., Regional synthesis of the Deep Sea Drilling results from Leg 22 in the Eastern Indian Ocean, *Init. Repts. DSDP, 22*, 815-831, 1974.

Sclater, J. G., and Fisher, R., The evolution of the east central Indian Ocean, with emphasis on the tectonic setting of the Ninetyeast Ridge, *Bull. Geol. Soc. Am., 85*, 683-702, 1974.

Ségoufin, J., and Patriat, P., Reconstitutions de l'Océan Indien occidental pour les époques des anomalies M21, M2 et 34. paléoposition de Madagascar, *Bull. Geol. Soc. France, 23*, 605-607, 1981.

Sengupta, S., Geological and geophysical studies in the western part of Bengal Basin, India, *Am. Assoc. Petrol. Geol. Bull., 50*, 1001-1017, 1966.

Simmons, G. R., Subsidence history of basement sites and sites along a carbonate dissolution profile, Leg 115, *Proc. ODP, Sci. Results, 115*, 123-128, 1990.

Storey, M., Saunders, A. D., Tarney, J., Leat, P., Thirwall, M. F., Thompson, R. N., Menzies, M. A., and Marriner, G. F., Geochemical evidence for plume-mantle interactions beneath Kerguelen and Heard Islands, Indian Ocean, *Nature, 336*, 371-374, 1988.

Storey, M., Kent, R. W., Saunders, A. D., Salters, V. J., Hergt, J., Whitechurch, H., Sevigny, J. H., Thirlwall, M. F., Leat, P., Ghose, N. C., and Gifford, M., Lower Cretaceous volcanic rocks on continental margins and their relationship to the Kerguelen Plateau, *Proc. ODP, Sci. Results, 120*, 33-53, 1992.

Tarduno, J. A., Sliter, W. V., Kroenke, L., Leckie, M., Mayer, H., Mahoney, J. J., Musgrave, R., Storey, M., and Winterer, E. L., Rapid formation of Ontong Java Plateau by Aptian mantle plume volcanism, *Science, 254*, 399-403, 1991.

ten Brink, U. S., and Brocher, T. M., Multichannel seismic evidence for a subcrustal intrusive complex under Oahu and a model for Hawaiian volcanism, *J. Geophys. Res., 92*, 13,687-13,707, 1987.

Vandamme, D., and Courtillot, V., Paleomagnetism of Leg 115 basement rocks and latitudinal evolution of the Réunion hotspot, *Proc. ODP, Sci. Results, 115*, 111-118, 1990.

Watkins, N. D., Gunn, B. M., Nougier, J., and Baksi, A. K., Kerguelen: continental fragment or oceanic island? *Geol. Soc. Am. Bull., 85*, 201-212, 1974.

White, R. S., and McKenzie, D. P., Magmatism at rift zones: the generation of volcanic continental margins and flood basalts, *J. Geophys. Res., 94*, 7685-7729, 1989.

Whitechurch, H., Montigny, R., Sevigny, J., Storey, M., and Salters, V., K-Ar and $^{40}Ar/^{39}Ar$ ages of Central Kerguelen Plateau basalts, *Proc. ODP, Sci. Results, 120*, 71-77, 1992.

Whitehead, J. A., and Luther, D. S., Dynamics of laboratory diapir and plume models, *J. Geophys. Res., 80*, 705-717, 1975.

Whitford, D. J., and Duncan, R. A., Origin of the Ninetyeast Ridge-Sr isotope and trace element evidence, in *Short Papers of the Fourth International Conference, Geochronology, Cosmochronology, Isotope Geology*, edited by R. E. Zartman, U.S. Geol. Survey, Open-file Report, 78-701, pp. 451-453, 1978.

Explosive Ocean Island Volcanism and Seamount Evolution in the Central Indian Ocean (Kerguelen Plateau)

P. R. BITSCHENE

Lab. Bioestratigrafia, Depto. Geologia, Universidad Nacional De La Patagonia SJB, Km 4, 9000 Comodoro Rivadavia, Argentina

J. DEHN, K. W. MEHL, AND H.-U. SCHMINCKE

Abteilung Vulkanologie und Petrologie, GEOMAR, Wischhofstr. 1-3, 2300 Kiel, Germany

Ocean island/seamount evolution on the Kerguelen Plateau occurred at about 80 Ma, when peridotite xenolith-bearing alkali basalt magmas (OIB) erupted onto the foundering basaltic basement. Erosion of the Kerguelen Plateau, a large igneous province topped with Cretaceous ocean islands/seamounts, including outcropping peridotite massifs, was the source for regional Cr and Ir anomalies in Cretaceous Indian Ocean marine sediments.

Explosive ocean island volcanism on the Kerguelen Plateau is manifested by marine ash layers in Oligocene to Recent sediments. Peaks of explosive ocean island volcanism occurred in the early Oligocene (basaltic, shield-building phase of the Kerguelen archipelago), middle to late Miocene (silicic, subaerial ocean island volcanism), Pliocene (silicic, subaerial to shallow marine ocean island volcanism), and late Quaternary (bimodal, subaerial to shallow marine ocean island volcanism). The marine ash layers coincide chemically and chronologically with the pulses of enhanced magmatic activity on Kerguelen Island. The basaltic ashes match transitional mid-ocean ridge basalt (T-MORB) compositions from the Kerguelen Plateau. K-rich alkali basaltic ash compositions also match K-rich ocean island basalts (OIB) from Kerguelen and/or Heard Island. The more silicic ashes compare closely with high-K silicic magmatic rocks from Kerguelen Island. Trachytic compositions may derive from the basaltic magmas by fractional crystallization. Alkali feldspar crystallization in the trachytic magmas led to the more silicic, rhyolitic compositions. Low-Ti and low-K basaltic to rhyolitic ash particles are products of calc-alkaline arc volcanism, most probably from the South Sandwich island arc.

INTRODUCTION

Many ocean islands and seamounts are the volcanic expressions of deep-seated, asthenospheric hotspots. Ocean islands are also important loci of explosive volcanism that leads to ocean wide volcanic ash distribution forming disseminated or discrete marine ash layers. Marine ash layers give an instant record of (a) the onset and timing of explosive, subaerial to shallow marine ocean island volcanic activity, (b) the magmatic and volcanologic evolution of ocean islands and their sources, and (c) the interaction of the hotspot/plume with the oceanic lithosphere [e. g., Kennett, 1981; Bitschene and Schmincke, 1990].

There are two especially long-lived hotspots in the Indian Ocean basin, the Kerguelen hotspot, active since at least 80 Ma, and the Réunion hotspot, active since about 66 Ma [Duncan et al., 1989]. The actual surface expressions of the Kerguelen hotspot are the islands of the Kerguelen archipelago (Figure 1). ODP Legs 119 and 120 [Barron, Larsen et al., 1989; Schlich, Wise et al., 1989] recovered marine ash layers from the Kerguelen Plateau, allowing the first tephrochronologic studies from this region [Bitschene et al., 1992a; Morche et al., 1991, 1992]. Leg 120 also drilled alkalic ocean island basalts (OIB) and

Synthesis of Results from Scientific
Drilling in the Indian Ocean
Geophysical Monograph 70

overlying Cr-rich glauconitic sediments at Site 748, indicative of Cretaceous seamount and plateau basalt erosion on the Kerguelen Plateau [Bitschene et al., 1992b].

This contribution reviews the occurrence, composition and significance of the ash layers related to Kerguelen Plateau explosive ocean island volcanism, and reconstructs the history of Late Cretaceous seamount evolution and erosion on the Kerguelen Plateau.

TEMPORAL AND SPATIAL DISTRIBUTION OF MARINE ASH LAYERS

ODP Legs 119 [Barron, Larsen et al., 1989], and 120 [Schlich, Wise et al., 1989] were designed to drill a transect along the Kerguelen Plateau (Figures 1 and 2), in order to reveal its nature and origin. Basaltic basement was drilled in five sites (Figure 2); volcaniclastic sediments were encountered in nearly all sites in the form of (a) volcanogenic, epiclastic clayey and conglomeratic sediments (Sites 748 and 750), (b) disseminated and discrete ash layers (Sites 736, 737, 745, 747, 749, 751), and (c) volcaniclastic mass flows (Sites 736 and 747).

The Danian volcaniclastic mass flows at Site 747 [Aubry and Berggren, 1989] were derived from erosion of the T-MORB basement of the Kerguelen Plateau, whereas Plio-Pleistocene debris flows at Site 736 (and probably also at Site 737) are part of the clastic apron of Kerguelen Island [Bitschene et al., 1992a]. There is also a noticeable background ash component in the Neogene sediments from

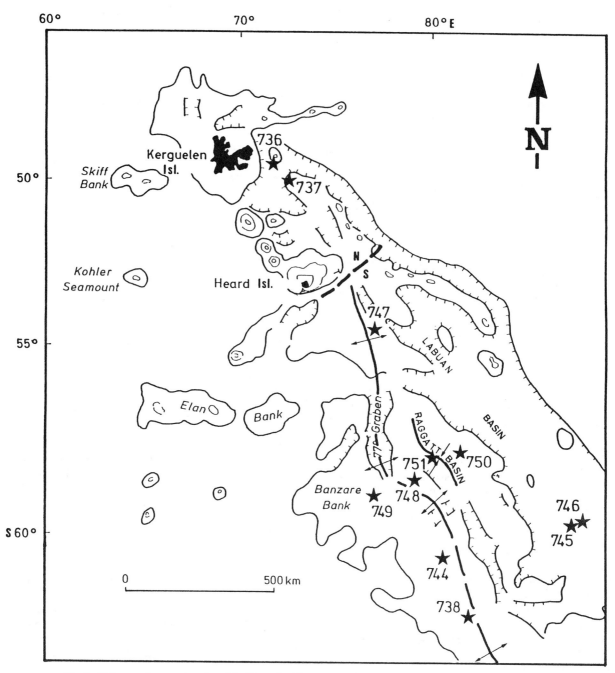

Fig. 1 Major morphotectonic units of the Kerguelen Plateau with locations of ODP Legs 119 and 120 drillsites.

the Kerguelen Plateau. This constant ash input represents subaerial fallout from mildly explosive volcanism on the Kerguelen Islands, and probably from ice rafting. Well-sorted, blocky (40 ± 20 microns), high-Ti basaltic tachylite and sideromelane, and clear, blocky and cuspate trachyphonolitic vitric shards with hedenbergite, Ti-magnetite, apatite and biotite phenocrysts are dispersed in Quaternary marine sediments and may have resulted from subglacial volcanic eruptions on Kerguelen Island. These volcanogenic "background" particles are responsible for the magnetic signal of the Quaternary sediments from Site 747 [Bitschene and Heider, 1991].

Marine ash layers occur in lower Oligocene (28 to 32 Ma), upper Miocene (8 Ma), Pliocene (1.7 to 4.5 Ma), and upper Quaternary (1 Ma to Recent) sediments (Figure 3). These stratigraphic intervals with explosive volcanism products coincide with identified epochs of enhanced magmatic activity on Kerguelen Island [Watkins et al., 1974; Nougier and Thomson, 1990].

PETROLOGIC EVOLUTION AND SOURCES

The results of more than a thousand major element

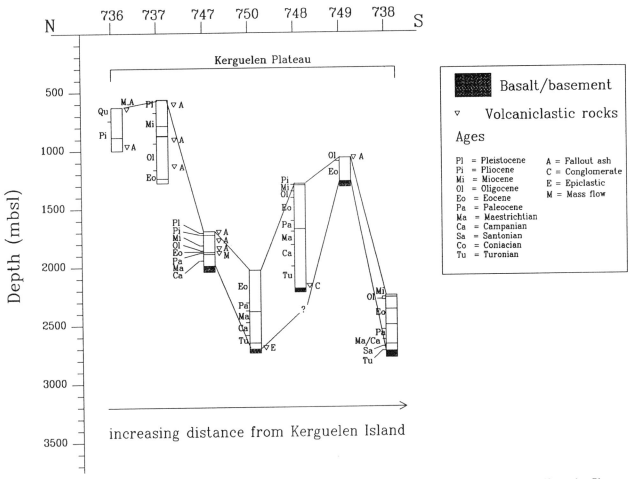

N   736   737   747   750   748   749   738   S

Kerguelen Plateau

increasing distance from Kerguelen Island

Fig. 2 Depth, relative geographic situation and stratigraphic position of basement rocks and volcaniclastic layers of Kerguelen Plateau basement drill holes.

microprobe analyses [Bitschene et al., 1992a; Morche et al., 1991, 1992] of single vitric shards and phenocrysts of marine ash layers from the Kerguelen Plateau are summarized as follows: The basaltic vitric shards are medium to high-K tholeiitic T-MORB with high Ti-concentrations (>3.2 %, high-Ti T-MORB; for convenience we use the term T-MORB to mean transitional basalts that are LREE-enriched, high-Ti tholeiitic MORB, thus clearly separating them from LREE-depleted, low-Ti N-MORB, and from silica-undersaturated, highly LREE-enriched OIB). Also, high-K alkali basalt compositions are reported by Morche et al. [1991a]. A few dispersed ash particles belong to a low-K and low-Ti basaltic series (Figure 4). Basaltic ashes contain rare olivine (?), augite, Fe- and Ti-oxides, and plagioclase as phenocrysts. The silicic ashes have high-K trachytic and rhyolitic compositions and contain Na- and Fe-rich clinopyroxene, alkali feldspar, apatite, and Ti-magnetite as phenocrysts, and subordinate biotite and zircon. Fractionation of the above mineral assemblages may well explain the major element evolution of the magmas from basaltic (T-MORB) to trachytic compositions (Figure 5). A few glass shards have intermediate compositions, thus corroborating a simple fractional crystallization history.

An alternative process to explain the generation of trachytic magmas would be partial melting of already existing basaltic oceanic crust. This process, however, would lead to highly silicic magmas, as evidenced in silicic magma compositions from Iceland. Mixing of basaltic and rhyolitic magma to produce the trachytic compositions can also be dismissed, because K_2O is highest in the trachytic magmas, and not in the rhyolitic magmas. Extensive alkali feldspar crystallization in the trachytic magmas is required to generate the more silicic, less K-rich rhyolitic magmas. Based on major element concentrations, the magmatic evolution of the Kerguelen explosive volcanism is controlled by long term, simple crystal fractionation differentiation, and not by extensive magma mixing or crustal melting.

The major element compositions of single vitric ash grains match bulk rock major element compositions from the Kerguelen archipelago [Bitschene et al., 1992a; Morche et al., 1991, 1992]. Cretaceous [Mehl et al., 1991; Salters et al., 1991], modern T-MORB basalts from Kerguelen Island [Nougier and Thomson, 1990], and T-MORB ash layers also have very similar major element concentrations and ratios, e. g. TiO_2/K_2O ratios of about 3.2. The low-K series ash compositions of some shards

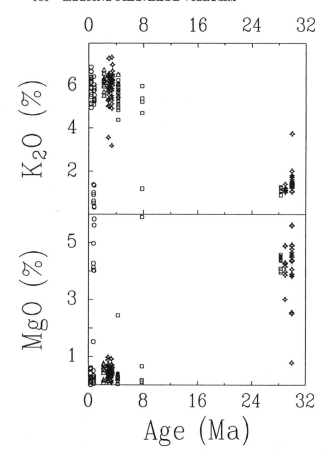

Fig. 3 K$_2$O and MgO concentrations of glass shards from discrete ash layers, and their approximate biostratigraphic age. Symbols as in Fig. 4, but without significance.

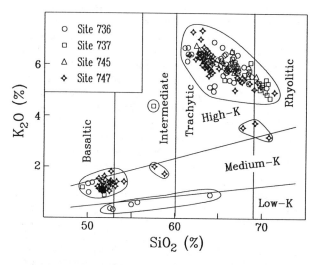

Fig. 4 Glass compositions, recalculated to 100%, in a K$_2$O vs. SiO$_2$ classification diagram.

detected in the ash layers (Figure 5) are most probably ice- or wind-rafted distal fallout products from the South Sandwich islands [Morche et al., 1991]. Admittedly, the low-K basaltic shards have rather high TiO$_2$ and alkali concentrations not common in a calc-alkaline, subduction related volcanic scenario. However, similar basalt compositions with higher than normal Ti and K-concentrations are known from Deception Island [Baker, 1990], and from the recent August 1991 Hudson eruption in the Southern Andes at 46° S latitude. The trachyandesitic ashes of the Hudson eruption spread within four days as far as Australia, using the same strong winds (roaring forties) that sweep the Kerguelen Plateau.

The massive alkali basalts recovered at Site 748 are silica undersaturated ocean island basalts (OIB) with low Zr/Nb ratios of 2.1, high (Ce/Yb)$_{cn}$ ratios of 24, very high Ba concentrations (1100 ppm), and no Ta and Nb depletion in N-MORB normalized trace element spidergrams (Figure 6). Isotopically they reflect the DUPAL-anomaly with ^{87}Sr/^{86}Sr$_i$ of 0.70516, and ^{208}Pb/^{204}Pb of 38.495 [Salters et al., 1992]. From Figure 6 it is also evident that simple mixing of an Indian Ocean OIB (Site 748 Kerguelen Plateau alkali basalt) and N-MORB (Site 758 Ninetyeast Ridge N-MORB) source (magma or protolith) does not produce Indian Ocean T-MORB. The alkali basalt from Site 748 represents the Late Cretaceous Kerguelen hotspot/plume signal and indicates the location of the Kerguelen hotspot at about 80 to 85 Ma [extrapolation of biostratigraphic and lithologic data in Schlich, Wise et al., 1989]. According to the paleolatitudes determined by Inokuchi and Heider [1992], the site of eruption was at about 45° S whereas the current position is at 59° S. A drift rate of 1.92 cm per year, or 19.2 km/m.y. can thus be calculated for the southern Kerguelen Plateau overriding the Site 748 hotspot, assumed to be in a fixed position ever since. This figure for the Mesozoic to Recent mean Kerguelen Plateau drift velocity - equivalent with a half spreading rate - coincides well with the spreading rate of the South East Indian Ridge since the Eocene, which is between 25 and 45 km/m.y. [Munschy et al., 1992].

OCEAN ISLAND/SEAMOUNT EROSION ON THE KERGUELEN PLATEAU

The erosion of young (Neogene to Recent) Kerguelen ocean islands is reflected in volcaniclastic debris flows drilled at Site 736 [Barron, Larsen et al., 1989]. These epiclastic volcanogenic sediments, which consist of basaltic and syenitic lithic clasts, and volcanic glasses and minerals, make up the clastic apron of the subaerially exposed and eroded Kerguelen Island volcanic centers [Bitschene et al., 1992a]. The input of volcanogenic material started in the late Pliocene [about 1.8 to 1.6 Ma; Barron, Larsen et al., 1989], but only reached its climax in the late Quaternary. This indicates major ocean island erosion roughly 30 million years after the onset of explosive ocean island volcanism on the Kerguelen Plateau.

The erosion of the only Cretaceous seamount so far known

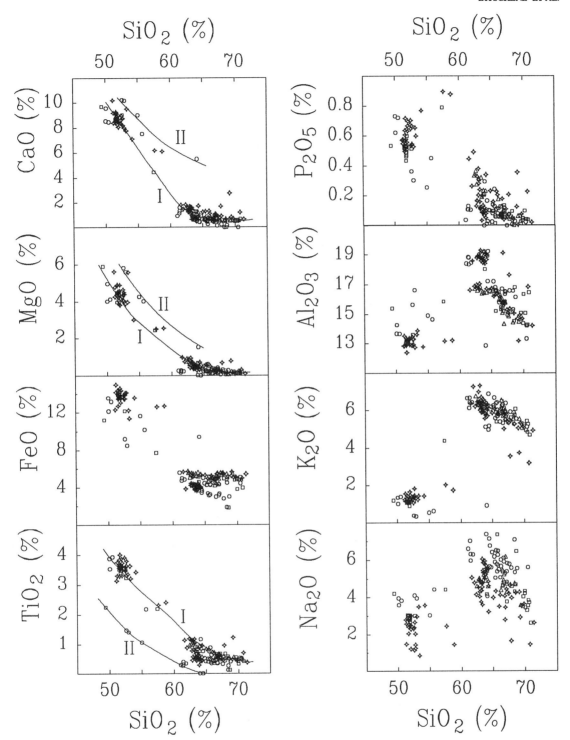

Fig. 5 Major element concentrations vs. SiO$_2$ as differentiation index. Line I is the trend for Kerguelen ashes, line II for the calc alkaline series.

from the Kerguelen Plateau is deduced from ODP Site 748, where an upper Cretaceous, xenolith-bearing alkali basalt (80 to 85 Ma) is overlain by 500 m of Turonian to Maestrichtian, Cr-rich glauconitic sediments [Bitschene et al., 1992b]. A Late Cretaceous volcanotectonic

reconstruction (Figure 7) from the Banzare Bank (Site 749) to the Raggatt Basin (Site 748) suggests that after termination of the basaltic, subaerial plateau volcanism (tholeiitic T-MORB) at about 114 to 100 Ma [Leclaire et al. 1987; Schlich, Wise et al., 1989], subaerial alteration,

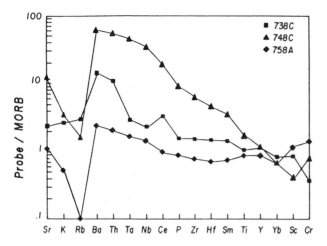

Fig. 6 N-MORB normalized spidergram of N-MORB from the Ninetyeast Ridge (filled diamonds), T-MORB (filled squares) and OIB (filled triangles) from the Southern Kerguelen Plateau.

subsidence, and erosion of the Southern Kerguelen Plateau led to the deposition of volcanogenic, clayey epiclastic sediments. A major tectonic event occurred at about 88 Ma, when the Eastern Raggatt Basin formed [Fritsch et al., 1992], and a magmatic event at about 80 Ma, when the tephritic alkali basalts at Site 748 were emplaced and the western Raggatt Basin developed.

The erosion of the Late Cretaceous seamount/ocean island and its underlying basaltic basement is monitored by the glauconitic sequence, which served as sinks for the solutions carrying Cr, Ir, Fe, K, Al, and other elements from the eroded ocean island and its T-MORB basaltic basement [Bitschene et al., 1992b]. Especially Cr (average of 600 ppm; peak of 860 ppm), and Ir (average of 30 ppt; peak of 110 ppt) show high concentrations in the glauconitic sediments (Table 1). If Sample 120-748C-30R-1, 7-10 cm is taken to represent background values (73 ppm Cr, 10 ppt Ir; Michel et al. [1990] report 4 to 20 ppt Ir as typical background values for marine carbonate sediments from ODP holes) for Cretaceous Kerguelen Plateau sediments, then the glauconitic sediments have a three- to eleven-fold Cr and Ir enrichment.

The high Cr-concentrations, abundant Cr-spinels, and high Ir-concentrations (Table 1) of the 500 m thick glauconitic sediments suggest a sustained Ir input either detritus bound, atomized or dissolved. Extraterrestrial bolide impact, basaltic volcanism or the erosion of major ultramafic bodies can be envisaged to carry along Ir and Cr. A bolide impact can be ruled out, as the glauconitic sediments encompass at least a 12 m.y. time span [Schlich, Wise et al., 1989]. Also, basaltic subaerial volcanism in this region virtually ceased with the underlying OIB and even older tholeiite basaltic basement. Peridotites, on the other hand crop out in several fracture zones of the Indian Ocean, and may have concentrations of several 1000 ppm of Cr and up to 400 to 4000 ppt of Ir [Keays, 1982].

A plot of Cr, Ba and Ir against depth (Figure 8) shows that

Cr and Ir do not correlate. Ir is highest in the basal conglomerate (Unit IIIC) from Site 748, which contains alkali basalt pebbles, detrital Cr-spinel, and also has high Ba concentrations (Table 1), as well as high Zr, Nb and REE concentrations [Bitschene et al., 1992b]. Cr, on the other hand, shows the highest concentrations in the pebble-free, glauconitic sand and siltstone (Unit IIIB). Ir is thus thought to be directly related to the erosion of a peridotitic source accompanying the alkali basalt, whereas the Cr-enrichment is caused by erosion and reworking of the Late Cretaceous OIB and the underlying T-MORB basement.

TABLE 1. Selected Trace Element Concentrations of Cr-rich Glauconitic Sediments from Hole 748C, ODP Leg 120

Core-Section, cm-interval	Cr (ppm)	Zr (ppm)	Ba (ppm)	Ir (ppt)
30R-1, 7-10 cm	73	10	55	10
56R-2, 137-144 cm	610	67	156	30
63R-1, 103-105 cm	730	97	101	40
65R-1, 43-44 cm	860	94	113	50
75R-1, 36-38 cm	570	137	250	30
75R-5, 50-52 cm	820	137	204	20
79R-4, 93-95 cm	214	402	1441	50
79R-4, 107-109 cm	228	502	1701	110

If redistributed within a thin, only 50-cm-thick layer, the total amount of Ir contained in the glauconitic sediments is sufficient to cover an area 1000 times larger than the area covered by the restricted, tectonically bound glauconite accumulation found in Site 748. Assuming 4000 km^2 areal extent of the glauconitic sediments, and loss of half of the Ir during redistribution processes, a 50-cm-thick sediment blanket with high Cr and Ir concentrations covering 2,000,000 km^2 of the Indian Ocean would be caused by erosion, reworking and redistribution of the Cr-rich glauconitic sediments or the basaltic mother rock. Any widespread Indian Ocean Cretaceous Ir anomaly could thus be produced by simple erosion of a large basaltic igneous province topped with oceanic islands and seamounts with associated ultramafic bodies. This plateau erosion and detritus reworking scenario could well explain the extraordinary Ir enrichment in an at least 1-m-thick sequence at the Cretaceous/Tertiary boundary at Hole 738C [Schmitz et al., 1991], not withstanding presentation of arguments in favor of an extraterrestrial bolide impact. In conclusion, the erosion and reworking of basaltic plateaus, and especially of ultramafic massifs associated with alkaline ocean island volcanism and/or deep-cutting transform faults, is a viable and reasonable way of explaining sustained, and sudden, Ir (and Cr and other elements) input into the marine environment. This process may be as important as the currently favored models of bolide impact or basaltic explosive volcanism in explaining Ir anomalies.

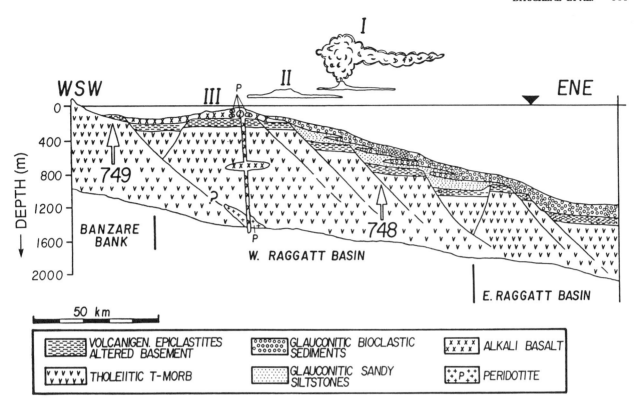

Fig. 7 Volcanotectonic evolution of the Southern Kerguelen Plateau (Banzare Bank and Raggatt Basin) during the Late Cretaceous. After cessation of T-MORB volcanism at about 114 to 100 Ma, volcanogenic epiclastic, clayey sediments formed on the foundering and weathering Southern Kerguelen Plateau basaltic basement. The Raggatt Basin then formed, receiving the terrigenic solutions and detritus from the still subaerially exposed Banzare Bank and environment. A major alkali basaltic episode occurred during this period (about 80 Ma), forming upper Cretaceous seamounts/ocean islands; peridotitic material was also exposed. After this ultimate pulse of magmatism, the Southern Kerguelen Plateau continued its subsidence, glauconitic sediments formed in a shallow-marine environment influenced by major input from the still exposed and deeply weathered Banzare Bank and the exposed seamount/ocean island with the peridotite cropping out. In a restricted, tectonically bound basin with synsedimentary faulting, the glauconitic sediments accumulated to 500 m thickness in about 12 m.y. (Cenomanian/Turonian to Maestrichtian). After that, the Southern Kerguelen Plateau subsided as a whole, pelagic carbonate sediments formed. During the Cretaceous and the Cenozoic, the Kerguelen Plateau moved southwards thus overriding the Kerguelen hotspot. This southward trend is manifested by a relative northward trend of the ocean island/seamount evolution, indicated here through the pulses III, II, and I, the latest being todays Kerguelen archipelago.

SEAMOUNT AND OCEAN ISLAND EVOLUTION AND EROSION

Seamount evolution began at least 80 m.y. ago, when alkali basaltic lavas were emplaced upon the foundering Kerguelen Plateau. Since then, a relative southward migration of the Kerguelen Plateau overriding the Kerguelen hotspot probably induced ocean island-type volcanism on the central and northern Kerguelen Plateau (Figure 7). Seamount evolution may have persisted during the Paleogene on the Central Kerguelen Plateau (pulse II), but no ash layers indicative of explosive ocean island volcanism are reported for that period, which was otherwise tectonically very active. The Late Cretaceous to Paleogene saw major tectonic events at 66 Ma, and the final separation of the Kerguelen Plateau from the Broken Ridge at about 42 to 45 Ma [e.g. Fritsch et al., 1992; Munschy et al., 1992]. Explosive ocean island activity started only after the final separation of Kerguelen Plateau and Broken Ridge, and after a Kerguelen Plateau-wide erosional event at about 38 Ma [Munschy et al., 1992]. Explosive ocean

island volcanism on the northern Kerguelen Plateau (Figure 7, pulse III) started with a shield building, basaltic stage in the early Oligocene, around 30 Ma. A major phase of silicic ocean island volcanism occurred in the late Miocene and Pliocene. Basaltic and silicic explosive activity resumed in the late Quaternary, perhaps after removal of thick glacial ice cover from the Kerguelen Archipelago.

Episodes of explosive ocean island volcanism on the Kerguelen Plateau occurred in the early Oligocene, late Miocene, Pliocene, and late Quaternary. These epochs with enhanced explosive ocean island volcanism in the Kerguelen archipelago do not correspond with the rather continuous explosive volcanism from the Iceland-Jan Mayen ocean island realm [Bitschene et al., 1989]. However, absolute dating of the Kerguelen Plateau and other ocean island-derived ash layers is still lacking, as well as correction for hiatuses. Ar-retentive minerals such as volcanic plagioclase, biotite, and especially sanidine are nearly always present in the Kerguelen Plateau ash

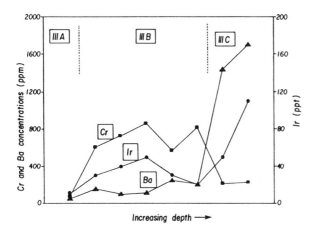

Fig. 8 Trace element (Ba, Cr, Ir) concentrations of whole-rock samples from the glauconitic sequence at ODP Site 748, and their relative stratigraphic relationship (Subunits IIIA, IIIB, IIIC).

layers, so one of the major future tasks is to establish an absolute time frame for oceanic sediments and marine ash layers world-wide using single grain Ar-Ar dating [Bitschene and Schmincke, 1990].

Acknowledgments. We thank C. Orth (Los Alamos/USA) and H. Hilbrecht (ETH Zürich/Switzerland) for making available and carrying out the Ir analysis. We especially thank J. Hertogen (Leuven/Belgium) who provided us with the REE and trace element analysis of the KP basalts. The Deutsche Forschungsgemeinschaft is thanked for financial support through grants Schm 250/37-1 to -4. Constructive comments from reviewers H. Ferriz and R. Mühe helped to clarify some results.

REFERENCES

Aubry, M.-P., Berggren, W. A., Age of the upper volcaniclastic debris flow at Site 747: a special study, in *Proc. ODP, Init. Repts.*, *120*, edited by R. Schlich, S. W. Wise Jr. et al., pp. 57-72, Ocean Drilling Program, College Station, TX, 1989.

Barron, J., Larsen, B. et al., *Proc. ODP, Init. Repts.*, *119*, Ocean Drilling Program, College Station, TX, 1989.

Baker, P. E., Deception Island, in *Volcanoes of the Antarctic Plate and Southern Oceans*, edited by W.E. LeMasurier and J.W. Thomson, 316-321, Am. Geophys. Union Antarctic Research Series, Washington, 1990.

Bitschene, P. R., Schmincke, H.-U., and Viereck, L., Cenozoic ash layers on the Voring Plateau (ODP Leg 104), in *Proc. ODP, Sci. Results*, *104*, edited by O. Eldholm, J. Thiede et al., 357-366, Ocean Drilling Program, College Station, TX, 1989.

Bitschene, P. R. and Schmincke, H.-U., Fallout tephra layers: composition and significance, in *Sediments and Environmental Geochemistry*, edited by D. Heling, P. Rothe, U. Forstner, and P. Stoffers, pp. 48-82, Springer, Heidelberg, 1990.

Bitschene, P. R., and Heider, F., Zusammensetzung, Herkunft und Bedeutung dispers verteilter vulkanigener Fallout Partikel in Nannofossilschlämmen des Kerguelen Plateaus, *Nat. Koll. DFG-Schwerpunkt ODP/DSDP*, 1 p (Abstr.), 1991.

Bitschene, P. R., Mehl, K. W., and Schmincke, H.-U., Composition and origin of marine ash layers and epiclastic rocks from the Kerguelen Plateau, southern Indian Ocean (Legs 119 and 120), in *Proc. ODP, Sci. Results, 120*, edited by R.

Schlich, S. W. Wise, Jr. et al., pp. 135-149, Ocean Drilling Program, College Station, TX, 1992a.

Bitschene, P. R., Holmes, M.-A., and Breza, J., Composition and origin of Cr-rich glauconitic sediments from the Southern Kerguelen Plateau (Site 748), in *Proc. ODP, Sci. Results*, *120*, edited by R. Schlich, S. W. Wise, Jr. et al., pp. 113-134, Ocean Drilling Program, College Station, TX, 1992b.

Duncan, R. A., Backman, J., and Peterson, L., Reunion hotspot activity through Tertiary time: Initial results from the Ocean Drilling Program, Leg 115, *J. Volcanol. Geotherm. Res., 36*, 193-198, 1989.

Fritsch, B., Schlich, R., Munschy, M., Fezga, F., and Coffin, M. F., Evolution of the Southern Kerguelen Plateau deduced from seismic stratigraphic studies and drilling at Sites 748 and 750, in *Proc. ODP. Sci. Results, 120*, edited by R. Schlich, S. W. Wise, Jr. et al., pp. 895-906, Ocean Drilling Program, College Station, TX, 1992.

Inokuchi, H. and Heider, F., Paleolatitude of the Southern Kerguelen Plateau inferred from the paleomagnetic study of Late Cretaceous basalts, in *Proc. ODP, Sci. Results*, *120*, edited by R. Schlich, S. W. Wise, Jr. et al., pp. 89-96, Ocean Drilling Program, College Station, TX, 1992.

Keays, R. R., Archean gold deposits and their source rocks: the upper mantle connection, in *The Geology, Geochemistry and Genesis of Gold Deposits*, edited by R. P. Foster, pp. 17-51, Balkema, Rotterdam, 1982.

Kennett, J. P., Marine tephrochronology, in *The Oceanic Lithosphere-The Sea*, edited by C. Emiliani, pp. 1373-1436, J. Wiley and Sons, New York, 1981.

Leclaire, L., Bassias, Y., Denis-Clochiatti, M., Davies, H., Gautier, I., Gensous, B., Giannesini, P. J., Patriat, P., Segoufin, J., Tesson, M., and Wanneson, J., Lower Cretaceous basalts and sediments from the Kerguelen Plateau, *Geo-Mar. Lett., 7*, 169-176, 1987.

Mehl, K.-W., Bitschene, P. R., Schmincke, H.-U., and Hertogen, J., Composition, alteration and origin of the basement lavas and volcaniclastic rocks at Site 738, Southern Kerguelen Plateau, in *Proc. ODP, Sci. Results, 119*, edited by J. Barron, B. Larsen et al., pp. 299-322, Ocean Drilling Program, College Station, TX, 1991.

Michel, H. V., Asaro, F., Alvarez, W. and Alvarez, L. W., Geochemical studies of the Cretaceous/Tertiary boundary in ODP Holes 689B and 690C, in *Proc. ODP, Sci. Results, 113*, edited by P. F. Barker, J. P. Kennett et al., pp. 159-169, Ocean Drilling Program, College Station, TX, 1990.

Morche, W., Hubberten, H.-W., Ehrmann, W. U., and Keller, J., Geochemical investigations of volcanic ash layers from Leg 119, Kerguelen Plateau, in *Proc. ODP, Sci. Results, 119*, edited by J. Barron, B. Larsen et al., pp. 323-344, Ocean Drilling Program, College Station, TX, 1991.

Morche, W., Hubberten, H.-W., Mackensen, A. and Keller, J., Geochemistry of Cenozoic ash layers from the Kerguelen Plateau (Leg 120): a first step toward a tephrostratigraphy of the Southern Indian Ocean, in *Proc. ODP, Sci. Results, 120*, edited by R. Schlich, S. W. Wise Jr. et al., pp. 151-160, Ocean Drilling Program, College Station, TX, 1992.

Munschy, M., Dyment, J., Boulanger, M. O., Boulanger, D., Tissot, J. D., Schlich, R., Rotstein, Y., and Coffin, M. F., Breakup and seafloor spreading between the Kerguelen Plateau-Labuan Basin and the Broken Ridge-Diamantina Zone, in *Proc. ODP, Sci. Results, 120*, edited by R. Schlich, S. W. Wise Jr. et al., pp. 931-944, Ocean Drilling Program, College Station, TX, 1992.

Nougier, J., and Thomson, J. W., Volcanoes of the Antarctic Plate and Southern Oceans - Iles Kerguelen, in *Volcanoes of the Antarctic Plate and Southern Oceans*, edited by W. E. LeMasurier and J.W. Thomson, pp. 429-434, Am. Geophys. Union Antarctic Research Series, Washington, 1990.

Salters, V., Storey, M., Sevigny, J. H., and Whitchurch, H., Trace element and isotopic characteristics of Kerguelen-Heard Plateau Basalts, in *Proc. ODP, Sci. Results, 120*, edited by R. Schlich, S. W. Wise Jr. et al., pp. 55-62, Ocean Drilling Program,

College Station, TX, 1992.

Schlich, R., Wise, S. W., et al., *Proc. ODP, Init. Repts, 120*, Ocean Drilling Program, College Station, TX, 1989.

Schmitz, B., Asaro, F., Michel, H. V., Thierstein, H. R., and Huber, B. T., Element stratigraphy across the Cretaceous/Tertiary boundary in Hole 738C, in *Proc. ODP, Sci. Results, 119*, edited by J. Barron, B. Larsen et al., pp. 719-730, Ocean Drilling Program, College Station, TX, 1991.

Watkins, N. D., Gunn, B. M., Nougier, J., and Baksi, A. K., Kerguelen: continental fragment or oceanic island?, *Geol. Soc. Am. Bull., 85*, 201-212, 1974.

Emplacement and Subsidence of
Indian Ocean Plateaus and Submarine Ridges

MILLARD F. COFFIN

*Institute for Geophysics, The University of Texas at Austin, 8701 Mopac Blvd., Austin, TX , 78759-8397,
USA*

Ocean Drilling Program, Deep Sea Drilling Project, and industrial borehole results from Indian Ocean
plateaus and submarine ridges help to constrain their subsidence histories. I use a simple Airy isostatic
model to calculate basement depths at ODP sites in the absence of sediment, and then backtrack these sites
using previously determined age-depth relationships for oceanic lithosphere to determine the original
depth or elevation of the sites. Resulting subsidence curves for each site were then checked by examining
sedimentologic and biostratigraphic evidence for when each site subsided below shelf depths. The analysis
suggests that thermal subsidence has been the dominant tectonic process affecting Indian Ocean plateaus
and submarine ridges following emplacement. I conclude that large portions of these features were
emplaced and began subsiding well above sea level, similar to large igneous provinces (LIPs) worldwide
today. This resulted in significant subaerial erosion and redeposition of volcanic material mixed with
biogenic sediment, and a gradual development of facies from terrestrial through terrigenous to shallow
water and pelagic, resulting in a sedimentary record with both continental and oceanic characteristics.

INTRODUCTION

Voluminous emplacements of mafic igneous rock
originating via processes other than "normal" seafloor
spreading include continental flood basalt and associated
intrusive rock; volcanic passive margins; oceanic plateaus;
submarine ridges; ocean basin flood basalts; and seamount
groups. These large igneous provinces (LIPs) share many
temporal, spatial, and compositional characteristics. The
early evolution of LIPs, especially their subsidence
history, is not well documented, and oceanic provinces
offer an opportunity to study this topic in the simplest
possible lithospheric setting.

The Indian Ocean contains a plethora of submarine ridges
and oceanic plateaus (Figure 1; Table 1), about which little
was known until the advent of deep sea drilling. Since
1972, however, seven of these features have been drilled,
and unequivocal igneous basement has been recovered and
dated from four - the Chagos-Laccadive Ridge, the
Kerguelen Plateau, the Mascarene Plateau, and Ninetyeast
Ridge. These drill cores have yielded much information on
the features' origin and evolution, especially from
stratigraphic, petrologic, geochemical, and
geochronologic studies, and they provide essential data for
examining their subsidence histories. In this paper I will
first briefly summarize our knowledge of four features for
which sufficient data are available to study their early
development, then analyze their subsidence history, and
finally discuss some geologic and geophysical
implications of the analysis.

Synthesis of Results from Scientific
Drilling in the Indian Ocean
Geophysical Monograph 70

OCEANIC PLATEAUS AND SUBMARINE RIDGES

Of fourteen known mafic plateaus and ridges in the Indian
Ocean (Figure 1; Table 1), reliable dates for drilled, dredged,
and subaerially sampled volcanic rock constituting
basement are available for six - Broken Ridge, Chagos-
Laccadive Ridge, Crozet Plateau, Kerguelen Plateau,
Mascarene Plateau, and Ninetyeast Ridge (Table 2). These
LIPs range in age from 110 (Kerguelen) to 0 (Crozet) Ma,
and geochronologic studies indicate that construction of
individual features took from <10 to as many as 44 m.y.
Three hotspots are apparently responsible for these six
features [Morgan, 1981], the Kerguelen hotspot for the
Kerguelen Plateau, Broken Ridge, and Ninetyeast Ridge;
the Réunion hotspot for Chagos-Laccadive Ridge and the
Mascarene Plateau; and the Crozet hotspot for the Crozet
Plateau. The complex tectonic evolution of the Indian
Ocean is responsible for one hotspot creating multiple
LIPs.

Although basement dates are available for Broken Ridge
and the Crozet Plateau (Table 2), the study of subsidence
immediately following emplacement of these features is
not yet possible. In the case of Broken Ridge, basement
samples were dredged, and are thus not reliably located with
regard to depth [Duncan, 1991]. Furthermore, Broken
Ridge experienced major tectonism prior to and during its
breakup with the northern Kerguelen Plateau in Eocene time
[Driscoll et al., 1991]. The only basement samples from
the Crozet Plateau are from islands, and lack of overlying
sediment precludes subsidence analysis.

The Chagos-Laccadive Ridge extends over ~2500 km west
and south of India, and its width is ~200 km. Numerous
atolls, reefs, banks, and shoals form the subaerial
expression of the feature, although no volcanic basement
is exposed. The ridge represents part of a hotspot track
which also includes the Deccan Traps in India and the
Mascarene Plateau [Figure 1; Morgan, 1981; Duncan,
1990]. ODP Leg 115 recovered the first basement samples

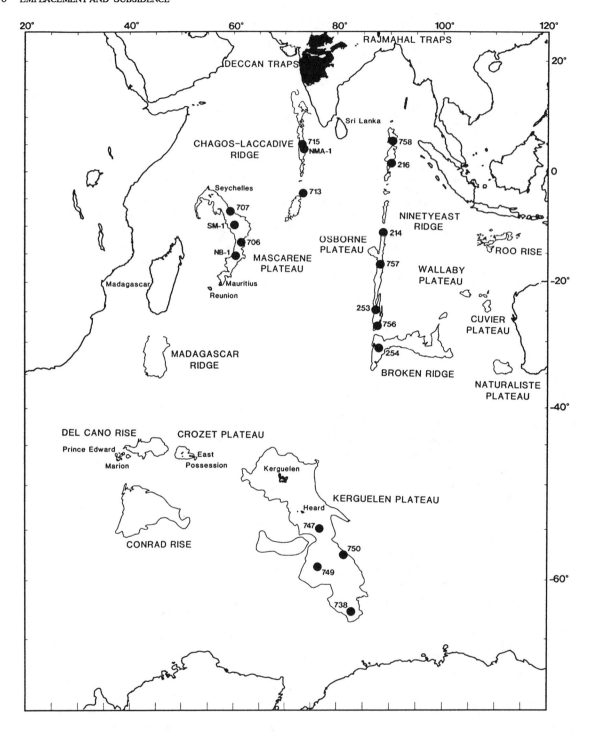

Fig. 1. Indian Ocean plateaus and submarine ridges, and the Deccan and Rajmahal Traps. Locations of DSDP, ODP, and industry volcanic basement sites are indicated by circles (Table 2).

TABLE 1. Indian Ocean Plateaus and Submarine Ridges: DSDP/ODP Sites

LIP	Type§	DSDP/ODP Leg(s)	DSDP/ODP Site(s)	Basement Reached?	Reference(s)
Broken Ridge	SR	26	255	no	Davies, Luyendyk et al., 1974
		121	752	no	Peirce, Weissel et al., 1989;
		121	753	no	Weissel, Peirce, Taylor, Alt et al., 1991
		121	754	no	ibid.
		121	755	no	ibid.
Chagos-Laccadive Ridge	SR	23	219	no	Whitmarsh, Weser, Ross et al., 1974
		115	712	no	Backman, Duncan et al., 1988;
		115	713	yes	Duncan, Backman, Peterson et al., 1990
		115	714	no	ibid.
		115	715	yes	ibid.
		115	716	no	ibid.
Conrad Rise	OP	-	-	-	-
Crozet Plateau	OP	-	-	-	-
Cuvier Plateau	OP	-	-	-	-
Del Caño Rise	OP	-	-	-	-
Kerguelen Plateau	OP	119	736	no	Barron, Larsen et al., 1989;
		119	737	no	Barron, Larsen et al., 1991
		119	738	yes	ibid.
		119	744	no	ibid.
		120	747	yes	Schlich, Wise et al., 1989;
		120	748	no	Wise, Schlich et al., 1992
		120	749	yes	ibid.
		120	750	yes	ibid.
		120	751	no	ibid.
Madagascar Ridge	SR	25	246	no	Simpson, Schlich et al., 1974
		25	247	no	ibid.
Mascarene Plateau	OP	24	237	no	Fisher, Bunce et al., 1974
		115	705	no	Backman, Duncan et al., 1988;
		115	706	yes	Duncan, Backman, Peterson et al., 1990
		115	707	yes	ibid.
Naturaliste Plateau	OP	26	258	no	Davies, Luyendyk et al., 1974
		28	264	?	Hayes, Frakes et al., 1975
Ninetyeast Ridge	SR	22	214	yes	von der Borch, Sclater et al., 1974
		22	215	yes	ibid.
		22	216	yes	ibid.
		22	217	no	ibid.
		26	253	yes	Davies, Luyendyk et al., 1974
		26	254	yes	ibid.
		121	756	yes	Peirce, Weissel et al., 1989;
		121	757	yes	Weissel, Peirce, Taylor, Alt et al., 1991
		121	758	yes	ibid.
Osborn Knoll	OP	-	-	-	-
Roo Rise	OP	-	-	-	-
Wallaby Plateau	OP	-	-	-	-

§OP, oceanic plateau; SR, submarine ridge

from the feature (Figure 1), and reliable dates from two sites are available (Table 2). From these dates, the minimum time for creation of the entire ridge is 8 m.y.

The Kerguelen Plateau, in the south-central Indian Ocean, is ~2500 km long and ~500 km wide. It encompasses an area in excess of 10^6 km², making it the second most voluminous LIP yet discovered [Coffin and Eldholm, 1991]. It is suggested to be part of a hotspot track which includes, in chronological order [Davies et al., 1989], the Bunbury Basalt (Australia), Naturaliste Plateau, Rajmahal Traps (India), Kerguelen/Broken Ridge, Ninetyeast Ridge, and the northernmost Kerguelen Plateau (Figure 1).

Structural interpretations [Houtz et al., 1977; Coffin et al., 1986] divide the Kerguelen Plateau into two distinct sectors, a northern and a southern, with the boundary at ~54°S (Figure 1), which have experienced differing histories of volcanism. Volcanic basement has been recovered and dated from four ODP sites and one dredge site on the southern Kerguelen Plateau (Figure 1; Tables 1, 2). It formed between 114 and 101 Ma [Whitechurch et al., 1992], with some subsequent alkalic volcanism. Kerguelen and Heard islands on the northern Kerguelen Plateau record volcanism over the past 39 m.y. [Giret and Lameyre, 1983; Clarke et al., 1983], and volcanic basement from its

TABLE 2. Ages of Indian Ocean Plateaus and Submarine Ridges

LIP	Type[1]	DSDP/ODP Leg (s)	Basement Site(s)	Basement Age (Ma)	Reference(s)
Broken Ridge	SR	26, 121	-	-	-
		-	Dredge	88	Duncan, 1991
Chagos-Laccadive Ridge	SR	23, 115	713	49	Duncan & Hargraves, 1990
		115	715	57	ibid.
		-	NMA-1	?	-
Conrad Rise	OP	-	-	?	-
Crozet Plateau	OP	-	Est	0-8	Lameyre & Nougier, 1982
		-	Possession	0-8	Chevallier et al., 1983
Cuvier Plateau	OP	-	-	?	-
Del Caño Rise	OP	-	Marion[2]	0	McDougall, 1971b
		-	Prince Edward[2]	0	ibid.
Kerguelen Plateau	OP	-	Dredge	114	Leclaire et al., 1987
		119	738	? (>91)	Barron, Larsen et al., 1989
		120	747	101-110	Whitechurch et al., 1992
		120	749	110	ibid.
		120	750	101	ibid.
		-	Heard	0-11	Clarke et al., 1983
		-	Kerguelen	0-39	Nougier et al., 1983; Giret & Lameyre, 1983
Madagascar Ridge	SR	25	-	?	Simpson, Schlich et al., 1974
Mascarene Plateau	OP	115	706	33	Duncan & Hargraves, 1990
		115	707	64	ibid.
		-	NB-1	31	ibid.
		-	SM-1	59[3]	ibid.; Meyerhoff & Kamen-Kaye, 1981
		-	Mauritius	1-7	McDougall, 1971a
		-	Reúnion	0-2	ibid.
Naturaliste Plateau	OP	26, 28	264	? (>91)	Hayes, Frakes et al., 1975
Ninetyeast Ridge	SR	22	214	59	Duncan, 1978
		22	216	81	ibid.
		26	253	? (>44)	Davies, Luyendyk et al., 1974
		26	254	38	Duncan, 1978
		121	756	43	Duncan, 1991
		121	757	58	ibid.
		121	758	82	ibid.
Osborn Knoll	OP	-	-	?	-
Roo Rise	OP	-	-	?	-
Wallaby Plateau	OP	-	-	?	-

[1]OP, oceanic plateau; SR, submarine ridge
[2]Marion and Prince Edward islands probably do not represent the age of the bulk of Del Caño Rise - see text for discussion.
[3]the integrated age of Duncan & Hargraves [1990] is used rather than their "best radiometric age estimate," because biostratigraphic evidence in Meyerhoff & Kamen-Kaye [1981] supports the former.

conjugate, Broken Ridge, is dated 88 Ma [Duncan, 1991].

The Mascarene Plateau, east and north of Madagascar, extends for ~2000 km from the continental Seychelles Bank [Baker and Miller, 1963] in the north to Réunion in the south (Figure 1). It consists of several discrete banks, with a maximum width of ~500 km. Volcanic basement crops out on the two youngest portions of the plateau, Réunion and Mauritius; these features formed over the past 7 m.y. (Table 2). Two industry and two ODP sites recovered volcanic basement from the plateau north of the two islands; dates from these rocks indicate that the feature has formed over the past 64 m.y. (Table 2).

Ninetyeast Ridge is one of the longest linear features on the Earth's surface, extending north-south for ~5000 km in the eastern Indian Ocean (Figure 1). It is 150 to 250 km wide. North of 7°S the ridge consists of several discrete blocks; to the south it is continuous to its termination at the western end of Broken Ridge. The feature lacks any

subaerial expression, but volcanic basement has been recovered and dated from six DSDP and ODP sites along the ridge (Table 2), documenting a 44 m.y. constructional history [Duncan, 1991].

SUBSIDENCE ANALYSIS

The general subsidence history of oceanic plateaus and submarine ridges has been the focus of only one study to date [Detrick et al., 1977]. A principal conclusion was that subsidence could be attributed to cooling and thickening of the lithospheric plate on which the plateaus and ridges were constructed at rates comparable with those of normal oceanic lithosphere, assuming that emplacement occurred close to sea level. Since that study the age-depth relationship of oceanic lithosphere has been documented further and refined [Parsons and Sclater, 1977; Hayes, 1988]. Below are described the methods and data used to analyze subsidence histories of drill sites on the Chagos-

Laccadive Ridge, Kerguelen Plateau, Mascarene Plateau, and Ninetyeast Ridge.

Methods

The procedure for calculating the subsidence history of oceanic crust has been thoroughly documented [e.g., Parsons and Sclater, 1977], and the most relevant study for the Indian Ocean is that of Hayes [1988], who analyzed data from the Southeast Indian Ocean. The level of emplacement of igneous basement is calculated by the equation,

$$D_0 = D_c - C \ age^{1/2} \tag{1}$$

in which D_0 is the original depth or elevation of emplacement of the crust in meters, D_c is the present corrected depth of the crust in meters, C is an empirical constant in meters, and age is in m.y.

Global averages of C have been determined to be 350 m [Parsons and Sclater, 1977] and 300 m [Hayes, 1988]. These averages were determined from Cenozoic age lithosphere; there is no reason to suspect, however, that Mesozoic lithosphere followed different rules of thermal subsidence. I employed the value of 300 m for the Chagos-Laccadive Ridge, Kerguelen Plateau, Mascarene Plateau, and Ninetyeast Ridge in the calculations.

D_c is obtained using to the equation of Crough [1983]:

$$D_c = d_w + t_s(\rho_s - \rho_m/\rho_w - \rho_m) \tag{2}$$

in which d_w is water depth in meters, t_s is sediment thickness in meters, ρ_s is average sediment density in gcm^{-3} (1.90), ρ_m is upper mantle density in gcm^{-3} (3.22), and ρ_w is water density in gcm^{-3} (1.03).

Detrick et al. [1977] suggest that it is valid to apply age-depth relationships for oceanic lithosphere to oceanic plateaus in the absence of major post-emplacement tectonism which may have resulted in thermal rejuvenation of the lithosphere and/or flexural uplift or subsidence. Evidence for such tectonism is lacking at the majority of drill sites examined [Davies, Luyendyk et al., 1974; von der Borch, Sclater et al., 1974; Backman, Duncan et al., 1988; Peirce, Weissel et al., 1989; Schlich, Wise et al., 1989].

To confirm or deny application of the thermal subsidence model to oceanic plateaus and submarine ridges, facies of the sediment at the various sites were examined, and it was assumed that the end of shallow-water sedimentation occurred at a depth of 200 m. Using basement ages determined by radiometric dating (Table 2), depth to igneous basement at the end of shelf deposition for each site was calculated by the equation:

$$D_{bes} = D_0 + C \ (age-age_{es})^{1/2} \tag{3}$$

in which D_{bes} is depth of igneous basement at the end of shelf deposition in meters, and age_{es} is age at the end of shelf deposition in m.y. (using time scale of Berggren et

al., 1985, and Kent and Gradstein, 1985]. Then depth to seafloor at the end of subaerial and shallow-water deposition was calculated by combining the equations of Hayes (1988) and Crough (1983) as follows:

$$D_{ses} = D_0 + C \ (age-age_{es})^{1/2} - t_{es}(\rho_s - \rho_m/\rho_w - \rho_m) \tag{4}$$

in which D_{ses} is depth of seafloor at the end of shelf deposition in meters, and t_{es} is sediment thickness at the end of shelf deposition in meters. Various equation parameters, and calculated basement and seafloor depths at the end of shelf deposition are shown in Tables 3 and 4, and calculated basement depths for the various sites are plotted (triangles) on Figure 2.

Another approach to predict levels of emplacement of Indian Ocean plateau and submarine ridge sites would be to use a plate model [e.g., Parker and Oldenburg, 1973; Davis and Lister, 1974; Crough, 1975; Parsons and Sclater, 1977]. Plate model equations, however, contain one variable - temperature at the base of the lithosphere - for which little agreement exists, especially in the case of LIPs such as oceanic plateaus and submarine ridges [e.g., White and McKenzie, 1989; Griffiths and Campbell, 1990; Hill, 1991]. Furthermore, this temperature may vary both spatially, over the $\sim10^5$ to $\sim10^6$ km dimensions, and temporally, over the $\sim10^5$ to $\sim10^6$ yr emplacement phase, of Indian Ocean plateaus and submarine ridges. Calculations of emplacement elevations are extremely sensitive to temperature, and consideration of a plate model in both forward and inverse modeling is beyond the scope of this study.

Data and Results

Fundamental data for this subsidence study are radiometric basement dates, and sedimentology and biostratigraphy of overlying sedimentary rock. Sedimentary facies of rocks provide information on environment of deposition, and macro- and microfossil assemblages can document the depth at which sediment was deposited. Biostratigraphy is used to date the sediment. All of these types of data and accompanying interpretations have errors associated with them, errors which generally increase with age. This study is limited to using the most consistent radiometric age dates (Table 2) for basement; sites for which no or inconsistent dates are available are not included. Combined sedimentary facies, benthic foraminiferal zonation, and biostratigraphic interpretations are used in most cases to estimate when a drill site descended beneath shelf (~200 m) depths; exceptions are noted (Table 4). Changes in sea level and sediment compaction were not considered. At times over the past 120 m.y., eustatic sea level appears to have been as much as ~250 m higher than present [Kominz, 1984; Haq et al., 1987]. This magnitude does not affect the results of the subsidence analysis. Sediment thicknesses at the oceanic plateau and submarine ridge sites examined herein are, with two exceptions, less than 700 m. This amount of overburden does not result in compaction significant enough to alter results of the

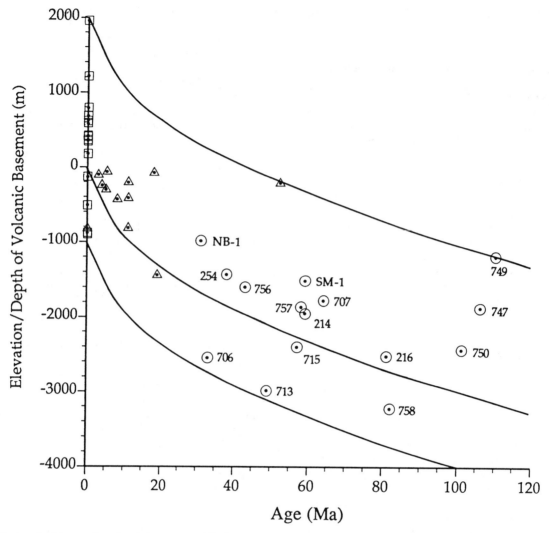

Fig. 2. Age-depth curves for volcanic basement at drill sites listed in Tables 3 and 4. Calculated basement depths (Table 3) for all sites are given for zero age (boxes) and the present (circles); where possible, calculated basement depths at the end of shelf deposition are indicated by triangles. See Table 4 for the corresponding depths to seafloor at the end of shelf deposition. Theoretical subsidence curves according to the equation of Hayes [1988] for crust emplaced 2000 m above sea level and at a depth of 1000 m (C=300) bound all points. Theoretical subsidence curve for crust emplaced at sea level indicates that basement at 11 of 15 sites (circles) was emplaced above sea level.

analysis [Sclater and Christie, 1980]. Below I address limitations of the data and results of the subsidence analysis.

Chagos-Laccadive Ridge. Basement samples from ODP Leg 115 Sites 713 and 715 on the ridge give consistent $^{40}Ar/^{39}Ar$ plateau and isochron ages [Duncan and Hargraves, 1990], and are consistent with plate motion models of the Indian Ocean [e.g., Duncan, 1990]. They are therefore included in the analysis (Tables 3, 4; Figure 2).

Sediment recovered at Site 713 indicates that the site was never shallow enough to experience a shelf depositional regime [Backman, Duncan et al., 1988]; therefore, only the depth of emplacement, 885 m, was calculated (Table 3). Site 715 appears to have been emplaced at shelf depths. The duration of a shallow-water depositional regime at this

site, however, is open to question because of a hiatus between upper Eocene shallow-water carbonate and lower Miocene deep-water ooze and chalk. Even using a maximum age for the end of shelf deposition (Table 4), the calculated seafloor depth of 1379 m at the end of shelf deposition is far too great, indicating that something is awry.

Kerguelen Plateau. $^{40}Ar/^{39}Ar$ basement dates from the southern Kerguelen Plateau (Figure 1; Table 2) show fairly wide scatter [Whitechurch et al., 1992]. Basalt from Site 747 yielded plausible dates of 101 to 110 Ma, and a value of 105.5 was used in the subsidence calculations (Tables 3, 4). Site 749 yielded the best documented date, 110 Ma. Plausible dates of 101 Ma were obtained for Site 750 basalt. The only other reported date, 114±1 Ma, was

TABLE 3. Equation parameters and elevations/depths of emplacement (D$_0$) of drill sites.

LIP	Site	d$_w$	t$_s$	D$_c$	Basement Age (Ma)	D$_0$
Chagos-Laccadive Ridge	713	-2920.3	107	-2985	49	-885
	715	-2272.8	211	-2400	57	-135
Kerguelen Plateau	747	-1695.2	297	-1874	105.5	1215
	749	-1069.5	202	-1191	110	1955
	750	-2030.5	672	-2435	101	580
Mascarene Plateau	706	-2518.0	48	-2547	33	-824
	707	-1551.9	376	-1779	64	621
	NB-1	-49.7	1556	-988	31	683
	SM-1	-50.9	2432	-1517	59	788
Ninetyeast Ridge	214	-1655	490	-1950	59	354
	216	-2247	457	-2522	81	178
	254	-1253	301	-1434	38	415
	756	-1513.1	145	-1600	43	367
	757	-1643.6	369	-1866	58	419
	758	-2923.6	499	-3224	82	-508

d$_w$ = water depth, in meters.
t$_s$ = sediment thickness, in meters.
D$_c$ = present corrected depth of crust, in meters.
D$_0$= original depth or elevation of crust emplacement, in meters.

TABLE 4. Equation parameters, calculated basement (D$_{bes}$), predicted seafloor depths (D$_{ses}$) , and duration of subaerial/shelf conditions of drill sites.

LIP	Site	age$_{es}$	D$_{bes}$	t$_{es}$	D$_{ses}$	Duration of Subaerial/Shelf Environments
Chagos-Laccadive Ridge	713[1]	-	-	-	-	0
	715	38[2]	-1443	105	-1379	19[10]
Kerguelen Plateau	747	87.5[3]	-76	1.5	-75	18[11]
	749	57.8[4]	-212	0	-212	52.2[11]
	750	90[5]	-415	52	-384	11[11]
Mascarene Plateau	706	33[6]	-824	0.5	-823	0[10]
	707	58.8[7]	-63	48	-34	5.2[11]
	NB-1[8]	-	-	-	-	31
	SM-1	48	-207	286	-35	11
Ninetyeast Ridge	214	55	-246	167	-145	4
	216	70	-817	109	-752	11
	254	30[9]	-434	92	-378	8
	756	38	-304	15	-295	5
	757	55	-101	157	-6	3
	758[1]	-	-	-1-	0	

age$_{es}$ = age at end of shelf deposition, in m.y.
t$_{es}$ = sediment thickness at end of shelf deposiiton , in meters.
[1]basaltic basement at these sites was emplaced well below shelf depths.
[2]maximum - a hiatus exists between upper Eocene shallow-water carbonate and lower Miocene deep-water ooze and chalk.
[3]minimum - a hiatus exists between 106 Ma basaltic basement and lower Santonian chalk and chert.
[4]minimum - a hiatus exists between 110 Ma basaltic basement and lower Eocene chalk and chert.
[5]minimum - a hiatus exists between Albian claystone and coal, and upper Turonian chalk and marl.
[6]maximum - 33 Ma basaltic basement is overlain by lower Oligocene shallow-water carbonate, which is in turn overlain by lower Oligocene ooze.
[7]minimum - a hiatus exists between upper Paleocene shallow-water and deep-water carbonate.
[8]this site experienced shallow water sediment deposition throughout its history.
[9]detailed age information for this well is not available.
[10]minimum.
[11]maximum.

obtained by the K-Ar method on a plagioclase from a dredged basalt [Leclaire et al., 1987], and no dates have been obtained from the Site 738 basalts. Dates for southern Kerguelen Plateau basalts appear less reliable than those for the other three features studied; more work is clearly needed on dating the cored basalts.

Petrologic analyses [Schlich, Wise et al., 1989] and subsidence calculations [Coffin, 1992] suggest that the three dated basement sites on the southern Kerguelen Plateau were constructed above sea level (Table 3; Figure 2). Sediment recovered from Sites 747 and 749 do not provide information on when the two descended below shelf depths [Schlich, Wise et al., 1989]. At the former, a hiatus exists between basement and lower Santonian chalk and chert; at the latter, between basement and lower Eocene chalk and chert. Using these dates as minimum values in the calculations, however, produces agreement with shelf depths predicted by the thermal subsidence model (Table 4; Figure 2). Site 750 contains the most complete sediment record for subsidence analysis, and again the calculated depths at the end of shelf deposition do not conflict with those interpreted from the sediment (Table 4). The southern Kerguelen Plateau, unlike its northern counterpart, shows no evidence of Tertiary or Quaternary volcanism. This observation and the subsidence analysis suggest that the southern portion has not been thermally rejuvenated. Major normal and possibly transform faults were active on the southern Kerguelen Plateau in Late Cretaceous time [Coffin et al., 1990], but do not appear to have affected the overall subsidence of the feature.

Mascarene Plateau. Two ODP and two industry boreholes on the Mascarene Plateau recovered volcanic basement which was subsequently dated (Table 2). Samples from Leg 115 Sites 706 and 707, and from well NB-1 on the plateau (Figure 1) give consistent $^{40}Ar/^{39}Ar$ plateau and isochron ages [Duncan and Hargraves, 1990], and are consistent with plate motion models of the Indian Ocean [e.g., Duncan, 1990]. The "best radiometric age estimate" of 45 Ma [Duncan and Hargraves, 1990] for volcanic basement at well SM-1, however, contradicts biostratigraphic evidence [Meyerhoff and Kamen-Kaye, 1981]; therefore the integrated age of Duncan and Hargraves [1990] is used (Tables 3, 4; Figure 2).

Petrologic studies [Backman, Duncan et al., 1988] and subsidence analysis indicate that igneous basement at Site 706 was emplaced below sea level. Lower Oligocene shallow-water carbonate overlies basement, and is in turn overlain by lower Oligocene deep-water ooze [Backman, Duncan et al., 1988]. These sedimentologic and biostratigraphic data suggest that the site remained in shallow water only briefly; it was probably emplaced there, and subsided rapidly into deeper water. Because thermal subsidence is of large magnitude immediately following volcanic emplacement, the discrepancy between the predicted seafloor depth at the end of shelf deposition (Table 4) and actual seafloor depth is probably not significant. Interbedded shallow-water limestone and

basalt at Site 707 demonstrate shallow water emplacement [Backman, Duncan et al., 1988], although subsidence calculations suggest construction high above sea level (Table 3; Figure 2). A hiatus between upper Paleocene shallow-water and deep-water carbonate suggests rapid initial subsidence. It thus appears that this site was kept elevated by dynamic forces, i.e., thermal or mechanical, after subsiding to deep water in late Paleocene time. Further evidence for this includes the calculated depth to seafloor at the end of shelf deposition, which is less than geologic evidence shows (Table 4). Calculations suggest that volcanic basement at wells NB-1 and SM-1 was constructed well above sea level (Table 3; Figure 2). The calculated depth at the end of shelf deposition for well SM-1 agrees quite well with that interpreted from sediment [Meyerhoff and Kamen-Kaye, 1981; Table 4].

Ninetyeast Ridge. Ninetyeast Ridge basement samples from three DSDP Leg 22 and 26 sites (214, 216, 254) provide K/Ar dates [Duncan, 1978], and samples from three ODP Leg 121 sites (756, 757, 758) on the ridge give consistent $^{40}Ar/^{39}Ar$ plateau and isochron ages [Duncan, 1991]. These dates appear reliable and are consistent with plate motion models of the Indian Ocean [e.g., Duncan, 1990], and are thus included in the analysis (Tables 3, 4; Figure 2).

All basement sites on the Ninetyeast Ridge with the exception of Site 758 backtrack to subaerial levels (Table 3; Figure 2). In all cases subaerial emplacement is consistent with sedimentologic, biostratigraphic, and petrologic interpretations of recovered core material [von der Borch, Sclater et al., 1974; Davies, Luyendyk et al., 1974; Peirce, Weissel et al., 1989]. Calculated depths at the end of shelf deposition for Sites 214, 756, and 757 correlate well with those interpreted from sediment [von der Borch, Sclater et al., 1974; Davies, Luyendyk et al., 1974; Peirce, Weissel et al., 1989]. Detailed biostratigraphic ages for Site 254 sediment are not available [Davies, Luyendyk et al., 1974], so it was impossible to check the calculated versus an observed depth at the end of shelf deposition for this site. Volcanic basement at Site 758, both from backtracking and from geologic evidence [Peirce, Weissel et al., 1989], was emplaced at depths greater than 200 m. Only at Site 216 does the calculated depth to basement at the end of shelf deposition (Table 4; Figure 2) differ markedly from that observed [von der Borch, Sclater et al., 1974]; this site was probably kept elevated by dynamic forces, i.e., thermal or mechanical, for a period after emplacement in the Campanian.

CONCLUDING DISCUSSION

The preceding subsidence analysis confirms that cooling of the lithosphere has been the main factor causing subsidence of the Chagos-Laccadive Ridge, Kerguelen Plateau, Mascarene Plateau, and Ninetyeast Ridge. It also reveals that these features have experienced significant portions of their development subaerially and in shallow water (Figure 3). Progressive development of dominantly

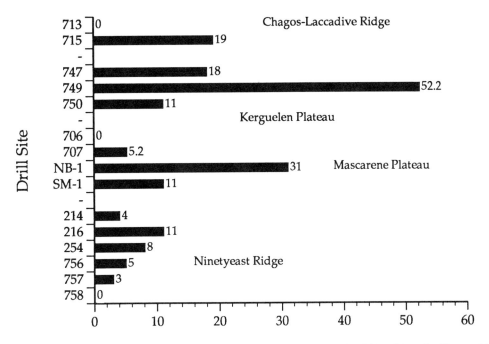

Fig. 3. Duration of subaerial and shelf environments at drill sites on the Chagos-Laccadive Ridge, Kerguelen Plateau, Mascarene Plateau, and Ninetyeast Ridge (Table 4).

terrestrial, then terrigenous, and eventually biogenic shallow-water and pelagic facies sediment in response primarily to thermal subsidence has produced a sedimentary record intermediate between continental margins and open marine.

Long hiatuses at many sites may be explained by their remaining above sea level for as much as ~50 m.y. following emplacement. Two other lines of evidence support this by suggesting that significant erosion of basement is possible. At Site 749 on the Kerguelen Plateau (Figure 1), compressional wave velocities are higher than those determined from samples from other plateau sites [Schlich, Wise et al., 1989], suggesting that deeper levels of the igneous crust were penetrated. At the same site, intermediate zeolite facies [Sevigny et al., 1992] were encountered, indicating higher temperatures and perhaps deeper crustal levels than zeolite facies at Sites 747 and 750.

Present-day, but smaller analogs to the oceanic plateaus and submarine ridges analyzed all suggest that subaerial emplacement and early evolution of such LIPs may be the rule rather than the exception. Table 5 summarizes maximum elevations of some of these provinces, all of which have been active in Quaternary time. These maximum elevations are probably comparable to the original setting of Site 749 on the Kerguelen Plateau (Figure 1), which is now situated on the shallowest part - Banzare Bank - of its southern sector, as well as to sites on other submarine LIPs. Samples obtained from the Naturaliste Plateau [Davies, Luyendyk et al., 1974; Hayes, Frakes et al., 1975; Coleman et al., 1982], Rio Grande Rise [Supko, Perch-Nielsen et al., 1977], and the Iceland-Faeroe

Ridge [Talwani, Udinstev et al., 1976] also show strong evidence of subaerial basalt extrusion. Furthermore, at least part of the basaltic seaward-dipping reflector sequences drilled on the Vøring Plateau [Talwani, Udinstev et al., 1976; Eldholm, Thiede, Taylor et al., 1987, 1989] and Rockall Bank [Roberts, Schnitker et al., 1984] was erupted and eroded subaerially.

Following the constructional phase of oceanic plateau and submarine ridge development, erosional processes become the dominant factor in altering plateau morphology if thermal rejuvenation does not keep the feature high-standing. If the plateau or ridge is submarine, erosion of basalt and associated extrusive and intrusive rocks is negligible. However, if the plateau or ridge is subaerial, erosion can be widespread, significantly altering the topography of the feature (e.g., Table 5). Rate of denudation of any rock type is a function of topography and climate. Wood recovered in basal sediment [Francis and Coffin, 1992] and clay mineralogy [Holmes, 1992] at Site 750 on the southern Kerguelen Plateau (Figure 1) suggest a temperate climate. Average denudation rates in temperate climates vary from 10 to 200 m/m.y., depending on the topographic relief [Saunders and Young, 1983]. Evidence from drilling suggests subaerial exposure for up to 50 m.y., allowing 500-10000 m of erosion. Drilling results do not support the latter figure, but several hundred to over a thousand meters of basalt could have been eroded from the Kerguelen Plateau based on compressional wave velocities [Schlich, Wise et al., 1989] and zeolite facies [Sevigny et al., 1992]. This erosional phase of development is marked by the deposition of terrestrial and terrigenous sediment. After the bulk of the feature subsides

TABLE 5. Present maximum elevations of Quaternary Indian Ocean hotspots, Iceland, Hawaii, and their tracks.

Hotspot Track (old→young)	Hotspot Source	Maximum Elevation (m)
Naturaliste Plateau-Kerguelen Plateau/Broken Ridge-Ninetyeast Ridge-northernmost Kerguelen Plateau	Heard Island Kerguelen Island	2745 1849
Chagos-Laccadive Ridge-Mascarene Plateau	Réunion Mauritius	3069 826
Del Caño Rise	Marion[1] Prince Edward[1]	1230 672
Crozet Plateau	Est Possession	1090 934
Iceland/Faeroe-Greenland Ridge	Iceland	2119
Hawaiian-Emperor Seamounts	Hawaii	4169

[1]Marion and Prince Edward islands probably do not represent the age of the bulk of Del Caño Rise - see text for discussion.

below sea level, biogenic shallow-water and pelagic sediment follow.

Why the oceanic plateaus and submarine ridges examined herein, as well as oceanic LIPs in general, are commonly emplaced above sea level and spend a significant portion of their history subaerially exposed or at shallow water depths is probably a combination of two factors. First, the mantle thermal anomaly responsible for LIP emplacement is greater than that for seafloor spreading centers [White and McKenzie, 1989], which results in more decompressional melting, greater thermal expansion of the lithosphere, and a lower surface area-to-volume ratio, meaning less efficient decay of the original thermal anomaly. More melt allows construction of many oceanic LIPs well above sea level, and hotter lithosphere enhances this. Lower surface area-to-volume ratios of LIPs and associated heated lithosphere results in slower decay of the thermal anomaly, i.e., subsidence at a lesser rate. Second, erosion of subaerial oceanic LIPs results in isostatic rebound, which serves to further prolong their anomalous elevation. Emplacement and subsidence analyses of LIPs globally, using both forward and inverse modeling, will help to further constrain and quantify these effects.

Acknowledgments. I thank the entire complements of the many Deep Sea Drilling Project and Ocean Drilling Program legs for obtaining the samples used in this study. A very special thanks to Kerry Kelts for stimulating the ideas presented here, and for providing an idyllic work environment in Kilchberg. Dave Pasta of Texaco generously provided information on industry wells on the Mascarene Plateau, and I thank Lucas Hottinger for checking biostragraphic age determinations from well SM-1 on the Mascarene Plateau. Bob Detrick and two anonymous reviewers provided constructive reviews, and the manuscript was carefully edited by Bob Duncan and Jeff Weissel. I am grateful to John Peirce and John Sclater for reviewing a preliminary version of this manuscript. Lisa Gahagan and Wayne Lloyd contributed to the preparation of Figure 1. This work was supported in part by the sponsors of PLATES, the global plate reconstruction project based at the University of Texas Institute for Geophysics (UTIG). UTIG contribution no. 905.

REFERENCES

Backman, J., Duncan, R. A., et al., *Proc. ODP, Init. Repts.,* 115, 1085 pp., 1988.

Baker, B. H., and Miller, J. A., Geology and geochronology of the Seychelles Islands and structures of the floor of the Arabian Sea, *Nature,* 199, 346-348, 1963.

Barron, J., Larsen, B. et al., *Proc. ODP, Init. Repts.,* 119, 942 pp., 1989.

Barron, J., Larsen, B. et al., *Proc. ODP, Sci. Results,* 119, 1003 pp., 1991.

Berggren, W. A., Kent, D. V., Flynn, J. J., and Van Couvering, J. A., Cenozoic geochronology, *Geol. Soc. Am. Bull.,* 96, 1407-1418, 1985.

Chevallier, L., Nougier, J., and Cantagrel, J. M., Volcanology of Possession Island, Crozet Archipelago, in *Antarctic Earth Science,* edited by R. L. Oliver, J. B. Jago, and P. R. James, pp. 652-658, Cambridge Univ. Press, Cambridge, 1983.

Clarke, I., McDougall, I., and Whitford, D. J., Volcanic evolution of Heard and McDonald islands, southern Indian Ocean, in *Antarctic Earth Science,* edited by R. L. Oliver, J. B. Jago, and P. R. James, pp. 631-635, Cambridge Univ. Press, Cambridge, 1983.

Coffin, M. F., Subsidence of the Kerguelen Plateau: the Atlantis concept, *Proc. ODP, Sci. Results,* 120, 945-949, 1992.

Coffin, M. F., Davies, H. L., and Haxby, W. F., Structure of the Kerguelen Plateau province from SEASAT altimetry and seismic reflection data, *Nature,* 324, 134-136, 1986.

Coffin, M. F., Munschy, M., Colwell, J. B., Schlich, R., Davies, H. L., and Li, Z. G., Seismic stratigraphy of the Raggatt Basin, southern Kerguelen Plateau: tectonic and paleoceanographic implications, *Geol. Soc. Am. Bull.,* 102, 563-579, 1990.

Coffin, M. F., and Eldholm, O., eds., Large Igneous Provinces: JOI/USSAC Workshop Report, *Univ. Texas at Austin Inst. for Geophys. Tech. Rept. 114,* 79 pp., 1991.

Coleman, P. J., Michael, P. J., and Mutter, J. C., The origin of the Naturaliste Plateau, SE Indian Ocean: implications from dredged basalts, *J. Geol. Soc. Australia, 29,* 457-468, 1982.

Crough, S. T., Thermal model of oceanic lithosphere, *Nature, 256,* 388-390, 1975.

Crough, S. T., The correction for sediment loading on the seafloor, *J. Geophys. Res., 88,* 6449-6454, 1983.

Davies, H. L., Sun, S.- S., Frey, F. A., Gautier, I., McCulloch, M. T., Price, R. C., Bassias, Y., Klootwijk, C. T., and Leclaire, L., Basalt basement from the Kerguelen Plateau and the trail of a Dupal plume, *Contrib. Mineral. Petrol., 103,* 457-469, 1989.

Davies, T. A., Luyendyk, B. P. et al., *Init. Repts. DSDP, 26,* 1129 pp., 1974.

Davis, E. E. and Lister, C. R. B., Fundamentals of ridge crest topography, *Earth Planet. Sci. Lett., 21,* 405-413, 1974.

Detrick, R. S., Sclater, J. G., and Thiede, J., The subsidence of aseismic ridges, *Earth Planet. Sci. Lett., 34,* 185-196, 1977.

Driscoll, N. W., Karner, G. D., and Weissel, J. K., Stratigraphic response of carbonate platforms and terrigenous margins to relative sea-level changes: are they really that different? *Proc. ODP, Sci. Results, 121,* 743-761, 1991.

Duncan, R. A., Geochronology of basalts from the Ninetyeast Ridge and continental dispersion in the eastern Indian Ocean, *J. Volcanol. Geotherm. Res., 4,* 283-305, 1978.

Duncan, R. A., The volcanic record of the Réunion hotspot, *Proc. ODP, Sci. Results, 115,* 3-10, 1990.

Duncan, R. A., Age distribution of volcanism along aseismic ridges in the eastern Indian Ocean, *Proc. ODP, Sci. Results, 121,* 507-517, 1991.

Duncan, R. A., and Hargraves, R. B., $^{40}Ar/^{39}Ar$ geochronology of basement rocks from the Mascarene Plateau, the Chagos Bank, and the Maldives Ridge, *Proc. ODP, Sci. Results, 115,* 43-51, 1990.

Duncan, R. A., Backman, J., Peterson, L. C. et al., *Proc. ODP, Sci. Results, 115,* 887 pp., 1990.

Eldholm, O., Thiede, J. Taylor, E. et al., *Proc. ODP, Init. Repts., 104,* 783 pp., 1987.

Eldholm, O., Thiede, J. Taylor, E. et al., *Proc. ODP, Sci. Results, 104,* 1141 pp., 1989.

Fisher, R. L., Bunce, E. T. et al., *Init. Repts. DSDP, 24,* 1183 pp., 1974.

Francis, J. E., and Coffin, M. F., Cretaceous fossil wood from the Raggatt Basin, southern Kerguelen Plateau (Site 750), *Proc. ODP, Sci. Results, 120,* 273-280, 1992.

Giret, A., and Lameyre, J., A study of Kerguelen plutonism: petrology, geochronology, and geological implications, in *Antarctic Earth Science*, edited by R. L. Oliver, P. R. James, and J. B. Jago, pp. 646-651, Cambridge Univ. Press, Cambridge, 1983.

Griffiths, R. W., and Campbell, I. H., Stirring and structure in mantle starting plumes, *Earth Planet. Sci. Lett., 99,* 66-78, 1990.

Haq., B. U., Hardenbol, J., and Vail, P. R., Chronology of fluctuating sea levels since the Triassic, *Science, 235,* 1156-1167, 1987.

Hayes, D. E., Age-depth relationships and depth anomalies in the Southeast Indian Ocean and South Atlantic Ocean, *J. Geophys. Res., 93,* 2937-2954, 1988.

Hayes, D. E., Frakes, L. A. et al., *Init. Repts. DSDP, 28,* 1017 pp., 1975.

Hill, R. I., Starting plumes and continental break-up, *Earth Planet. Sci. Lett., 104,* 398-416, 1991.

Holmes, M. A., Cretaceous subtropical weathering followed by cooling at 60°S latitude: the mineral composition of southern Kerguelen Plateau sediment, Leg 120, *Proc. ODP, Sci. Results, 120,* 99-111, 1992.

Houtz, R. E., Hayes, D. E., and Markl, R. G., Kerguelen Plateau bathymetry, sediment distribution, and crustal structure, *Mar. Geol., 25,* 95-130, 1977.

Kent, D. V., and Gradstein, F. M., A Cretaceous and Jurassic geochronology, *Bull. Geol. Soc. Am., 96,* 1419-1427, 1985.

Kominz, M. A., Oceanic ridge volumes and sea-level change - an error analysis, in *Interregional Unconformities and Hydrocarbon Accumulation,* edited by J. S. Schlee, pp. 109-127, Am. Assoc. Petrol. Geol. Mem. 36, Tulsa, OK, 1984.

Lameyre, J., and Nougier, J., Geology of Ile de l'Est, Crozet Archipelago (TAAF), in *Antarctic Geoscience,* edited by C. Craddock, pp. 767-770, Univ. Wisconsin Press, Madison, 1982.

Leclaire, L., Bassias, Y. Denis-Clocchiatti, M. Davies, H. Gautier, I. Gensous, B. Giannesini, P.-J. Patriat, P. Ségoufin, J. Tesson, M., and Wannesson, J., Lower Cretaceous basalt and sediments from the Kerguelen Plateau, *Geomar. Lett., 7,* 169-176, 1987.

McDougall, I., The geochronology and evolution of the young oceanic island of Réunion, Indian Ocean, *Geochim. Cosmochim. Acta, 35,* 261-270, 1971a.

McDougall, I., Geochronology, in *Marion and Prince Edward Islands, Report on the South African Biological and Geological Expedition 1965/66,* edited by E. M. van Zinderen Bakker, J. M. Winterbottom, and R. A. Dyer, pp. 72-77, A.A. Balkema, Cape Town, 1971b.

Meyerhoff, A. A., and Kamen-Kaye, M., Petroleum prospects of Saya de Malha and Nazareth Banks, Indian Ocean, *Am. Assoc. Petrol. Geol. Bull., 65,* 1344-1347, 1981.

Morgan, W. J., Hotspot tracks and the opening of the Atlantic and Indian oceans, in *The Sea, Volume 7, The Oceanic Lithosphere,* edited by C. Emiliani, pp. 443-487, Wiley, New York, 1981.

Nougier, J., Pawlowski, D., and Cantagrel, J. M., Chrono-spatial evolution of the volcanic activity in southeastern Kerguelen (T.A.A.F.), in *Antarctic Earth Science,* edited by R. L. Oliver, J. B. Jago, and P. R. James, pp. 640-645, Cambridge Univ. Press, Cambridge, 1983.

Parker, R. L., and Oldenburg, D. W., Thermal model of ocean ridges, *Nature, 242,* 137-139, 1973.

Parsons, B., and Sclater, J. G., An analysis of the variation of ocean floor bathymetry and heat flow with age, *J. Geophys. Res., 82,* 803-827, 1977.

Peirce, J. W., Weissel, J. K. et al., *Proc. ODP, Init. Repts., 121,* 1000 pp., 1989.

Roberts, D. G., Schnitker, D. et al., *Init. Repts. DSDP 81,* 923 pp., 1984.

Saunders, I., and Young, A., Rates of surface processes on slopes, slope retreat, and denudation, *Earth Surface Processes Landforms, 8,* 473-501, 1983.

Schlich, R., Wise, S. W., Jr. et al., *Proc. ODP, Init. Repts., 120,* 648 pp., 1989.

Sclater, J. G., and Christie, P. A. F., Continental stretching: an explanation of the post-mid-Cretaceous subsidence of the central North Sea basin, *J. Geophys. Res., 85,* 3711-3739, 1980.

Sevigny, J. H., Whitechurch, H. Storey, M. and Salters, V. J. M., Zeolite-facies metamorphism of central Kerguelen Plateau basalts, *Proc. ODP, Sci. Results, 120,* 63-69, 1992.

Simpson, E. S. W., Schlich, R. et al., *Init. Repts. DSDP, 25,* 884 pp., 1974.

Supko, P. R., Perch-Nielsen, K. et al., *Init. Repts. DSDP, 39,* 1139 pp., 1977.

Talwani, M., Udinstev, G. et al., *Init. Repts. DSDP, 38,* 1256 pp., 1976.

von der Borch, C., Sclater, J. G. et al., *Init. Repts. DSDP, 22,* 890 pp., 1974.

Weissel, J., Peirce, J. Taylor, E. Alt, J. et al., *Proc. ODP, Sci. Results, 121,* 990 pp., 1991.

White, R., and McKenzie, D., Magmatism at rift zones: the generation of volcanic continental margins and flood basalts, *J. Geophys. Res., 94,* 7685-7729, 1989.

Whitechurch, H., Montigny, R. Sevigny, J. Storey, M., and Salters, V., K-Ar and $^{40}Ar^{-39}Ar$ ages of central Kerguelen Plateau basalts, *Proc. ODP, Sci. Results, 120,* 71-77, 1992.

Whitmarsh, R. B., Weser, O. E. Ross, D. A. et al., *Init Repts. DSDP 23,* 1180 pp., 1974.

Wise, S. W., Jr., Schlich, R. et al., *Proc. ODP, Sci. Results, 120,* 1155 pp., 1992.

Extensional and Compressional Deformation of the Lithosphere in the Light of ODP Drilling in the Indian Ocean

JEFFREY K. WEISSEL , VICKI A. CHILDERS[1], AND GARRY D. KARNER

Lamont-Doherty Geological Observatory of Columbia University, Palisades, NY, 10964, USA
[1] Also at: Department of Geological Sciences, Columbia University

Simple kinematic and isostatic models of lithospheric deformation (extension and compression) are often employed to gain insight into the mechanical behavior and thermal properties of the lithosphere. Predictions from such modeling efforts (topography, gravity, crustal structure, and time-line stratigraphy) are matched against observed seismic reflection, refraction, gravity, and topography data. The timing and history of deformational events, however, can often be extracted from sedimentary successions deposited over deformed crust and lithosphere. Because the mechanical properties of oceanic lithosphere depend to a large extent on the lithosphere's temperature structure and therefore on its age, information relating to deformation timing and depositional environment obtained through drilling of sediment sequences provides important constraints on the model predictions. In this paper, we highlight the contribution of ODP drilling in the Indian Ocean to the understanding of the processes of lithospheric extension and compression. In particular, we focus attention on: 1) Broken Ridge, a rift flank flexurally uplifted in response to middle Eocene rifting (drilled during ODP Leg 121), 2) Owen Ridge, part of a fracture zone uplifted under transpression in the early Miocene (drilled during Leg 117), and 3) the Central Indian Ocean basin, a broad zone of lithospheric shortening which began in the late Miocene (drilled during Leg 116).

We summarize modeling studies of Broken Ridge, Owen Ridge, and the Central Indian Ocean basin to show that the topographic and gravity expressions of lithospheric extension and compression are controlled in a fundamental way by lithospheric "strength", as measured by flexural rigidity or equivalently by T_e, the thickness of an elastic plate which serves as a mechanical analogue for the lithosphere. For example, the uplift of >2 km which occurred at Broken Ridge during the rifting event can be explained as flexural isostatic rebound that occurs because the lithosphere is mechanically unloaded during extension. We show, however, that the flexural strength of the lithosphere makes it difficult to distinguish between extensional and compressional deformation in the absence of other definitive information (such as seismicity) when that deformation is accomplished by finite slip on one or a small number of discrete faults. T_e for the lithosphere of the three deformational features has been estimated by matching predicted against observed topography and gravity. We then compare the resulting T_e estimates against the values expected on the basis of lithospheric age at the time the deformational events occurred. That critical information on the age of the deformation relative to the age of the oceanic lithosphere is best obtained by drilling and dating the sedimentary sequences preserved over the deformed lithosphere.

INTRODUCTION

The Indian Ocean region contains many geological features of global interest such as broad flood basalt provinces (Karoo, Deccan, and Rajmahal) and submarine plateaus (Mascarene, Kerguelen-Heard, Broken Ridge) with associated hotspot trails (Chagos-Laccadive Ridge; Ninetyeast Ridge); two of the world's largest deep-sea fans (the Indus and Bengal); passive continental margins with rifting ages ranging from Jurassic to Neogene; and zones of active plate collision and convergence along the Himalayan mountain chain and the Indonesian arc-trench system.

The seafloor of the Indian Ocean is particularly noteworthy for the apparently simple examples of lithospheric extension and compression which occur there, and form the focus of this report. Simple kinematic (or dynamic) and isostatic models for extension and compression of the lithosphere are often employed to gain insight into the mechanical behavior and thermal properties of the lithosphere. Seismic reflection and

refraction, gravity, and topography data provide essential constraints in such modeling efforts. In the oceans, however, drilling of sedimentary successions deposited over deformed seafloor allows us to extract information on the timing and history of the deformational process that is unobtainable by other means. Chronostratigraphy in fact, provides a "tape recording" of the deformational process analogous to the record of seafloor spreading contained in magnetic lineation patterns on mid-ocean ridge flanks.

The purpose of this paper is to highlight the contribution of ODP drilling in the Indian Ocean to our understanding of the mechanical properties and behavior of the oceanic lithosphere obtained through studies of three examples of lithospheric deformation. We will discuss: 1) the Central Indian Ocean basin (Figure 1), where a broad region of oceanic lithosphere has been shortened under N-S compression since the late Miocene; 2) Owen Ridge, part of a fracture zone in the northwest Indian Ocean (Figure 1), that was uplifted under transpression starting in the early Miocene; 3) Broken Ridge (Figure 1), formerly contiguous with the northern part of the Kerguelen-Heard Plateau, which was uplifted as a result of rifting between the two plateaus in the middle Eocene. These three features were drilled during Ocean Drilling Program (ODP) Legs 116,

Synthesis of Results from Scientific
Drilling in the Indian Ocean
Geophysical Monograph 70

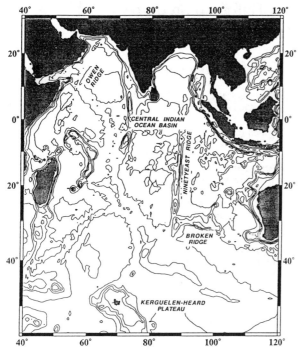

Fig. 1. Location map for the Indian Ocean showing generalized bathymetry at a contour interval of 1000 m, and features of interest for this study.

117, and 121 respectively. We acknowledge at the outset the contributions to the present work made by the shipboard scientists on these legs. This synthesis distills the analyses and interpretations made by many individuals. We refer the reader to the Leg 116, 117, and 121 *Initial Reports* and *Scientific Results* volumes for more detailed discussion of the drilling and logging results [Cochran, Stow et al., 1989, 1990; Prell, Niitsuma et al., 1989, 1991; Peirce, Weissel et al., 1989; Weissel, Peirce, Taylor, Alt et al., 1991].

BROKEN RIDGE

Background

The Kerguelen-Heard Plateau, Broken Ridge and Ninetyeast Ridge (Figure 1) are features of thickened oceanic crust originally derived from magmatism associated with the Kerguelen-Rajmahal hotspot [see reviews by Duncan and Storey, this volume; Weis et al., this volume]. Broken Ridge and the Kerguelen-Heard Plateau are conjugate rifted fragments of a large oceanic plateau which likely formed from hotspot volcanism in early or mid-Cretaceous time [e.g., Morgan, 1981]. The basement rocks of both features are basaltic, based on the recovery to date from limited dredging [Leclaire et al., 1987; Duncan, 1991], and ODP Leg 119 and 120 drilling on the Kerguelen-Heard Plateau [Barron, Larsen, et al., 1989; Schlich, Wise, et al., 1989]. The present separation of Broken Ridge and the northern part of the Kerguelen-Heard Plateau (Figure 1) results from an episode of lithospheric extension (i.e. rifting), followed by seafloor spreading which began at about anomaly 18 time [~42 Ma,

Berggren et al., 1985; Mutter and Cande, 1983] and has continued to the present day. Since middle Eocene time, Broken Ridge has moved north by about 20° of latitude as part of the Indo-Australian plate, while the Kerguelen-Heard Plateau, as part of the Antarctic Plate, has remained almost stationary.

Broken Ridge (Figure 2) is noteworthy because the effects of rifting are particularly evident in its morphology and seismic stratigraphy. A marked angular unconformity, which is developed widely over the crest of Broken Ridge, is observed in seismic reflection profiles including the one crossing the ODP Leg 121 drill site locations (Figure 3). This unconformity most likely reflects rift-related uplift and resulting exposure of the crest of Broken Ridge to wave-base and possibly subaerial erosion. Four seismic tectono-stratigraphic units defined from seismic reflection data over Broken Ridge are discussed in detail elsewhere [Driscoll et al., 1989; 1991a,b]. The two older units which are clearly truncated by the erosional unconformity (Figure 3), comprise a north-dipping sedimentary sequence in which the upper, more highly-stratified unit thins to the north and downlaps on to the older, underlying unit. A thin, subhorizontal seismic unit capping the truncation surface (Figure 3) is most likely a post-rift sequence marking the resubmergence of Broken Ridge after the middle Eocene rifting event. The fourth seismic stratigraphic unit which onlaps the gentle northern slope of Broken Ridge (Figure 3), is possibly the downslope-transported and reworked products of stratigraphically-lower parts of the section that were uplifted during rifting and eroded. Because this fourth seismic stratigraphic unit has never been drilled, its age and origin can only be inferred indirectly from the seismic stratigraphy.

Earlier drilling at Deep Sea Drilling Project (DSDP) Site 255 (Figure 2) revealed that Broken Ridge has remained a fairly shallow-water carbonate platform throughout most of its history (Davies, Luyendyk et al., 1974). Thus, the stratigraphic section at Broken Ridge was expected to preserve a record of the vertical motions of Broken Ridge as it responded to the rifting process. ODP Sites 752-755 (Figures 2 and 3) were planned to sample the sediments preserved above and below the prominent truncation surface in a series of stratigraphically overlapping, single-bit holes. The overall aim for drilling at Broken Ridge was to extract information from the preserved sedimentary record about how the lithosphere responds to extension. In particular, ODP drilling sought answers to the following three questions:

(1) What is the role of the sublithospheric mantle in initiating lithospheric extension?

(2) What is the magnitude of vertical motion of rift flanks during (and after) extension?

(3) What is the implication of such vertical motions for the mechanical strength of extended lithosphere?

Leg 121 drilling on Broken Ridge

Drilling results (Figure 4) which bear on these three questions can be summarized as follows:

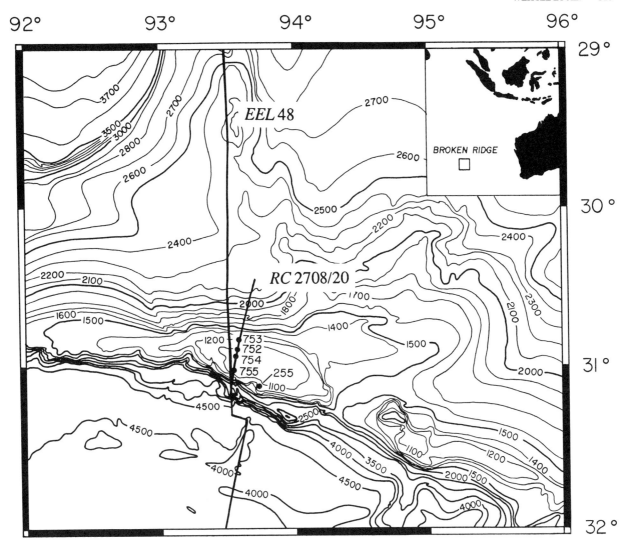

Fig. 2. Bathymetry map of the western part of Broken Ridge [Driscoll et al, 1989]. Contour interval is 100 m except along the southern escarpment of Broken Ridge where some contours are omitted for clarity. ODP Sites 752-755 drilled during Leg 121 and DSDP Site 255 are located by the small dots.

(1) The youngest sediments below the unconformity are chalks of middle Eocene age (P11/CP13c) at Site 753 (Figure 4). Thus, rifting between Broken Ridge and the northern part of the Kerguelen-Heard Plateau which led to the uplift and erosion of Broken Ridge and the ultimate separation of the two plateaus, began at about 45 Ma.

(2) The oldest, biostratigraphically-dated sediments above the erosional surface at Sites 752 and 754 (Figure 4) are upper Eocene oozes (P15 and CP15). About half the microfossil assemblage, however, consists of older reworked forms, some of which are from a shallow-water, high-energy environment. Clastic debris (shell and bryozoan fragments, pebbles of limestone and chert mixed with sand from the uplifted sediments), much of it iron-stained, was shed from the exposed crest of Broken Ridge, reworked into gravel and sand layers, and deposited as Broken Ridge sank below wave-base in the late Eocene.

(3) Water paleodepths estimated from benthic foraminiferal assemblages indicate that Broken Ridge subsided from upper to lower bathyal depths between the late Cretaceous and the middle Eocene before rifting began (Figure 5). After rifting, Broken Ridge subsided rapidly from upper bathyal depths in the late Eocene, reaching its present depth of slightly more than 1000 m (Figure 5).

These basic drilling results provide answers to the three questions posed above. First, because depositional depths were increasing rather than decreasing at Broken Ridge prior to rifting, we conclude that lithospheric extension was initiated by far-field horizontal stress, rather than by a mantle convection process which should lead to thermal uplift and shallowing of the lithosphere prior to rifting. Second, the water paleodepth estimates (Figure 5), together with the geometry of the pre-rift stratigraphic section (Figure 3), indicate that Broken Ridge was uplifted by

N

S

LINE 20

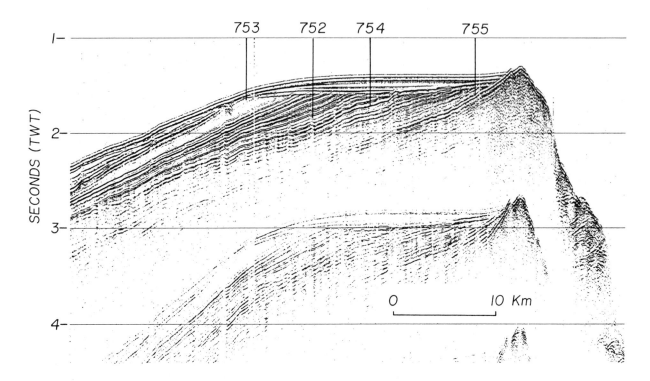

Fig. 3. Single-channel seismic reflection profile (*Conrad* cruise 2708, Line 20, located in Figure 2) passing through the ODP Leg 121 drill sites. Approximate penetration depths are indicated for the four sites.

>2 km during the rifting event. Third, the rifting event affecting Broken Ridge and the northern part of the Kerguelen-Heard Plateau was of short duration. It apparently began after deposition of middle Eocene chalks sampled at Site 753 (Figure 4), and was completed by magnetic anomaly 18 time [~42 Ma, Berggren et al., 1985] as this magnetic lineation marks the start of seafloor spreading between the two plateau fragments [e.g., Mutter and Cande, 1983].

Both the brevity of the rifting event (about 5 m.y.) and the low present-day heatflow [approximately 45 mW/m^2; Peirce, Weissel et al., 1989; Anderson et al., 1977] are consistent with a mechanical, rather than thermal explanation for the uplift of Broken Ridge. Rift flank uplift, like that found at Broken Ridge, can be attributed to flexural isostatic rebound following mechanical unloading of the lithosphere during extension, as proposed by Weissel and Karner [1989]. If this explanation is correct, the lithosphere must retain finite mechanical strength (or flexural rigidity) during extension. This implication is controversial; many workers contend that extended lithosphere has little or no mechanical strength [e.g., Barton and Wood, 1984; Fowler and McKenzie, 1989; Watts, 1988], a condition attributed to mechanical causes (pervasive faulting in the upper part of the lithosphere),

and to thermal causes (elevating the geotherm by extending and thinning the lithosphere).

Models for lithospheric extension: Constraints from Broken Ridge

Weissel and Karner [1989] proposed that rift flank uplift results from mechanical unloading of the lithosphere during extension and consequent flexural isostatic rebound. This mechanism is viewed as an alternative to explanations for rift flank uplift involving thermal or dynamic processes, and magmatic thickening of the crust. The flexural uplift hypothesis provides a better explanation for the topography and free air gravity anomaly across Broken Ridge than does an explanation involving crustal thickness variations, as shown below.

The hypothesis incorporates several simplifying assumptions about how lithospheric extension actually occurs. First, it is assumed that the kinematics of extension can be "decoupled" from the resulting isostatic problem. This way, the resulting topography is simply the sum of the topography resulting from the kinematics of extension and the topography produced by isostatic restoring stresses. Second, lithospheric extension is assumed to occur "instantaneously". Third, it is assumed that the lithosphere retains finite mechanical strength or flexural

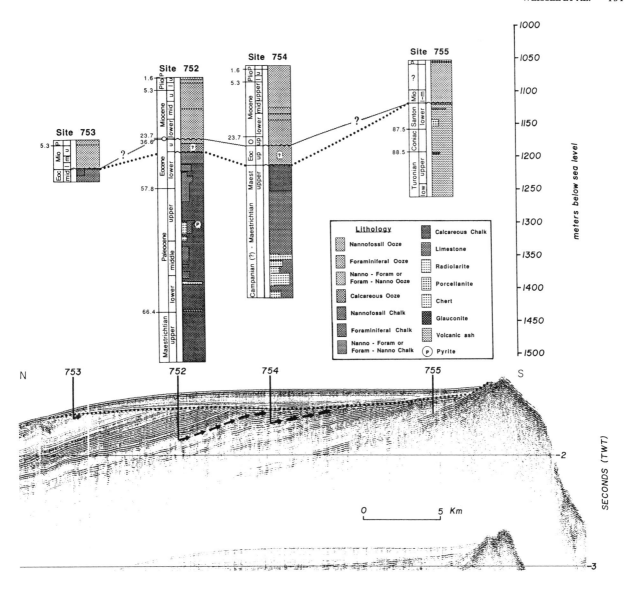

Fig. 4. Summary lithostratigraphic columns for Sites 752-755 on Broken Ridge [after Driscoll et al., 1991a,b]. The dotted line represents the middle Eocene erosional unconformity, and the wavy line denotes the Oligocene hiatus found in the post-rifting pelagic cap [Peirce, Weissel et al., 1989]. The arrows on the seismic profile show the deepest horizon penetrated at each of the drill sites.

rigidity during extension and responds to isostatic restoring stresses by flexure of a continuous elastic plate overlying a fluid substrate, despite the fact that the upper part of the lithosphere might be pervasively faulted during extension.

A simple kinematic model for lithospheric extension involves finite slip on an initially planar, dipping normal fault cutting through the entire lithosphere (Figure 6). The kinematics of extension are parameterized in terms of fault dip γ and amount of extension (or fault heave) X_O. We now determine the isostatic consequences of this kinematic model, and compare the resulting model topography and free-air gravity effect to those observed over Broken Ridge. Finite slip on a normal fault transforms an originally undeformed lithosphere in isostatic equilibrium to a configuration which is not in isostatic equilibrium (Figure

6). Thus, isostatic restoring stresses will immediately act in order to regain isostatic equilibrium. The magnitude and lateral distribution of these restoring stresses are found by requiring the mass/unit area in columns above a depth of isostatic compensation to be the same before and after extension. As usual in isostatic calculations, we define the depth of compensation as the level below which there are no lateral density variations. Inspection of Figure 6 shows that this surface lies at a depth of $a + X_o \tan \gamma$, where a, the thermal thickness of the lithosphere, is the depth where temperature reaches T_m, the temperature of the underlying asthenosphere [Weissel and Karner, 1989].

In Figure 7 we show the distribution of isostatic restoring stresses resulting from finite slip on a planar normal fault, as depicted in the kinematic model (Figure 6). The vertical displacements of the lithosphere which are required to

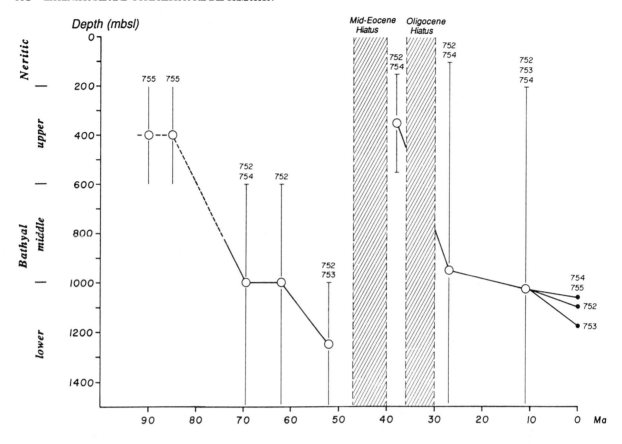

Fig. 5. Paleodepth estimates for Broken Ridge based on benthic foraminiferal assemblages recovered by ODP drilling [Peirce, Weissel, et al., 1989]. The open circles denote the paleodepth estimate from the most common species in the sample, and the vertical lines show the paleodepth range using all benthic foraminifers present in the sample. The general subsidence curve for Broken Ridge as taken as the line joining the open circles.

balance these isostatic restoring stresses, can be calculated by assuming that the lithosphere responds to the stress distribution by flexure of a continuous elastic plate with thickness T_e. As applied to the Earth's lithosphere, T_e is termed an effective elastic thickness because it reflects the mechanical strength of the lithosphere in a depth-averaged sense. Studies of lithospheric flexure at trenches and beneath seamounts suggest that T_e for oceanic lithosphere is related to lithospheric temperature structure (i.e., lithospheric age) at the time of deformation [Watts et al., 1980; Bodine et al., 1981; McNutt, 1984]. The resulting surface topography at the time of rifting is the sum of two components: (1) the topography introduced by the kinematics of extension (Figure 6), and (2) the topography which balances the isostatic restoring stresses engendered by the kinematics of extension. Additional subsidence or uplift of the lithospheric surface will occur following rifting in response to the decay of lithospheric temperature perturbations, filling of basinal areas with sediment, or erosion of emergent parts of rift flank topography. We refer the reader to earlier studies [Weissel and Karner, 1989; Ebinger et al., 1991] where the isostatic effects of these additional "loads" are explicitly treated.

We now employ the modeling approach described above to explain the topography and free air gravity observed at Broken Ridge. The north-south topography profile shown in Figure 8 clearly shows that the effects of the mid-Eocene rifting episode are evident in the present-day morphology of the feature. Broken Ridge displays uplift along its southern margin, which is delineated by major southward-dipping normal faults (see also Figure 3) with a cumulative throw of >5 km and an overall dip of about 20°. Depths shoal to about 1000 m adjacent to the south-facing escarpment, and the topography deepens northwards to a morphologic saddle at about 2500 m depth over a distance between 80-150 km from the escarpment (Figures 2 and 8). As stated before, the water paleodepth estimates (Figure 5) together with the geometry of the pre-rift seismic stratigraphy (Figure 3) indicate that Broken Ridge was uplifted by >2 km during the rifting event. Maximum uplift, which presumably occurred adjacent to the southern escarpment of Broken Ridge, is imprecisely known because the crest of the uplifted topography was removed by wavebase and possibly subaerial erosion. The preserved structural and stratigraphic evidence indicates that Broken Ridge represents the footwall of an extensional system,

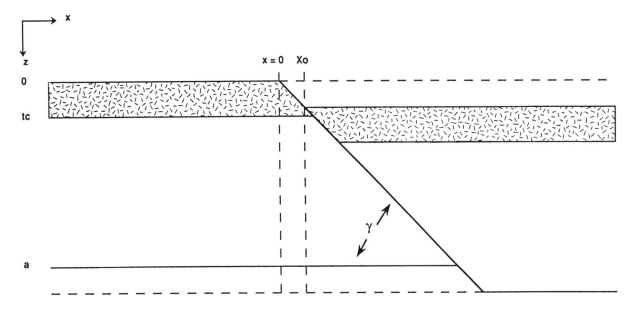

Fig. 6. Kinematic description of instantaneous slip along a plane, dipping normal fault cutting the entire lithosphere [after Weissel and Karner, 1989]. The model is parameterized by γ the fault dip, X_o the heave on the fault, the initial thickness t_c for the crust (stippled), and the initial thickness of the lithosphere a (defined as the depth where temperature reaches T_m, the asthenosphere temperature).

Isostatic Restoring Stress

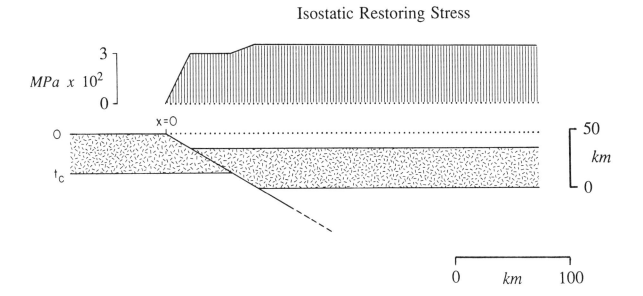

Fig. 7. Magnitude and distribution of isostatic restoring stress (vertically hatched area) resulting from instantaneous, finite slip on a normal fault as shown in Figure 6. We used a fault dip $\gamma = 30°$, a heave $X_o = 20$ km, a crustal thickness $t_c = 32.5$ km and a lithospheric thickness $a = 125$ km [after Weissel and Karner, 1989].

BROKEN RIDGE

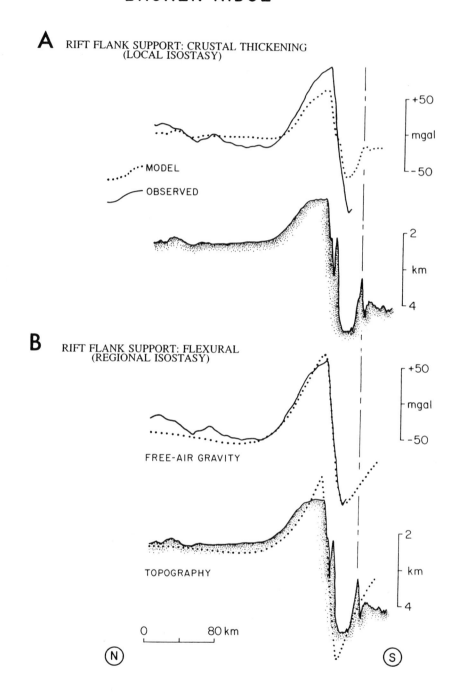

A RIFT FLANK SUPPORT: CRUSTAL THICKENING
(LOCAL ISOSTASY)

MODEL

OBSERVED

B RIFT FLANK SUPPORT: FLEXURAL
(REGIONAL ISOSTASY)

FREE-AIR GRAVITY

TOPOGRAPHY

0   80 km

Fig. 8. Comparison of gravity and topography observed along profile EEL48 across Broken Ridge (see Figure 2) with model calculations.

a) Predicted gravity effect (dots) for the situation in which the uplifted topography of Broken Ridge is supported by crustal thickness variations in a local isostatic manner. Note that the model gravity anomaly is much smaller in amplitude compared to the observed (solid line).

b) Predicted topography and gravity effect (dots) using the model for the isostatic effects of slip on a normal fault cutting the entire lithosphere [Figure 6; Weissel and Karner, 1989]. We assume that seawater overlies the lithosphere in these calculations.

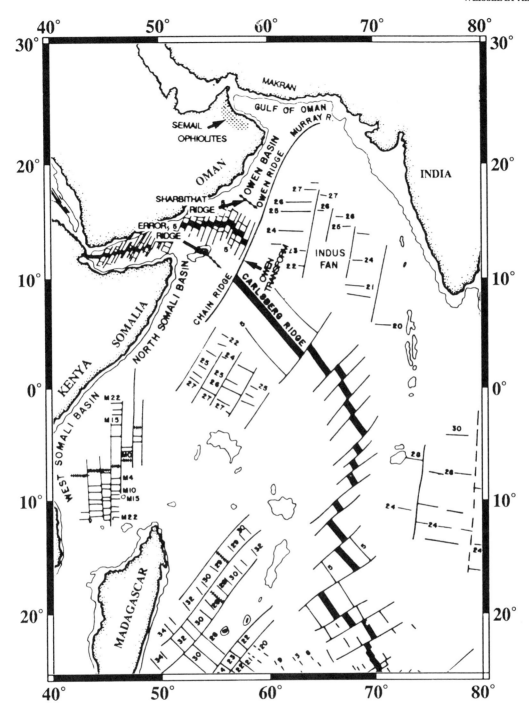

Fig. 9. Location map for the northwestern Indian Ocean showing the main physiographic features and marine magnetic lineations [from Mountain and Prell, 1990].

and that the south-facing escarpment represents the main border fault of the system.

The modeling results in Figure 8b show that the observed topography and gravity anomaly across Broken Ridge are explained very effectively by the simple normal faulting model when we use a fault dip of 20°, an amount of extension (heave) X_o of 18 km, crustal thickness t_c of 18 km, and an elastic plate thickness T_e of 15-20 km. The crustal thickness used in the modeling is based on the 18-20 km crustal thickness determined for Broken Ridge from re-analysis of early seismic refraction measurements [MacKenzie, 1984]. The value of T_e required in the modeling corresponds to a thermal thickness a for the lithosphere of 45-60 km at the time of rifting, if we assume

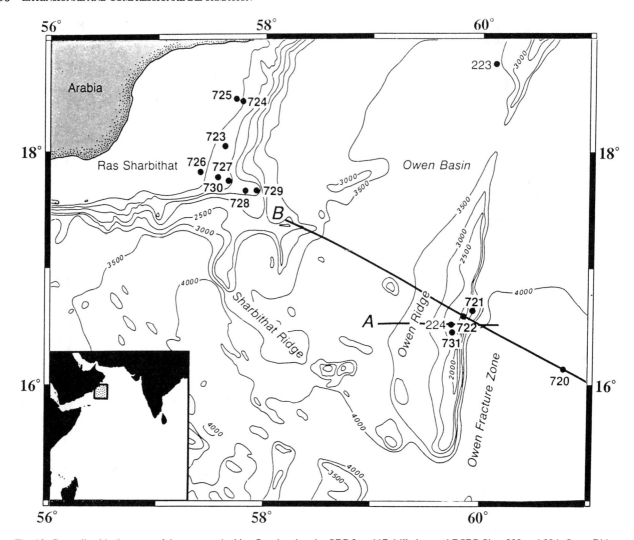

Fig. 10. Generalized bathymetry of the western Arabian Sea showing the ODP Leg 117 drill sites and DSDP Sites 223 and 224. Owen Ridge is depicted by the dotted pattern. The location of seismic profile C2704 line 25 (Figure 11) is also shown.

that T_e equates to the depth of the 450°C isotherm in cooling oceanic lithosphere [Parker and Oldenburg, 1973]. The magnitude of flexural uplift in the model topography uplift is over 2000 m at the south-facing escarpment, agreeing with the uplift inferred from the drilling results and the seismic stratigraphy after allowing for the material eroded from the crest of Broken Ridge in the middle to late Eocene.

The calculated gravity in Figure 8b explains the free-air gravity observed over Broken Ridge very well. Notice that the ratio of free-air gravity anomaly to topography over the crest of Broken Ridge is large (about 90 mgal/km). This large gravity/topography ratio provides good evidence that the crust mantle boundary (Moho) and the surface topography are uplifted by similar magnitudes, as would be predicted by the flexural rebound model. The large gravity anomalies over the crest of Broken Ridge, in fact, preclude explanations for support of the uplifted crest of Broken Ridge from thermal causes or by crustal thickening as shown by Weissel and Karner [1989] and below.

In Figure 8a, we test the hypothesis that the topography observed at Broken Ridge is supported by thickening of the crust, in a manner analogous to an Airy scheme of local isostatic compensation. Such crustal thickening can potentially be attributed to magmatic underplating of the crust by extension-induced partial melting of the mantle [White et al., 1987; Mutter et al., 1988; McKenzie and Bickle, 1988]. To model the free-air gravity anomaly over Broken Ridge in the case of crustal thickening, we must first determine the shape of the compensating Moho topography. The variation of crustal thickness across Broken Ridge can be determined from the bathymetry by assuming that the crust is in local isostatic equilibrium with a "standard" column of oceanic crust. Driscoll et al. [1989] defined the standard column in this region as 6 km-thick oceanic crust lying at a depth of 4.5 km, similar to late Eocene oceanic crust to the south of Broken Ridge (Figure 2). Because we use the observed topography of Broken Ridge to estimate its crustal thickness, the objective for the modeling in this case is to match the

OWEN RIDGE

W   E

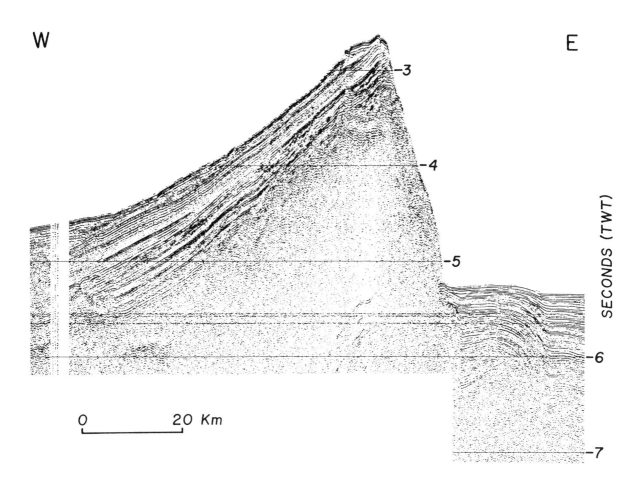

Fig. 11. Single-channel seismic reflection profile C2704 line 25 across Owen Ridge (profile A, Figure 10).

observed gravity, rather than the topography (cf. Figure 8b). Figure 8a shows that the gravity anomaly predicted assuming local isostatic compensation of the topography by crustal thickening is significantly smaller than that observed. We conclude, therefore, that the observed topography of Broken Ridge is not supported by crustal thickness variations (e.g., by magmatic thickening) in a local isostatic manner.

In summary, the kinematic and flexural isostatic model developed by Weissel and Karner [1989] involving finite simple slip on a planar normal fault cutting the lithosphere (Figure 6) explains both the topography and gravity anomaly observed over Broken Ridge (Figure 8b). Broken Ridge provides a good example of an uplifted rift flank where the cause of the uplift can be attributed to flexural rebound following mechanical unloading of the lithosphere during extension.

OWEN RIDGE

Background

Owen Ridge, along with Chain Ridge and Murray Ridge, constitutes a continuous line of bathymetric features extending from the Pakistan margin to the Somali Basin in the northwest Indian Ocean (Figure 9). The line of ridges and associated troughs is thought to mark a fracture zone because of (1) its continuity with the Owen transform fault, which right-laterally offsets the East Sheba Ridge in the Gulf of Aden and the Carlsberg Ridge by about 280 km, (2) its orientation perpendicular to post-anomaly 29 magnetic lineations in the Arabian Sea (Figure 9), and (3) the existence of an age contrast across the ridge, separating early Tertiary crust to the east (Figure 9) from unknown, but presumably older crust to the west in the Owen Basin [Whitmarsh, 1979; Stein and Cochran, 1985; Mountain and Prell, 1990].

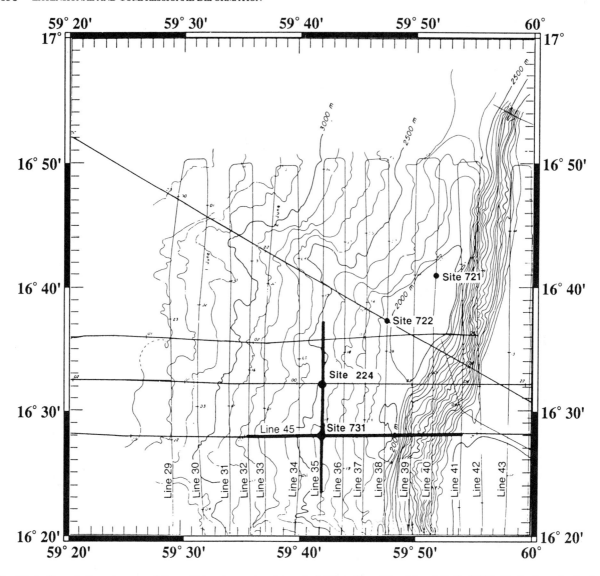

Fig. 12. SeaBeam bathymetry map of Owen Ridge at a contour interval of 100 m [from Mountain and Prell, 1990]. Single channel seismic coverage is shown by the solid track-lines. The strike- and dip-lines passing through Site 731 (Figure 13) are emphasized.

Owen Ridge is an asymmetric feature with a gently-dipping western side (about 2°) and a steep (about 15°) east-facing escarpment (Figures 10-12). Owen Ridge borders the Owen Basin to the west, and abuts the Indus Fan to the east. The ridge itself extends from about 15°N to 21°N, at about 60°E longitude (Figure 10). In this study, we will focus on the segment of the ridge that extends from 15°N to 18°N. Sediment supply to the ridge is heavy from either side (Figure 11).

The age of the crust on the east side of the ridge is well-constrained by marine magnetic anomalies 25 through 27 [Mountain and Prell, 1990]. The age of the crust in Owen Basin is less well known. Identification of M-series magnetic anomalies in the Mozambique, West Somali and North Somali basins (Figure 9) has linked the formation of these basins to the breakup of Gondwanaland in Late Jurassic/Early Cretaceous time [Segoufin and Patriat, 1980;

Patriat and Segoufin, 1988]. A similar history might be inferred for Owen Basin, the northernmost of these small basins which lie along the eastern coast of Africa and Arabia. Although seismic refraction studies in the southern part of Owen Basin suggest that it is underlain by typical oceanic crust, seafloor spreading magnetic anomalies have not been identified [Whitmarsh, 1979]. Low amplitude (~150 nT), ENE-trending magnetic lineations have been mapped, but the trend is not consistent with any known tectonic features in the basin [Whitmarsh, 1979].

Without identifiable magnetic anomalies, the age of the Owen Basin crust must be estimated in other ways. Heat flow data are consistent with a Late Cretaceous age, as are basement depths [Stein and Cochran, 1985]. This crustal age is supported by DSDP drilling of the Owen Ridge at Site 224 (Figure 10), which reached a lamprophyric igneous unit overlain by middle to lower Eocene sediments [Whitmarsh, Weser, Ross et al., 1974].

DSDP Site 224 drilling results provided only a sketch of the uplift history of Owen Ridge, because poor core recovery hampered precise timing of the event. Upper Oligocene to lower Miocene clastic turbidite beds which were initially flat-lying were tilted westward as Owen Ridge began to uplift. DSDP Site 224 was uplifted above the water depths associated with turbidite deposition in late Oligocene to early Miocene time. The uplift is marked by the thinning of the turbidite units, the interfingering of turbiditic and pelagic layers, and then the transition to pelagic deposition.

Seismicity marks present-day left-lateral motion along the Owen transform, which separates the Carlsberg Ridge from the East Sheba Ridge in the Gulf of Aden. The line of ridges and associated troughs north of the Owen transform is marked by sparse, infrequent shallow earthquake activity. Normal faulting is occurring in the Dalrymple Trough, and right-lateral strike-slip motion is occurring along the Owen Ridge north of 20°N, and southwest of the study area along lineaments that parallel the Owen Ridge [Gordon and DeMets, 1989; White, 1984; Sykes, 1970]. Within the study area south of 17°N, Owen Ridge is seismically quiet. Studies of present day plate motions suggest that if Owen Ridge represents a plate boundary between the Indian and Arabian plates, relative motion is very slow, about 2 mm/yr [Gordon and DeMets, 1989] and probably reflects slight spreading rate differences between the East Sheba and Carlsberg ridges.

Previous workers have attributed the uplift of Owen and Murray ridges to a component of compression which has occurred along the fracture zone as a result of the change in spreading directions when Arabia rifted away from Africa in the Neogene [Whitmarsh, 1979; White, 1984; Mountain and Prell, 1990]. The fact that Owen Ridge is a ridge instead of a trough is not in itself diagnostic of a compressional or transpressional origin, because as shown earlier, Broken Ridge which is a rift flank uplifted by mechanical unloading during extension, has a similar magnitude of topographic relief [Figure 8; Weissel and Karner, 1989]. Nonetheless, a compressional or transpressional origin seems most likely. Sediments to the east at the foot of Owen Ridge (Figure 11) show evidence of uplift attributable only to compression. If extension were occurring, we would expect to see diverging reflectors within the sediment column east of the ridge as space is made available for sediments. We know that a small amount of strike-slip motion is occurring along the Owen Ridge today, and presumably has occurred for some time. A component of compression along this feature could have developed simply as a consequence of strike-slip motion along a section of the fault which does not exactly parallel the slip direction.

ODP Leg 117 Drilling on Owen Ridge

The western flank of Owen Ridge is characterized by numerous and deep slump scars between sediment ridges

(Figures 12 and 13). Leg 117 Sites 721, 722, and 731 (Figures 10 and 12) were drilled on top of sediment ridges near the crest of Owen Ridge to obtain an undisturbed Neogene and Paleogene sedimentary section to study Milankovich cycles, the onset of the monsoon, and the tectonic uplift of Owen Ridge. The most complete section was recovered at Site 731 (Figure 12), and a lithostratigraphic summary of the four holes drilled at Site 731 is shown in Figure 14.

The oldest lithologic unit sampled at Owen Ridge is designated Unit IV, which comprises alternating coarse-grained sand and silt turbidites and mud turbidites with no large scale fining-upward trend. Unit IV was deposited in the late Oligocene to late early Miocene. The thickest undisturbed section extends from 320 to 994 meters below sea floor (mbsf) in Hole 731C (Figure 14). The uppermost 70-80 m of this unit (lower Miocene) consist of interbedded fine-grained turbidites which fine upwards and are interbedded with nannofossil chalks toward the top of the unit. The increase in pelagic sediment upsection marks the uplift of Owen Ridge out of the regime of turbidite deposition. Planktonic foraminifer preservation is poor, however, and benthic foraminifers from upper and middle bathyal depths are absent.

Unit III is a nannofossil chalk of early to middle Miocene age. The chalk is 80-90% calcium carbonate, and shows the dominance of pelagic sedimentation after Owen Ridge was uplifted beyond the level of turbidite deposition. Unit II is a siliceous nannofossil chalk of late Miocene age and Unit I is a nannofossil ooze of late Miocene to Holocene age. Both units show the continuing dominance of pelagic sedimentation (Figure 14).

The timing of the uplift of Owen Ridge can be determined from the Leg 117 drilling results. Uplift of the ridge is marked by the transition within Unit IV from turbiditic to pelagic facies in the late early Miocene (Figure 14), and by phases of nondeposition and erosion. Ages for the tops of the transitional lithofacies are approximately 15-16 Ma at each site on Owen Ridge. Uplift, therefore, was not a local event, as Sites 721 and 731 are separated by about 30 km along the strike of Owen Ridge (Figure 12). This time links the uplift event with the initiation of rifting of Arabia away from Africa, and the first emplacement of oceanic crust in the Red Sea.

Modeling Lithospheric Shortening: Constraints from Owen Ridge

The success in explaining the morphology and the gravity over Broken Ridge by treating extension as instantaneous slip along a single normal fault (Figure 6) and calculating the expected isostatic response [Weissel and Karner, 1989] led us to ask if compressional features might be equally well explained simply by reversing the sense of displacement on the fault (Figure 15). We assume a finite amount of slip occurs instantaneously along a single reverse fault which cuts the lithosphere, and that the kinematics can be completely specified by fault dip γ and amount of shortening or heave X_O. The tip of the

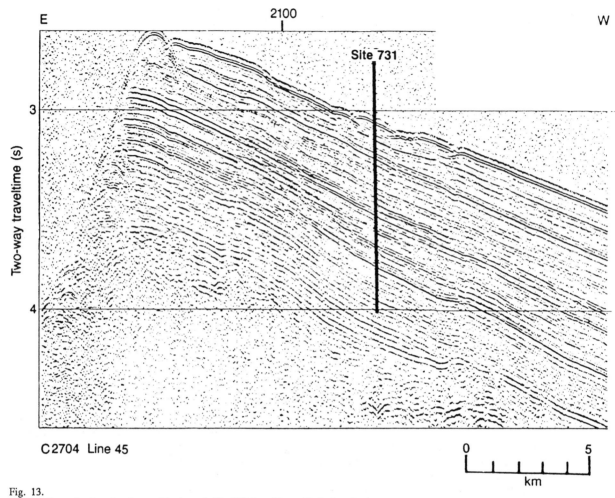

Fig. 13.
  a) Dip-line seismic reflection profile through Site 731 (see Figure 12 for location).
  b) (next page) Strike-line seismic reflection profile through Site 731 (see Figure 12 for location).

overthrust block is allowed to collapse (without extension) down onto the underthrusting block (Figure 15), such as would occur if the tip fails locally under the influence of gravitational forces. If enough shortening occurs or if the crust is thin enough, mantle material can be included in the collapsed tip.

Thrust faulting locally thickens the crust and lithosphere (Figure 15) and emplaces a "load" on the lithosphere. As with the extensional modeling, we assume that the lithosphere responds to the loading as would a continuous elastic plate on a fluid foundation. Isostatic restoring stresses engendered by the faulting cause the lithosphere to deflect in order to regain isostatic equilibrium. Additional vertical displacements occur as the faulted lithosphere thermally re-equilibrates with time following the compressional event. The free-air gravity effect over the feature is then calculated using the line integral technique of Tanner [1967].

Modeling of Owen Ridge is more complicated than for Broken Ridge because we must include the loading effect of sediment emplaced since shortening occurred in the early

Miocene. The loading effect of sedimentation is approximated by infilling the modeled topography with sediment of assumed constant density to a specified level on either side of the ridge. The isostatic consequence of sediment infilling is additional lithospheric flexure. The sediment loading effects are calculated iteratively for each time increment since the time of deformation.

Observations constraining the modeling results are (1) bathymetry, single-channel seismic reflection data, and gravity measurements from shipboard surveys of Owen Ridge, and (2) information of the timing of the uplift of Owen Ridge obtained from the ODP Leg 117 drilling results. Known parameters include (1) the age of the underthrust lithosphere [anomaly 25-26, about 60 Ma; Mountain and Prell, 1990], (2) the time of the deformation (about 15 Ma, from ODP Leg 117 drilling) and therefore the age of the underthrust lithosphere at the time of faulting (about 45 Ma), and (3) the crustal thickness, which is assumed to be 6 km, typical of normal oceanic crust. An age difference across Owen Ridge probably exists between the overthrust and the underthrust lithosphere, as the

S   N

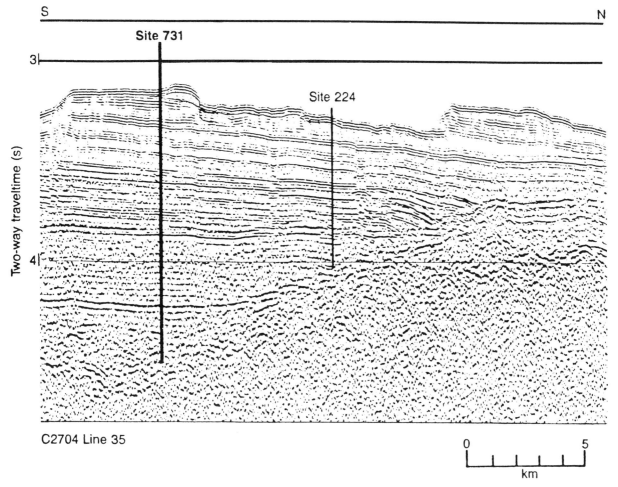

C2704 Line 35

0 |_|_|_|_|_| 5
km

underthrust lithosphere is early Tertiary, and the Owen Basin lithosphere is thought to be Late Cretaceous in age. If the age difference is small (on the order of 10 m.y., as seems likely), it makes little difference in the calculations and can be neglected. Parameters that must be determined by the modeling are: (1) the amount of shortening or heave X_o, (2) the fault dip γ, and (3) the thickness of sediment infill either side of Owen Ridge.

Figure 16 shows the best model fit to the seafloor topography and the free-air gravity, with 20 km of shortening occurring across a fault dipping 20° to the west. The results were obtained using a density of 2200 kg/m³ for the infilling sediment, and allowing sediment to fill to a level 300 m higher west of the ridge than to the east, as shown in the single-channel seismic profiles (e.g., Figure 11). The crustal structure and total sediment thickness from the best-fitting model are shown in Figure 17. The best-fitting elastic plate thickness T_e was found to be equivalent to the depth of the 600°C isotherm in 45 m.y.-old oceanic lithosphere.

CENTRAL INDIAN OCEAN BASIN

Background

Probably the clearest example worldwide of "diffuse" or continuum-style compressional deformation of oceanic lithosphere is found in the Central Indian Ocean basin south of India and Sri Lanka. Deformation is distributed over an unusually broad zone, between about 5°N and 10°S and extending from about 75°E longitude eastwards toward the Sumatra - Andaman Trench (Figure 18). The properties of the deformed lithosphere, and the implications for global plate motions and the mechanical behavior of oceanic lithosphere have intrigued Earth scientists for more than two decades [Sykes, 1970; Eittreim and Ewing, 1972; Stein and Okal, 1978; Weissel et al., 1980; Bergman and Solomon, 1980, 1985; Geller et al., 1983; Wiens and Stein, 1983; McAdoo and Sandwell, 1985; Wiens, 1986; Zuber, 1987; Neprochnov et al., 1988; Petroy and Wiens, 1989; Stein et al., 1990; Stein and Weissel, 1990; Karner and Weissel, 1990a,b]. For example, it has been proposed that the zone of compressional deformation and tectonic activity is part of a "diffuse" plate boundary between an Australian and Indian plate extending across the equatorial Indian Ocean from the Central Indian Ridge to the Andaman Trench [Figure 18; Wiens et al., 1985; DeMets et al., 1988, 1990a,b; Gordon et al., 1990].

The deformed region exhibits an anomalously high level of seismicity, the larger of these events providing evidence for widespread thrust and strike slip faulting in the oceanic mantle [e.g., Figure 18; Bergman and Solomon, 1985;

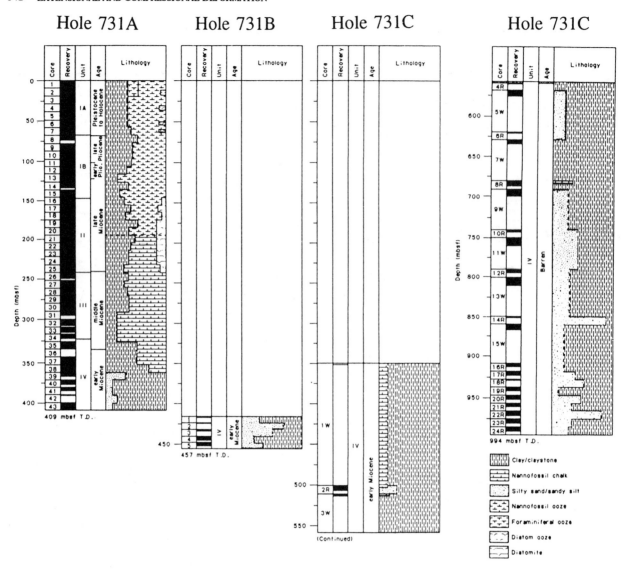

Fig. 14. Summary lithostratigraphic columns for Holes 731 (A-C) drilled on Owen Ridge during Leg 117 [Prell, Niitsuma et al., 1989].

Wiens and Stein, 1983; Petroy and Wiens, 1989]. Between 75°E and the Ninetyeast Ridge, the deformation takes the form of dramatic E-W trending undulations in the oceanic basement, with wavelengths of 150-300 km and peak-to-trough amplitudes up to 2 km [Figures 19 and 20; Weissel et al., 1980; Geller et al., 1983; McAdoo and Sandwell, 1985; Neprochnov et al., 1988]. Throughout the deformed region, the surface of oceanic crust is broken into fault blocks bounded by high-angle reverse faults (e.g., Figure 21) at an average spacing of about 8 km [Bull, 1990]. The faults, which strike roughly E-W, have a maximum throw of several hundred meters [Weissel et al., 1980; Geller et al., 1983; Neprochnov et al., 1988]. As described in the modeling section below, we favor an hypothesis that the long wavelength undulations are primarily a flexural response of the lithosphere to approximately N-S oriented compression, as described recently by Karner and Weissel

[1990a,b]. We also subscribe to the view that the reverse faulting is a brittle response of the upper lithosphere to compression, and that the faults themselves represent selective reactivation of a population of inward- and outward-facing faults created in the vicinity of the ridge crest during crustal formation [Weissel et al., 1980; Bull, 1990].

Until recently, our observational knowledge of the deformation was based largely on historical earthquake data, and "conventional" shipboard geophysical measurements such as gravity, bathymetry, single-channel seismics, and heat flow (see references cited above). In the last few years, however, both ODP drilling and MCS data have provided important new insights into the time history and the nature of the compressional deformation. In 1987, Bull and Scrutton [1990a,b] acquired 12-channel analogue MCS data using a modest seismic source array, and for the

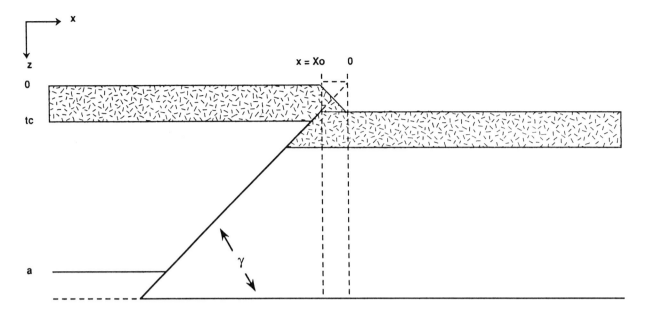

Fig. 15. Simple kinematic model for instantaneous reverse faulting of the entire lithosphere. Fault dip is γ, and the shortening or heave is X_o.

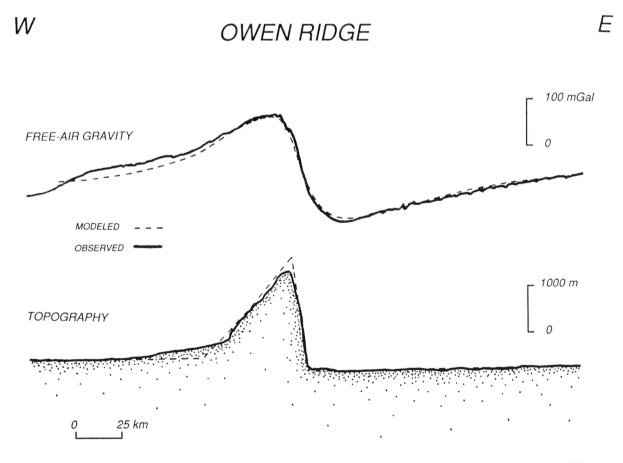

Fig. 16. Calculated topography and gravity profiles across Owen Ridge are compared with observed values from Conrad 2704 line 1 (profile B, Figure 10) projected perpendicularly to the trend of the ridge. Conrad 2704 line 1 passes through Site 722 in Figure 12.

Owen Ridge

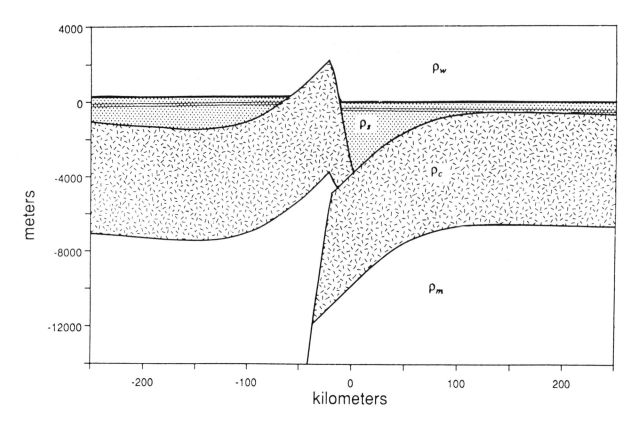

Fig. 17. Modeled crustal structure of Owen Ridge. The hatchered area is crust of density ρ_c, the stippled region is sediment of density ρ_s, above is seawater of density ρ_w, and below the crust is mantle of density ρ_m. Sediment is infilled to 0 m on the right side of the ridge, and to +300 m on the left. The lines extending through the sediment layer represent the top of the layer at given times during the evolution of the feature.

first time imaged reverse faults in the oceanic crust dipping 30°- 40° to the north (Figure 21). In 1991, an international expedition involving French, U.S., and Dutch scientists collected 96-channel MCS data across the deformed region. The new data reveal south-dipping faults in the crust for the first time, and provide evidence that the faults penetrate through the crust down into the mantle [Chamot-Rooke, de Voogd et al., 1991].

In 1987, three sites were drilled during ODP Leg 116 on the distal Bengal Fan about 800 km south of Sri Lanka near 1°S, 81.4°E (Figure 19). A north-south seismic reflection profile through these sites (717-719) is shown in Figure 21. The sites are located on two adjacent faults blocks bounded by the north-dipping crustal faults clearly observed in the seismic section. Lithospheric shortening by reverse displacement on the faults is recorded in the seismic stratigraphy of the overlying sediments. In essence, a pre-deformation sedimentary section marked by reflectors subparallel to oceanic basement gives way upward to stratal packages which pinch out toward the "noses" of the fault blocks and onlap the underlying pre-deformation section. These packages denote episodic

movement on the faults during the deformation. The top of the pre-deformation sedimentary section is therefore marked by an angular unconformity at the nose of the fault blocks passing into a correlative conformity in the trough areas of the rotated blocks (i.e., at Site 717). A main drilling objective for Leg 116 was to probe these seismic stratigraphic units to determine the time that faulting began at this particular location within the deformed region. Site 718, because of its thin syn-deformation section, provided an opportunity to drill deeply into the pre-deformation section in order to document the early history of Bengal Fan sedimentation. It was hoped that the deformational events recorded in the sedimentary section at the drill sites could be related to main phases of Himalayan orogenesis inferred from from onland geological, thermochronologic, and paleomagnetic studies.

Leg 116 Drilling Results

Turbidites predominate in the sedimentary succession encountered during Leg 116 drilling, as shown in the summary lithostratigraphic columns for the three sites (Figure 20). The sedimentary section has been divided into

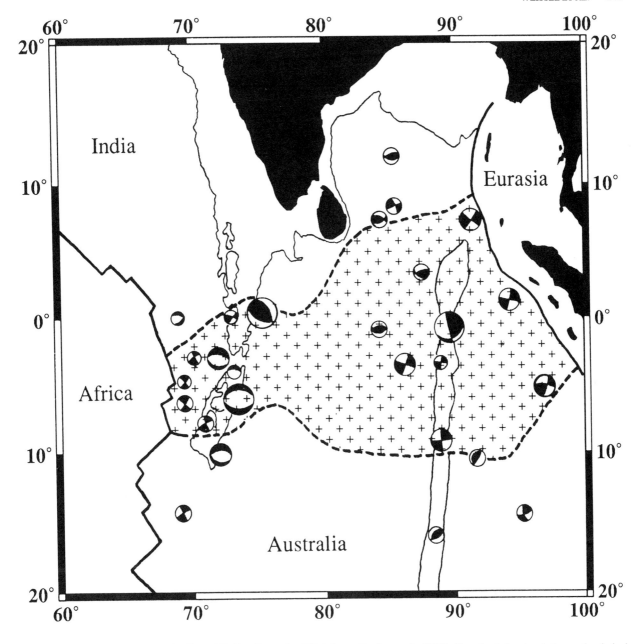

Fig. 18. Plate tectonic setting of the Central Indian Ocean [modified from Gordon et al., 1990]. The stippled region between the dashed lines represents a diffuse plate boundary between otherwise rigid Indian and Australian plates, according to Wiens et al. [1985], DeMets et al. [1988, 1990a,b], and Gordon et al. [1990]. Representative focal mechanisms of shallow earthquakes are shown.

5 units on the basis of dominant lithologies [Cochran, Stow et al., 1989]. Unit I, which is only about 2-5.5 m thick, consists of calcareous clay with thin mud turbidite deposited during the Holocene and perhaps the latest Pleistocene. Unit II (Figure 20), which is a sequence of silty turbidites 98-147.5 m thick, was deposited rapidly during the early Pleistocene.

Units III and IV are primarily finer-grained mud turbidites of latest Miocene through Late Pliocene age. These two units can be differentiated by the presence of distinctive green and white biogenic turbidites in Unit III. Sediment

accumulation rates at Site 717 average 70 m/m.y. through the Pliocene. Accumulation rates decrease towards the top of Unit III, however, and the lower Pleistocene and uppermost Pliocene parts of the section appear to be condensed or perhaps missing.

The deepest lithostratigraphic unit penetrated during Leg 116 drilling (Unit V) is a thick sequence of grey silty mud turbidites (Figure 20), which range in age from early to late Miocene. Accumulation rates averaged 75 m/m.y. over that interval of time. Mud turbidites, biogenic turbidites and

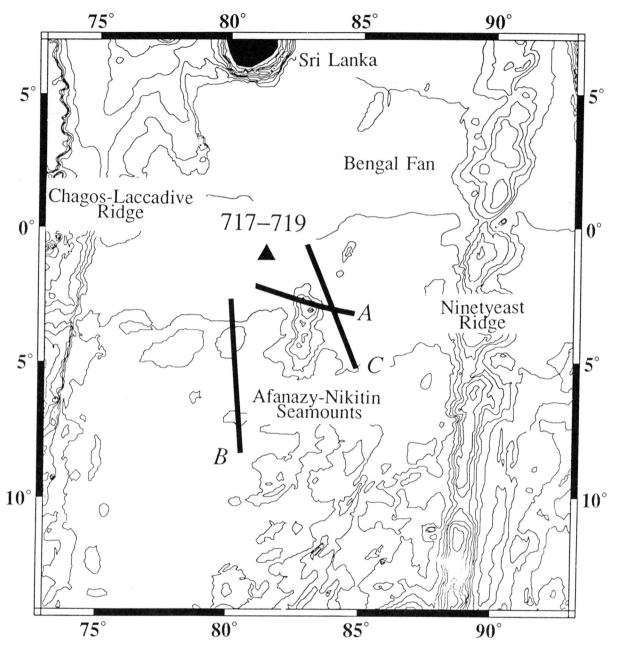

Fig. 19. General bathymetry of the Central Indian Ocean showing the major bathymetric features discussed in the text. Contour interval is 500 m. The Afanazy-Nikitin seamount group is delineated with the 4500 m isobath. The solid triangle denotes the location of ODP Leg 116 drilling (Sites 717-719). Heavy solid lines labeled A, B, and C locate the seismic reflection profiles shown in Figure 20.

pelagic clays, up to 20 m thick, were encountered sporadically below a depth of 605 mbsf at Site 718.

Two main conclusions can be drawn from Leg 116 drilling of the sediments overlying the deformed lithosphere in the Central Indian Ocean basin. First, the beginning of lithospheric shortening, marked by the unconformity (correlative conformity, at Site 717) observed in seismic reflection profiles (e.g., Figure 21), is dated as late Miocene (7.5 - 8.0 Ma). This major tectonic event does not correspond to any change in the nature of the

sediments. Second, the drilling did not define the onset of Bengal Fan sedimentation at the latitude of the Leg 116 drill sites. Drilling was terminated at Site 718 while still in fan turbidite lithologies (Figure 22). The 17 - 17.5 Ma age assigned to the oldest sediments sampled is inferred from sporadically recovered nannofossils [Gartner, 1990]. Cochran [1990] states that "although the base of the fan was not penetrated, a number of lines of evidence suggest it was approached and that fan sedimentation began in the early Miocene". Fan sedimentation of early Miocene age as

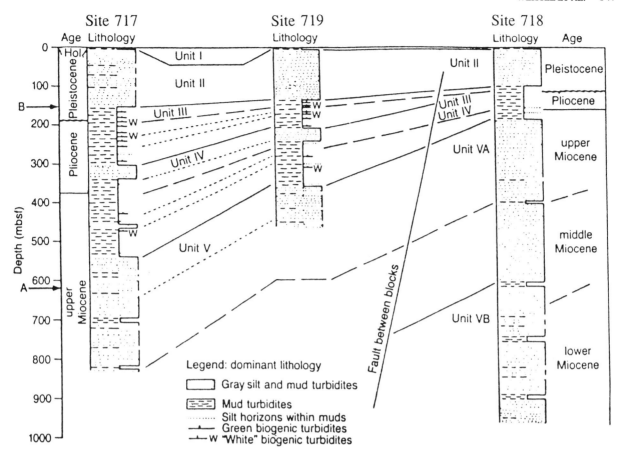

Fig. 20. Summary lithostratigraphic columns for Sites 717-719, drilled during ODP Leg 116 [from Cochran, Stow et al., 1989].

revealed through drilling implies that the Himalayan mountains were high enough at that time to provide a significant source of clastic sedimentation to the Central Indian Ocean basin. This inference about the early Miocene topography of the Himalayan - Tibetan region agrees with thermochronologic data from southern Tibet which are interpreted as showing large and rapid erosional denudation and (by implication) uplift beginning at about 20 Ma [Harrison et al., 1992].

Provenance studies based on detrital clay mineralogy [Bouquillon et al., 1990; Brass and Raman, 1990] and heavy mineral assemblages [Amano and Taira, 1992] from Leg 116 samples have shed light on possible phases of tectonism in the Himalayan - southern Tibetan region. The clay studies point to a Himalayan source for the clastics of Units V and II (Figure 20), while sediments derived from the Indian subcontinent are prominent in Units III and IV. These results are echoed in the heavy minerals study. Amano and Taira [1992] note that sediments shed from the high Himalaya abruptly decreased in the interval 7.5 to 6.5 Ma. Between 6.5 and 0.9 Ma, sediments derived from the Indian subcontinent and Sri Lanka became prominent. Since 0.9 Ma, the Himalayas have been the primary source of clastic material delivered to the distal Bengal Fan. Amano and Tairo [1992] conclude from the heavy mineral

provenance study that there have been two primary phases of Himalayan uplift since the early Miocene, the first one culminating between 11 and 7.5 Ma, and a later phase which started in the early Pleistocene. Cochran [1990] cautions against an exclusively tectonic explanation for the results of the provenance studies. He suggests that changes in sea level, and the change in the period of glacial cycles, might also explain the variations in clay mineralogy and heavy mineral compositions found in the Neogene section at the Leg 116 drill sites.

Although it is not obvious why compressional deformation should have started 7 - 8 million years ago in the Central Indian Ocean basin, speculation has centered on whether this age marks the time when the Tibetan Plateau reached its maximum elevation and crustal thickness [e.g., Harrison et al., 1992]. Several lines of evidence can be cited to support this view. The approximately N-S trending graben in southern Tibet probably started to form at this time, possibly indicating that maximum elevation of the plateau had been reached and that "gravitational collapse" by E-W extension had commenced. The intensification of the Indian Ocean monsoon occurred at about 8 Ma, according to Leg 117 drilling results [Kroon et al., 1991; Prell et al., this volume]. The intensification is thought to be directly related to the emergence of an extensive region

717   719   718   6 s

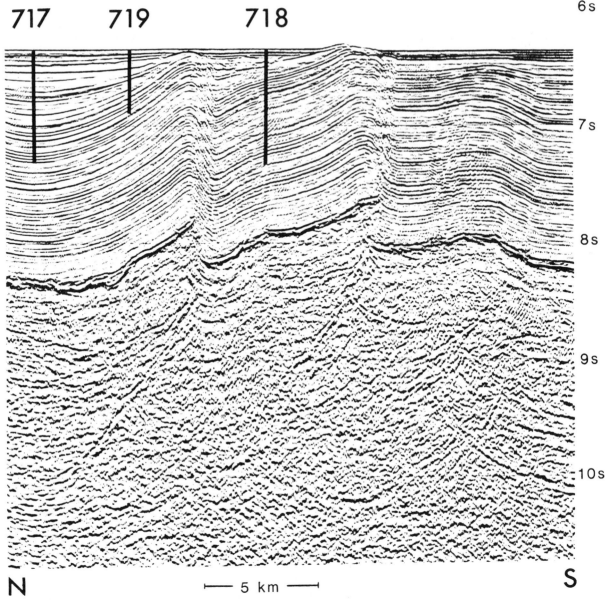

N   ⊢— 5 km —⊣   S

Fig. 21. 12-channel MCS reflection profile across the Leg 116 drill sites [Bull and Scrutton, 1990a,b]. Note the fault surface reflectors in the oceanic crust dipping 30° - 40° to the north. The approximate drilling penetration at each of the Sites 717-719 is indicated.

of high elevation in Tibet by the late Miocene. A speculative conclusion [e.g., Harrison et al., 1992] is that the surface of the Tibetan Plateau reached its maximum supportable elevation 7 - 8 million years ago, and that part of the continued convergence between India and Asia has been taken up by the shortening observed in the Central Indian Ocean basin.

Modeling Lithospheric Shortening: Constraints from the Central Indian Ocean basin

The continuum style of compressional deformation of oceanic lithosphere expressed in the Central Indian Ocean basin is very unusual on a world-wide basis. Theoretical investigations of the deformation have primarily

concentrated on explaining the long wavelength pattern of basement highs and lows (e.g., Figure 22) and associated gravity anomalies. The main point of modeling studies is to learn directly from the deformation about the mechanical properties of the oceanic lithosphere, the level of compressive tectonic force acting on the lithosphere, and why deformation is so prominent in the Central Indian Ocean basin.

Weissel et al. [1980] suggested that the long-wavelength basement undulations (e.g., Figure 22) might be analogous to the buckling of a thin elastic plate on a fluid foundation. They found that to buckle an elastic plate at the observed wavelength of about 160 km requires a buckling stress of 2.4 GPa (24 kbar) and an effective elastic plate thickness T_e

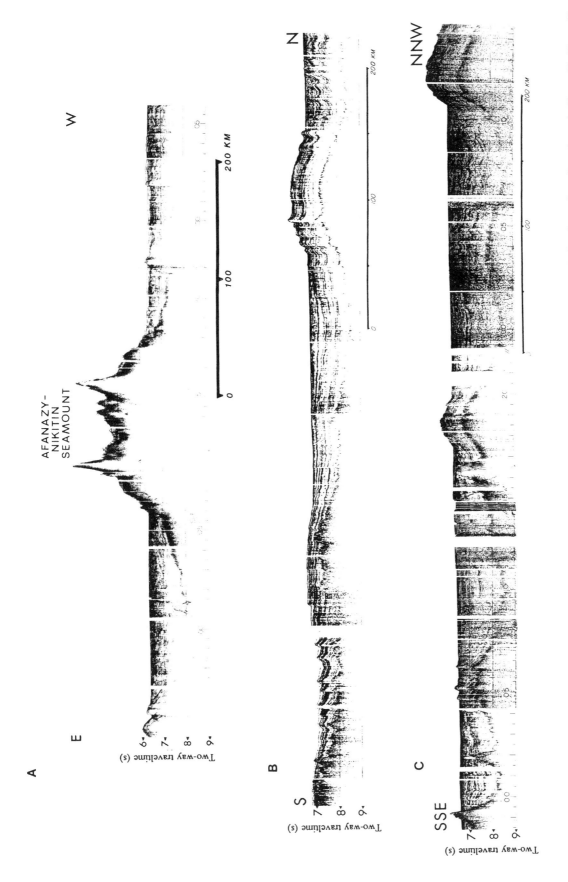

Fig. 22. Single-channel seismic reflection profiles from region of compressional deformation of oceanic lithosphere in the Central Indian Ocean. Locations shown in Figure 19. Figure 20a shows an east-west profile across the Afanazy-Nikitin seamounts. Figures 20b and 20c clearly indicate the long wavelength (about 200 km) and large amplitude (about 2 s two-way time) undulations in the oceanic basement of the deformed zone. Bengal Fan turbidites can be seen infilling the topographic lows created by the deformation, completely covering the deformation in places.

OBSERVED FREE-AIR GRAVITY   MODELED FREE-AIR GRAVITY

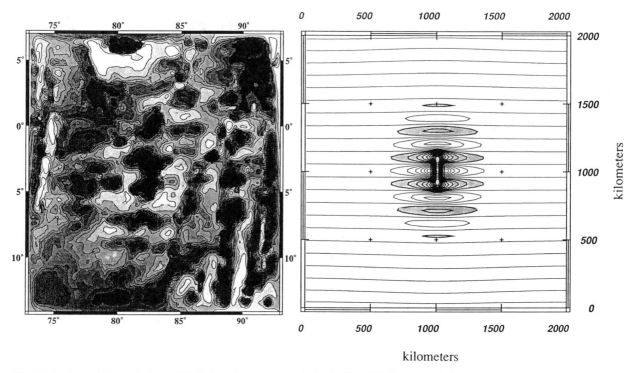

Fig. 23. Predicted (right) and observed (left) free-air gravity anomaly in the Central Indian Ocean basin [Karner and Weissel, 1990 a,b]. The gravity grey scale runs from most positive (black) to most negative (white). Note that the modeled gravity accounts for hourglass planform pattern of the positive gravity anomaly over the Afanazy-Nikitin seamounts and the characteristic 200 km deformation wavelength.

of about 12 km. Weissel et al. [1980] concluded that such high stresses are unrealistic, and that non-elastic processes must be occurring. McAdoo and Sandwell [1985] pointed out that because the Central Indian Ocean lithosphere was more than 55 m.y. old at the time compression began, a T_e of 46 km would be expected. A 46 km-thick elastic plate would buckle at a wavelength of 450 km under an average horizontal compressive stress of 4.9 GPa (49 kbar). This wavelength is more than twice the average wavelength (estimated to be 190 km by McAdoo and Sandwell) of the basement deformation, and the stress level would far exceed the failure strength of lithospheric materials. McAdoo and Sandwell [1985] applied an elastic-plastic yield strength envelope [YSE; Brace and Kohstedt, 1980; Goetze and Evans, 1979] model for lithospheric strength, and calculated that buckling with a characteristic wavelength of about 190 km would result under a compressive stress of 600 MPa (6 kbar), which lithospheric material could probably support.

Karner and Weissel [1990 a,b] questioned whether the deformation actually represents buckling behavior. They noted that the Afanazy-Nikitin seamounts (Figures 19 and 22) are centrally located in the area where the long-wavelength basement undulations (Figure 22) and associated gravity anomalies are best developed. Karner and Weissel suggested that certain wavelength components of an initial plate deflection, caused by the emplacement of the Afanazy-Nikitin seamounts on the Indian plate in the

Late Cretaceous or Early Tertiary, have been preferentially amplified since the lithosphere came under N-S compression in the late Miocene. They regarded the deformation as amplification of a pre-existing deformation rather than as buckling. By matching both the amplitude and wavelengths of the basement undulations and the gravity anomalies, Karner and Weissel determined that a NS-directed compressive force/unit length of magnitude 1.5 -2.0 x 10^{13} N m^{-1} could explain the deformation observed in the Central Indian Ocean. They also found that Bengal Fan sedimentation has played an important role in enhancing the magnitude of deformation caused by in-plane compression. Figure 23 compares observed gravity anomalies over the Central Indian Ocean basin with those calculated by Karner and Weissel from their best fitting model.

The level of force/unit length which Karner and Weissel [1990a,b] determine in their forward modeling of the deformation is indicated by the shaded area under the yield strength envelope (YSE) in Figure 24. Note that this level of compressive force agrees with that implied in the elastic-plastic plate buckling study of McAdoo and Sandwell [1985]. Moreover, a force/unit of magnitude 2.0 x 10^{13} N m^{-1} is consistent with state of stress models for the Indo-Australian Plate [Cloetingh and Wortel, 1985, 1986], which predict typical horizontal compressive stresses in the 3-5 kbar range for the northern Indian

Ocean, supported over a nominal plate thickness of 100 km.

The modeling studies summarized above provide important predictions about the nature of the deformation, and the implications for the mechanical properties and behavior of the oceanic lithosphere:

(1) The calculated level of tectonic force when applied to a YSE concept for the lithosphere (Figure 24) implies that the upper part of the lithosphere (15-20 km) has yielded and that brittle deformational behavior should be manifest. This is consistent with the seismic images of the reverse faulting in the crust (e.g., Figure 21), and with the areal and depth distribution of earthquakes [although some hypocentral depths are as great as 39 km, Bergman and Solomon, 1985].

(2) The YSE analysis (Figure 24) predicts a 10-15 km thick "core" region in the lithosphere that has not yielded and where, presumably, compressional strain (on the order of 1%) is stored elastically.

DISCUSSION

For each of the three deformation examples considered in this paper, the mechanical properties of the lithosphere play a major role in determining the resulting morphology and gravity anomalies that we observe in shipboard geophysical data. If we plot seismic reflection data from Broken Ridge and Owen Ridge at similar vertical and horizontal scales (Figure 25), we find a strong resemblance between the morphology of both features, despite the fact that Broken Ridge has been affected by extension, and Owen Ridge by transpression. The flexural rigidity of the lithosphere makes it difficult to distinguish between extensional and compressional deformation in the absence of other definitive information, such as seismicity. We reinforce this point by calculating the topography resulting from the kinematic models for extension and shortening (Figure 6 and 15, respectively) in which all model parameters are the same except for the sense of displacement on the fault. We see in Figure 26 that the predicted topographies for extension and shortening are quite similar, and resemble the observed morphologies of Broken Ridge and Owen Ridge (Figure 25). The major difference is that the area under the curve (i.e. the horizontal integral of the topography) is positive for lithospheric shortening and negative for lithospheric extension. Thus, topography by itself might not be diagnostic of the type of deformation that has occurred.

Compressional deformation of the lithosphere does not always have the same expression. Lithospheric shortening at Owen Ridge is localized along 1 or 2 major faults, whereas in the Central Indian Ocean basin, shortening is accommodated over a broad region, on the order of 1000 km in the north-south direction. Although it is not entirely clear why one style of deformation is favored in a given locale, we suspect that several factors are important in determining how the deformation is expressed. In the case of the Central Indian Ocean basin, emplacement of the Afanazy-Nikitin seamounts in the Late Cretaceous-Early

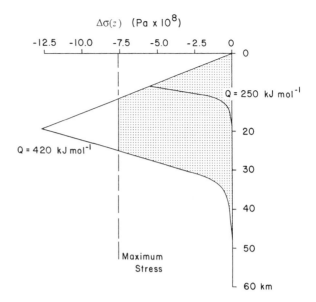

Fig. 24. Yield-stress envelope (YSE) versus depth constructed for oceanic lithosphere of thermal age 65 Ma and a compressional strain rate of $10^{-15}s^{-1}$. The activation energy of the lithosphere Q was chosen to be 420 kJ mole^{-1} in agreement with laboratory measurements on "wet" olivine [Chopra and Paterson, 1981]. The magnitude of horizontal compressive force/unit length required to reproduce the amplitude and wavelength characteristics of the broad basement undulations is the shaded area under the YSE. The maximum compressive stress supported in a strong "core" region of the lithosphere some 15 km thick, is denoted by the broken vertical line.

Tertiary provided a pre-existing lithospheric deflection, and certain wavelength components of this early deflection were preferentially amplified by compression in the late Miocene. Perhaps an important factor is the orientation of pre-existing tectonic fabrics of the lithosphere relative to the applied compression. In the Central Indian Ocean, a population of originally ridge-crest parallel normal faults appears to have been "reactivated" as reverse faults during the deformation [Weissel et al., 1980; Bull, 1990]. There is no evidence in seismic reflection data, however, for reactivation of fracture zones by faulting parallel to their traces. This behavior might be attributed to the approximately north-south orientation of the principal compressive stress in the Central Indian Ocean as revealed for instance, by earthquake focal mechanism solutions [e.g., Figure 18; Bergman and Solomon, 1985; Wiens and Stein, 1983; Petroy and Wiens, 1989]. This direction is sub-parallel to the fracture zones, and is roughly orthogonal to the magnetic lineation trends and, therefore, to the original ridge-crest parallel normal faults. For Owen Ridge, on the other hand, the Owen fracture zone provided a zone of weakness oriented favorably to the direction of compression.

We noted earlier when discussing the modeling procedures that the effective elastic thickness of oceanic lithosphere T_e reflects the temperature structure of the lithosphere (i.e., lithospheric age) at the time of deformation [Watts et al., 1980; Bodine et al., 1981; McNutt, 1984]. In general, T_e

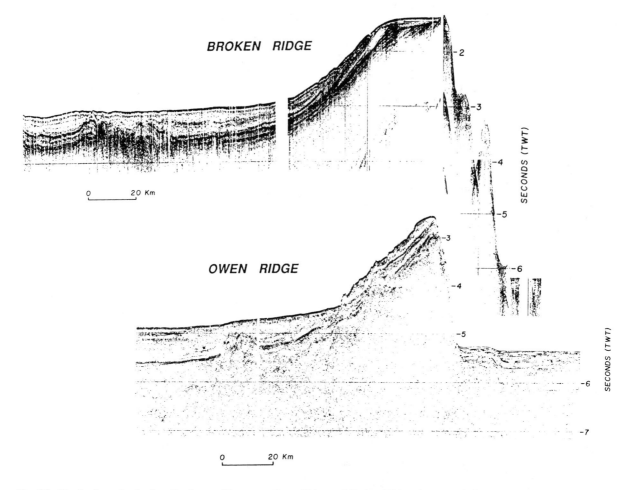

Fig. 25. Single-channel seismic reflection profiles across Owen Ridge and Broken Ridge shown at similar horizontal and vertical scales. Note the similarity in general morphology of these features despite their different tectonic origins.

appears to follow the depth of a characteristic isotherm between 300° - 600° C in lithosphere cooling toward an asymptotic thickness of about 125 km [Parsons and Sclater, 1977]. This behavior is thought to reflect a fundamental dependence of rock strength on temperature. Drilling at Broken Ridge, Owen Ridge, and in the Central Indian Ocean has provided age constraints for the deformational events at those locations from biostratigraphically dated sediment samples. The geophysical models, on the other hand, provide estimates of T_e of the lithosphere involved in the deformation, and the level of compression in the case of the Central Indian Ocean.

Broken Ridge, which was more than 50 m.y. old when rifting began, and the Central Indian Ocean basin, which was probably more than 60 m.y. old when compression began, require T_e values in the modeling which are considerably smaller than expected from the empirical relation between T_e and lithospheric age derived from studies of flexure at seamounts and seaward of trenches [Watts et al., 1980; Bodine et al., 1981; McNutt, 1984]. For the Central Indian Ocean basin, the observed T_e value

of 12-15 km (a factor of 2 less than the value expected on the basis of age) probably reflects the reduction in thickness of the "mechanically-strong" part of the lithosphere by yielding at the top and bottom of the plate under compression, as depicted in the YSE diagram in Figure 24. As stated before, this situation is consistent with observed widespread reverse faulting in the crust, and with the areal and depth distribution of earthquakes.

For Broken Ridge, however, the small value of T_e in the range 15-20 km (compared to the expected value of 15-30 km) probably has a thermal or compositional explanation, rather than a mechanical cause as for the Central Indian Ocean basin. First, Broken Ridge formed as a result of hotspot volcanism and its crust is thick (18-20 km) compared to that of "normal" oceanic lithosphere (6-8 km). If crustal rocks are mechanically weaker than mantle rocks (as in the case of continental lithosphere), we would expect the T_e value for Broken Ridge to be lower than for comparably-aged, normal oceanic lithosphere. Second, the water paleodepth estimates obtained from Leg 121 drilling on Broken Ridge (Figure 5) lead us to suspect a thermal cause for the low value of T_e found for Broken Ridge

TOPOGRAPHY (at $t=0$)

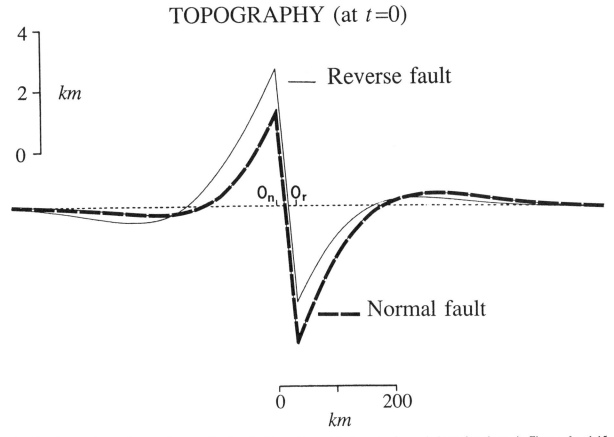

Fig. 26. Topography resulting at time $t = 0$ using the kinematic models for extension and shortening shown in Figures 6 and 15, respectively. Note that the heave on normal fault and the reverse fault is the same, X_o, but the sense of displacement is opposite. The topography generated by the two models is similar, except that the horizontal integral of the topography is positive for the reverse fault, and negative for the normal fault.

(Figure 8b). The paleodepth information coupled with the seismic stratigraphy indicates that Broken Ridge lay in water depths of about 1.0-1.5 km just before the mid-Eocene rifting event occurred. Had Broken Ridge been subsiding as oceanic lithosphere cooling towards an asymptotic thermal thickness of 125 km, it should have been at a depth of at least 2000 m after 50 m.y. of cooling and the acquisition of a sediment cover 1500 m thick. Further, had the lithosphere at Broken Ridge continued to cool towards a 125 km thermal "plate" thickness after the rifting episode, no more than about 300 m of subsidence should have occurred during the past 40 m.y. Yet we know that Broken Ridge reached its present depth of more than 1000 m quite rapidly after rifting (Figure 5).

We suggest that before rifting, temperatures in the mantle beneath the Broken Ridge-Kerguelen-Heard plateau were elevated because of the proximity of the Kerguelen-Rajmahal hotspot to the region. This situation might explain the "anomalously slow" subsidence of Broken Ridge before rifting, the "anomalously fast" subsidence after rifting, and the low value for T_e at Broken Ridge. The Late Cretaceous - Early Tertiary section at Broken Ridge contains abundant volcanic ash layers whose composition are consistent with hotspot volcanism from a nearby

eruptive center [Shipboard Scientific Party, 1989]. Higher than normal mantle temperatures would reduce the amount of subsidence of Broken Ridge compared to that expected from a cooling, 125 km-thick lithosphere, essentially because the bottom of the lithosphere, defined as the depth where temperature reaches that of the underlying asthenosphere T_m, would lie at a shallower depth (45-60 km, say) in a mantle affected by a hot spot. The depth-integrated mechanical strength of the lithosphere (including T_e) would be reduced under these circumstances, for the same reason. After rifting, the seafloor spreading process removed the Broken Ridge from the influence of the Kerguelen-Rajmahal hotspot, which has remained beneath the Kerguelen-Heard plateau during the last 40 m.y. Broken Ridge was then able to cool towards an asymptotic thickness of 125 km. Subsidence following rifting was rapid because the lower part of the lithosphere "buffered" against heat loss by the presence of the hotspot before rifting, was now able to cool.

Flexural coupling between Broken Ridge and the younger oceanic lithosphere to the south could help explain the "anomalous" large amount of subsidence of Broken Ridge since rifting. Essentially, the new oceanic lithosphere being hotter than the lithosphere beneath Broken Ridge, tends to lose heat faster and subside more rapidly. If no differential vertical slip occurs between the rifted lithosphere and the newly accreted oceanic lithosphere after seafloor spreading starts, the rift flank (i.e., Broken Ridge) will be dragged down flexurally by the rapidly-subsiding new oceanic lithosphere to the south, while the new oceanic lithosphere is flexurally held up by the slowly subsiding Broken Ridge lithosphere. Karner et al. [1991] present a model for the tectonic evolution of Broken Ridge which includes the effects of flexural coupling between Broken Ridge and the younger Southeast Indian Ridge oceanic lithosphere. Note, however, that while flexural coupling provides a simple explanation for the large amount of subsidence of Broken Ridge since rifting, it cannot explain the anomalously small amount of subsidence for the pre-rift period, nor the small value for T_e found at Broken Ridge.

CONCLUSIONS

Stratigraphic successions over deformed lithosphere potentially contain information on the timing and duration of the deformational event. Insights into tectonic uplift or subsidence, as reflected in changing depositional environments, might also be found from sediment lithofacies and microfossil assemblages. Such information can only be unlocked by deep scientific drilling. The Indian Ocean region contains three notable examples of deformation of the oceanic lithosphere: Broken Ridge, which records the effects of a middle Eocene rifting episode in its structure and stratigraphy; Owen Ridge, part of the Owen fracture zone uplifted under transpression in the early Miocene; and the Central Indian Ocean basin which has been deformed under approximately N-S compression since the late Miocene. These deformational features were the targets of ODP drilling during Legs 121, 117 and 116, respectively.

Drilling results from the three deformational features yield important constraints for theoretical studies of lithospheric deformation which employ simplifying assumptions on the mechanical properties and behavior of the lithosphere. The modeling studies of Broken Ridge, Owen Ridge, and the Central Indian Ocean basin, summarized in this paper, all show that the topographic and gravity expressions of lithospheric extension and compression are controlled in a fundamental way by lithospheric "strength", as measured by flexural rigidity or equivalently by T_e, the effective elastic plate thickness. In particular, the uplift of >2 km at Broken Ridge during the middle Eocene rifting event can be explained as flexural isostatic rebound that occurs because the lithosphere is mechanically unloaded during extension. For each of the three deformational features, T_e has been estimated by matching predicted against observed topography and gravity. The resulting T_e estimates can be compared to the values expected on the basis of lithospheric age at the time the deformational events occurred. That critical information on time is best obtained by drilling and dating

the sedimentary sequences preserved over the deformed lithosphere.

What are the prospects for future ODP drilling with the objective of better understanding how the Earth responds to extensional and compressional processes? Based on the knowledge gained from drilling the deformational features in the Indian Ocean, the answer depends how well sedimentary sections record the changes in depth and depositional environment that occur in response to the deformational process. An ideal sedimentary section would have good biostratigraphic resolution (microfossils should be plentiful and well-preserved), moderate sediment accumulation rates (particularly if dilution of pelagic material by clastic sedimentation occurs), and a means of recording depth variations (benthic foraminifer assemblages, or well-developed hiatuses such as erosional unconformities). The success of future attempts to investigate the mechanical behavior and properties of the oceanic lithosphere by deep scientific drilling will depend in large measure on how well these criteria are met.

Acknowledgments. We are indeed grateful to the many scientists involved in ODP Legs 116, 117, and 121, whose efforts made these drilling campaigns such a success. Many hours of discussion with Neal Driscoll have helped put the drilling results and the modeling studies into proper scientific perspective. This manuscript was reviewed by Jim Cochran, Don Forsyth, Keith Louden, and Greg Mountain, and their comments are appreciated. Support for the research was provided through National Science Foundation awards OCE 85-11980, 85-16918, and 89-18035; Office of Naval Research awards N00014-87-K-204 and N00014-89-J-1148, and JOI/USSAC awards TAMRF 20239 and 20258. This is Lamont-Doherty Geological Observatory contribution no. 4967.

REFERENCES

Amano, K. and Taira, A., Two phase uplift of the high Himalayas since 17 Ma, *Geology, 20,* 391-394, 1992.

Anderson, R. N., Langseth, M. G., and Sclater, J. G., Mechanisms of heat transfer through the floor of the Indian Ocean, *J. Geophys. Res., 82,* 3391-3409, 1977.

Barron, J., Larsen, B. et al., *Proc. ODP, Init. Repts., 119,* Ocean Drilling Program, College Station, TX, 1989.

Barton, P. and Wood, R., Tectonic evolution of the North Sea basin: crustal stretching and subsidence, *Geophys. J. R. Astr. Soc., 79,* 987-1022, 1984.

Berggren, W. A., Kent, D. V., Flynn, J. J., and Van Couvering, W. J., Cenozoic geochronology, *Geol. Soc. Am. Bull., 96,* 1407-1418, 1985.

Bergman, E. A. and Solomon, S. C., Oceanic intraplate earthquakes: Implications for local and regional intraplate stress, *J. Geophys. Res., 85,* 5389-5410, 1980.

Bergman, E. A. and Solomon, S. C., Earthquake source mechanisms from body-waveform inversion and intraplate tectonics in the northern Indian Ocean, *Phys. Earth Planet. Inter., 40,* 1-23, 1985.

Bodine, J. H., Steckler, M. S., and Watts, A. B., Observations of flexure and the rheology of the oceanic lithosphere, *J. Geophys. Res., 86,* 3695-3707, 1981.

Bouquillon, A., France-Lanord, C., Michard, A., and Tiercelin, J.-J., Sedimentology and isotope chemistry of Bengal Fan sediments: The denudation of the Himalaya, in *Proc. ODP, Sci.*

Results, 116, edited by J. R. Cochran, D. A. V. Stow et al., 43-48, Ocean Drilling Program, College Station, TX, 1990.

Brace, W. F. and Kohlstedt, D. L., Limits on lithospheric stress imposed by laboratory experiments, *J. Geophys. Res., 85,* 6248-6252, 1980.

Brass G. W. and Raman, C. V., Clay mineralogy of sediments of the Bengal Fan, in *Proc. ODP, Sci. Results, 116,* edited by J. R. Cochran, D. A. V. Stow et al., 35-42, Ocean Drilling Program, College Station, TX, 1990.

Bull, J. M., Structural style of intra-plate deformation, Central Indian Ocean basin: Evidence for the role of fracture zones, *Tectonophys., 184,* 213-228, 1990.

Bull, J. M. and Scrutton, R. A., Fault reactivation in the Central Indian Ocean and the rheology of oceanic lithosphere, *Nature, 344,* 85-858, 1990a.

Bull, J.M. and Scrutton, R. A., Sediment velocities and deep structure from wide-angle reflection data around Leg 116 sites, in *Proc. ODP, Sci. Results, 116,* edited by J. R. Cochran, D. A. V. Stow et al., 311-316, Ocean Drilling Program, College Station, TX, 1990b.

Chamot-Rooke, N., de Voogd B., and PHEDRE Shipboard Participants, Seismic reflection profiling across the central Indian Ocean deformed lithosphere, Abstract with Programs, AGU Fall Meeting 1991, p.488, 1991.

Chopra, P. N. and Paterson, M. S., The rheology of dunite and the influence of water, *Tectonophys., 78,* 453-473, 1981.

Cloetingh, S. and Wortel, R., Regional stress field of the Indian Plate, *Geophys. Res. Letts., 12,* 77-80, 1985.

Cloetingh, S. and Wortel, R., Stress in the Indo-Australian plate, *Tectonophysics, 132,* 49-67, 1986.

Cochran, J. R., Himalayan uplift, sea level, and the record of Bengal Fan sedimentation at the ODP Leg 116 sites, in *Proc. ODP, Sci. Results, 116,* edited by J. R. Cochran, D. A. V. Stow et al., 397-414, Ocean Drilling Program, College Station, TX, 1990 .

Cochran, J. R., Stow, D. A. V. et al., *Proc. ODP, Init. Repts., 116,* Ocean Drilling Program, College Station, TX , 1989.

Cochran, J. R., Stow, D. A. V. et al., *Proc. ODP, Sci. Results, 116,* Ocean Drilling Program, College Station, TX, 1990.

Coffin, M. F., Rabinowitz, P. D., and Houtz, R. E., Crustal structure in the western Somali basin, *Geophys. J. R. Astr. Soc., 86,* 331-369, 1986.

Curray, J. R., and Moore, D. G., Growth of the Bengal deep-sea fan and denudation of the Himalayas, *Bull. Geol. Soc. Am., 82,* 563-572, 1971.

Davies, T. A., Luyendyk, B. P. et al., *Init. Repts. DSDP, 26,* 281-294, (U.S. Govt. Printing Office), Washington, D.C., 1974.

DeMets, C., Gordon, R. G. and Argus, D. F., Intraplate deformation and closure of the Australian-Antarctica-Africa plate circuit, *J. Geophys. Res., 93,* 11,877-11,898, 1988.

DeMets, C., Gordon, R. G., and Vogt, P. R., Seafloor deformation between the Indian and Australian plates near the Central Indian Ridge and Great Chagos Bank: results from an aeromagnetic study of the Carlsberg and Central Indian Ridges, *EOS, Trans. AGU, 71,* p. 1598, 1990a.

DeMets, C., Gordon, R. G., Argus, D. F., and Stein, S., Current plate motions, *Geophys. J. Int., 101,* 425-478, 1990b.

Driscoll, N. W., Karner, G. D., and Weissel, J. K., Stratigraphic response of carbonate platforms and terrigenous margins to relative sealevel changes: are they really that different? in *Proc. ODP, Sci. Results, 121,* edited by J. K. Weissel, J. W. Peirce, E. Taylor, J. Alt et al., 743-762, Ocean Drilling Program, College Station, TX, 1991a.

Driscoll, N. W., Weissel, J. K., Karner, G. D., and Mountain, G. S., Stratigraphic response of a carbonate platform to relative sea level changes: Broken Ridge, southeast Indian Ocean, *Am. Assoc. Pet. Geol. Bull., 75,* 808-831, 1991b.

Driscoll, N. W., Karner, G. D., Weissel, J. K., and Shipboard Scientific Party, Stratigraphic and tectonic evolution of Broken Ridge from seismic stratigraphy and Leg 121 drilling, in *Proc. ODP, Init. Repts., 121,* edited by J. W. Peirce, J. K.

Weissel et al., 71-92, Ocean Drilling Program, College Station, TX, 1989.

Duncan, R. A., Age distribution of volcanism along aseismic ridges in the eastern Indian Ocean, in *Proc. ODP, Sci. Results, 121,* edited by J. K. Weissel, J. W. Peirce, E. Taylor, J. Alt et al., 507-518, Ocean Drilling Program, College Station, TX, 1991.

Ebinger, C. J., Karner, G. D., and Weissel, J. K., Mechanical strength of extended continental lithosphere: constraints from the western rift system, East Africa, *Tectonics, 10,* 1239-1256, 1991.

Eittreim, S. K., and Ewing, J. I., Midplate tectonics in the Indian Ocean, *J. Geophys. Res., 77,* 6413-6421, 1972.

Fowler, S. and McKenzie, D. P., Flexural studies of the Exmouth and Rockall Plateaux using SEASAT altimetry, *Basin Res., 2,* 27-34, 1989.

Gartner, S., Neogene calcareous nannofossil biostratigraphy, Leg 116 (Indian Ocean), in *Proc. ODP, Sci. Results, 116,* edited by J. R. Cochran, D. A. V. Stow et al., 165-188, Ocean Drilling Program, College Station, TX, 1990.

Geller, C. A., Weissel, J. K., and Anderson, R. N., Heat transfer and intraplate deformation in the Central Indian Ocean, *J. Geophys. Res., 88,* 1018-1032, 1983.

Goetze, C., and Evans, B., Stress and temperature in the bending lithosphere as constrained by experimental rock mechanics, *Geophys. J. R. Astr. Soc., 59,* 463-478, 1979.

Gordon, R. G. and DeMets, C., Owen fracture zone and Dalrymple trough in the Arabian Sea, *J. Geophys. Res., 94,* 5560-5570, 1989.

Gordon, R. G., DeMets, C., and Argus, D. F., Kinematic constraints on distributed deformation in the equatorial Indian Ocean from present motion between the Australian and Indian plates, *Tectonics, 9,* 409-422, 1990.

Harrison, T. M., Copeland, P., Kidd, W. S. F., and Yin, A., Raising Tibet, *Science, 255,* 1663-1670, 1992.

Karner, G. D., and Weissel, J. K., Factors controlling the location of compressional deformation of oceanic lithosphere in the Central Indian Ocean, *J. Geophys. Res., 95,* 19,795-19,810, 1990a.

Karner, G. D., and Weissel, J. K., Compressional deformation of oceanic lithosphere in the Central Indian Ocean: Why it is where it is, in *Proc. ODP, Sci. Res., 116,* edited by J. R. Cochran, D. A. V. Stow et al., 279-290, Ocean Drilling Program, College Station, TX, 1990b.

Karner, G. D., Driscoll, N. W., and Peirce, J.W., Gravity and magnetic signature of Broken Ridge, southeast Indian Ocean, in *Proc. ODP, Sci. Results, 121,* edited by J. K. Weissel, J. W. Peirce, E. Taylor, J. Alt et al., 681-696, Ocean Drilling Program, College Station, TX, 1991.

Kroon, D., Steens, T. N. F., and Troelstra, S. R., Onset of monsoonal related upwelling in the western Arabian Sea as revealed by planktonic foraminifers, in *Proc. ODP, Sci. Results, 117,* edited by W. L. Prell, N. Niitsuma et al., 257-264, Ocean Drilling Program, College Station, TX, 1991.

Leclaire, L., Bassias, Y., Denis-Clocchiatti, M., Davies, H. L., Gautier, I., Gensous, B., Giannesini, P.-J., Morand, F., Patriat, P., Segoufin, J., Tesson, M., and Wanneson, J., Lower Cretaceous basalt and sediments from the Kerguelen Plateau, *Geo-Mar. Letts., 7,* 169-176, 1987.

MacKenzie, K., Crustal stratigraphy and realistic seismic data, Ph.D. Thesis, Univ. of California, San Diego (unpublished), 121 pp, 1984.

McAdoo, D. C., and Sandwell, D. T., Folding of oceanic lithosphere, *J. Geophys. Res., 90,* 8563-8568, 1985.

McKenzie, D. P., and Bickle, M. J., The volume and composition of melt generated by extension of the lithosphere, *J. Petrol., 29,* 625-679, 1988.

McNutt, M. K., Lithospheric flexure and thermal anomalies, *J. Geophys. Res., 89,* 11,180-11,194, 1984.

Morgan, W. J., Hotspot tracks and the opening of the Atlantic and Indian Oceans, in *The Sea, Vol. 7: The oceanic lithosphere,* edited by O. Miliani, John Wiley and Sons, New York, 443-487, 1981.

Mountain, G. S., and Prell, W. L., A multiphase plate tectonic history of the southeast continental margin of Oman, in *The Geology and Tectonics of the Oman Region*, edited by A. H. F. Robertson, M. P. Searle, and A. C. Ries, 725-743, Geol. Soc. Amer., Sp. Pub. No. 49, 1990.

Mutter, J. C., and Cande, S. C., The early opening between Broken Ridge and Kerguelen Plateau, *Earth Planet. Sci. Lett.*, 65, 369-376, 1983.

Mutter, J. C., Buck, W. R., and Zehnder, C. M., Convective partial melting, 1: a model for the formation of thick basaltic sequences during the initiation of spreading, *J. Geophys. Res.*, 93, 1031-1048, 1988.

Neprochnov, Y. P., Levchenko, O. V., Merklin, L. R., and Sedov, V. V., The structure and tectonics of the intraplate deformation area in the Indian Ocean, *Tectonophys.*, 156, 89-106, 1988.

Parker, R. L., and Oldenburg, D. W., Thermal model of ocean ridges, *Nature*, 242, 137-139, 1973.

Parsons, B., and Sclater, J. G., An analysis of the variation of ocean floor bathymetry and heat flow with age, *J. Geophys. Res.*, 82, 803-827, 1977.

Patriat, P., and Segoufin, J., Reconstruction of the Central Indian Ocean, *Tectonophys.*, 155, 211-234, 1988.

Peirce J. W., Weissel J. K. et al., *Proc. ODP, Init. Repts., 121*, Ocean Drilling Program, College Station, TX, 1989.

Petroy, D. E., and Wiens, D. A., Historical seismicity and implications for diffuse plate convergence in the northeast Indian Ocean, *J. Geophys. Res.*, 94, 12,301-12,326, 1989.

Prell, W.L., Niitsuma, N. et al., *Proc. ODP, Init. Repts., 117*, Ocean Drilling Program, College Station, TX, 1989.

Prell, W. L., Niitsuma, N. et al., *Proc. ODP, Sci. Results, 117*, Ocean Drilling Program, College Station, TX, 1991.

Schlich, R., Wise, S. W., Jr., et al., *Proc. ODP, Init. Repts., 120*, Ocean Drilling Program, College Station, TX, 1989.

Segoufin, J. and Patriat, P., Existence d'anomalies mesozoiques dans le basin de Somalie. Implications pour les relations Afrique-Antarctique-Madagascar, *C. R. Acad. Sci.*, 91, 85-88, Paris, 1980.

Shipboard Scientific Party, Sites 721, 722, 731, in *Proc. ODP, Init. Repts., 117*, Ocean Drilling Program, College Station, TX, 1989a.

Shipboard Scientific Party, Broken Ridge Summary, in *Proc. ODP, Init. Repts., 121*, Ocean Drilling Program, College Station, TX, 1989b.

Stein, C. A., and Cochran, J. R., The transition between the Sheba Ridge and Owen Basin: rifting of old oceanic lithosphere, *Geophys. J. R. Astr. Soc. 81*, 47-74, 1985.

Stein, C. A., and Weissel, J. K., Constraints on Central Indian Basin thermal structure from heat flow, seismicity and bathymetry, *Tectonophys,.* 176, 315-332, 1990.

Stein, C.A., Cloetingh, S., and Wortel, R., Kinematics and mechanics of the Indian Ocean diffuse boundary zone, in *Proc. ODP, Sci. Results, 116*, edited by J. R. Cochran, D. A. V. Stow et al., 261-278, Ocean Drilling Program, College Station, TX, 1990.

Stein, S., and Okal, E. A., Seismicity and tectonics of the Ninetyeast Ridge area: evidence for internal deformation of the Indian plate, *J. Geophys. Res.*, 83, 2233-2245, 1978.

Sykes, L. R., Seismicity of the Indian Ocean and a possible nascent island arc between Ceylon and Australia, *J. Geophys. Res.*, 75, 5041-5055, 1970.

Tanner, J. G., An automated method of gravity interpretation, *Geophys. J. R. Astr. Soc.*, 13, 1967.

Watts, A. B., Gravity anomalies, crustal structure and flexure of the lithosphere at the Baltimore Canyon Trough, *Earth Planet. Sci. Letts.*, 89, 221-238, 1988.

Watts, A. B., Bodine, J. H., and Steckler, M. S., Observations of flexure and the state of stress in the oceanic lithosphere, *J. Geophys. Res.*, 85, 6369-6376, 1980.

Weissel, J. K. and Karner, G. D., Flexural uplift of rift flanks due to mechanical unloading of the lithosphere during rifting, *J. Geophys. Res.*, 94, 13,919-13,950, 1989.

Weissel, J. K., Anderson, R. N., and Geller, C. A., Deformation of the Indo-Australian plate, *Nature*, 287, 284-291, 1980.

Weissel, J. K., Peirce, J. W., Taylor, E. Alt, J. et al., *Proc. ODP, Sci. Results, 121*, Ocean Drilling Program, College Station, TX, 1991.

White, R. S., Active and passive plate boundaries around the Gulf of Oman, *Deep Sea Res.*, 31, 731-745, 1984.

White, R. S., Spence, G. D., Fowler, S. R., McKenzie, D. P., Westbrook, G. K., and Bowen, A. N., Magmatism at rifted continental margins, *Nature*, 330, 439-444, 1987.

Whitmarsh, R. B., The Owen Basin off the south-east margin of Arabia and the evolution of the Owen Fracture Zone, *Geophys. J. R. Astr. Soc.*, 58, 441-470, 1979.

Whitmarsh, R. B., Weser, O. E., Ross, D. A. et al., *Init. Repts. DSDP, 23*, (U.S. Government Printing Offic), Washington, D.C., 291-420, 1974.

Wiens, D. A., and Stein, S., Age dependence of oceanic intraplate seismicity and implications for lithospheric evolution, *J. Geophys. Res.*, 88, 6455-6468, 1983.

Wiens, D. A., DeMets, C., Gordon, R. G., Stein, S., Argus, D., Engeln, J. F., Lundgren, P., Quible, D. Stein, C. Weinstein, S., and Woods, D. F., A diffuse plate boundary model for Indian Ocean tectonics, *Geophys. Res. Letts.*, 12, 429-432, 1985.

Wiens, D. A., Historical seismicity near Chagos: a complex deformation zone in the equatorial Indian Ocean, *Earth Planet. Sci. Letts.*, 76, 350-360, 1986.

Zuber, M. T., Compression of oceanic lithosphere: An analysis of intraplate deformation in the central Indian Basin, *J. Geophys. Res.*, 92, 4817-4826, 1987.

Mesozoic Paleoenvironment of the Rifted Margin off NW Australia (ODP Legs 122/123)

ULRICH VON RAD

Bundesanstalt für Geowissenschaffen und Rohstoffe, Postfach 510153, D-3000, Hannover 51, Germany

NEVILLE F. EXON

Bureau of Mineral Resources, Geology and Geophysics, GPO Box 378, Canberra, A.C.T. 2604, Australia

RON BOYD

Geology Department, Dalhousie University, Halifax, Nova Scotia B3H 3J5, Canada

BILAL U. HAQ

Marine Geology and Geophysics, National Science Foundation, Washington, DC, 20550, USA

During early Mesozoic time Northwest Australia was a passive margin of eastern Gondwana facing the southern Tethys Sea. ODP Legs 122 and 123 drilled a complete transect of eight boreholes across the old, sediment-starved passive margin of the Exmouth Plateau down to the nearby abyssal plains: (1) the Wombat Plateau (Sites 759-761, 764) - Argo Abyssal Plain transect (Site 765) and (2) the central Exmouth Plateau (Sites 762, 763) - Gascoyne Abyssal Plain (Site 766) transect.

Upper Triassic (Carnian-Rhaetian) fluviodeltaic to carbonate platform deposits of the Wombat Plateau document the synrift phase. The Rhaetian carbonate platform of lagoonal and reefal facies is characterized by sequence boundaries (211 and 215 Ma) and correlates with coeval facies changes in the western Tethys (Alps). During earliest Jurassic times, Wombat Plateau underwent platform drowning and early-rift volcanism, followed by a major blockfaulting episode (Callovian/Oxfordian). This resulted in uplift, rift flank tilting, subaerial erosion of the horst, and the formation of a "post-rift unconformity".

A precursor of the Indian Ocean formed at the Argo Abyssal Plain after the Callovian/Oxfordian breakup of eastern Gondwana. At Wombat Plateau, this was followed by rapid subsidence and a Berriasian transgression with nearshore to hemipelagic "juvenile-ocean" sediments, characterized by a condensed section of terrigenous littoral sands and belemnite-rich sandy muds. Calcisphere-nannofossil chalks with a low-diversity opportunistic flora indicate a stressful environment. Bentonites suggest early post-breakup volcanism.

The Albian/Cenomanian represents a gradual transition from hemipelagic ("juvenile ocean") to pelagic ("mature ocean") conditions. Purely pelagic conditions started after the Cenomanian/Turonian anoxic boundary event and led to declining subsidence and sedimentation rates.

On the central Exmouth Plateau, "failed" breakup during late Jurassic times was followed by major uplift of the southern hinterland. This uplift was associated with renewed rifting attempts south of the Exmouth Plateau and resulted in erosion and northward progradation of a major Berriasian shelf-margin clastic wedge (Barrow Group). This late synrift sequence is overlain by a condensed Valanginian section, followed by late Valanginian/early Hauterivian final breakup between Australia and Greater India and an early Aptian transgressive "juvenile ocean" section.

The Mesozoic depositional sequences are strongly affected by multistage rift tectonics, climatic changes due to plate motions, cyclic deltaic sediment input, and global sea level fluctuations. Eustasy played a significant role in generating the observed depositional sequences, especially in the Rhaetian of the Wombat Plateau and the Early Cretaceous of the central Exmouth Plateau.

INTRODUCTION

Before Ocean Drilling Program (ODP) Legs 122 and 123, the early Mesozoic tectonic and paleoceanographic history of the world's oceans received little attention through scientific drilling activity. The stratigraphic records of old

Synthesis of Results from Scientific
Drilling in the Indian Ocean
Geophysical Monograph 70
Copyright 1992 American Geophysical Union

sediment-starved rifted margins, like the Exmouth Plateau off Northwest Australia, contain important clues about the thermal and structural processes that lead to rifting and continental breakup, and the related paleoenvironmental changes, prior to and after the initiation of ocean basins. Legs 122 and 123 were the first of the Indian Ocean drilling program to core such an old stratigraphic sequence since *JOIDES Resolution* drilled the Galicia margin off Spain [Boillot, Winterer et al., 1987].

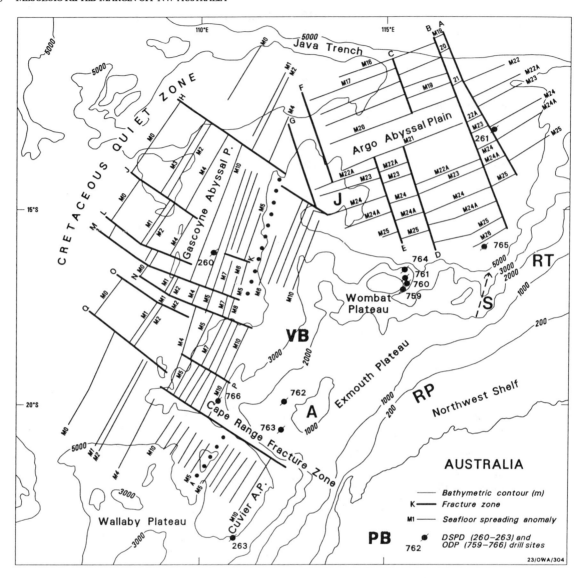

Fig. 1. The NW Australian continental margin with ODP Sites 759-766 (Legs 122 and 123), DSDP Sites 260, 261 and 263, and magnetic anomalies after Fullerton et al. [1989]. A = Exmouth Plateau Arch, J = Joey Rise, PB = Pilbara Block, RP = Rankin Platform, RT = Rowley Terrace, S = Swan Canyon, VB = volcanic basement [Exon and Buffler, 1992].

During ODP Legs 122 and 123 we drilled two transects across the continent/ ocean boundary from the Exmouth Plateau to the adjacent abyssal plains: the first transect comprised Wombat Plateau (N Exmouth Plateau) Sites 759, 760, 761 and 764, and the Argo Abyssal Plain Site 765, whereas the second transect consists of the central Exmouth Plateau Sites 762 and 763 together with Site 766 in the easternmost Gascoyne Abyssal Plain (Figure 1). Within easy reach of the drill-bit, this margin provided an opportunity

1. to study the Triassic to Cretaceous biostratigraphic, magnetostratigraphic, and paleoenvironmental history of the eastern Tethys ocean off northeastern Gondwana in a well-preserved southern- hemisphere, mid- to low-latitude setting;

2. to investigate the tectonic and magmatic rifting history of the Northwest Australian margin;

3. to study the Jurassic (to Valanginian) rift-drift transition and the early Cretaceous juvenile ocean phase, as well as the late Cretaceous to Cenozoic mature ocean phase;

4. to date the breakup of the Exmouth Plateau margins and the formation of some of the oldest oceanic basins of the Indian Ocean;

5. to test sequence-stratigraphic and eustatic models in order to discern the effects of vertical tectonics, sediment supply, climate, and global sea-level changes on depositional patterns [Haq et al., 1987; Wilgus et al., 1988].

The Triassic intracratonic early rift stage was studied at Wombat Plateau Sites 759-761 and 764. This sequence is

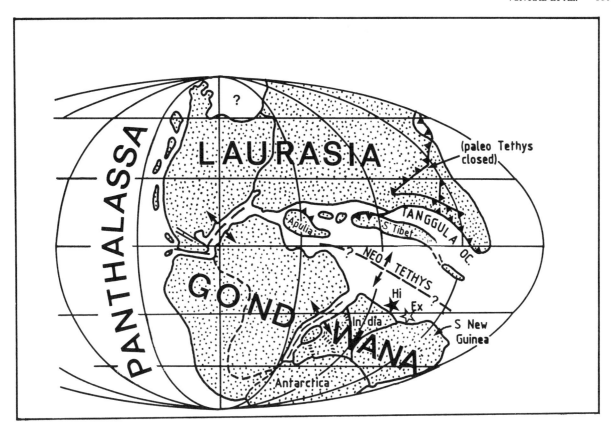

Fig. 2. Reconstruction of continents and oceans during Late Jurassic times. Note that the Tethys Himalaya (Hi) and Exmouth Plateau (Ex) were adjacent margins bordering NE Gondwana. After Sengör [1985] and other sources.

capped by a major regional "post-rift unconformity" spanning the Callovian to Oxfordian interval which is overlain by Lower Cretaceous, hemipelagic, "juvenile ocean"-stage sediments (Figures 1,5). The central Exmouth Plateau Sites 762 and 763 concentrated on delineating the depositional sequences associated with a Berriasian-Valanginian clastic wedge related to pre-breakup uplift.

The post-breakup stages can be studied in the different paleoenvironments of the Wombat Plateau, Central Exmouth Plateau (Sites 762, 763), and the Argo and Gascoyne Abyssal Plains (Sites 765-766).

The Northwest Australian margin was the site of repeated rifting, whereby successive continental terranes were ripped off the East Gondwana continent and carried northward by spreading oceans to collide with the accreting Asian continent [Görür and Sengör, 1992].

According to Sengör [1985], Audley-Charles [1988] and Veevers [1988], there were three major late Paleozoic-Mesozoic rifting events between Gondwana and Laurasia:

(1) Latest Carboniferous to mid-Permian rifting removed continental blocks ("Sibumasu" = China, W Thailand and E Burma) from Australian Gondwana, created "Tethys II" and initiated subsidence of the "Westralian superbasin".

(2) Jurassic (Oxfordian, M25/26) rifting split the Argo Landmass or South Tibet/Lhasa Block (including Mount

Victoria Land in Burma) from eastern Gondwana, and created a southeastern arm of Tethys ("Tethys III" or Argo Basin) between Australia and the Argo Landmass (Figure 2).

(3) Final breakup between Australia (central Exmouth Plateau) and "Greater India" at late Valanginian/early Hauterivian times (M-11) formed Gascoyne/Cuvier Abyssal Plains and initiated the modern Indian Ocean [Gradstein, Ludden et al., 1990; Boyd et al., 1992].

During Triassic to Middle Jurassic times, the Exmouth Plateau was part of a South-Tethyan passive margin, at the northern border of the East Gondwanan continent; this margin was in the vicinity of the northeastern rim of "Greater India" (Tethys Himalaya) to the west and of South Tibet, Timor-Burma to the northeast [Figure 2; Audley-Charles, 1988; Gradstein et al., 1992]. The demise of Tethys III between Greater India and Tibet by the Himalayan collision during Late Cretaceous to Paleogene times was a compensation for the Late Jurassic disassembly of Gondwana and the opening of the Indian Ocean [Sengör, 1985].

Unless otherwise stated, we use the time scale and Mesozoic-Cenozoic "cycle chart" of Haq et al. [1987] in this paper. The reader is referred to following papers for further details of the geological evolution of Wombat and Exmouth Plateaus: von Rad et al. [1992], Haq et al. [1992],

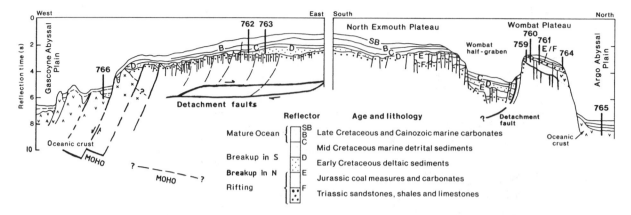

Fig. 3. Schematic E-W and N-S sections across Exmouth Plateau and adjacent abyssal plains with approximate locations of ODP Sites 759-766. The marked post-rift unconformity is underlain by block-faulted Triassic to Jurassic sediments and overlain by thin post-breakup sediments.

Boyd et al. [1992], Exon and Buffler [1992], and von Rad and Bralower [1992].

STRATIGRAPHIC/STRUCTURAL SETTING

Northern Exmouth Plateau/Wombat Plateau

The northern Exmouth Plateau is a north-facing part of the Australian continental margin, with the oceanic crust of the Argo Abyssal Plain bordering it immediately towards the north (Figure 1). The Exmouth Plateau consists of thinned and deeply subsided continental crust, with a Phanerozoic sedimentary sequence in excess of 10 km thick [Exon and Willcox, 1978]. The known sequences along the northern margin range in age from Triassic to Neogene. The margin is cut by major east-northeast- trending, steeply dipping faults into a number of horsts (including the Wombat Plateau), grabens, and half-grabens. Under the shelf, continental basement of the Pilbara Block and pre-Permian and Permian detrital sediments are overlain by Triassic sediments that can be traced westward along the Exmouth Plateau margin.

Seasat altimetry data of the Exmouth Plateau area are very similar to the free-air gravity map of Exon and Willcox [1978]. Gravity anomalies over Exmouth Plateau as a whole are low-amplitude and broad, suggesting that the plateau overall is in local isostatic equilibrium. However, Wombat Plateau, the volcanic northwestern extension of the plateau (Joey Rise), the culmination of Exmouth Plateau arch, and the basement high mapped by Exon and Buffler [1992] are marked by gravity anomalies of larger amplitude. This suggests that all these features are in regional isostatic equilibrium and supported by the fluxural strength of the lithosphere, as shown by the strong uplift and erosion of the Wombat Plateau horst compared to the nearby Swan Canyon area (Figure 1).

Sites 759, 760, 761 and 764 were drilled across the Wombat Plateau to recover a composite Late Triassic synrift to Cretaceous/Cenozoic post-rift record of sediments (Figure 6). The oldest sediments recovered are of mid-Carnian age. The Wombat Plateau area experienced various Permian to late Triassic rift phases. An important

period of uplift and erosion in the late Middle Jurassic is documented by the major angular post-rift "E" unconformity (Figure 4). The minimum stratigraphic extent of the hiatus on the northwest Australian margin spans the early Oxfordian, but in many areas the entire Upper Jurassic sequence is missing. We believe that the lack of the Jurassic sequence on the eastern Wombat Plateau is a result of Callovian uplift and subsequent erosion. This Callovian unconformity has been called "breakup unconformity", or "main unconformity" [Falvey, 1974]. Because a time gap of several million years is observed between the formation of this erosional unconformity at the rifted margin (Callovian) and the initiation of sea floor spreading in the associated Argo Abyssal Plain (Oxfordian), we prefer the term "post-rift unconformity" which avoids a direct correlation with the onset of drifting (i.e. "breakup").

During Berriasian times subsidence following the breakup of the Argo Abyssal Plain resulted in the transgression of the sea and in the deposition of a thin veneer of Cretaceous and Cenozoic hemipelagic to eupelagic sediments (see also Figure 5). This Berriasian marine transgression is documented in the Delambre No. 1 well 200 km SSW of the Wombat Plateau [von Rad et al., 1992], in Site 761, and possibly in Site 760. A post-Berriasian "D" unconformity at this time [Bradshaw et al., 1988] is documented in Delambre No 1 well. The change in the Turonian from marls to limestones marks the regional "C" seismic reflector. The Cretaceous sequence is rarely thicker than 200 m. The northern Exmouth Plateau margin subsided rapidly to bathyal depths in the early Cretaceous, whereas shallow-water sedimentation characterizes the Australian Northwest Shelf during the Hauterivian [Islam, 1988]. The Tertiary sequence on the northern Exmouth Plateau consists of thin bathyal chalks and oozes. The "B" unconformity encompasses the Cretaceous/Tertiary boundary, and the "A" unconformity is of early Oligocene age (Figures 4,5,6).

Central Exmouth Plateau

In Paleozoic times the Exmouth Plateau was an integral part of Gondwana and surrounded by continental crust. In

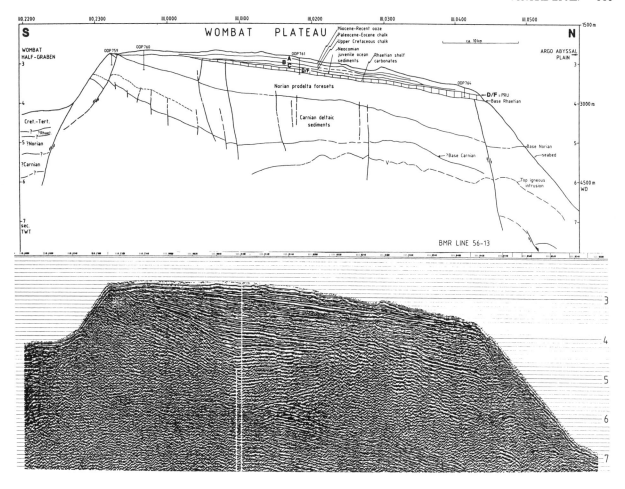

Fig. 4. Bottom: Multichannel seismic line BMR-56-13, N-S across Wombat Plateau. Top: Line drawing showing sediment packages and major reflectors [reflector nomenclature after Exon and Willcox, 1978], with locations of ODP Sites 759-761 and 764 (all sites are projected). D/F is post-rift unconformity (PRU).

the Permian, the Tethys Ocean started to form to the north and there was considerable stretching and thinning of Exmouth Plateau crust [Mutter et al., 1989; Williamson et al., 1990]. This led to rapid subsidence, and the deposition of a thick section of largely fluviodeltaic Triassic sediments. These sediments were cut by numerous steeply-dipping faults during latest Triassic to Late Jurassic rifting. There was little subsidence or deposition on the central and western plateau in the Jurassic.

Breakup between the northern Exmouth Plateau and the "Argo Landmass" in the Late Jurassic [Gradstein, Ludden et al., 1990], may have been related to the final rift movements on the major faults on most of the Central Exmouth Plateau. Uplift along the southern margin of the present Exmouth Plateau in the Berriasian to early Valanginian led to progradation of a massive delta across the southern plateau [Exon et al., 1992a]. Breakup of the western and southern Exmouth Plateau margins, which began in the late Valanginian, formed the Gascoyne and Cuvier abyssal plains, and the flow of detrital sediments to the plateau decreased greatly. The plateau started to subside

during the Aptian and its later history is one of slow sedimentation with decreasing detrital input.

Argo and Gascoyne Abyssal Plains

The Argo Abyssal Plain is flat and about 5.7 km deep. On the north it is bounded by the Java Trench (Figure 1). It is underlain by the oldest oceanic crust known in the Indian Ocean. This crust has slowly been consumed by the convergence of Australia and the Sunda Arc during the Cenozoic. At Site 765 the sequence of oceanic sediments is only 935 m thick [Gradstein, Ludden et al., 1990]. The area was starved of sediment due to the low relief of the Australian craton and the generally arid climate.

In the Argo Abyssal Plain, Larson [1975], Heirtzler et al. [1978], Veevers et al. [1985], and Fullerton et al. [1989] delineated the location of marine magnetic anomalies, generally trending N70 E (Figure 1). The presence of M26 to M16 marine magnetic lineations and isochrons is indicated. This suggests that at Site 765 basement may be of M26 or late Oxfordian age. Site 765 is less than 75 km from a prominent positive magnetic anomaly that lies

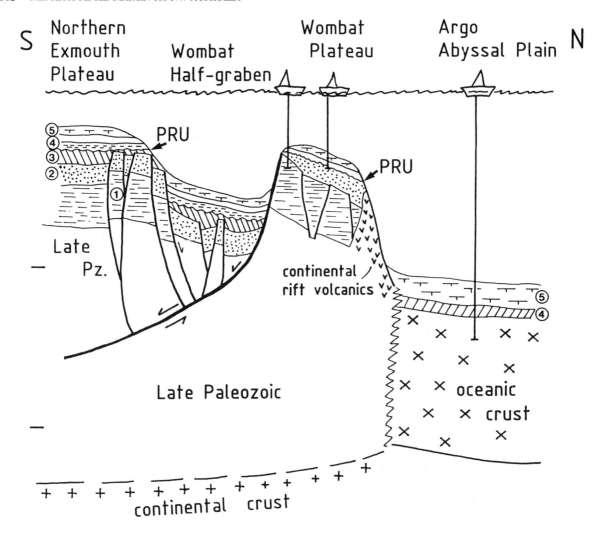

Fig. 5. Schematic N-S cross-section from N Exmouth Plateau to Argo Abyssal Plain, showing geological setting of Wombat Plateau: 1 = Permo-Triassic fluviodeltaic sediments, 2 = Upper Triassic fluviodeltaic to marginal marine deposits, 3 = Jurassic coal measure sequence and shelf carbonates, 4 = juvenile ocean sediments (Early Cretaceous), 5 = mature ocean sediments (Late Cretaceous-Cenozoic).

along the continent/ocean boundary [Veevers et al., 1985].

DSDP Site 261 [Veevers, Heirtzler et al., 1974], drilled during DSDP Leg 27, is located approximately 320 km to the north of Site 765. Magneto-stratigraphic interpretations placed DSDP Site 261 at M24, in the middle Kimmeridgian. Tholeiitic basalts drilled at the base of this hole are typical mid-ocean ridge basalts. The oldest sediments of that site are of late Kimmeridgian age [Dumoulin and Bown, 1992].

Site 766 is located on the western limit of the Exmouth Plateau, at the foot of the continental slope, leading down to the Gascoyne Abyssal Plain (Figures 1 and 3). The seafloor represents an erosional surface that may expose strata as old as Cretaceous. The sedimentary section is less than 0.5 s (500-700 m) of two-way traveltime on seismic sections. This section overlies a basement, which may represent the top of volcanics intermediate between oceanic and continental basement (Figure 7). The

overlying 500- to 700-m-thick sedimentary sequence consists of a late Valanginian and younger post-rift sequence directly on oceanic basement or an extensive sediment/volcanic complex underlain by continental crust [Exon and Buffer, 1992]. Seafloor spreading in the Gascoyne region is constrained by the Hauterivian to Barremian age of sediments overlying tholeiitic basalt at DSDP Site 260. Seafloor spreading is thought to have started in M10 time at the Valanginian/Hauterivian boundary, with a spreading jump seaward between M5 and M4 time (early Barremian).

EARLY-RIFT EVOLUTION (PERMO-TRIASSIC): THE INITIAL SPLIT OF GONDWANA

Permian to mid-Triassic evolution

In the early Mesozoic, the present northwest Australian margin was part of a continental rift zone on northeast Gondwana that bordered the Tethys Sea to the north. The

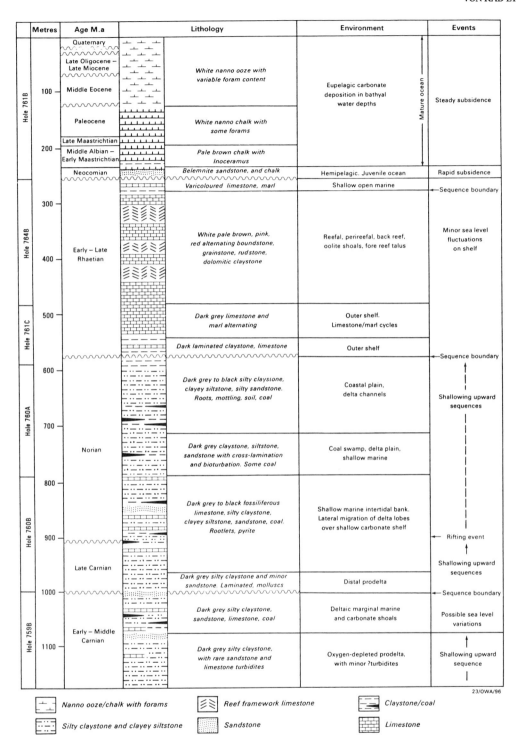

Metres	Age M.a	Lithology	Environment	Events
Hole 761B	Quaternary			
100	Late Oligocene – Late Miocene / Middle Eocene	White nanno ooze with variable foram content	Eupelagic carbonate deposition in bathyal water depths	Steady subsidence
200	Paleocene	White nanno chalk with some forams		
	Late Maastrichtian			
	Middle Albian – Early Maastrichtian	Pale brown chalk with Inoceramus		
	Neocomian	Belemnite sandstone, and chalk	Hemipelagic. Juvenile ocean	Rapid subsidence
300		Varicoloured limestone, marl	Shallow open marine	← Sequence boundary
400	Early – Late Rhaetian	White pale brown, pink, red alternating boundstone, grainstone, rudstone, dolomitic claystone	Reefal, perireefal, back reef, oolite shoals, fore reef talus	Minor sea level fluctuations on shelf
500		Dark grey limestone and marl alternating	Outer shelf. Limestone/marl cycles	
		Dark laminated claystone, limestone	Outer shelf	← Sequence boundary
600		Dark grey to black silty claystone, clayey siltstone, silty sandstone. Roots, mottling, soil, coal	Coastal plain, delta channels	Shallowing upward sequences
700	Norian	Dark grey claystone, siltstone, sandstone with cross-lamination and bioturbation. Some coal	Coal swamp, delta plain, shallow marine	
800 900		Dark grey to black fossiliferous limestone, silty claystone, clayey siltstone, sandstone, coal. Rootlets, pyrite	Shallow marine intertidal bank. Lateral migration of delta lobes over shallow carbonate shelf	← Rifting event
	Late Carnian			Shallowing upward sequences
1000		Dark grey silty claystone and minor sandstone. Laminated, molluscs	Distal prodelta	← Sequence boundary
	Early – Middle Carnian	Dark grey silty claystone, sandstone, limestone, coal	Deltaic marginal marine and carbonate shoals	Possible sea level variations
1100		Dark grey silty claystone, with rare sandstone and limestone turbidites	Oxygen-depleted prodelta, with minor ?turbidites	Shallowing upward sequence

23/OWA/96

Legend:
- Nanno ooze/chalk with forams
- Silty claystone and clayey siltstone
- Reef framework limestone
- Sandstone
- Claystone/coal
- Limestone

Fig. 6. Composite stratigraphic diagram from the Wombat Plateau Sites 759, 760, 761, and 764 showing maximum thicknesses and representative facies. Note post-rift unconformity (PRU) separating upper Triassic from overlying Berriasian sediments.

Exmouth Plateau is underlain by continental crust with a late Paleozoic to Triassic sedimentary section which is about 10 km thick [Exon and Willcox, 1978] and overlies early Paleozoic metamorphic and Precambrian plutonic rocks.

By Triassic time eastern Gondwana had drifted significantly northward and a temperate and humid climate prevailed during subsidence and progradation of a thick Triassic depositional wedge along the Western Australian margin [Boote and Kirk, 1989].

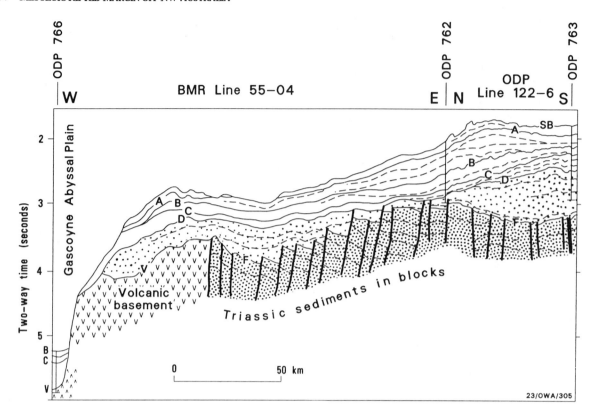

Fig. 7. Line drawing from seismic profiles ODP 122-6 and BMR 55-04 linking Sites 762, 763, and 766.

Upper Triassic (Carnian-Norian) fluviodeltaic and marginal-marine environment

At the Carnian/Norian boundary a broad delta plain graded into a proximal marine delta, and farther seaward into a distal prodelta with some shelf carbonates drilled at Sites 759 and 760. The region of the present Wombat Plateau represents the distal part of this delta system. Along the western margin of the present Exmouth Plateau, there was a major elongate island chain, characterized by rift volcanics [Exon and Buffler, 1992] which shed its debris into adjacent areas.

The 600-m-thick Carnian to Norian, mainly siliciclastic succession, was penetrated in Sites 759 and 760 (Figure 6) and can be correlated, using palynomorph biostratigraphy [Brenner, 1992a], wireline logs, microfacies and lithostratigraphy. This section can be subdivided by two sequence boundaries separating three upward-shoaling third-order cycles (Figure 8): (1) middle to upper Carnian prodelta mudstone to delta plain deposits; (2) uppermost Carnian to lower Norian marginal-marine to fluviodeltaic sequence topped by a prominent unconformity; and (3) middle to upper Norian proximal marine deltaic to non-marine delta plain deposits. These cycles of sedimentation consist of a large number of smaller (10-m-scale) shoaling-upward parasequences [Röhl et al., 1992], and document the common facies shift within a delta-dominated, marginal-marine environment.

The siliciclastic-dominated Carnian/Norian succession includes several shallow-water carbonate intercalations; they increase in frequency up-section, due to the general shallowing of the environment from prodelta to lagoon to delta plain. The limestone intercalations contain large proportions of terrigenous debris and are partly interpreted as storm deposits (tempestites), derived from carbonate shoals. Partly, they are bioclastic sands, washed together in front of river deltas, and intertidal to lagoonal limestones, deposited in restricted areas within a tidal flat and/or interdistributary bay.

Rhaetian transgression: lagoonal and reefal carbonates

The Rhaetian shelf limestone recovered in the Wombat Plateau sites, especially the reefal facies of Site 764, represents the first discovery of upper Triassic platform carbonates anywhere in the Australian Northwest Shelf region. This has potentially opened up a new target for petroleum exploration in the region [Williamson et al., 1989].

The Rhaetian was characterized by a major southward transgression of Tethys. In the Wombat Plateau area, the delta plain paleoenvironment (Site 760) changed into a shallow-marine shelfal carbonate setting. Due to a northward drift of Gondwana, Northwest Australia and northern Greater India reached a subtropical latitude [25°-30°S, Ogg et al., 1992] and faced the open Tethys sea. Thus

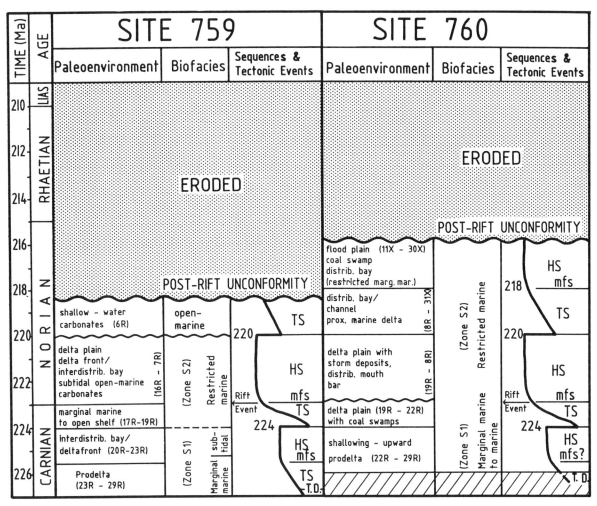

Fig. 8. Paleoenvironment, biofacies, tectonics and sequence stratigraphy of the Carnian-Norian sections of Sites 759 and 760. After von Rad et al. [1992]. Abbreviations see Fig. 9.

it is not surprising that coral, sponge, and hydrozoan reefs flourished near the outer edge of a broad carbonate platform with lagoons, patch reefs, and tidal flats, which in turn lay seaward of a delta system similar to that of the Norian stage [Dumont and Röhl, this volume].

The Rhaetian sediments can be subdivided using high-resolution seismic profiles into reefal and lagoonal facies [Williamson et al., 1989]. Criteria for the recognition of reefs include a zone of low reflectivity over the reef core, blanking of seismic energy beneath the core; and onlapping seismic facies around the reef core, corresponding to wedges of reef-derived detritus. In general, the reefs appear to be patch reefs, concentrated on structurally high blocks, particularly in the north of the plateau [Exon and Ramsay, 1990].

Williamson [1992] identified three seismic sequences within the Rhaetian strata. The entire succession is unconformably overlain by Cretaceous sediments, the unconformity being clearly angular in the south and almost subparallel in the north. Seismic data (Figures 4A, B), show the progressively younger nature of the terminations

of dipping Rhaetian strata below the main (post-rift) unconformity, and indicate its angular and erosive character.

The "Rhaetian" age assignment in Sites 761 and 764 is based on a correlation of foraminiferal, calcareous nannoplankton, ostracod, and especially palynological biostratigraphies [Brenner et al., 1992c]. The base of the Rhaetian is defined by the first appearance of the dinoflagellate *Rhaetogonyaulax rhaetica* [Brenner, 1992a]. The Tethyan ammonite zonation, based on the Alpine Koessen Beds [Figure 10; Tozer, 1990], is not correlated with the Australian dinoflagellate zonation [Helby et al., 1987; Brenner, 1992a], reflecting provinciality and ecological control.

Detailed microfacies analysis allowed clear characterization of the depositional environment. Based on the kind and amount of the major biogenic and abiogenic allochems, Röhl et al. [1991] distinguished 25 main microfacies types for the Carnian/Norian and Rhaetian carbonates. Based on microfacies analysis, gamma spectroscopy logs, and the interpretation of paracycles,

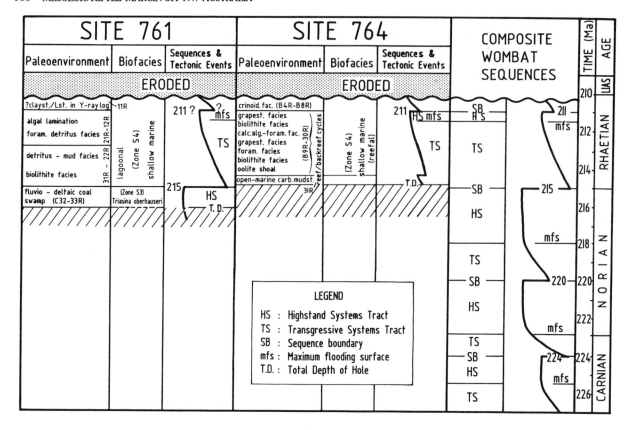

Fig. 9. Norian-Rhaetian paleoenvironment, biofacies, tectonics and sequence stratigraphy of Sites 761 and 764, plus composite sequence stratigraphy of all four Wombat Plateau sites. Age in Ma after Haq et al. [1987]. HS = highstand systems tract, TS = transgressive systems tract, mfs= maximum flooding surface. After von Rad et al. [1992].

the Rhaetian series of the Wombat Plateau can be correlated between the "lagoonal" Site 761 and the "reefal" Site 764 [Röhl et al., 1992; Sarti et al., 1992; Dumont and Röhl, this volume]. In general, the Rhaetian (except for its latest part) was a tectonically quiescent period of moderate platform subsidence compensated by carbonate buildup. It is characterized by two sequence boundaries, one at the Norian/Rhaetian boundary and one near the top of the Rhaetian (Figure 9). The two shallowing-upward sequences correspond to third-order depositional cycles of the global cycle chart [Haq et al., 1987].

Site 761 is characterized by an intertidal to shallow-subtidal lagoonal environment with many deepening-upward limestone/marl parasequences in a transgressive systems tract. Upwards they grade into a sand shoal facies with patch reefs in a lagoonal setting (highstand systems tract). Site 764 shows a similar development, but with typical features of reef development with higher carbonate productivity, better ventilated subtidal oolite shoals, and several reefal/lagoonal cycles.

Using a sequence-stratigraphic approach, we can subdivide the Rhaetian section into four systems tracts (Figure 9). There is no lowstand systems tract between the sequence boundary and the transgressive system tracts, since the tops of carbonate platforms typically lack lowstand wedges, which can only be found off-bank [Haq, 1991]. The four paracycles are:

1) The lower Rhaetian transgressive systems tract between the lower sequence boundary ("215 Ma") and the maximum flooding surface ("211.5 Ma")(in Site 761 this represents the beginning of carbonate buildup with deepening-upward limestone/marl cycles; in Site 764 this sequence is characterized by reef growth).

2) The "middle Rhaetian" highstand systems tract between the maximum flooding surface ("211.5 Ma") and the upper sequence boundary ("211 Ma").We observed at Site 761 many shallowing-upward (regressive) cycles consisting of pure wackestones with *Megalodon* and/or *Triasina* at the base of each cycle and less pure, supratidal algal laminite at the top. In Site 764, this paracycle is characterized by patch to pinnacle to back reefs.

3) The "upper Rhaetian" transgressive systems tract overlies the upper sequence boundary ("211 Ma"). It is represented by a deepening-upward sequence of open-marine shelf marl and claystone at Site 764 and a limestone/claystone cycle, interpreted from the gamma-ray wireline logs at Site 761.

4) An "uppermost Rhaetian" deepening-upward sequence of open-marine crinoidal grainstone and packstone, overlying an unconformity (Site 764).

Continuing tectonic subsidence during the Rhaetian resulted in a slow relative sea-level rise. The reefal facies

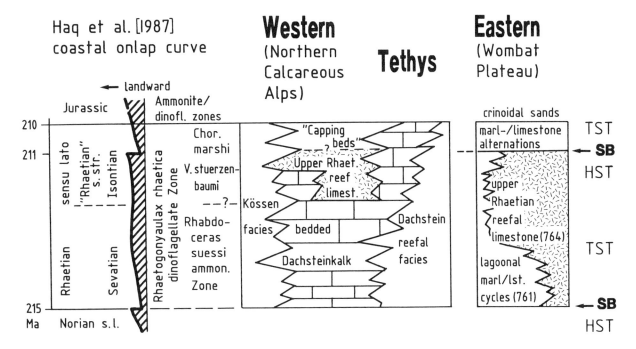

Fig 10. Comparison of western and eastern Tethyan facies and sequence stratigraphy of the Rhaetian [modified from Röhl et al., 1991]. Note good correlation of general facies and global eustatic events. SB = sequence boundary, TST = transgressive systems tract, HST = highstand systems tract. Note: Tethyan ammonite zones [Tozer, 1990] are not correlated with Australian *Rhaetogonyaulax rhaetica* dinoflagellate zone.

show first "keep-up" deposits that grade upward into "catch-up" lagoonal to reefal cycles. Drowning of the carbonate platform during a major latest Rhaetian relative sea-level rise, which was probably caused by accentuated tectonic subsidence and eustatic sea level rise, killed the reefs (= "give-up") and terminated carbonate buildup (Figures 6 and 9).

Two global late Triassic (Rhaetian) sea-level events (sequence boundaries) are documented in Sites 761 and 764 by litho- and biofacies, and by log character (Figure 9). Detailed comparisons between the structural and stratigraphic settings of the Alps and the Australian margin by Dumont [1992], Röhl et al. [1991] and Dumont and Röhl [this volume], show a sequence boundary inside the upper part of the Australian Rhaetian (Figure 10). Apart from this tectono-eustatic sea-level change, which was probably caused by major plate-tectonic reorganizations, the Rhaetian facies of the Wombat Plateau shows a striking similarity to that of the western Tethys. As in the Wombat Plateau area, the base of the Rhaetian in the western, southern, and northern Alps is characterized by significant upward-decreasing terrigenous input which may be explained by global sea-level changes [Dumont, 1992]. In the Northern Calcareous Alps the Kössen marls (limestone and marly shale facies) can be compared to the marl/limestone alternations within Site 761. Also the northern Alpine "Upper Rhaetian reefal limestone" (Oberrhät-Riffkalk) can be compared to the reefal series in Site 764 (Figure 10). In the Northern Calcareous Alps two sequence boundaries were discovered in the Upper Triassic

[R. Brandner, personal communication, 1992], one at the base of the Rhaetian Kössen Beds, and one near the top of the Rhaetian at the base of the "capping beds" [Figure 10; Stanton and Flügel, 1989].

The geodynamic evolution of the east Gondwanan Tethyan margin around the Triassic/Jurassic boundary is also very similar to the synchronous evolution of the European margin of the Ligurian Tethys exposed in the Western Alps of France [Dumont, 1992].

JURASSIC RIFT-DRIFT TRANSITION

In NW Australia, a prolonged phase of rifting began in the latest Triassic and continued until final breakup occurred in the Early Cretaceous [Audley-Charles, 1988; Boote and Kirk, 1989]. Initial rifting began along the margin north from the Exmouth Plateau and resulted in the Late Jurassic breakup during which a now-lost continental fragment separated, and moved away to form the Argo Abyssal Plain [Figure 11; Veevers, 1988]. Prior to and concurrent with this rifting event, a thick Jurassic synrift sedimentary succession accumulated in the restricted Barrow/Dampier rift basins [Veenstra, 1985]. Barber [1988] documented mid-Jurassic erosion over much of the central Exmouth Plateau in response to the rifting, and this region only began to accumulate thin marine muds after the Callovian breakup event. Wright and Wheatly [1979] interpreted the main angular unconformity seen on seismic profiles over much of the plateau to be mid-Jurassic in age. A thick Oxfordian to Tithonian marine sequence filled the Barrow-Dampier rift and marine shales up to 2000 m thick

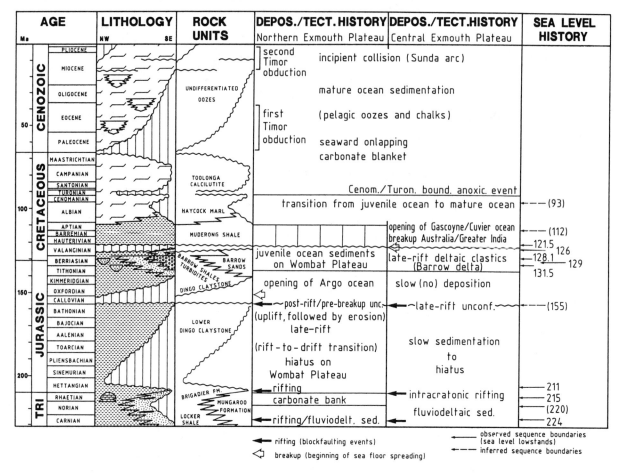

AGE	LITHOLOGY	ROCK UNITS	DEPOS./TECT. HISTORY Northern Exmouth Plateau	DEPOS./TECT.HISTORY Central Exmouth Plateau	SEA LEVEL HISTORY

Fig. 11. Stratigraphic summary of the central Exmouth Plateau and Wombat Plateau [modified after Boyd et al., 1992]. Lithology and rock units represent mainly central Exmouth Plateau, whereas the depositional, tectonic and sea level history is also shown for the northern Exmouth Plateau/Wombat Plateau area. Chronostratigraphy and ages of sequence boundaries after Haq et al. [1987].

prograded into the southern Kangaroo Trough [Barber, 1988].

On the central and western Exmouth Plateau the Jurassic sequence is condensed and intermittent in occurrence, and when Cretaceous deposition began it was onto (or virtually onto) Upper Triassic sediments, with substantial fault blocks forming an irregular surface with highs and lows trending northeast [Exon and Willcox, 1978; Barber, 1988; Exon and Buffler, 1992].

Early Jurassic paleogeography

During Early Jurassic times, the Exmouth Plateau was still within subtropical latitudes [25°-30°S, Ogg et al., this volume]. The major change since the Rhaetian was a further deepening of the depositional environment, possibly caused by blockfaulting. Outer shelf to upper slope limestones, dredged from the northern Wombat escarpment and the Swan Canyon area (Figure 1), include crinoidal packstone and a mud- to wackestone with redeposited large wood fragments and occasional quartz grains [von Rad and Exon, 1983; von Rad et al., 1990]. The microfacies of these hemipelagic to possibly turbiditic Rhaetian to

lowermost Jurassic limestones suggests a differentiation of the carbonate platform into several swells and deeper basins. A similar situation prevailed in the western Tethys, where the upper Rhaetian reef limestones are overlain by Lower Jurassic crinoidal limestones (Hierlatzkalk), red ammonite-bearing marlstone (Adneter Kalk) or open-marine marls (Figure 10).

Rhaetian-Early Jurassic rift tectonics and synrift volcanism

The Wombat Plateau was affected by major Early Jurassic volcano-tectonic events [Dumont, 1992; Figure 11]. The crustal extension caused (1) small volcanic eruptions from shallow magma chambers with diverse composition [cf. White and McKenzie, 1989; Hooper, 1990], and (2) a drastic change in the subsidence pattern of the margin. Nearshore rift basins, such as the Barrow-Dampier Basin, were rapidly filled and marginal plateaus, such as the Exmouth and Scott Plateaus, uplifted and truncated by a latest Pliensbachian unconformity [Hocking et al., 1987]. Similar synchronous events are recorded on the European passive margin of the Ligurian Tethys in the Western Alps. This suggests Tethys-wide plate-tectonic reorganization

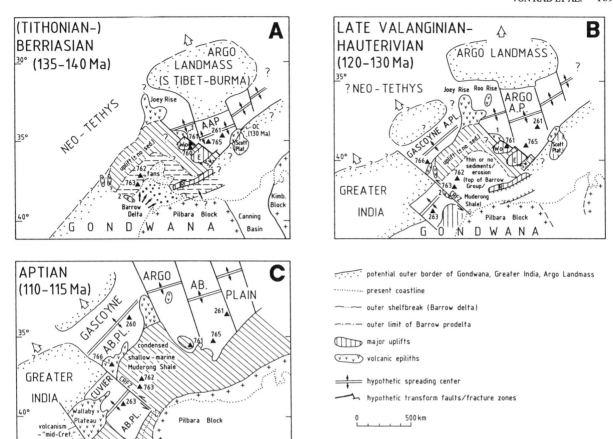

Fig. 12. Paleogeography of the Northwest Australian margin and vicinity during Tithonian to Aptian times. Modified after Barber [1988]; Exon et al. [1992b], Kopsen and McGann [1989]; Gradstein et al. [1992]; and Boyd et al. [1992]. Paleolatitudes after Ogg et al. [in press]. A. Tithonian to Berriasian. B. Late Valanginian to Hauterivian. C. Aptian. Abbreviations: AAP = Argo Abyssal Plain, CRFZ = Cape Range. Fracture Zone, E = Emu Plateau, OC = Oates Canyon (western escarpment of Scott Plateau), R = Rankin Platform, Sw = Swan Canyon graben, Wo = Wombat Plateau (transgressive Berriasian sand and calcisphere/bentonite series): 1 = Rhaetian-early Jurassic older volcanics (northern Wombat escarpment), 2 = mid-Berriasian intrusion.

causing changes of subsidence patterns and individualization of nearshore basins and marginal plateaus at the Triassic/Jurassic boundary [Dumont, 1992].

During a late Rhaetian (K-Ar age: 213 Ma) to earliest Jurassic (K-Ar age: 190-206 Ma) intracratonic rift phase, a suite of highly differentiated K-rich rhyolitic to trachytic rocks erupted along the northern Wombat Plateau, probably under subaerial to very shallow-marine conditions [von Rad and Exon, 1983; von Rad et al., 1990]. Volcanic activity in a "rift-valley situation" between the "Argo landmass" or "Mount Victoria Land block" (now located in Burma) and Exmouth Plateau is indicated by intermediate volcanics in Burma [Görür and Sengör, 1992], by the lowermost Jurassic volcanics dredged from the "contiguous" margin of the Wombat Plateau, by earliest Jurassic basaltic flows alternating with Hettangian shallow-water limestone at the Scott Plateau well Scott Reef No. 1, and by the strong magnetic anomaly found landward of the continent/ocean boundary south of the Argo Abyssal Plain [Veevers, 1988].

Exon and Buffler [1992] found that reflector-free acoustic basement along the outer western Exmouth Plateau, whose rough and irregular top is approximately level with the F unconformity, forms a transition zone between Triassic sediments and the continent/ocean boundary. Seismic and dredging evidence indicate that this acoustic basement consists of volcanic rocks intruded and extruded during the Triassic/Jurassic phases of rifting.

Short-lived volcanic activity accompanied the major plate-tectonic reorganization at or near the Rhaetian/ Hettangian boundary in the western Alps, eastern and western Pyrenees, Ebro Basin, Baleares, Betic zone, Sicily, Morocco, US East coast, and Northwest Australia [Dumont, 1992]. The synchroneity of these magmatic and tectonic events with global eustasy is conspicuous and suggests that during this time eustasy was tectonically forced.

Middle Jurassic coal measure succession

During the middle Jurassic, the Exmouth Plateau and the adjacent Tethyan Himalaya (Nepal) were at relatively higher (about 35°S) latitudes compared to the late Triassic/Early Jurassic; this resulted in increased siliciclastic input and reduced carbonate deposition [Gradstein et al., 1992]. The Exmouth Plateau was extensively structured by northeast-southwest trending faults. The central Exmouth Plateau and Rankin Platform became emergent, whereas very thick (up to 2000 m) nonmarine to marginal-marine siliciclastic sediments were deposited in troughs and basins located on both sides of this platform [Exon et al., 1992].

Studies on dredge material by von Stackelberg et al. [1980], von Rad and Exon [1983], and von Rad et al. [1990] show that there is a thick paralic Middle Jurassic coal-measure succession in the Swan Canyon area of the northern Exmouth Plateau, consisting largely of mudstone, siltstone, and sandstone (Figure 1). The Bathonian to Bajocian sequence in Delambre No. 1 well is 1007 m thick. Middle Jurassic sequences are, however, thin to absent on the central Exmouth Plateau and on the Wombat Plateau, which appear to have both been high at that time.

The "coal measure succession" documents a relative Middle Jurassic sea level lowstand with deltaic sedimentation (floodplain claystones, channel sands, delta foresets, and coal swamps) alternating with marginal marine sedimentation [von Rad et al., 1990]. Impregnation of many of these sediments with goethite cement was probably a post-depositional process during lateritic weathering after the Callovian/Oxfordian uplift.

Middle/Late Jurassic extensional tectonics (blockfaulting)

Callovian/Oxfordian late-rift tectonics caused major uplift of large parts of the Exmouth Plateau which lead to a post-rift unconformity. A trough persisted southeast of the Rankin Platform. The Tethys Sea invaded the southern Exmouth Plateau area, where the Dingo Claystone, a marine claystone and marl, was deposited. The northern Exmouth Plateau was flooded from the north, as documented by dredged paralic claystones and shelf carbonates in the Swan Canyon area and along the Rowley Terrace [Figure 1; Exon and Ramsey, 1990]. A major high along the western edge of the present Exmouth Plateau can be assumed. Upper Jurassic sediments were only locally deposited on the Exmouth Plateau as thin, condensed marine claystones ("Upper Dingo Claystone") or hemipelagic marls and micritic limestones (Swan Canyon dredges).

Regional geological evidence from seismic profiles and commercial wells indicates that the Wombat Plateau experienced uplift and northward tilting during a major Callovian/Oxfordian rift phase [von Rad et al., 1992]. Isostatic (flexural) rebound might be the reason for the tilting, because the adjacent Wombat half-graben subsided at the same time as the horst was uplifted along a major normal (listric ?) fault with a throw of several kilometers (Figures 4 and 5). Another hypothesis to explain the post-rift unconformity is the strong volcanic influence during the rift-drift transition: the uplift of the plateau might be due to a plume caused by the upwelling of hot asthenosphere and underplating during the late stages of

rifting and initiation of sea floor spreading [White, 1989; Mutter et al., 1988]. The blockfaulting event coincided with a widespread plate-reorganization during Callovian times and the resulting major sea level fall which was followed by a global early Oxfordian transgression caused by accelerated sea floor spreading [Gradstein et al., 1992]. After uplift above sea level, the Wombat Plateau was strongly eroded, forming the major "post-rift unconformity". Evidence for the post-depositional uplift and erosion of Wombat Plateau during Jurassic times includes (1) regional geological and seismic data, (2) post-depositional, late-diagenetic overprinting of the Rhaetian carbonates, and (3) the composition of the Berriasian to Valanginian transgressive sand at Site 761.

Seismic and regional geological evidence [Dumont, 1992; Exon et al., 1982; Veevers, 1988; Williamson, 1992; Exon and Ramsay, 1990] show lower to middle Jurassic sediments over most of the northern Exmouth Plateau (e.g., Swan Canyon area). This Callovian post-rift unconformity is commonly overlain by a thin (upper Jurassic to Neocomian) "juvenile ocean" sequence.

Rhaetian diagenetic features on the Wombat Plateau were strongly overprinted by a Jurassic late-diagenetic phase which affected the whole upper Triassic sediment pile during the post-Rhaetian uplift of the Wombat Plateau above sea level. Röhl et al. [1992] identified an upward increase in moldic porosity and dedolomitization features, due to a presumed upward increase of freshwater influence in the upper Rhaetian sections of Sites 761 and 764. Fe-oxide laminations in Site 764 indicate a reduction/oxidation boundary and were probably due to a fossil groundwater horizon.

INITIAL (LATE JURASSIC) BREAKUP NORTH OF THE EXMOUTH PLATEAU

Tectonics and volcanism

Although there was an extended history of Triassic/Jurassic rifting and subsidence prior to breakup, the final episode of breakup occurred in response to a thermal plume that caused widespread volcanism related to passive rifting of the margin [White and McKenzie, 1989; Ludden and Dionne, 1992]. This volcanism produced a series of volcanic plateaus, such as the Scott Plateau, the Joey Rise north of the Exmouth Plateau, and the Wallaby Plateau south of Exmouth Plateau [Figure 12; von Rad and Exon, 1983; von Rad and Thurow, 1992]. This widespread volcanism was connected to the late Jurassic uplift and erosional truncation of Exmouth Plateau.

Argo Abyssal Plain evolution

The start of sea-floor spreading in Argo Abyssal Plain lead to the formation of a new offspring of Tethys, "Tethys III" (Figure 12). This event was dated by sea floor magnetic anomalies (Figure 1), calibrated by the age of the overlying oldest sediments or of the oceanic basement, recovered in two boreholes, DSDP Site 261 and ODP Site 765 [Gradstein, Ludden et al., 1990].

The "basement ages" of both Argo Abyssal Plain sites

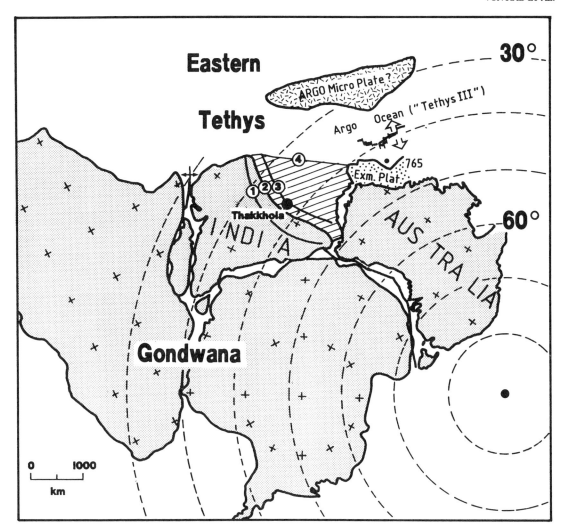

Fig. 13. Latest Jurassic (Kimmeridgian-Tithonian) paleogeographic reconstruction of east Gondwana and southeastern Tethys [modified after Ogg et al., this volume]. 1 = Main boundary fault of Himalaya (present); 2 = Indus-Tsangpo Suture; 3 = minimum northward extent of India in Gondwana after unwinding the doubled thickness of the crust south of the Indus-Tsangpo Suture; 4 = postulated northern edge of Greater India. N = Nepal [Thakkhola section; Gradstein et al., 1992]; cross-hatched: Himalayan Tethys sediments.

agree reasonably well with the magnetic anomalies. At about late Oxfordian times (M26, 155 Ma), sea floor spreading started in Argo Abyssal Plain between Northwest Australia (Gondwana) and the "Argo microplate" (see Figures 12 and 13). This southward jump of the spreading center from the Neo-Tethys in the north eventually led to the closure of Mesozoic Tethys. The hypothetical "Argo Landmass", located north of this proto-Indian Ocean drifted northward and was possibly accreted to the Laurasian continent as "Sikuleh Natal fragment" (now in Sumatra), or as "North Victoria Land" [now in Burma; Veevers, 1988], or as "Lhasa Block" (Figure 13). Alternatively, all of these blocks may be slivers of a single landmass.

For Site 765, drilled on magnetic anomaly M26, a K/Ar age of the celadonite, cementing a basaltic hyaloclastite directly overlying basaltic basement and underlying the oldest sediments in Core 123-765C-62R, is 155.3 ± 3.4 Ma [Ludden and Dionne, 1992]. This minimum basement

age corresponds to "Callovian" [Haq et al., 1987] or "late Oxfordian to early Kimmeridgian" according to the Kent and Gradstein [1985] scale. The oldest datable sediments in Core 123-765C-62R are of Tithonian age on the basis of nannofossils [Kaminski et al., 1992]. However, there is a 1.5 m interval of sediments barren of age-diagnostic fossils between the lower-most dated nannofossil samples and oceanic basement, suggesting a 10 to 15-m.y. hiatus or an extremely condensed sedimentation between basement emplacement (155 Ma) and Tithonian times (Figure 6). The oldest sediments recovered in Site 261 (on magnetic anomaly M24) are of late Kimmeridgian to early Tithonian age based on nannofossils [Dumoulin and Bown, 1992]. These sediments document a history of starved sedimentation punctuated by periodic influx of calcareous pelagic turbidites, especially during times of eustatic lowstands and depressed sea level, such as at latest Berriasian, early Valanginian, and early Aptian times

[Dumoulin and Bown, 1992]. Additional evidence for an extremely condensed sedimentation between Callovian/Oxfordian and Tithonian times comes from exploration wells in the Timor Sea area [Bradshaw, 1991].

The lowermost (Tithonian to Berriasian) sediments are red, brown and green claystones interbedded with bentonitic clays; they were deposited very slowly (2 m/m.y.) at or just below the CCD at around 2800 m paleodepth [Gradstein, 1992] and contain winnowed concentrations of inoceramid prisms and nannofossils, redeposited layers rich in calcispheres, Mn nodules and volcanogenic debris [Dumoulin and Bown, 1992]. They are overlain by Valanginian to Aptian calcareous claystone and by Upper Cretaceous to Tertiary claystones with turbidites (Figure 7). The Argo region apparently occupied a position at the southern limit of the Tethys nannofloral realm, thus yielding both Tethyan and Austral biogeographic features [Dumoulin and Bown, 1992].

Wombat Plateau juvenile ocean evolution

During Tithonian to Valanginian times, both the young Argo ocean basin and the adjacent passive margins experienced rapid tectonic subsidence. Extension lowered the flanks of the margin and caused a landward migration of the shoreline (a typical onlap pattern). The "juvenile ocean stage" is characterized by (1) hemipelagic transgressive sediments with a strong, upward-decreasing terrigenous influence (condensed sections); (2) restricted marginal-marine flora and fauna, indicating a "stressful" environment; and (3) post-breakup volcanism, indicated by altered ash layers (bentonites).

During latest Jurassic times eastern Gondwana had drifted southward placing Northwest Australia in temperate mid-latitudes [about 35°S according to Ogg et al., in press b]. Figure 13 shows the plate-tectonic setting at about 134 Ma: Eastern Gondwana is still intact with Antarctica-Australia-Greater India connected. In this reconstruction the northern margin of "Greater India" (India and Tethyan Himalaya) lies close to and southwest of the Exmouth Plateau [Gradstein and von Rad, 1991].

Tithonian-Berriasian paleogeography is shown in Figure 12A. Continental breakup was followed by sea floor spreading and rapid subsidence of the Argo Abyssal Plain. This "birth" of a narrow, east-west trending Argo (or proto-Indian) Ocean was marked by the deposition of a peculiar suite of hemipelagic "juvenile ocean" sediments of Berriasian to Valanginian age. Rapid tectonic subsidence caused a major relative sea level rise at the outer continental margin and the deposition of a unique record of an incipient ocean basin.

On the Wombat-Plateau (Site 761) a 20 m-thick, condensed, fining-upward (transgressive) nearshore-siliciclastic to hemipelagic "juvenile ocean" margin section of Berriasian to Valanginian age was recovered. It contained conspicuous lithofacies, such as ferruginous arkosic sand and belemnite-rich sandy-silty clay, overlain by calcisphere nannofossil chalks with intercalations of six thick layers of bentonite [altered ash turbidites; Figure

14; von Rad and Thurow, 1992; von Rad et al., 1992; von Rad and Bralower, 1992]. Considerably enhanced igneous activity can be expected, if the asthenosphere under the Exmouth Plateau was comparatively hot and the overlying lithosphere stretched and thinned during rifting and after the onset of oceanic spreading [White, 1989]. This resulted in explosive volcanic activity of dacitic composition during continental breakup and the following rapid tectonic subsidence at the outer continental margin.

NEOCOMIAN LATE-RIFT DELTAIC SEDIMENTATION AT THE CENTRAL EXMOUTH PLATEAU AND THE FINAL SEPARATION BETWEEN AUSTRALIA AND GREATER INDIA

A thick Neocomian wedge of clastic sediments was penetrated in Sites 763 (more proximal prodelta) and 762 (distal prodelta). This late-rift prograding continental margin wedge was deposited during Berriasian times, when there was a sudden progradation of deltaic sediments of the Barrow Group from the south [Exon et al., 1992], reaching a peak sedimentation rate of around 300 m/m.y. This progradation terminated suddenly in the early Valanginian (Figure 12).

The proximal marine delta east of Site 763 consists largely of sandstone and siltstone, is up to 1500 m thick and built toward the north. The maximum extent of the delta front [Erskine and Vail, 1988; Boyd et al., 1992] lies just south of Site 763 (Figure 7). In the adjacent Vinck No. 1 well, 677 m of prodelta mudstones with some turbidite sandstones were drilled. Site 762 is toward the outer limit of the distal delta. In the adjacent Eeendracht No. 1 well the distal deltaic succession, comprised of prodelta and basin floor mudstones, is only 149 m thick. Water depths during deposition of the Barrow Group are estimated to be 200-400 m at the two ODP sites.

Veevers and Powell [1979] and Exon and Buffler [1992] provide indirect evidence that a large part of the Barrow Group sediment may have been provided by a major ridge, probably of the order of two kilometers high, which formed by thermal doming above the future Cape Range Fracture Zone Exmouth Plateau (Figure 1). This ridge provided reworked sediments for the prograding delta, and collapsed immediately after breakup, thus cutting off the sediment supply. An eastern sediment source in the Pilbara Block may also have been important, at least in the Barrow Sub-basin and on eastern Exmouth Plateau [Boote and Kirk, 1989]. Boyd et al. [1992] make a strong case for another major sediment source in the uplifted area where the CRFZ met the north-south spreading center which led to the formation of the Cuvier Abyssal Plain. Although Greater India lay to the southwest, the ridge between it and the plateau prevented it from providing sediment to the delta. Volcanic highs in the southwest, probably of latest Triassic to earliest Jurassic age [Exon and Buffler, 1992], formed islands and were planated by wave action in the Berriasian, shedding sediment which formed local sand bodies in the process.

Well north of the delta, which loaded and depressed the Triassic surface, the water was apparently shallow, and thin

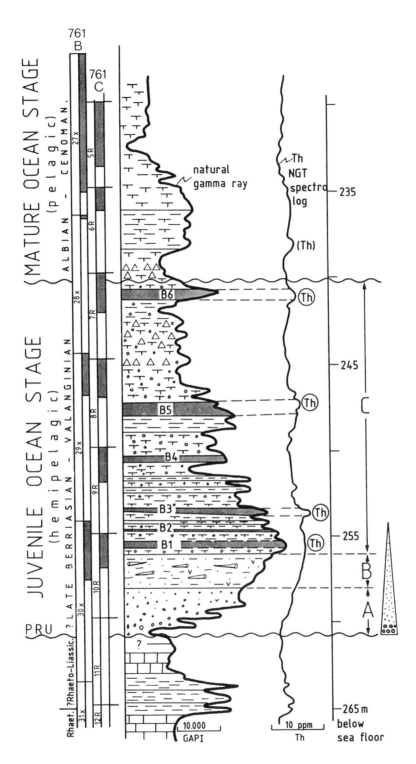

Fig. 14. Lithostratigraphy of the Lower Cretaceous section in Site 761 (Wombat Plateau) correlated with wireline logs [from von Rad and Thurow, 1992]. Note the good correlation of the six bentonite layers (B1 to B6) with natural gamma-ray maxima and high Th contents. Due to the limited core recovery (left-hand column) the lithostratigraphic column is largely based on interpretation of the natural gamma-ray tool (NGT) record. A,B,C = lithostratigraphic units of "juvenile ocean section", PRU = post-rift unconformity.

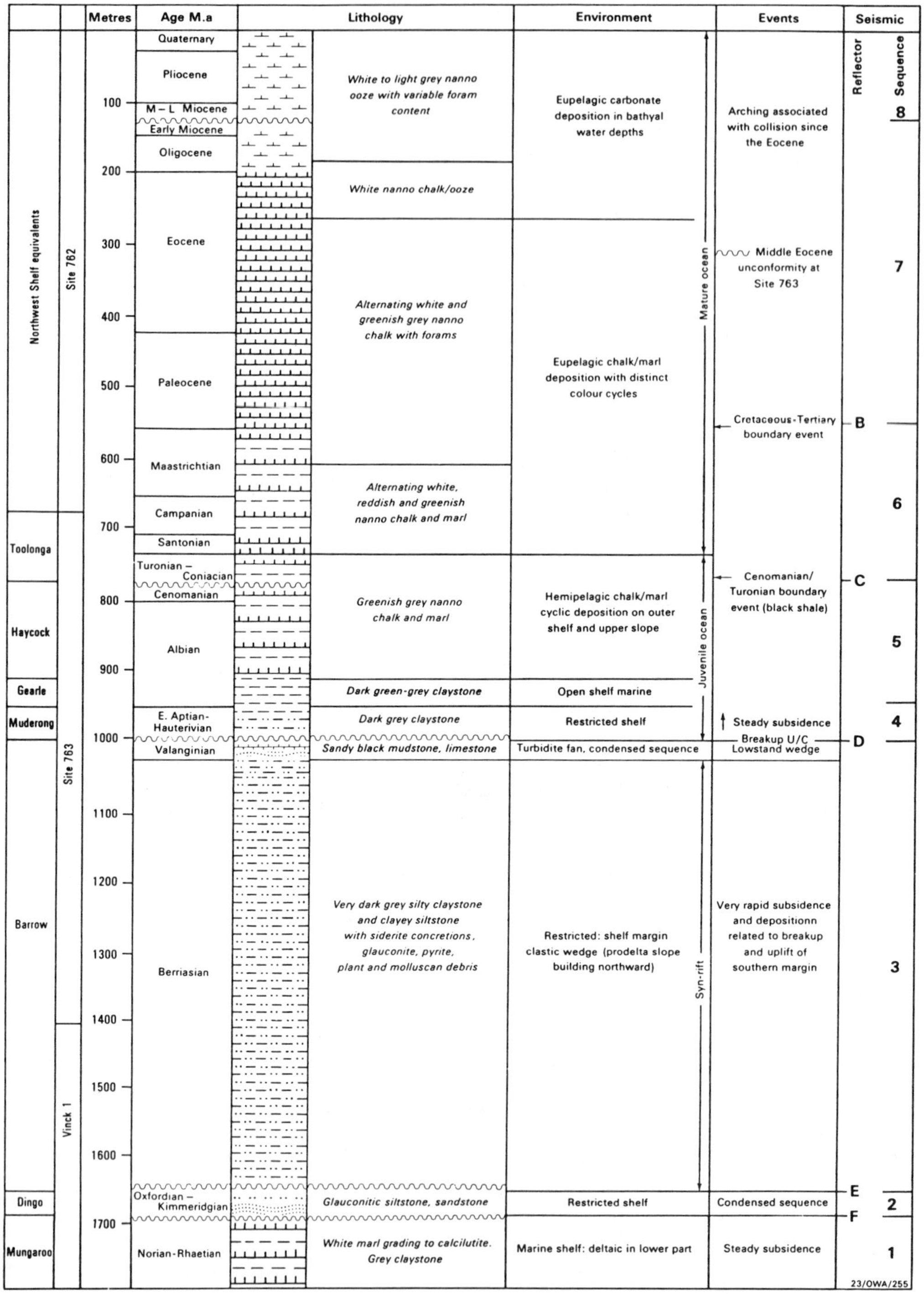

Metres	Age M.a	Lithology	Environment	Events	Seismic Reflector	Sequence
	Quaternary					
	Pliocene	White to light grey nanno ooze with variable foram content	Eupelagic carbonate deposition in bathyal water depths	Arching associated with collision since the Eocene		8
100	M – L Miocene					
	Early Miocene					
	Oligocene					
200		White nanno chalk/ooze				
300	Eocene			Middle Eocene unconformity at Site 763		7
400		Alternating white and greenish grey nanno chalk with forams	Eupelagic chalk/marl deposition with distinct colour cycles			
500	Paleocene			Cretaceous-Tertiary boundary event	B	
600	Maastrichtian					
		Alternating white, reddish and greenish nanno chalk and marl				6
700	Campanian					
	Santonian			Cenomanian/ Turonian boundary event (black shale)	C	
800	Turonian – Coniacian	Greenish grey nanno chalk and marl	Hemipelagic chalk/marl cyclic deposition on outer shelf and upper slope			
	Cenomanian					5
900	Albian					
	Gearle	Dark green-grey claystone	Open shelf marine			
1000	E. Aptian-Hauterivian	Dark grey claystone	Restricted shelf	Steady subsidence		4
	Valanginian	Sandy black mudstone, limestone	Turbidite fan, condensed sequence	Breakup U/C Lowstand wedge	D	
1100						
1200		Very dark grey silty claystone and clayey siltstone with siderite concretions, glauconite, pyrite, plant and molluscan debris	Restricted: shelf margin clastic wedge (prodelta slope building northward)	Very rapid subsidence and deposition related to breakup and uplift of southern margin		3
1300	Berriasian					
1400						
1500						
1600						
1700	Oxfordian – Kimmeridgian	Glauconitic siltstone, sandstone	Restricted shelf	Condensed sequence	E / F	2
	Norian-Rhaetian	White marl grading to calcilutite. Grey claystone	Marine shelf: deltaic in lower part	Steady subsidence		1

Left column groups: Northwest Shelf equivalents; Site 762; Site 763; Toolonga; Haycock; Gearle; Muderong; Barrow; Vinck 1; Dingo; Mungaroo. Right side labels: Mature ocean; Juvenile ocean; Syn-rift.

23/OWA/255

Fig. 15. Composite stratigraphic diagram from central Exmouth Plateau Sites 762 and 763 with maximum thicknesses and representative facies.

bathyal to shallow marine successions were laid down. Still further north on the plateau virtually no sediments are preserved, and that area may have been emergent [Exon et al., 1992a].

The dating of the opening of Gascoyne Abyssal Plain at the northwestern margin of Exmouth Plateau is documented by Site 766 (Figure 1). According to Fullerton et al. [1989] the site is on anomaly M10 (latest Valanginian to early Hauterivian), and breakup was most likely somewhat earlier. Site 766 drilled a "basement" of basaltic sills with thin interbeds of latest Valanginian sediments, and overlain by glauconitic bathyal sandstone and siltstone dated by palynology as late Valanginian and Hauterivian with a paleodepth of 900 m predicted from backtracking calculation [Gradstein, Ludden et al., 1990]. Therefore breakup along the northwestern margin of the Exmouth Plateau occurred in the late Valanginian.

The lack of late Valanginian and early Hauterivian sediments on the Exmouth Plateau contrasts with the situation in Site 766. Sites 762 and 763 both contain Berriasian to lower Valanginian deltaic mudstone of the Barrow Group, overlain disconformably by middle Hauterivian to Barremian shallow-marine mudstones. Three dinoflagellate subzones are missing, representing about 7 m.y. There are no major time breaks in the overlying Barremian, Aptian, or Albian sequences [Brenner, 1992b].

LATE APTIAN-ALBIAN EVOLUTION

During the late Aptian-Albian(-Cenomanian), the Exmouth Plateau experienced a series of major transgressions. At Site 761, early Albian transgression followed a 20-m.y.-long hiatus (Figure 14). A similar hiatus was observed in the central Exmouth Plateau Sites 762 and 763 where continental breakup was dated as late Valanginian to early Hauterivian [Brenner, 1992b].

During the Aptian, i.e., after the Greater India-Australia separation, bathyal chalk was deposited on Wombat Plateau, whereas a thin transgressive shallow-marine mudstone succession was deposited on the central Exmouth Plateau and on the Northwest Shelf [Exon et al., 1992a].

In the Albian, the Windalia Radiolarite, the Gearle Siltstone and the lower Haycock Marl were deposited in various combinations across most of he region [Hocking et al., 1987]. In general, mudstone was deposited on the inner shelf, marly sediments on the outer shelf, hemipelagic claystone, chalk and calcilutite in lower neritic and upper bathyal depths, and pelagic claystone on the abyssal plains [Exon et al., 1992a].

In Sites 762 and 763 the Gearle Siltstone consists of mudstone, marl and limestone, is relatively thin, and is confined to the early Albian; it rests unconformably on the lower Aptian or Muderong Shale [Exon et al., 1992a]. The Haycock Marl was deposited on the Gearle Siltstone, and it marks the onset of a steady increase in the proportion of pelagic carbonate and decrease in mud, which culminated in the Late Cretaceous with the deposition of chalks (Figure 15). According to wireline log data, the major transition from mudstone to marl and chalk occurs above the Aptian Muderong shale.

During the Albian [Exon et al., 1992a], another major transgression produced a very thin chalk sequence on the Wombat Plateau. At Site 761, nannofossils indicate that Albian to Cenomanian is present [Bralower and Siesser, 1992]. Radiolarian blooms suggest high fertility (?upwelling), but the deposited radiolarian chalks were later transformed to porcellanites and might be partly equivalent to the Windalia Radiolarite. The Albian chalks were probably deposited at a water depth of 1000 m and indicate

the transition from the hemipelagic "juvenile ocean" to the pelagic "mature ocean" stage.

In Site 766 (Figure 8), located on Valanginian oceanic crust at the foot of the western scarp (Figure 1), all Cretaceous sediments are bathyal [Gradstein, Ludden et al., 1990]. Chalks are present in the Aptian, earlier than on the plateau. The Barremian/ Aptian boundary represents a marked change in sedimentation from rapidly deposited clayey and sandy submarine fan deposits to hemipelagic/pelagic nannofossil oozes [Gradstein, 1992]. This implies a global tectonic and/or eustatic event.

MATURE OCEAN EVOLUTION (LATE CRETACEOUS AND YOUNGER)

The "mature ocean" stage is characterized by the lack of terrigenous influx and unrestricted thermohaline circulation resulting in pelagic sedimentation in a relatively deep marine environment.

By Turonian-Coniacian time, the Wombat Plateau lay in bathyal depths (500-1000 m), accumulating pure foraminiferal nannofossil chalks. Subsidence continued at a steady pace throughout the late Cretaceous and early Paleogene, decreasing somewhat in the late early Eocene (around 50 Ma) either due to diminishing heat loss from the oceanic lithosphere or in response to the initial collision of India with Asia which effected a slowing down of seafloor spreading rates throughout the Indian Ocean. In general, the Late Cretaceous-Paleogene represents an extended period of tectonic quiescence without any record of volcanic activity on the Wombat Plateau, which slowly subsided to its present water depth (2000-2500 m). The Australian continent separated from Antarctica during late Cretaceous/ Paleocene times and drifted northward from about 40°S (latitude of NW Australia during Albian) to its present latitude (17°S).

After the Albian the Wombat Plateau had apparently subsided sufficiently so as not to show a direct response to sea-level fluctuations. The Upper Cretaceous section is quite condensed (60 m at Site 761) and disrupted by four unconformities [Bralower and Siesser, 1992].

The light and dark cycles in the Campanian to Maestrichtian pelagic sediments depend on clay content and probably on climatic changes driven by Milankovich cycles. Dark clay-rich layers were deposited in wetter and warmer climatic phases with high insolation and terrestrial runoff and low bioproductivity. The broad dating of the sequence suggests that the most likely controlling mechanisms, at least in the Turonian to Maestrichtian interval are the Milankovitch 19-23,000 years precession cycle and the 41,000 year obliquity cycle [Huang et al., 1992]. The change in Site 762, from dominantly reduced (green) sediments in the Albian, to dominantly red or multicolored sediments in the younger portion of the formation, must reflect a change in oxidation state and a decline in enclosed organic matter. This coincides with the disappearance of plant debris and pyrite after the Albian.

The Cenomanian-Turonian boundary event, an ocean-wide anoxic episode [Schlanger et al., 1987] is well documented

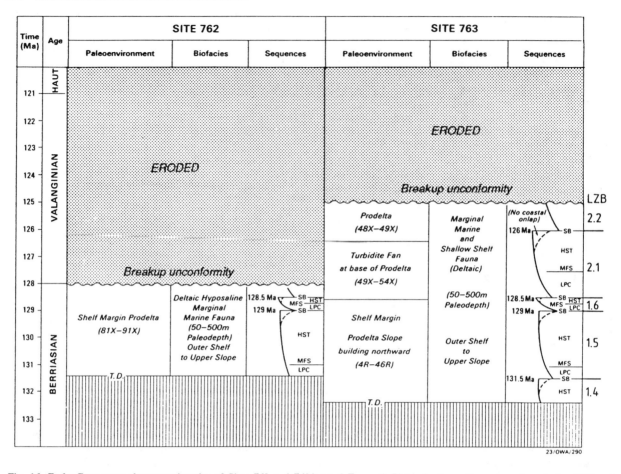

Fig. 16. Early Cretaceous chronostratigraphy of Sites 762 and 763(central Exmouth Plateau). Ages of sequence boundaries (SB) and maximum flooding surfaces (MFS) from Haq et al. [1987]. HST = highstand systems tract, LPC = lowstand prograding complex. After Haq et al. [1992]. 1.4-1.6, 2.1-2.2 = third-order cycles after Haq et al. [1987].

in Site 763 but is missing at Site 761 due to a major hiatus between the upper Turonian and lower Albian. At Site 763, a carbonate-free black shale layer has up to 25% total organic carbon, zeolite-replaced radiolarians, and a lack of bioturbation. At Site 762, the black claystone generally lacks calcareous foraminifers, but contains arenaceous foraminifers and radiolarians which are commonly pyritized or zeolitized. This global anoxic event was due to a major sea level rise at the beginning of the Turonian which caused a strong decrease in terrigenous input and a peak of bioproductivity. High production or preservation of planktonic organic matter resulted in a rising calcite compensation depth and an expanded mid-ocean O_2-minimum layer impinging on the upper continental slope [Thurow et al., this volume].

An apparently uninterrupted Cretaceous/Tertiary (K/T) boundary section was recovered from Hole 761C. An iridium anomaly has been recorded in close coincidence with the planktonic turnover at this site which is also associated with abundance of chromium and Mg-Al-Cr-Ni-rich magnetites [Rocchia et al., 1992]. These authors suggest that the terrestrial material was possibly accreted

during a number of impacts with multiple small bodies dispersing most of their mass in the atmosphere. The K/T boundary section is biostratigraphically complete [Pospichal and Bralower, 1992]: mid-latitude (about 36°S) Cretaceous nannofossils are replaced at the boundary by opportunistic "survivor" species, which in turn are replaced by newly evolved Danian taxa.

DISCUSSION

Sequence Stratigraphy

Wombat Plateau. The upper Triassic depositional sequences recovered from the four Wombat Plateau sites are a composite record of regional tectonics, changes in subsidence rates, changes in the source and rate of sediment supply, climate, and eustatic sea-level fluctuations [von Rad et al., 1992].

The Carnian-Norian section at Sites 759 and 760 (Figure 8) is characterized by three third-order sequences [UAA 3.2, UAA 4.1a and b of Haq et al., 1987], separated by two sequence boundaries (SB). The lower SB (late Carnian) is correlated with the 224 Ma SB of the Haq et al. [1987] chart; the upper SB, of mid-Norian age (approximately 220

Ma), was not included in the global cycle chart of Haq et al. [1987]. In general, we observe an upward-shallowing trend with an evolution from prodelta to delta front, or from subtidal lagoon to delta plain, or from distributary bay to flood plain and coal swamp.

The Rhaetian at Sites 761 and 764 (Figure 9) is characterized by two sequence boundaries (SB), one at the Rhaetian/ Norian boundary ("215 Ma"), and one in the uppermost Rhaetian ("211 Ma"). The lower SB is underlain by Norian flood plain deposits (cycle UAB 4.1b) and overlain by "lower" Rhaetian transgressive systems tract sediments (cycle UAB-1) which are lagoonal at Site 761. The "middle" Rhaetian represents lagoonal/reefal cycles which belong to the highstand systems tract; again they are lagoonal-dominated (sub- to intertidal) at Site 761 and reef-dominated (shallow subtidal) at Site 764. The "uppermost" Rhaetian SB ("211 Ma") documents the first drowning of the Rhaetian carbonate platform, and the overlying more open-marine (subtidal) sediments represent the next transgressive systems tract of cycle UAB 2.1b.

Central Exmouth Plateau. The Cretaceous strata on the Exmouth Plateau rest on a condensed sequence of Jurassic shelf sediments ("Dingo Claystone equivalent") or directly on Triassic paralic to fluviodeltaic non-marine sequences. A sequence stratigraphic interpretation is shown in Figure 16 [Haq et al., 1992] and the lithostratigraphy and seismic stratigraphy in Figure 15.

Sequence-stratigraphic analysis of the 700 m-thick Berriasian-Valanginian progradational prodelta sediments recovered in Sites 762 and 763 is facilitated by the availability of extensive seismic data and well logs in the vicinity [Erskine and Vail, 1988; Boyd et al., 1992]. Five third-order cycles [LZB 1.5, 1.6, 2.1, 2.2, 2.3 after Haq et al., 1987], separated by four sequence boundaries ("131.5 Ma", "129 Ma", "128.5 Ma", "126 Ma") can be distinguished in the Berriasian-Valanginian pre-breakup section of Sites 762 and 763 (Figure 16).

In Site 763 the highstand systems tract of third-order cycle LZB 1.4 is overlain by two complete sequences (LZB 1.5 and 1.6) and a condensed sequence (2.1) [Haq et al., 1992]. The highest sequence boundary ("126 Ma") is closely followed by the breakup unconformity marking a nearly 10-m.y. (late Valanginian to late Barremian) hiatus associated with the final breakup of Exmouth Plateau. This breakup unconformity is overlain by uppermost Barremian to Aptian claystone (Muderong Shale equivalent) deposited in a shelf environment following a major transgression and accelerated deepening as the rate of basin subsidence increased. Site 762 contains a less complete sequence-stratigraphic record of the Early Cretaceous (Figure 16) with only two sequence boundaries.

Structural/stratigraphic evolution and subsidence history

Figures 3 and 5 show schematic cross-sections across Exmouth Plateau to the adjacent abyssal plains and the stratigraphic development of this passive margin sequence from an intracratonic "early rift" stage (late Triassic) to

"late rift" or rift-drift transition (Jurassic). The marked post-rift unconformity (E) is underlain by block-faulted Triassic (and partly lower to middle Jurassic) synrift sediments and overlain by a thin cover of Cretaceous to Cenozoic post-breakup sediments. The post-rift phase starts with a hemipelagic "juvenile ocean stage" (Early Cretaceous) which grades into a purely pelagic "mature ocean stage".

Figure 17 is a graphic summary of the structural and stratigraphic evolution of Wombat Plateau during the past 230 m.y. shown in seven time slices. During Carnian/Norian times (Figure 17A) we observe the northward progradation of seaward-thickening fluviodeltaic sediments from the Gondwanan continent with thin intercalations of lagoonal carbonates, rift-related block-tectonic movements, and volcanism (aborted "early rift" phase). The Rhaetian (Figure 17B) is characterized by an extensive southward transgression of the Tethys sea. This caused a major buildup of reefal and lagoonal carbonates. The carbonate platform was drowned for the first time during latest Rhaetian times ("211 Ma"). At the Callovian/Oxfordian boundary (Figure 17C), the whole area of the northern Exmouth Plateau was uplifted, tilted, and eroded during a major late-rift block-faulting event which lead to breakup of the northern margin in the Late Jurassic. This produced the marked post-rift unconformity with non-deposition and/or erosion of Jurassic strata on the uplifted horst.

Dumont [1992] suggests two different, independent rift phases: (1) a "Tethyan" rift phase with tectonic pulses in the Carnian and near the Rhaetian/Hettangian boundary, responsible for the "Jurassic breakup of northern Gondwanan continental blocks", and (2) an Indo-Australian rift phase (after Callovian?) causing renewed uplift and erosion when true oceanic crust was formed in Argo Abyssal Plain. During Berriasian times (Figure 17D-1) the newly formed Argo Abyssal Plain was flexurally coupled with Wombat Plateau. This resulted in a major transgression and the deposition of a condensed, deepening-upward succession of terrigenous to hemipelagic "juvenile ocean" sediments: nearshore arkosic sands, overlain by belemnite-rich sandy muds (deeper shelf) and calcisphere-nannofossil chalks (outer shelf to upper slope), intercalated with thick bentonites. During Late Cretaceous (post-Albian) times (Figure 17E), Wombat Plateau had subsided to bathyal depths resulting in the slow deposition of eupelagic chalks and oozes ("mature ocean stage").

A geohistory diagram of a composite Wombat Plateau section shows the paleobathymetry, decompacted burial history (subsidence) and the sedimentation rates for the past 200 m.y. (Figure 18). During the Carnian-Norian we observe several shallowing-upward cycles (prodelta to coal swamp); at the Rhaetian/Norian boundary we see evidence of a global transgression which resulted in carbonate buildup. Subsidence and sedimentation rates were high during the late Triassic aborted rifting stage. Drowning of the carbonate platform was followed by subsidence to

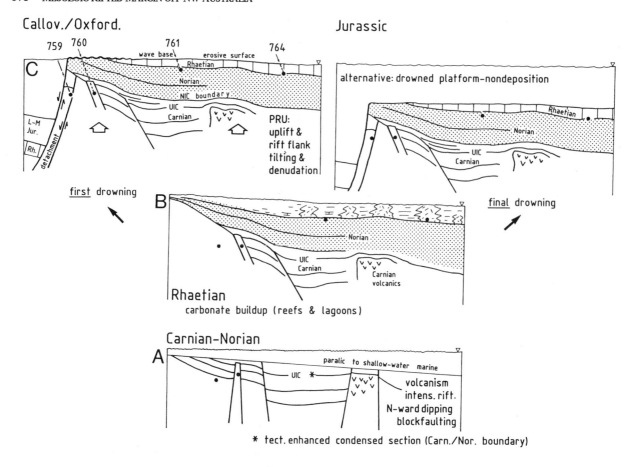

Fig. 17. Structural evolution of Wombat Plateau (and Exmouth Plateau) during the past 230 Ma. A. Carnian/Norian boundary (fluviodeltaic siliciclastics). B. Rhaetian (carbonate buildup). C-1. Callovian/Oxfordian (post-rift unconformity). C-2. Jurassic [alternative explanation after Sarti et al., 1992]. D-1. Neocomian "juvenile ocean" stage of Wombat Plateau. D-2. Berriasian [alternative explanation after Sarti et al., 1992]. E. Late Cretaceous ("mature ocean" stage).

slightly greater water depths during Early Jurassic times (based on dredge information). After uplift during Callovian/ Oxfordian times and the formation of the post-rift unconformity, the plateau experienced rapid tectonic subsidence submerging it during Early Cretaceous times. All commercial Exmouth Plateau wells show a similar subsidence history with rapid subsidence during late Triassic to early Jurassic, a phase of uplift (truncation) or negligible subsidence in the middle to late Jurassic, followed by renewed rapid tectonic subsidence after breakup during latest Jurassic to early Cretaceous times [Cloetingh et al., 1992].

The central Exmouth Plateau experienced the following evolution [Boyd et al., 1992; Figure 11]:

(1) Early-rift stage of intracratonic sedimentation with a seaward-thickening sequence of marine clastics during the Norian and Rhaetian;

(2) Late-rift stage with continued rifting and block-faulting during Jurassic to Berriasian/Valanginian times. The initial breakup north of the Exmouth Plateau was accompanied by rifting in the central plateau area ("failed breakup event"). This resulted in infilling of the nearshore rift-valley troughs (Barrow-Dampier Basin, Kangaroo Trough). Major uplift southeast and south of Exmouth Plateau during latest Jurassic to Berriasian times caused subsidence and deposition of 1-2 km of deltaic marine clastics on the central Exmouth Plateau (Barrow Group).

(3) Initiation of sea floor spreading northwest and south of the Exmouth Plateau (late Valanginian). The rifting, heating and extension shifted seaward to the Gascoyne and Cuvier rifts and southward to the Perth Basin, as Greater India separated from Gondwana.

According to a model for the geological development of the southwestern margin of the Exmouth Plateau by Exon and Buffler [1992], the plateau was firmly linked to Greater India in the Triassic and Jurassic. During the Jurassic the central plateau was high, perhaps because of thermal effects associated with rifting. In the latest Jurassic and Berriasian, there was major uplift along the length of the future Cape Range Fracture Zone, along normal faults produced by thermal doming immediately prior to breakup. Erosion stripped Triassic and Permian sediments from it,

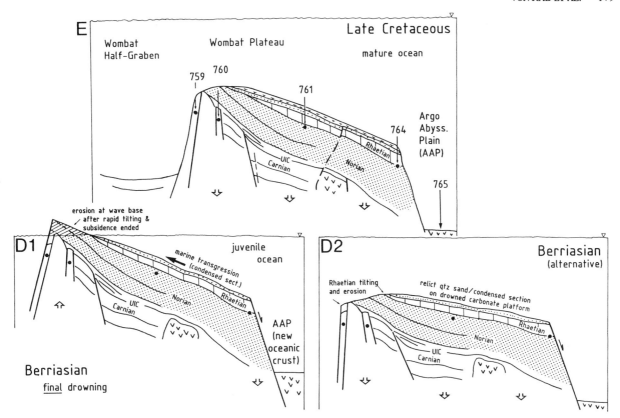

and deposited them rapidly across the southern plateau as part of the Berriasian to early Valanginian Barrow Delta. Late Valanginian breakup along the Cape Range Fracture Zone by west-northwest oriented transcurrent faulting removed Greater India and left behind the basalts in the Cuvier Abyssal Plain. The outermost margin of the plateau subsided along pre-existing and new faults, in harmony with the cooling abyssal plain basalts. However, the inner part of the Cape Range Fracture Zone uplift was preserved as a high, presumably because of igneous buttressing at depth, and the margin as a whole formed an anticlinorium.

The breakup of the northwestern margin of the Exmouth Plateau occurred at the same time (late Valanginian) as the southwestern margin, as attested to by seismic studies of Exon and Buffler [1992], the magnetic anomaly study of Fullerton et al. [1989], and the age of the sediments overlying basement in Site 766 (Figure 8). However, the northwestern margin formed by rifting, with the magnetic anomalies sub-parallel to the margin (Figure 1). Breakup produced a margin bounded by normal faults and fracture zones. Immediately after breakup, fairly deep water (ca. 800-1000 m) bounded the margin to the west, as shown by the basal sediments in Site 766. A gentle oceanward tilt of the volcanic basement developed, as the oceanic crust cooled and sank, pulling the continental margin down.

CONCLUSIONS

1. The Mesozoic depositional sequences recovered from the sediment-starved old, rifted margin at Wombat and Exmouth Plateau (Figure 19) provide a case study of the dynamic relationship between vertical tectonics (uplift during blockfaulting, differential subsidence), climatic changes, changes in the source and rate of siliciclastic input, volcanic activity (early rift and post-breakup), and eustatic fluctuations. The recovered section provides an excellent example of the interaction of these factors and the depositional response.

2. We have shown the important influence of tectonics during the rift stage and rift-drift transition. From regional geology, seismic stratigraphy, and ODP drilling we know that at least five major tectonic events influenced the evolution of this margin:

(a) a rift event near the Carnian/Norian boundary, caused by block faulting (uplift, rapid erosion, and redeposition);

(b) further rifting near the Triassic/Jurassic boundary, leading to widespread intermediate volcanism;

(c) a major post-rift unconformity at Callovian/ Oxfordian times resulting from uplift of the outer margin, tilting of the block, and subaerial erosion; this caused a major Jurassic hiatus on the Wombat Plateau horst;

(d) an initial breakup during Late Jurassic (Oxfordian) times north of Exmouth Plateau, resulting in the opening of the Argo Abyssal Plain, and

(e) a final breakup during late Valanginian/Hauterivian (M-10) times northwest and south of Exmouth Plateau, following intense uplift of the southern hinterland and

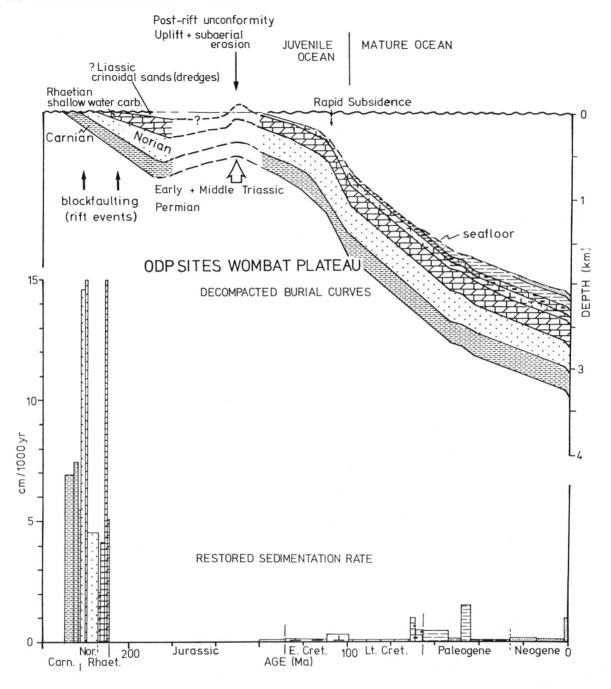

Fig. 18. Geohistory diagram showing decompacted burial curves and paleobathymetry, as well as restored sedimentation rates of the Wombat Plateau sites during the past 130 m.y. After uplift (post-rift unconformity) in the Middle Jurassic, rapid tectonic subsidence submerged the plateau in the Early Cretaceous. Note asymptotically decreasing thermal subsidence rates during Late Cretaceous to Cenozoic times. Sedimentation rates are comparatively high during the Late Triassic early-rift stage and exponentially decrease during the accumulation of the pelagic Cretaceous to Cenozoic sediments.

resulting in the formation of Gascoyne and Cuvier Abyssal Plains between Australia and "Greater India" by movement away to the west.

3. Climatic changes, such as the northward drift of the Northwest Australian area to subtropical latitudes during Rhaetian times, in addition to the generally warm and arid climate [Frakes, 1979], favored the buildup of reefal

carbonates and terminated the deposition of fluviodeltaic siliciclastics at the outer margin.

4. Changes in the source and rate of siliciclastic sediment input were especially important during rapid fluviodeltaic early-rift sedimentation in the Carnian/Norian. On the central Exmouth Plateau, a 1500 m-thick-sequence of Berriasian-lower Valanginian deltaic sediments prograded

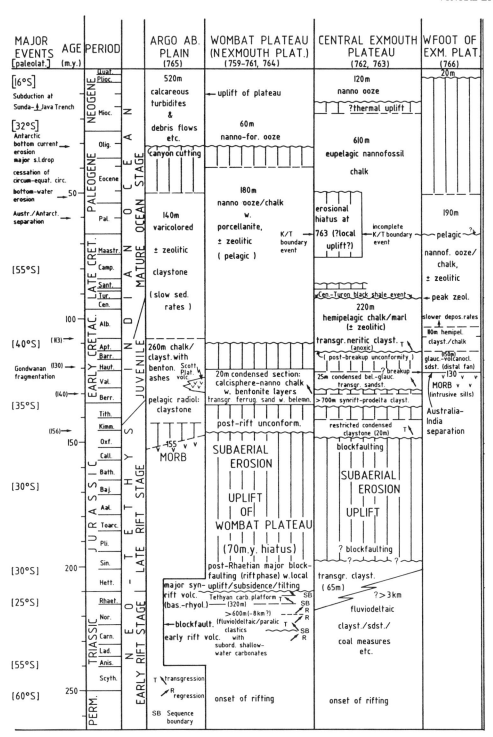

Fig. 19. Summary of structural and paleoenvironmental evolution of Leg 122 and 123 sites during the past 250 m.y. Paleolatitudes after Ogg et al. [1992].

to the north-northwest away from the present-day Cuvier Abyssal Plain. If there is an excess of siliciclastic sediment supply, this factor may override the eustatic signal [Boyd et al., 1992].

   5. The latest Triassic/Hettangian aborted early rift phase

and the Late Jurassic breakup was accompanied and followed by major pulses of volcanic and volcaniclastic activity of felsic to intermediate composition [von Rad and Exon, 1983]. The volcanism can be explained by the upwelling of hot asthenosphere during the stretching and thinning of the overlying lithosphere. A broad band of uppermost Triassic/lowermost Jurassic volcanics forms the basement at the western and northern Exmouth Plateau. Wave erosion of highs in the Triassic/Jurassic rift volcanics along the northwestern margin provided sediment for a seismically recognized, large shallow-marine body of lower Cretaceous sediment, separate from but equivalent to the Barrow Group. This prograded from the west, and filled a new depression caused by fault movements associated with breakup.

6. Eustacy seems to have played a significant role in generating the depositional sequences in the Late Triassic (Figure 9A) and Early Cretaceous (Figure 16). The Upper Triassic shallow-water carbonates are thought to bear a strong eustatic signal. However, separating the tectonic signal from eustatic events is not straightforward, especially if the tectonic development is as complex as at the Wombat Plateau during late Triassic to Jurassic times. A good argument in favor of global sea-level fluctuations is the excellent correlation of two major sequence boundaries (SB) and of the resulting conspicuous facies changes between the eastern Tethys (Northwest Australia) and the well-studied western Tethyan outcrops in the western, northern, and southern Alps. We found a SB close to the Norian/Rhaetian boundary. A major event of the same age, the "early Rhaetian transgression", is known from the Alps. This is an argument in favour of a global eustatic origin for this event which fits well with the 215 Ma SB of the Haq cycle chart. Also the "latest Rhaetian drowning event" which is also observed in the Alps can be correlated with the sea level rise following the 212 Ma SB of Haq et al. [1987].

Global synchronism of third-order cycles can be also explained by intraplate deformation reflecting high levels of regional stress during major plate reorganizations [Cloetingh et al., 1992]. Although the ultimate cause is obscure, we favour eustasy as the main responsible factor to produce the late Triassic and earliest Cretaceous sequence boundaries. At other times in the late Triassic the tectonic signal is mixed with the eustatic signal to produce the preserved stratigraphic record.

Acknowledgments. We are grateful to all members of the shipboard scientific parties of ODP Legs 122 and 123 for the excellent cooperation on board, the free exchange of data after the cruises, and their scientific contribution to the success of both legs. We appreciate especially the detailed discussions with Tim Bralower (Chapel Hill), Wolfram Brenner (Kiel), Dick Buffler (Austin), Felix M. Gradstein (Halifax), Thierry Dumont (Grenoble), Ursula Röhl (Hannover), and Paul Williamson (Canberra, Australia). Technical assistance (drafting and typing) was given by H. Karmann and E. Müller. Some of the drafting was carried out in the BMR drawing office. U. von Rad acknowledges financial support by the Deutsche Forschungsgemeinschaft (Bonn; DFG grants Ra 191-10/2-4). N. Exon publishes with the permission of the Director, Bureau of Mineral Resources, Canberra. We are very grateful for a great number of helpful critical comments, especially by Jim Ogg (West Lafayette, Indiana), Marita Bradshaw (Canberra, Australia), Rainer Brandner (Insbruck, Austria), Jeff Weissel (Lamont, New York), and an anonymous reviewer.

REFERENCES

Audley-Charles, M., Evolution of the southern margin of Tethys (North Australian region) from early Permian to late Cretaceous, in *Tethys and Gondwana*, edited by M. G. Audley-Charles, and A. Hallam, 79-100, Geol. Soc. Spec. Publ., 37, Oxford, 1988.

Barber, P. M., The Exmouth Plateau deep water frontier: a case history, in *The North West Shelf, Australia,* edited by P. G. Purcell, and R. R. Purcell, 173-187, Proc. Petroleum Exploration Soc. Australia Symposium, Perth, 1988.

Boillot, G., Winterer, E. L. et al., *Proc. ODP, Init. Repts., 103,* Ocean Drilling Program, College Station, TX, 1987.

Boote, D. R. D. and Kirk, R. B., Depositional wedge cycles on evolving plate margin, western and northwestern Australia, *Am. Assoc. Pet. Geol. Bull., 73 (2),* 216-243, 1989.

Boyd, R., Williamson, P. E. and Haq, B. U., Seismic stratigraphic and passive margin evolution of the southern Exmouth Plateau, in *Proc. ODP, Sci. Res., 122,* edited by U. von Rad, B. U. Haq et al., 39-59, Ocean Drilling Program, College Station, TX, 1992.

Bradshaw, J., Geological cross-section across the western Timor Sea, Bonaparte Basin, *Bureau of Min. Res., Record 1991/07,* Canberra, Australia, 1991.

Bradshaw, M.T., Yeates, A.N., Beynon, R.M., Brakel, A.T., Langford, R.P., Totterdell, J.M. and Yeung, M., Palaeogeographic evolution of the North West Shelf region, in *The North West Shelf, Australia,* edited by P. C. Purcell, and R. R. Purcell, 29-54, Proc. Petroleum Exploration Soc. Australia Symposium, Perth, 1988.

Bralower, T. J. and Siesser, W. G., Cretaceous calcareous nannofossil biostratigraphy of Sites 761, and 763, Exmouth and Wombat Plateaus, northwest Australia, in *Proc. ODP, Sci. Res.,* 122, edited by U. von Rad, B. U. Haq et al., 529-556, Ocean Drilling Program, College Station, TX, 1992.

Brenner, W., Bown, P. R., Bralower, T. J., Crasquin Soleau, S., Depeche, F., Dumont, T., Martin, R., Siesser, W. G., and Zaninetti, L., Correlation of Carnian to Rhaetian palynological, foraminiferal, calcareous nannofossil, and ostracode biostratigraphy, Wombat Plateau, in *Proc. ODP, Sci. Results, 122,* edited by U. von Rad, B. U. Haq et al., 487-495, Ocean Drilling Program, College Station, TX, 1992.

Brenner, W., First results of late Triassic palynology of the Wombat Plateau, northwestern Australia, in *Proc. ODP, Sci. Res., 122,* edited by U. von Rad, B. U. Haq et al., 413-426, Ocean Drilling Program, College Station, TX, 1992a.

Brenner, W., Dinoflagellate cyst stratigraphy of the lower Cretaceous sequence at Sites 762 and 763, Exmouth Plateau, northwest Australia, in *Proc. ODP, Sci. Res., 122,* edited by U. von Rad, B. U. Haq et al., 511-528, Ocean Drilling Program, College Station, TX, 1992b.

Cloetingh, S., Stein, C., Reemst, P., Gradstein, F. M., Williamson, P. E., Exon, N. F. and von Rad, U., The relationship between continental margin stratigraphy, deformation, and intraplate stresses for the Indo-Australian region, in *Proc. ODP, Sci. Results, 123,* edited by F. M. Gradstein, J. N. Ludden et al., Ocean Drilling Program, College Station, TX. in press, 1992.

Dumont, T., Upper Triassic (Rhaetian) sequences of the northwestern Australian Shelf recovered on Leg 122: Sea-level changes, Tethyan rifting, and overprint of Indo-Australian breakup, in *Proc. ODP, Sci. Res., 122,* edited by U. von Rad, B. U. Haq et al., 197-211, Ocean Drilling Program, College Station, TX, 1992.

Dumoulin, J. A. and Bown, P. R. Depositional history,

nannofossil biostratigraphy, and correlation of Argo Abyssal Plain Sites 765 and 261, in *Proc. ODP, Sci. Results, 123*, edited by F. M. Gradstein, J. N. Ludden et al., Ocean Drilling Project, College Station, TX, in press, 1992.

Erskine, R. and Vail, P. R., Seismic stratigraphy of the Exmouth Plateau, in *Atlas of Seismic Stratigraphy, 2, AAPG Studies in Geology, 27*, edited by A. W. Balley, 163-173, 1988.

Exon, N. F., Borella, P. and Ito, M., Sedimentology of marine Cretaceous sequences in the central Exmouth Plateau (northwest Australia), in *Proc. ODP, Sci. Res., 122*, edited by U. von Rad, B. U. Haq et al., 233-257, Ocean Drilling Program, College Station, TX, 1992a.

Exon, N. F. and Buffler, R. T., Mesozoic seismic stratigraphy and tectonic evolution of the western Exmouth Plateau, in *Proc. ODP, Sci. Res., 122*, edited by U. von Rad, B. U. Haq et al., 61-81, Ocean Drilling Program, College Station, TX, 1992.

Exon, N. F., Haq, B. U. and von Rad, U., Exmouth Plateau revisited: scientific drilling and geological framework, in *Proc. ODP, Sci. Res., 122*, edited by U. von Rad, B. U. Haq et al., 3-20, Ocean Drilling Program, College Station, TX, 1992b.

Exon, N. F. and Ramsay, D. C., BMR-Cruise 95, Triassic and Jurassic sequences of the northern Exmouth Plateau and offshore Caning Basin, *Postcruise Report: Bureau Min. Res.*, Canberra, 1990.

Exon, N. F., von Rad, U. and von Stackelberg, U., The geological development of the passive margins of the Exmouth Plateau off northwest Australia, *Mar. Geol., 47*, 131-152, 1982.

Exon, N. F. and Willcox, J. B. Geology and petroleum potential of the Exmouth Plateau area off Western Australia, *Am. Assoc. Pet. Geol. Bull. 62 (1)*, 40-72, 1978.

Falvey, D.A., The development of continental margins in plate tectonic theory, *APEA (Austral. Pet. Explor. Assoc.) Journal, 14*, 95-106, 1974.

Frakes, L. A., *Climates throughout geological time*, Elsevier, 310 pp., Amsterdam, 1979.

Fullerton, L. G., Sager, W. W. and Handschumacher, D. W., Late Jurassic-Early Cretaceous evolution of the eastern Indian Ocean adjacent to Northwest Australia, *J. Geophys. Res. 94*, 2937-2953, 1989.

Görür, N. and Sengör, A. M. C., Paleogeography and tectonic evolution of the eastern Tethysides: implications for the northwest Australian margin breakup history, in *Proc. ODP, Sci. Res., 122*, edited by U. von Rad, B. U. Haq et al., 83-106, Ocean Drilling Program, College Station, TX, 1992.

Gradstein, F. M., Legs 122 and 123, NW Australian margin - a stratigraphic and paleogeographic summary, in *Proc. ODP, Sci. Res., 123*, edited by F. M. Gradstein, J. N. Ludden et al., 0-0, Ocean Drilling Program, College Station TX, in press.

Gradstein, F. M., von Rad, U., Gibling, M. R., Jansa, L. F., Kaminski, M. A., Kristiansen, I. L., Ogg, J. G., Röhl, U., Sarti, M., Thurow, J. W., Westermann, G. E. G. and Wiedmann, J., Stratigraphy and depositional history of the Mesozoic continental margin of central Nepal, *Geol. Jb., Reihe B, 77*, 3-141, 1992.

Gradstein, F. M., and von Rad, U., Stratigraphic evolution of Mesozoic continental margin and oceanic sequences: Northwest Australia and northern Himalayas, in *Evolution of Mesozoic and Cenozoic Continental Margins, Mar. Geol., 102*, edited by A. W. Meyer, T. A. Davies, and S. W. Wise Jr., 131-173, 1991.

Gradstein, F. M. Ludden, J. N. et al., *Proc. ODP, Init. Repts., 123*, Ocean Drilling Program, College Station, TX, 1990.

Haq, B. U., Sequence stratigraphy, sea level change, and significance for the deep sea, *Internat. Assoc. Sediment., Spec. Publ. 12*, 3-39, 1991.

Haq, B. U., Boyd, R., Exon, N. F. and von Rad, U., Evolution of the central Exmouth Plateau: A post-drilling perspective. In von Rad, U., Haq, B.U. et al., *Proc. ODP, Sci. Res., 122*, 801-816, Ocean Drilling Program, College Station, TX, 1992.

Haq, B. U., Hardenbol, J. and Vail, P., Chronology of fluctuating sea levels since the Triassic (250 million years ago to present), *Science, 235*, 1156-1166, 1987.

Haq, B. U., ⌐ . Rad, U., O'Connell S. et al., *Proc. ODP, Init. Repts., 122*, Ocean Drilling Program, College Station, TX, 1990.

Heirtzler, J. R., Cameron, P., Cook, P. J., Powell, T., Roeser, H. A., Sukardi, S. and Veevers, J. J., The Argo Abyssal Plain, *Earth Planet. Sci. Lett., 41*, 21-31, 1978.

Helby, R., Morgan, R. and Partridge, A., A palynological zonation of the Australian Mesozoic, *Assoc. Australasian Paleont. Mem. 4*, 1-94, 1987.

Hocking, R. M., Moors, M. T. and van der Graaff, W. J. E., Geology of the Carnarvon Basin, Western Australia, *Geol. Surv. Western Australia Bull., 133*, 289, 1987.

Hooper, P. R., The timing of crustal extension and the eruption of continental flood basalts, *Nature, 345*, 246-249, 1990.

Huang, Z., Boyd, R. and O'Connell, S., Upper Cretaceous cyclic sediments from Hole 762C, Exmouth Plateau, northwest Australia, in *Proc. ODP, Sci. Res., 122*, edited by U. von Rad, B. U. Haq et al., 259-277, Ocean Drilling Program, College Station, TX, 1992.

Islam, M. A., Palynological age-dating of seismic horizons D, E and F in the Beagle and Dampier Sub-basin, in *The Northwest Shelf, Australia, Proc. Petrol. Expl. Soc. Australia, Symposium, Perth*, edited by P. G. Purcell, and R. R. Purcell, 599-604, 1988.

Kaminski, M. A., Baumgartner, P. O., Bown, P. R., Haig, D. W., McMinn, A., Moran, M. J., Mutterlose, J. and Ogg, J. G., Magnetobiostratigraphic synthesis of Ocean Drilling Progam Leg 123: Site 765 and 766 (Argo Abyssal Plain and lower Exmouth Plateau), in *Proc. ODP, Sci. Results, 123*, edited by F. M. Gradstein, J. N. Ludden et al., in press, 1992.

Kent, D. V. and Gradstein, F. M., A Cretaceous and Jurassic geochronology, *Geol. Soc. Am. Bull., 96*, 1419-1429, 1987.

Kopsen, E. and McGann, G., A review of the hydrocarbon habitat of the eastern and central Barrow/Dampier sub-basin, Western Australia, *APEA J. (Austr. Pet. Explor. Assoc.), 25*, 154-176, 1985.

Larson, R. L., Late Jurassic sea-floor spreading in the eastern Indian Ocean, *Geology, 3*, 69-71, 1975.

Ludden, J. N. and Dionne, B., The geochemistry of oceanic crust at the onset of rifting in the Indian Ocean, in *Proc. ODP, Sci. Res., 123*, edited by F. M. Gradstein, J. N. Ludden et al., 0-0, Ocean Drilling Program, College Station, TX, 1992.

Mutter, J. C., Buck, W. R. and Zehnder, C. M., Convective partial melting, 1. A model for the formation of thick basaltic sequences during the initiation of spreading, *J. Geophys. Res., 93 (2)*, 1031-1048, 1989.

Mutter, J. C., Larson, R. L. and the NW Australia Study Group, Extension of the Exmouth Plateau: Deep seismic reflection/ refraction evidence of simple and pure shear mechanisms, *Geology, 17*, 15-18, 1989.

Ogg, J. G., Kodama, K. and Wallick, B., in *Proc. ODP, Sci. Results, 123*, edited by F. M. Gradstein, J. N. Ludden et al., Ocean Drilling Program, College Station, TX, 1992.

Pospichal, J. J. and Bralower, T. J., Calcareous nannofossils across the Cretaceous/Tertiary boundary, Site 761, northwest Australian margin, in *Proc. ODP, Sci. Res., 122*, edited by U. von Rad, B. U. Haq et al., 735-751, Ocean Drilling Program, College Station, TX, 1992.

Rocchia, R., Boclet, D. Bonté, P., Froget, L., Galbrun, B., Jéhanno, C. and Robin, E., Iridium and other element distributions, mineralogy, and magnetostratigraphy near the Cretaceous/Tertiary boundary in Hole 761C, in *Proc. ODP, Sci. Res., 122*, edited by U. von Rad, B. U. Haq et al., 753-762, Ocean Drilling Program, College Station, TX, 1992.

Röhl, U., Dumont, T., von Rad, U., Martini, R. and Zaninetti, L., Upper Triassic Tethyan carbonates off Northwest Australia (Wombat Plateau, ODP Leg 122), *Facies, 25*, 211-252, Erlangen, 1991.

Röhl, U., von Rad, U. and Wirsing, G., Microfacies, paleoenvironment, and facies-dependent carbonate diagenesis in upper Triassic platform carbonates off northwest Australia, in *Proc. ODP, Sci. Res., 122*, edited by U. von Rad, B. U. Haq et al., 129-159, Ocean Drilling Program, College Station, TX, 1992 .

Sarti, M., Russo, A. and Bosellini, F., Rhaetian strata, Wombat plateau: analysis of fossil communities as a key to paleoenvironment changes, in *Proc. ODP, Sci. Res., 122*, edited by U. von Rad, B. U. Haq et al., 181-195, Ocean Drilling Program, College Station, TX, 1992.

Schlanger, S. O., Arthur, M. A., Jenkyns, H. C. and Scholle, P. A., The Cenomanian-Turonian oceanic anoxic event, I. Stratigraphy and distribution of organic carbon-rich beds and the marine $\delta^{13}C$ excursion, in *Marine Petroleum Source Rocks*, edited by J. Brooks, and A. Fleet, 347-375, Geol. Soc. London, Spec. Publ., 24, 1987.

Sengör, A. M. C., The story of Tethys: how many wives did Okeanos have? *Episodes, 8*, 3-12, 1985.

Stanton, R. J. and Flügel, E., Problems with reef models: the Late Triassic Steinplatte "Reef" (Northern Alps, Salzburg/Tyrol, Austria), *Facies, 20*, 1-138, Erlangen, 1985.

Tozer, E. T., How many Rhaetians? *Albertiana, 8*, 10-14, 1990.

Veenstra, E., Rift and drift in the Dampier Sub-basin, a seismic and structural interpretation, *APEA J. (Austral. Petrol. Explor. Assoc.), 18 (1)*, 177-189, 1985.

Veevers, J. J., Morphotectonics of Australia's northwestern margin, in *The North West Shelf Australia*, edited by P. G. Purcell, and R. R. Purcell, 19-27, Proc. Petroleum Exploration Soc. Australia Symposium, Perth, 1988.

Veevers, J. J., Heirtzler, J. R. et al., *Init. Repts. DSDP, 27*, U.S. Government Printing Office) Washington, 1974.

Veevers, J. J., Tayton, J. W., Johnson, B. D. and Hansen, L., Magnetic expression of the continent-ocean boundary between the western margin of Australia and the eastern Indian Ocean, *J. Geophys., 56*, 106-120, 1985.

Veevers, J. J., and Powell, C. McA., Sedimentary wedge progradation from transform-faulted continental rim: southern Exmouth Plateau, western Australia, *Am. Assoc. Pet. Geol. Bull., 63(11)*, 2088-2096, 1979.

von Rad, U. and Bralower, T., Unique record of an incipient ocean basin: lower Cretaceous sediments from the southern margin of Tethys, *Geology, 20*, 551-555, 1992.

von Rad, U. and Exon, N. F., Mesozoic-Cenozoic sedimentary and volcanic evolution of the starved passive margin off northwest Australia, *Am. Assoc. Pet. Geol. Memoir, 34*, 253-281, 1983.

von Rad, U., Exon, N. F. and Haq, B. U., Rift-to-drift history of the Wombat Plateau, northwest Australia: Triassic to Tertiary Leg 122 results, in *Proc. ODP, Sci. Res., 122*, edited by U. von Rad, B. U. Haq et al., 765-800, Ocean Drilling Program, College Station, TX, 1992.

von Rad, U., Schott, M., Exon, N. F., Mutterlose, J., Quilty, P. G. and Thurow, J., Mesozoic sedimentary and volcanic rocks dredged from the northern Exmouth Plateau: petrography and microfacies, *BMR (Bur. Min. Res.) J. Australian Geol. Geophys., 11*, 449-476, 1990.

von Rad, U. and Thurow, J., Bentonitic clays as indicators of early Neocomian post-breakup volcanism off northwest Australia, in *Proc. ODP, Sci. Res., 122*, edited by U. von Rad, B. U. Haq et al., 213-232, Ocean Drilling Program, College Station, TX, 1992.

von Stackelberg, U., Exon, N. F., von Rad, U., Quilty, P., Shafik, S., Beiersdorf, H., Seibertz, E. and Veevers, J. J., Geology of the Exmouth and Wallaby Plateaus off northwest Australia: sampling of seismic sequences, *BMR (Bur. Min. Res.) J. Australian Geol. Geophys., 5*, 13-140, 1980.

White, R. S., Volcanism and igneous underplating in sedimentary basins and at rifted continental margins, in *Origin and Evolution of Sedimentary Basins and Their Energy and Mineral Resources, IUGG Geophys. Monogr. 48 3*, edited by R. A. Price, 125-127, 1989.

White, R. S. and McKenzie, D. P., Magmatism at rift zones: the generation of volcanic continental margins and flood basalts, *J. Geophys. Res., 94*, 7685-7729, 1989.

Wilgus, C. K., Hastings, B. S., Ross, C. A., Posamentier, H., Van Wagoner, J., and Kendall, C. G. St. (editors), Sea level changes: an integrated approach, *Soc. Econ. Pal. Min., Spec. Publ. 42*, 407pp., Tulsa OK, 1988.

Williamson, P.E., Seismic sequence analysis of a late Triassic carbonate reefal platform on the Wombat Plateau, Australia, in *Proc. ODP, Sci. Res., 122*, edited by U. von Rad, B. U. Haq et al, 23-37, Ocean Drilling Program, College Station, TX, 1992.

Williamson, P. E., Swift, M. G., Kravis, S. P., Falvey, D. A., and Brassil, F., Permo-Carboniferous rifting of the Exmouth Plateau region, Australia: an intermediate plate model, in *The Potential of Deep Seismic Profiling for Hydrocarbon Exploration, Arles 1989*, Edition Technip., 1990.

Wright, A. J., and Weatley, T. J., Trapping mechanisms and the hydrocarbon potential of the Exmouth Plateau, *J. Austr. Pet. Expl. Assoc. 19 (1)*, 19-29, 1979.

Tectonics and Sea-Level Changes Recorded in Late Triassic Sequences at Rifted Margins of Eastern and Western Tethys (Northwest Australia, Leg 122; Western Europe)

THIERRY DUMONT

Laboratoire de Géodynamique des Chaines Alpines, Institut Dolomieu, 15 rue M. Gignoux, 38031 Grenoble, France

URSULA RÖHL

Bundesanstalt für Geowissenschaften und Rohstoffe, Postfach 510153, D-3000 Hannover 51, Germany

During ODP Leg 122 upper Triassic (Carnian to Rhaetian) sediments were recovered at the sediment-starved passive continental margin off Northwest Australia [Haq et al., 1990]. The early-rift series of the Wombat Plateau, a northern sub-plateau of the Exmouth Plateau, consists of upper Triassic fluviodeltaics and shallow-marine carbonates including reefal facies. These sequences are capped by an erosional "post-rift unconformity" with a 70 m.y. hiatus during the Jurassic. The Wombat Plateau bears only a thin pelagic post-rift sedimentary cover of Cretaceous to Cenozoic age.

Detailed investigations of microfacies, wireline logs and high-resolution seismics allow the reconstruction of paleoenvironments and identification of the depositional sequences. The Rhaetian paleoenvironments and facies are very similar to those of the Western Tethys (Western and Eastern Alps). Some of the Australian sequences fit well with the sequences of the Haq et al. [1987] global cycle chart, and with the sequences in Western Europe. Their boundaries seem to correspond to global events (especially the Norian/Rhaetian boundary), suggesting that the most important causal factor was eustasy. However, short-lived tectonic or volcanic events have affected the Australian and European continental margins at the same time (Carnian, Triassic/Jurassic boundary). This unexpected synchronism suggests that these short tectonic events are not confined to the Australian margin and may have played a significant role in the globally synchronous deposition of third-order sequences in addition to eustasy.

The erosional "post-rift unconformity" which overlies the Rhaetian series corresponds to several superimposed events related to the multi-stage rifting and opening of the oceanic domains surrounding the Northwest Australian shelf. These events began with a tectonic reorganization around the Triassic/Jurassic boundary, which also affected the Exmouth Plateau and the Rankin Platform, and which is known from other plate margins (Western Tethys, Central Atlantic).

INTRODUCTION

Sites ODP 759, 760, 761, and 764 are located on the Wombat Plateau (Fig. 1 and 2), a small subplateau, which is separated from the northern Exmouth Plateau by a deep canyon superimposed on a half-graben [Exon et al., 1982]. Our main objectives were (1) a better understanding of the early rift history of this margin during the late Triassic (extensional tectonics affecting fluviodeltaic siliciclastics and shallow-water carbonate environments with reefal buildups) and (2) the paleoenvironmental evolution of this northeastern Gondwanian margin bordering the southern Tethys sea [Dumont, 1992; Röhl et al., 1991; Röhl et al., 1992]. A composite 900-m-thick section of Carnian to late Rhaetian age was reconstructed using the data from the four drillsites. Sites 759, 760 and the lowermost part of Site 761 contain siliciclastic-dominated deltaic sequences of Carnian-Norian age with some interbedded shallow-water carbonates. Sites 761 and 764 consist of Rhaetian shallow-water carbonates with intertidal to shallow subtidal

Synthesis of Results from Scientific
Drilling in the Indian Ocean
Geophysical Monograph 70
Copyright 1992 American Geophysical Union

sediments in Site 761 and shallow subtidal (lagoonal and reefal) carbonates in Site 764.

LITHOSTRATIGRAPHY AND PALEOENVIRONMENTS
Carnian and Norian (Sites 759 and 760)

Claystones, siltstones, sandstones and several shallow-water carbonate intercalations characterize the series of both sites. These Carnian and Norian delta-dominated sequences were dated by palynomorphs, nannofossils, ostracodes and foraminifers [Brenner et al., 1992].

From bottom to top three main lithological parts can be distinguished (Fig. 3, "Lithology Correlation" column):

-Lower part: lowermost Carnian claystones, siltstones and minor sandstones, which contain siderite nodules and layers. These series are interpreted as deposits of a prodelta environment [Ito et al., 1992].

-Middle part: Carnian to Norian siliciclastics (claystones, siltstones, sandstones) with intercalated limestones. The frequency of limestone beds and percentage of allochems in the limestones increase upwards, together with an upward shallowing of the depositional environment which is documented by the vertical microfacies evolution [Röhl et al., 1991]. Three types of carbonate facies are found in succession in this part, that is, from base to top: (1)

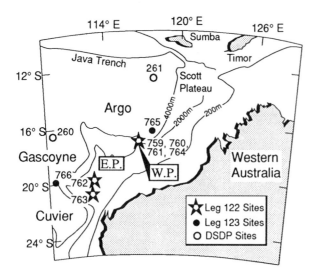

Fig. 1. Location of the ODP Sites 759, 760, 761 and 764 on the Wombat Plateau, north from the Exmouth Plateau, northwestern Australian shelf (E.P.: Exmouth Plateau; W.P.: Wombat Plateau).

Storm deposits showing erosive bases, grading of allochems, and often containing a large percent of quartz grains, pelecypod shells and crinoid fragments. (2) Bioclastic sands (grainstones), whose components (fragments of calcareous algae, oncoids, ooids) were deposited on migrating carbonate sand shoals. (3) Algal mats and patch reefs occurring within restricted areas protected from siliciclastic dilution.

-Upper part: Norian sandstones and fine siliciclastics, showing a shallowing-upward trend and interpreted as fluviodeltaics [Ito et al., 1992].

Rhaetian (Sites 761 and 764)

"Rhaetian" is a controversial stage name, since it is considered either as part of the late Norian [Decade of North American Geology-Geological Society of America, (DNAG-GSA) time scale], or as a stage properly so called [Haq et al., 1987; see discussion in Röhl et al., 1991]. In this paper, the Rhaetian is considered as a stage because (1) the biostratigraphic content is different from the Norian one and allows location of the Norian/Rhaetian boundary, and (2) as shown in the following, this boundary corresponds to a major, probably global, sequence boundary.

This 180- to 240-m-thick limestone-dominated series is dated as Rhaetian by palynomorphs and dinoflagellates [Brenner et al., 1992], by nannofossils [Bralower et al., 1991], by ostracodes [Depeche and Crasquin-Soleau, 1992], and by foraminifers [Zaninetti et al., 1992]. At the base of Site 761, the Rhaetian beds overly 10 m of coal-seam bearing claystone and siltstone of Norian age [Brenner et al., 1992], which compare well with the regressive upper Norian shallow deltaics of Site 760. The change from fluviodeltaic claystones to marine marl/limestone alternations in Site 761 (Unit VB or

member A, Figure 4) documents the early Rhaetian relative sea-level rise.

Rhaetian beds can be divided into three parts using the lithofacies, microfacies facies analysis and log interpretation [Dumont, 1992; Röhl et al., 1992; Figure 4]: a marly lower part is overlain by a carbonate-dominated middle part in both sites, which is overlain by a marl-dominated upper part in Site 764 only.

- Lower part [members A and C of Dumont, 1992]: Site 761 contains crinoid-rich marl/limestone alternations [Unit VB of Röhl et al., 1992] of shallow open-marine environments, overlain by more restricted marl/limestone alternations [Unit VA, Röhl et al., 1992]. The episodic occurrence of grainstone, packstone and floatstone with pelecypod shells, foraminifers, coated grains and calcisponge fragments in the marl/limestone alternations of Unit VA (Figure 4) suggests shallow subtidal water depths. In Site 764, the lowermost part (Unit VII, Figure 4) consists of an alternation of highly bioturbated marl, dark clayey carbonate mudstone, and mud- to wackestone, comparable to those of Unit VB in Site 761. But contrary to Site 761, it is overlain at Site 764 (member C) by lagoonal carbonate cycles with few interbedded marly layers. The environments at Site 764 were shallower and better ventilated than at Site 761 during the deposition of member C and upper member A.

- The middle, limestone-dominated part of the Rhaetian sequence corresponds in Site 761 [member B of Dumont, 1992] to sand shoal facies associated with calcisponge patch reefs in a lagoonal setting. They grade upwards into intertidal carbonate flat to shelf-lagoon environments. At Site 764, typical features of reef development with several lagoonal/reefal cycles are observed [see final section on "Facies model", and Röhl et al., 1991].

- The upper, marly-dominated part was not recovered in Site 761, but was recovered in Site 764. Marl/limestone alternations characterize the base of this part (lower two-thirds of member F, Figure 4). The limestone types are skeletal packstone with crinoid and brachiopod fragments, mudstone with sponge spicules, and peloidal, bioturbated wackestone with foraminifers. *Triasina hantkeni* Mazjon and crinoid fragments are very abundant in the carbonate-dominated upper half (upper third of member F), which is capped by a major erosional surface, the so-called "post-rift unconformity" (PRU). The PRU corresponds to a sedimentary gap of 70 m.y. with an erosion visible on the seismic profiles. During this time interval, an unknown part of upper Rhaetian and possibly Liassic section has been removed.

Just below the PRU, the top of the Rhaetian sequence at Site 764 was affected by late diagenesis: we found a thin horizon with iron and manganese oxide lamination which postdates hardening, increased porosity and fresh water calcite precipitation, chalkification and dedolomitization at the top [Röhl et al., 1992]. Geochemical data supporting late diagenesis in meteoric-phreatic environments are provided by Sarti and Kälin [1992], but they are not

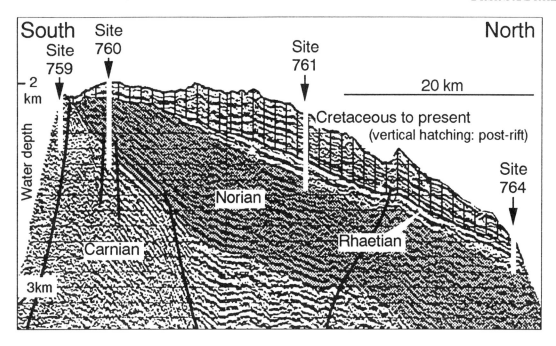

Fig. 2. NS seismic profile across the Wombat Plateau [line BMR 56-13, modified from Williamson et al., 1989]. Note the faulting of the Carnian sequence prior to the deposition of the Norian wedge, and the angular unconformity corresponding to the erosional post-rift unconformity (at the base of the hatched series), which truncates the Triassic beds in the central and southern part of the plateau.

conclusive. However, the assumption of a post-Rhaetian fossil water table/groundwater level produced by a fresh-water lens, and hence the emersion of the Wombat Plateau during the Jurassic, looks reasonable.

FACIES MODEL [AFTER RÖHL ET AL., 1991]

The recovered carbonates have been affected by strong diagenetic alteration. However, by studying all transitions between preserved and totally disturbed biogenic structures and sedimentary textures, 25 main types of microfacies were distinguished [Röhl et al., 1991], which have been documented in 354 samples. These 25 types of microfacies are listed and grouped into several facies units in Table1. Similar facies types were found in western and eastern Tethyan paleoenvironments [see following section, "Comparison with European Triassic", and Röhl et al., 1991].

Carnian/Norian Carbonate Facies

From microfacies analysis [Röhl et al., 1992], the carbonate layers of Sites 759 and 760 belong to the following types: (1) calcareous algae-rich facies with oolites (415 to 465 meters below sea floor in Site 760; 165 to 210 mbsf in Site 759), (2) oolitic/oncolitic-rich, partly dolomitized facies with coral/calcisponge floatstones (370 to 390 mbsf in Site 760; 100 to 145 mbsf in Site 759), and (3) oolitic/algal stromatolitic facies in the upper part (315 to 370 mbsf in Site 760; 55 to 95 mbsf in Site 759). In each case, the limestones of Site 760 indicate better oxygenated environments than those of Site 759. This shows a more landward position of Site 759 compared to Site 760 during Triassic times [Röhl et al., 1991]. The

limestone intercalations within the fluviodeltaic to marginal- marine Carnian and Norian series occur in between thin and evenly laminated siltstones. Some limestones represent storm deposits, reworked from nearby shallow-water carbonate shoals. Autochthonous and parauthochthonous carbonate sands were either deposited in restricted areas within interdistributary bays as migrating sand waves, or might have developed on the top of barrier island arcs in front of an abandoned delta lobe.

Rhaetian carbonate buildup

Changes of the paleolatitude of the Wombat Plateau area [Ogg et al., this volume] combined with a global sea-level rise were responsible for the development of a carbonate platform overlying the Carnian/Norian fluviodeltaic to shallow-marine sequences. The first sediments deposited after the flooding of these sequences were marl/limestone alternations, which occur at the base and top of the Rhaetian. The components within the limestones (e.g., crinoid fragments) reflect the subtidal, ventilated environment. Bioclastic sands, which developed on uplifted fault blocks [Williamson et al., 1989], were redistributed by currents. They preceded the development of oolitic shoals, which occur mainly in the lower part of the section, when reefal development had not yet started to shelter the lagoons from the open sea. Overlying the oolitic to skeletal substrate, calcisponge-hydrozoan patch reefs first developed in both Sites 761 and 764, but were drowned at Site 761.

The biolithite facies of the central reef areas (Site 764) shows a characteristic association of microfacies types. Ostracode-rich, echinoderm-rich and foraminifer-rich

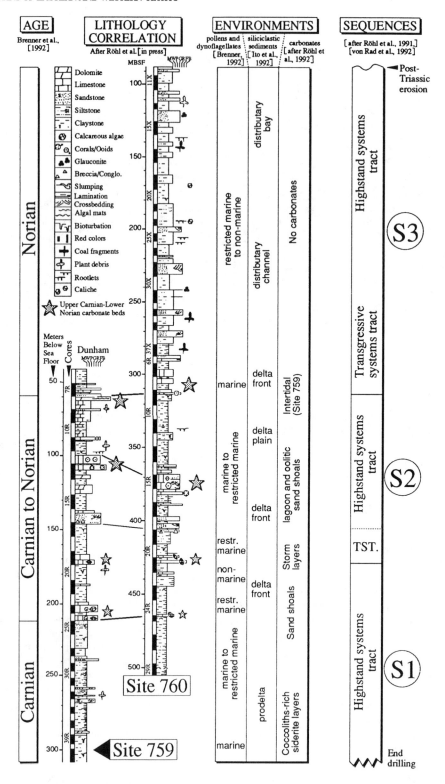

Fig. 3. Carnian and Norian series drilled at ODP Sites 759 and 760. A sequence stratigraphic interpretation is proposed, based on the vertical evolution of paleoenvironments observed in siliciclastic and in carbonate sediments, which are interbedded. The lower part of sequence S1 has not been penetrated, and the upper part of sequence S3 is lacking in both sites due to post-Triassic erosion (post-rift unconformity).

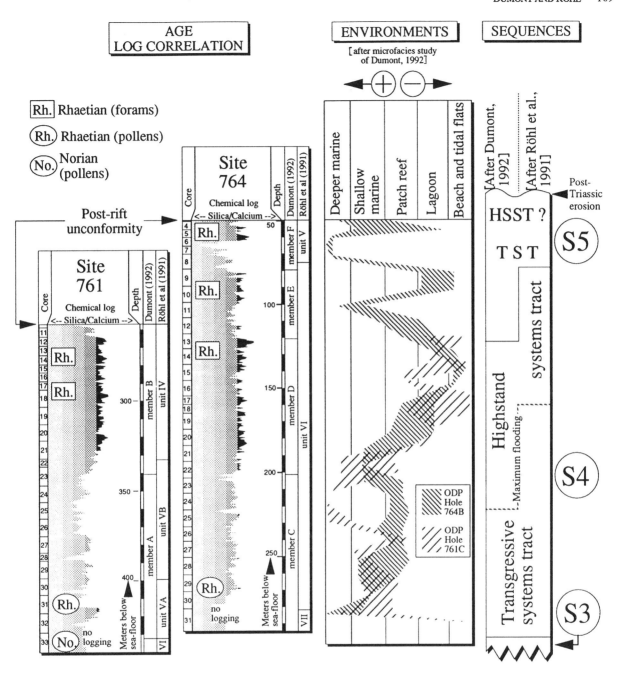

Fig. 4. Upper Norian and Rhaetian series drilled at ODP Sites 761 and 764. Two sequence stratigraphic interpretations are proposed, after Dumont [1992] and Röhl et al. [1992]. The S3/S4 sequence boundary was penetrated at Site 761 only. The lower part of sequence S5 and possibly the upper part of S4 [after Röhl et al., 1992] are lacking at Site 761 due to post-Triassic erosion. The upper part of S5 was also eroded at Site 764.

limestones are intercalated between the individual reef structures. The reefal facies is characterized by boundstones with *Retiophyllia* and *Astraeomorpha*-type corals [Sarti et al., 1992], calcisponges and hydrozoans (Table 1). Peloidal packstones to grainstones are typical fillings of reef-growth cavities. In the Rhaetian reefal platform

carbonates of the Wombat Plateau, five reef cycles are identified. They were probably produced by minor relative sea-level changes and correlative carbonate production fluctuations.

The lagoonal area (Site 761) is characterized by a relatively more restricted setting, beginning with the

Detritus-Mud Facies	1. Mudstone	(Estuarine) Lagoonal facies (shelf)

Foraminifers-Detritus Facies	2- Bioturbated wackestone
	3- Foraminiferal wackestone:
	  a) Triasina predominated
	  b) Involutinidae predominated
	  c) Duostominidae predominated
	4- Wackestone *sensu lato*
	  a) peloidal wackestone
	  b) coated grain wackestone
	5- Ostracod-rich wackestone
	6- Echinodermal wackestone
	7- Skeletal wackestone	

Foraminifers-Calcareous algae and Crinoid Facies	8- Foraminiferal packstone
	  a) Involutinidae predominated
	  b) Triasina predominated
	9- Peloidal packstone
		Lagoon to open marine facies

	10- Skeletal packstone
	  (to grainstone partly
	  grapestone facies)
	11- Echinodermal packstone
	12- Codiacean wacke- to packstone	

	13- Skeletal floatstone 14- Coral/sponge floatstone	Reefal debris
Oncolitic-Oolitic Facies	15- Foraminiferal grainstone 16- Coated grain grainstone 17- Oolitic grainstone 18- Oncolitic grainstone 19- Dasycladacean grainstone 20- Codiacean grainstone 21- Peloidal grainstone 22- Skeletal grainstone s.l. to rudstone	Carbonate sand shoal facies
Biolithite Facies	23- Algal bindstone	Reefal facies

	24- Boundstones
	  a) Coral boundstone
	  b) Sponge boundstone
	  c) Sponge-hydrozoan boundstone
	25- Boundstones and framestones
	  (sponges and corals)	

Table 1. Microfacies types of the late Triassic on Wombat Plateau, after Röhl et al. [1992].

deposition of marl/limestone alternations between several patch reefs. An algal-foraminiferal detritus facies was deposited in a quiet lower reef slope to lagoon setting with a wide variety of microfacies types including reef debris that decreases towards the lagoonal setting. Regressive cycles of lagoonal foraminifer- and Megalodon-bearing wackestone to algal bindstone (stromatolites) occur in the upper part, and indicate the progradation of coastal environments.

The vertical facies evolution of the Rhaetian series on Wombat Plateau at Sites 761 and 764 represents an upward gradient from "catch-up" sedimentation [Sarg, 1988] as the initial response to relative sea-level rise, to "keep-up" carbonate sedimentation with patch-reef development in the seaward areas, and finally a rapid regression followed by renewed transgression which killed the reefs ("give-up", only preserved at Site 764).

Our representation of facies distribution of the Wombat Plateau during late Triassic (Figure 5) shows the SE-NW seaward transition from the fluviodeltaic setting to marginal-marine deltaic environments, and then to the carbonate platform environments. The latter setting was

dominant in the Wombat Plateau area at the end of the Triassic (Rhaetian) due to both landward migration of marine environments and subtropical paleolatitudes [Ogg et al., this volume].

VERTICAL EVOLUTION: SEQUENCE STRATIGRAPHY

Sequence stratigraphy and the testing of eustasy-driven depositional models was one of the major objectives of Leg 122. In particular, an attempt was made to test the validity of the cycle chart of Haq et al. [1987] for long-distance correlations, and of the concept of global sea-level changes in the Triassic. Concerning the Triassic sequences of the Wombat Plateau, several interpretations have been proposed [Dumont, 1992; Röhl et al., 1992; von Rad et al., 1992].

Depositional sequences and sequence boundaries

Five depositional sequences have been identified [von Rad et al., 1992 and this volume; Röhl et al., 1991; Röhl et al., 1992; Dumont, 1992]. Sedimentologic and biostratigraphic criteria (transgressive or regressive trends, depositional gaps and erosion surfaces), together with log interpretation, were used for identification of the sequences. The microfacies types were situated on a theoretical platform profile, and their vertical evolution was used for documenting the retrogradational or progradational character of each lithologic unit. Paleodepths were not quantified since most of the facies, either fully marine or restricted, are still shallow and a paleodepth curve would be too speculative. Detailed interpretations and discussions are not presented in this paper, which refers to the papers mentioned above.

Three sequences were found in the Carnian-Norian at Sites 759 and 760 (S1, S2 and S3; Figure 3). A fourth sequence (S4) and the base of a fifth one (S5) were found in the Rhaetian at Sites 761 and 764 (Figure 4):

S1: The oldest sediments penetrated (Carnian; upward-shallowing prodelta shales and siltstones) represent the highstand systems tract (HST) of sequence S1. The transgressive systems tract (TST) and the maximum flooding surface were not penetrated.

S2: This sequence contains delta front to delta plain siliciclastics interfingered with shallow carbonates. The carbonate layers are more frequent in the upper part of the sequence, and they have recorded a shallowing-upward evolution which characterize the highstand systems tract. The spacial distribution of terrigenous sediments is probably controlled by local factors [laterally shifting delta lobes; Borella et al., 1992]. On the contrary, the carbonate layers, which can be correlated between Sites 759 and 760 and are represented by continuous seismic reflectors (Figure 3), are widespread. Their location and development do not correspond to a depositional model involving randomly distributed deltaic bodies which created appropriate shallow environments for carbonate development. The carbonates were more likely developed during particular time intervals, owing to favorable paleoecological and climatic setting and a lowering in

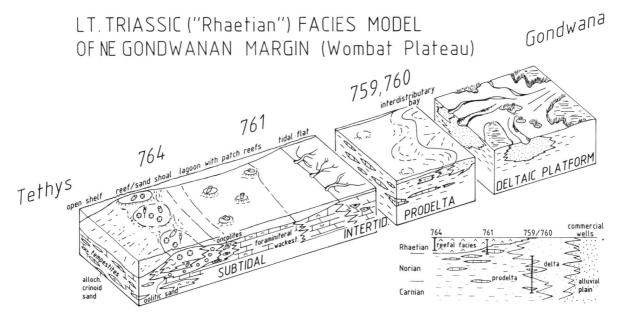

Fig. 5. Reconstruction of the late Triassic paleoenvironments and facies distribution across the northwestern Australian margin using the sedimentological data of Leg 122 [from Röhl et al., 1991].

terrigenous supply. We suspect carbonate development was driven by relative sea-level changes together with climatic influences.

S3: This sequence has a transgressive base marked by a shift into an open-marine paleoenvironment as shown by dinoflagellate evolution [Brenner, 1992]. The sequence boundary is overlain by an aggradational wedge (transgressive systems tract: distributary channel with sand bars), followed by a markedly regressive, terrigenous-dominated wedge (highstand systems tract: restricted and low-energy environments with episodic development of paleosoils). The seismic data show that the uppermost part of the Triassic series in Site 760 does not overlap the lowermost sediments penetrated at Site 761, which are thus younger. But the paleontological and palynological assemblages of both are similar [Brenner, 1992] and so are the facies. Therefore we consider the base of Site 761 to belong to the same systems tract as the top of the Triassic series at Site 760, i.e., the upper highstand systems tract of S3.

S4 and base of S5: The S3/S4 sequence boundary was penetrated at the base of Site 761 (Figure 4). It corresponds to a sharp transgressive and erosional surface overlying delta plain siliciclastic sediments (late highstand systems tract of S3), and overlain by open marine marls and crinoidal limestones. The S3/S4 sequence boundary is in fact superimposed on the transgressive surface of S4. There is no lowstand wedge or shelf margin wedge in sequence S4, because this site was too far from the platform margin during the late Triassic. The marly, lower part of S4 is retrogradational (transgressive systems tract, Figure 4), and the upper part consists of an aggradational to progradational carbonate wedge with a markedly regressive

top (highstand systems tract). Within the TST, the reefal series at Site 764 is more calcareous and shallower than at Site 761, although the latter site is supposed to be in a more internal position. This is because the reefal buildups, which initially developed at Sites 761 and 764, were later drowned at Site 761. During this time interval, most of the platform (except the reefs) underwent unfavorable conditions for carbonate production. The potential carbonate production of reefs in an appropriate ecological setting gives theoretical growth rates much higher than any sea-level rise. As shown by Schlager [1981] and Gildner and Cisne [1989], reef drowning can only be explained by modifications in paleoecological setting (temperature, salinity, oxygen content, and terrigenous pollution). We suggest that these modifications, which restricted high carbonate production to the patch-reefs and drowned the reef at Site 761, correspond to the maximum rate of relative sea-level rise.

As shown in Figure 4, the subbottom depth of the maximum flooding surface of S4 and the S4/S5 sequence boundary is higher in the interpretation of Dumont [1992] than in Röhl et al. [1992]. This is due to minor discrepancies in the interpretation of paleoenvironments, and especially because those of the lower part of sequence S4 at Site 761 (clayey interval corresponding to Cores 761-23R to 761-26R) are thought to be deeper in the first interpretation than in the second one:

- According to Dumont [1992], this clayey interval represents the drowning of the underlying reef-bearing carbonate parasequences (Cores 761-27R to 761-32R), and the thick overlying lagoonal carbonate wedge (Cores 761-12R to 761-22R) is regressive. Thus the clayey interval of Cores 761-23R to 761-26R is interpreted to contain the maximum flooding surface.

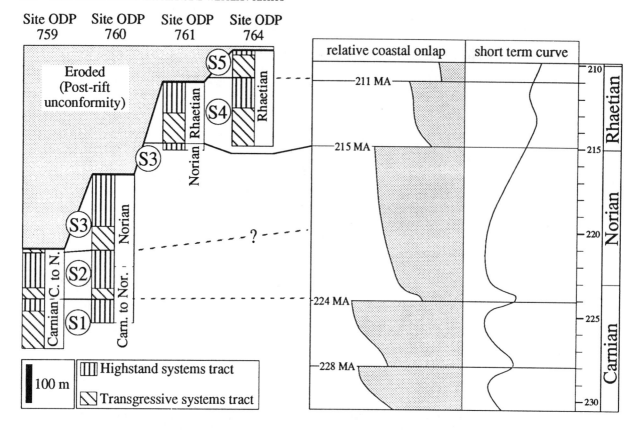

Fig. 6. Comparison between the sequences identified during Leg 122 and the global cycle chart of Haq et al. [1987]. One sequence boundary corresponds to the chart [215 Ma: S3/S4; Norian/ Rhaetian boundary, Brenner et al., 1992]. Two others are probably represented in the chart (S1/S2: 224 Ma; S4/S5: 211 Ma). The last one (S2/S3: 220 ?Ma) is unknown, and may be related to local relative changes driven by tectonics or sedimentary supply.

- According to Röhl et al. [1992], the same clayey interval is interpreted as a shallow lagoonal deposit developed in a restricted back-reef setting included in the transgressive systems tract.

In both interpretations there is a sequence boundary within the upper Rhaetian (S4/S5), and the marly interval of Cores 764-6R to 764-8R (Figure 4) belongs to the TST of sequence 5. The base of this sequence was only recovered at Site 764, whereas it was eroded at Site 761. The uppermost Rhaetian beds recovered at Site 764 do possibly represent a part of the highstand systems tract of S5, since they are made of shallow carbonate facies quite similar to some facies of the HST of sequence S4 (lagoonal facies with *Triasina hantkeni*, isolated corals), but it is not possible to confirm this due to the truncation of S5 by the erosional PRU.

The five depositional sequences S1 to S5 could have been induced by fluctuations in accommodation space [Posamentier and Vail, 1988]. Their identification does not document global sea-level changes, but only relative sea-level changes which result from the interplay between eustasy, tectonic subsidence, and sediment supply (terrigenous supply and carbonate production).

Fit With Known Global Events

The sequences observed in the Triassic series of the Wombat Plateau are compared to the most commonly used chart of sea-level fluctuations [Haq et al., 1987; Figure 6]. The S1/S2 sequence boundary is assumed to correspond to the 224 Ma sequence boundary of the chart, but the paleontological support is weak. The S2/S3 sequence boundary is Norian, and it is unknown in the chart. S3/S4 is well dated [Norian below, Rhaetian above; Brenner, 1992], and corresponds to the 215 Ma sequence boundary. S4/S5 is within the upper Rhaetian [Bralower et al., 1991] and is a good candidate to correspond to the 211 Ma sequence boundary of the chart, which occurred 1 m.y. before the Triassic/Jurassic boundary.

To summarize, the 215 Ma sequence boundary has undoubtedly been found in NW Australia. Two other sequence boundaries may correspond to the 224 Ma and 211 Ma sequence boundaries, but their ages are poorly constrained. Another sequence boundary which is not in the cycle chart has been found within the Norian.

TRIASSIC TECTONIC AND VOLCANIC EVOLUTION IN THE WOMBAT PLATEAU AREA

The best evidence of extensional tectonics in the Wombat Plateau area consists of late Carnian normal faults which are sealed by depositional sequences S2 and S3, as shown by seismic profiles (Figure 7; BMR line 56-13). These faults have offsets of approximately 100m and accommodate the thickening of the series to the north or northwest. A coeval tectonic and volcanic event is documented by the occurrence of volcanic clasts resedimented in some conglomerate layers of sequence S2 [von Rad et al., 1990].

A few normal faults, of a smaller order of magnitude, cut through the uppermost Norian [Williamson et al., 1989]. The hanging wall of these faults may have helped to initiate the Rhaetian reefal buildups which settled on bioclastic and oolitic accumulations on small topographic highs (for example at Site 764). They are however hardly visible on seismic profiles because of the masking effect of the reefs. These faults were no longer active during reef growth.

A volcanic event occurred close to the Triassic/Jurassic boundary. The lavas were not recovered during the Leg 122, but they were dredged along the northern Wombat Plateau escarpment [von Rad and Exon, 1982; von Rad et al., 1990]. The most reliable absolute age is 213 ± 3 Ma [Rhaetian; Haq et al., 1987]; other good age determinations give 206±4 Ma (Hettangian) and 193±4 Ma (Sinemurien). On Scott Plateau, volcaniclastics and basaltic flows are interbedded in the Hettangian series [Scott Reef 1 well; Bint, 1988].

At about the same time (Triassic/Jurassic boundary), the NW Australian margin experienced a major geodynamic reorganization: the thick Triassic series of the Exmouth Plateau was dissected by numerous normal faults [Exon and Willcox, 1980], and overlain by a thin, condensed Jurassic series with local erosion and non-deposition [Barber, 1988]. The Barrow and Dampier basins experienced a different evolution starting from the Triassic/Jurassic boundary [thick Jurassic series; Barber, 1988; Boote and Kirk, 1989], which led to a "failed Jurassic rift" between the Australian continent and the Exmouth marginal plateau.

COMPARISON WITH EUROPEAN TRIASSIC SERIES

The late Triassic sequences of the NW Australian shelf have many similarities with those of Western Europe:

Facies

The Carnian and Norian series are deltaic siliciclastics with minor neritic carbonate interbedded. In the Western Alps, the coeval sequences are subtidal to tidal-flat carbonates [mainly dolomites; Megard-Galli and Baud, 1977]. The sediments are different, due to a higher paleolatitude for NW Australia [Ogg et al., this volume], but the total thickness of the series is comparable [about 1 km; Megard-Galli and Faure, 1988; Kirk, 1985].

The Rhaetian facies are much more similar between western and eastern Tethys, as pointed out by Röhl et al.

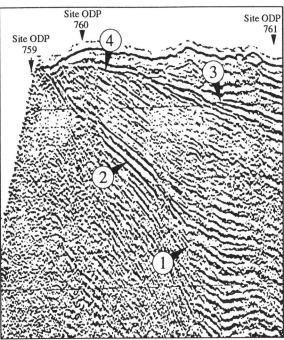

Seismic line BMR 56-13, southern Wombat Plateau, NW Australia:
1- Carnian synsedimentary fault.
2- Upper Carnian-Lower Norian carbonate layers (see fig. 3).
3- Southwards truncation of Rhaetian beds by the PRU.
4- Post-rift unconformity (PRU).

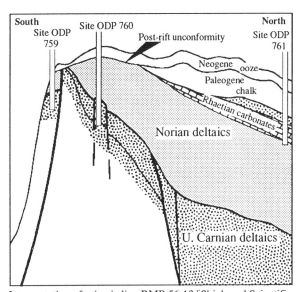

Interpretation of seismic line BMR 56-13 [Shipboard Scientific Party, 1990]

Fig. 7. Seismic profile across the southern Wombat Plateau, showing the strong angular unconformity between the upper Triassic beds and the Tertiary post-rift sediments [seismic line BMR 56-13, interpretation after Shipboard Scientific Party, 1990a]. Several upper Carnian normal faults are visible.

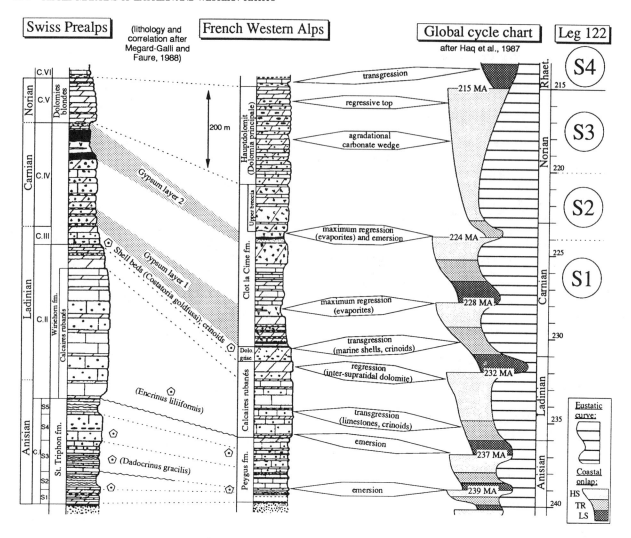

Fig. 8. Triassic sequences in the Western Alps compared to the global cycle chart of Haq et al. [1987] and to the sequences identified during the Leg 122.

(1991), with a major development of lagoonal or reefal carbonates in the Wombat Plateau area, due to the decreasing paleolatitude [Ogg et al., this volume]. The fauna (Megalodons, *Triasina hantkeni*, corals, calcisponges and hydrozoans), and the patch-reef/lagoon organization of the carbonate platform (Figure 7) are similar to the faunas and facies distribution observed in the Eastern Alps [Flügel, 1981] or other late Triassic platforms [i.e. Sicily: Di Stefano and Senowbari-Daryan, 1987].

Sequence Stratigraphy

Sequence boundaries. Two synthetic lithostratigraphic columns of Triassic sequences from the Western Alps are shown in Figure 8 (after Megard-Galli and Faure, 1988). These series are usually split into six lithological cycles (I to VI), which are not depositional sequences of Vail et al. [1977]. However, they have recorded some regional sedimentological events which are meaningful in terms of

relative sea-level changes: i.e., the transgression of the so-called "*Costatoria goldfussi* beds" during the late Ladinian-early Carnian, the two Carnian regressions with development of evaporites, and the late Norian regression followed by the early Rhaetian transgression. Based on the age of the lithologic cycles, a correlation with the depositional sequences of the Haq et al. [1987] chart is proposed (Figure 8). A sequence stratigraphic analysis concerning the Southern Alps has been published recently by Doglioni et al. [1990] (Figure 9A).

In the Western and Southern Alps, the 224 Ma and the 215 Ma sequence boundaries, which may correspond respectively to the S1/S2 and the S3/S4 boundaries of Leg 122, are documented, but there is no equivalent of the S2/S3 boundary. In the Western Alps, there is a prominent sedimentological boundary which roughly coincides with the Triassic/Jurassic boundary [Megard-Galli and Baud, 1977; Dumont, 1988]. This boundary is a sequence

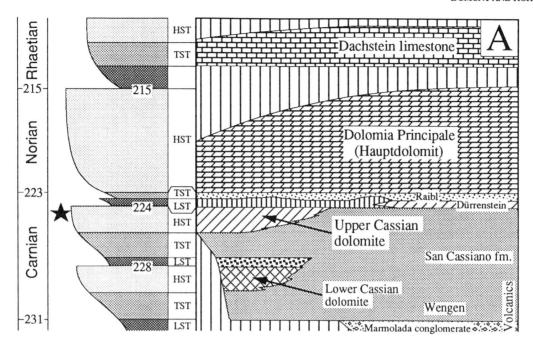

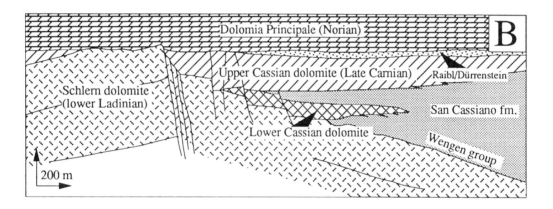

Fig. 9A. Triassic sequence stratigraphy in the Dolomites (Southern Alps), after Doglioni et al. [1990]. The upper Cassian dolomite is a highstand wedge deposited during the late Carnian. B. Carnian synsedimentary tectonics in the Southern Alps after Fois and Gaetani [1981]. The major tectonic activity occurred during the deposition of the upper Cassian dolomitic wedge.

boundary superimposed on a transgressive surface, which may correspond to the 211 Ma sequence boundary of the Haq et al. [1987] cycle chart, and to the S4/S5 boundary of the Wombat Plateau sites.

To summarize, the 215Ma sequence boundary is found both in NW Australia and in Western Europe, the 224Ma and 211Ma sequence boundary may also be present in both places, whereas the S2/S3 boundary (220 Ma?) is not found in Europe.

Rhaetian sequences. In Western Europe, the Rhaetian series show the same facies development as those recovered during Leg 122: a sharp transgressive base coinciding with

the Norian/Rhaetian boundary, a siliciclastic-rich lower part and a shallower, carbonate upper part with patch reef development [Dumont, 1988; Röhl et al., 1992]. This sequence, which is about 100m to 200m thick in the internal Western Alps, is very similar to S4. The age of the base of the overlying sequence is still a matter of discussion in Western Europe [late Rhaetian or early Hettangian; Hallam, 1990]. It may be of Rhaetian age as in NW Australia.

Synsedimentary Tectonics

Most of the Triassic synsedimentary tectonic activity in the Western Alps occurred during the Carnian. The

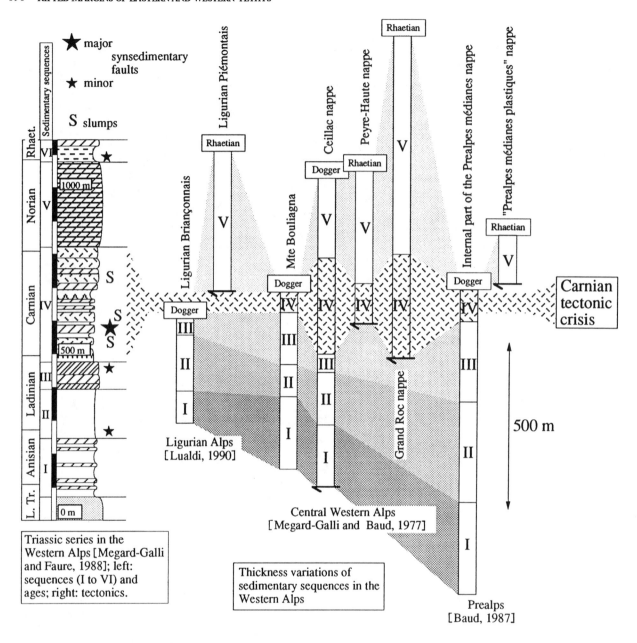

Fig. 10. The Carnian tectonic crisis in the internal Western Alpine series. The pre-Carnian sequences (I to III) are deposited and preserved in most areas of the domain, although they thicken eastwards. During and after the Carnian, a strong differential subsidence is shown by the highly variable thickness of sequences IV and V. Together with the observed faulting and local erosion, this documents a tectonic change during the Carnian.

sediments were highly disturbed [breccias, normal faults and erosion surfaces; Megard-Galli and Faure, 1988]. During and after the Carnian event, the platform organization and the subsidence pattern changed dramatically (Figure 10). In the Western Alps, many breccias are found in the upper part of the Carnian, at the base of the so-called Hauptdolomit formation.

Some tectonic activity is also recorded in the Carnian sediments of the southern margin of the Ligurian Tethys: Figure 9B shows a Carnian tilted block in the Southern Alps [after Fois and Gaetani, 1981], which is sealed by the Norian Hauptdolomit formation. Figure 9A [Doglioni et al., 1990] shows that the main tectonic activity is late Carnian in age (during deposition of the upper Cassian dolomitic wedge). Carnian megabreccias are also found in the sequences bordering the intracontinental Triassic basins of the Southern Appenines and Calabria [Boni et al., 1990].

In NW Australia, the major normal faults which are shown on seismic profile BMR 56-13 (Figure 7) also affected the upper Carnian sediments. Thus, during the late Carnian time interval, important tectonic activity occurred both in Western Europe and in the Wombat Plateau area.

Volcanism

Intraplate volcanism occurred in the Western Alps (1) during the late middle Triassic [ash layers; Caby and Galli, 1964], and (2) at the Triassic/Jurassic boundary [Lemoine et al., 1986; Dumont, 1988]. The former event is coeval from one of the major volcanic events of the Dolomites. The latter event is younger and is the most important one in the Western Alps. In the Dauphiné external zone, subaerial basaltic flows less than 100m thick overlie the late Triassic dolomites (Norian p.p.) and are onlapped by lower Hettangian neritic limestones. This volcanism is not widespread in the Western Alps and the total volume of lava is low. But similar events are also found in some other late Triassic-early Liassic Tethyan series:
- Pyrenees: late Rhaetian to early Hettangian [Curnelle and Cabanis, 1989];
- Iberian chain and Mallorca: Triassic/Liassic boundary [Pocovi et al., 1989];
- Sicily: Hettangian [Patacca et al., 1979];
- Morocco and Central Atlantic: early Hettangian (Manspeizer, 1988) to late Hettangian [Sutter, 1989].

This rift volcanic event, which was short although the affected domain is widespread, is younger and has a different meaning than the Ladinian-early Carnian volcanism which affected the southern Alps [De Zanche, 1990]. In the eastern American Central Atlantic realm, this former event may coincide with a plate-tectonic reorganization around the Triassic/Jurassic boundary [Klitgord et al., 1988].The age of this event is surprisingly consistent with both the age of the lavas dredged on the northern edge of the Wombat Plateau [213±3 Ma, 206±4 Ma and 193±4 Ma; von Rad and Exon, 1982] and the Hettangian age of the volcaniclastics interbedded in the Scott Reef 1 well on Scott Plateau [Bint, 1988; Bint, personal communication, 1989].

Triassic/Jurassic Tectonic Reorganization

In the Western Alps, there was a drastic change in the distribution of subsidence around the Triassic/Jurassic boundary [Lemoine et al., 1986; Dumont, 1988; Rudckiewicz, 1988]. Before this time, the domains which now crop out in the external zone (Dauphiné) were not subsiding, and those of the internal zones experienced a significant subsidence which was balanced by sedimentation (about 1 km of Triassic series in the Briançonnais domain). After the Triassic/Jurassic boundary, the situation was reversed. The external domains underwent strong differential subsidence related to extensional tectonics with block faulting which started in

the Hettangian [Dumont, 1988], whereas the internal Briançonnais domain became a marginal plateau which was emerged and eroded during the Liassic [Lemoine et al., 1986; Faure et Megard-Galli, 1988].

The NW Australian margin also experienced a major reorganization around the Triassic/Jurassic boundary. The Triassic series are thick all over the NW Australian margin: 1 km after well data [Kirk, 1985], and up to 3 km.predicted by seismic data [Exon and Willcox, 1980]. The Jurassic series are very thick in the Barrow-Dampier basins (about 1 km) and very condensed or absent on the Rankin platform and the Exmouth and Wombat Plateaus [Exon et al., 1982; Barber, 1988]. Uplift of the marginal plateau and sediment starvation started in the early Liassic [Barber, 1988; Boote and Kirk, 1989], and increased in the late Liassic [so-called rift onset unconformity; Barber, 1988; Kirk, 1985]. This history is similar and synchronous to the evolution of the European margin of the Ligurian Tethys in the Western Alps, which recorded (1) a major change around the Triassic/Jurassic boundary, and (2) the uplift of the Briançonnais marginal plateau during the Liassic. This suggests a major (global?) plate-tectonic reorganization around the Triassic/Jurassic boundary. Additionally, many other domains were affected by major changes in subsidence pattern or plates motion at the same time, for example:

(1) Both the European and the Apulian Tethyan margins in the Eastern and Southern Alps recorded in many places a general drowning of the Rhaetian reefs, and became the locus of hemipelagic sedimentation. The non-parallelism of Triassic rift directions with that of the later breakup is typical of the Western Tethys [Laubscher and Bernoulli, 1977], and the northwestern Australian margin also shows the same feature.

(2) In the western margins of Central Atlantic, Klitgord et al. [1988] document two rifting events (one late Triassic and the other early Jurassic) with different directions of extension. A major cusp in the apparent polar wander (APW) path and a change in the rate of APW is documented in the Newark Basin close to the Triassic/Jurassic boundary by Witte and Kent [1990].

DISCUSSION AND CONCLUSIONS

Depositional sequences: Causal factors

The main causal factors for the deposition of sedimentary sequences are global eustasy, tectonic subsidence (or uplift), and fluctuations in sediment supply.

Fluctuations in sediment supply are an "autocyclic" control. They cannot explain any global similarities with Western Europe, or any far-reaching correlation since the paleolatitudes of NW Australian margin [Ogg et al., this volume] and of Europe, and hence their climate and paleogeography, were fundamentally different.

Tectonic subsidence or uplift is commonly assumed to be a local factor, contrary to eustasy which is global. If this were true, the synchronism between the NW Australian sequences and other sequences in the world would document global sea-level changes. However, this assumption is a matter of discussion: according to Cloetingh [1986], tectonic subsidence on passive margins is related to intraplate stresses, which can be transmitted from one plate to another. This model involves tectonic coupling of different margins, especially during plate-tectonic reorganizations, and provides an explanation for the synchronism of short-term subsidence or uplift on several plate margins. As explained below, the results of our comparison between northwest Australia and Western Alps would be consistent with such a model.

Tectonics and/or Eustasy ?

The whole NW Australian margin underwent subsidence during the late Triassic, which produced accommodation for the deposition of a thick siliciclastic and carbonate series. This subsidence was probably a combination of thermal subsidence which followed the late Permian breakup [Boote and Kirk, 1989], and of tectonic subsidence coeval with the Triassic stretching of the lithosphere [Cloetingh et al., in press]. Short-term variations, which produced the sedimentary sequences S1 to S5, are superimposed upon this long-term subsidence. The short-term signal may also be due to rift tectonics, or to eustasy, and this signal may be local or global. These two questions have to be considered separately. Concerning the second question, the data from Leg 122 and their comparison with Western Europe show that the short-term signal is not only local. At least one sequence boundary (215 Ma) is synchronous in NW Australia, in Western Europe, and in other places [i.e. Grand Banks, off Newfoundland, after Hubbard, 1988, and in the Canadian Arctic, after Embry, 1988], and is thus generated by a global event. Two other sequence boundaries were also identified both in NW Australia and in Western Europe, although their synchronism is poorly defined. Whether these global events are of eustatic or of tectonic origin is more controversial. The area drilled during Leg 122 was affected by a short late Carnian extensional tectonic pulse and by a later volcanic event, phenomena which are also recorded in the Western European series. Furthermore, a tectonic change around the Carnian/Norian boundary is also recorded in the sediments of the northern Indian margin, which was adjacent to the northwestern Australian margin during late Triassic [Gaetani and Garzanti, 1991] This would support the concept of the transmission of tectonic activity in several contiguous plates and plate margins by way of intraplate stress propagation [Cloetingh, 1986]. In the studied area, it is likely that this short Carnian tectonic pulse has generated or enhanced a sequence boundary synchronously with other plate margins. It is thus very speculative to pick out a global eustatic signal from the observed sequences. However, considering the fit with the global cycle chart for the late Triassic [Haq et al., 1987], a causal linkage can be assumed between short-term tectonic pulses and global eustasy, as proposed by Dewey [1988].

The periodic assembly and disruption of plates or strips of continental lithosphere, such as the "Argo landmass" [Audley Charles, 1988] or other North Gondwanian

microcontinents, certainly had a major impact on sea-level changes, especially if several plate margins were moving at the same time. For example, two peak transgressions of the NW Australian shelf, during the latest middle Triassic-early late Triassic and during the Pliensbachian [Bradshaw et al., 1988], respectively correspond to the Carnian tectonic event (this paper) and to the Pliensbachian rift onset unconformity and tectonic change [Kirk, 1985; Barber, 1988].

Multi-stage rifting

It is essential to consider the Mesozoic disintegration of the Gondwanian margin as a discontinuous history. Unfortunately, its sequential character is difficult to decipher using data from the Wombat Plateau alone, owing to the very long sedimentary gap encountered. Figure 11 shows the main regional events that may have affected the Wombat Plateau during the Jurassic-earliest Cretaceous time interval, which corresponds to the gap [Shipboard Scientific Party, 1990a,b]. At least two rifting/breakup cycles occurred: the neo-Tethyan one and the Indian Ocean one. The neo-Tethyan cycle itself is polyphased, since several continental fragments broke off the northern Gondwanian margin during the Triassic and the Jurassic [Audley-Charles, 1988]. The concept of multi-stage rifting linked to the periodic disintegration of the northern Gondwanian margin is also documented on the northern Indian margin, which was the western continuation of the northwestern Australian margin and whose remnants are now incorporated in the "Tethys Himalaya" [Gaetani and Garzanti, 1991]. The northern Exmouth Plateau area itself was affected by at least two breakup events which overlapped in time and space: the Argo breakup related to the neo-Tethyan cycle, and the Gascoyne breakup related to the Indian Ocean cycle (Figure 11). The Wombat Plateau lies in a particular position, near the junction of the two oceanic domains (Figure 1). It is likely that the sedimentary gap represented by the "post-rift unconformity" (Figure 7) is very long due to the superimposed effects of the Argo breakup and of the Gascoyne breakup [Dumont, 1992]. The late Triassic sedimentary and tectonic history of the Wombat Plateau was related to the Neo-Tethyan history.

However, with the available data, it is not possible to state that the location of the Argo breakup north from the Wombat Plateau was predetermined during the Triassic because (1) the Triassic paleogeographic trends do not parallel the future breakup, and (2) the Triassic sediments and the geometry of the coeval sedimentary bodies do not represent a platform margin which could indicate a Triassic hinge-line. Our study points to a northwestward-facing broad siliciclastic, then carbonate platform, which was periodically flooded by Tethyan transgressions, and which was stamped out obliquely by the Argo breakup 55 m.y. later [Ludden and Dionne, in press].

In our opinion, the neo-Tethyan "syn-rift" stage which was successfully followed by the opening of the Argo ocean did not start before the Triassic/Jurassic boundary,

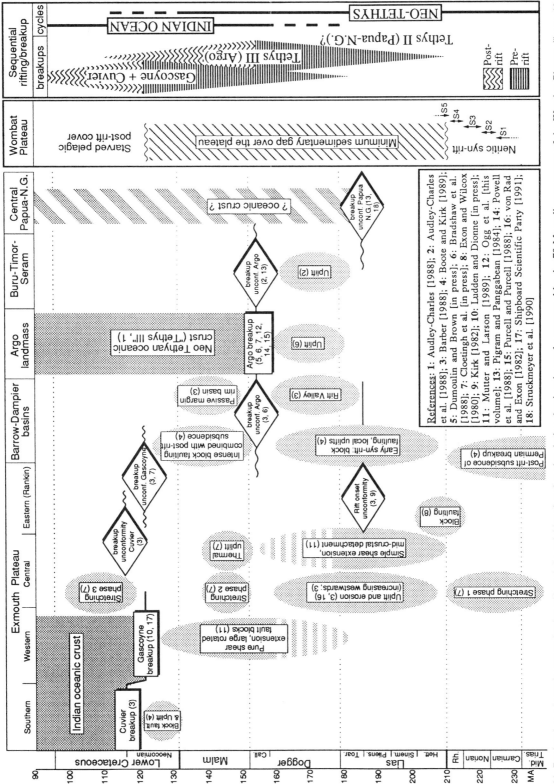

Fig. 11. Regional geodynamic events having affected the neighbouring domains during the time interval represented by the 70 Ma sedimentary gap of the Wombat Plateau ("post-rift unconformity"). To the right are shown the two proposed rifting/breakup cycles (the Argo/neo-Tethyan cycle, and the Gascoyne/Indian ocean cycle), whose effects are superimposed in the Wombat Plateau area.

after the plate-tectonic reorganization which was described above. This "Argo rifting stage" caused renewed subsidence in the Barrow-Dampier "rift-valley" system, and uplift , sediment starvation and local erosion on the Exmouth and Wombat plateaus. The late Triassic sequences of the Wombat Plateau however were deposited in a different geodynamic setting, with tectonic subsidence and faulting probably related to an earlier neo-Tethyan breakup of a North Gondwanian fragment which was not the Argo Landmass (possibly Central Papua-New Guinea?; Figure 11).

The very long time interval of non-deposition and erosion observed in the series of the Wombat Plateau (late Triassic to early Cretaceous), which is represented by the "post-rift unconformity" capping the late Triassic series, could be explained by the superimposed effects of the late Jurassic Argo breakup and of the early Cretaceous Gascoyne breakup. Starvation and erosion started with the "Argo rifting stage". The Wombat Plateau was flooded again by pelagic sediments after the Gascoyne breakup, with the development of thermal subsidence following the initial formation of the Indian Ocean. All these events took place in a rejuvenated geodynamic setting, different from the late Triassic one.

Acknowledgments. The authors are very grateful to the Ocean Drilling Program (Texas A & M University, College Station, Texas, U.S.A.) for inviting one of us (T.D.) to participate in ODP Leg 122 and for giving us access to Leg 122 samples, and to the ODP personnel and the Sedco crew which made Leg 122 a success. We thank the Deutsche Forschungsgemeinschaft for its financial support of the ODP research project (Ra 191/1, U. Röhl). T. Dumont was funded by ODP France (A.I.P. "Rifting mésozoïque polyphasé"). We are grateful for critical comments by P. C de Graciansky, H. C. Jenkyns, J. Marcoux and especially U. von Rad.

REFERENCES

Audley-Charles, M. G., Evolution of the Southern margin of Tethys (North Australian region) from early Permian to late Cretaceous, in *Gondwana and Tethys, Geological Soc. of London Special Publication, 37*, edited by M. G. Audley-Charles, and A. Hallam, 79-100, 1988.

Audley-Charles, M. G., Ballantyne, P. D., and Hall, R., Mesozoic-Cenozoic rift-drift sequence of Asian fragments from Gondwanaland, *Tectonophysics, 155*, 317-330, 1988.

Barber, P. M., The Exmouth Plateau deep water frontier: A case history, in *The North West Shelf Australia, Proceedings of the North West Shelf Symposium,* edited by P. G. Purcell, and R. R. Purcell, 173-187, Perth, 1988.

Baud A., Stratigraphie et sédimentologie des calcaires de Saint-Triphon (Trias, Préalpes, Suisse et France), *Mémoires de Géologie, 1*, 322p, Lausanne, 1987.

Bint, A. N., Gas fields of the Browse Basin, in *The North West Shelf Australia, Proceedings of the North West Shelf Symposium,* edited by P. G. Purcell, and R. R. Purcell, 413-418, Perth, 1988.

Boni, M., Torre, M., Zamparelli, V., Il Trias medio-superiore dell'unita' di S. Donato (Appenino Meridionale, Calabria): risultati preliminari, *Rend. Soc. Geol. It., 13*, 89-92, 1990.

Boote, D. R., and Kirk, R. D., Depositional Wedge Cycles on Evolving Plate Margin, Western and Northwestern Australia, *Am. Assoc. Petroleum Geol. Bulletin, 73/2*, 216-243, 1989.

Borella, P., Görür, N., Dumont, T., Sarti, M., Stefani, A., and Lewis, T., Upper Triassic (Rhaetian) carbonate environments, Wombat Plateau, Northwest Australian Shelf, in *Proc. ODP, Sci. Results, 122*, edited by U. von Rad, B. U. Haq et al., 161-180, College Station, TX, 1992.

Bradshaw, M. T., Yeates, A. N., Beynon, R. M., Brakel, A. T., Langford, R. P., Totterdell, J. M., and Yeung M., Paleogeographic Evolution of the North West Shelf region, in *The North West Shelf Australia, Proceedings of the North West Shelf Symposium,* edited by P. G. Purcell, and R. R. Purcell, 29-53, Perth, 1988.

Bralower, T., Bown, P. R., and Siesser, W. G., Significance of Upper Triassic nannofossils from the Southern hemisphere (ODP Leg 122, Wombat Plateau, N.W. Australia), *Mar. Micropaleontol., 17*, 119-154, 1991.

Brenner, W., Bown, P. R., Bralower, T. J., Crasquin-Soleau, S., Depeche, F., Martini, R., Siesser, W. G., and Zaninetti, L., Correlation of Carnian to Rhaetian palynological, foraminiferal, calcareous nannofossil, and ostracode biostratigraphy, Wombat Plateau, in *Proc. ODP, Sci. Results, 122*, edited by U. von Rad, B. U. Haq et al., 487-496, College Station, TX, 1992.

Brenner, W., First results of Late Triassic palynology of the Wombat Plateau, northwestern Australia, in *Proc. ODP, Sci. Results, 122*, edited by U. von Rad, B. U. Haq et al., 413-426, College Station, TX, 1992.

Caby, R. and Galli, J., Existence de cinérites et tufs volcaniques dans le Trias moyen de la zone briançonnaise, *Comptes rendus Acad. Sc. Paris, 259*, 417-420, 1964.

Cloetingh, S., Intraplate stresses: a new tectonic mechanism for fluctuations of relative sea level, *Geology, 14*, 617-620, 1986.

Cloetingh, S., Stein, C., Reemst, P., Gradstein, F., Williamson, P., Exon, N., von Rad, U., Continental margin stratigraphy, deformation, and intraplate stresses for the Indo-Australian region, *Proc. ODP, Sci. Results, 123*, in press.

Curnelle, R. and Cabanis, B, Relations entre le magmatisme "triasique" et le volcanisma infra-liasique des Pyrénées et de l'Aquitaine; Apports de la géochimie des éléments en traces, *Bull. Centres Rech. Explor.-Prod. Elf Aquitaine, Pau, 13*, 347-375, 1989.

Depeche, F., and Crasquin-Soleau, S., Triassic marine ostracodes of the Australian margin (Holes 759B, 760B, 761C, 764A and 764B), in *Proc. ODP, Sci. Results, 122*, edited by U. von Rad, B. U. Haq et al., 453-462, College Station, TX, 1992.

Dewey, J. F., Lithospheric stress, deformation, and tectonic cycles: the disruption of Pangea and the closure of Tethys, in *Gondwana and Tethys, Geological Soc. of London Special Publication, 37*, edited by M. G. Audley-Charles, and A. Hallam, 23-40, 1988.

De Zanche, V., A review of Triassic stratigraphy and paleogeography in the Eastern and Southern Alps, *Boll. Soc. Geol. Italiana, 109*, 59-71, 1990.

Di Stefano P. and Senowbari-Daryan B., Upper Triassic dasycladales (green algae) from the Palermo Mountains (Sicily, Italy), *Geol. Rom., 24*, 189-220, 1987.

Doglioni, C., Bosellini, A. and Vail, P. R., Stratal patterns: a proposal of classification and examples from the Dolomites, *Basin Research, 2*, 83-95, 1990.

Dumont, T., Late Triassic-early Jurassic evolution of the western Alps and of their European foreland: initiation of the Tethyan rifting, *Bull. Soc. Géol. France, 4/4*, 601-612, 1988.

Dumont, T., The Late Triassic (Rhaetian) sequences of the North-Western Australian shelf recovered during ODP Leg 122: sea-level changes, Tethyan rifting and overprint of Indo-Australian breakup, in *Proc. ODP, Sci. Results, 122*, edited by U. von Rad, B. U. Haq et al., 197-212, College Station, TX, 1992.

Dumoulin, J.A., and Bown, P.R., Depositional history, nannofossil biostratigraphy, and correlations of Argo Abyssal

Plain Sites 765 and 261, in *Proc. ODP, Sci. Results, 123*, edited by F. M. Gradstein, J. N. Ludden et al., College Station, TX, in press.

Embry, A. F., Triassic Sea-Level Changes: evidences from the Canadian Arctic Archipelago, in *Soc. Econ. Pal. Mineral. Special Publication N° 42*, Sea-level changes: an integrated approach, C. K. Wilgus, B. S. Hastings, C. G. Kendall, H. W. Posamienter, C. A. Ross, and J. C. Van Wagoner, 299-260, Tulsa, 1988.

Exon, N. F., von Rad, U., and von Stackelberg, U., The geological development of the passive margins of the Exmouth Plateau off northwest Australia, *Mar. Geology, 47*, 131-152, 1982.

Exon, N. F., and Willcox, J. B., The Exmouth Plateau; Stratigraphy, structure, and petroleum potential, *Bureau of Min. Res. Bulletin, 199*, 52 p, 1980.

Faure, J.L., and Megard-Galli, J., L'évolution jurassique en Briançonnais: sédimentation continentale et fracturation distensive, *Bull. Soc. Géol. France, 4/4*, 681-692, 1988.

Fois, E., and Gaetani, M., The Northern margin of the Civetta buildup. Evolution during the Ladinian and the Carnian, *Riv. Ital. Paleont., 86/3*, 469-542, 1981.

Flügel E., Paleoecology and facies of Upper Triassic reefs in the northern Calcareous Alps, in *European Fossil Reef Models, Soc. Econ. Pal. Mineral. Special Publication, 30*, edited by D. F. Toomey, 291-360, 1981.

Gaetani, M., and Garzanti, E., Multicyclic history of the Northern India Continental Margin (Northwestern Himalaya), *Am. Assoc. Petroleum Geol. Bulletin, 75/9*, p. 1427-1446, 1991.

Gildner, R. F., and Cisne, J. L., Quantitative modelling of carbonate stratigraphy and water-depth history using depth-dependent sediment accumulation function, in *Quantitative Dynamic Stratigraphy*, edited by T. A. Cross, Prentice Hall, p. 417-432, 1989.

Hallam, A, Correlation of the Triassic-Jurassic boundary in England and Austria, *J. Geol. Soc., 147*, 421-424, London, 1990.

Haq, B. U., Hardenbold, J., and Vail, P. R., Chronology of fluctuating sea-levels since the Triassic, *Science, 235*, 1156-1167, 1987.

Haq, B.U., von Rad, U., O'Connell, S., and Shipboard Scientific Party of Leg 122, *Proc. ODP, Init. Repts., 122*, College Station, TX, 1990.

Hubbard, R. J., Age and signifiance of Sequence Boundaries on Jurassic and Early Cretaceous Rifted Continental Margins, *AAPG Bulletin, 72/1*, 49-72, 1988.

Ito, M., O'Connell, S, Stefani, A., and Borella, P., Fluviodeltaic successions at the Wombat Plateau: upper Triassic siliciclastic-carbonate cycles, in *Proc. ODP, Sci. Results, 122*, edited by U. von Rad, B. U. Haq et al., 109-128, College Station, TX, 1992.

Kirk, R. B., A Seismic Stratigraphic Case History in the Eastern Barrow Subbasin, North West Shelf, Australia, in *Seismic Stratigraphy II, an integrated approach*, edited by O. R. Berg, and D. G. Woolverton, 183-208, *Am. Assoc. Petroleum Geol. Memoir, 39*, 1985.

Klitgord K.D., Hutchinson D.R., and Schouten H., U.S. Atlantic Continental Margin; Structural and tectonic framework, in *The Atlantic Continental Margin: U.S. Geol. Soc. of America, The Geology of North America, I-2*, edited by R. E. Sheridan, and J. A. Grow, 19-52, 1988.

Laubscher, H.P., and Bernoulli, D., Mediterranean and Tethys, in *The Ocean Basins and Margins, 4A*, edited by A. Nairn, W. Kanes, and F. Stehli, 1-27, Plenum Press, New York, 1977.

Lemoine, M., Bas, T., Arnaud-Vanneau, A., Arnaud, H., Dumont, T., Gidon, M, Bourbon, M., de Graciansky, P. C., Rudckiewicz, J. L., Megard-Galli, J., and Tricart, P., The continental margin of the Mesozoic Tethys in the western Alps, *Mar. Petrol. Geol., 3*, 179-200, London, 1986.

Lualdi A., An outline on the Triassic in the Ligurian Alps (Briançonnais and Prepiemont domains), *Boll. Soc. Geol. It., 109*, 51-58, 1990.

Ludden, J.N., and Dionne, B. The geochemistry of oceanic crust at the onset of rifting in the Indian Ocean, in *Proc. ODP, Sci. Results, 123*, edited by F. M. Gradstein, J. N. Ludden et al., College Station, TX, in press.

Manspeizer, W., Triassic-Jurassic rifting and opening of the Atlantic: an overview, in *Triassic-Jurassic rifting: Continental breakup and the origin of the Atlantic Ocean and passive margins*, edited by W. Manspeizer, 41-79, Elsevier, 1988.

Megard-Galli, J., and Baud, A., Le Trias moyen et supérieur des Alpes nord-occidentales: données nouvelles et corrélations stratigraphiques, *Bulletin BRGM, 4/3*, 233-250, Orléans, 1977.

Megard-Galli, J., and Faure, J. L., Tectonique distensive et sédimentation au Ladinien supérieur-Carnien dans la zone briançonnaise, *Bull. Soc. Géol. France, 4/5*, 705-716, 1988.

Mutter, J. C., and Larson, R. L., Extension of the Exmouth Plateau, offshore northwestern Australia: Deep seismic reflection/refraction evidence for simple and pure shear mechanisms, *Geology, 17*, 15-18, 1979.

Patacca E., Scandone P., Giunta G. and Liguori V., Mesozoic paleotectonic evolution of the Ragusa zone (Southeastern Sicily), *Geol. Rom., 18*, 331-369, 1979.

Pigram, C. J., and Panggabean, H., Rifting of the Northwestern margin of the Australian continent and the origin of some microcontinents in Eastern Indonesia, *Tectonophysics, 107*, 331-353, 1984.

Pocovi, A., Lago, M., Bastida, J., Enrique, P., Zachmann, D., and Vaquer, R., The Triassic-Liassic alkaline magmatism of the Iberian Chain, Tarragona, and Majorca (Spain): petrological and geochemical features and setting conditions, *Terra Cognita, Abstracts, 1*, 287, 1989.

Posamentier, H. W., and Vail, P. R., Eustatic control on clastic deposition II- Sequence and systems tracts models, in *Sea-level changes: an integrated approach: SEPM Special Publication N° 42*, edited by C. K. Wilgus, B. S. Hastings, C. G. Kendall, H. W. Posamentier, C. A. Ross, and J. C.Van Wagoner, 125-154, Tulsa, 1988.

Powell, C. M., Roots, S. R., and Veevers, J. J., Pre-breakup continental extension in East Gondwanaland and the early opening of the eastern Indian Ocean, *Tectonophysics, 155*, 261-283, 1988.

Purcell, P. G., and Purcell, R. R., The Northwest Shelf Australia. An introduction, in*The North West Shelf Australia, Proceedings of the North West Shelf Symposium*, edited by P. G. Purcell, and R. R. Purcell, 1-17, Perth, 1988.

Röhl, U., Dumont, T., von Rad, U., Martini, R. and Zaninetti, L., Upper Triassic Tethyan carbonates off Northwest Australia (Wombat Plateau, ODP, Leg 122), *Facies, 25*, Erlangen, 1991.

Röhl, U., von Rad, U. and Wirsing, G., Microfacies, paleoenvironment and facies-dependent carbonate diagenesis in Upper Triassic platform carbonates off Northwest Australia, in *Proc. ODP, Sci. Results, 122*, edited by U. von Rad, B. U. Haq et al., 129-160, College Station, TX, 1992.

Rudckiewicz, J. L., Structure et subsidence de la marge téthysienne entre Grenoble et Briançon au Lias et au Dogger, PhD Thesis, *ENSM*, Paris, 126 p, 1988.

Sarg, J. F. Carbonate sequence stratigraphy, in *Sea-level changes: an integrated approach: Soc. Econ. Pal. Mineral. Special Publication N° 42*, edited by C. K. Wilgus, B. S. Hastings, C. G. Kendall, H. W. Posamentier, C. A. Ross, and J. C. Van Wagoner, 155-182, Tulsa, 1988

Sarti, M., and Kälin, O., Stable carbon and oxygen isotopic composition of Rhaetian shelf carbonates, Wombat Plateau, northwest Australia, in *Proc. ODP, Sci. Results, 122*, edited by U. von Rad, B. U., Haq B.U. et al., 839-850, College Station, TX,1992.

Sarti, M., Russo, A., and Bosellini, F., Rhaetian strata, Wombat Plateau: Analysis of fossil communities as a key to paleoenvironmental changes, in *Proc. ODP, Sci. Results, 122*, edited by U. von Rad, B. U., Haq et al., 181-196, College Station, TX,1992.

Schlager, W., The paradox of drowned reefs and carbonate platforms, *Bull. Geol. Soc. Am., 92*, 197-211, 1981.

Shipboard Scientific Party, Site 761, in *Proc. ODP, Init. Repts., 122*, edited by B. U. Haq, U. von Rad, S. O'Connell et al., 161-212, 1990a.

Shipboard Scientific Party, Site 764, in *Proc. ODP, Init. Repts., 122*, edited by B.U. Haq, U. von Rad, S. O'Connell et al., 353-386, 1990b.

Shipboard Scientific Party, Sites 765 and 766, in *Proc. ODP, Init. Repts., 123*, edited by F. Gradstein, J. Ludden et al., 1991.

Struckmeyer, H. I., Yeung, M., and Bradshaw, M. T., Mesozoic Paleogeography of the Northern margin of the Australian Plate and its implications for hydrocarbon exploration, in *Petroleum Exploration in Papua New Guinea, Proc. of the First PNG Petroleum Convention*, edited by G. J. Carman and Z. Carman, 137-152, Port Moresby, 1990.

Sutter J. F., Innovative approaches to the dating of igneous events in the Early Mesozoic basins of the eastern United States, *U.S.G.S Bulletin, 1776*, 194-200, 1989.

Vail, P.E., Mitchum, R. M. Jr., Todd R. G., Widmier, J. M., Thompson, S., Sangree, J. B., Bubb, J. N., and Hatlelid, W. G., Seismic stratigraphy and global changes of sea-level, in *Am. Assoc. Petroleum Geol. Memoir 26*, 49-212, 1977.

von Rad, U., and Exon, N. F., Mesozoic-Cenozoic Sedimentary and Voicanic Evolution of the Starved Passive Continental Margin off Northwest Australia, in *Studies in Continental Margin Geology. Am. Assoc. Petroleum Geol. Memoir 34*, edited by J. S. Watkins and C. L. Drake, 253-281, 1982.

von Rad , U., Haq, B.U., and Exon, N. F., Rift-to-drift history of the Wombat Plateau, northwest Australia: Triassic to Tertiary Leg 122 results, in *Proc. ODP, Sci. Results, 122*, edited by U. von Rad, B. U. Haq et al., 765-800, College Station, TX, 1992.

von Rad, U., Schott, M., Exon, N. F., Mutterlose, J., Quilty, P. G., and Thurow, J., Mesozoic sedimentary and volcanic rocks dredged from the northern Exmouth Plateau: petrography and microfacies, *BMR J. Australian Geol. Geophys., 11*, 449-472, 1990.

Williamson, P. E., Exon, N. F., Haq, B. U., von Rad, U., and Leg 122 Shipboard Scientific Party, A North West shelf Triassic reef play: results from ODP Leg 122, *The APEA Journal, 29*, 328-344, 1989.

Witte W.K., and Kent D.V., The Paleomagnetism of Red Beds and Basalts of the Hettangian Extrusive Zone, Newark Basin, New Jersey, *J. Geophys. Res., 95/B11*, 17533-17545, 1990.

Zaninetti, L., Martini, R. and Dumont, T., The Triassic foraminifera from Sites 761 and 764, Wombat Plateau, NW Australia, in *Proc. ODP, Sci. Results, 122*, edited by U. von Rad, B. U. Haq et al., 427-436, College Station, TX, 1992.

Sedimentary History of the Tethyan Margins of Eastern Gondwana during the Mesozoic

JAMES G. OGG

Department of Earth and Atmospheric Sciences, Purdue University, West Lafayette, IN, 47907, USA

FELIX M. GRADSTEIN

Atlantic Geoscience Centre, Bedford Institute of Oceanography, Dartmouth, Nova Scotia B2Y 4A2 Canada

JULIE A. DUMOULIN

U.S. Geological Survey, 4200 University Drive, Anchorage, AK, 99508-4667, USA

MASSIMO SARTI

Dipartimento di Scienze della Terra, Universita della Calabria, 1-87030 Castiglione Cosentino Scalo, Italy

PAUL BOWN

Department of Geological Sciences, University College London, London, WC1E 6BY, UK

A composite Mesozoic geological history for the Gondwana margins to the Eastern Tethys Ocean can be assembled from stratigraphic successions on the Australian and Himalayan margins and from drill sites of Ocean Drilling Program Legs 122 and 123. During the Triassic, this region drifted northwards, entering tropical paleolatitudes during the Late Triassic-Early Jurassic, then returned to mid-latitudes for the Middle Jurassic through Early Cretaceous. Shallow-water carbonates are restricted to the tropical-latitude interval; at other times, the margins are dominated by clastic deposition. Episodes of deltaic sandstone progradation over the shelves are caused by eustatic sealevel fluctuations, by wet climatic conditions within the source regions and by local tectonic activity. A major hiatus between Callovian shallow-water shelf deposits and Oxfordian deep-water sediments is an ubiquitous feature, which may be related to a widespread plate tectonic reorganization and the cascading effects of associated sealevel rise and elevated carbon dioxide levels. Off Northwest Australia, this Callovian/Oxfordian event also coincides with an episode of block faulting. Marginal sediments deposited during the Late Jurassic are mainly marine claystone containing abundant terrigenous organic matter. Shallow depths of carbonate compensation (CCD) during the Late Jurassic through Early Cretaceous prevented the preservation of carbonate over most of the Argo basin off Northwest Australia, and these deep-sea sediments consist mainly of condensed, oxygenated radiolarian-rich claystone. During the late Kimmeridgian-early Tithonian, a downward excursion in the CCD enabled limited preservation of some larger nannofossils and mollusc fragments within the pelagic deposits, a feature also recorded in coeval deposits in the Atlantic. Explosive volcanism accompanied the final stages of rifting between India and Australia during the late Berriasian and Valanginian, producing volcaniclastic debris washing into the deltas and widespread ash deposits. The late Barremian and Aptian sediments indicate a rise in the CCD, accompanied by warming of the region and an increased delivery of organic-rich claystone into the basins.

Introduction

The present Indian Ocean was created during the Early Cretaceous through Tertiary when the former supercontinent of Gondwana shattered and dispersed its fragments in the form of the continents of Africa, India, Antarctica and Australia. Prior to this disruption, a triangular extension of the Pacific between the northern shores of Gondwana and the southern margin of Eurasia formed the Tethys Ocean (Figure 1). The eastern portion of this Tethys Ocean had a complex history of seafloor renewal, as slivers of northern Gondwana split from the supercontinent, traveled across the Eastern Tethys basin leaving young seafloor in their wake, and collided with southern Asia forming the composite terrains of Indochina, Tibet, the Middle East and Turkey [e.g., Sengör, 1987]. Ophiolites within the sutures between these accreted microplates preserve a few fragments of the former Tethys oceanic crust and deep-sea sediments. During the Late Jurassic, the separation of a microplate from Gondwana

Synthesis of Results from Scientific
Drilling in the Indian Ocean
Geophysical Monograph 70

Eastern Tethys

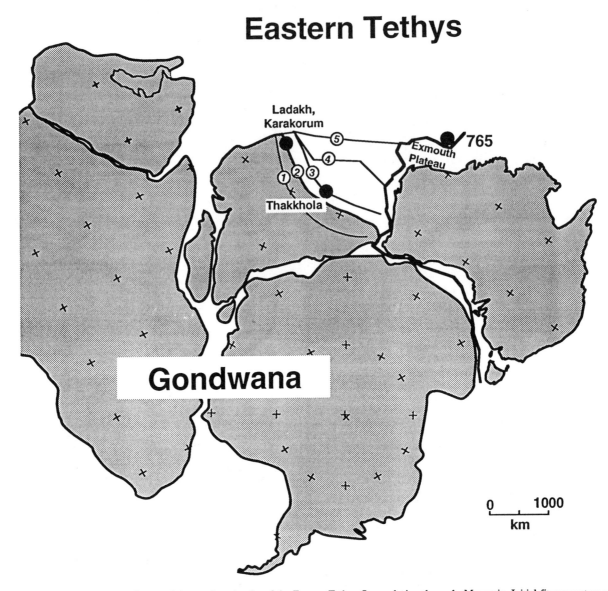

Fig. 1. Gondwana supercontinent and the southern border of the Eastern Tethys Ocean during the early Mesozoic. Initial fit parameters are described in the text. For the Indian plate, line 1 is present Main Boundary Fault of the Himalayas, line 2 is Indus-Tsangpo Suture, line 3 is minimum northward extent of India in East Gondwanaland after extending the doubled thickness of the crust south of the Indus-Tsangpo Suture, line 4 is the present Kun Lun-southern Tsaidam mountain front, and line 5 is postulated northern edge of Greater India [from Klootwijk and Bingham, 1980; Powell et al., 1988]. Position of central Nepal shelf (Thakkhola region) is partially adjusted for crustal shortening. In this reconstruction, each continent is drawn according to a centered mercator projection. The Northwest Australian and Himalayan shelves are adjacent Gondwana margins facing Eastern Tethys.

formed the present passive margin of Northwest Australia and the oceanic crust of the adjoining Argo Abyssal Plain. The Argo portion of Eastern Tethys was extended during the Early Cretaceous when Greater India separated from Australia.

In 1972, Deep Sea Drilling Project Leg 27 drilled the Argo Abyssal Plain at Site 261 and recovered a condensed section of reddish calcareous claystone of Late Jurassic age [Veevers, Heirtzler et al., 1974] (Figure 2). In 1988, Ocean Drilling Program Leg 123 returned to this region to drill Site 765, a deep penetration of the Jurassic oceanic crust to

enable improvement of Mesozoic bio- and lithostratigraphy of the mid-latitude Eastern Tethys and to serve as a "geochemical reference site" for use in global mass-balance models [Ludden, Gradstein et al., 1990; Gradstein, Ludden et al., 1992]. These sites provide our only in-situ stratigraphy of the vanished Eastern Tethys ocean.

Prior to the Cretaceous break-up of Eastern Gondwana, the southern margin of the Tethys encompassed portions of Northwest Australia and the Indian subcontinent (Figure 1). The Northwest Australian passive margin has been

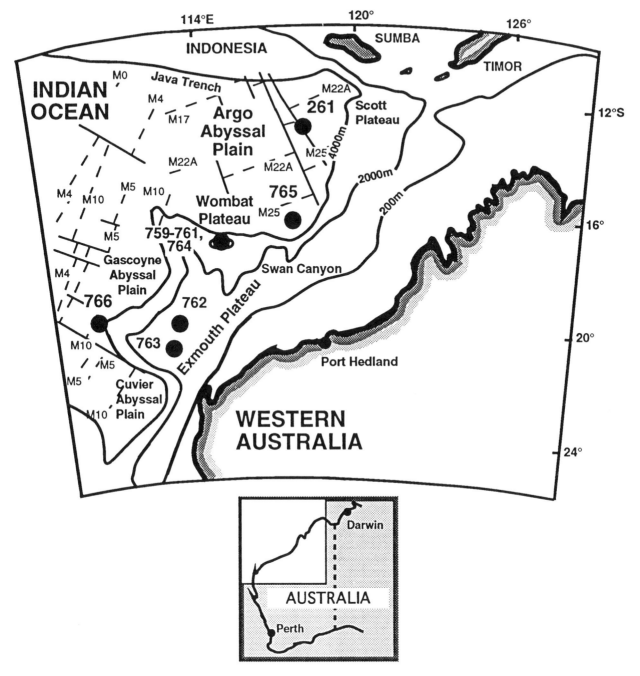

Fig. 2. Northwest Australian margin showing selected magnetic anomalies [Sager et al., 1992], ODP Leg 122 and 123 drill sites, and DSDP Site 261.

extensively investigated through seismic stratigraphy tied to various drilling sites, especially the Exmouth and Wombat Plateaus at the southern margin of Argo Abyssal Plain [e.g., Exon and Willcox, 1978; von Stackelberg et al., 1980; Exon et al., 1982; von Rad and Exon, 1983; Erskine and Vail, 1988]. This region was further investigated at a series of drill sites during ODP Leg 122 [Haq, von Rad et al., 1990; von Rad, Haq et al., 1992] and Leg 123 [Ludden, Gradstein et al., 1990; Gradstein, Ludden et al., 1992]. The Mesozoic tectonic and sedimentation

history of the Northwest Australian margin is reviewed by von Rad et al. [this volume]. The former Tethyan margin of India is exposed in truncated slices in the Himalayan collision belt [Gansser, 1964], especially in the Thakkhola region north of the Annapurna range in Central Nepal [e.g., Bodenhausen et al., 1964; Bordet et al., 1971] and in the Ladakh region of India [e.g., Thakur, 1981; Gaetani et al., 1986; Garzanti et al., 1987]. In conjunction with ODP Legs 122-123, the Mesozoic stratigraphy of the Thakkhola region of the Central Nepal margin was investigated by the Lost Ocean Expeditions of 1988 [Gradstein et al., 1989; Gradstein et al., 1991; Gradstein et al., 1992] and of 1991 [Ogg, Sarti et al., unpublished manuscript, 1992].

When assembled in a paleogeographic framework and compared to coeval patterns in other ocean basins, the Mesozoic successions of the pelagic and marginal-marine facies in these different pieces of the Eastern Tethys begin to form a coherent picture. Global paleoceanographic factors of changing atmospheric CO_2 levels and carbonate dissolution, evolutionary trends of marine plankton, sealevel fluctuations and plate tectonic reorganizations are superimposed upon various local factors including subsidence, paleolatitude drift, climate, terrigenous and volcanic influx, and productivity. We will summarize the major features of the facies patterns and possible paleoceanographic interpretations for the Late Triassic through Early Cretaceous of this eastern portion of Tethys.

PALEOGEOGRAPHIC FRAMEWORK

Two paleogeographic factors, important for understanding the relationships of sedimentary facies between different regions, are the configuration of continents and the paleolatitudes at the time of deposition. For Gondwana, the paleolatitude framework is partially dependent upon the choice of the initial pre-rifting configuration.

The Late Cretaceous through Tertiary paleogeography of the Indian Ocean is constrained by magnetic anomaly lineations, fracture patterns, hot-spot traces, and paleomagnetic measurements of continental and oceanic sediments [reviewed by Royer et al., this volume].

In contrast, continental reconstructions prior to the mid-Cretaceous are poorly constrained for several reasons: (1) The mid-Cretaceous Magnetic Quiet Zone masks the transition between the initial stages of Gondwana breakup and the position of the continents at Late Cretaceous magnetic anomaly C34. (2) The collision of India with Asia resulted in considerable shortening and underthrusting of its former northern extent. (3) Flood basalts and thick Tertiary submarine-fan deposits of the Bay of Bengal have concealed any underlying magnetic anomalies. (4) Antarctica consists of two plates, East and West Antarctica, which had relative displacements of unknown distances during the Mesozoic. (5) The paleomagnetic database for continental orientations through the Jurassic and Early Cretaceous time is extremely limited. (6) The timing of rifting and the fate of microplates that separated from the Gondwana margins is poorly known.

The Australian and Himalayan margins facing the Eastern Tethys behaved as passive margins throughout the Mesozoic, but experienced a progressive rifting of different continental slivers. The timing of the most recent rift events off Northwest Australia are constrained by magnetic anomalies as late Oxfordian-Kimmeridgian in the Argo Abyssal Plain and as late Valanginian near the Exmouth Plateau [Fullerton et al., 1989; Sager et al., 1992]. Identification of the conjugate Late Jurassic "Argo Landmass" is debated. Published possibilities have

included the Tarim block of Asia [Larson, 1975], South Tibet [Audley-Charles, 1983; Audley-Charles et al., 1988], Tibet-Burma [Görür and Sengör, 1992], Shillong Plateau to the north of Bangladesh [Ricou et al., 1990], a series of continental terrains accreted to Sumatra [Metcalfe, 1990] and Mount Victoria Land in western Burma plus Sumba Island in Indonesia [Metcalfe, 1991]. Most Gondwana reconstructions implicate that the Exmouth Plateau faced a vague "Greater India", with rifting initiated in late Valanginian (Figure 1). However, this portion of "Greater India" is generally assumed to have been completely lost during the Tertiary collision with Tibet. A similar identification problem exists for the rifted blocks that once faced the Himalayan and New Guinea margins. Facies comparisons of suspect terrains, while providing constraints on tectonic scenarios, do not necessarily imply paleogeographic continuity. Rifting episodes prior to the mid-Jurassic for the Himalayan and Australian margins are inferred from the subsidence history, from pulses of volcanic activity, or from suturing of continental fragments to Asia [e.g., Gaetani et al., 1986, 1990; Sengör, 1987; Audley-Charles, 1988; Görür and Sengör, 1992; von Rad et al., this volume], but all such evidence is indirect. Therefore in our synthesis, we mainly examine the sedimentation history of the known north-facing Gondwana margins towards Eastern Tethys of Australia and the Himalayas.

The pre-rift assembly of the main Gondwana continents, especially of India, is primarily based upon matching of continental outlines. Most initial fits resemble the reconstruction published by du Toit [1937]. Seafloor spreading between India and Australia in the late Valanginian and Hauterivian [Ludden, Gradstein et al., 1990] left a northwest-trending set of marine magnetic anomalies M10 through M0 off the western coast of Australia [Fullerton et al., 1989]. There is no corresponding set of magnetic anomalies identified off the eastern coast of India, which suggests that the associated Early Cretaceous oceanic crust was consumed during the Himalayan collision. Therefore, the corresponding fit of India to Australia, hence the relationship of Tethyan sequences in the Himalayas to coeval deposits off Northwest Australia, is based upon independent fits of (1) Australia to Antarctica and (2) India to Antarctica. Most reconstructions imply that the Indian plate and its attached seafloor rotated about 30° counterclockwise with respect to Australia during the mid-Cretaceous magnetic quiet zone, followed by rapid northward drift towards its collision with Tibet. This mid-Cretaceous pirouetting was avoided by Veevers et al. [1971], who placed the eastern margin of India directly against southwestern Australia. This reconstruction was not used by Veevers in later publications [e.g., Johnson et al., 1976; Veevers, 1984], but has been resurrected by Ricou et al. [1990] and by Baumgartner [1992].

Faced with the dilemma between the lack of Early Cretaceous magnetic anomalies off India and the pivoting India in the mid-Cretaceous, we have reluctantly selected

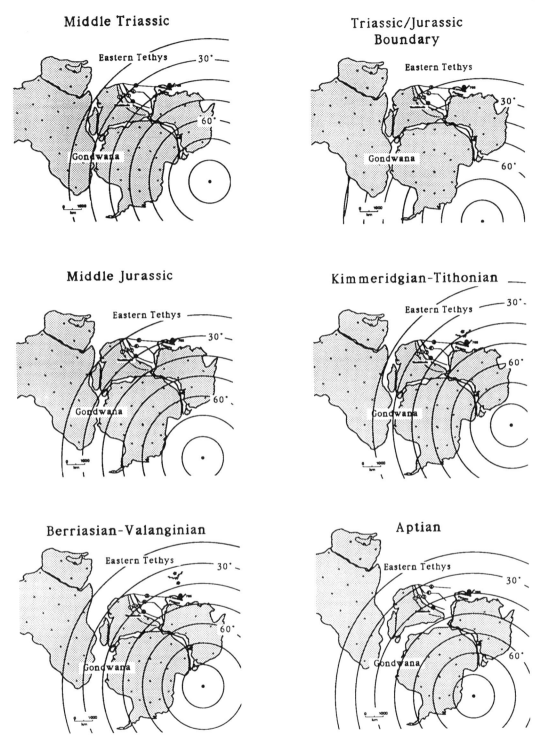

Fig. 3. Paleolatitude history of the eastern Gondwana continents during the Triassic and Jurassic. Paleolatitude grid is based upon the composite apparent polar wander path of the South Pole (Table 1) and the initial fits of Australia-Antarctica-India-Africa. In this reconstruction, each continent is drawn according to a centered mercator projection; therefore, there is a slight distortion of the paleolatitude grid at latitudes close to the equator. Separation of India and Africa from Gondwana occurred in the Early Cretaceous.

TABLE 1: Mean Apparent Polar Wander Path of Gondwana and Paleolatitudes of Northwest Australia and Central Nepal

Mean Age	South Pole	Argo Basin (Site 765) Paleolatitude, Declination	Thakkhola, Nepal Paleolatitude, Declination
Aptian	55°S, 160°E	39°S, 330°	40°S, 315° (est.)
Berr.-Valanginian	45°S, 160°E	44°S, 318°	43°S, 304°
Kimm.-Tith.	35°S, 165°E	44°S, 304°	37°S, 292°
Middle Jurassic	50°S, 180°E	30°S, 319°	30°S, 313°
Early Jurassic	55°S, 190°E	23°S, 324°	27°S, 320°
end-Triassic	55°S, 190°E	23°S, 324°	27°S, 320°
Middle Triassic	40°S, 170°E	39°S, 309°	35°S, 300°

South Pole latitude-longitude coordinates are relative to Australia. Paleolatitudes are computed for the Argo Abyssal Plain off Northwest Australia (Site 765; 15.98°S, 117.57°E) and for the Thakkhola region of Central Nepal (28.8°N, 83.85°E). To compensate for compressional shortening during the Himalayan collision [Klootwijk and Bingham, 1980], the Thakkhola region was first rotated 15° counterclockwise about a pole near Ladakh (34.6°N, 76.1°E).

the "traditional" method of refitting Gondwana. Placing the eastern coast of India against Antarctica also appears to be consistent with the paleomagnetic data from Australia and India [Klootwijk, 1984]. The three components to our selected fit are (1) Australia against Antarctica following Royer and Sandwell [1989], (2) India against Antarctica following Powell et al. [1988], and (3) Africa against India following Lawver and Scotese [1987] and Coffin and Rabinowitz [1988]. Based upon these poles of rotation for the initial fit of Gondwana, selected Triassic/Jurassic paleomagnetic poles from these continents were projected into Australian coordinates. From this compilation of apparent polar wander paths, mean poles were estimated to the nearest 5° for different geological periods from Middle Triassic to Aptian (Table 1). These mean poles enable computation of the paleolatitudes associated with the sedimentary histories for the different Gondwana margins (Table 1; Figure 3).

The Argo Basin is predicted to have been at 39° to 44°S latitude during the Early Cretaceous. This position is only slightly southward of the measured paleolatitudes of approximately 35° to 37°S (±2°) at DSDP Site 261, and ODP Sites 765 and 766 [Ogg et al., 1992a]. The paleomagnetic declinations measured at Site 765 of a Berriasian orientation of 307° to an Aptian orientation of 323° [Ogg et al., 1992a] are also consistent with this mean apparent polar wander path.

The Late Triassic-earliest Jurassic paleolatitude from Thakkhola is 28°S [Klootwijk and Bingham, 1980], which is identical to the predicted 27°S from the mean Gondwana path. The 38°S paleolatitude of possible Berriasian age from Thakkhola agrees with the predicted 43° paleolatitude of "Berriasian-Valanginian". The main discrepancy is the 38°S measured in the Middle Jurassic Bagung Formation, compared with a predicted 30°S from the Gondwana path. However, it is possible that one data point or the other is in error.

The consistency among Gondwana continent apparent polar wander paths and agreement with paleomagnetic measurements in Tethyan sediments indicate three main trends are important during the Mesozoic: (1) Rapid northward movement from mid-latitudes into tropical latitudes occurred during the Triassic with maximum tropical settings during the latest Triassic-Early Jurassic. (2) During the Middle Jurassic to Berriasian, there was a gradual southward drift to higher latitudes. (3) After the separation of India from Australia in the late Valanginian, India experienced rapid northward movement during the remainder of the Cretaceous. Australia remained at approximately the same latitude until the onset of rapid northward drift in the late Tertiary [Embleton, 1984].

LATE TRIASSIC TO MIDDLE JURASSIC FACIES
Northwest Shelf of Australia and Exmouth Plateau
The various basins of the broad shelf off Northwest Australia display a fairly consistent Mesozoic stratigraphy [e.g., Butcher, 1988; Bradshaw et al., 1988; Purcell and Purcell, eds., 1988]. Deposition of the Mesozoic succession commenced in Early Triassic with transgressive marine claystones and minor siltstones overlying a thin layer of shelf limestone (Figure 4). During the Middle and Late Triassic, the clastic Mungaroo Formation was deposited by prograding deltas, floodplains and braided river systems. Sandstones within the Mungaroo Formation form the most extensive oil and gas reservoirs on the shelf, and the succession displays cycles of marine incursions [Crostella and Barter, 1980].

The Triassic/Jurassic boundary is marked by a widespread hiatus, commonly encompassing the Hettangian [Barber, 1982]. Lower Jurassic carbonates and sandstone are followed by thick transgressive deposits of the marine lower Dingo Claystone (Figure 4). The dark-colored, slightly calcareous Dingo Claystone grades laterally into prodelta or deltaic facies, with local progressive upward-shallowing during the Bathonian and earliest Callovian.

During mid-Carnian, portions of the modern Northwest Shelf of Australia underwent block faulting, creating offshore plateaus and troughs [von Rad et al., this volume]. The Exmouth Plateau is an England-sized block of thinned and deeply subsided continental crust, bounded by steep 2-3 km escarpments to the Argo Abyssal Plain in the north and to the Gascoyne and Cuvier Abyssal Plains to the west

Mesozoic Stratigraphy of Himalayan and Northwest Australian Margins

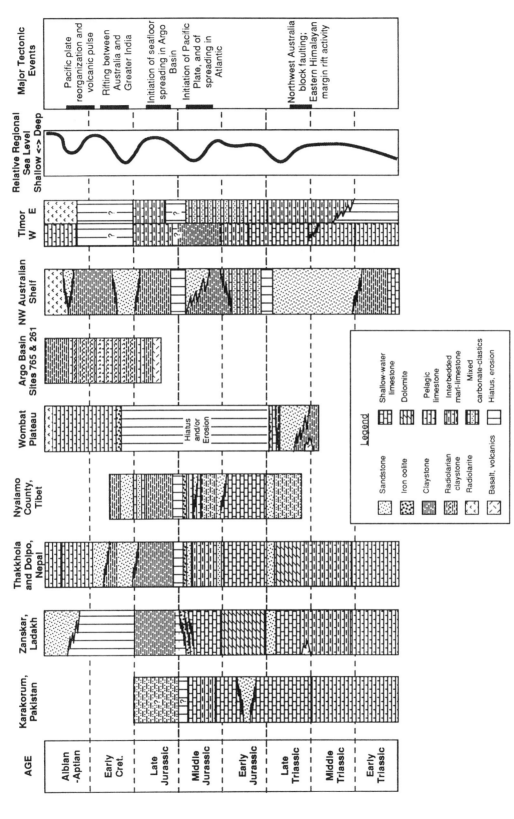

Fig. 4. Stratigraphic summary of Himalaya and Northwest Australian margins facing the Eastern Tethys and of selected Leg 122-123 sites. Most preserved successions are marginal marine shelf sequences; deep water facies are mainly known from sites in the Argo Abyssal Plain and from Timor. Age-facies compilations are schematic representations; see text for descriptions. Regional sea-level changes are generalized from transgressive marine and prograding continental facies observed on these margins, but the curve is similar to other global eustatic sea-level compilations.

and south, respectively (Figure 2). The Exmouth Plateau received deltaic sediments during the Middle and Late Triassic, but after the end-Triassic hiatus, the main Exmouth Plateau received only localized condensed sediments of Dingo Claystone [Exon et al., 1982].

Wombat Plateau (ODP Leg 122)

The Wombat Plateau is a northward-tilted horst isolated between the main Exmouth Plateau and the Argo Abyssal Plain. ODP Leg 122 drilled a transect of Sites 759, 760, 761 and 764 into the Upper Triassic succession [Haq, von Rad et al., 1990].

Crustal extension continued through the Carnian and formed a system of half-graben depocenters. Carnian prodeltaic claystone, exceeding 270 m thickness, grades upward into a Norian deltaic facies (Figure 4). The thick (300-1000 m) succession of Norian deltaic complex also contains interbeds of intradeltaic molluscan algal-oolitic limestone and bioclastic storm beds. The Norian delta was terminated in the early Rhaetian[1]. Lower Rhaetian dark laminated claystone interbedded with intraclast- and crinoid-rich, bioclastic limestone and graded quartzose sandstone represents a relatively restricted environment punctuated by storm deposits [Röhl et al., 1992] and/or by redeposition from localized eroding highs following a major block faulting and tilting episode [Sarti et al., 1992]. The upper Rhaetian consists of intertidal carbonate flat to shelf-lagoon facies, possibly with a fringing barrier reef in the north (Site 764) and patch reefs to the south [Röhl et al., 1992]. An overall shallowing-upward trend is subdivided by minor transgressions [Dumont, 1992]. The Rhaetian facies and microfacies of open-shelf and overlying lagoon-reefal platform resemble coeval tropical to subtropical deposits in the Alps of the western Tethys [Dumont and Röhl, this volume].

The Rhaetian reef is terminated by approximately 20 m of an open-marine facies of alternating calcareous claystone and bioturbated limestone. The uppermost meters consist of a reddish-colored, Triasina-rich, crinoidal grainstone and biomicrite containing Fe-oxide micronodules and stringers. This transgression and condensation may reflect an end-Triassic global eustatic sea-level rise [e.g., Vail et al., 1977; Haq et al., 1987], or may partially represent a diachronous breakup of the southern Tethys margin, similar to the situation in the Alps. Trachyte and rhyolite volcanics dredged from the northern escarpment of the plateau are interpreted to be early syn-rift volcanics (210 Ma) erupted at the Triassic/Jurassic boundary [von Stackelberg et al., 1980; von Rad and Exon, 1983].

Following the Rhaetian drowning and truncation of Triassic facies, the Wombat Plateau either remained as a sediment-starved submarine horst throughout the Jurassic [Sarti et al., 1992], or received sediments which were later removed by erosion during a postulated uplift and tectonic

---

[1] The usage of "Rhaetian" as a distinct stage name for the latest Triassic is controversial, and some Triassic time scales incorporate the "Rhaetian" into the late Norian stage.

tilting event during the Callovian/Oxfordian [von Rad, Haq et al., 1992; von Rad et al., this volume]. A submarine sediment-starved Wombat Plateau is suggested by marine-derived isotope signatures of carbonate cements [Sarti and Kälin, 1992], by the lack of karst development on the Rhaetian surface, and by the absence of Jurassic pelagic sediments. An uplift-erosion event is suggested by pervasive dolomitization and dissolution of the Rhaetian facies [Röhl et al., 1992], by the seismic and drilling evidence of progressive truncation of deeper Norian strata to the south, and by the immature character of the overlying Berriasian(?) belemnite-rich sand which may incorporate reworked siliciclastics from the underlying Norian.

Timor

Triassic through Lower Jurassic strata generally are a basinal facies in East Timor and a shallow-water facies in West Timor (Figure 4). Middle to Late Triassic strata (Aitutu Formation) in the autochthonous units of East Timor consist of approximately 1000 m of purple to dark blue shale and marl and of fine-grained radiolaria-rich limestones interbedded with shale, bioclastic limestone, quartzose sandstone, and radiolarite [Audley-Charles, 1968]. A Ladinian to Norian age is indicated by Daonella and Halobia bivalves [Grunau, 1956], but locally there is a basal conglomerate rich in bivalves and Carnian ammonites. The Aitutu Formation is generally unconformable over the Permian strata of the Cribas Formation, and Lower Triassic strata, if present, are only 2-m thick [Umbgrove, 1938]. The upper Aitutu Formation has Norian ammonites (Halorites sp., Arcestes sp.) and is conformable to the overlying Lower Jurassic Wai Luli Formation. The 800- to 1000-m thick Wai Luli Formation comprises ammonite-bearing blue-gray marl, limestone and micaceous shale, with intervals in the upper portion of quartz sandstones with plant debris and conglomerates [Audley-Charles, 1968].

In West Timor, the resemblance of the shallow-water facies to the succession of the Italian-Austrian Alps led to the application of Alpine facies nomenclature. Lower Triassic limestone, rich in ammonites (Meekoceras, Owenites, Anasibirites), is conformable over Permian marine strata [Wanner, 1931] and is overlain by Anisian ammonite- and brachiopod-rich "Hallstätter Fazies" limestone. The Ladinian-Carnian suite includes (1) red-brown, Mn-rich, crinoidal limestone with ammonites, (2) brown, thinly bedded, radiolarian limestone and radiolarite, rich in Halobia and Daonella bivalves (corresponding to the Aitutu Formation of East Timor), and (3) clastic "flysch" facies including sandstones, bioclastic calcarenites and limestone breccias. The Norian "timorische Korallenkalk" and "timorische Dachsteinkalkfazies" comprise oolitic limestone and coral boundstones with stromatoporoids and sponges [Wanner, 1931]. These shallow-water carbonates are similar in both lithofacies and fossil assemblages to the "Rhaetian" strata of the Wombat Plateau [Sarti et al., 1992]. The overlying

Rhaetian consists of organic-rich, thinly bedded marlstone with ammonites and molluscs, similar to the Kössen beds of the Northern Calcareous Alps. Middle Lower Jurassic gray shallow-water limestone contains a fauna of large molluscs (e.g, *Lithiotis, Mytilus, Pachymegalodus*). Middle Jurassic strata consist of ferruginous shales with siderite [Wanner, 1931]. Reddish-purple marly limestone with Hettangian ammonites and gray ammonite-bearing marly limestone of Lower and Middle Jurassic punctuate the series in both West and East Timor [Wanner, 1931], but the stratigraphic relationships are uncertain due to intense tectonic disruption. In the Kekneno region of West Timor, Jurassic strata are missing [Bird and Cook, 1990].

Himalayan Margins

Triassic. Throughout the Himalayan region, the Triassic displays a general shallowing-upward trend, beginning with Lower Triassic pelagic nodular limestones and terminating in the Norian-Rhaetian with fluvial-deltaic quartzites (Nepal and Tibet) or shallow-water carbonates (Ladakh and Pakistan) (Figure 4).

The Triassic Tethyan succession of Thakkhola in Central Nepal begins with a condensed, Lower Triassic to Anisian, ammonite-rich nodular limestone overlain by up to 800 m of ?Ladinian to Norian calcareous to silty shales, probably representing open-shelf to tidal-shelf environments [Bordet et al., 1971]. The upper Norian-Rhaetian portion of the Thini Formation in Thakkhola consists predominantly of thick-bedded, fine- to coarse-grained, quartzose sandstone, up to 130-m thick, with planar and trough stratification and intervals with root structures [Gradstein et al., 1989, 1992]. Biomicritic limestone, dolomite and grey shale are locally abundant. A conspicuous white quartzite, 22-m thick, is present at the top of the formation, followed by a transition to Lower Jurassic shallow-water carbonates. The paleoflow pattern suggests flow reversals (tidal influence) within northward-flowing river channels.

In the Dolpo region (west of Thakkhola) and in Western Nepal, the Lower Triassic consists of 15 to 30 m of grayish ammonite-rich nodular limestone and shale [Fuchs, 1977], similar to the Chocolate Series of Kumaon in northern India [Heim and Gansser, 1939]. This nodular limestone is overlain by the Anisian to Carnian Mukut Limestone Formation, consisting of 50 to 300 m of alternating blue limestone, marl and black shale, typically with *Daonella* and *Halobia* bivalves and ammonites [Fuchs, 1977]; a facies equivalent to the Kalapani Limestone Formation of Kumaon [Heim and Gansser, 1939]. Lower Norian shale and sandstone (Tarap Shales in Dolpo), as much as 400-m thick, are equivalent to Kuti Shales of Kumaon. The formation is capped by a 50-m thick Quartzite Bed and overlain by Lower Jurassic limestone of the Kioto Formation.

In South Tibet close to the Nepal border (Nyalamo County), the clastic Derirong Formation, supposedly Triassic [Xu et al., 1989], terminates in approximately 50 m of white fine-grained quartzite with large-scale cross-laminations [M. Sarti and W. Wise, unpublished data, 1992], similar to the top of the Thini Formation in Thakkhola. There appears to be a gradual transition to the overlying Lower Jurassic shallow-water carbonates (Pupuga Formation) consisting of oolitic limestone containing abundant quartz.

The Zanskar region in Ladakh exhibits a gradual upward-shallowing trend consisting of Lower Triassic to Anisian condensed pelagic nodular limestone (Tamba Kurkur Formation), of thick deposits of Ladinian-Carnian marly limestone (Hanse Formation), and of Carnian-Norian tidal carbonates with planar stromatolites (Zozar Formation), and culminating in the Norian-Rhaetian deltaic inner-shelf facies of the Quartzite Series [Gaetani et al., 1986]. The Quartzite Series, correlative with the upper part of the Thini Formation of Thakkhola in Nepal, shows a similar polymodal paleoflow pattern with a dominant NE direction. Episodes of pillow lavas and reddish-colored limestone in the Carnian may indicate rift-related submarine volcanism. From the Triassic through the Cretaceous, the Zanskar shelf shed calcareous and terrigenous turbidites into the adjacent continental slope and rift basin of the Lamayuru region of northern Zanskar [Bassoullet et al., 1980, 1984; Fuchs, 1982].

The north Karakorum region of Pakistan is interpreted as a microplate that rifted from Gondwana in the Permian and developed a Triassic/Jurassic series very similar to the Gondwana passive margin [Gaetani et al., 1990]. The Triassic is also represented by an overall shallowing-upward succession: Lower to Middle Triassic dark platy limestones with chert bands and redeposited shallow-water carbonates ("Borom Formation") are overlain by thick deposits of Upper Triassic-Lower Jurassic black limestone and light-colored dolomite with cyclothems (Aghil Formation). Upper Triassic shallow-water carbonates also occur in the adjacent southeast Pamir region [Dronov et al., 1982].

Lower Jurassic. The Jomosom Formation of Thakkhola region of Central Nepal consists of more than 250 m of thin-bedded bioclastic and micritic limestone alternating with oolitic limestone [Gradstein et al., 1989, 1991, 1992]. Distinctive features of the formation include coquina layers with wave-rippled tops, oncolites, intensive bioturbation (including *Chondrites*), cross-stratification, intraclastic conglomerates, and hummocky cross-stratification. This Lower Jurassic limestone was deposited on a shallow subtidal carbonate platform subject to periodic storm waves. Periodic shoaling took place from calcareous shales upwards into wave-influenced oolitic-shoal environments.

The Lower Jurassic in the Ladakh region is also characterized by shallow-water limestone (Figure 4). The Kioto Formation, 600-m thick, potentially includes uppermost Triassic to pre-Callovian. The lower of the two main facies units consists of dolomitized bio-intraclast packstones, which lacks age-diagnostic fauna [Gaetani et al., 1986]. We consider this lower facies to be correlative to the Lower Jurassic Jomosom Formation of Central Nepal.

In the north Karakorum region of Pakistan, the lowermost Jurassic portion of the Aghil Formation consists of limestone and dolomitic cyclothems. A reddish sandstone interval, the Yashkuk Formation, occurs in the middle Lower Jurassic and is interpreted as deposits from meandering streams and a tide-influenced delta prograding to the northwest over a shallow-water carbonate shelf [Gaetani et al., 1990]. The sands are rich in rock fragments of various volcanic, sedimentary and metamorphic material, suggesting a distant episode of tectonic uplift and dissection, although Karakorum has been interpreted to be part of a former microplate [Gaetani et al., 1990]. The sandstone unit is overlain by limestones of restricted lagoonal to intertidal facies.

Middle Jurassic (pre-Callovian). The Bagung Formation of Thakkhola region of Central Nepal, a mixed facies of clastics and carbonate, spans the Early Bajocian (or older) to Early Callovian [Bordet et al., 1971; Gradstein et al., 1989, 1991, 1992]. The lower portion of this 250-m thick formation is dominated by biomicritic silty limestone interbedded with cross-bedded sandstone lenses deposited in a marginal marine setting, perhaps at the edge of a deltaic or estuarine complex. The diverse marine fauna indicates more temperate conditions than in the underlying Jomosom Limestone Formation and includes coquina beds of intact *Praeexogyra* oyster shells. The upper Bagung Formation is rich in calcareous shale, with thin beds of micritic limestone, of sand-rich coquina and of cross-stratified to hummocky- and wave-rippled quartzose sandstone and siltstone. This upper unit is interpreted as a middle shelf environment influenced by storm waves, and terminates in a lower Callovian ferruginous oolite bed (see below). In contrast to the underlying Jomosom Formation, the Bagung Formation is dominated by terrigenous influx and lacks tropical carbonates and coincides with a southward drift from tropical to mid-latitudes (Table 1; Figure 3).

In southern Tibet, the Early Jurassic Pupuga Formation of shallow-water carbonates is overlain by coastal facies of 300 to 350 m of quartzose sandstones of the lower (Bajocian) part of the Niehnieh Hsiungla Formation [Wan et al., 1982; Wang and Sun, 1983, Westermann and Wang, 1988, Xu et al., 1989]. A general Bathonian through early Callovian deepening in the shelfal facies of over 500 m of dark shales and sandstones within the upper Niehnieh Hsiungla and Lalung La Formations is interrupted by thick-bedded, white quartzite, representing a coastal sand deposited during a short regression [Wang and Sun, 1983].

The upper portion of the Kioto Limestone Formation of Ladakh consists of alternating mudstone and packstone limestones, representing an inner shelf environment, and abruptly terminates in a Callovian sandy ferruginous oolite [Gaetani et al., 1986].

In the Karakorum region of Pakistan, intertidal to subtidal cyclic carbonates, containing bioclastic grainstone lag

deposits, sand bars and stromatolites, transgressively overlie Lower Jurassic sandstones [Gaetani et al., 1990]. Transitional sabkha sediments are in the south, and open-lagoonal limestones are in the north. The tectonically overlying upper Middle Jurassic strata consist of two units. The 300-m thick Reshit Formation unit is dark gray micrite to clayey biomicrite. The other unnamed unit is biointraclastic packstone to grainstone alternating with thin-bedded basinal limestone containing Callovian ammonites.

Summary of Triassic to Middle Jurassic Facies

Three main factors governed the patterns of Triassic through Middle Jurassic facies along the eastern Tethyan margin of Gondwana -- paleolatitude, relative sealevel, and regional tectonics.

Northwest Australia and the adjacent Himalayan margin lay at about 40°S during the Middle Triassic, drifted steadily northward into tropical latitudes to reach 20°-25°S in the Rhaetian to Early Jurassic, then again returned to temperate latitudes in the Middle Jurassic (Table 1; Figure 3). As a result, shallow-water carbonate platforms are favored during the Rhaetian and Lower Jurassic -- Aghil Formation in Karakorum, Kioto Formation in Ladakh, Jomosom Formation in Central Nepal, Pupuga Formation in southern Tibet, reef limestone on the Wombat Plateau and in Timor. Such shallow-water carbonate facies are rare during the Middle Triassic-Carnian-early Norian or during the Middle Jurassic when most of these regions were in temperate latitudes. In addition, the influx of terrigenous clastics was also governed by the climatic regime, with increased clay input favored under tropical chemical weathering conditions and increased detrital components favored under subtropical monsoonal or seasonal temperate climatic conditions where physical weathering is important.

Throughout this drift in paleolatitudes, Northwest Australia was always further from the equator than the Himalayan margins, and the Karakorum and Ladakh regions were in the most tropical latitudes (Figure 3). As a result, the facies are generally more calcareous on the Himalayan margins than on the Northwest Australian margin, and shallow-water carbonate deposition may continue into the Middle Jurassic in the more tropical latitudes (e.g., Ladakh and Pakistan).

Following a rapid deepening in the basal Triassic, the margins display a progressive shallowing culminating with deltas prograding over Middle Triassic to Carnian mudstone on the Northwest Australian shelf and over Norian shallow-shelf sediments on the Himalayan margin (Figure 4). A mid- to late-Carnian episode of tectonics is indicated by the formation of fault-bound basins on the Australian margin [von Rad et al., this volume] and by volcanics in the Ladakh region, and may have contributed to the increased influx of terrigenous clastics during the Norian. A Rhaetian isolation of the Wombat Plateau is suggested by the termination of the Norian delta complex and establishment of a Rhaetian reef complex, in contrast to the continued deltaic and floodplain deposition elsewhere along the Northwest Australian margin. There may have been similar Rhaetian (and possibly also Lower Jurassic) reef buildups at other locations along the Northwest Australian margin [(N. Exon, personal communication, 1992]. Some distal rift-blocks on the margin, such as the Wombat Plateau and portions of Timor, may have undergone permanent drowning and/or sediment starvation at the end of the Triassic; however, the absence of Jurassic sediments on these blocks may also be due to mid-Jurassic uplift and subaerial erosion.

Marine calcareous claystone deposits are characteristic of the mid-Middle Jurassic in most regions, with a shallowing episode possibly beginning in late Bathonian. On the Northwest Australian margin, a late Early Jurassic transgression produced widespread deposition of marine claystone and possible localized deposition of shallow-water carbonate; this highstand was followed by a late Bathonian to Callovian shallowing.

CALLOVIAN AND LATE JURASSIC FACIES
Northwest Shelf of Australia ("Breakup Unconformity")

A widespread hiatus and sharp facies change occurs on the Northwest Australian shelf during the Callovian/Oxfordian. The shallowing-upward to emergent clastic succession of the Bathonian to Early Callovian is truncated and overlain by marine facies of the Upper Dingo Claystone [Butcher, 1988]. The Exmouth Plateau became isolated from the mainland by the formation of the "Barrow-Dampier Trough" and other rift-fault-bounded basins, and therefore experienced widespread sediment starvation during the Jurassic [Exon et al., 1982].

The Wombat Plateau also did not preserve Jurassic sediments due to submarine or subaerial erosion and/or sediment starvation. Drill sites on the Wombat Plateau [Haq, von Rad et al., 1990] recovered a thin interval of ferruginous feldspathic sand associated with Mn-oxide crusts and micronodules. Belemnite rostra are disseminated within this sand at Site 761. Overlying the reworked relict sand is a hemipelagic calcisphere-nannofossil chalk of late Berriasian to Valanginian age [von Rad and Bralower, in press].

According to Falvey [1974] and Braun and Beaumont [1989], the "breakup unconformity" forms a major hiatus in continental margins resulting from rift flank uplift and subaerial erosion during the initiation of seafloor spreading. A supporting example for the concept of "breakup unconformity" was the apparent coincidence of a widespread Callovian/Oxfordian hiatus and tectonic activity on the Northwest Australian margin with the reported Oxfordian age for the oldest oceanic crust in the Argo Abyssal Plain [Veevers, Heirtzler et al., 1974]. The age spanned by the unconformity was considered to be correlative with the oldest age of the neighboring oceanic crust, therefore the "breakup unconformity" model has been applied to various margins for timing the transition from the rift stage to the active spreading stage [e.g., Falvey, 1974; Montadert et al., 1979; Hubbard et al., 1985; Grow et al., 1988; Hubbard, 1988].

However, oceanic drilling and correlation of magnetic anomalies have shown that the proposed "break-up unconformities" identified on seismic profiles do not always correspond to the actual onsets of seafloor spreading, for example between Australia and Antarctica [Cande and Mutter, 1982; Powell et al., 1988] or between Iberia and Newfoundland [Boillot, Winterer et al., 1988]. In the case of the Argo Abyssal Plain, the onset of seafloor spreading was approximately 2 million years prior to late-Oxfordian magnetic polarity chron M26 [Fullerton et al., 1989; Sager et al., 1992; see next section], and hence began during the mid-Oxfordian. Therefore, the pronounced Callovian/Oxfordian hiatus and unconformity on the Northwest Australian margin predates the onset of seafloor spreading by several million years. This Callovian/Oxfordian unconformity in Northwest Australia has also been called a "post-rift unconformity" [von Rad et al., this volume; Dumont, 1992], without a specific linkage to the initiation of seafloor spreading.

Upper Jurassic sediments on the Northwest Australian margin mainly consist of restricted marine claystones infilling the rifted basins. Erosion of adjacent uplifted regions locally produced deep-water submarine-fan turbidites and overlapping sandy fan lobes [Butcher, 1988]. Shallow-marine sands were deposited over widespread areas during the Tithonian, followed by an end-Jurassic disconformity.

Argo Abyssal Plain (DSDP Site 261 and ODP Site 765)
Basement Age and Character. Marine magnetic anomalies striking N20°W to N30°W were identified by Falvey [1972] and correlated to the Jurassic portion of the M-sequence by Larson [1975]. The magnetic anomaly pattern from M26 to M16 and intervening fracture zone offsets has been correlated throughout the basin to the edge of the subduction zone [Fullerton et al., 1989; Sager et al., 1992, and references therein]. According to magnetic anomaly ages [Ogg et al., 1984; Steiner et al., 1986; Ogg et al., 1991] and the absolute time scale of Kent and Gradstein [1985], the Argo oceanic ridge was spreading continuously from late Oxfordian (158 Ma) until at least late Berriasian (142 Ma) time. Spreading rates averaged about 40 km/m.y. during the Late Jurassic, slowing to about 20 km/m.y. during the Berriasian [recomputed from Sager et al., 1992].

ODP Site 765 was drilled on oceanic crust between magnetic anomalies M25A and M26 of latest Oxfordian age (Figure 2), and is less than 75 km seaward from a prominent positive magnetic anomaly that lies along the ocean/continent boundary [Veevers et al., 1985]. The basal sediments are barren of microfossils, but the crustal age of latest Oxfordian is supported by two K/Ar radiometric dates on celadonite and glass samples from the oceanic basalts, which yield an age of 155.3±3.4 Ma [Ludden, 1992].

DSDP Site 261 was drilled between magnetic anomalies M24A and M24B of early Kimmeridgian age (Figure 2). The basal sediments at Site 261 contain Kimmeridgian to early Tithonian nannofossil assemblages, characterized by the abundance of *Watznaueria britannica*, and *W. barnesae*, occurrence of *Stephanolithion bigotii*, *Ethmorhabdus gallicus* and *Axopodorhabdus cylindratus*, and overlying lowest occurrence of *Conusphaera mexicana minor* [Dumoulin and Bown, 1992].

The sediments at both sites show a common stratigraphic succession (Figure 4), and variations between sites are primarily due to differences in proximity to the continental margin (Dumoulin and Bown, 1992). These sediments provide the only in-situ record of deep-ocean conditions in the Eastern Tethys; we therefore present a detailed description of the facies.

Kimmeridgian-earliest Tithonian. The oldest sediment recovered in the Argo Abyssal Plain is probably a 7-cm interval of firm, dusky-red claystone interstitial to and overlying basalt hyaloclastite at Site 765 (Core 123-62R-4, interval 21-28 cm). The overlying 1.4 m of reddish-brown, homogeneous, non-calcareous silty claystone contains disseminated FeMn micronodules, angular quartz and some feldspar grains, and volcanic lithic clasts [Dumoulin and Bown, 1992]. The only microfossils are abundant agglutinated benthic foraminifers with taxa that are not age-diagnostic. A pre-Tithonian age is assigned based on the age of overlying units and the facies correlation to Site 261. The absence of carbonate indicates sedimentation below the carbonate compensation depth (CCD). Enrichment in oxidized Fe and Mn oxides may reflect slow sedimentation under oxygenated bottom water and proximity to metalliferous hydrothermal emissions from the active spreading ridge. The spreading ridge was also the source of the volcanic fragments and may have contributed to the formation of the Fe-smectite clay. The abundant quartz silt was possibly wind-transported from the adjacent mainland, although bioturbation may have obscured any turbidite structures.

Between basalt pillows at Site 261 is pale pinkish-gray, recrystallized, fine-grained limestone containing calcispheres, inoceramid mollusc fragments, calcareous foraminifers, coccoliths and radiolarians. Directly overlying the Kimmeridgian basalt is red nannofossil-rich (35%) claystone, but only 25 cm were recovered at the top of Core 261-33R. Rare thin lenses of coarser material, rich in calcispheres, are probably current lag deposits.

The good preservation of upper Kimmeridgian nannofossils and other calcareous fauna at Site 261 is in sharp contrast to the absence of carbonate in the coeval claystone at Site 765 on older, and presumably deeper, oceanic crust. If the oceanic crust formed at an initial depth of 2700 m, then this contrast may imply that the late Kimmeridgian CCD was at approximately 3000 m depth within the Argo Basin. The CCD in the equatorial Pacific was also at depths shallower than 3000 m during the Late Jurassic [Ogg et al., 1992b]. In the present northeastern Indian Ocean, the CCD is more than 5000 m deep, but shallows to less than 3500 m near the continental margins [Berger and Winterer, 1974]. A similar shallowing of the Jurassic CCD toward the margin probably was a factor in the poor preservation of carbonate at Site 765.

Tithonian. Brown to reddish-brown calcareous claystone with layers containing abundant nannofossils and prismatic fragments of inoceramid mollusc shells comprises a 5- to 10-m interval at both Sites 765 and 261 [Dumoulin and Bown, 1992]. The non-carbonate fraction consists mainly of smectite clay, FeMn nodules, and grains of quartz and volcanic lithic fragments. Mottling may reflect pervasive bioturbation, but discrete burrows are rare. At Site 765, the facies also contains some bentonitic ash layers [Thurow and von Rad, 1992]. Nannofossils, especially the large robust coccolith *Watznaueria manivitae*, dominate the carbonate. In the lower part of the unit, nannofossil- and clay-rich layers alternate with coarser-grained inoceramid-rich layers and lenses on the scale of millimeters to centimeters. These inoceramid-rich layers decrease in thickness and abundance upward, as does the general carbonate content of the unit. Calcispheres are sparse, except for concentrations in light red or pale green layers in the upper part of the unit. The distinct calcisphere-rich layers may be the result of unusual, hence ecologically stressful, oceanographic conditions resulting in blooms by this microfossil [Thierstein, 1981]. However, calcisphere-rich pelagic limestones of Tithonian-Early Cretaceous age are common along the former outer margin of northern Australia (see below), so these calcisphere-rich layers in the Argo basin may represent redeposition events.

Inoceramids are epibenthic bivalves adapted to soft mud substrates at all depths and have a global distribution, but occur most frequently on continental shelves [Thiede and Dinkelman, 1977; MacLeod and Hoppe, 1992]. Many of the inoceramid fragments in the Argo basin contain pits, typically 10-20 microns in diameter and 20-50 microns deep, which are probably fungal or sponge microborings. Although only a few whole inoceramid shells were recovered from the Tithonian calcareous claystone, the abundance and size of the prismatic fragments suggest a local origin. Oxygenated bottom waters probably enhanced the disintegration of the inoceramid shells [E. Kauffman, in Dumoulin and Bown, 1992]. Black manganese nodules and agglutinated foraminifers are most abundant in inoceramid-rich intervals. The absence of any indications of turbidity-current structures suggests that the local concentrations of inoceramid debris are probably the result of current winnowing; however, turbidite transport followed by thorough bioturbation is also a possibility. The greater abundance of coarse inoceramid prisms at Site 765 relative to Site 261 reflects the closer proximity of the continental shelf. Another type of fine-scale winnowing was observed at Site 261, where unusual millimeter-thick layers of stacked *Watznaueria manivitae* coccolith plates, essentially without other interstitial material, alternate with clay-rich layers. These concentrations of stacked nannofossil plates at Site 261 are interpreted as the fine-grained equivalents of the winnowed inoceramid concentrations at Site 765 [Dumoulin and Bown, 1992].

Deposition of this Tithonian claystone unit took place above the CCD. However, the poor preservation of the nannofossil assemblages and pervasive etching of the remaining robust forms indicate bottom conditions within the carbonate lysocline. A greater degree of dissolution and diversity impoverishment is noted at Site 765, on older and presumably deeper oceanic crust. A slow rate of sediment accumulation and the proximity to the active spreading ridge probably contributed to the enrichment in FeMn nodules. Mean accumulation rates, including possible redeposited layers, are only on the order of 1 to 2 m/m.y.

These sites indicate a Tithonian CCD at approximately 3000-3500 m, a deeper level than during the Kimmeridgian, but rising again at the end of the Jurassic causing the disappearance of in-situ pelagic carbonate. This late Kimmeridgian to early Tithonian downward excursion in CCD parallels the observed aragonite compensation depth history in portions of the Western Tethys and in the Central Atlantic, where the upper Kimmeridgian and lower Tithonian red calcareous claystone preserved aragonite fragments at paleodepths of approximately 3500 m, in contrast to upper Tithonian and younger carbonates which did not preserve aragonite [Shipboard Scientific Party, 1983; Ogg et al., 1983]. In contrast, the CCD in the equatorial Pacific remained above 3000 m until the Berriasian [Ogg et al., 1992b].

Timor and New Guinea

The deeper portions of the former northern margin of Australia have facies similar to those of the Argo Abyssal Plain (Figure 4). On Buton Island, a possible displaced sliver of the New Guinea margin, Kimmeridgian brown siliceous mudstone (Rumu Formation) contains redeposited layers of red calcareous mudstone and clayey limestone [Smith and Silver, 1991]. The uppermost Jurassic-Lower Cretaceous portion of the overlying Tobelo Formation consists of radiolarian limestone interbedded with calcisphere-rich pelagic limestone containing abundant calcite prisms from disaggregated inoceramid shells. Calcispheres are also abundant in Upper Jurassic or Lower Cretaceous pelagic limestones found on Timor [Wanner, 1940]. No sediments of Oxfordian-Kimmeridgian age are known from the autochthonous units on Timor, and uppermost Jurassic shales are overlain by Lower Cretaceous pelagic calcareous claystones and radiolarite [Audley-Charles, 1968; 1991]. In West Timor, the Ofu series of marls and marly limestone with radiolarite horizons spans the upper lower Oxfordian through Tithonian [Wanner, 1931], whereas in East Timor, the Oxfordian-Kimmeridgian consists of light greenish marls.

Facies patterns on the former northern shelf of Australia are distorted by tectonic rearrangements and accretion of terranes during the Cenozoic [e.g., Pigram and Davies, 1987; Metcalfe, 1988; Audley-Charles, 1991], but appear to reflect a succession of clastic influxes and marine

transgressions similar to those of Northwest Australia. The Kimmeridgian-Tithonian Imburu Mudstone Formation lies above the Bathonian-?Oxfordian Koi Iange Sandstone Formation in western Papua New Guinea [Pigram and Panggabean, 1984].

Himalayan Margin

Callovian Ferruginous Layer. In the Thakkhola region of Central Nepal, the upper Bagung Formation represents a middle shelf environment with calcareous claystone interbedded with sandy bioclastic storm beds [Gradstein et al., 1989, 1991, 1992]. This unit is capped by a resistant-weathering unit of 5 to 8 meters of reddish-brown, bioclastic-rich quartzose sandstone, and a clay-rich uppermost meter containing scattered Fe-oolites and phosphatic ooids. This "Ferruginous Oolite" unit of Bordet et al. [1971] contains abundant bivalves, ammonites, belemnites, brachiopods, gastropods, *Chondrites*, bored intraclasts, and Fe-oxide-coated crinoid ossicles. An age of earliest Callovian is well-established by the ammonite fauna. The main sandstone is cross-stratified to hummocky-bedded, and is interpreted as being deposited and repeated reworked on a shallow shelf by periodic high-energy episodes.

A similar "Ferruginous Oolite Formation" lies unconformably on the Kioto Limestone Formation of the Ladakh region [Gaetani et al., 1986; Garzanti et al., 1989]. Above and below an interval of bioclastic-rich Fe-oolite-bearing sandstone are cross-bedded bioclastic quartzites, which locally display graded bedding and hummocky cross-stratification indicating northeastwards paleoflow. A varicolored shale member is generally present between the lower Fe-oolite and the quartzite. Thicknesses of each member varies from 2 to 20 m. The general upward-coarsening succession is interpreted as progradation of clastics on a storm-dominated shelf, followed by capping by a Fe-oolite "roof bed" [Gaetani et al., 1986]. The unit is age transgressive from ?Bathonian-early Callovian in the west to middle or late Callovian in the east.

In south Tibet, the Ferruginous Oolite member of early Callovian age is a 2 m layer of iron- and manganese-oxide-impregnated skeletal calcarenite with ironstone crusts and iron oolites, and sharply overlies the dark micaceous shale of the upper Lalung La Formation [Westermann and Wang, 1988].

Fe-oolite beds preferentially occur at the base of transgressive events on shallow shelves. The origin of Fe-oolite horizons is debated [e.g., studies in Young and Taylor (eds) 1989], but is generally ascribed to redeposited and winnowed products from lateritic soils on adjacent well-vegetated coastal plains, where Fe has been concentrated by intensive leaching under a warm humid climate [e.g., Hallam, 1975; Hallam and Bradshaw, 1979; Siehl and Thein, 1989]. Fe-oolite-rich condensed strata are unusually common within Callovian shelf sequences in Europe, occurring in the Swiss Jura, Paris Basin, Poland, Spain, Portugal and elsewhere. Typically, marine shales pass into sandstone, which is capped by thin ironstone having reworked Fe-oolites. The coeval facies on the Himalayan margins are a local expression of a widespread event during the Callovian.

Oxfordian-Tithonian Organic-Rich Shales.

In Thakkhola region of Central Nepal, a major stratigraphic break, encompassing the middle Callovian through early Oxfordian, occurs between the ferruginous sands and overlying marine shales of the "Nupra Formation" [Bordet et al., 1971; Gradstein et al., 1989, 1991, 1992; Figure 4]. The Nupra Formation consists of at least 250 meters of organic-rich shale, which averages 1.7% total organic carbon [Heroux et al., 1979]. Escaping methane gas is observed as eternal "holy flames" in the Hindu temple at Muktinath. Macrofossil-bearing concretions, locally called "saligrams" and considered sacred to Vishnu, are compositionally zoned, indicating successive precipitation by silica, accompanied and followed by pyrite and locally ankerite, and finally by siderite. The shale contains abundant ammonites, belemnites, agglutinated foraminifers, and vascular plant fragments. The diverse ammonite fauna indicates an age span from middle Oxfordian (Transversarium Zone) through late Tithonian. As throughout the Himalayan region, Kimmeridgian ammonites are poorly represented [Krishna et al., 1982; Westermann and Wang, 1988]. The diverse assemblage of agglutinated foraminifers in the lower Nupra Formation are more typical of slope rather than shelfal conditions. The topmost Nupra foraminiferal assemblage is less diverse and contains calcareous foraminifers, suggesting shallower, less restricted conditions.

The Nupra Formation is interpreted as a deep shelf to upper slope facies deposited under low oxygen bottom conditions. A Septal Strength Index [Westermann, 1973] of 12+ for *Belemnopsis* from the upper Nupra suggests a potential implosion depth of about 300 m. There is no evidence of storm-wave effects. The predominance of terrestrial organic matter in the shales and overlying prograding deltaic clastics suggests that much of the shale may have originated as prodelta clays. Calcareous microfossils appear to have undergone diagenetic dissolution, followed by reprecipitation in the form of gypsum and siderite. The transition to the overlying Lower Cretaceous deltaic complex is gradational, but encompasses only a few meters between the lowest thin beds of siltstone and the lowest thick sand bodies.

The Nupra Formation is lithologically similar to and correlates with the ammonite-bearing Spiti Shale of Ladakh and southern Tibet [Krishna et al., 1982; Krishna, 1983; Gaetani et al., 1986; Westermann and Wang, 1988]. In the Zanskar (Ladakh) region, the Spiti Formation consists of 20 to 60 meters of dark shales with intercalations of marls and calcareous quartzose siltstone and with rare lags of macrofossils. The depositional environment of Spiti Shale in this region is interpreted as a quiet mid-outer shelf with episodic major storm events [Gaetani et al., 1986]. In this region, strata of Early Cretaceous age are absent, and the succession continues with prograding deltas of late Aptian-

early Albian age. The Spite Shale (Membu or Menkatum Formation) in south Tibet is at least 450 m thick and spans the late Oxfordian through early Tithonian [Westermann and Wang, 1988].

In the Karakorum region, the Upper Jurassic overlying Callovian carbonate shelf deposits also consist of thick black slates interbedded with hummocky- or parallel-laminated, quartzose sandstones [Gaetani et al., 1990]. These deposits contain shallow-water fauna, but lack age-diagnostic fossils. A storm-dominated shelfal environment is indicated, with episodes of turbiditic sandstone deposition and rare basaltic layers.

The occurrence of dark shales over a distance in excess of 1500 km along the Himalayan margin and the apparent monotonous deposition encompassing most of the Late Jurassic suggest a low-relief hinterland and relatively uniform depositional conditions offshore.

Summary of Callovian-Upper Jurassic Facies

The Late Jurassic along the Northwest Australian shelf and Himalayan margins is characterized by dark marine shales deposited in restricted basins or in outer shelf-slope settings (Figure 4). These organic-rich shales are unconformable upon Callovian shallow-shelf deposits, and a major hiatus encompasses the late Callovian and early Oxfordian. In some regions, such as Central Nepal, this hiatus also includes most of the Callovian.

The Callovian/Oxfordian boundary is a worldwide hiatus or condensed horizon in both shelf and deep-sea deposits. Even pelagic radiolarites on Pacific oceanic crust display an unconformity and stratigraphic gap at this level [Shipboard Scientific Party, 1990; Ogg et al., 1992b]. A major eustatic sealevel rise occurred between the middle Callovian and the middle Oxfordian [Haq et al., 1987]. As on these eastern margins, many of the Western Tethys shelves and epicontinental embayments display a facies change from Callovian ferruginous or shallow-shelf sediments to Oxfordian black clays (Figure 4). This phenomenon includes the classical sections of England, where the organic-rich Oxford Clay overlies lower Callovian carbonate shelf facies of the Cornbrash beds [e.g., Arkell, 1933; House, 1989]. On the Northwest Australian shelf, this Callovian/Oxfordian boundary hiatus and deepening coincides with rift block-faulting events, but not with the actual onset of active Argo seafloor spreading in early late Oxfordian.

The Callovian/Oxfordian boundary event and associated eustatic sealevel rise may represent a widespread plate reorganization episode within the Pacific-Tethys system. Changes in the direction and rate of Pacific plate spreading, such as occurred during the early Aptian or the Eocene (40 Ma Emperor-Hawaiian bend), coincide with tectonic and volcanic disruptions along many of the Pacific margins and may have resulted in sympathetic effects in other ocean basins, although cause-effect relationships are ambiguous. The early Aptian and Eocene disruptions in the Pacific are associated with increased oceanic volcanic activity, resulting in elevated atmospheric carbon dioxide and global warming [Sheridan, 1983; Arthur et al., 1985; Owen and Rea, 1985; Larson, 1991a,b]. A change in Pacific-Eastern Tethys spreading orientations in the late Callovian may have changed the stresses on the Australian and other Tethys margins, thereby causing the local "breakup unconformity" ("post-rift unconformity") and setting the stage for the Oxfordian initiation of seafloor spreading in the Mozambique-Somali basins between Africa and Madagasar-India [Coffin and Rabinowitz, 1988] and in the Argo basin. Unfortunately, our record of Callovian/Oxfordian events in the Pacific-Tethys ocean basins is largely erased through subduction. However, the hypothesized Callovian/Oxfordian plate reorganization in the Pacific would be consistent with the observed rise in global sealevel, elevated CCD in the oceans, enhanced preservation of siliceous microfossils, precipitation of iron oolites on shelves, and increased deposition of organic-rich deposits during the late Callovian through Oxfordian [Ogg et al., 1992b].

The Northwest Australian and Himalayan margins underwent a rapid southward drift into mid-latitudes, changing from approximately 30°S in the Middle Jurassic to 40°S in the Kimmeridgian-Tithonian (Table 1). This paleolatitude drift may have contributed to the dominance by clastics on these shelves during the Late Jurassic. In contrast, the carbonate-rich shelf sediments of the Tethys margins further toward Africa were deposited under subtropical paleolatitudes. The widespread Spiti-Nupra-Dingo Shale deposition throughout the Late Jurassic suggests that coarser clastics were either trapped on broad coastal floodplains, or broken down through chemical weathering. The Nupra Shale is a mixture of illite and kaolinite clay and enriched in terrigenous organic material [Gradstein et al., 1991, 1992], which supports the latter possibility. In either case, a low relief in the hinterland source regions is implied, despite the local occurrence of Callovian/Oxfordian rift tectonics.

In the new Argo Basin, the basal condensed red claystones are primarily smectite-illite, with carbonate being preserved only during a deepening of the CCD during the late Kimmeridgian-Tithonian. Steady subsidence of the oceanic crust, perhaps coupled with a minor rise in CCD, resulted in a progressive decline in carbonate preservation during the Tithonian.

LOWER CRETACEOUS FACIES
Northwest Australian Margin and Exmouth Plateau

A disconformity separates the Tithonian upper Dingo Claystone from the overlying sandy "Barrow Group" of Berriasian-Valanginian age [Butcher, 1988; Figure 4]. The Barrow Group is a series of clastic facies, from submarine fan upwards into coastal plain or wave-dominated delta environments, which prograded towards the northwest as far as the central part of the Exmouth Plateau. Sites 762 and 763 can be tied into the seismic stratigraphic framework of the central Exmouth Plateau and reveal that over 1700 m of clastic sediments were deposited during the latest Tithonian through early Valanginian as a 300-km

wide prodelta to delta wedge, with at least 5 depositional episodes [Exon and Buffler, 1992; von Rad et al., this volume]. However, an excess of sediment supply during synrift subsidence may obscure the eustatic signal [Boyd et al., 1992].

A major transgression during the late Valanginian led to deposition of the Muderong Shale, representing widespread restricted marine conditions that continued on the Northwest Australian shelf until the latest Aptian [Butcher, 1988; Exon et al., 1992]. The transgression coincides with the onset of spreading between India and Australia at magnetic chron M11-M10, as documented by marine magnetic anomalies [Larson, 1977; Larson, et al., 1979; Johnson et al., 1980; Powell and Luyendyk, 1982; Fullerton et al., 1989] and by the basal age of Site 766 drilled on transitional crust at the edge of the Gascoyne Abyssal Plain. In this case, this transgressive unconformity is nearly coincident with the "break-up". The transgressive base of the Muderong Shale is a glauconitic quartz sandstone. The equivalent of the Muderong Shale at the Exmouth Plateau sites consists of Hauterivian to early Aptian, organic-rich claystone deposited in a neritic shelf environment. A glauconitic sandstone, the Windalia Sandstone Member, terminates the Muderong Shale during the latest Aptian and earliest Albian, and marks a minor regressive phase and erosional episode associated with continued separation of India [Butcher, 1988]. The Albian transgression led to widespread deposition of the Windalia Radiolarite.

Site 766 off the western escarpment of this Exmouth Plateau continued through the Barremian to receive dark greenish-gray turbidites of glauconite-, bioclastic- and volcaniclastic-rich siltstones and sandstones [Ludden, Gradstein et al., 1990]. Upward coarsening of the sequence indicates progradation of a submarine fan derived from either the Exmouth Plateau or an adjacent shelf of Greater India. Thin waxy clay laminae in the upper Valanginian-lower Hauterivian are interpreted as bentonitic ash beds. The sharp contact between the middle Barremian distal submarine fan deposits of dark claystone and the Aptian-Albian tan-colored siliceous chalk with zeolites suggests a transgressive event.

A similar pattern of clastic deposition occurred on the former northern shelf of Australia. At the end of the Tithonian, the Toro Sandstone Formation prograded over the Upper Jurassic Imburu Mudstone Formation in western Papua New Guinea [Pigram and Panggabean, 1984]. This was followed in the mid-Early Cretaceous by a marine transgression and deposition of claystone facies.

Argo Abyssal Plain (DSDP Site 261 and ODP Site 765)

Berriasian through lower Aptian sedimentation on the Argo Abyssal Plain consists mainly of reddish-brown claystone [Leg 123 Shipboard Scientific Party, 1988]. This series at Site 765 was subdivided into units and subunits by the presence of bentonitic ash beds [Berriasian-Valanginian], nannofossil chalk turbidite beds (Hauterivian), horizons rich in radiolarians (Barremian), and levels of enrichment in rhodochrosite concretions (lower Aptian) [Ludden, Gradstein et al., 1990]. We will use a more general stratigraphic framework for the Argo Basin [Dumoulin and Bown, 1992].

At Sites 261 and 765, the Berriasian through lower Barremian unit spans 60-70 meters; however, much of this unit consists of redeposition events. Numerous nannofossil-rich turbidites occur within the non-calcareous claystone. These carbonate turbidites include mixtures of radiolarians, calcispheres, inoceramid prisms, calcareous and agglutinated foraminifers, quartz, glauconite and claystone clasts [Dumoulin, 1992]. In general, turbidites are finer-grained and thinner at Site 261, reflecting the greater distance from the continental margin. Thin beds of dark claystone (slightly enriched in organic carbon) and silty laminae of angular quartz are also interpreted as turbidite beds. Turbidites at both sites were derived from the outer shelf or slope environments above the CCD but below the photic zone, and include a volcanic component. Calcareous turbidites are not present in the lower half of the unit (Berriasian-lower Valanginian) at Site 765. This lower interval of Site 765 has an irregular FeMn-oxide crust near the base and contains the highest abundance of bentonite layers. Ash beds are also present as smectite to bentonite clay layers within the upper Berriasian-lower Valanginian calcisphere-rich nannofossil chalk on the adjacent isolated Wombat Plateau [von Rad and Thurow, 1992].

The host reddish-brown claystone is homogeneous. Greenish-gray mottling indicates that the depositional environment was oxidizing, allowing hematite precipitation near the sediment/water interface followed by localized post-depositional reduction around burrow fillings or concentrations of organic matter. Rhodochrosite micronodules and rhombs are concentrated into white layers, lenses or ovoids a few millimeters wide, or less commonly, are dispersed in the claystone. These rhodochrosite micronodules are generally replacements and overgrowths on radiolarians and perhaps calcispheres [Dumoulin and Bown, 1992].

Radiolarians are abundant, except within the Berriasian strata, and are generally recrystallized to microcrystalline quartz. Radiolarians occur in two distinct situations -- dispersed in the claystone and concentrated in numerous lenses and layers from a few millimeters to several centimeters thick. Some light-colored radiolarite layers display parallel- and cross-lamination and were probably deposited by turbidity currents. Other radiolarian concentrations are discontinuous wispy laminae within a clay-rich matrix and are probably pelagic accumulations winnowed by bottom currents [Dumoulin and Bown, 1992]. The claystone fauna has a low diversity, and the assemblages are interpreted as reflecting restricted oceanic conditions. In contrast, the radiolarian-enriched layers are dominated by a different fauna, which may have been derived from an Antarctic provenance [Baumgartner, 1992]. Baumgartner [1992] suggests that "cold" Antarctic-India seaway water may have intermittently entered the Argo Basin via the rift channel between India and Australia where

upwelling along the Northwest Australian margin resulted in mass mortality of the transported radiolarian population. However, the Early Cretaceous paleolatitudes of the proposed Antarctica-India seaway and of the Argo Basin are both about 40°S (Figure 3). An alternative possibility is that these radiolarian enrichments may have resulted from periodic high fertility conditions within the Argo Basin, perhaps governed by Milankovitch climatic cycles. Milankovitch-governed cycles of Early Cretaceous productivity produced interlayering of claystone and radiolarite in the equatorial Pacific [Molinie and Ogg, 1992; Ogg et al., 1992b] and periodic enrichments of radiolarians and nannofossils within the Atlantic and Western Tethys pelagic sediments [Huang, 1991].

A change in sediment coloration from the reddish brown upwards to predominantly greenish gray in the upper Barremian and Aptian corresponds to a dramatic increase in sedimentation rates. The upper Barremian-Aptian unit spans over 200 meters. Except for sediment color and turbidite composition, this unit is similar to the underlying reddish-brown unit. Carbonate-rich turbidites are rare, except for an upper Aptian interval at Site 765, suggesting that the source regions experienced a rise in the CCD or reduced nannofossil productivity. Bioturbation produces a streaky texture, with varying degrees of homogenization. The change in coloration and organic content is probably mainly a reflection of increasing burial rates [e.g., Habib, 1983], rather than the oxygen levels of bottom waters. Intervals within the Aptian of black organic-rich claystone are broadly correlative between sites, and may be a local expression of mid-Cretaceous "anoxic events" [e.g., Jenkyns, 1986] or organic-rich influx from the continents. Except for some of the radiolarian-rich layers, redeposited sediment is not evident through most of the unit. Nevertheless, Dumoulin and Bown [1992] postulate that fine-grained mud turbidites are a component of the increased sedimentation rate.

Concurrently during the Barremian and early Aptian, there is a gradual increase in abundance and diversity of Tethyan radiolarian fauna diluting the endemic "circum-Antarctic" assemblages. This Tethyan influence may reflect a more globally "equitable climate" coupled with a more open exchange between the Argo Basin and the main Tethys Ocean [Baumgartner, 1992]. A contributing factor may have been the subsidence to the northwest of the Argo Basin of the Scott Plateau and Ashmore Platform in the late Early Cretaceous, thereby allowing increased oceanic circulation [Butcher, 1988].

Timor

The Lower Cretaceous is poorly documented in Timor, largely due to tectonic disruption or unconformities. In East Timor, the Wai Bua Formation consists primarily of Aptian-lower Albian radiolarites, radiolarian marls and shales [Audley-Charles, 1968] and corresponds to the Windalia Sandstone and Windalia Radiolarite of the Northwest Australian margin. Middle Cretaceous limestone and marls rich in planktonic foraminifers occur in both Rotti and Timor [Wanner, 1931].

Himalayan Margin

During the Berriasian, quartzose sandstones of the Chukh Unit deltaic complex in the Thakkhola region of Central Nepal, prograded across the marine clays of the Late Jurassic shelf [Gradstein et al., 1989, 1991, 1992; Ogg, Sarti et al., unpublished manuscript, 1992]. Within this 75- to 100-m unit, numerous channel sands are imbedded into silty shales and overlain by paleosols and display north to northwestward paleoflows. This quartzose unit is followed by another delta series composed of 75 m of greenish sands with volcaniclastic (alkali basalt) lithic grains, glauconite, quartzite fragments, and abundant pieces of gymnosperm plants [Bordet et al., 1971]. Lower Cretaceous basalt flows and quartzite-volcaniclastic sands exposed in the Lesser Himalayan thrust sheet near the Indian craton are considered to represent more proximal equivalents to this episode of volcaniclastic-rich fluvial-deltas and may coincide with the Valanginian rifting between India and Australia [Sakai, 1983, 1989] and/or to the Aptian-Albian Rajmahal-Bengal trap events in Bangladesh [Baksi et al., 1987]. Major and minor element chemistry of the volcaniclastic grains in these two Lower Cretaceous deposits appear to indicate sources associated with continental rift volcanism [S. Dürr, personal communication, 1992].

Lower Aptian dark shales, 100-m thick, with thin beds of glauconitic siltstone probably represent a deep, poorly oxygenated shelf environment [Gradstein et al., 1989, 1991, 1992]. The overlying Albian Dzong Unit consists of planktonic foraminifera-bearing marls interbedded with calcareous glauconitic sandstones, and probably represents a continental slope environment with periodic turbidite influx [Gradstein et al., 1989; Gradstein and von Rad, 1991].

In south Tibet, the dark lower Tithonian shales are also overlain by a 100-m thick Berriasian cross-bedded quartzose sandstone (Gucuo Formation), but with a thick (over 300 m) upper Tithonian transitional interval of a prograding carbonate-siliciclastic shelfal facies (Shumo Formation) consisting of calcareous mudstone and wackestone, oolitic and bioclastic limestone, and quartzose sandstone followed by dark shales [Xu et al., 1989]. The Berriasian sandstone unit is overlain by dark shales and silty shales containing upper Berriasian ammonites.

In the Ladakh region, the Albian-age Giumal Sandstone Formation of interbedded dark gray sandstone (glauconitic to volcaniclastic) and bioturbated shales is interpreted as multiple progradations of a delta onto a shelf influenced by storm waves or tidal currents [Gaetani et al., 1986]. There appears to be a major stratigraphic gap between this unit and the underlying Tithonian Spiti Shale, which may be related to regional uplift associated with the northward drift of the Indian Craton [Mathur, 1990]. The Albian deltaic sands are overlain by upper Cenomanian pelagic chalks [Fuchs, 1987].

Summary of Lower Cretaceous Facies

Following the Late Jurassic dominance by marine clay deposition on the Tethyan margins of Gondwana (upper Dingo Claystone, Spiti Shale, Nupra Formation, etc.), the Early Cretaceous began with rapid progradation of deltaic complexes (Barrow Group, Toro Sandstone, Chukh Unit, etc.) typical of the worldwide "Wealden" facies (Figure 4). At the start of the Cretaceous, a cooling episode occurred on the adjacent Australian continent, ending the long Jurassic period of warm and humid conditions and replacing tropical flora with cool-temperate assemblages [White, 1990a,b]. The shift from tropical to temperate conditions, coupled with possible tectonic uplifts in the regions adjacent to the developing rift between India and Australia, was probably the cause of this pulse of clastic deposition. Paleolatitudes of the Northwest Australian and Himalayan margins were about 40°-45°S (Table 1), only slightly more temperate than during the Late Jurassic.

Volcanic eruptions accompanying the late Berriasian through Valanginian birth of the new Indian Ocean deposited ash beds across the Exmouth Plateau and Argo Basin. Related volcanic activity may have shed volcaniclastic debris into deltaic systems in the Himalayan margin and into the deepening India-Australia rift (volcanic-rich sediments at Site 766). Localized explosive volcanism continued with lesser intensity into the Hauterivian-Aptian, depositing volcaniclastics and ash beds at Site 766 and perhaps contributing to the Lower Cretaceous of Nepal.

After the late Valanginian initiation of seafloor spreading between India and Australia, the new "passive" margins subsided. Deltaic deposits were replaced by marine claystones along the Australian-Himalayan margin (Figure 4). A major global rise in sealevel from late Valanginian to early Aptian [Haq et al., 1987] contributed to this flooding. Climatic conditions on the continents warmed [White, 1990a,b] in response to the Early Cretaceous drift into lower latitudes (Table 1). Contributing to this warmth was an increase in atmospheric CO_2 levels during the mid-Cretaceous [Arthur et al., 1985].

Within the Argo Basin, Lower Cretaceous claystone and radiolarians were slowly deposited under oxygenated bottom conditions below the CCD. Periodic changes in surface fertility or in exchange of waters between basins resulted in pulses of radiolarian influx, and overall radiolarian productivity increased following the rifting between India and Australia. Surrounding continental margins shed nannofossil-rich turbidites into the basin, especially after the drowning of the shelves in the late Valanginian. The abundance of these silica-rich calcareous turbidites coincide with times of postulated short-term eustatic lowstands in the Valanginian and late Aptian [Haq et al., 1987], and inferred associated lowering of CCD levels on the continental slopes. During these episodes, the accumulating pelagic carbonate material on the steep lower slopes of the Australian margin may have been susceptible to displacement by gravity flows [Dumoulin, 1992]. This model for the abundance of pelagic calcareous

turbidites during sea-level lowstands contrasts with the typical association of increased shallow-water carbonate detritus being shed into basins during highstands [e.g., Ogg et al., 1983; Droxler and Schlager, 1985; Mullins et al., 1988].

During the late Barremian and Aptian, the Argo Basin received an increased influx of clays from the margins, resulting in increased burial rates and preservation of organic carbon. At this time, there was increased dissolution of carbonate -- a local reflection of a global rise in CCD as atmospheric CO_2 levels increased [Sheridan, 1983; Arthur et al., 1985]. Tethyan radiolarian assemblages become more important as the waters in the Argo Basin warmed and had greater interchange with the shrinking Eastern Tethys and the growing Indian Ocean.

CONCLUSIONS

The complex sedimentation history of the margins of Gondwana and of the Argo basin portion of the Eastern Tethys reflects the interplay of global processes, regional tectonics and climate. These sedimentary deposits were primarily deposited in temperate latitudes and include abundant clastic influx from the adjacent landmasses. In contrast, the coeval "typical Western Tethys facies" were characteristic of tropical latitudes, and also reflect a partial isolation of those shelves and basins from terrigenous influx. Only during the Late Triassic-Early Jurassic were these eastern Gondwana margins within the tropical belt. During this period, the margins were subject to significant shallow-water carbonate deposition, and some local Norian-Rhaetian deposits resemble the coeval "Dachstein-Kössen-Oberrhätkalk" facies of the Northern Calcareous Alps.

A hiatus and stratigraphic break between Callovian shallow-water strata and Oxfordian deep-water sediments is a curious ubiquitous feature observed both on most margins and in the oceanic sediments. This event is not fully understood, but may have been caused by a global plate tectonic reorganization and the associated cascading effects of rising eustatic sea levels and increasing CO_2 levels. Another global tectonic reorganization caused similar effects during the late Barremian and early Aptian.

Clastic influx is dramatically exhibited by the deposition of hundreds of meters of organic-rich claystone on these shelves during the Late Jurassic (e.g., Spiti-Nupra Shale in the Himalayas, Muderong Shale in Northwest Australia). In contrast, Upper Jurassic sedimentation on Western Tethys margins are characterized by condensed successions of radiolarian chert and ammonite-rich marl. The organic-rich claystone contains ammonites and radiolarians encased within siliceous nodules, but these pelagic remains are dispersed within the outflow from the humid continents. A change to more temperate climatic conditions coincided the Berriasian-Valanginian rifting of Australia and Greater India. As a result, large delta complexes prograded across these shelves, and numerous volcanic eruptions created widespread volcaniclastics and ash beds. After the onset of seafloor spreading between these continents in the late Valanginian, the new passive margins subsided and transgressive fine-grained deposits are common. Claystone accumulation was further enhanced by the return to warm humid conditions in the hinterlands.

The reddish-brown claystone facies of the Argo Basin resemble their Upper Jurassic and Lower Cretaceous clayey radiolarite equivalents in the tropical Pacific [Shipboard Scientific Party, 1990; Ogg et al., 1992b], but reflect a relatively lower productivity. A downward excursion of the carbonate compensation depth (CCD) during the late Kimmeridgian-early Tithonian is indicated in the Argo Basin by limited preservation of nannofossils and mollusc fragments. This event correlates to a similar CCD excursion noted in the Atlantic. Otherwise, as within the Pacific basin, the elevated CCD during the Late Jurassic prevented the preservation of carbonate at oceanic depths.

There are many features of the Eastern Tethys paleoceanography and tectonic history that are not yet understood. Further field work within the few remaining Mesozoic exposures that have not been swallowed by subduction will help fill gaps, but much of the history may remain forever veiled.

Acknowledgments. This project was an outgrowth of our involvement with ODP Legs 122 and 123 off Northwest Australia. We thank our colleagues particularly Ulrich von Rad, Peter Baumgartner, Thierry Dumont, Nevell Exon, Martin Gibling, Lubomir Jansa, Mike Kaminski, Inger Lise Kristiansen, Ursula Röhl, William Sager, Jürgen Thurow, Gerd Westermann, and Jost Wiedmann on these ventures and our comrades on the Lost Ocean Expeditions into the Himalayas for providing much of the knowledge summarized in this report, and apologize for any mistakes in interpreting their careful observations. Participation on the ODP legs, support for post-cruise research, and involvement with this Indian Ocean Workshop was funded by the U.S. Science Advisory Committee (USSAC), the Geological Survey of Canada, and the Consiglio Nazionale delle Ricerche. Earlier manuscripts benefited greatly from reviews by Gabi Ogg (who also drafted some key figures), Ulrich von Rad, Hugh Jenkyns, David Boote, and Georges Mascle.

REFERENCES

Arkell, W. J., *The Jurassic System in Great Britain*, 681 pp., Clarendon Press, Oxford, 1933.

Arthur, M. A., W. E. Dean, and S. O. Schlanger, Variations in the global carbon cycle during the Cretaceous related to climate, volcanism, and changes in atmospheric CO_2, in *The Carbon Cycle and Atmospheric CO_2: Natural Variations Archean to Present, Am. Geophys. Union Monograph, 32*, 504-529, 1985.

Audley-Charles, M. G., The geology of Portuguese Timor, *Mem. Geol. Soc. London, 4*, 76 pp, 1968.

Audley-Charles, M. G., Reconstruction of eastern Gondwanaland, *Nature, 310*, 165-166, 1983.

Audley-Charles, M. G., Ballantyne, P. D., and Hall, R., Mesozoic-Cenozoic rift-drift sequence of Asian fragments from Gondwanaland, In Mesozoic and Cenozoic Plate Reconstructions, edited by C. R. Scotese, and W. W. Sager, *Tectonophysics, 155*, 317-330, 1988.

Audley-Charles, M. G., Evolution of the southern margin of Tethys (North Australian region) from early Permian to late Cretaceous, in *Gondwana and Tethys. Geol. Soc. London, Spec. Publ., 37*, edited by M. G. Audley-Charles, and A. Hallam, 79-100, 1988.

Audley-Charles, M. G., Tectonics of the New Guinea area, *Annu. Rev. Earth Planet. Sci., 19*, 17-41, 1991.

Baksi, A. Y., Barman, T. R., Paul, D. K., and E. Farrer, Widespread Early Cretaceous flood basalt volcanism in eastern India: geochemical data from the Rajmahal-Bengal-Sylhet Traps, *Chem. Geol., 63*, 133-141, 1987.

Barber, P. M., Palaeotectonic evolution and hydrocarbon genesis of the Central Exmouth Plateau, *APEA Journal, 22*, 131-144, 1982.

Bassoullet, J. P., Colchen, M., Marcoux, J., and Mascle, G., Les masses calcaires du flysch triasico-jurassique de Lamayuru (zone de la suture de l'Indus, Himalaya du Ladakh); klippes sédimentaires et éléments de plate-forme remanies, *Riv. Ital. Paleontol. Stratigr. 86*, 825-844, 1980.

Bassoullet, J. P., Colchen, M., Gilbert, E., Marcoux, J., Mascle, G., Suture, E., and van Haver, T., L'orogenè Himalayan au Crétacé, *Mém. Soc. Géol. France, Nouvelle Série 147*, 9-20, 1984.

Baumgartner, P. O., Lower Cretaceous radiolarian biostratigraphy and biogeography off Northwest Australia (ODP Sites 765, 766 and DSDP Site 261, Argo Abyssal Plain and lower Exmouth Plateau), in *Proc. ODP, Sci. Results, 123*, edited by F. M. Gradstein, J. N. Ludden et al., Ocean Drilling Program, College Station, TX, 1992, in press.

Berger, W. H., and Winterer, E. L., Plate stratigraphy and the fluctuating carbonate line, in *Pelagic Sediments: On Land and Under the Sea, Int. Assoc. Sedimentol. Spec. Pub. 1*, edited by K. J. Hsü, and H. C. Jenkyns, 11-48, 1974.

Bird, P., and Cook, S. E., Permo-Triassic succession of the Kekneno area, West Timor: implications for palaeogeography and basin evolution, in *Orogenesis in Action*, edited by R. Hall, G. Nichols, and C. Rangin, abstract volume, 1990.

Bodenhausen, J. W. A., de Booy, T., Egeler, C. G. and Nijhuis, H. J., On the Geology of Central West Nepal - a preliminary note, *Report, 22th International Geological Congress*, part XI, India (Dehli), 101-122, 1964.

Boillot, G., Winterer, E. L. et al., *Proc. ODP, Sci. Results, 103*, Ocean Drilling Program, College Station, TX, 1988.

Bordet, P., Colchen, M., Krummenacher, D., Le Fort, P., Mouterde,R. and Remy, M., *Recherches Geologiques dans L'Himalaya du Nepal, Region de la Thakkhola*: Editions Centre National Recherches Scientifique, Paris, 279 p, 1971.

Boyd, R., Williamson, P. and Haq, B., Seismic stratigraphy and passive margin evolution of southern Exmouth Plateau, in von Rad, U, Haq, B. et al., *Proc. ODP, Sci. Results, 122*, Ocean Drilling Program, College Station, TX, 1992, in press.

Bradshaw, M. T., Yeates, A. N., Beynon, R. M., Brakel, A. T., Langford, R. P., Totterdell, J. M., and Yeung, M., Paleogeographic evolution of the Northwest shelf region, in *The Northwest Shelf of Australia*, edited by P. G. Purcell, and R. R. Purcell, 29-59, Pet. Expl. Soc. Australia Symp., Perth, 1988.

Braun, J., and Beaumont, C., A physical explanation of the relation between flank uplifts and the breakup unconformity at rifted continental margins, *Geology, 17*, 760-764, 1989.

Butcher, B. P., Northwest shelf of Australia, in *Divergent/Passive Margin Basins, Am. Assoc. Petrol. Geol. Mem. 48*, edited by J. D. Edwards, and P .A. Santogrossi, 81-115, 1988.

Cande, S. C. and Mutter, J. C., A revised identification of the oldest sea-floor spreading anomalies between Australia and Antarctica, *Earth Planet. Sci. Lett. 58*, 151-160, 1982.

Coffin, M. F., and Rabinowitz, P. D., Evolution of the conjugate East African-Madagascan margins and the western Somali Basin, *Geol. Soc. Amer. Special Paper 226*, 78 pp, 1988.

Crostella, A., and Barter, T., Triassic/Jurassic depositional history of the Dampier and Beagle subbasins, Northwest shelf of Australia, *Australian Pet. Explor. Assoc. J., 20*, 25-33, 1980.

Dronov, V. I., Gazdzicki, A., and Melnikova, G. K., Die triadischen Riffe im südöstlichem Pamir, *Facies, 6*, 107-128, 1982.

Droxler, A. W., and Schlager, W., Glacial versus interglacial sedimentation rates and turbidite frequencies in the Bahamas, *Geology, 13*, 799-802, 1985.

du Toit, A. L., *Our Wandering Continents, an Hypothesis of Continental Drifting*, 366 pp, Oliver and Boyd Edinburgh, 1937.

Dumoulin, J. A., Lower Cretaceous smarl turbidites of the Argo Abyssal Plain, Indian Ocean, in *Proc. ODP, Sci. Results, 123*, edited by F. M. Gradstein, J. N. Ludden et al., Ocean Drilling Program, College Station, TX, 1992, in press.

Dumoulin, J. A., and Bown, P. R., Depositional history, nannofossil biostratigraphy, and correlation of Argo Abyssal Plain Sites 765 and 261, in *Proc. ODP, Sci. Results, 123*, edited by F. M. Gradstein, J. N. Ludden et al., Ocean Drilling Program, College Station, TX, 1992, in press.

Dumont, T., The Late Triassic (Rhaetian) sequences of the Northwestern Australian shelf recovered during ODP Leg 122: sea-level changes, Tethyan rifting and overprint of Indo-Australian breakup, in *Proc. ODP, Sci. Results, 122*, edited by U. von Rad, B. U. Haq et al., Ocean Drilling Program, College Station, TX , 1992, in press.

Embleton, B. J. J., Continental paleomagnetism, in *Phanerozoic Earth History of Australia*, edited by J. J. Veevers, p.11-16, Oxford Geological Science Series #2, Oxford University Press, New York, 1984.

Erskine, R. D. and Vail, P. R., Seismic stratigraphy of the Exmouth Plateau, in *Atlas of Seismic Stratigraphy, AAPG Studies in Geology, 27*, edited by A. W. Bally, 163-173, 1988.

Exon, N. F. and Willcox, J. B., Geology and petroleum potential of the Exmouth Plateau area off Western Australia, *Am. Assoc. Petrol. Geol. Bull, 62*, 40-72, 1978.

Exon, N. F., von Rad, U. and von Stackelberg, U., The geological development of the passive margins of the Exmouth Plateau off northwest Australia, *Mar. Geol. 47*, 131-152, 1982.

Exon, N. F., and Buffler, R. F., Mesozoic seismic stratigraphy and tectonic evolution of the western Exmouth Plateau, in *Proc. ODP, Sci. Results, 122*, edited by U. von Rad, B. U. Haq et al., Ocean Drilling Program, College Station, TX, 1992, in press.

Exon, N. F., Borella, P., and Ito, M., Sedimentology of marine Cretaceous sequences in the central Exmouth Plateau (NW Australia), in *Proc. ODP, Sci. Results, 122*, edited by U. von Rad, B. U. Haq et al., Ocean Drilling Program, College Station, TX, 1992, in press.

Falvey, D. A., Sea-floor spreading in the Wharton Basin (northeast Indian Ocean) and the break-up of eastern Gondwanaland, *Australian Petroleum Exploration Association Journal, 12*, 86-88, 1972.

Falvey, D. A., The development of continental margins in plate tectonic theory, *Australian Petroleum Exploration Association Journal, 14*, 95-106, 1974.

Fuchs, G., The geology of the Karnali and Dolpo regions, Western Nepal, *Jahrbuch der Geologischen Bundesanstalt (Jahrb. Geol. B.-A.), 120*, 165-217, 1977.

Fuchs, G., The geology of western Zanskar, *Geol. Jahrbuch 125*, 1-50, 1982.

Fuchs, G., The geology of the southern Zanskar (Ladakh) -- evidence for the autochthony of the Tethys Zone of the Himalaya, *Geol. Jahrbuch 130*, 465-491, 1987.

Fullerton, L. G., Sager, W. W. and Handschumacher, D. W., Late Jurassic-Early Cretaceous evolution of the eastern Indian Ocean adjacent to Northwest Australia, *J. Geophys. Res, 94*, 2937-2953, 1989.

Gaetani, M., Casnedi, R., Fois, E., Garzanti, E., Jadoul, F., Nicora, A. and Tintori, A., Stratigraphy of the Tethys Himalaya in Zanskar, Ladakh -- Initial report, *Riv. It. Paleont. Strat., 91* , 443-478, 1986.

Gaetani, M., Garzanti, E., Jadoul, F., Nicora, A., Tintori, A., Pasini, M., and Ali Khan, K. S., The north Karakorum side of the Central Asia geopuzzle, *Geol. Soc. Amer. Bull., 102,* 54-62, 1990.

Gansser, A., *Geology of the Himalayas,* 289 pp, J. Wiley Publ., 1964.

Garzanti, E., Haas, R. and Jadoul, F., Ironstones in the Mesozoic passive margin sequence of the Tethys Himalaya (Zanskar, Northern India): sedimentology and metamorphism, in *Phanerozoic Ironstones, Geol. Soc. London Spec. Publ., 46,* edited by T. P. Young, and W. E. G. Taylor, 229-244, 1989.

Garzanti, E., Baud, A. and Mascle, G., Sedimentary record of the northward flight of India and its collision with Eurasia (Ladakh Himalaya, India), *Geodynamica Acta, 1,* 87-102, 1987.

Görür,N., and Sengör, A. M. C., Tethyan evolution of the NW Australian margin: Implications for the evolution of the Exmouth Plateau, in *Proc. ODP, Sci. Results, 122,* edited by U. von Rad, B. U. Haq et al., Ocean Drilling Program, College Station, TX , 1992, in press.

Gradstein, F.M., and von Rad, U., Stratigraphic evolution of Mesozoic continental margin and oceanic sequences: Northwest Australia and northern Himalayas, in *Evolution of Mesozoic and Cenozoic Continental Margins. Mar. Geology, 102,* edited by A. W. Meyer, T. A. Davies, and S. W. Wise, Jr., 131-173, 1991.

Gradstein, F. M., Gibling, M. R., Jansa, L. F., Kaminski, M. A., Ogg, J. G., Sarti, M., Thurow, J. W., von Rad, U. and Westermann, G. E. G., *Mesozoic stratigraphy of Thakkhola, Central Nepal, Special Report 1,* Centre for Marine Geology, Dalhousie Univ., Halifax, Nova Scotia, 115 pp, 1989.

Gradstein, F. M., Gibling, M. R., Sarti, M., von Rad, U., Thurow, J. W., Ogg, J. G., Jansa, L. F., Kaminski, M. A., and Westermann, G. E. G., Mesozoic Tethyan strata of Thakkhola, Nepal: evidence for the drift and breakup of Gondwana, *Palaeogeography, Palaeoclimatology, Palaeoecology* 88, 193-218, 1991.

Gradstein, F. M., von Rad, U., Gibling, M. R., Jansa, L. F., Kaminski, M. A., Kristiansen, I. L., Ogg, J. G., Röhl, U., Sarti, M., Thurow, J. W., Westermann, G. E. G., and Wiedmann, J., Stratigraphy and depositional history of the Mesozoic continental margin of Central Nepal, *Geol. Jahrbuch, Reihe B77,* 3-142, 1992.

Gradstein, F.M., Ludden, J., et al., *Proc. ODP, Sci. Results, 123,* Ocean Drilling Program, College Station, TX , 992, in press.

Grow, J. A., Klitgord, K. D., and Schlee, J. S., Structure and evolution of Baltimore Canyon Trough, in *The Atlantic Continental Margin: U.S., The Geology of North America, I-2,* edited by R. E. Sheridan, and J. A. Grow, Geol. Soc. Amer., Boulder, CO, 269-2901, 1988.

Grunau, H. R., Zur Geology von Portugiesisch Ost-Timor, *Mitt. naturf. Ges. Bern, 13,* 11-18, 1956.

Habib, D., Sedimentation-rate-dependent distribution of organic matter in the North Atlantic Jurassic-Cretaceous, in *Init. Repts. DSDP, 76,* edited by R. E. Sheridan, F. M. Gradstein et al., 781-794, U.S. Government Printing Office, Washington, D.C., 1983.

Hallam, A., *Jurassic Environments,* Cambridge University Press, Cambridge, U.K., 269 pp, 1975.

Hallam, A. and Bradshaw, M.J., Bituminous shales and oolitic ironstones as indicators of transgressions and regressions, *J. Geol. Soc. London, 136,* 157-164, 1979.

Haq, B. U., Hardenbol, J., and Vail, P., Chronology of fluctuating sea levels since the Triassic (250 million years ago to present), *Science, 235,* 1156-1166, 1987.

Haq, B. U., von Rad, U., O'Connell, S. et al., *Proc. ODP, Init. Repts., 122,* 826 pp, Ocean Drilling Program, College Station, TX, 1990.

Heim, A., and Gansser, A., Central Himalaya. Geological observations of the Swiss expedition 936, *Mém. Soc. Helv. Sc. Nat., 73,* Mém. 1, 1-245, 1939.

Heroux, Y., Chagnon, A. and Bertrand, R., Compilation and correlation of major thermal maturation indicators, *Am. Assoc. Petrol. Geol. Bull., 63,* 2128-2144, 1979.

House, M., *Geology of the Dorset Coast, Geologists' Association Guide,* Geologists' Association, London, 162 pp, 1989.

Huang, Z., Periodicity in Cretaceous pelagic sequences. PhD thesis, Dalhousie University, Halifax, 171 pp, 1991.

Hubbard, R. J., Age and significance of sequence boundaries on Jurassic and Early Cretaceous rifted continental margins, *Amer. Assoc. Petrol. Geol. Bull., 72,* 49-72, 1988.

Hubbard, R. J., Pape, J., and Roberts, D. G., Depositional sequence mapping to illustrate the evolution of a passive continental margin, in *Seismic Stratigraphy II: An Integrated Approach to Hydrocarbon Exploration, Amer. Assoc. Petrol. Geol. Mem. 39,* edited by O. R. Berg, and D. G. Woolverton, 93-116, 1985.

Jenkyns, H.C., Pelagic environments, in *Sedimentary Environments and Facies (2nd ed.),* edited by H. G. Reading, Blackwell Scientific, Oxford, 343-397, 1986.

Johnson, B. D., Powell, C. McA. and Veevers, J. J., Spreading history of the eastern Indian Ocean and Greater India's northward flight from Antarctica and Australia, *Geol. Soc. Amer. Bull., 87,* 1560-1566, 1976.

Johnson, B. D., Powell, C. McA., and Veevers, J. J., Early spreading history of the Indian Ocean between India and Australia, *Earth Planet. Sci. Lett., 47,* 131-143, 1980.

Kent, D. V. and Gradstein, F. M., A Cretaceous and Jurassic geochronology, *Geol. Soc. Amer. Bull., 96,* 1419-1427, 1985.

Klootwijk, C. T., A review of Indian Phanerozoic palaeomagnetism: implications for the India-Asia collision, *Tectonophysics, 105,* 331-353, 1984.

Klootwijk, C. T. and Bingham, D. K., The extent of Greater India, III. Paleomagnetic data from the Tibetan sedimentary series, Thakkhola region, Nepal Himalaya, *Earth Planet. Sci. Lett., 51,* 381-405, 1980.

Krishna, J., Callovian-Albian ammonoid stratigraphy and paleobiogeography in the Indian sub-continent, with special reference to the Tethys Himalaya, *Himalayan Geology, 11,* 43-72, 1983.

Krishna, J., Kumar, V. S. and Singh, I. B., Ammonoid stratigraphy of the Spiti Shale (Upper Jurassic), Tethys Himalaya, India, *Neues Jahrbuch für Geologie und Paläontologie, Monatshefte,* 580-592, 1982.

Larson, R. L., Late Jurassic seafloor spreading in the eastern Indian Ocean, *Geology, 3,* 69-71, 1975.

Larson, R. L., Early Cretaceous breakup of Gondwanaland off western Australia, *Geology, 5,* 57-60, 1977.

Larson, R. L., Latest pulse of Earth: Evidence for a mid-Cretaceous superplume, *Geology, 19,* 547-550, 1991a.

Larson, R. L., Geological consequences of superplumes, *Geology, 19,* 963-966, 1991b.

Larson, R. L. Mutter, J. C., Diebold, J. B., Carpenter, G. B., and Symonds, P., Cuvier Basin: a product of ocean crust formation by Early Cretaceous rifting off western Australia, *Earth Planet. Sci. Lett., 45,* 105-114, 1979.

Lawver, L. A., and Scotese, C. R., A revised reconstruction of Gondwanaland, in *Gondwana Six: Structure, Tectonics and Geophysics, Amer. Geophys. Union, Geophysical Monograph Ser., 40,* edited by G. D. McKenzie, 17-23, AGU, Washington, D.C., 1987.

Leg 123 Shipboard Scientific Party, Sedimentology of the Argo and Gascoyne abyssal plains, NW Australia: report on Ocean Drilling Program Leg 123 (Sept.1-Nov.1, 1988), *Carbonates and Evaporites, 3,* 201-212, 1988.

Ludden, J. N., Geochemistry and radiometric ages on Late Jurassic oceanic basalt, ODP Site 765, in *Proc. ODP, Sci. Results, 123,* edited by F. M. Gradstein, J. N. Ludden et al., Ocean Drilling Program, College Station TX, 1992, in press.

Ludden, J. N., Gradstein, F. M., et al., *Proc. ODP, Init. Repts., 123,* 716 pp, Ocean Drilling Program, College Station, TX, 1990.

MacLeod, K. G., and Hoppe, K. A., Evidence that inoceramid bivalves were benthic and harbored chemosynthetic symbionts, *Geology, 20,* 117-120, 1992.

Mathur, N. S., Tethyan Cretaceous sediments in the Northwest Himalaya, *Cretaceous Research 11*, 289-305, 1990.

Metcalfe, I., Origin and assembly of Southeast Asian continental terranes, in *Gondwana and Tethys. Geol. Soc. London, Spec. Publ. 37*, edited by M. G. Audley-Charles, and A. Hallam, 101-118, 1988.

Metcalfe, I., Allochthonous terrane processes in Southeast Asia, *Phil. Trans. R. Soc. Lond., A331*, 625-640, 1990.

Metcalfe, I., Late Palaeozoic and Mesozoic palaeogeography of Southeast Asia, *Palaeogeography, Palaeoclimatology, Palaeoecology, 87*, 211-221, 1991.

Molinie, A. J., and Ogg, J. G., Milankovitch Cycles in Upper Jurassic and Lower Cretaceous Radiolarites of the Equatorial Pacific (ODP Site 801) -- Spectral Analysis and Sedimentation Rate Curves, in *Proc. ODP, Sci. Results, 123*, edited by R. L. Larson, Y. Lancelot et al., Ocean Drilling Program, College Station TX, in press, 1992.

Montadert, L., de Charpal, O., Roberts, D., Guennx, P., and Sibuet, J.S., Northeast Atlantic passive continental margins: Rifting and subsidence processes, in *Deep Drilling Results in the Atlantic Ocean; Continental Margins and Paleo-Environment*, edited by M. Talwani, W. Hay, and W. B. F. Ryan, Amer. Geophys. Union, Maurice Ewing Series 3, 154-186, 1979.

Mullins, H. T., Gradulski, A. F., Hine, A. C., Melillo, A. J., Wise, S. W., Jr., and Applegate, J., Three-dimensional sedimentary framework of the carbonate ramp slope of central west Florida: a sequential seismic stratigraphic perspective, *Geol. Soc. Am. Bull., 100*, 514-533, 1988.

Ogg, J. G., Robertson, A. H. F., and Jansa, L. F., 1983. Jurassic sedimentation history of Site 534 (western North Atlantic) and of the Atlantic-Tethys seaway, in *Init. Repts. DSDP, 76*, edited by R. E. Sheridan, F. M. Gradstein et al., 829-884, U.S. Government Printing Office, Washington, D.C., 1983.

Ogg, J. G., Steiner, M. B., Oloriz, F., and Tavera, J. M., Jurassic magnetostratigraphy, 1. Kimmeridgian-Tithonian of Sierra Gorda and Carcabuey, southern Spain, *Earth Planet. Sci. Lett., 71*, 147-162, 1984.

Ogg, J. G., Hasenyager, R. W., Wimbledon, W. A., Channell, J. E. T., and Bralower, T. J., Magnetostratigraphy of the Jurassic-Cretaceous boundary interval -- Tethyan and English faunal realms, *Cretaceous Research, 12*, 455-482, 1991.

Ogg, J. G., Kodama, K. and Wallick, B. P., Lower Cretaceous magnetostratigraphy and paleolatitudes off NW Australia (ODP Site 765 and DSDP Site 261, Argo Abyssal Plain, and ODP Site 766, Gascoyne Abyssal Plain), in *Proc. ODP, Sci. Results, 123*, edited by F. M. Gradstein, J. N. Ludden et al., Ocean Drilling Program, College Station TX, in press, 1992a.

Ogg, J. G., Karl, S. M., and Behl, R., Jurassic through Early Cretaceous sedimentation history of the central Equatorial Pacific and of ODP Sites 800 and 801, in *Proc. ODP, Sci. Results, 129*, edited by R. L. Larson, Y. Lancelot et al., Ocean Drilling Program, College Station TX, in press, 1992b

Owen, R. M. and Rea, D. K., Sea-floor hydrothermal activity links climate to tectonics; The Eocene carbon dioxide greenhouse, *Science, 227*, 166-169, 1985.

Pigram, C. J., and Davies, H. L., Terranes and the accretion history of the New Guinea orogen, *BMR Journal of Australian Geology and Geophysics, 10*, 193-211, 1987.

Pigram, C. J., and Panggabean, H., Rifting of the northern margin of the Australian continent and the origin of some microcontinents in eastern Indonesia, *Tectonophysics, 107*, 331-353, 1984.

Powell, T. S., and Luyendyk, B. P., The sea-floor spreading history of the eastern Indian Ocean, *Mar. Geophys. Res., 5*, 225-247, 1982.

Powell, C. McA., Roots, S. R., and Veevers, J. J., Pre-breakup continental extension in East Gondwanaland and the early opening of the eastern Indian Ocean, *Tectonophysics, 155*, 261-283, 1988.

Purcell, P. G., and Purcell, R. R. (eds), *The Northwest Shelf of Australia*, Pet. Expl. Soc. Australia Symp., Perth, 1988.

Ricou, L.-E., Besse, J., Marcoux, J., and Patriat, P., Un fit Gondwanien révisée à partir données pluridisciplinaires, *C. R. Acad. Sci., 311*, 463-469, 1990.

Röhl, U., von Rad, U., and Wirsing, G., Microfacies, paleoenvironment and facies-dependent carbonate diagenesis in Upper Triassic platform carbonates off Northwest Australia, in *Proc. ODP, Sci. Results, 122*, edited by U. von Rad, B. U. Haq et al., Ocean Drilling Program, College Station, TX, in press, 1992.

Royer, J.-Y. and Sandwell, D. T., Evolution of the eastern Indian Ocean since the Late Cretaceous: Constraints from Geosat altimetry, *J. Geophys. Res., 94*, 13755-13782, 1989.

Sager, W. W., Fullerton, L. G., Buffler, R. T., and Handschumacher, D. W., Argo Abyssal Plain magnetic lineations revisited: Implications for the onset of seafloor spreading and tectonic evolution of the eastern Indian Ocean, in *Proc. ODP, Sci. Results, 123*, edited by F. M. Gradstein, J. N. Ludden et al., Ocean Drilling Program, College Station TX, in press, 1992.

Sakai, H., Geology of the Tansen Group of the Lesser Himalaya in Nepal, *Memoir Fac. Sci., Kyushu Univ., D, 25*, 27-74, 1983.

Sakai, H., Rifting of the Gondwanaland and uplifting of the Himalayas recorded in Mesozoic and Tertiary fluvial sediments in the Nepal Himalayas, in *Sedimentary Facies in the Active Plate Margin*, edited by A. Taira, and F. Masuda, Terra Scientific Publ. Co., 723-732, 1989.

Sarti, M., Russo, F. R., and Bosellini, A., Rhaetian strata, Wombat Plateau: analysis of fossil communities as a key to paleoenvironmental change, in *Proc. ODP, Sci. Results, 122*, edited by U. von Rad, B. U. Haq et al., Ocean Drilling Program, College Station, TX, in press, 1992.

Sarti, M., and Kälin, O., Stable carbon and oxygen isotopic composition of Rhaetian shelf carbonates, Wombat Plateau, Northwest Australia, in *Proc. ODP, Sci. Results, 122*, edited by U. von Rad, B. U. Haq et al., Ocean Drilling Program, College Station, TX, in press, 1992.

Sengör, A. M., Tectonics of Tethysides: orogenic collage development in a collisional setting, *Ann. Rev. Earth Planet. Sci., 15*, 213-244, 1987.

Sheridan, R. E., Phenomena of pulsation tectonics related to the breakup of the eastern North American continental margin, in *Init. Repts. DSDP, 76*, edited by R. E. Sheridan, F. M. Gradstein et al., 897-909, U.S. Government Printing Office, Washington, D.C., 1983.

Shipboard Scientific Party, Site 534: Blake-Bahama Basin, in *Init. Repts. DSDP, 76*, edited by R. E. Sheridan, F. M. Gradstein et al., 141-340, U.S. Government Printing Office, Washington, D.C., 1983.

Shipboard Scientific Party, Site 801, in *Proc. ODP, Init. Repts., 129*, edited by Y. Lancelot, R. Larson et al., 91-170 and 303-418, Ocean Drilling Program, College Station, TX, 1990.

Siehl, A., and Thein, J., Minette-type ironstones, in *Phanerozoic Ironstones. Geol. Soc. Special Publ. 46*, edited by T. P. Young, and W. E. G. Taylor, 175-193, 1989.

Smith, R. B., and Silver, E. A., Geology of a Miocene collision complex of Buton, eastern Indonesia, *Geol. Soc. Amer. Bull., 103*, 660-678, 1991.

Steiner, M. B., Ogg, J. G. Melendez, G. and Sequieros, L., Jurassic magnetostratigraphy, 2, Middle-Late Oxfordian of Aguilon, Iberian Cordillera, northern Spain, *Earth Planet. Sci. Lett., 76*, 151-166, 1986.

Thakur, V. C., Regional framework and geodynamic evolution of the Indus-Tsangpo suture zone in Ladakh Himalaya, *Trans. R. Soc. Edinburgh, Earth Sciences 72*, 89-97, 1981.

Thierstein, H. R., Late Cretaceous nannoplankton and change at the Cretaceous-Tertiary boundary, in *The Deep Sea Drilling Project: A Decade of Progress, Soc. Econ. Paleont. Mineral. Spec. Pub., 32*, edited by J. E. Warme, R. G. Douglas, and E. L. Winterer, 355-394, 1981.

Thiede, J., and Dinkelman, M. G., Occurrence of Inoceramus remains in late Mesozoic pelagic and hemipelagic sediments, in *Init. Repts. DSDP, 39*, edited by P. R. Supko, K. Perch-

Nielsen et al., 899-910, U.S. Government Printing Office, Washington, D.C., 1977.

Thurow, J., and von Rad, U., Early Neocomian bentonites of ODP Leg 123 Sites, in *Proc. ODP, Sci. Results, 123*, edited by F. M. Gradstein, J. N. Ludden et al., Ocean Drilling Program, College Station TX, in press, 1992.

Umbgrove, J. H. F., Geological history of the East Indies, *Am. Assoc. Petrol. Geol. Bull., 22*, 1-70, 1938.

Vail, P. R., Mitchum, R. M., Jr., Todd, R. G., Widmier, J. M., Thomson, S. III, Sangree, J. B., Bubb, J. N., and Hatleid, W. G., Seismic stratigraphy and global changes of sea level, part 1-11: Seismic stratigraphy and global cycles of relative changes of sea level, in *Seismic Stratigraphy -- Applications to Hydrocarbon Exploration, Am. Assoc. Petrol. Geol. Mem., 26*, edited by C. E. Payton, 49-212, 1977.

Veevers, J. J. (ed.), *Phanerozoic Earth History of Australia*, Oxford Geological Science Series #2, Oxford University Press, New York, 1984

Veevers, J. J., Jones, J. G., and Talent, J. A., Indo-Australian stratigraphy and the configuration and dispersal of Gondwanaland, *Nature, 229*, 383-388, 1971.

Veevers, J. J., Heirtzler, J. R. et al., *Init. Rept. DSDP, 27*, (U.S. Govt. Printing Office), Washington, 1060 pp, 1974.

Veevers, J. J., Tayton, J. W. and Johnson, B. D., Prominent magnetic anomaly along the continent-ocean boundary between the northwestern margin of Australia (Exmouth and Scott Plateaus) and the Argo Abyssal Plain, *Earth Planet. Sci. Lett., 72*, 415-426, 1985.

von Rad, U., and Exon, N. F., Mesozoic-Cenozoic sedimentary and volcanic evolution of the starved passive margin off northwest Australia, *Am. Assoc. Petrol. Geol. Mem., 34*, 253-281, 1983.

von Rad, U., and Bralower, T. J., Unique record of an incipient ocean basin: Lower Cretaceous sediments from the southern margin of Tethys, *Geology*, in press, 1992.

von Rad, U., Haq, B. U., et al., *Proc. ODP, Sci. Results, 122*, Ocean Drilling Program, College Station, TX, in press, 1991

von Rad, U., and Thurow, J., Bentonitic clays as indicators of Early Neocomian post-breakup volcanism off Northwest

Xu, Y., Wan, X., Gou, Z. and Zhang, Q., *Biostratigraphy of Xizang (Tibet) in the Jurassic, Cretaceous and Tertiary Periods*, China University of Geosciences, 1989.

Young, T. P., and Taylor, W. E. G. (eds), *Phanerozoic Ironstones, Geol. Soc. Special Publ. 46*, 251 pp, 1989.

Australia (ODP Leg 122), in *Proc. ODP, Sci. Results, 122*, edited by U. von Rad, B. U., Haq et al., Ocean Drilling Program, College Station, TX, in press, 1992.

von Stackelberg, U., Exon, N. F., von Rad, U., Quilty, P., Shafik, S., Beiersdorf, H., Seibertz, E. and Veevers, J. J., Geology of the Exmouth and Walaby Plateaus off northwest Australia: sampling of seismic sequences, *BMR (Bur. Min. Res.) J. Australian Geol. Geophys., 5*, 113-140, 1980.

Wan, Z., G. Li, Y. Cao, Q. Gu, X. Zhou, S. Zhang, Q. Wu and X. Yuan, *Tectonics of Yarlung Zangbo Suture Zone Xizang (Tibet) -- Guide to Geological Excursion*, Geological Bureau of Xizang (China), 49 pp., 1982.

Wang, Y.-G., and Sun, D.-L. Triassic and Jurassic paleogeography and evolution of the Qinghai-Xizang (Tibet) Plateau, *Circum-Pacific Jurassic Special Paper #2*, compiled by G. E. G. Westermann, McMaster Univ., I.G.C.P. Project #171, 19 pp, 1983.

Wanner, J., De stratigraphie van Nederlandsch Oost Indie, Mesozoicum, *Leidsche Geologische Mededeelingen, 5*, 567-610, 1931.

Wanner, J., Gesteinsbildende Foraminiferen aus Malm und Unterkreide des östlichen Ostindischen Archipels, *Palaeontologische Zeitschrift, 22*, 75-99, 1940.

Westermann, G. E. G., Strength of concave septa and depth limits of fossil cephalopods, *Lethaia, 6*, 383-403, 1973.

Westermann, G. E. G. and Wang Yi-Gang, Middle Jurassic ammonites of Tibet and the lower Spiti shales, *Palaeontology, 31*, 295-335, 1988.

White, M. E., Plant life between two Ice Ages Down Under, *Amer. Sci., 78*, 253-262, 1990a.

White, M. E., *The Flowering of Gondwana*, Princeton Univ. Press, Princeton, N.J., 1990b.

Middle and Late Cretaceous History of the Indian Ocean

MARY ANNE HOLMES AND DAVID K. WATKINS

Department of Geology, University of Nebraska, Lincoln, NE 68588-0340, USA

By the Late Jurassic the Somali and Mozambique basins opened to form a restricted west Indian Ocean and rifting of fragments from Gondwana formed the Argo Basin to the north. The east Indian Ocean opened during the Early Cretaceous when India separated from Australo-Antarctica to form the Wharton Basin where hotspot activity formed the Naturaliste Plateau and Kerguelen Plateau-Broken Ridge (KPBR). By the Albian, east and west Indian Oceans were small arms of the Tethys. Fluvio-deltaic systems developed on continental margins. KPBR was a volcanic archipelago forested by a mild and wet climate-loving climax forest of podocarpaceaen conifers with an understory of tree and seed ferns. Elevated kaolinite at several sites and gibbsite on Kerguelen Plateau further indicate an Albian warming. A mid-Cretaceous marine transgression marked by widespread black shale/radiolarite deposition also brought marine sediment to western Australian basins. The warm climate persisted through the Cenomanian as reefs formed on the southeast margin of India, but began to decline during the Turonian as kaolinite disappeared from southern sites. The Wharton Basin sank to sub-CCD abyssal depths by this time. The western KPBR formed a shallow shelf in a poorly circulated region where thick greensands accumulated over the slowly subsiding Kerguelen Plateau. During the Turonian abundant ash and tuff accumulated at Broken Ridge sites prior to the onset of Ninetyeast Ridge hotspot activity as India drifted northward. A second major Cretaceous transgression occurred in the Santonian and allowed chalk deposition over marginal western Australia. Beginning in the Santonian and persisting into the Paleocene chert-rich nannofossil chalk formed on the elevated plateaus, indicating upwelling. Circulation over KPBR improved and a diverse bryozoan-dominated benthonic fauna and flora developed on KPBR. The mid-Campanian is marked by a nearly ocean-wide disconformity which corresponds to the development of marked provinciality of the calcareous plankton. This divergence of tropical and austral plankton communities continues through the upper Campanian and most of the Maestrichtian and probably resulted from the onset of shallow circum-Antarctic circulation. Subsidence of Ninetyeast Ridge began during the late Campanian, as shallow-water carbonates, some dolomitized by seepage refluxion, became buried by deep-water pelagic carbonates. Another transgression in the late Maestrichtian brought pelagic marls to southwest Australian Perth Basin.

INTRODUCTION

Five Ocean Drilling Program (ODP) cruises recovered Cretaceous sediment during 1987-1989. The information derived from these cruises, together with that from the seven Deep Sea Drilling Project (DSDP) cruises in 1972-1973 and data from the continental margins of India, southern Africa and Australia (Figure 1) allow for a more detailed paleoceanographic reconstruction of the early Indian Ocean than was previously possible. We attempt here to bring together all of the ocean drilling data available to date combined with recent data on the marginal basins and plate motions to reconstruct sedimentation patterns, planktonic distributions and paleoclimate for the post-Neocomian Cretaceous Indian Ocean.

Pre-Aptian History

The northeast margin of Gondwana began to break apart in the late Permian, opening an oceanic basin between a rifted fragment and the remainder of Gondwana [Gaetani and Garzanti, 1991; Audley-Charles et al., 1988]. The rifted fragment may have comprised northern Tibet and other microcontinents [Audley-Charles et al., 1988], or may have comprised the southern Tibet-Lhasa block [Gaetani and Garzanti, 1991]. The oceanic basin opened at this time ['Neo-Tethys' of Gaetani and Garzanti, 1991; 'Tethys II' of

Synthesis of Results from Scientific
Drilling in the Indian Ocean
Geophysical Monograph 70

Audley-Charles et al., 1988; 'Paleotethys' of Gradstein and von Rad, 1991] has largely disappeared down the subduction zone north of India and along the Java Trench, but its marginal deposits are still found in the Himalaya region [Gradstein and von Rad, 1991; Gaetani and Garzanti, 1991] and along the northwest margin of Australia [e.g., Gradstein and von Rad, 1991]. Permo-Triassic rifting extended into the interior of Gondwana and formed marine basins in northern Indian-Nepal and in northwest Australia [Gradstein and von Rad, 1991; Gaetani and Garzanti, 1991].

A second major rifting event observed in these regions began in the mid-Jurassic and rifted southern Tibet/Lhasa, Burma, Malaysia, Timor, Papua-New Guinea, and other microcontinents ['Sundaland' of Audley-Charles et al., 1988; 'Argo Land Mass'? of Powell et al., 1988] from Gondwana [Audley-Charles et al., 1988; Powell et al., 1988; Gaetani and Garzanti, 1991; Gradstein and von Rad, 1991]. The ocean thus formed ['Paleotethys III' of Audley-Charles et al., 1988] is the earliest northeast Indian Ocean to form (Figure 2). The oldest magnetic anomaly of this ocean is identified adjacent to northwest Australia as M10 and dated as 133 Ma [Powell et al., 1988; Larson et al., 1979]. Powell et al. [1988] infer that breakup began in the Cuvier Abyssal Plain at 132 Ma (Figure 1). Rifting progressed southward along what is now the west coast of Australia and split India from western Australia around 130 Ma, the age of the oldest anomaly in the Perth Abyssal Plain [Powell et al., 1988]. Rifting of India from Australia formed the Wharton Basin (Figure 2).

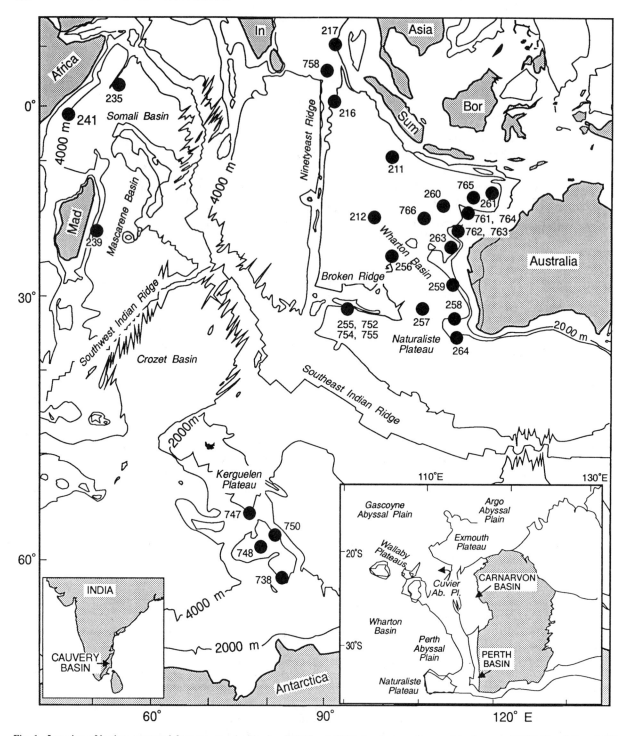

Fig. 1. Location of basins, structural features, marginal basins, DSDP and ODP sites that are discussed in the text. DSDP Sites 249 and 250 are off scale and illustrated in Figure 2. Contours in meters. In = India; Mad = Madagascar; Sum = Sumatra; Bor = Borneo.

The western Indian Ocean had opened by the Callovian (late mid-Jurassic), with the Mozambique and Somali basins formed by 157 Ma [Ségoufin and Patriat, 1981]. However, the oldest sediment recovered to date in either basin is Santonian.

We focus on the Indian Ocean which opened by the

Aptian. We begin with a summary of the sediment deposited in and around this basin and subsequent basins and plateaus, including the latest available paleontologic data.

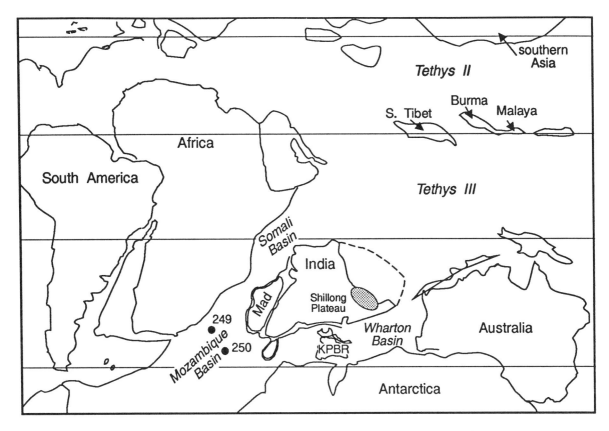

Fig. 2. Indian Ocean in relation to continents and Tethys Ocean, 119 Ma [after Scotese et al., 1988]. Position of continental fragments of southern Tibet, Burma and Malaya after Audley-Charles et al. [1988]. Shillong Plateau is shaded area. Mad. = Madagascar; KPBR = Kerguelen Plateau-Broken Ridge: a hotspot volcanic platform at this time.

STRATIGRAPHY OF THE INDIAN OCEAN AND ITS MARGINAL BASINS

Marginal Cratonic Basins

Carnarvon Basin, Australia. The Neocomian-Aptian Nanutarra Formation rests unconformably on Precambrian basement or conformably on the Yarraloola Conglomerate in the northeastern parts of the Carnarvon Basin [Figure 3; Tait and Smith, 1987]. This unit consists of interbedded silt and immature sandstone and ranges in thickness from 24 to 45 m. The lower part is fluvial in proximal areas and is increasingly marine both offshore and with younger age. The unit partly underlies and partly interfingers with the Birdrong Sandstone, the lowermost member of the Winning Group [Tait and Smith, 1987].

The Winning Group consists of 6 formations which form two depositional sequences. The lower, transgressive sequence ranges from Neocomian to Aptian at the base to late Aptian/early Albian at the top and comprises the lowermost Birdrong Sandstone, the Muderong Shale and the Windalia Radiolarite. These are diachronous units which partly interfinger and grade from sandstone to radiolarite offshore [Figure 3; Tait and Smith, 1987]. The lower, cross-bedded sands are fluvial and underlie highly bioturbated glauconitic sands, indicating rising sea level and a change to marine conditions. The Muderong Shale is a clayey siltstone with thin lenses of sand and becomes more calcareous offshore [Tait and Smith, 1987; Playford et al., 1975]. The Windalia Radiolarite grades from a rippled siliceous siltstone onshore to glauconitic and pyritic offshore where it is indistinguishable from the Muderong Shale [Tait and Smith, 1987]. Its deposition was at maximum transgression for the lower Winning Group [Tait and Smith, 1987].

The upper Winning Group consists of the Gearle Siltstone and its lateral equivalents, the Haycock Marl and the Alinga Formation, all deposited on a quiet marine shelf [Tait and Smith, 1987]. The Gearle Siltstone is an Albian to Turonian interval of glauconitic, carbonaceous, pyritic, clay-rich siltstones with interbeds of radiolarite and ranges in thickness from 120 to 170 m in outcrop to over 700 m offshore [Tait and Smith, 1987; Playford et al., 1975]. Nannofossils are sporadic and poorly preserved in the Gearle and Alinga formations [Shafik, 1990]. The Haycock Marl formed northeast of the locus of Gearle Siltstone deposition and comprises clayey chalk and marl [Tait and Smith, 1987].

The upper Winning Group is overlain unconformably by the Toolonga Calcilutite (Figure 3), a Santonian-lower Campanian chalk, partly glauconitic, which represents another transgressive phase for the Carnarvon Basin. It is enveloped by thin phosphatic deposits and ranges from 50 to 100 m thick in outcrop to 600 m in offshore wells [Tait

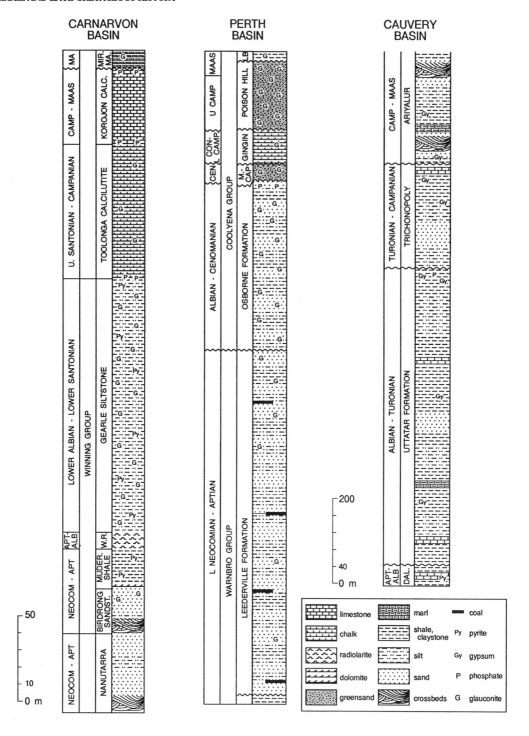

Fig. 3. Composite section of Aptian-Maestrichtian sediment in the Carnarvon, Perth, and Cauvery basins. Location of basins are given in insets of Figure 1. Note the change of scale for Cauvery Basin. The Ariyalur Formation is over 2700 m thick. Jurassic marine sediment underlies lower Cretaceous sediment in all basins, although the Dalmiapuram Formation may overlie Precambrian basement. Left column = stage; right column = formation; center column, if present, = group. Straight lines = conformable contacts; wavy lines = unconformable contacts. Neocom. = Neocomian. Apt. = Aptian; Alb. = Albian; Cen. = Cenomanian; Con. = Coniacian; Camp. = Campanian; Ma and Mas. = Maestrichtian. For Carnarvon Basin: Mir. Ma. = Miria Marl; Calc. = calcarenite; W. R. = Windalia Radiolarite; Muder. = Muderong. For Perth Basin: LB = Lancelin Beds; M.Cap = Molecap Greensand. For Cauvery Basin: Dal. = Dalmiapuram Formation.

and Smith, 1987]. Calcareous nannofossils are abundant and generally well preserved in this unit. The presence of *Biscutum coronum* suggests cool surface waters during the early Campanian [Shafik, 1990]. Overlying the Toolonga Calcilutite is the 45 m-thick Korojon Calcarenite. Large, intact pieces of inoceramids indicate deposition over a quiet outer shelf [Tait and Smith, 1987]. Nannofossils in this interval include rare *Cribrosphaerella daniae* and *Monomarginatus quaternarius*, suggesting moderately cool surface waters [Shafik, 1990].

The Miria Marl unconformably overlies the Korojon Calcarenite and is a one to 30 m thick unit of marl with glauconitic claystone, becoming more argillaceous offshore. The base of the unit contains phosphatic nodules and phosphatized fossils [Tait and Smith, 1987]. McGowran [1974] characterized the foraminiferal assemblages as "extratropical" indicating the transitional nature of the fauna. Nannofossils from this interval are apparently transitional in nature, with the co-occurrence of *Micula murus*, *Nephrolithus frequens*, *Ceratolithoides aculeus*, and *Petrarhabdus copulatus* (among others) indicating affinities to both tropical and cooler, austral nannofloras [Shafik, 1990].

Perth Basin, Australia. The late Neocomian-Aptian Leederville Formation is an interbedded sequence of coarse to fine fluvio-deltaic siliciclastics (Figure 3). Glauconite and pyrite occur locally and there is coal in the southern Perth Basin. Around Perth the formation is 250 m thick, and thickens offshore to a maximum of 545 m [Playford et al., 1976]. Spores, pollen, and ostracods poorly date this unit. The Osborne Formation unconformably overlies the Leederville Fm. (Figure 3). It comprises glauconitic sandstone with minor beds of dark gray to black shale and thickens offshore from 60 m to 200 m [Playford et al., 1976]. This sequence was deposited in nearshore marine to paralic environments, probably during the Albian [Playford et al., 1976].

The Cenomanian Molecap Greensand disconformably overlies the Osborne Formation [Shafik, 1990; Playford et al., 1976]. Over its 9 to 12 m thickness it comprises mainly glauconite pellets and contains fish, reptile parts, a few bivalves and belemnites [Playford et al., 1976], but is devoid of calcareous microfossils [Shafik, 1990]. In the basin center the Molecap has thin phosphatic beds at base and top. The Gingin Chalk is a 20 m-thick unit unconformably overlying the Molecap Greensand [Playford et al., 1976; Shafik, 1990]. It is partly glauconitic and contains a diverse assemblage of coccoliths, foraminifers, bivalves, sponges, brachiopods, ammonites, ostracods, echinoids, crinoids, cirriped and annelid worms [Playford et al., 1976]. The Gingin Chalk has been dated as Santonian based on crinoids [*Marsupites and Uintacrinus*; Withers, 1924, 1926] and foraminifers [Belford, 1960], although recent work by Shafik [1990] indicates that the upper Gingin is early Campanian in age. It correlates to the Toolonga Calcilutite of the Carnarvon Basin.

The Poison Hill Greensand and the correlative Lancelin Beds (Figure 3) of the Perth Basin are time-correlative to the Korojon Calcarenite of the Carnarvon Basin [Shafik, 1990]. The Poison Hill Greensand ranges from 23 to 43 m thick, coarsens upward and is conglomeratic near the top. The Lancelin Beds comprise glauconitic marls with abundant inoceramid debris and are known only from boreholes in the Lancelin area [Playford et al., 1976]. The Lancelin Beds are late Campanian in age based on the occurrence of *Aspidolithus parcus*, *Reinhardtites levis*, and *Reinhardtites anthophorus* [Shafik, 1990]. The common occurrence of high latitude species (e.g. *Monomarginatus* spp., *Biscutum magnum*) and the sparsity of tropical species (e.g. *Ceratolithoides aculeus*, *Quadrum* spp.) indicate cold surface water conditions [Shafik, 1990].

Overlying this unit is the Breton Marl of Shafik [1990]. This unit is less than 6 m thick and known only from boreholes in the Lancelin area. The Breton Marl is upper Maestrichtian based on the occurrence of *Nephrolithus frequens*, *Micula murus*, and *Lithraphidites quadratus*. It is correlative with the Miria Marl of the Carnarvon Basin [Shafik, 1990]. The occurrence of these three species, as well as the nature of the rest of the assemblage, suggests transitional surface water conditions [Shafik, 1990].

Cauvery Basin, India. The Upper Jurassic-Aptian Sivaganga Beds overlie Archean basement in the Cauvery Basin [Sastri et al., 1973; Sastri et al., 1981; Kumar, 1985]. This unit comprises terrestrial to paralic siliciclastics which thicken offshore from 350 to over 1090 m [Sastri et al., 1973; Sastri et al., 1981]. Unconformably overlying is the Aptian-Albian Dalmiapuram Formation, a 400 m-thick reefoidal limestone which grades offshore to marine pyritic shales, sandstone, and subordinate limestone [Figure 3; Sastri et al., 1973; Sastri et al., 1981; Kumar, 1985]. The overlying Uttatar Formation has a probable Cenomanian age and comprises black carbonaceous shales near the base followed by reefoidal limestone with minor sandstones [Sastri et al., 1973] and gypsiferous shales and phosphates near the top (Kumar, 1985). Overlying the Uttatar is the Trichonopoly Formation, a 300 to 600 m-thick Turonian-Lower Senonian conglomeratic sandstone with minor limestone and gypsiferous claystone and marlstone [Sastri et al., 1973; Sastri et al., 1981; Kumar, 1985]. Kumar [1985] reports tree trunks from the upper part and describes the environment for the Trichonopoly as shallow neritic with littoral and mud flat facies.

The Campanian-Maestrichtian Ariyalur Formation unconformably overlies the Trichonopoly Formation [Sastri et al., 1973; Sastri et al., 1981; Kumar, 1985]. It is informally divided into a lower unit of marine variegated or gypsiferous claystones with increased bioclastic debris upsection. The upper Ariyalur Formation comprises fluvial sandstone with interbeds of lacustrine claystone and biostromal limestone [Sastri et al., 1973; Sastri et al., 1981; Kumar, 1985]. The entire unit thins from over 2700 m in the central Cauvery Basin to 730 m offshore.

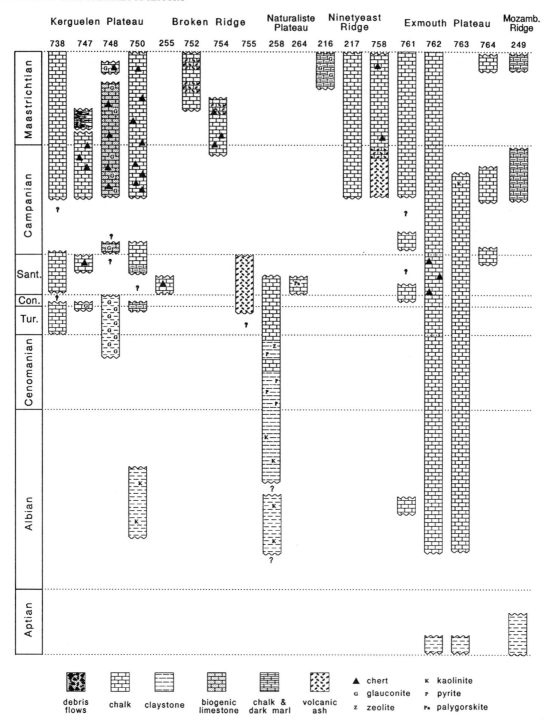

Fig. 4. Age assignments for Aptian-Maestrichtian sediment recovered by ODP and DSDP for shallow sites: continental margins and oceanic plateaus.

Shillong Plateau, India. Along the southern slopes of the Shillong Plateau (Figure 2), a series of shelf-deposited conglomerates and glauconitic sandstones of Senonian age comprise the Mahadek Formation which overlies Jurassic basalt or Precambrian basement [Kumar, 1985]. The Mahadek unconformably underlies the Danian Langpar

Formation, a series of calcareous shales and sandy limestones [Kumar, 1985].

OCEANIC PLATEAU AND OUTER CONTINENTAL SHELF SITES
Kerguelen Plateau. The oldest age determined for the freshest basalt on the Kerguelen Plateau is 110-100 Ma (K-

Ar and ^{40}Ar/^{39}Ar, Whitechurch et al., 1992). The oldest sediment is fluvial cross-bedded soft pebble conglomerate interbedded with soft, kaolinitic clay at Site 750 [Figure 4; Schlich, Wise et al., 1989]. The pebbles are composed of soft kaolinitic clay or highly altered volcaniclasts and contain gibbsite. Examination of the palynomorph [Mohr and Gee, 1992a] and macrofloral remains [Francis and Coffin, 1992] in this sequence has yielded a detailed description of land plant colonization and succession on the volcanic islands of the early Kerguelen Plateau. The lowest samples yield sporomorphs with probable affinities to the gleicheniaceaen and dicksoniaceaen pteridophytes, the pioneer colonizers of nutrient-poor soils developed on exposed basalt. Successively higher samples indicate the rise to dominance of podocarpaceaen conifers. These trees probably formed a canopy over 30 m tall, sheltering understory tree ferns, seed ferns, and smaller podocarps. This climax forest and clay mineral composition suggests mild, wet climatic conditions during the Albian [Watkins et al., 1992; Holmes, 1992].

Site 750 was spot-cored and the contact with the overlying unit was not recovered. Above the terrestrial sediment, black, smectitic, marly chalk of Turonian age was recovered [Schlich, Wise et al., 1989]. This is overlain by Santonian-Maestrichtian partly silicified chalk with chert nodules and faint dissolution laminae with concentrations of clinoptilolite [Schlich, Wise et al., 1989]. Nannofossils within the Turonian and Santonian-lower Campanian interval are abundant and moderately preserved. The presence of *Seribisctum primitivum*, *Thiersteinia ecclesiastica*, and *Repagulum parvidentatum* suggests austral affinities for these assemblages. This interval is separated from the overlying upper Campanian-Maestrichtian section by a significant disconformity spanning much of the mid-Campanian. The uppermost Campanian-Maestrichtian interval is characterized by rich nannofloras dominated by high latitude taxa (e.g. *Monomarginatus* spp., *Nephrolithus* spp.) indicating a strongly austral nature for these assemblages [Watkins, 1992]. Foraminifers are generally sparse and poorly preserved except in the Maestrichtian interval. In general, the foraminifers also indicate an austral affinity except for in the upper Maestrichtian, where there are indications of somewhat transitional conditions [Watkins et al., 1992]. Benthic foraminifers indicate an upper bathyal paleodepth throughout the Upper Cretaceous [Watkins et al., 1992].

Site 748 was drilled west of Site 750 in the Raggatt Basin (the central basin of the southern Kerguelen Plateau). A 3.2 m alkali basalt flow capped by highly rounded alkali basalt cobbles interspersed with Cenomanian glauconitic, mollusc-laden calciclastic sediment was recovered as the basal unit [Figure 4; Schlich, Wise et al., 1989]. Kaolinitic drilling chips corresponding to the terrestrial sediment of Site 750 were the only material recovered below this. The basalt cobbles formed at the shore by sedimentary rounding or as large lava droplets when the flow reached the sea [Schlich, Wise et al., 1989]. Overlying the basalt is 155 m of outer shelf-deposited greensand of Cenomanian-Coniacian age which contains abundant fossil wood, mollusk fragments, siderite nodules up to 30 cm thick, pyrite, clinoptilolite and diagenetic silica [Schlich, Wise et al., 1989; Bitschene et al., 1992]. Sedimentary features other than burrows are sparse in this section except for a 10 m interval where cross-laminae and graded beds occur [Schlich, Wise et al., 1989]. This corresponds to a distinct increase in marine character of the kerogen in the sediment, indicating increased communication with marine waters in what must have been a somewhat restricted marine setting [Watkins et al., 1992]. Calcite lines slickensided normal faults in Turonian sediment which corresponds with extensional activity in the Raggatt Basin [Fritsch et al., 1992; Munschy et al., 1992]. Palynological investigation indicates that the greensands contain diverse assemblages of marine dinoflagellate cysts as well as terrestrially derived pollen and spores. Biostratigraphic analyses of the palynomorph assemblages indicates that this sequence spans the upper Cenomanian through Coniacian [Mohr and Gee, 1992b].

Overlying the greensand is 335 m of glauconitic biohermal grainstones, packstones, and wackestones of early Campanian through late (but not latest) Maestrichtian age [Schlich, Wise et al., 1989]. The principal components are bryozoans and inoceramid prisms and pieces, with lesser amounts of thin-shelled molluscs, crinoid columnals, sponge spicules and (near the middle of the unit) red algal fragments. An entire *Micraster* echinoid indicates the sediment underwent little transport prior to deposition [Schlich, Wise et al., 1989]. Bioturbation is intense throughout. With the exception of 40 m near the middle of this unit, where the presence of red algae indicates very shallow water depths, this sequence contains a fairly diverse assemblage of calcareous nannofossils, dinocysts, and calcareous benthic foraminifers as well as rare hedbergellid and globigerinelloidid planktonic foraminifers. Biostratigraphy indicates a significant mid-Campanian disconformity within the lower portion of this sequence [Watkins et al., 1992]. Abruptly overlying this unit is upper Paleocene chalk with chert, indicating foundering of Site 748 during the late Maestrichtian-early Paleocene, associated with extensional activity in the Raggatt Basin [Munschy et al., 1992].

Lower Turonian shallow-water calciclastic limestone with basalt pebbles was recovered in the core catcher over basalt at ODP Site 738 [Barron, Larsen et al., 1989]. Overlying the shallow-water limestone is 61 m of lower Turonian-Campanian silicified limestone, alternately laminated and bioturbated. The limestone contains chert and anastomosing dissolution laminae similar to the section at Site 750 [Barron, Larsen et al., 1989]. Also present are microfractures indicating an extensional tectonic regime through the Campanian. Shallow-water limestone is overlain by 164 m of gray Campanian-Eocene bioturbated chalk with chert and anastomosing dissolution laminae similar to that recovered at Site 750 [Barron, Larsen et al., 1989]. Microfossil biostratigraphy suggests at least three disconformities (lower Santonian, mid-upper Campanian, and mid-Maestrichtian) above the upper Turonian (total depth; Figure 4). However, microfossil preservation was generally poor and the resolution of the biostratigraphy is low. The presence of high latitude forms indicates austral affinities for the nannofossils (Turonian-Maestrichtian) and Maestrichtian planktonic foraminifers [Barron, Larsen et al., 1989].

Upper Turonian-Coniacian shallow-water calciclastic limestone with glauconite and basalt granules, similar to that recovered at Site 738, overlies basalt at ODP Site 747 [Schlich, Wise et al., 1989]. This is overlain by 75 m of nannofossil chalk with inoceramid fragments and chert nodules, similar to that described for Sites 738 and 750 [Schlich, Wise et al., 1989]. Nannofossils within this sequence contain numerous high latitude taxa but lack all tropical ones, indicating strongly austral affinities. Overlying the chalk is 30 m of volcaniclastic debris flows with at least three separate flows interbedded with chalk and ooze [Schlich, Wise et al., 1989]. The clasts are dominantly fresh to highly altered volcaniclasts and also include pieces of older chalk, angular chert fragments, and cemented limestone. Crude lamination between clay-rich and conglomerate beds occur, but grading is not evident, indicating deposition from concentrated debris flows rather than from turbidity currents. The debris flows denote uplift of the Kerguelen Plateau and exposure of previously deposited sediment, perhaps even basement, during the early Maestrichtian [Munschy et al., 1992]. Powell and others [1988] proposed that a leaky strike-slip fault which eventually became the Southeast Indian Ridge became active by 95 Ma, and separation of Australia and Antarctica (and extension of the southern Kerguelen Plateau) began during Chron 34 (80 to 84 Ma, early Campanian).

Broken Ridge. Gray cherty limestone of Santonian age, dipping at approximately 20°, was recovered as DSDP Site 255 [Figure 4; Davies, Luyendyk et al., 1974]. Foraminifers and radiolarians have pyritic, chalcedonic, or glauconitic fill. This pelagic unit is truncated by erosion and overlain by Eocene-age littoral calcarenite and gravel, a result of a 2500 m uplift of the Ridge prior to separation from the Kerguelen Plateau during the middle Eocene [Davies, Luyendyk et al., 1974; Rea et al., 1990]. Common but poorly preserved (overgrown) nannofossils indicate a Santonian (*Marthasterites furcatus* Zone) age for this limestone [Thierstein, 1974].

At Site 752, mid- to upper Maestrichtian sediment comprises 76 m of mottled chalk with ash and porcellanite [Figure 4; Peirce, Weissel et al., 1989]. Nannofossil assemblages contain numerous high latitude taxa (e.g. *Nephrolithus corystus, Monomarginatus* spp.) throughout this interval, indicating cool surface waters [Resiwati, 1991]. A biostratigraphically complete Cretaceous/ Tertiary boundary was recovered at this site. Nannofossil assemblages exhibit a typical Southern Ocean boundary sequence including very high abundances of the diminutive *Prediscosphaera stoveri* just below the boundary [Pospichal, 1991]. At the top of the Cretaceous is an ash layer with elevated Ir levels [Michel et al., 1991].

Limestone with chert and ash layers form the 204 m-thick lower Maestrichtian section at ODP Site 754 [Figure 4; Peirce, Weissel et al., 1989]. Dolomite replacement occurs in the lower 65 m. Nannofossil assemblages from this section include numerous high latitude taxa (e.g. *Nephrolithus corystus, Monomarginatus* spp.) but lack significant tropical taxa (e.g. *Ceratolithoides aculeus, Quadrum* spp.), indicating austral surface water masses [Resiwati, 1991]. Planktonic foraminiferal assemblages are dominated by *Archaeoglobigerina australis*, and *Globigerinelloides* and *Heterohelix* spp. [van Eijden and Smit, 1991], also suggesting austral affinities.

A 143 m section of tuff spans the Turonian-late Santonian at ODP Site 755 [Figure 4; Peirce, Weissel et al., 1989]. The lower 20 m includes turbidity current-derived glauconitic intervals with sharp basal contacts and fining upwards sequences while the middle 48 m has abundant glauconite dispersed throughout [Peirce, Weissel et al., 1989]. Calcareous microplankton fossils are generally rare and poorly preserved through this interval, although they occur in sufficient quantity and quality to provide biostratigraphic age assignments.

Ninetyeast Ridge. The oldest sediment recovered from the Ninetyeast Ridge is probably Campanian from DSDP Site 217 [Figure 4; Pimm, 1974]. It comprises a sequence of approximately 65 m of redeposited dolarenite and shelly chalk interbedded with chert and claystone [von der Borch, Sclater et al., 1974]. The partially dolomitized shelly material includes bivalves (inoceramids and ostreids), scleractine corals, and calcareous algae, indicating a shallow, off-reef depositional environment which underwent dolomitization by seepage refluxion from an intermittently desiccating lagoon [von der Borch and Trueman, 1974]. These deposits are overlain by 130 m of clay-rich Maestrichtian chalk, the lower part of which is rich in inoceramids and ostreids. The upward decrease in bivalve content [von der Borch, Sclater et al., 1974] and the progression of benthonic foraminifers [McGowran, 1974] suggests progressive deepening of the site through the Campanian and early Maestrichtian.

Approximately 130 m of Maestrichtian calcareous clay interbedded with glauconitic, shelly chalk and ash beds was recovered at DSDP Site 216 [Figure 4; von der Borch, Sclater et al., 1974]. The shallow water facies is overlain by pelagic chalk. Late Maestrichtian planktonic foraminiferal assemblages from Site 217 are characterized by abundant, diverse keeled and large heterohelicid forms, whereas coeval assemblages from the southern Site 216 contain fewer, less diverse keeled and large heterohelicid forms [McGowran, 1974]. This suggests cooler waters at the southern Site 216 during the late Maestrichtian.

The trend from shallow to deep water facies found at Sites 216 and 217 is also seen at ODP Site 758, located between DSDP Sites 216 and 217 (Figure 1). The Campanian-Maestrichtian section recovered here is 377 m thick [Figure 4; Peirce, Weissel et al., 1989]. The lower 169 m is upper

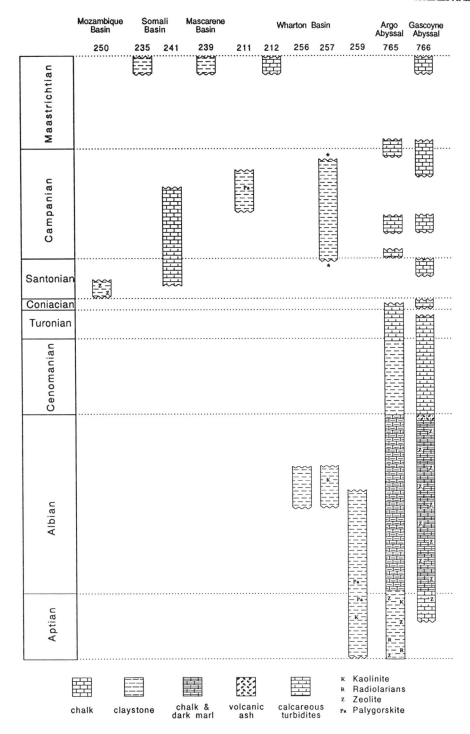

Fig. 5. Age assignments for Aptian-Maestrichtian sediments recovered by ODP and DSDP for deep sites: oceanic basins.

Campanian (*Q. trifidum* Zone) and comprises three units: a lower 67 m of tuff with shelly chalk interbeds, a middle 64 m of volcanic clay with nannofossils, foraminifers, and chert, and an upper 38 m of chalk with porcellanite, chert and ash similar to that recovered from the Kerguelen Plateau. Nannofossil assemblages within this interval contain only temperate or tropical forms (such as *Quadrum*

trifidum, Q. sissinghii, and *Ceratolithoides aculeus*) and lack high latitude forms (such as *Biscutum coronum* and *B. dissimilis*). In the overlying Maestrichtian chalk, alternation between assemblages containing tropical species and high latitude taxa suggests repeated lateral migrations in the position of cool and warm surface water masses over the site [Resiwati, 1991].

Naturaliste Plateau. The oldest sediment recovered at DSDP Site 258 is brown silty mudstone with pyrite and siderite nodules and glauconitic, volcaniclastic sandstone [Figure 4; Davies, Luyendyk et al., 1974]. The clay fraction is dominantly kaolinitic, a warm climate indicator [Cook et al., 1974b; Millot, 1970]. The unit contains abundant, moderately preserved (etched) nannofossil assemblages characteristic of the middle Albian to lower Cenomanian [*Prediscosphaera columnata* and *Eiffellithus turriseiffelii* Zones; Thierstein, 1974]. Common *Seribiscutum primitivum* throughout this interval indicate austral affinities for these assemblages. Foraminiferal assemblages are sparse within this interval, but suggest ages similar to that based on nannofossils [Herb, 1974]. It is time-correlative with the glauconitic Osborne Formation in the adjacent Perth Basin (Figure 3). Nannofossil content increases upwards in the sequence, with upper Albian to lower Cenomanian claystones changing to middle Cenomanian zeolite- and clay-rich chalks [Davies, Luyendyk et al., 1974]. Nannofossils are abundant, though etched, in this interval. Abundant *S. primitivum* indicate austral affinities. Foraminifers are sparse and poorly preserved. The Cenomanian equivalent in the Perth Basin is the Molecap Greensand.

Overlying is Turonian-Santonian interbedded chalk and siliceous chalk with chert nodules, similar to that described for Kerguelen Plateau and Broken Ridge. The Santonian section is truncated by erosion with upper Miocene chalks overlying [Davies, Luyendyk et al., 1974]. The thick Turonian and Coniacian biogenic sedimentary sequence indicates high surface water productivity during this time. Calcareous nannofossils are abundant throughout this interval [Thierstein, 1974]. The presence of common *Thiersteinia ecclesiastica* and *Seribiscutum primitivum* in the Coniacian-Santonian suggest cool surface waters over the plateau. Nannofossil preservation is moderate in the Turonian but significantly improves in the Coniacian and Santonian. There is a concomitant increase in the abundance of planktonic foraminifers, with the assemblages dominated by small globigerine and heterohelicid taxa [Herb, 1974]. Keeled forms are a minor component of these assemblages, which also suggests cooler surface waters. The Coniacian-Santonian chalk of Site 258 is equivalent to the Gingin Chalk of the Perth Basin. However, the Gingin Chalk lacks the siliceous cement and nodules of the chalks recovered on the Naturaliste, Broken Ridge, and Kerguelen plateaus.

On the southern edge of the Naturaliste Plateau at DSDP Site 264, 1 m of gray, clay-rich nannofossil chalk overlies basalt conglomerate [Figure 4; Hayes, Frakes et al., 1975]. The clay is composed of palygorskite and smectite [Cook et al., 1975]. These clays may form by alteration of volcanic ash or may be derived from terrestrial sources [Jones and Galan, 1988]. Vallier and Kidd [1977] attributed their presence at Site 264 to volcanic contributions. The age of the sediment corresponds to tuffs at Site 755 on Broken Ridge. Shafik [1990] assigned this unit to lower Campanian, equivalent to the Gingin Chalk (Perth

Basin) and the Toolonga Calcilutite (Carnarvon Basin). The abundance of *Biscutum coronum* and the presence of *Biscutum dissimilis* indicates relatively cool (austral) surface water at this site during the Early Campanian.

Exmouth Plateau. A 41.3 m thick lower-middle Albian to upper Maestrichtian pale brown nannofossil chalk with inoceramid debris unconformably overlies Valanginian sediment at ODP Site 761 [Figure 4; Haq, von Rad et al., 1990]. The chalk contains rhythmic alternations of bioturbated and faintly laminated zones. Nannofossil biostratigraphy indicates at least four significant disconformities in this sequence, with only mid-Albian, mid-Turonian, lower Santonian, and upper Santonian through Campanian represented by thin chalk units (Figure 4). This is overlain by 22.8 m of white to greenish white chalk with bioturbated/ laminated alternations and interbeds of radiolarian porcellanite and chert nodules of mid- to late Maestrichtian age [Haq, von Rad et al., 1990].

The Muderong Shale of the Carnarvon Basin extends offshore to ODP Site 762 where 10 m of lower Aptian black shale was recovered [Haq, von Rad et al., 1990]. Pelagic chalk deposition began at this site during the Albian with 288 m of Albian-Upper Maestrichtian nannofossil chalk or clayey nannofossil chalk alternating with calcareous clay at 5 to 100 cm-thick intervals [Haq, von Rad et al., 1990]. The contacts of the clay-rich units are gradational, indicating they are not turbiditic. Chert occurs in the Santonian interval only. The dominant clay mineral of the Albian through Coniacian clay is smectite followed by illite [Haq, von Rad et al., 1990]. Traces of chlorite, a cool climate indicator, occur in the lower Campanian through lower Maestrichtian [Haq, von Rad et al., 1990]. The Albian through lower Turonian section appears biostratigraphically complete but is only approximately 22 m thick, indicating a somewhat condensed sequence. The middle Turonian through Maestrichtian section is apparently complete (Figure 4), although this is difficult to verify in light of the absence of some nannofossil marker taxa and reported differences in the stratigraphic ranges of other taxa relative to European stratotypes [Haq, von Rad et al., 1990].

Beginning in the Albian and persisting into the Santonian, Site 763 received increasing amounts of pelagic carbonate and decreasing amounts of felsic terrestrial detritus, as did the adjacent Carnarvon Basin (Figures 3 and 4). This sequence begins with a 52.5 m-thick fining upwards Muderong Shale-equivalent with glauconite and bentonites in the lower part which alternates between bioturbated and laminated intervals. Mineral composition varies from dominantly smectitic to illitic [Haq, von Rad et al., 1990]. The shale is overlain by 38 m of Albian-Cenomanian claystones and calcareous claystones with hard carbonate interbeds altered to siderite or dolomite, equivalent to the Gearle Siltstone. Calcareous microfossils within this interval are generally abundant and well preserved. This, in turn, is overlain by 14.6 m of alternating upper Cenomanian nannofossil claystone with zeolites and darker, clay-rich intervals [Haq, von Rad et al., 1990]. The boundaries between carbonate-rich and -poor intervals are gradational and suggest 20,000 to 40,000 year alternations of pelagic carbonate input [Haq, von Rad et al., 1990]. Calcareous microfossils within this interval are virtually absent but plant debris is present. Overlying this unit is a 139 m-thick section of nannofossil chalk with foraminifers and clay-enriched intervals [Haq, von Rad et al., 1990], equivalent to the Toolonga Calcilutite of the Carnarvon Basin. Kaolinite was detected as a minor component near the top of the Campanian; otherwise illite dominates. Calcareous microfossils within this interval are abundant and well-preserved, although some of the marker taxa typical of tropical-temperate nannofossil assemblages of this age are absent [Haq, von Rad et al., 1990]. This unit is truncated and overlain by Eocene sediment, apparently subject to an erosive event which occurred sometime during the late Campanian to Eocene.

A highly condensed, 8 m-thick section of Coniacian to upper Maestrichtian nannofossil chalk with some clay-rich intervals was recovered at Site 764 (Figure 4). The unit contains pyrite, carbonate bioclastic debris, dolomite rhombs, and minor plant debris and is characterized by poorly preserved nannofossils and mixed Upper Cretaceous foraminiferal assemblages [Haq, von Rad et al., 1990]. It overlies Triassic reefal limestone and underlies Eocene nannofossil ooze [Haq, von Rad et al., 1990].

Mozambique Ridge. Approximately 110 m of medium gray to olive black silty claystone to volcanic clayey siltstone overlie basalt at Site 249 on the Mozambique Ridge [Figure 4; Simpson, Schlich et al., 1974]. At the base of this sequence, poorly preserved (highly dissolved) nannofossil assemblages suggest a Barremian to Aptian age [Bukry, 1974]. The foraminiferal assemblages are dominated by benthonic forms (mostly Lagenidae), with only rare indeterminate globigerine forms representing the planktonic foraminifers. Ostracods are common although not age-diagnostic [Simpson, Schlich et al., 1974]. Kaolinite was detected in the basal unit but otherwise smectite dominates this interval [Matti et al., 1974]. The middle of this unit is barren of diagnostic microfossils. At the top of the unit, planktonic foraminifers suggest a late Albian to Cenomanian age. Radiolarians become relatively common in this upper interval, although nannofossils are sparse and poorly preserved [Simpson, Schlich et al., 1974]. The volcanic siltstones are disconformably overlain by approximately 110 m of Campanian through Maestrichtian clayey chalk. Both nannofossils and planktonic foraminifers are abundant and well preserved [Figure 4; Simpson, Schlich et al., 1974]. A disconformity (between sections 17-3 and 17-4), encompassing most of the lower Maestrichtian, is indicated by the nannofossil stratigraphy [Bukry, 1974]. X-ray data indicate smectite dominates the fine fraction, with a palygorskite-rich interval in the mid-Campanian and kaolinite appearing in late Maestrichtian sediment [Matti et al., 1974; Marchig and Vallier, 1974].

Continental Rise and Abyssal Plain Sites
Wharton Basin. The oldest sediment in the Wharton Basin was recovered overlying basalt from Site 259 and comprises 140 m of greenish gray zeolite-bearing claystone with dolomite nodules at the base [Veevers, Heirtzler et al., 1974]. The clay is dominated by smectite but contains kaolinite-rich intervals [Cook et al., 1974a]. This interval is largely devoid of siliceous and calcareous microfossils, although some radiolarians are present at mid-section. Organic-walled microfossils indicate that this sequence spans the Aptian (Figure 5), with marine palynomorphs (dinoflagellates) comprising 70-80% of the palynological assemblages [Wiseman and Williams, 1974]. In the middle part of the unit, organic-walled microfossils disappear and are replaced by diverse assemblages of calcareous nannofossils, foraminifers, and radiolarians. The co-occurrence of the nannofossils *Prediscosphaera columnata, Hayestites albiensis,* and *Axopodorhabdus albianus* [Proto Decima, 1974] indicate the lower part of the *Prediscosphaera columnata* Zone of middle to early late Albian age. Planktonic foraminifers in this interval are dominated by hedbergellids [Veevers, Heirtzler et al., 1974]. Benthic foraminifers within this interval are diverse and similar to those found in the Uttatar Group of the Cauvery Basin. They indicate an upper slope depth for the site during the early late Albian [Scheibernová, 1974]. The sediment in this interval consists of 100 m of zeolite-rich nannofossil clay and clayey ooze with palygorskite, cristobalite, and pyritized radiolarians [Veevers, Heirtzler et al., 1974]. The zeolitic nannofossil clay sequence is overlain by approximately 25 m of yellowish brown zeolitic claystone which is barren of calcareous and organic-walled microplankton. Sparse radiolarians suggest that this unit is Upper Cretaceous. The unit is overlain by upper Paleocene nannofossil ooze [Veevers, Heirtzler et al., 1974].

Approximately 11 m of upper Albian nannofossil-rich, zeolitic, brown clay occurs at DSDP Site 256 (Figure 5). Nannofossils are abundant and moderately preserved (etched) throughout the interval [Thierstein, 1974]. The presence of *S. primitivum* suggests cooler surface waters. Planktonic foraminifers occur only in the lower 2 m of this sequence. These strongly etched assemblages are composed almost solely of *Hedbergella planispira* [Herb, 1974]. A similar unit was recovered at Site 257 in the southern Wharton Basin. Two cores over a 57 m interval contain pelagic zeolitic clay with traces of glauconite [Davies, Luyendyk et al., 1974]. Nannofossils increase downhole in the middle Albian section and foraminifers become increasingly corroded [Davies, Luyendyk et al., 1974]. The co-occurrence of the nannofossils *Prediscosphaera columnata, Hayestites albiensis,* and *Axopodorhabdus albianus* indicate the lower part of the *Prediscosphaera columnata* Zone of middle to early late Albian age [Thierstein, 1974]. Planktonic foraminiferal assemblages are dominated by *Hedbergella planispira* and show indications of increasing dissolution downwards in the middle Albian section [Herb, 1974].

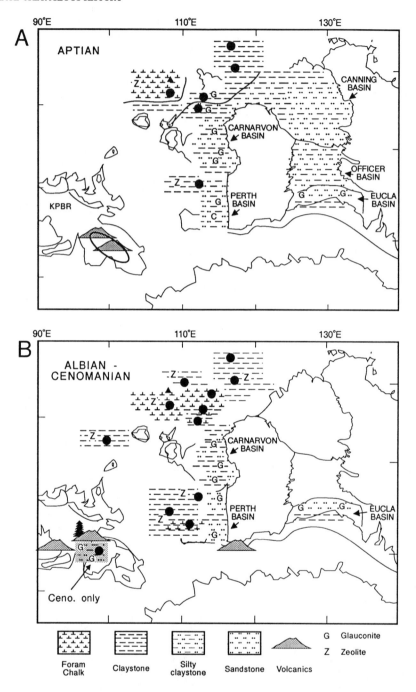

Fig. 6.   Summary of deposition in the Indian Ocean at various stages for the mid- and Late Cretaceous. Reconstructions of Royer and Sandwell [1989]. KPBR = Kerguelen Plateau-Broken Ridge. Heavy line outlines Raggatt Basin of the southern Kerguelen Plateau. A. Aptian. B. Albian-Cenomanian. C. Coniacian-Santonian. D. late Campanian-early Maestrichtian.

Site 211 in the northern part of the Wharton Basin received 10 m of nannofossil ooze during the Campanian because of its topographically elevated position. This unit also contains thin interbeds of iron-rich ash [Figure 5; von der Borch, Sclater et al., 1974]. Venkatarathnam [1974] attributed the presence of palygorskite to alteration of volcanic detritus. Near the contact with the underlying basalt, the nannofossil assemblages are depauperate and contain only solution-resistant taxa, indicating deposition near the CCD [Gartner, 1974]. The presence of *Quadrum trifidum* and *Quadrum sissinghii* indicate warm surface water over the site during the early late Campanian.

A 60 m-thick unit of zeolitic brown clay interbedded with calciclastic turbidites overlies basalt at DSDP Site 212

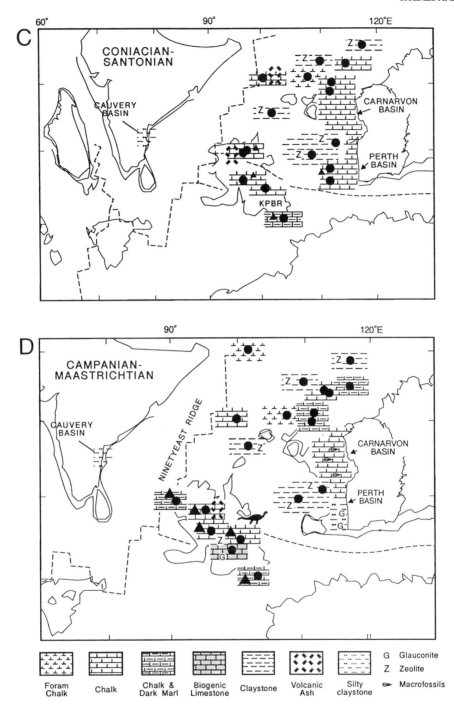

[Figure 5; von der Borch, Sclater et al., 1974]. The brown clay is dominated by smectite and contains clinoptilolite and volcanic ash. The turbiditic clayey chalks contain mixed assemblages of calcareous microplankton fossils. The most consistent age for this material is based on the sporadic occurrence of the latest Maestrichtian nannofossil *Nephrolithus frequens* mixed with older nannofossils. The late Maestrichtian calcareous turbidites are overlain by 15

m of undated brown zeolitic claystone followed by middle Eocene chalk.

Argo Abyssal Plain. The entire Cretaceous section at ODP Site 765 is hemipelagic, with graded calcareous or siliceous (or both) turbidites interbedded with pelagic clay and nannofossil chalk [Figure 5; Gradstein, Ludden et al., 1990]. The lower 65 m of the Aptian section has rhodochrosite-replaced radiolarians and microconcretions [Gradstein, Ludden et al., 1990]. Clay minerals are dominated by mixed-layer illite/smectite, probably terrestrially derived. Kaolinite increases as a minor component upsection, reaching a maximum in upper Aptian sediment. Thin radiolarites in the upper Aptian section correlate with the Windalia Radiolarite of the Carnarvon Basin. Microfossil assemblages through the turbiditic interval are generally sparse and poorly preserved. Highly etched nannofossils and the virtual absence of calcareous foraminifers indicate deposition near or below the CCD during much of the Aptian. Abundance and preservation is sporadically better in the Albian and Upper Cretaceous, although the effects of dissolution are still evident. Age assignments (Figure 5) indicate sporadic deposition of sediment throughout the Aptian-Campanian. Benthic foraminifers suggest abyssal depths throughout this interval [Gradstein, Ludden et al., 1990].

Gascoyne Abyssal Plain. A 48 m-thick Aptian sequence of nannofossil chalks and graded calciclastic sediments was recovered at ODP Site 766 [Figure 5; Gradstein, Ludden et al., 1990]. This is overlain by 54 m of Albian-Cenomanian chalk interbedded with bioturbated, zeolitic, pelagic clay capped by a thin, immature, glauconitic sandstone with acidic volcaniclasts and zeolites [Gradstein, Ludden et al., 1990]. Overlying is 22 m of Turonian-lower Campanian variegated zeolitic pelagic clay interbedded with thin (<15 cm) graded calciclastic sediment with volcanic lithic fragments [Gradstein, Ludden et al., 1990]. Thirty-two m of lower Campanian-lower Paleocene nannofossil ooze overlies the clay and interbeds of graded carbonates reappear along with polymictic conglomerates with calcareous matrices [Gradstein, Ludden et al., 1990]. Microfossil biostratigraphy (Figure 5) indicates relatively continuous sedimentation for the Aptian through Cenomanian giving way to discontinuous sediment preservation through most of the Upper Cretaceous. Nannofossil assemblages are generally abundant and moderately preserved through this interval. Planktonic foraminifers generally show evidence of strong corrosion [Gradstein, Ludden et al., 1990].

Somali Basin. DSDP Site 235 contains pods of metamorphosed orange pink clay within basalt flows recovered at the base of the hole [Fisher, Bunce et al., 1974]. Roth [1974] identified a few, poorly preserved calcareous nannofossils from this sediment indicating a latest Maestrichtian (*Micula murus* Zone) age (Figure 5).

Mozambique Basin. Zeolitic claystones at Site 250 contain highly etched nannofossil assemblages of the *Marthasterites furcatus* Zone, indicating a late Coniacian to Santonian age [Figure 5; Thierstein, 1974]. Foraminiferal assemblages in this interval consist only of sparse arenaceous benthonic forms [Simpson, Schlich et al., 1974].

Mascarene Basin. Less than 45 cm of nannofossil-rich brown clay of late Maestrichtian age overlies basalt at Site 239 (Figure 5). Nannofossil assemblages are strongly etched, containing only solution-resistant forms. The presence of *Micula murus* indicates the latest Maestrichtian [Bukry, 1974]. The foraminiferal assemblages are composed exclusively of benthonic forms [Simpson, Schlich et al., 1974].

MID- AND LATE CRETACEOUS HISTORY BY STAGE
Aptian

The Indian Ocean by the Aptian consisted of a narrow arm through the Mozambique Basin to the west between southern Africa and Madagascar-India, a narrow arm through the newly opened Wharton Basin in the east between India and Australia, and the northern margins of India and Australia (Figure 2). The latter was actually a southern shore of the Tethys. Between India, Australia and Antarctica, Kerguelen Plateau-Broken Ridge was forming by hotspot activity. Additional volcanic activity near the northwest Australian margin is indicated by palygorskite at Site 261 and ash at Site 763.

The only Aptian sediment recovered in the Indian Ocean to date comes from two localities: 1) the northwest margin of Australia (Exmouth Plateau, Argo and Gascoyne Abyssal Plains) and 2) the Wharton Basin (Figures 1 and 6a). The deposits off northwest Australia and the adjacent northern Indian margin are considered by Gradstein and von Rad [1991] to be those of a juvenile northern Indian Ocean wherein siliciclastics give way to pelagic deposits (see von Rad et al., this volume]. They suggest that chalk deposition at Site 765 prior to the Aptian may have corresponded to a deepening of the CCD to around 2800 m. The Wharton Basin was also in a juvenile stage, with the paucity of fossils at Site 259 indicating deposition below the CCD. Site 259 lies on extended continental crust and its paleodepth cannot be calculated.

The Aptian-Albian anoxic event left black shales in the Carnarvon and Cauvery basins, at sites off northwest Australia, and in the Mozambique and Wharton Basins. The Perth Basin may have been shallower than the oxygen minimum zone or anoxic bottom waters which caused the event, as there are no black shales reported from there. However, offshore on the Naturaliste Plateau black shales did accumulate.

A transgression occurred in the nascent Indian Ocean which peaked in the late Aptian-early Albian. The evidence for this includes the transition from fluvial to marine sediment and the appearance of greensands in the Perth and Carnarvon basins [Playford et al., 1975], and the occurrence of marine deposits in the Australian interior. The Aptian-Albian Windalia Radiolarite represents maximum transgression. Transgression was accompanied by a decline in denudation rates for the western Australian craton from 5 m/m.y. to 1.5 to 2 m/m.y [van de Graaf, 1981].

The dominant clay mineral is smectite or a mixed-layer illite/smectite. The only kaolinite reported is a minor component at Site 259. At the more northerly Site 765, kaolinite increases upsection through upper Aptian-lower Albian sediment but remains a minor component. These low kaolinite levels suggest a rather cool to temperate climate.

Albian

A distinct warming trend occurred for the Albian, as indicated by abundant kaolinite on the Naturaliste and Kerguelen Plateaus and off the northwest Australian shelf (Site 263; Figures 6a and 7). The climax forest of podocarpaceaen conifers and tree and seed ferns on the Kerguelen Plateau required a warm, wet climate [Mohr and Gee, 1992a]. Further, gibbsite occurs in Kerguelen Plateau fluvial deposits, indicating high rainfall there [Holmes, 1992]. High rainfall could result from location near the subpolar low or may have been orographically induced. In support of the latter hypothesis, Coffin [1992; this volume] has estimated that over 1000 m, and perhaps as much as 1850 m, of relief existed on the southern Kerguelen Plateau when the last basalts were emplaced by the Albian. Kaolinite is absent from most sites on the northwest Australian margin. Smectite and/or illite predominate in pelagic sediment there with illite indicating a stronger terrestrial input.

Low diversity foraminifer assemblages in the Indian Ocean as well as on the Falkland Plateau led Sliter [1976] to propose that austral waters were cool from the Albian to Maestrichtian. Some endemism among foraminifers is evident by the similarity between the assemblages of Wharton and Cauvery basins and as noted by Sliter [1976] for mid- to Upper Cretaceous sediment from several southern ocean sites. Nevertheless, some reef carbonates began to form in the Cauvery Basin, located then at about 50°S (Figure 2). The CCD deepened enough to allow carbonate accumulation at Site 259. Luyendyk and Davies [1974] estimated the CCD at approximately 2700 m in the Albian Wharton Basin. The low diversity foraminifer assemblages identified by Sliter [1976] on the Falkland Plateau were deposited at depths estimated at 100 to 400 m.

High kaolinite levels and low diversity foraminifer assemblages in Indian Ocean sediment appear to have contradictory climatic indications for the Albian. Possibly some other factors limited foraminifer diversity or kaolinite may have been generated by very high rainfall.

Cenomanian

The proposed warming trend of the Albian persisted into the Cenomanian as indicated by kaolinite persisting as an important sediment component on the Kerguelen and Naturaliste plateaus and its appearance in Argo Abyssal Plain sediment. In addition, reefal limestones (the Uttatar Formation) persisted in the Cauvery Basin. The major tectonic event was the beginning of movement between Australia and Antarctica, and in the Raggatt Basin of Kerguelen Plateau [Figure 7; Munschy et al., 1992; Scotese et al., 1988; Powell et al., 1988]. This movement coincides with the sinking of the eastern Raggatt Basin

(Site 750) from a forest-covered volcanic archipelago to a shallow marine shelf with poor circulation where pelagic marl accumulated (Figures 6b and 7). Volcanic activity persisted in the western Raggatt Basin (Site 748), and continued off the northwest coast of Australia as volcaniclastic sediment accumulated at Site 766. Renewed terrestrial input occurred in the Carnarvon Basin, where the Gearle Siltstone began to accumulate.

In northwest India on the Tethyan coast, the foraminiferal Chikkim Limestone, an outer carbonate shelf deposit with rudist fragments, began to accumulate above glauconitic arenites as the clastic shelf drowned [Gaetani and Garzanti, 1991]. As this unit is correlative to the Molecap Greensand, another transgression is indicated for the Cenomanian. The Wharton Basin dropped below the CCD prior to the Cenomanian and remained there throughout the rest of the Cretaceous. Circulation over Kerguelen Plateau was poor, as indicated by organic-rich, fossil-poor glauconite and marl. The abundance of glauconite deposition during this time may be a result of sea level fluctuations in a small, newly-formed seaway which did not receive large amounts of sediment. The glauconites are associated with hiatuses and indications of sealevel rise or, in the case of the Kerguelen Plateau, of slow subsidence (Figures 3 to 5). In addition, glauconite deposits are associated with enhanced sediment organic content and are usually found in outer shelf regions not far from large rivers [McRae, 1972; Odin and Fullagar, 1988]. Glauconite deposition is also widespread along the margins of the North Atlantic-Tethys during the mid and Late Cretaceous: in the mid-Atlantic United States, southern England and northern Germany. The Atlantic was also a young ocean at this time. The margins of young oceans may be ideal places for glauconite to form when terrestrial input is low. Iron is abundant in the form of rifted flood basalts. Nutrient supply is probably high and promotes planktonic as well as benthonic productivity. This enhances organic matter accumulation in sediment and promotes the reducing microenvironments necessary for ferrous iron mobility.

Turonian-Coniacian

The climate cooled in the southern Indian Ocean during the Turonian as indicated by the disappearance of kaolinite on the Kerguelen and Naturaliste plateaus. However, kaolinite remains an important component in the more northerly Site 765 turbidites. The southern coast of India had moved to approximately 30° S which promoted evaporitic claystone deposition within the Cauvery Basin. Volcanic activity is evident on Broken Ridge with the deposition of some 90 m of tuff with glauconite and micrite (Figures 4, 6c, and 7). The unaltered ash persists as an important sediment component through Paleocene sediment. Ash was also observed in volcaniclastic sediment from Site 766. Sedimentation rates were low all over the Indian Ocean, generally less than 5 m/m.y., with the western Raggatt Basin as the only exception. This area subsided slowly and was relatively closed with poor bottom circulation up to the Coniacian, as indicated by rates of organic-rich

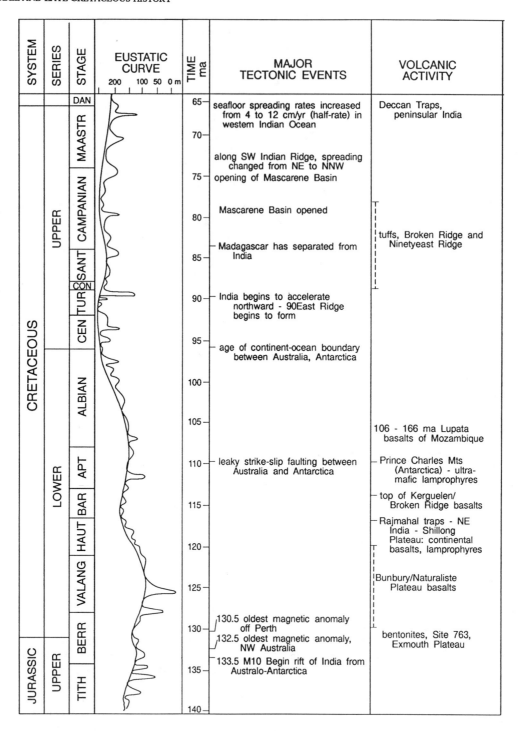

Fig. 7. Summary of major tectonic, volcanic, and sedimentary events in the Indian Ocean during the Cretaceous, compared with the eustatic sea-level curve of Haq et al. [1988]. Sources of information include all DSDP and ODP *Initial Reports* volumes and Schlich et al. [1974]; Förster [1975]; Davies and Kidd [1977]; Sclater and Heirtzler [1977]; Sclater et al. [1977]; Kidd and Davies [1978]; Dingle et al. [1983]; Audley-Charles et al. [1988]; Powell et al. [1988]; Scotese et al. [1988]; Royer and Sandwell [1989]; Gradstein and von Rad [1991]; Munschy et al. [1992]; and Storey et al. [1992] .

STAGE	TIME ma	OCEANIC BASINS	OCEANIC PLATEAUS, CONTINENTAL SHELVES	CONTINENTS
DAN	65			In general, high sea level stand deposits: peak transgression in SW Africa; Mahadek Fm., Shillong Plateau Zanskar-Spiti: marls, foram ooze Perth Basin: greensand, chalks Carnarvon Basin: calcarenite, marl Cauvery Basin: neritic sand mudflats give way to continental deposits in Maas.
MAASTR	70		Silicified limestones, chalk with chert nodules; glauconite on Kerguelen Plateau and 90East R.	
CAMPANIAN	75–80	brown pelagic clays; calcareous turbidites off NW Australia		
SANT	85	pelagic clays Mozambique B.; nanno chalk in Somali B. and off NW Australia		transgression: Perth and Carnarvon Basins: Tur - Sant Gingin Chalk; Toolonga Calcilutite SW Africa: Con - Sant St. Lucia Fm. W. Peninsular India: Con Abur Fm.
CON				
TUR	90			
CEN	95		Raggatt Basin (Kerg. Pl.) subsided to receive pelagic chalk Begin glauconitic sedimentation, Kerguelen Plateau	minor evaporites and reefs, (Uttatar Fm) in Cauvery Basin, SW peninsular India
ALBIAN	100–105	nanno-bearing pelagic clay, Wharton Basin smectitic claystone turbidites and zeolitic pelagic clays, NW Australian margin	terrestrial, kaolinitic sediment, Kerguelen Plat. kao.-rich detrital clay and glauc. sandstones, Naturaliste Plateau pelagic chalk, Exmouth Plateau	
APT	110	oldest sediment in Wharton Basin: sparsely fossilferous gray zeo-bearing claystone	black shales off NW Australia	mid-upper Aptian: SW Africa: regressive conglomerates; Carnarvon Basin: mid-Apt unconformity overlies Windalia Radiolarite (max transgression) Zanskar-Spiti region: black shales (anoxic event) Perth Basin: unconformity
BAR	115		volcaniclastic sediment Mozambique Ridge	Largely fluvio-deltaic and paralic siliciclastics which range from early Jurassic through Cenomanian;
HAUT	120			Ironstones at base, marls towards the top; Glauconite marks transgressions; marls, thin limestones mark regressions
VALANG	125			Perth Basin: Leederville and Osborne Formations Carnarvon Basin: Barrow and Winning Groups
BERR	130			Zanskar-Spiti region, Himalaya: Giumal Sandstone Thakkhola region, Himalaya: Chukh Group Cauvery Basin: Sivaganga Beds and Dalmiapuram Fm.
TITH	135–140			southern Africa: Lower Uitenhage Group (to Berr/Val)

greensand deposition of 63 m/m.y. or more at Site 748. Renewed transgression is apparent in western peninsular India, where deposits of the Bagh and Lameta Beds are estuarine to fluvial and contain abundant mollusc and reptilian fossils [Kumar, 1985].

Santonian

A greenish gray chalk with chert first appears in Santonian sediment at Broken Ridge, Naturaliste Plateau, and Kerguelen Plateau (Figures 4 and 6c). This chalk persists into the Paleocene on Broken Ridge and Kerguelen Plateau. The chert contains the remains of radiolarians and suggests nutrient-rich waters over these plateaus, perhaps as a result of enhanced circulation and even some upwelling. A major transgression is indicated by deposition of Gingin Chalk in the Perth Basin and the

Toolonga Calcilutite in the Carnarvon Basin. Deposition is preserved on the Shillong Plateau with glauconitic sandstones and conglomerates of the Mahadek Formation [Kumar, 1985]. Deposition is first recorded for the Somali Basin at Site 241 with a series of siliciclastic turbidites. Smectite dominates the clay mineral suites of all Indian Ocean sites with minor kaolinite in Site 765 turbidites only, indicating the climate remained cool to temperate.

Campanian

The major tectonic event is the formation of and deposition on the Ninetyeast Ridge. The oldest sediment on the ridge is shallow-water reefal limestone at Site 217 (Figure 4). Paleolatitude for this site is about 40-45°S [Scotese et al., 1988]. Downfaulting occurred along the northwest Australian margin as polymictic conglomerates accumulated at Site 766. An important oceanographic event for the Indian Ocean is a mid-Campanian major disconformity in all the deep Australian marginal basins, with the possible exception of sites on the distal Exmouth Plateau (Figures 4 and 5). This disconformity appears to coincide with the development of strong biogeographic separation of the southern and northern Indian Ocean [Sliter, 1976; Huber and Watkins, 1992]. Huber and Watkins [1992] document the post-mid-Campanian rise of the austral nannoflora and planktonic foraminiferal faunas that characterize the Southern Ocean during the late Campanian and Maestrichtian. This same provincialism is evident in the sediments of the Western Australian margin, with the Perth Basin dominated by austral forms whereas the Carnarvon Basin was characterized by microfossil assemblages of transitional (mixed austral and tropical) forms [Shafik, 1990].

A bryozoan-echinoid-mollusc bioherm flourished over the Raggatt Basin (Site 748) as circulation was enhanced after a Santonian hiatus. Chert-rich chalk throughout the Indian Ocean indicates enhanced planktonic productivity. Sedimentation rates were high on Broken Ridge where tuffaceous limestones and chalk accumulated [Rea et al., 1990].

Climates remained cool at all southern sites, as indicated by the appearance of highly austral planktonic assemblages and the continued dominance of smectite in sediment. Kaolinite occurs only along the northwest Australian margin at Sites 763 and 765. Chlorite appears in the suite of nearby Site 762, an indicator of cool and/or dry climate.

Maestrichtian

Major structural changes occurred in the Indian Ocean during the Maestrichtian (Figures 6d and 7). The Ninetyeast Ridge lengthened as India drifted northward at an accelerated pace. As the ridge moved northward it subsided and deposition changed from shallow water, dolomitized reefal limestones to pelagic chalk (Figure 4). Kerguelen Plateau underwent more rifting which caused downfaulting of the Raggatt Basin and emplacement of a series of debris flows with basaltic clasts at Site 747. At Site 748, shallow-water biohermal limestones are abruptly overlain by pelagic chalks. The Somali Basin's oldest sediment consists of Maestrichtian red nannofossil clay intercalated in MORB basalt flows at Site 235.

The early Maestrichtian was regressive, with renewed siliciclastics, the Chikkim Shale, on the northwest Indian margin [Gaetani and Garzanti, 1991], and marine siliciclastics replaced by lacustrine deposits in the upper Ariyalur Formation of the Cauvery Basin. A late Maestrichtian transgression deposited the chalks of the Breton Beds in the Perth Basin. Chert-rich chalk, deposited on the plateaus since the Santonian, persisted through the Maestrichtian on Ninetyeast Ridge and the Kerguelen Plateau. Chert is also present in sediment at Sites 761 and 762 on the Exmouth Plateau, a further indication of enhanced radiolarian productivity. Ash is found in the calciclastic turbidites of Site 212 in the northern Wharton Basin. Enhanced sedimentation rates at this site and at Site 766 allowed preservation of carbonates at these sites, while other deep sites received only pelagic clay. Both halves of the Indian Ocean had opened by now, and the basaltic plateaus formed early in Indian Ocean history had subsided from sealevel during the mid-Cretaceous into pelagic depths, some 2000 m deep, by the Late Maestrichtian.

The Cretaceous/Tertiary boundary event is discussed by Pospichal and Huber [this volume].

SUMMARY

The north Indian Ocean opened with rifting of fragments from Gondwana forming the area now northwest of Australia during the Jurassic. The west opened during the Jurassic with movement of Africa from India, forming the Mozambique and Somali basins. The east Indian Ocean opened with India rifting from Australo-Antarctica to form the Wharton Basin during the Early Cretaceous. Hotspot activity during the Early Cretaceous formed Naturaliste Plateau and Kerguelen Plateau-Broken Ridge. As India moved northward hotspot activity formed the Ninetyeast Ridge, beginning in the Santonian-early Campanian. Prolonged volcanic activity has left abundant ash-rich sediment on Broken Ridge, the northwest Australian margin, and in the western deep basins.

During the Early Cretaceous, fluvio-deltaic sedimentation dominated the marginal basins and continental shelves, but high sealevel stands trapped sediment on the shelves and led to starved deep ocean basins. The mid-Cretaceous anoxic event left black shales in all deep and marginal basins with the exception of the Perth Basin, which may have been elevated above the anoxic water mass which caused the event. Sealevel rises were accompanied by extensive greensand deposition, particularly in the mid Cretaceous. During the mid and late Cretaceous high sealevel stand, pelagic chalk and marl formed on the shelves and in the western Australian and northwest Indian basins, particularly during the Santonian and late Campanian-Maestrichtian. During the Santonian particularly nutrient-rich waters promoted radiolarian growth and led to abundant chert formation within chalks on the plateaus and shelves. A major unconformity occurs in the early Campanian which Huber and Watkins [1992] attribute to the next phase of Gondwana breakup: the initiation of extensional movement which would lead to seafloor spreading between Australia, New Zealand, and Antarctica, and the northward drift of South America accompanied by high sea level allowed for circum-polar shallow marine communication. This led to the segregation of an austral water mass as reflected in the high degree of provincialism exhibited by calcareous micro- and nannofossils in late Campanian and Maestrichtian sediment.

Cool or temperate Aptian climate gave way to an Albian warming as indicated by the climax forest of podocarpaceaen conifers with tree and seed fern understory on Kerguelen Plateau and greatly increased kaolinite levels in sediment on the plateaus and in the deep basins. Climatic deterioration began in the Turonian and led ultimately to the development of a separate, cooler austral watermass by the late Campanian. A brief warming less than 500,000 years before the Cretaceous/Tertiary extinction is suggested by a negative $\delta^{18}O$ excursion on the Maud Rise and the migration of warm water calcareous planktonic taxa southward [Huber and Watkins, 1992].

Acknowledgments. We wish to thank Brian McGowran, Isabella Premoli-Silva and David Rea for constructive criticism of the text. James E. Thomas provided helpful editorial assistance. The Ocean Drilling Program provided funds for partial completion of this review. We also wish to acknowledge all of the people who have worked for the Deep Sea Drilling Project and the Ocean Drilling Program to provide the wealth of data now available on the Indian Ocean.

REFERENCES

Audley-Charles, M. G., Ballantyne, P. D., and Hall, R., Mesozoic-Cenozoic rift-drift sequences of Asian fragments from Gondwanaland, *Tectonophys., 155*, 317-330, 1988.

Barron, J., Larsen, B. et al., *Proc. ODP, Init. Repts., 119*, Ocean Drilling Program, College Station, TX, 1989.

Belford, D.J., Upper Cretaceous foraminifera from the Toolonga Calcilutite and Gingin Chalk, Western Australia, *Bureau of Mineral Resources, Australia, Bulletin 57*, 1-198, 1960.

Bitschene, P. R., Holmes, M. A., and Breza, J. R., Composition and origin of Cr-rich glauconitic sediments from the southern Kerguelen Plateau (Site 748), *Proc. ODP Sci. Results, 120*, 113-134, 1992.

Bukry, D., Phytoplankton stratigraphy, offshore East Africa, Deep Sea Drilling Project Leg 25, *Init. Repts. DSDP, 25*, 635-646, 1974.

Cook, H. E., Zemmels, I., and Matti, J. C., X-ray mineralogy data, eastern Indian Ocean - Leg 27, Deep Sea Drilling Project, *Init. Repts. DSDP, 27*, 535-548, 1974a.

Cook, H. E., Zemmels, I., and Matti, J. C., X-ray mineralogy data, southern Indian Ocean - Leg 26, Deep Sea Drilling Project, *Init. Repts. DSDP, 26*, 573-592, 1974b.

Cook, H. E., Zemmels, I., and Smith, J. C., X-ray mineralogy data, Austral-Antarctic region, Leg 28, Deep Sea Drilling Project, *Init. Repts. DSDP, 28*, 981-998, 1975.

Coffin, M. F., Subsidence of the Kerguelen Plateau: the Atlantis Concept, *Proc. ODP Sci. Results, 120*, 945-949, 1992.

Davies, T. A., and Kidd, R. B., Sedimentation in the Indian Ocean through time, in *Indian Ocean Geology and Biostratigraphy*, edited by J. R. Heirtzler, H. M. Bolli, T. A. Davies, J. B. Saunders, and J. G. Sclater, pp. 61-85, Am. Geophys. Union, Washington, 1977.

Davies, T. A., Luyendyk, B. P. et al., *Init. Repts. DSDP, 26*, U.S. Government Printing Office, Washington, D.C., 1974.

Dingle, R. V., Siesser, W. B., and Newton, A. R., *Mesozoic and Tertiary geology of southern Africa*, 375 pp., A. A. Balkema, Rotterdam, 1983.

Fisher, R. L., Bunce, E. T. et al., *Init. Repts. DSDP, 24*, U.S. Government Printing Office, Washington, D.C., 1974.

Förster, R., The geological history of the sedimentary basin of southern Mozambique, and some aspects of the origin of the Mozambique Channel, *Palaeogeog., Palaeoclim., Palaeoecol., 17*, 267-287, 1975.

Francis, J. E., and Coffin, M. F., Cretaceous fossil wood from the Raggatt Basin, southern Kerguelen Plateau (Site 750), *Proc. ODP Sci. Results, 120*, 273-280, 1992.

Fritsch, B., Schlich, R., Munschy, M., Fezga, F., and Coffin, M. F., Evolution of the Southern Kerguelen Plateau deduced from seismic stratigraphic studies and drilling at Sites 748 and 750, *Proc. ODP Sci. Results, 120*, 895-906, 1992.

Gaetani, M., and Garzanti, E., Multicyclic history of northern India continental margin (northwestern Himalya), *Amer. Assoc. Petrol. Geol. Bull., 75*, 1427-1446, 1991.

Gartner, S., Jr., Nannofossil biostratigraphy, Leg 22, Deep Sea Drilling Project, *Init. Repts. DSDP, 22*, 577-599, 1974.

Gradstein, F. M., Ludden, J. N., et al., *Proc. ODP, Init. Repts., 123*, Ocean Drilling Program, College Station, TX, 1990.

Gradstein, F. M., and von Rad, U., Stratigraphic evolution of Mesozoic continental margin and oceanic sequences: northwest Australia and northern Himalayas, *Mar. Geol., 102*, 131-173, 1991.

Haq, B. U., Hardenbol, J., and Vail, P. R., Mesozoic and Cenozoic chronostratigraphy and cycles of sea-level change, in *Sea-Level Changes and Integrated Approach*, edited by C. K. Wilgus, B. S. Hastings, C. G. St. C. Kendall, H. W. Posamentier, C. A. Ross, and J. C. Van Wagoner, pp. 71-108, Soc. Econ. Paleont. Mineral., Spec. Publ. No. 42, Tulsa,1988.

Haq, B. U., von Rad, U., O'Connell, S. et al., *Proc. ODP, Init. Repts., 122*, Ocean Drilling Program, College Station, TX, 1990.

Hayes, D. E., Frakes, L. A. et al., 1975, *Init. Repts. DSDP, 28*, U.S. Government Printing Office, Washington, D.C., 1975.

Herb, R., Cretaceous planktonic foraminifera from the Eastern Indian Ocean, *Init. Repts. DSDP, 26*, 745-769, 1974.

Holmes, M. A., Cretaceous subtropical weathering followed by cooling at 60°S latitude: the mineral composition of southern Kerguelen Plateau sediment, Leg 120, *Proc. ODP Sci. Results, 120*, 99-111, 1992.

Huber, B. T., and Watkins, D. K., *Biogeography of Campanian-Maestrichtian calcareous plankton in the region of the southern Ocean: paleogeographic and paleoclimatic implications*, in press, 1992.

Jones, B. F., and Galan, E., Sepiolite and palygorskite, in Hydrous phyllosilicates exclusive of micas, edited by S. W. Bailey, *Rev. Mineral., 19*, 631-674, 1988.

Kidd, R. B., and Davies, T. A., Indian Ocean sediment distribution since the Late Jurassic, *Mar. Geol., 26*, 49-70, 1978.

Kumar, R., Fundamentals of historic geology and stratigraphy of India, pp. 147-175, Halsted Press, New Delhi, 1985.

Larson, R. L., Mutter, J. C., Diebold, J. B., and Carpenter, G. B., Cuvier Basin: a product of ocean crust formation by early Cretaceous rifting off western Australia, *Earth Planet. Sci. Lett., 45*, 105-114, 1979.

Luyendyk, B. P., and Davies, T. A., Results of DSDP Leg 26 and the geologic history of the southern Indian Ocean, *Init. Repts. DSDP, 26*, 909-943, 1974.

Marchig, V., and Vallier, T. L., Geochemical studies of sediment and interstitial water, Sites 248 and 249, Leg 25, Deep Sea Drilling Project, *Init. Repts. DSDP, 25*, 405-415, 1974.

Matti, J. C., Zemmels, I., and Cook, H. E., *Init. Repts. DSDP, 25*, 843-861, 1974.

McRae, S. G., Glauconite, *Earth-Sci. Rev., 8*, 397-440, 1972.

McGowran, B., Foraminifera, *Init. Repts. DSDP, 22*, 609-627, 1974.

Michel, H.V., Asaro, F., and Alvarez, W., Geochemical study of the Cretaceous-Tertiary boundary region at Hole 752B, *Proc. ODP, Sci. Results, 121*, 415-422, 1991.

Millot, G., Geology of clays (Engl. trans.), 430 pp., Springer-Verlag, New York, 1970.

Mohr, B. A. R., and Gee, C. T., An early Albian palynoflora from the Kerguelen Plateau, Southern Indian Ocean (Leg 120), *Proc. ODP Sci. Results, 120*, 255-271, 1992a.

Mohr, B. A. R., and Gee, C. T., Late Cretaceous palynofloras (sporomorphs and dinocysts) from the Kerguelen Plateau, Southern Indian Ocean (Sites 748 and 750), *Proc. ODP Sci. Results, 120,* 281-306, 1992b.

Munschy, M., Dyment, J., Boulanger, M. O., Boulanger, D., Tissot, J. D., Schlich, R., Rotstein, Y., and Coffin, M. F., Breakup and seafloor spreading between the Kerguelen Plateau-Labuan Basin and Broken Ridge-Diamantina Zone, *Proc. ODP Sci. Results, 120,* 931-944, 1992.

Odin, G. S., and Fullagar, P. D., Geological significance of the glaucony facies, in *Green Marine Clays,* edited by G. S. Odin, pp. 295-332, Elsevier, Amsterdam, 1988.

Pimm, A. C., Sedimentology and history of the northeastern Indian Ocean from the Late Cretaceous to Recent, *Init. Repts. DSDP, 22,* 469-476, 1974.

Peirce, J., Weissel, J., et al., *Proc. ODP, Init. Repts., 121,* Ocean Drilling Program, College Station, TX, 1989.

Playford, P. E., Cockbain, A. E., and Low, G. H., Geology of the Perth Basin, *West. Aust. Geol. Survey, Bull. 124,* 311p., 1976.

Playford, P. E., Cope, R. N., Cockbain, A. E., Low, G. H., and Lowry, D. C., Phanerozoic, in *Geology of Western Australia, West. Aust. Geol. Survey, Mem. 2,* 223-433, 1975.

Pospichal, J. J., Calcareous nannofossils across the Cretaceous/Tertiary boundary at Site 752, Eastern Indian Ocean, *Proc. ODP, Sci. Results, 121,* 395-413, 1991.

Powell, C. M., Roots, S. R., and Veevers, J. J., Pre-breakup continental extension in East Gondwanaland and the early opening of the eastern Indian Ocean, *Tectonophys., 155,* 261-283, 1988.

Proto Decima, F., Leg 27 calcareous nannoplankton, *Init. Repts. DSDP, 27,* 589-621, 1974.

Rea, D. K., Dehn, J., Driscoll, N. W., Farrell, J. W., Janecek, T. R., Owen, R. M., Pospichal, J. J., Resiwati, P., and the ODP Leg 121 Scientific Party, Paleoceanography of the eastern Indian Ocean from ODP Leg 121 drilling on Broken Ridge, *Geol. Soc. Amer. Bull., 102,* 679-690, 1990.

Resiwati, P., Upper Cretaceous calcareous nannofossils from Broken Ridge and Ninetyeast Ridge, Indian Ocean, ODP Leg 121, *Proc. ODP Sci. Results, 121,* 141-170, 1991.

Roth, P. H., Calcareous nannofossil from the northwestern Indian Ocean, Leg 24, Deep Sea Drilling Project, *Init. Repts. DSDP, 24,* 651-766, 1974.

Royer , J.-Y., and Sandwell, D., Evolution of the Eastern Indian Ocean since the Late Cretaceous: constraints from Geosat altimetry, *J. Geophys. Res., 94,* 13,755-13,782, 1989.

Sastri, V. V., Sinha, R. N., Singh, G., and Murti, K. V. S., Stratigraphy and tectonics of sedimentary basins on east coast of peninsular India, *Amer. Assoc. Petrol. Geol. Bull., 57,* 655-678, 1973.

Sastri, V. V., Venkatachala, B. S., and Narayanan, V., The evolution of the east coast of India, *Palaeogeog., Palaeoclim., Palaeoecol., 36,* 23-54, 1981.

Scheibnerová, V., Aptian-Albian benthonic foraminifera from DSDP Leg 27, Sites 259, 260, and 263, Eastern Indian Ocean, *Init. Repts. DSDP, 27,* 697-718, 1974.

Schlich, R., Simpson, E. S. W., and Vallier, T. L., Regional aspects of Deep Sea Drilling in the western Indian Ocean, Leg 25, DSDP, *Init. Repts. DSDP, 25,* 743-759, 1974.

Schlich, R., Wise, S. W. Jr. et al., *Proc. ODP, Init. Repts., 120,* Ocean Drilling Program, College Station, TX, 1989.

Sclater, J. G., Abbott, D., and Thiede, J., Paleobathymetry and sediments of the Indian Ocean, in *Indian Ocean Geology and Biostratigraphy,* edited by J. R. Heirtzler, H.M. Bolli, T. A. Davies, J. B. Saunders, and J. G. Sclater. *Amer. Geophys. Union,* pp. 25-59, Washington, 1977.

Sclater, J. G., and Heirtzler, J. R., An introduction to Deep Sea Drilling in the Indian Ocean, in *Indian Ocean Geology and Biostratigraphy,* edited by J. R. Heirtzler, H.M. Bolli, T. A. Davies, J. B. Saunders, and J. G. Sclater, *Amer. Geophys. Union,* Washington, pp. 1-24, 1977.

Scotese, C. R., Gahagan, L. M., and Larson, R. L., Plate tectonic reconstructions of the Cretaceous and Cenozoic ocean basins, *Tectonophys. 155,* 27-48, 1988.

Ségoufin, J., and Patriat, P., Reconstructions de l'océan Indien occidental pour les époques des anomalies M21, M2 et 34, Paléoposition de Madagascar, *Bull. Soc. Géol. France, XXIII,* 603-607, 1981.

Shafik, S., Late Cretaceous nannofossil biostratigraphy and biogeography of the Australian western margin, *Bureau Mineral Resources, Geology Geophysics Report 295,* 164 p, Australian Government Publ. Service, Canberra, 1990.

Simpson, E.S.W., Schlich, R. et al., *Init. Repts. DSDP, 25,* 884 p., U.S. Govternment Printing Office, Washington, D.C., 1974.

Sliter, W. V., Cretaceous foraminifers from the southwestern Atlantic Ocean, Leg 36, Deep Sea Drilling Project, *Init. Repts. DSDP, 36,* 519-573, 1976.

Storey, M., Kent, R., Saunders, A. D., Salters, V. J., Hergt, J., Whitechurch, H., Sevigny, J. H., Thirlwall, M. F., Leat, P., Ghose, N. C. and Gifford, M., Lower Cretaceous volcanic rocks on continental margins and their relationship to the Kerguelen Plateau, *Proc. ODP, Sci. Res., 120,* 33-53, 1992.

Tait, A. M., and Smith, D. N., Cretaceous, in *Geology of the Carnarvon Basin - Western Australia, West. Austr. Geol. Bull. 133,* edited by R. M. Hocking, H. T. Moors, and J. E. van de Graaf, pp. 135-161, Department of Mines, Perth, 1987.

Thierstein, H. R., Calcareous nannoplankton - Leg 26, Deep Sea Drilling Project, *Init. Repts. DSDP 26,* 619-667, 1974.

Vallier, T. L., and Kidd, R. B., Volcanogenic sediments in the Indian Ocean, in *Indian Ocean Geology and Biostratigraphy,* edited by J. R. Heirtzler, H. M. Bolli, T. A. Davies, J. B. Saunders, and J. G. Sclater, pp. 87-118, Amer. Geophys. Union, Washington, 1977.

van de Graaf, W. J. E., Paleogeographic evolution of a rifted cratonic margin: S.W. Australia - Discussion, *Palaeogeog., Palaeoclim., Palaeoecol., 34,* 163-172, 1981.

van Eijden, A. J. M., and Smit, J., Eastern Indian Ocean Cretaceous and Paleogene quantitative biostratigraphy, *Proc. ODP, Sci. Results, 121,* 1991.

Veevers, J. J., Heirtzler, J. R. et al., *Init. Repts. DSDP, 27,* 77-123, 1974.

Venkatarathnam, K., Mineralogical data from Sites 211, 212, 213, 214, and 215 of Deep-Sea Drilling Project Leg 22 and origin of noncarbonate sediments in the equatorial Indian Ocean, *Init. Repts. DSDP, 22,* 489-501, 1974.

von der Borch, C. C., Sclater, J. G. et al., *Init. Repts. DSDP, 22,* U.S. Government Printing Office, Washington, D.C., 1974.

von der Borch, C. C., and Trueman, N. A., Dolomitic basal sediments from northern end of Ninetyeast Ridge, *Init. Repts. DSDP, 22,* 477-483, 1974.

Watkins, D. K., Upper Cretaceous nannofossils from ODP Leg 120, Kerguelen Plateau, Southern Ocean, *Proc. ODP Sci. Results, 120,* 343-370, 1992.

Watkins, D. K., Quilty, P. G., Mohr, B. A. R., Mao, S., Francis, J. E., Gee, C. T., and Coffin, M. F., Cretaceous Paleontology of the central Kerguelen Plateau, *Proc. ODP Sci. Results, 120,* 951-960, 1992.

Whitechurch, H., Montigny, R., Sevigny, J., Storey, M., and Salters, V. J. M., K-Ar and $^{40}Ar/^{39}Ar$ ages of central Kerguelen Plateau Basalts, *Proc. ODP Sci. Results, 120,* 71-77, 1992.

Wiseman, J. F., and Williams, A. J., Palynological Investigation of samples from Sites 259, 261, and 263, Leg 27, Deep Sea Drilling Project, *Init. Repts. DSDP, 27,* 915-924, 1974.

Withers, A. T., The occurrence of the crinoid Uintacrinus in Australia, *J. Royal Soc. Western Australia, 11,* 15-18, 1924.

Withers, A. T., The crinoid Marsupites in the Upper Cretaceous of Western Australia, *J. Royal Soc. Western Australia, 12,* 97-100, 1926.

Aptian-Albian Calcareous Nannonfossil Biostratigraphy of ODP Site 763 and the Correlation Between High- and Low-Latitude Zonations

TIMOTHY J. BRALOWER

Department of Geology, University of North Carolina, Chapel Hill, NC, 27599-3315, USA

Recent studies of expanded Aptian-Albian sections from a range of latitudes and from the Indian, and North and South Atlantic Ocean basins have produced higher resolution calcareous nannofossil zonation schemes for this interval. Integration of high- and low-latitude calcareous nannofossil biostratigraphies is proposed based on the study of ODP Site 763 located on the central Exmouth Plateau. This site was situated in temperate latitudes in the Aptian-Albian ages.

INTRODUCTION

Original calcareous nannofossil zonations for the Aptian and Albian stages were developed 20 years ago in sections from Tethyan regions [e.g., Manivit, 1971; Thierstein, 1971]. Although nannofossil assemblages in high-latitude Aptian-Albian sections are very different from those in the Tethyan Realm [e.g., Black, 1973; Wise and Wind, 1977], zonal markers are present in practical abundances. These original biostratigraphic schemes do not, however, offer sufficient stratigraphic resolution to help solve the detailed paleoceanographic and paleoenvironmental problems which abound in the Aptian-Albian interval. Recent studies have therefore attempted to establish a series of subzonal markers and biohorizons to provide additional resolution [e.g., Wise, 1983; Applegate and Bergen, 1988; T. J. Bralower et al., unpublished manuscript, 1992]. At this scale, however, many stratigraphically useful taxa are restricted to high- or low-latitude areas. The Aptian and Albian ages lasted approximately 27 m.y. and thus account for over one third of the Cretaceous Period [Harland et al., 1990]. Because of the duration of this interval, the absence of magnetic reversals for correlation, and the host of paleoceanographic problems which exist, at this time when the earth was close to maximum "greenhouse" state, the availability of accurate, high-resolution biostratigraphy is extremely important.

An expanded, apparently complete upper Aptian-Albian section was recovered at Site 763, drilled on the central Exmouth Plateau during ODP Leg 122. This site, which was located between 40° and 50° S in the Aptian-Albian, contains a highly diverse temperate nannoflora with a mixture of high and low-latitude markers, and therefore allows an accurate correlation of the ranges of subzonal and other markers between the two areas.

MATERIAL AND METHODS

This investigation is based upon compilation of previous calcareous nannofossil biostratigraphies as well as

Synthesis of Results from Scientific
Drilling in the Indian Ocean
Geophysical Monograph 70

selective reinvestigation of parts of particular sections. The centerpiece of this study is the Aptian-Albian section at ODP Site 763 situated on the central Exmouth Plateau (Figure 1). A thick upper Aptian-Albian sequence was recovered in Hole 763B corresponding to the Gearle Siltstone unit onshore Australia. This section is predominantly composed of green-grey calcareous claystone or marl, and although recovery was excellent (87%), the material is biscuited (regular horizons disturbed by drilling) and thus not all of it is useful. Sedimentation rates average 10 m/m.y. and therefore the section is suitable for higher resolution biostratigraphic studies. Calcareous nannofossil preservation is dominantly excellent, but decreases to moderate in a few intervals. Detailed biostratigraphy of this section is described in Bralower and Siesser [1992].

The biostratigraphy of Site 763 is compared with several other expanded and relatively complete upper Aptian-Albian sequences from high and low-latitude DSDP sites (Figure 1). An expanded upper Aptian-Albian section was recovered at DSDP Site 545 on the continental slope seaward of the Mazagan Escarpment [Hinz, Winterer et al., 1984]. This section consists predominantly of green nannofossil claystone with very minor thin intervals of conglomerate and black shale. The sediments recovered are largely homogeneous but contain slump folds, microfaults and low-angle sliding surfaces which have led to some stratigraphic repetition and omission. A fairly major unconformity exists towards the base of Core 40 corresponding to part of the middle-upper Albian [Leckie, 1984; T. J. Bralower et al., unpublished manuscript, 1992]. Several minor slump blocks occur in Cores 38 and 39 but these are more observable in planktonic foraminiferal than calcareous nannofossil biostratigraphy. Recovery in the upper Aptian-Albian part of Site 545 averaged 73%. Moderate and well preserved calcareous nannofossils occur throughout the section [Wiegand, 1984]. If extreme care is taken with structural problems in this hole, the preservation and high sedimentation rates (10-20 m/m.y.) allow detailed investigation of the relative order of events in a low-latitude setting.

LATE ALBIAN

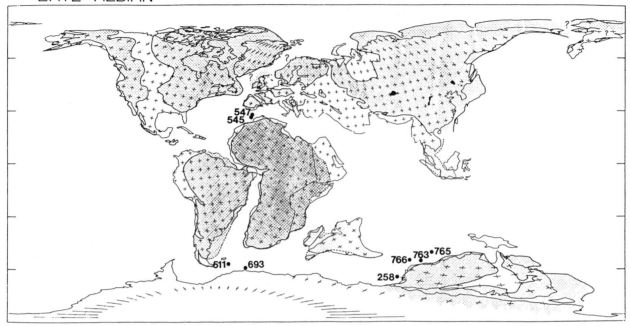

Fig. 1. Location of sections investigated and others mentioned in the text. Paleogeographic reconstruction of Barron et al. [1981].

A more continuous upper Albian section was recovered at DSDP Site 547, located further down the Mazagan Slope. This section contains claystone with frequent intervals of conglomerate, and was deposited at higher sedimentation rates (>30 m/m.y.) than Site 545. Study of this section complements that of Site 545 in that it contains complete intervals correlating to the omitted and repeated upper Albian parts of the latter site. Two holes were drilled at Site 547. The lower part of the upper Albian was recovered in Hole 547B, and the overlying, remainder of the upper Albian was obtained in Hole 547A. Recovery in these holes averaged 56%. Nannofossil biostratigraphy at Sites 545 and 547 was studied by Wiegand [1984]. The calcareous nannofossils in these sites have been reinvestigated paying detailed attention to the ranges of non-zonal markers.

DSDP Site 511 was drilled on the Falkland Plateau where an Aptian-Albian section deposited at high (~65° S) latitude was recovered [Ludwig, Krasheninnikov et al., 1983]. The section consists predominantly of brown, green, red and grey calcareous claystone and chalk. Minor intervals of black shale are interbedded in the section. This sequence appears to be continuous and had high recovery (averaging 88%). A detailed calcareous nannofossil investigation of Site 511 was carried out by Wise [1983]. The present study has concentrated on the ranges of selected nannofossil marker taxa.

Smear slides were prepared of all samples using standard preparation methods. Because nannofossils are the major component of most samples, no concentration techniques were employed. All samples were inspected briefly in the light microscope and the best preserved sample per section

was chosen for detailed biostratigraphic investigations. The taxonomy of most species observed is standard and described in more detail by various authors [e.g., Thierstein, 1973; Wise and Wind, 1977; Perch-Nielsen, 1985; Covington and Wise, 1987; Crux, 1989].

Thickly concentrated smear slides were prepared so that even the rarest of occurrences could be noted. Preservation of nannofossils ranges from poor to excellent in the investigated sections. Special attention was paid to samples in critical stratigraphic intervals, especially near the ends of species ranges, where slides were observed for several hours. Range charts for the sections investigated are tabulated in Bralower and Siesser [1992, Site 763] and Bralower et al. [unpublished manuscript, 1992, other sites].

APTIAN-ALBIAN CALCAREOUS NANNOFOSSIL BIOSTRATIGRAPHY OF ODP SITE 763

A highly diverse calcareous nannoflora was observed in upper Aptian-Albian samples from Hole 763B. Biostratigraphic results indicate that the entire Albian stage was recovered at this site, corresponding to over 80 m of section. The number of species averages 30-35 per sample and is fairly constant through the section. Although detailed studies of nannofossil assemblages have not been conducted, there do not appear to be any dramatic fluctuations in the relative abundances of taxa with Tethyan and Austral affinities. This indicates that Site 763 was located in a fairly stable part of the temperate realm throughout the late Aptian-Albian, providing an ideal opportunity to calibrate the ranges of high and low latitude markers.

TABLE 1. Sub-bottom Depths (m) of Nannofossil Events in DSDP/ODP Sites

NANNOFOSSIL EVENT	Site 511	Site 545	Site 547	Site 763
base *Corollithion kennedyi*			621.96	424.29
base *Gartnerago nanum*				457.80
top *Hayesites albiensis*		360.83	715.87	475.30
top *Parhabdolithus infinitus*		377.18	755.05	440.30
base *Eiffellithus turriseiffelii*	436.42	360.83	757.81	457.80
base *Gartnerago striatum*		374.18	768.10	462.30
top *Rucinolithus irregularis*		370.12		
base *Cribrosphaerella ehrenbergii*		376.18		462.30
base *Eiffellithus trabeculatus*		374.18		465.25
base *Eiffellithus* cf. *E. eximius*		376.18		478.30
base *Broinsonia signata*				484.75
base *Braarudosphaera stenorheta*		376.18		
base *Axopodorhabdus albianus*		376.18		486.32
base *Percivalia hauxtonensis*		379.83		486.32
top *Sollasites falklandensis*	444.85			484.50
base *Tranolithus orionatus*	447.43	374.18		494.28
base *Corollithion signum*		376.18		
base *Hayesites albiensis*	506.64	417.87		517.80
base *Sollasites falklandensis*	481.30			527.27
base *Prediscosphaera columnata*	490.80	441.17		528.75
base *Seribiscutum primitivum*	471.80			543.70
base *Prediscosphaera spinosa*	506.64	446.61		524.28
base *Parhabd. achlyostaurion*		495.13		527.27

Potentially useful markers from this interval have been compiled in Table 1. These include species which have abundance peaks in high- and/or temperate latitude regions: *Seribiscutum primitivum, Sollasites falklandensis, Corollithion kennedyi, Eiffellithus* cf. *E. eximius* and *Percivalia hauxtonensis*. These taxa provide both first and last occurrence events in the upper Aptian-Albian. The ranges of *S. primitivum* and *E.* cf. *E. eximius* are by far the most continuous in Hole 763B; the other species having patchier distributions toward the ends of their ranges. Other temperate species with less clear biostratigraphic potential include: *Gartnerago striatum* (base), and *Octocyclus magnus* (top). Potentially useful species with a Tethyan or cosmopolitan distribution include the zonal markers, *Prediscosphaera columnata* and *Eiffellithus turriseiffelii* (both bases) and also *Axopodorhabdus albianus, Tranolithus orionatus* (both bases) and *Hayesites albiensis* (base and top) (Table 1). Other Tethyan/Cosmopolitan events with less demonstrated potential include *Cribrosphaerella ehrenbergii* (base), *Eiffellithus trabeculatus* (base), *Corollithion signum* (base), *Rucinolithus irregularis* (top), *Broinsonia signata* (base), *Prediscosphaera spinosa* (base) and *Parhabdolithus achlyostaurion* (base).

Nannofossil assemblages in Aptian-Albian sediments from Hole 763B are characterized by overall unusually small size. The significance of coccolith size is not understood in modern assemblages, therefore the origin of size variability in the ancient record cannot be interpreted. A second interesting facet of the biostratigraphic results from Hole 763B is that there appears to have been an interval of rapid nannofossil turnover in the middle and late

Albian (Table 1, Figure 2). This was predominantly a period of diversification with the evolution of several new, important taxa, some of which became major components of Late Cretaceous assemblages.

THE APTIAN-ALBIAN RECORD AT OTHER DSDP SITES
DSDP Site 511

Calcareous nannofossil biostratigraphy of Site 511 was studied by Wise [1983]. The upper Aptian-Albian part of this section has been reinvestigated in detail (Table 1) and remarkably similar ranges to those of Wise [1983] have been found with the large majority of events lying within two meters of his reported results. The ranges of the few marker species which differ from the results given by Wise [1983] by more than two sections (approximately three meters) include the first occurrence of *Seribiscutum primitivum* (Section 511-55-6 in our study versus 511-55-3 in Wise [1983]). In addition, Wise [1983] did not distinguish *Prediscosphaera spinosa* or *Hayesites albiensis*, the first occurrences of which were determined to lie in Section 511-57-6, over fifteen meters below the first occurrence of *Prediscosphaera columnata* (*P. cretacea* in Wise [1983]). This interval is largely barren, thus accounting for the rather large discrepancy.

DSDP Site 545

An entire upper Aptian-Albian section was recovered in DSDP Site 545 except for a significant unconformity in the middle Albian near the base of Core 545-40 (Table 1). Several disparities occur between the results of this reinvestigation of the Albian portion of Sites 545 and 547 and the ranges given by Wiegand [1984]. In particular,

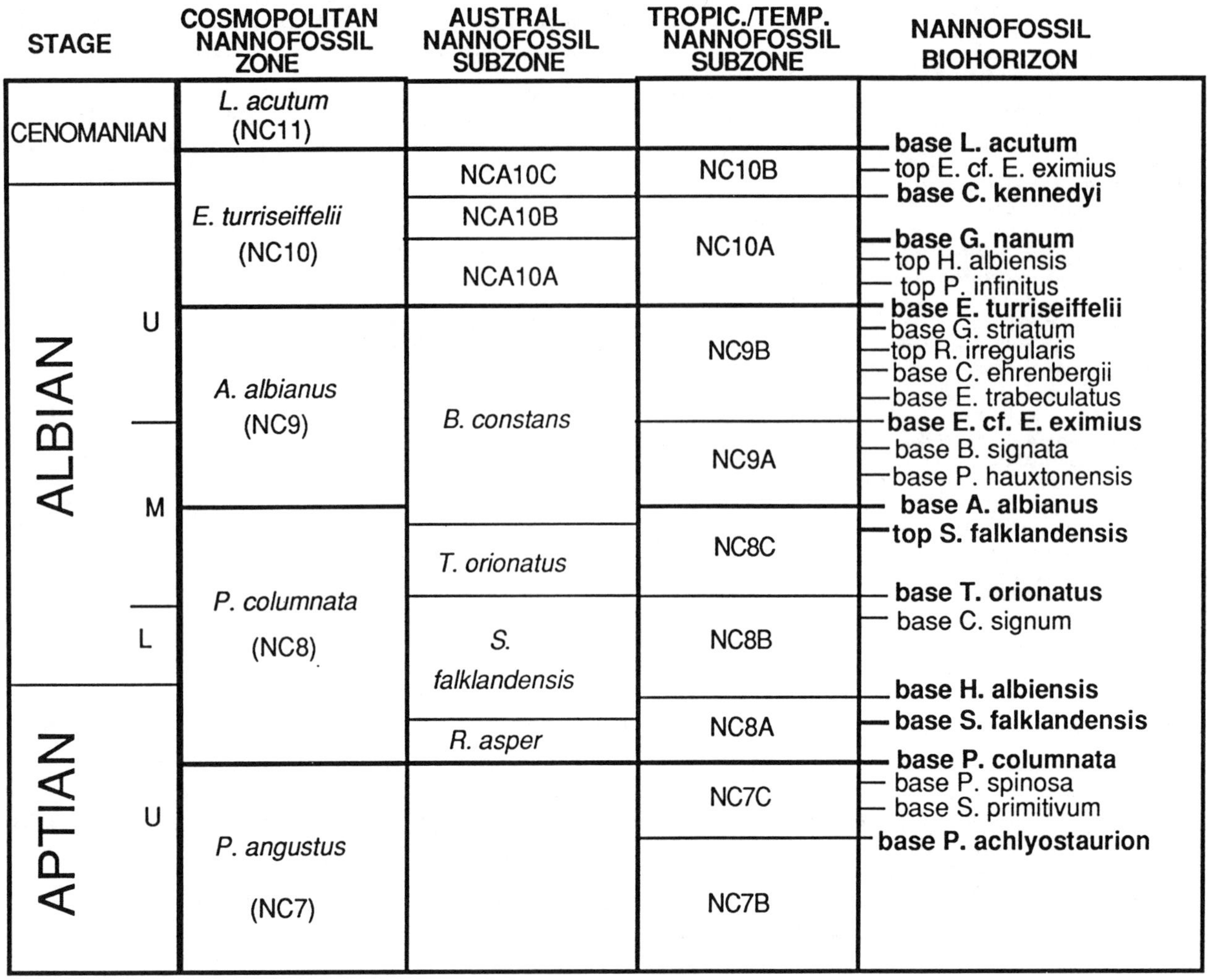

STAGE			COSMOPOLITAN NANNOFOSSIL ZONE	AUSTRAL NANNOFOSSIL SUBZONE	TROPIC./TEMP. NANNOFOSSIL SUBZONE	NANNOFOSSIL BIOHORIZON
CENOMANIAN			*L. acutum* (NC11)			— **base L. acutum**
ALBIAN	U		*E. turriseiffelii* (NC10)	NCA10C	NC10B	— top E. cf. E. eximius — **base C. kennedyi**
				NCA10B	NC10A	— **base G. nanum** — top H. albiensis
				NCA10A		— top P. infinitus
			A. albianus (NC9)	*B. constans*	NC9B	— **base E. turriseiffelii** — base G. striatum — top R. irregularis — base C. ehrenbergii — base E. trabeculatus
	M				NC9A	— **base E. cf. E. eximius** — base B. signata — base P. hauxtonensis
						— **base A. albianus**
			P. columnata (NC8)	*T. orionatus*	NC8C	— **top S. falklandensis**
	L					— **base T. orionatus** — base C. signum
				S. falklandensis	NC8B	— **base H. albiensis**
				R. asper	NC8A	— **base S. falklandensis**
APTIAN	U		*P. angustus* (NC7)		NC7C	— **base P. columnata** — base P. spinosa — base S. primitivum
						— **base P. achlyostaurion**
					NC7B	

Fig. 2. Proposed late Aptian-Albian calcareous nannofossil biostratigraphy [after T. J. Bralower et al., unpublished manuscript, 1992] and correlation with Austral subzonal scheme of Wise and Wind [1977] and Wise [1983]. Events shown in bold are zonal/subzonal markers. Relative durations of substages are not drawn to scale.

Wiegand [1984] recognized sporadic occurrences of *Eiffellithus turriseiffelii* in Core 545-40, whereas I have not observed this species below Section 545-39-1, similar to the continuous range of Wiegand [1984]. Wiegand [1984] found rare occurrences of *Tranolithus orionatus* down to Section 545-44-1, whereas I have not observed this species below Section 545-40-4. Significant differences were also observed in the events of the following rare taxa: last occurrence of *Rucinolithus irregularis* (Section 545-40-1 here, Section 545-38-3 in Wiegand [1984]) and the first occurrence of *Braarudosphaera stenorheta* (Section 545-40-5 here, Section 545-39-4 in Wiegand [1984]). Minor differences exist between the levels of several other events determined here and in Wiegand [1984]. These include the first occurrence of *Hayesites albiensis* (Section 545-45-1 here, 45-4 in Wiegand [1984]), and the last occurrence of *H.*

albiensis (Section 545-39-1 here, 38-3 in Wiegand [1984]). In addition, Wiegand [1984] did not report the following taxa: *Prediscosphaera spinosa, Corollithion kennedyi, Parhabdolithus achlyostaurion, Eiffellithus* cf. *E. eximius* and *Percivalia hauxtonensis.* Some very similar ranges exist between the two studies. These include the first occurrences of *Prediscosphaera columnata, Axopodorhabdus albianus, Cribrosphaerella ehrenbergii,* and *Corollithion signum.* All of these events lie within two sections (3 meters) in the two investigations.

DSDP Site 547

An entire upper Albian sequence was recovered in combined Holes 547A and 547B. This lies above the interval in which there is rapid turnover of nannofossil species. However, there are significant differences in the levels of those events which do occur between the results obtained by Wiegand [1984] and those found here. In particular, the first occurrence of *Eiffellithus turriseiffelii* was found to lie in Section 547B-4-4, whereas Wiegand [1984] found this event to lie in Section 5476B-6-1. In addition, the last occurrence of *Hayesites albiensis* was found to lie in Section 547A-63-1 by Wiegand [1984] and in Section 547B-3-2 in this investigation. I have also determined the meter levels of events which were not previously detected: the first occurrence of *Corollithion kennedyi* (Section 547A-60-1, although there is a gradual transition between this species and *C. signum,* in Cores 60-64), and the last occurrences of *Eiffellithus* cf. *E. eximius* (Section 547A-60-1) and *Percivalia hauxtonensis* (Section 547B-3-1).

OTHER APTIAN-ALBIAN SEQUENCES FROM THE INDIAN OCEAN

A summary of Aptian-Albian sediments recovered by early legs of DSDP in the Indian Ocean was given by Thierstein [1977]. Albian sediments were cored in DSDP Sites 256 and 257 from the Wharton Basin. These sections are not useful from a biostratigraphic sense as the recovered Albian section is very thin. A reasonably thick middle and upper Albian sequence was spot cored at DSDP Site 258 on the Naturaliste Plateau. Assemblages at this site show a mixture of temperate and high-latitude affinities comparable to those from the Exmouth and Falkland Plateaus [e.g., Wise and Wind, 1977; Bralower and Siesser, 1992]. Although the poor sediment recovery at Site 258 renders it inadequate for detailed biostratigraphic studies, the order of first and last occurrences of both high- and low-latitude zonal and subzonal markers is similar to that obtained at other investigated sites. Apparently contemporaneous sections were recovered from Sites 260 and 263 from the Wharton Basin, and preliminary biostratigraphic studies were carried out by Proto Decima [1974].

A lower-middle Albian section was recovered in ODP Site 765 in the Argo Abyssal Plain [Mutterlose, 1992]. This section is stratigraphically below the middle Albian interval of major nannofossil turnover and thus is not useful from a biostratigraphic viewpoint. A contemporaneous Albian sequence was recovered at Site 766 at the base of the Exmouth Plateau. The biostratigraphy of this site [Mutterlose, 1992] as well as Southern Ocean Site 693 [Mutterlose and Wise, 1990] indicates that the first occurrence of *Seribiscutum primitivum* lies below the first occurrence of *Prediscosphaera columnata,* in a similar order to Site 763, but the reverse order from Site 511, in which the first occurrence of *S. primitivum* lies several meters above that of *P. columnata* [Wise, 1983; T. J. Bralower et al., unpublished manuscript, 1992].

DISCUSSION

Aptian-Albian Calcareous Nannofossil Biostratigraphy

Calcareous nannofossil zonations have developed over the last twenty years, however, the early schemes of Thierstein [1971, 1973], Manivit et al. [1977], Sissingh

[1977] and Roth [1978] are still widely applied. The Sissingh [1977] zonation combines zonal units from other schemes in this interval and will not be discussed any further. The schemes of Thierstein [1971, 1973] and Manivit et al. [1977] contain the upper Aptian-upper Albian *Prediscosphaera cretacea* (*P. columnata*) Zone (defined as the interval between the first occurrences of the nominate taxon and *Eiffellithus turriseiffelii*) and the upper Albian and lower Cenomanian *Eiffellithus turriseiffelii* Zone (defined as the interval between the first occurrences of the nominate taxon and *Lithraphidites acutum*). The Roth [1978] scheme is similar but contains an additional zone in the middle and lower upper Albian, between the *P. cretacea* and *E. turriseiffelii* Zones, the *Axopodorhabdus albianus* Zone, the lower boundary of which corresponds to the first occurrence of the nominate species (Figure 2). Erba [1988] proposed a zonation scheme based on studies of sections from the Umbrian Apennines of Italy, including two new zones: the *Nannoconus regularis* Zone, in between the *Parhabdolithus angustus* and *Prediscosphaera columnata* Zones, the lower boundary of which is based on the first occurrence of the nominate taxon, and the *Parhabdolithus achlyostaurion* Zone between the *P. columnata* and *E. turriseiffelii* Zones, the base of which is also defined by the first occurrence of the nominate species.

Aptian-Albian nannofossil zones have been demonstrated to be widely applicable units and have been used in numerous subsequent studies of land, DSDP and ODP sections. Wise and Wind [1977] and Wise [1983] have proposed several subzonal divisions of Aptian-Albian nannofossil zones for high latitude sections based on results from Falkland Plateau Sites 327 and 511. These units include division of the *P. cretacea* (*P. columnata*) Zone into four subzones: the *Rhagodiscus asper* Subzone, defined as the interval between the first occurrences of *P. cretacea* (*P. columnata*) and *Sollasites falklandensis*; the *S. falklandensis* Subzone, corresponding to the interval between the first occurrences of the nominate species and *Tranolithus orionatus*; the *T. orionatus* Subzone, which is the interval between the first occurrence of the nominate taxon and the last occurrence of *S. falklandensis* and the *Biscutum constans* Subzone which corresponds to the interval between the last occurrence of *S. falklandensis* and the first occurrence of *E. turriseiffelii* (Figure 2).

Manivit et al. [1977] and Applegate and Bergen [1988] proposed subzonal units for tropical/temperate sites based on results from European/North African land sections and Site 641 on the Galicia Margin, respectively. Manivit et al. [1977] divided the *E. turriseiffelii* Zone into the *Hayesites albiensis* and *Prediscosphaera spinosa* Subzones based on the last occurrence of *H. albiensis*. Applegate and Bergen [1988] divided the *P. cretacea* (*P. columnata*) Zone, defined as in Thierstein [1971] into two subzones: the *Braarudosphaera africana* Subzone and the *T. phacelosus* Subzone, by the first occurrence of *T. phacelosus* (*T. orionatus*). These authors and Wiegand [1984] divided the *E. turriseiffelii* Zone in a similar fashion to Manivit et al. [1977]. Results obtained here indicate that the last occurrence of *H. albiensis* may be somewhat time-transgressive as this event lies below the first occurrence of *E. turriseiffelii* in Hole 763B, but above this event in Hole 547A.

The advantage of defining subzonal units is that it provides additional resolution to zonation in a formal fashion while being entirely consistent with previously defined zonal units. Here, these previous subzonal schemes are analyzed based upon studies of several sites. Formal definitions of new temperate-tropical subzones (Figure 2) are given in T. J. Bralower et al. [unpublished manuscript, 1992]. The Austral subzones proposed by Wise and Wind [1977] and Wise [1983] are not altered. The basis of the Roth [1978] zonation is adopted, including the *P. columnata* (NC8), the *A. albianus* (NC9) and the *E. turriseiffelii* (NC10) Zones. The *T. phacelosus* Subzone of Applegate and Bergen [1988] is divided into two subzones by the first occurrence of *H. albiensis*. Division of the *A. albianus* Zone into two subzones is proposed based on the first occurrence of *E.* cf. *E. eximius* and the *E. turriseiffelii* Zone into two subzones based on the first occurrence of *Corollithion kennedyi*. The correlation of Austral and temperate/tropical subzones proposed in Figure 2 is based largely on the biostratigraphic results from Hole 763B in which all of these units can be identified. Several of the marker species utilized are very rare or absent in high-latitude sites. These include *H. albiensis*, *A. albianus* and *E.* cf. *E. eximius*.

There is little agreement as to the stratigraphic order of the numerous subsidiary events which occur in the Albian. An objective of this investigation was to see whether these events could be ordered in a consistent fashion and therefore be applied in a higher resolution stratigraphy. In the following, the biostratigraphic potential of 12 biohorizons in this interval as determined by this investigation is discussed (Figure 2). These biohorizons include: the first occurrences of *Parhabdolithus achlyostaurion*, *Seribiscutum primitivum* and *Prediscosphaera spinosa* (all upper Aptian), *Broinsonia signata*, *Gartnerago striatum*, *Corollithion signum*, *Percivalia hauxtonensis* (lower and middle Albian), *Eiffellithus trabeculatus* and *Cribrosphaerella ehrenbergii* (both upper Albian) and the last occurrences of *Rucinolithus irregularis* and *Parhabdolithus infinitus* (both upper Albian).

Several of these markers (e.g. *E. trabeculatus*, *B. signata* and *P. infinitus*) occur too sporadically towards the end of their ranges to be potentially useful. Most of the other biohorizons of interest are largely restricted to either tropical/temperate or to high-latitude areas. The upper Albian first occurrence of *G. striatum*, a species restricted to temperate and tropical areas, has been observed in Hole 763B and Site 545. *S. primitivum* is largely restricted to temperate and high-latitude areas. The earlier occurrence of this taxon with respect to *P. columnata* in temperate sites investigated indicates that this event may be time-transgressive. *P. hauxtonensis* is largely restricted to temperate and Boreal areas, with rare occurrences in

Tethyan sections, and has not been observed on the Falkland Plateau or other Austral sites (e.g. DSDP Site 258). The level of first occurrence of this species has only been observed in detail in Hole 763B, thus the usefulness of this event has not been fully demonstrated. The first occurrence of *P. spinosa*, slightly precedes that of *P. columnata* in a few locations including Sites 511 and 545, but interestingly, not in Hole 763B. The other events, including the first occurrences of *P. achlyostaurion*, *C. ehrenbergii* and *C. signum* and the last occurrences of *R. irregularis*, species which are either very rare or absent in Austral sites, are potentially useful in temperate/tropical sequences, and their proposed order (Figure 2) is based on the consistent results from Sites 763, 545 and 547. The first occurrence of *P. achlyostaurion* occurs somewhat below the level proposed by Erba [1988] but is in a variable position in Hole 763B and Site 545.

Accurate detection of stage boundaries is very difficult in DSDP/ODP sites due to the absence of standard macrofossil taxa used in definitions. I therefore rely on indirect correlations based on microfossil biostratigraphies of stratotype sections. The Aptian/Albian boundary is very difficult to define using calcareous nannofossil biostratigraphy since there appears to be no significant events in this time interval. Thierstein [1971, 1973] placed the base of the *Prediscosphaera cretacea* (*P. columnata*) Zone in the lower Albian, based on studies of stratotype and parastratotype sections. Bréhéret et al. [1986] and Delamette et al. [1986] have subsequently modified this event to lie within the uppermost Aptian based on the study of more complete parastratotype sections. The age of this event, and therefore the nannofossil biostratigraphic definition of the Aptian/Albian boundary are still somewhat in question. In terms of the three basic nannofossil zones utilized here, the lower-middle Albian boundary lies within the *P. columnata* Zone (probably within the NC8B subzone), and the middle-upper Albian boundary lies close to the base of the *Eiffellithus turriseiffelii* Zone [Thierstein, 1973]. The Albian/Cenomanian boundary is also difficult to define using nannofossil biostratigraphy. Perhaps the closest event to this boundary is the first occurrence of *Corollithion kennedyi*, which lies in the lowest ammonite zone of the Cenomanian in Northern Europe [Crux, 1982].

An interesting observation which can be made based on the results of this investigation is that the middle and early late Albian appears to have been a time of rapid diversification of calcareous nannoplankton, some of the highest evolutionary rates in the middle Cretaceous.

APTIAN-ALBIAN PALEOENVIRONMENTS IN THE INDIAN OCEAN

There are two general conclusions which can be made regarding the Aptian-Albian paleoenvironment of the Indian Ocean. The first is that biogeographic gradients in calcareous nannofossil assemblages existed and appear to be within the same range of magnitude as those of the Late Cretaceous. Biogeography of mid-Cretaceous nannofossil assemblages from the Indian Ocean was investigated by

Roth and Krumbach [1986]. However, in their analysis, they only included two sites, 258 and 259 from the Naturaliste Plateau and Wharton Basin, respectively, with very little latitudinal contrast. With the current set of Indian and Southern Ocean sites, much greater latitudinal variability exists. Several species in this interval are largely endemic to temperate and high latitude (ie >40°) sites. These include *Seribiscutum primitivum* and *Sollasites falklandensis*. The former taxon can compose up to 25% of the assemblage of some samples [Roth and Krumbach, 1986; Mutterlose and Wise, 1990]. Conversely, numerous taxa exist which are endemic to low and temperate latitudes, including *Rucinolithus irregularis* and *Eiffellithus* cf. *E. eximius*. Clearly some of the highest diversities observed, in terms of the number of species, are in temperate-latitude sites such as Site 763. There are distinct differences between high-latitude Boreal and Austral floras in the Aptian-Albian. Certain taxa, common in the Boreal Aptian-Albian, such as *Nannoconus* and *Phanulithus anfractus*, have not been found nearly as commonly in the Austral realm. Conversely, Austral species such as *Biscutum dissimilis* and *S. falklandensis*, have not been observed in the Boreal realm [J. Crux, personal communication, 1990].

The second interesting observation which can be made is that there is a general absence of organic carbon-rich sediments in the late Aptian-Albian of the Indian Ocean. The highest organic carbon contents of 2-3% occur in middle and upper Albian sediments of DSDP Site 258. However, there are no organic-rich (>1% TOC) sediments in the entire late Aptian-Albian sequence of ODP Site 763, or in the lower Albian portions of ODP Sites 765 and 766 [Haq, von Rad et al., 1990; Gradstein, Ludden et al., 1990]. A possible reason for this is that none of the Indian Ocean sites were in an upwelling setting in this interval. ODP Site 763 was on the continental shelf, several hundred kilometers from land and at a latitude in which upwelling is rare. Sites 765 and 766 were below the continental shelf. The generally modest organic carbon contents at Site 258, located at a more likely upwelling latitude than the other sites further to the south, are also perplexing.

The Aptian-Albian is an interval in which deep sea drilling sites from the Atlantic and Pacific Oceans, and land sections from several areas are characterized by considerable thicknesses of organic carbon-rich sediments [e.g., Schlanger and Jenkyns, 1976]. At several sites, these sediments cluster in the early Albian and late middle Albian interval [T. J. Bralower et al., unpublished manuscript, 1992]. The general lack of organic carbon-rich late Aptian-Albian sediments in the Indian Ocean is certainly an interesting phenomenon, one which will benefit from future drilling in this region.

Acknowledgments. The author is grateful to W. Siesser for shipboard biostratigraphic collaboration. I acknowledge thoughtful comments of K. von Salis and an anonymous reviewer. I thank C. Botelho, C. Carr and L. Sierra for lab assistance. This research was supported by United States Science Advisory Committee.

REFERENCES

Applegate, J. L. and Bergen, J. A., Cretaceous calcareous nannofossil biostratigraphy of sediments recovered from the Galicia Margin, ODP Leg 103, *Proc. ODP, Init. Repts.*, *103*, 293-319, 1988.

Barron, E. J., Harrison, C. G., Sloan, J. L. and Hay, W. W., Paleogeography, 180 million years to the present, *Eclogae Geol. Helv.*, *74*, 443-470, 1981.

Black, M., British Lower Cretaceous Coccoliths. 1. Gault Clay, Part 2. *Palaeont. Soc. Monograph*, 49-112, 1973.

Bralower, T. J. and Siesser, W. G., Cretaceous calcareous nannofossil biostratigraphy of ODP Leg 122 Sites 761, 762 and 763, Exmouth and Wombat Plateaus, N.W. Australia, *Proc. ODP, Init. Repts.*, *122*, 529-556, 1992.

Bréhéret, J. G., Caron, M. and Delamette, M., Niveaux riches en matière organique dans l'Albien vocontien; quelques caractères du paléoenvironment; essai d'interprétation génétique. *Doc. Bur. Rech. Géol. Min.*, *110*, 141-191, 1986.

Covington, J. M. and Wise, S. W., Jr., Calcareous nannofossil biostratigraphy of a Lower Cretaceous deep sea fan complex: DSDP Leg 93, Site 603, lower continental rise off Cape Hatteras, U.S.A., *Init. Repts. DSDP, 93*, 617-660, 1987.

Crux, J. A., Upper Cretaceous (Cenomanian to Campanian) calcareous nannofossils, in *A Stratigraphical Index of Calcareous Nannofossils*, edited by A. R. Lord, pp. 81-135, Ellis Horwood, Chichester, 1982.

Crux, J. A., Biostratigraphy and paleogeographical applications of Lower Cretaceous nannofossils from north-western Europe, in *Nannofossils and Their Applications*, edited by J. A. Crux and S. E. van Heck, pp. 143-211, Ellis Horwood, Chichester, 1989.

Delamette, M., Caron, M., and Bréhéret, J. G., Essai d'interprétation génétique des faciès euxiniques de l'Eo-Albien du Bassin Vocontien (SE France) sur la base de données macro- et microfauniques, *C.R. Acad. Sc. Paris*, *302*, 1085-1090, 1986.

Erba, E., Aptian-Albian calcareous nannofossil biostratigraphy of the Scisti a Fucoidi cored at Piobbico (Central Italy), *Riv. It. Paleont. Strat.*, *94*, 249-284, 1988.

Gradstein, F. M., Ludden, J. et al., *Proc. ODP, Init. Repts.*, *123*, Ocean Drilling Program, College Station, TX, 1990.

Haq, B. U., von Rad, U., O'Connell, S. et al., *Proc. ODP, Init. Repts., DSDP, 122*, Ocean Drilling Program, College Station, TX, 1990.

Harland, W. B., Armstrong, R. L., Cox, A. V., Craig, L. E., Smith, A. G., and Smith, D. G., *A Geologic Time Scale*, 263pp., Cambridge University Press, 1990.

Hinz, K., Winterer, E. L. et al., Site 545, *Init Repts. DSDP, 79*, 1984.

Krasheninnikov, V. A. and Basov, I. A., Stratigraphy of Cretaceous sediments of the Falkland Plateau based on planktonic foraminifers, Deep Sea Drilling Project, Leg 71, *Init. Repts. DSDP, 71*, 789-820, 1983.

Leckie, M., Mid-Cretaceous planktonic foraminiferal biostratigraphy off central Morocco, Deep-Sea Drilling Project Leg 79, Sites 545-547, *Init. Repts. DSDP, 79*, 1984.

Ludwig, W. J., Krasheninnikov, V. et al., Site 511 Summary, *Init. Repts. DSDP, 71*, 21-109, 1983.

Manivit, H., Nannofossiles calcaires du Crétacé Francais (Aptien-Maestrichtian). Essai de biozonation appuyé sur les stratotypes, *Fac. Sci. d'Orsay, Thèse Doctorate d'État*, 187 pp, 1971.

Manivit, H., Perch-Nielsen, K., Prins, B., and Verbeek, J. W., Mid Cretaceous calcareous nannofossil biostratigraphy, *Kon. Ned. Akad. Wet. Proc. B*, *80*, 169-181, 1977.

Mutterlose, J., Lower Cretaceous nannofossil biostratigraphy of ODP Leg 123 off Northwest Australia, E. Indian Ocean, *Proc. ODP, Sci. Results*, *123*, 1992.

Mutterlose, J. and Wise, S. W., Jr. Lower Cretaceous nannofossil biostratigraphy of ODP Leg 113 Holes 692B and 693A, Continental Slope off East Antarctica, Weddell Sea, *Proc. ODP, Sci. Results*, *113*, 325-351, 1990.

Perch-Nielsen, K., Mesozoic calcareous nannofossils, in *Plankton Stratigraphy,* edited by H. M. Bolli, J. B. Saunders, and K. Perch-Nielsen, pp. 329-426, Cambridge University Press, 1985.

Proto Decima, F., Leg 27 calcareous nannoplankton, *Init. Repts. DSDP, 27,* 589-622, 1974.

Roth, P. H., Cretaceous nannoplankton biostratigraphy and oceanography of the northwestern Atlantic Ocean, *Init. Repts. DSDP, 44,* 731-759, 1978.

Roth, P. H., and Krumbach, K. P. Middle Cretaceous calcareous nannofossil biogeography and preservation in the Atlantic and Indian Oceans: implications for paleoceanography, *Mar. Micropaleontol., 10,* 235-266, 1986.

Schlanger, S. O., and Jenkyns, H.C., Cretaceous oceanic anoxic events: causes and consequences, *Geol. en Mijnbouw, 55,* 179-184, 1976.

Sissingh, W., Biostratigraphy of Cretaceous calcareous nannoplankton, *Geol. Mijnbouw, 56,* 37-65, 1977.

Thierstein, H. R., Tentative Lower Cretaceous calcareous nannoplankton zonation, *Eclog. Geol. Helv., 64,* 459-488, 1971.

Thierstein, H. R., Lower Cretaceous calcareous nannoplankton biostratigraphy, *Abhandlung. Geolog. Bund., 29,* 3-53, 1973.

Thierstein, H. R., Mesozoic calcareous nannoplankton biostratigraphy of marine sediments, *Mar. Micropaleontol., 1,* 325-362, 1976.

Thierstein, H. R., Mesozoic calcareous nannofossils from the Indian Ocean, DSDP Legs 22 to 27, in *Indian Ocean Geology and Biostratigraphy,* edited by J. R. Heirtzler et al., pp. 339-352, 1977.

Wiegand, G. E., Cretaceous nannofossils from the Northwest African margin, Deep Sea Drilling Project Leg 79, *Init. Repts. DSDP, 79,* 563-578, 1984.

Wise, S. W., Jr., Mesozoic and Cenozoic calcareous nannofossils recovered by Deep Sea Drilling Project Leg 71 in the Falkland Plateau Region, Southwest Atlantic Ocean, *Init. Repts. DSDP, 71,* 481-550, 1983.

Wise, S. W. Jr. and Wind, F. H., Mesozoic and Cenozoic calcareous nannofossils recovered by DSDP Leg 36 drilling on the Falkland Plateau, Southwest Atlantic sector of the Southern Ocean, *Init. Repts. DSDP, 36,* 269-492, 1977.

The Cenomanian/Turonian Boundary Event in the Indian Ocean - a Key to Understand the Global Picture

J. THUROW

Geol. Dept., Ruhr-Univ. Bochum, P.O. Box 102148, 4630 Bochum, FRG

H.-J. BRUMSACK AND J. RULLKÖTTER

ICBM, Univ. Oldenburg, 2900 Oldenburg, FRG

R. LITTKE

KFA Jülich GmbH, 5170 Jülich, FRG

P. MEYERS

Dept. Geological Sciences, Univ. Michigan, Ann Arbor, MI, 48109-1063, USA

The Mesozoic pelagic realm is a favorable environment for the preservation and detection of paleoceanographic events. These events, one of the most prominent expressions of which are black shales, contribute important information about the sensitivity of the ocean/atmosphere system to outside influence. They represent unstable states of the oceans and disruptions of major sedimentary cycles on various scales of intensity, distribution, and time.

The globally recorded Cenomanian/Turonian Boundary Event is the most distinct and best studied Mesozoic event. It has a very pronounced record in shallow- to deep-water settings drilled at a transect along the Northwest Australian continental margin [Ocean Drilling Program (ODP) Legs 122/123]. The sedimentological and geochemical data clearly indicate that the depositional environment was characterized by an intensified and probably expanded oxygen minimum zone with oxic conditions prevailing at greater water depth. This is especially demonstrated by the presence of organic carbon- and trace metal-rich black shales at shelf sites and parallel Mn-accumulation at deep-sea sites. Samples from the Atlantic/Western Tethys realm, by contrast, display trace metal patterns which are more compatible with severe oxygen depletion or even anoxia in the deep waters during the same time interval.

It is suggested that two factors are the major controls on the occurrence of this event: (1) oceanic circulation (restricted marginal basins with anoxia in the water column vs. open basins with a distinct oxygen minimum zone), and (2) sealevel fluctuations paralleling a general transgressive trend. Events correlate with maximum flooding within a 3rd order cycle. High-amplitude 5th order cycles, which may result from short-term climatic changes and/or changes in the pattern of intraplate stress, gave rise to rhythmic nutrient flux to the oceanic reservoir. Because the lowland areas were increasingly flooded, the nutrient supply never ceased at this time. An enormous amount of nutrients could be progressively leached from land and transported to accumulate in the oceans. Resulting high plankton productivity led to strong intensification and finally expansion of the oxygen minimum zone both horizontally and vertically in marginal seas, but only to weak oxygen minimum zone intensification along margins bordering the open ocean. The record from Northwest Australia is the best example of the latter process known so far.

The attainment of the modern deep-water circulation system with its highly oxygenated deep-water masses, together with the end of continuous nutrient flux to the basins (lower amplitude of transgressions), terminated global black shale sedimentation from the latest Cretaceous to the present.

INTRODUCTION

The Northwest Australian margin is one of the rare examples of an undeformed Tethyan passive continental margin preserved in the oceans. Rifted margins are a promising setting for detection of the Cenomanian/ Turonian Boundary Event (CTBE). Unfortunately, most of

these Tethyan margins were later accreted or subducted especially along the rim of the Pacific realm, and thus the global record is poor. The Northwest Australian margin is a sediment-starved margin with a relatively thin (<1 km) cover of post-breakup sediment. Thus, it is a good target for drilling Cretaceous sedimentary sequences. Three sites from ODP Legs 122 and 123 [von Rad et al. 1989; Haq, von Rad et al., 1990; Gradstein, Ludden et al., 1990, 1992,; von Rad, Haq et al., 1992] are of interest for the study presented here (Figure 1):

- Site 763 on the Exmouth Plateau,

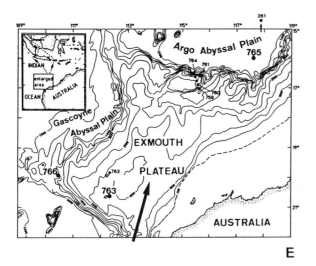

Fig. 1. Bathymetric map and geographic location of Northwest Australian sites on the Exmouth Plateau and adjacent Argo and Gascoyne Abyssal plains. Contour interval 500 m.

- Site 765 in the Argo Abyssal Plain, a region of Jurassic ocean floor overlain by about 1 km of sediments,

- Site 766 in the Gascoyne Abyssal Plain at the foot of an escarpment forming the boundary between continental and oceanic crust.

All three globally important mid-Cretaceous paleoceanographic events - the Selli Event of early -middle Aptian age, the CTBE, and the early Campanian biosiliceous event [LCE, Thurow, 1988a] - have been detected and can be studied at least at one of these drillsites.

Scientific ocean drilling has identified several Cretaceous episodes when organic-carbon-rich sediments ("black shales") were deposited in the world oceans. Some of these episodes are also represented by well-exposed, uplifted sections on land. "Black shales" in the deeper-marine environment throughout the late Cenomanian to early Turonian (ca. 91 Ma) are especially notable for their high concentrations of organic matter (OM), for the pronounced marine character of this material, for the lack of benthic life, for their particularly widespread occurrence, and for the short duration of their deposition. A distinct positive $\partial^{13}C$ spike marks the onset of the CTBE [e.g. Schlanger and Jenkyns, 1976; Scholle and Arthur, 1980; Pratt and Threlkeld, 1984; Arthur et al., 1987; Schlanger et al., 1987; Thurow et al., 1988], and trace metal enrichment in the deep-water environment is exceptionally high [Brumsack, 1980; Brumsack and Thurow, 1986]. The isotopic excursion is also recognized in sequences which lack other geochemical or sedimentary characteristics of the CTBE. This main "anoxic event" [Schlanger and Jenkyns, 1976] is commonly associated with a major sealevel rise and is coeval on the scale of ammonite, inoceramid, planktonic foraminiferal and radiolarian zonations. Refined isotope curves confirm the synchroneity of the CTBE on a global scale with an absolute timespan of about 0.5 m.y.

The CTBE occurs within a wide range of paleo-water depths and paleogeographic settings [Schlanger and Jenkyns, 1976; Wiedmann et al., 1978, 1982; Arthur and Schlanger, 1979; Jenkyns, 1980; Schlanger and Cita, 1982; Einsele and Wiedmann 1982; Thurow et al., 1982; Herbin et al., 1986a,b; Thurow and Kuhnt, 1986a; Arthur et al., 1987; Schlanger et al., 1987]. This boundary event was originally named Cenomanian/Turonian Oceanic Anoxic Event (OAE); (Schlanger and Jenkyns, 1976). Later it was named Bonarelli-Event, Niveau-Thomel, Oceanic Anoxic Event 2 (OAE 2), and "Evénement 2". The Cenomanian/Turonian Boundary Event (CTBE) has turned out to be one of the most important oceanic events in the Mesozoic and has been intensively studied.

PRESENT KNOWLEDGE ABOUT BLACK SHALE OCCURRENCES AND CHARACTERISTICS, AND DEPOSITIONAL MODELS

Three major types of CTBE occurrences are generally found: A) Deep-water settings of the North Atlantic-type (DSDP/ODP sites) with CTBE-sediment thickness of a few centimeters to a few meters. The depositional environment was anoxic and the black shales are true sapropels. B) Shallow-water settings along the Northwest African margin with CTBE-sediment thickness of several tens of meters. The depositional environment was characterized by high surface productivity and a strongly oxygen-depleted/anoxic water column. C) Shallow- to deep-water settings of the Tethyan realm (e.g. Northwest Australia, Legs 122 and 123 sites) with CTBE-sediment thickness of a few centimeters. The depositional environment was characterized by an oxygen minimum zone (OMZ) with oxic conditions prevailing below this zone.

Based on these results, it is obvious that the first two paleosettings (A and B) are similar with respect to the event-features described above, despite the completely different environments. They correspond to: (1) Continental margin to basin transects in marginal basins - basins which are insufficiently coupled with the open ocean circulation, the proto-Atlantic and the western Tethys being examples of this type. The literature about this type of environment is extensive, and comprehensive studies of the Cenomanian/Turonian Boundary Event problem began in this area [Arthur and Premoli Silva, 1982; Thurow et al., 1982]. (2) The other areas are parts of the mid-Cretaceous proto-Atlantic, which can be studied from the enormous record of DSDP and ODP cores. From the various studies available it is clear that all the CTBE occurrences in these environments are similar to each other to a large extent. The continental margin to basin transects bordering the open ocean (paleosetting C) are different from the environments described above. For the mid-Cretaceous occurrences one has to refer to the circum-Tethyan margins. Most of the record, however, has been subducted, and the remains of passive margins, which are the environments most suitable for this type of deposits, are rare. One of the exceptions is the continental margin transect drilled during ODP Legs 122 and 123 off Northwest Australia.

All CTBE occurrences have in common: (1) onset/end of 'event-characteristic' lithologies (i.e. black shale, radiolarite), (2) abrupt faunal changes or distinct extinction/evolution patterns, and (3) sudden changes in sedimentary composition and geochemical patterns. In general, sedimentation rates are extremely low (evidence from biostratigraphy), even if redeposition from shallower sites accounts for part of the sediment accumulation.

Apart from the distinct and unique black shale sedimentation with characteristic geochemical signatures, the most obvious effect of the CTBE in open-marine environments is a 'pulse' in radiolaria persisting into the middle Turonian [Thurow et al., 1982; Thurow, 1987; Thurow, 1988a]. The consequent high content of biogenic silica leads to common zeolite or chert formation in CTBE sediments.

Marine strata deposited during the CTBE display lithologic, faunal, and geochemical characteristics which indicate that significant parts of the world's ocean were periodically oxygen deficient. In particular, the black laminated 'anoxic' shales appear to be restricted to an expanded OMZ or an anoxic water column [Arthur et al., 1987; de Graciansky et al., 1984; Schlanger et al., 1987; Thurow et al., 1988].

The existence of anoxic or oxic conditions around mid-Cretaceous paleoceanographic events remains controversial [e.g., Brass et al., 1982; Sarmiento et al., 1988; Kempe, 1990; Pedersen and Calvert, 1990; Herbert and Sarmiento, 1991]. In general, it has been proposed that deep-water formation in the oceans would be quite different during geologic intervals with reduced equator-to-pole temperature gradients. Exchange of deep-water masses should be considerably slower with an overall warm climate, as assumed for the mid-Cretaceous. Salinity, rather than temperature differences, may have driven the deep-ocean circulation [Brass et al., 1982]. Saline waters tend to form at subtropical latitudes where evaporation exceeds precipitation. Herbert and Sarmiento [1991] point out a likely consequence of warm saline bottom-water formation on ocean chemistry, i.e. the tendency to drive the ocean toward anoxia. Pederson and Calvert [1990], however, challenge the idea that anoxia is responsible for the distinct black shale sedimentation in the mid-Cretaceous. They suggest that the flux of organic matter to the seafloor is the major factor. Another idea is that transgressive flooding of lowlands increased evaporation substantially and produced more saline waters. When these waters reach the deeper basin levels they may initiate density stratification, leading to anoxia. This is an intriguing idea, but the lack of similar processes in modern or subrecent (interglacial) times makes this process unlikely.

Considering all these interpretations it is obvious that all authors have to refer to modern oceanographic processes. Because not much is known about Cretaceous climate and circulation in detail, there are a lot of ambiguities in these models.

The tendency to drive the ocean towards anoxia is evident in the mid-Cretaceous. This effect is not due to lower oxygen solubility with increasing ocean temperature in the mid-Cretaceous, but rather to the increased efficiency by which plankton will extract nutrients from convecting waters at low latitudes [Herbert and Sarmiento, 1991]. The present ocean nutrient content is sufficient to induce deep-water anoxia, if the circulation pattern changes, although occurrences of anoxia are rarely recorded. The simple ocean chemical model of Herbert and Sarmiento [1991] defines the balance between mean ocean nutrient content and the circulation parameters that can resolve the controversy of an oxygenated vs. anoxic ocean since the mid-Cretaceous.

The global CTBE has been related to various causes, i.e., transgression, high productivity or enhanced preservation of organic matter, global upwelling, cool climate/warm climate, bolide impacts, (volcanic) extrusions or a combination of several of these phenomena. However, an ultimately satisfying explanation and model is lacking so far. Here, we hope to contribute new data and new arguments towards a global modelling of the CTBE.

NEW RESULTS FROM THE NORTHWEST AUSTRALIAN CONTINENTAL MARGIN TRANSECT
Environment and sedimentology

The section at Site 763 (the shallowest site, located on the Exmouth Plateau, Figure 1) displays the typical features of a pelagic deep-marine CTBE interval and can thus be compared with North Atlantic deep-water CTBE intervals [e.g. Site 641 at the Galicia Margin, Thurow et al., 1988]: green calcareous nannoclaystone below the Cenomanian/Turonian boundary overlain by carbonate-free green zeolitic clays and zeolitic black shale rich in barite and OM of marine origin. A pelagic depositional environment formed after the subsidence of the Barrow Delta. The paleo-water depth was approximately 1000m [based on backtracking calculations, Haq et al., 1992], thus it was well above the local carbonate compensation depth (CCD). At this site black shale forms two layers with up to 26% total organic carbon (TOC) consisting of marine kerogen [Rullkötter et al., 1992]. These layers are rich in clinoptilolite and opal-CT [Thurow, 1992; Meyers and Dickens, this volume]. The clinoptilolite is a diagenetic product of precursor radiolarian tests [e.g. Thurow, 1988b].

Common accessory minerals are framboidal pyrite and tabular barite, the barite suggesting high surface productivity and sediment starvation or condensation [Schmitz, 1987; Bishop, 1988]. Mn has not been detected. This mineral assemblage is typical for deep-water (below CCD) deposits of CTBE sediments.

Under the SEM we frequently observed oval forms (CaF-apatite) which resemble bacteria in shape and size [Thurow 1992, his plate 8, Figures B-E]. They were phosphatized and predate silica diagenesis (zeolites growing out of "bacteria"). These forms are definitely fossil bacteria [Cl. Monty, written communication, 1992; J. Lucas, oral communication, 1992]. Similar forms are common in phosphate nodules, but were never described from mid-Cretaceous black shales. Therefore, the Northwest Australian occurrence is the first description of in situ

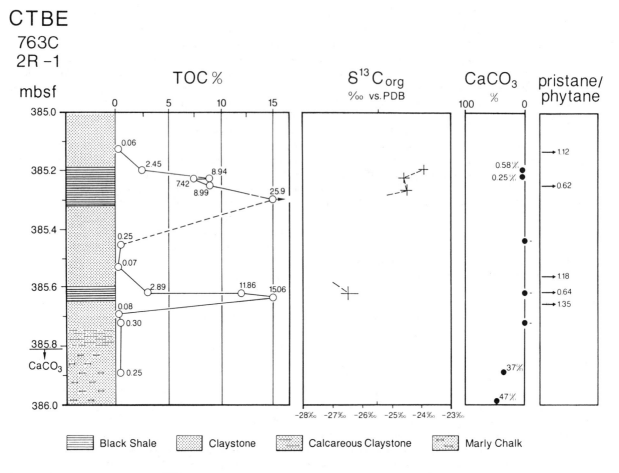

Fig. 2A. Sedimentary facies, TOC, ∂13TOC, CaCO3, and pristane/phytane ratios for the CTBE interval at Hole 763C. B. Site 763 Rock Eval analyses plotted in a modified van Krevelen diagram. Trend I corresponds to strictly algal-OM, trend II represents mixed terrestrial and marine OM, and trend III represents terrestrial OM.

anaerobic bacteria in CTBE black shales. Because opal-CT and clinoptilolite form during early diagenesis, the age of the bacteria coincides approximately with that of the black shale deposition.

Site 766 is located at the base of the steep western margin of the Exmouth Plateau (Figure 1). The oldest sediment penetrated is uppermost Valanginian sandstone and siltstone, interbedded with inclined basaltic sills. Site 766 appears to have remained close to or above the CCD throughout its history [Shipboard Scientific Party, 1988, Gradstein et al., 1989].

The Barremian/Aptian boundary represents a marked change in sedimentation, from clayey and sandy submarine fan deposits to a condensed Aptian to lower Paleocene sequence of pelagic and hemipelagic nannofossil oozes. The transition from rapidly deposited clastic sediments of Valanginian to Barremian age to slowly deposited Aptian to Paleocene pelagic and hemipelagic ooze appears to mark the transition from juvenile to mature oceanic depositional conditions. Similar reduction of sedimentation rates at Site 765 and at abyssal Atlantic sites during the late Cretaceous implies global tectonic and/or sea level processes.

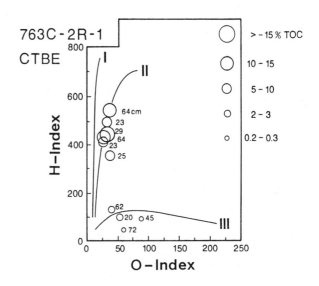

Cherty limestone occurs locally in the Aptian to lower Albian turbidites. Foraminifers in these rocks are commonly silicified; associated corroded radiolarian "ghosts" may have been an important source for the silica which replaced many of the foraminifers. Upper Cretaceous sediments were deposited close to the CCD [Gradstein et al., 1989], backtracking suggests a paleodepth of approximately 3000 m, and display alternating light-colored and dark to varicolored nannofossil-rich claystones.

The CTBE section at Site 766, the deep water equivalent of Site 763, is rather different: Mn-rich clays, devoid of any OM, are the temporal equivalent of the black shales. The only feature in common is the occurrence of radiolarian sands at the CTBE. The depositional environment was fully pelagic.

The CTBE interval overlies a prominent debris flow [Thurow, 1992; his plate 6, Figure D] and is characterized by an overall decrease in carbonate and an upward increase in dark claystone. Light-colored claystones alternate with Mn-rich claystones with intercalations of radiolarian-rich/zeolite-rich dark brown to black claystones. Radiolarian skeletons are often replaced by manganese oxides [Thurow, 1992; his plate 7, Figures A, B]. At this site oxic conditions dominated throughout the CTBE. Although these sediments look similar to black shales, they are devoid of autochthonous OM.

The dark layers are faintly laminated and their mineralogical composition is dominated by smectite and illite. No traces of benthic burrowing were observed. Mn is confined to spherical or "rose"-like todorokite aggregates. This special type of todorokite is common on the lower surface of diagenetic Mn-nodules [von Stackelberg, 1987], and the presence of this mineral may be taken as an indication of highly diminished sedimentation rates [Chester, 1990].

Site 765 is located in the Argo Abyssal Plain, northwest of Australia (Figure 1). Upper Jurassic/lower Cretaceous claystones directly overlie oceanic crust. Late Jurassic break-up along the northeastern rim of Gondwanaland quickly led to an oceanic rift-valley north of the Exmouth Plateau and formed the present Argo Abyssal Plain. From earliest Cretaceous time onward, less than 1000 m of clayey sediments accumulated in this abyssal environment. This ancient ocean floor is being subducted beneath the Banda/Sunda Arc along the Timor Trough to the north [Barber, 1988]. To the south lies the North West Shelf of Australia. Drilling at Site 765 was performed to study the sedimentation history and crustal/sediment geochemistry of what was believed to be one of the oldest sections of present-day sea floor [Veevers et al., 1974, Heirtzler et al., 1978].

The lower Cretaceous section is dominated by red, brown, and green claystone; dark-gray OM-rich claystone also occurs in the early Aptian, probably the local expression of the Tethyan Selli-Event (OAE 1a). Some calcareous and mixed-sediment turbidites are dominant in some intervals, but rare in event sections. Minor lithologies include

rhodochrosite-rich sediments, radiolarites, and bentonites in the lower part of the sequence.

The LCE can be detected in a sequence of fine-grained radiolarian-rich calciturbidites alternating with brown deep-sea clays. The clays are finely bedded and are characterized by a low, but significant Mn content. They contain well-crystallized illite and lesser amounts of smectite/mixed layer clays [Gradstein et al., 1989]. Very little organic carbon was detected. Radiolarian occurrences are confined to two calcareous turbidites intercalated in the brown clays. The radiolarians were transported from a shallower fertile zone downslope to the Argo Abyssal Plain. The brown clays are barren and contain no remains of biogenic silica.

The CTBE section at Site 765 with a paleo-water depth of approximately 5000 m - the deepest CTBE-interval ever drilled - is again rich in manganese and devoid of biological debris (even radiolaria). However, in contrast to the clays above and below, fine lamination is evident. This indicates that benthic life was at least impoverished due to the lack of food supply and/or due to oxygen depletion.

The sedimentary sequence displays alternating brown-reddish and grey claystone, with intercalations of black laminated claystone. The lamination is distinct, lacking any bioturbation. The dominant minerals are smectite and illite. Diagenesis is less advanced and the illite is preserved in its original form [Thurow, 1992]. The rare occurrence of framboidal pyrite points to sulphate reduction of precursor OM, an indication that at least some OM arrived at the sea floor. Fluctuations in the O_2 content close to the sediment/water interface may explain the sedimentary structures and the changing mineral content.

The mineralogy of "event sediments" from all sites does not differ significantly from the bulk mineralogy of Cretaceous sediments in this part of the Indian Ocean [Gradstein et al., 1990]. Furthermore, no coarse terrigenous detrital components or biogenic debris were found in the event intervals, as is the case below and above. XRD studies indicate that no particular volcanic influences and/or climatic changes, as demonstrated by the common occurrence of palygorskite in latest Cretaceous and younger sediments [Gradstein et al., 1990], were active.

Organic Geochemistry

Dark-colored layers of Cretaceous rocks with relatively high concentrations of OM were found at numerous DSDP sites, as well as in many on-shore sections. The distribution of their occurrence in the North Atlantic Ocean was discussed by Arthur [1979],Waples [1983], Herbin et al. [1986a], Meyers et al. [1986], Stein et al. [1986], Thurow and Kuhnt [1986a], Dean and Arthur [1987], Thurow et al. [1988], with the objective to identify the paleoceanographic factors involved in the formation of 'black shales'. Improved preservation of OM, increased contribution of continental OM to oceanic basins, and enhanced production of marine OM are some of the factors which have been suggested to be most important.

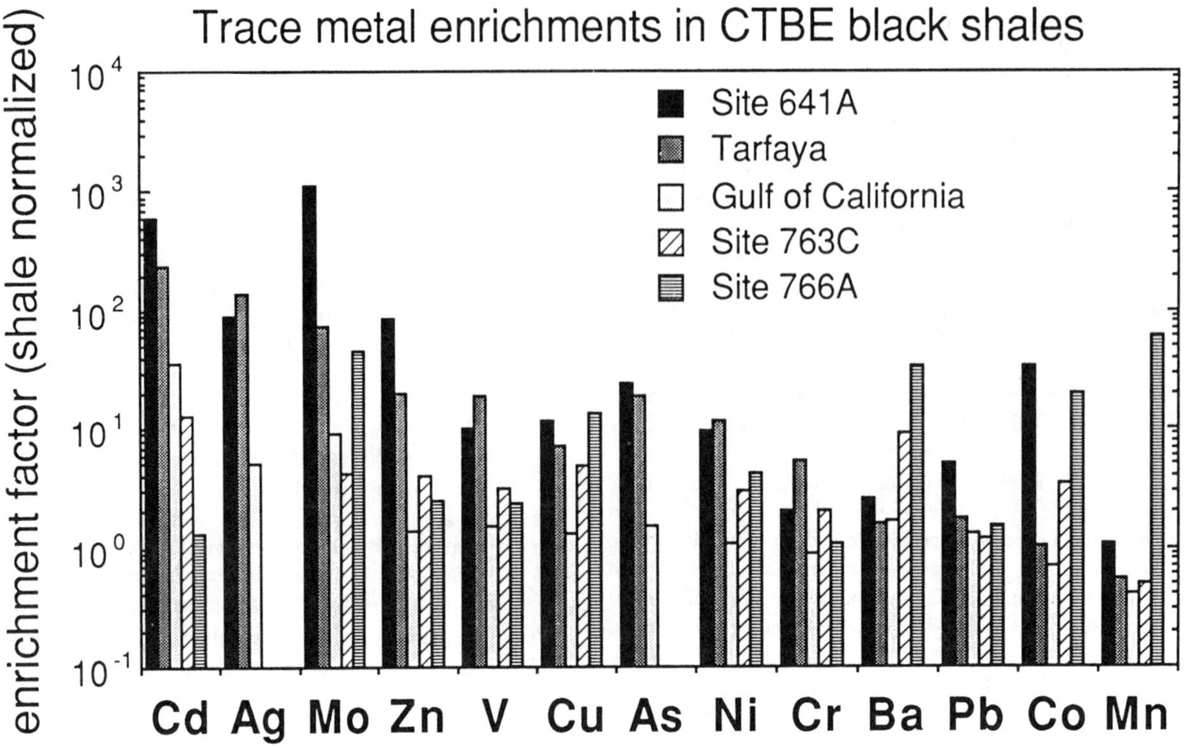

Fig. 3. Trace element enrichment in various CTBE sections

Hole 641C	CTBE-sapropel
Tarfaya	Coastal Basin/CTBE-upwelling
Gulf of California	Recent upwelling
Hole 763C	CTBE-"sapropel"
	OMZ - expanded and intensified
Hole 766A	"Oxic", Mn-rich shales below OMZ

Varying proportions of marine and terrigenous organic constituents are found in sediments deposited at different times and locations in the Cretaceous Atlantic Ocean, but black shales of the CTBE in the open ocean environment are always characterized by a dominance of marine OM [Meyers et al., 1984; Herbin et al, 1986a,b; Meyers et al., 1986; Stein et al., 1986; Thurow, 1987; Zimmerman et al., 1987].

Black shale at Site 763, the only site off Northwest Australia with sediments rich in OM at the CTBE interval, occurs in two distinct layers 5 cm and 15 cm thick which are separated by 25 cm of grey-green claystone, resembling the sediment below (Figure 2). The results of organic carbon determination and Rock Eval pyrolysis are shown in Figure 2. They demonstrate the strong enrichment in organic carbon (up to 26%) in the CTBE black shale at this particular site, while in the green claystone above, below, and between the black shale layers the content of OM is low.

Kerogen microscopy revealed that the OM at this site is of dominantly marine origin [Rullkötter et al., 1992]. Yet, the hydrogen indices are surprisingly low (between 350 and 550) with an average around 450 (Figure 2). This implies that the primary OM was degraded to some extent, an observation which is supported by the SEM evidence of microbial activity (fossil bacteria). Bituminite (i.e.

particles with a shape not corresponding to cellular structures) and phytoplankton-derived alginite and liptodetrinite are the dominant marine macerals in the black shales [Rullkötter et al., 1992]. In the less OM-rich black shale samples with organic carbon contents between 2 and 3% which are represented by the lower, thinner layer (Figure 2) and the uppermost part of the thicker upper layer, alginite and liptodetrinite are more abundant than bituminite. Bituminite predominates in the central and lower parts of the upper black shale.

Bituminite enrichment has also been found in CTBE black shales from the deep Atlantic Ocean at Site 603 off Cape Hatteras [Rullkötter et al., 1987] and Site 641 [Stein et al., 1988; Thurow et al., 1988]. In both cases, the organic carbon content exceeds 10%. Like at the Exmouth Plateau, the abundance of bituminite, representing degraded marine phytoplankton, goes along with moderate hydrogen index values. In contrast to this, OM in black shales consisting of alginite from (shallow-water) algal mats has hydrogen indices much higher than those observed for the CTBE black shales (Figure 2B). The bituminite may largely derive from zooplankton fecal pellets, and thus the relatively low hydrogen indices may correspond to partial oxidation of the OM in the food chain. However, extensive anaerobic microbial reworking of OM in the sediment after deposition will have a similar effect on the hydrogen budget.

Deposition of the black shales may have occurred by rapid sinking of fecal pellets from a surface-water zone with reasonably high productivity through partly oxygenated water masses which prevented the preservation of major amounts of primary OM particles (alginite) into an oxygen-depleted bottom water mass. This allowed preservation of high amounts of OM even at the generally low sedimentation rates during the Cretaceous at this site [Haq et al., 1990]. Strongly reducing conditions at least within the black shale sediments are indicated on a molecular level by very low pristane/phytane ratios (Figure 2).

Selected samples from Sites 765 and 766 in the Argo Abyssal Plain and Gascoyne Abyssal Plain (Figure 1) were investigated to find out if black shale deposition extended from the continental margin to farther off shore. The CTBE samples turned out to be typical open-marine deposits with very low organic carbon contents and were not suitable for further organic geochemical studies. Since the depositional sequences are continuous at both sites and their sedimentary facies are characteristic of the CTBE, including spikes rich in radiolarians, the absence of OM-rich layers can be explained in several related ways. The surface productivity may have been relatively low, because the deposition is unlikely to have occurred near a high-productivity area. As a result, the lateral and vertical extension of an OMZ near the shelf edge was insufficient to affect OM preservation in the Argo Abyssal Plain like in the shallower environment at Site 763. The split of the CTBE black shale into two layers, interrupted by grey to green claystone (background sedimentation) at Site 763, suggests a paleoposition of Site 763 near the base of a fluctuating OMZ. Hence we infer that the OMZ reached from the outer shelf to a water depth of approximately 1000m. In greater water depths below the OMZ no OM was preserved (Site 766). A similar situation applies to Site 765 (Argo Abyssal Plain). Because of its considerable distance from the shelf edge, it is also possible that this site was not overlain by a well developed OMZ.

Inorganic Geochemistry

The CTBE interval recovered off the Northwest Australian continental margin (Sites 763, 765, 766) has provided some important inorganic geochemical information regarding the paleoenvironment during Cenomanian/Turonian time. Generally, the use of trace metals as paleoenvironmental indicators is based on the assumption that the accumulation of specific metals is not a diagenetic but rather a syngenetic phenomenon [e.g., Brumsack, 1980].

Diagenetic remobilization is evident for Mn, which is mobile as Mn^{2+} in the pore waters of reducing sediments and diffuses along concentration gradients towards the sites of Mn removal. Mn may be incorporated as Mn^{2+} into diagenetically formed carbonate minerals, oxidized to Mn^{4+} and trapped as oxi/hydroxide at the sediment/seawater interface, or even dispersed as Mn^{2+} into the (oxygen-depleted) water column. Since certain trace metals are bound to oxi/hydroxide coatings, like in deep-sea clays, these phases represent a potential metal source.

Many black shales are intercalated with deep-sea TOC-poor clays. Trace metals may then be leached from these oxic sediments and fixed in TOC-rich intervals. Simple mass balance calculations have been performed to check this possibility [Brumsack, 1980; Brumsack and Thurow, 1986; Arthur et al., 1990]. These considerations reveal that for elements like Ag, As, Cd, Mo, V, and Zn a sedimentary sequence more than 10 times as thick as the TOC-rich layer has to be leached of its metal content, which seems rather unrealistic. The extreme enrichment of these metals in CTBE black shales consequently must result from other processes, probably during sedimentation.

Trace metals are removed from seawater, another potential metal source, by various processes. These include (1) bioproductivity and OM deposition, (2) adsorption to settling particles, and (3) chemical precipitation as sulphides or oxides. The type of host sediment also plays an important role for the quantity of major and minor elements present. For example, changes in the ratio of carbonate versus clay minerals are reflected in parallel variations in trace element abundances. For this reason the following discussion is based on element/Al ratios, which compensate for dilution effects by carbonate, biogenic silica, or OM and therefore allow to compare the trace metal patterns of different sedimentary environments. We use "average shale" data [Wedepohl, 1970] for comparison because this material most likely represents the detrital background sedimentation chemistry .

During growth, marine biota, assimilate several trace metals in addition to nutrient elements. This fact is reflected in the low concentration ranges of certain elements, like Cd and Zn, in surface seawater [see compilation by Bruland, 1980]. Nevertheless plankton do not seem to concentrate these elements to high enough levels to explain the trace metal content of black shales by rapid deposition of this primary material [Brumsack, 1980]. Since under present-day conditions in the oceans elements are involved in nutrient-type recycling processes and are therefore regenerated within the water column, it is unlikely that plankton are the only source for metals in black shales, especially in deep-water settings.

Marine particulate matter, as collected by sediment traps, may represent another source for metal enrichments in TOC-rich sediments. Even though OM can effectively absorb and concentrate trace metals [Chester, 1990], chemically these particulates are characterized by Ag/TOC, Cd/TOC, V/TOC, and Zn/TOC ratios an order of magnitude lower than required for Cretaceous black shales. Under modern, oxic conditions in the ocean these ratios are rather constant or even seem to drop further with depth due to the preferential remineralization of labile biogenic material. Therefore, the sedimentation of this material also does not seem to explain the chemical composition of Cretaceous black shales [Brumsack and Thurow, 1986].

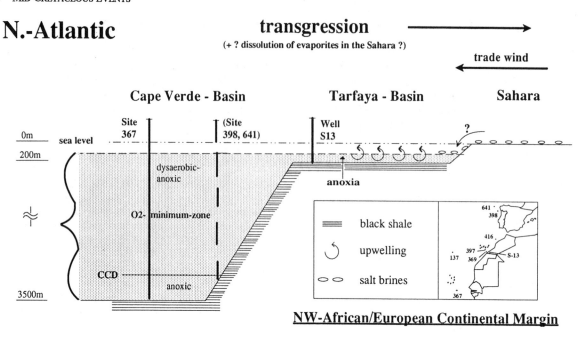

Fig. 4A. Hypothetical transect through the Northeast Atlantic margin showing the setting of coastal basins (transgressive basins), like the Tarfaya basin, and deep-sea drill sites like the Galicia margin (Site 641) or the Cape Verde Basin (Site 367). B. Transect of the Northwest Australian margin (simplified): Paleodepth, paleoceanography and sedimentary facies at Sites 763, 766, and 765 for Cenomanian/Turonian times. The oxidation state is indicated.

If a significant accumulation of trace metals and TOC is not conceivable under oxic water column conditions, the possibility of an anoxic or severely oxygen-depleted system should be evaluated. Lengthy periods are required to accumulate trace metals from the water column, since seawater concentrations at present are and probably in the geological past were very low. Furthermore, the preservation of OM in the water column has to be enhanced by orders of magnitude, compared to present-day conditions in the open ocean. An anoxic water column would be an ideal trap for redox sensitive and sulphide-forming trace metals. If preservation dominates over production, long periods of time are available for accumulating specific trace metals from seawater and producing trace metal anomalies in the sediment.

A compilation of trace metal enrichments from ODP Site 641 (Galicia margin), a very condensed sapropelic band representing the most extreme anoxic conditions in the proto-Atlantic during Cenomanian/Turonian time, the Tarfaya Basin, a Cretaceous upwelling system off Northwest Africa, the Recent Gulf of California upwelling system, Site 763 within the expanded OMZ during Cenomanian/Turonian time, and Site 766 below the OMZ is shown in Figure 3. Element enrichment factors of the TOC-rich CTBE sediments from Site 641 and Tarfaya are generally higher than those from Recent upwelling sediments. This may result from trapping of elements in anoxic waters of restricted basins or severely depleted oxygen levels in an expanded OMZ.

Samples from Site 763 are characterized by metal enrichments higher than those in Recent upwelling sediments from the Gulf of California, except for Cd and Mo. The relatively high enrichment of Ba at the Australian locations (Figure 3) may result from increased biological productivity [Schmitz, 1987; Dymond et al., 1992] and/or very low sediment accumulation rates and corresponding Ba-enrichment due to the lack of diluting detrital material. In view of the following arguments, the latter explanation seems more likely.

All TOC-rich CTBE locations are depleted in Mn. As has been mentioned earlier, these environments provide the source for Mn-rich suboxic or anoxic waters [Brumsack, 1991]. The OMZ, which acts like a transport medium, "exports" this element in the Mn^{2+} oxidation state from the continental margin towards the deep-sea [Klinkhammer and Bender, 1980; Martin and Knauer, 1984]. There, Mn may get oxidized and finally accumulated in deep-sea clays. In contrast to Site 763, where black shales are encountered and Mn is depleted in the CTBE sediments, the Mn concentrations in the sediments at Site 766 (Figure 3) off the Northwest Australian coast are extremely high. Regarding the paleodepth of approximately 3000 m, this site possibly was situated below an expanded OMZ and may have collected Mn-oxi/hydroxides originating from this zone, as is depicted in Figure 4. The Mn enrichment at Site 766 and, less pronounced, at Site 765 is also indicative of very low sediment accumulation rates in this deep sea

setting, because Mn concentrations are much higher than those in Recent deep-sea clays.

A comparison of the Mn and trace metal enrichments (expressed as element/Al weight-ratio) reveals that Co, Cu, Mo, and V seem to be scavenged from the water column during the formation and descent of the Mn-rich particulates or during extended residence time at the sediment/seawater interface (Figure 5). The chemistry of these Mn-rich layers seems to be partly analogous to Fe/Mn crusts reported from seamounts off Hawaii [De Carlo et al., 1987], which represent hydrogenous particulate material originating from the OMZ. Therefore these samples may be regarded as "extreme" deep-sea clays. In contrast to Co, Cu, Mo, and V, Fe is only slightly accumulated in the Mn-rich layers, even though the Fe/Mn crusts are enriched in this element. This may be an indication for the selective removal of Fe in anoxic CTBE sediments or even in anoxic waters as pyrite. For comparison also the data point for CTBE black shales with more than 3 % TOC is shown. This demonstrates, that elements like Co, Cu, Mo, and V are accumulating under oxic and anoxic conditions and that the intensity of metal signals strongly seems to depend upon time, i.e. sediment accumulation rate.

Coeval Cyprus umbers, i.e. Mn- and Fe-rich metalliferous sediments devoid of any OM, which are interpreted to be of hydrothermal origin [Robertson and Hudson, 1973], are also included in Figure 3. These umbers were deposited on oceanic crust (Troodos Ophiolite - a part of the suite of ophiolite-hosted sulphide deposits of Force et al. [1983] contemporaneous to the Mn-rich sediments off Northwest Australia based on the occurrence of the CTBE- and LCE-radiolarian spike, respectively [Thurow, 1992].

In summary, the chemical data clearly indicate the presence of an intensified and probably expanded OMZ and oxygen containing deep-waters off the Northwest Australian margin throughout Cenomanian/Turonian time. In contrast to this paleoenvironmental setting, samples from the Atlantic/Western Tethys realm display trace metal patterns which are more compatible with severe oxygen depletion or even anoxia in the deep-waters during the same time interval [e.g. Brumsack and Thurow, 1986].

Manganese-rich Sediments and Deposits

The intimate stratigraphic association of Cenomanian/Turonian Mn-rich sediments with organic carbon-rich and generally Mn-depleted black shales suggests a genetic connection between both environments.

An extensive column of anoxic or euxinic water either developed on or was carried onto the shelf during rapid relative rise in sea-level. These waters are characterized by large amounts of dissolved Mn^{2+}. We assume that the Mn originates predominantly from shelf sediments, as is demonstrated by the low Mn/Al-ratios of TOC-rich, reducing sediments. Mn may subsequently be transported within the OMZ and precipitated as Mn^{2+}-oxy/hydroxides in shallower environments above the OMZ (e.g., by upwelling) or at deeper settings below the OMZ [e.g.

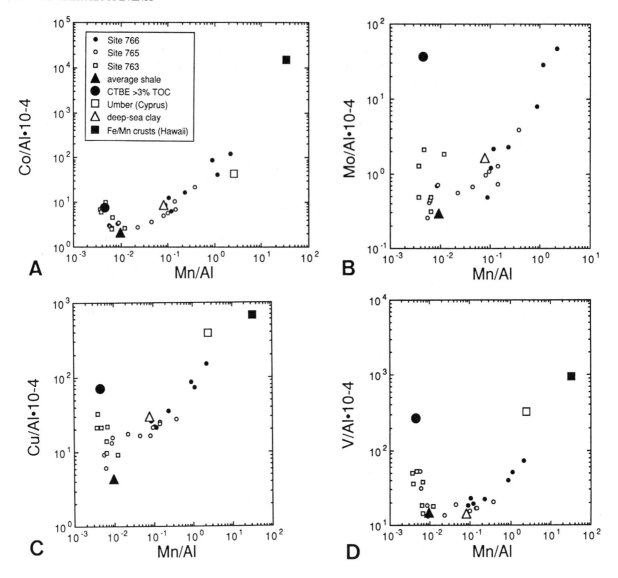

Fig. 5. Mn ratio (Al-normalized) for CTBE-black shales (Northwest Australia and others). Comparison of Mn and trace metal enrichments (element/Al weight ratio)

Jenkyns et al., 1991; Brumsack, 1991]. Such a model accounts for most large sedimentary Mn deposits of this type.

Recognition of anoxia characterized by the development of a well-developed OMZ off Australia indicates that manganiferous deposits can be related to mid-water transport of manganese, following the processes described above. The CTBE sediments at Site 766 and at Site 765 are characterized by high Mn concentrations. This Mn seems to originate from the reducing sediments, is transported by the OMZ and finally accumulates in the deep-sea. Since the pelagic sedimentation rates are strongly lowered during this time interval, the hydrogenous Mn-rich particulates form a major component of the sediments. A shallow water equivalent in this area, probably genetically related, is the

giant deposits of Groote Eylandt in Northern Australia dated as Albian-Cenomanian [Schlanger, 1986].

DISCUSSION

Production, Accumulation, and preservation of organic matter

To augment the evidence from OM, trace metal studies and faunal/floral patterns, other methods have been used to identify the potential productivity vs. preservation signal in a given environment.

As a puzzling outcome of comparative studies of modern and Cretaceous OM-rich deposits, Bralower and Thierstein [1984, 1987] concluded that Cretaceous black shale sequences do not indicate an increase in surface productivity. In contrast, they appear to represent a decrease compared to modern marine deposits which lack

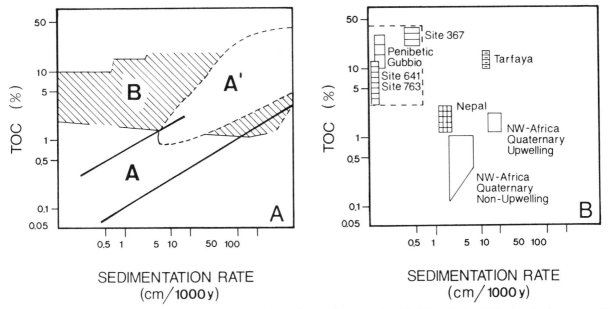

Fig. 6A. Plot of sedimentation rate vs. TOC for characteristic modern environments [modified from Stein, 1986]; A: oxic deep-water, A': high productivity, B: anoxic deep-water. B: Selected CTBE intervals in comparison to Nepal (middle Jurassic to lower Cretaceous Spiti Shale Facies) and to Quaternary environments off Northwest Africa [data for Quaternary sediments from Stein, 1991].

significant OM-enrichment, although underlying high productivity areas. To explain the striking black shale sequences deposited especially in the North Atlantic and adjacent areas Bralower and Thierstein [1984, 1987 argues that enhanced preservation in an oxygen-depleted environment is the dominant factor for the formation of these deposits.

For the bulk of mid-Cretaceous OM-rich deposits this is a convincing model, although it is not always supported by $\partial^{13}C$ and paleontological data. It is not an entirely satisfying explanation for those distinct black shale layers which parallel biosiliceous events, show geochemical anomalies, and can be globally correlated. Because productivity and preservation are typically coupled in today's ocean [Emerson and Hedges, 1988], it seems probable that productivity in the mid-Cretaceous oceans was only relatively higher (compared to before and after), and that preservation increased strongly coinciding with an increase in global temperature [e.g. Fischer, 1982].

The sedimentation rates in modern high-productivity areas are significantly higher than in most mid-Cretaceous environments with black shale sedimentation. Simple calculations of TOC content and sedimentation rate show that the mass accumulation rate of OM is generally higher in modern deposits than in mid-Cretaceous black shale deposits. Stein [1986] used the organic carbon/sedimentation rate relationship (Figure 6A) to compare depositional environments of the Mesozoic with modern and Quaternary data. Field A (Figure 6A) corresponds to oxic deep-water conditions, where TOC is positively correlated with sedimentation rate. Modern high productivity (i.e. upwelling) areas are characterized by high sedimentation rates and high TOC values (field A'). No correlation between sedimentation rates and TOC values is

obvious in anoxic deep-water environments as, for example, the modern Black Sea [field B, Degens and Ross, 1974, Stein, 1991]. Figure 6B shows the organic carbon/sedimentation rate relationship for selected CTBE intervals compared to the Quaternary high-productivity/low-productivity areas of Stein [1991] and to Jurassic OM-rich delta front sediments [taken as a proxy for a typical source rock environment, e.g. Nupra Shales, Nepal; Gradstein et al., 1989; Thurow and Gibling, 1989]. A decompaction of 30% for Mesozoic sediments was included. Where possible the time span of the *W. archaeocretacea* foraminiferal Zone was used, otherwise the duration of the CTBE was calculated as 0.5 m.y. [Thurow, 1992]. The intervals studied can be clearly divided in two depositional environments:

1. Equatorial zones of enhanced OM accumulation, mainly situated in outer shelf environments, which exhibit medium to high TOC values and comparatively high sedimentation rates.

2. Continental margin basins located above the CCD (Penibetic, Gubbio) and deep-sea environments (North Atlantic, Northwest Australia: Site 763) generally characterized by very high TOC values but extremely low sedimentation rates. Here a latitudinal factor is obvious in sedimentation/preservation rates of OM, with clearly distinguishable equatorial (Cape Verde Basin, DSDP Site 367) and more northerly (Galicia margin: ODP Site 641) end members.

OM-accumulation values have been used to estimate paleoproductivity of surface waters [Müller and Suess, 1979; Betzer et al., 1984 in oxic paleoenvironments and Bralower and Thierstein, 1984, 1987, in anoxic pelagic environments].

CENOMANIAN

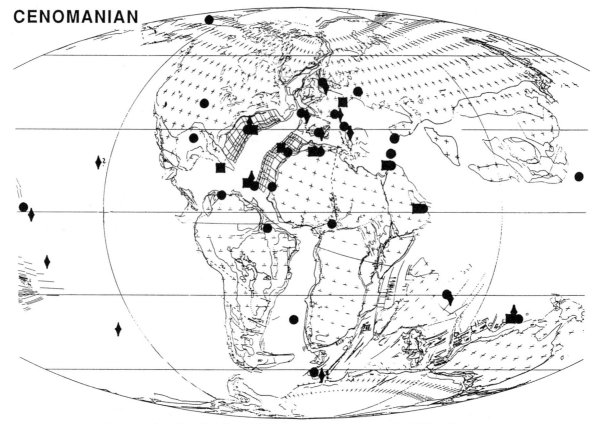

Geographic distribution of known occurences of middle Cretaceous paleoceanographic events.

References for high latitude areas are rare due to lack of knowledge.

■ LCE (Lower Campanian Event)

● CTBE (Cenomanian/Turonian Boundary Event)

♦ SELLI Event (Lower Aptian)

Fig. 7. Map of global occurrence of the Selli-Event, CTBE, and LCE.

Under anoxic conditions, the preservation of OM is distinctly higher than in oxic environments [Demaison and Moore, 1980; Bralower and Thierstein, 1984]. The latter authors suggest, from comparison of accumulation rates of marine organic carbon and primary production rates in Recent anoxic environments, that at least 2% of the primary organic carbon is preserved in the sediment. During the Cenomanian/Turonian boundary interval, anoxic water conditions prevailed, and paleoproductivity was about 50 times higher than the OM-accumulation rates [e.g. Stein, 1986].

For the deep-water settings the organic carbon accumulation rates are difficult to calculate because of poor stratigraphic resolution, but they seem to be low and fairly similar on a global scale, taking North Atlantic/Western Tethys and Northwest Australian deep-water environments as proxies. Values around $0.1 g/m^2/yr$ have been calculated.

Such a pattern is supported by the results of a similar study [Kuhnt et al., 1990], based on a much broader data base for the Western Mediterranean/Eastern North Atlantic. High organic carbon accumulation rates are confined to outer shelf/high productivity areas ($15 g/m^2/yr$), while other depositional environments even with striking black shale sedimentation have low accumulation rates (0.1-$0.5 g/m^2/yr$). However, uncertainty remains in correlating time with sedimentation rates around the CTBE until better stratigraphic resolution becomes available.

Radiolarian productivity and events

Radiolarians are the most striking planktonic group around the CTBE, and they show mass abundance and high diversity in all open marine deposits.

The mid-Cretaceous radiolarian record off Northwest Australia is poor in general, but there are sufficient faunas

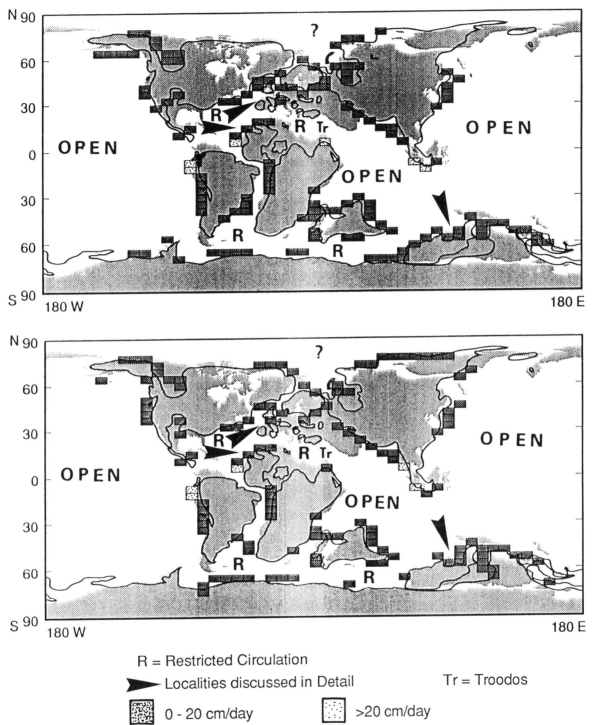

Fig. 8. Global distribution of mid-Cretaceous coastal upwelling [from Barron, 1990, unpublished atlas]. Upper part: simulated mean annual coastal upwelling; lower part: simulated mean annual coastal upwelling with quadrupled atmospheric CO_2-pressure.

R = Restricted Circulation

Localities discussed in Detail

Tr = Troodos

0 - 20 cm/day

>20 cm/day

to prove the presence of the most important Cretaceous global biosiliceous event, the CTBE [Thurow, 1988a, 1992]. The faunal assemblages are peculiar compared to those from the North Atlantic/Western Tethys.

Radiolarians are scarce in mid-Cretaceous samples off Northwest Australia (DSDP and ODP samples, dredge

samples), but there are exceptions. Radiolarians or their diagenetic remains have been found in the early Campanian (LCE), at the Cenomanian/Turonian boundary (CTBE), and in the late Albian and early Aptian (Selli). They show the same pattern of faunal composition/preservation as in the North Atlantic and Western Tethys. The Campanian and

Aptian faunas have a low diversity, and the faunal composition is different from that of typical Atlantic-Tethyan faunas [Thurow, 1988a, unpublished manuscript, 1992]. We found high-diversity faunas of cosmopolitan taxa in samples dated as late Albian and from the CTBE interval at Site 766. Faunal diversity is lower than in well studied low-latitude occurrences. This cannot be explained by poor preservation, because not only are fragile tests missing, but also some important markers which have a robust skeleton (i.e. *Crucella cachensis*).

In the coeval interval at Site 763 primary radiolarian skeletons in the black shale/green claystone are replaced by clinoptilolite similar to many other CTBE-intervals [Thurow, 1988b].

The cosmopolitan radiolarian record is evidence for an unrestricted water exchange with the Tethys at least at this time. Peculiarities in the faunal assemblage may be explained by the relatively high paleolatitude of this site (35-40°S).

Environment of the Cenomanian/Turonian boundary event - clues from Northwest Australia

The record from Northwest Australia provides new understanding of the paleoceanographic processes responsible for the formation of OM-rich facies connected with the CTBE and their pattern of distribution. Cenomanian/Turonian black shales, other than those in the Atlantic Ocean (Figure 4A), were only formed in the continental margin sedimentary sequence offshore Australia, but not in the Argo Abyssal Plain. Even at the continental margin they are restricted to the shallower Site 763, and were not found at the deeper Site 766 (Figure 4B). The composition of OM, at least in the two richest samples from Site 763, is comparable to those in Atlantic Ocean Cenomanian/Turonian black shales, i.e. there is a high concentration of structurally degraded OM (bituminite) with only a moderate hydrogen index. This may indicate sedimentation of the OM from a fairly prolific zone of bioproductivity through a significant vertical distance of oxygenated water mass into an anoxic environment at the sediment/water interface where the conditions favored OM preservation (anoxic bottom water mass?).

Documented by the thicker (upper) Cenomanian/ Turonian black shale layer, black shale formation started more or less abruptly and later ceased more gradually. This is obvious from the higher organic carbon contents and higher hydrogen indices of the basal part of this black shale layer and a decrease of organic carbon contents and hydrogen indices towards the top. Sediments between the two black shale layers are typical background sediments with respect to their OM characteristics, as observed in the underlying lower Cretaceous section. There is no indication that an increase in terrigenous OM supply had an influence on marine OM preservation as a contributing factor for black shale formation.

The kerogen microscopy results thus indicate that the Cenomanian/Turonian black shales in the North Atlantic and off Northwest Australia were deposited under strictly anoxic conditions with a strongly oxygen-depleted water mass overlying the sediment/water interface. The abundance of bituminite in the CTBE black shales, representing degradation products of marine phytoplankton, is consistent with the moderate hydrogen index values (around 500 mg hydrocarbons/g TOC) at the sites studied. OM in black shales rich in alginite from (shallow-water) algal mats would have led to HI values much higher than 500, for example as in the case of the Northwest African Coastal Basins (Figure 4A).

To recapitulate, the Northwest Australian margin is characterized by low productivity during mid-Cretaceous times. Although waters were not very well oxygenated, the formation of OM-rich facies is an extreme exception and only the CTBE-interval at Site 763 is comparable with similar deposits in the North Atlantic/West Tethys. The likely paleosituation off Northwest Australia might be limited nutrient supply, too small to produce normal or high fertility surface waters, and subsequent storage of larger amounts of OM in the sediment. The radiolarian-rich layers found at Site 766 were probably redeposited from an upslope location having an environment comparable to that of Site 763. Local upwelling may account for the distinct radiolarian-rich intervals.

The connection of the Cretaceous Northwest Australian margin with the open Tethys is still debated. Radiolarian evidence exists for an early Cretaceous connection to the southern ocean (proto-Antarctic Ocean, southernmost proto-Atlantic Ocean), based on the occurrence of endemic faunas. Around the CTBE, however, faunas are cosmopolitan and similar to those observed in the Tethyan and Atlantic sites. Because faunal exchange occurred, the paleoceanography off Northwest Australia and the adjacent Tethys must have been similar. Northwest Australia was part of the southern margin of Tethys at this time.

NORTHWEST AUSTRALIAN RECORDS AND THEIR SIGNIFICANCE FOR RE-CONSTRUCTION OF THE GLOBAL PALEOENVIRONMENTAL HISTORY OF MID-CRETACEOUS BLACK SHALE FORMATION

From the Indian Ocean drilling data, we conclude that the occurrence of the CTBE is not caused by globally enhanced fertility or world-wide stagnation of the oceans, but that in marginal seas like the North Atlantic/Mediterranean area oceanic circulation in the central Tethys was restricted and the OMZ was intensified and expanded. These seas were less than about 3000 m deep and thus became anoxic. In the open ocean the abyssal environments (>3000 m) remained oxic (Mn-rich/carbon-free shale). Evidence and dating of the CTBE in the deep environment is given by the prominent radiolarian spike containing the typical cosmopolitan CTBE fauna, but not by OM enrichments. From the data presented so far, a first attempt towards explaining the global distribution of CTBE deposits is possible. Figure 4A shows a continental margin transect displaying an intensified OMZ, expanded down to the seafloor. The margin is bordered by a marginal ocean basin (North Atlantic around the CTBE). OM-rich deposits are

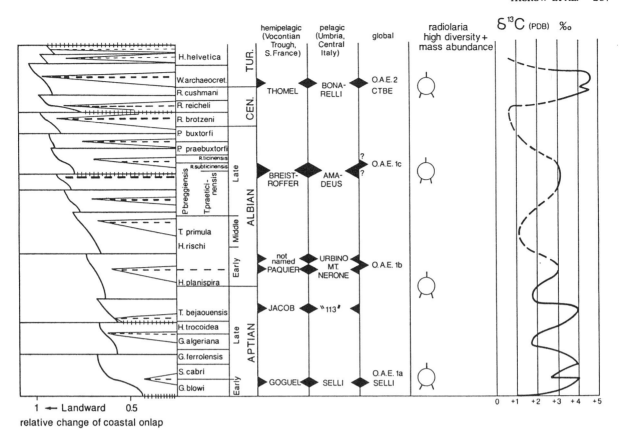

Fig. 9. Cretaceous oceanic anoxic events (OAE) according to Schlanger and Jenkyns [1976], and Jenkyns [1991] plotted vs. regional anoxic events in the Vocontian Trough [S. France, Bréhéret, 1988] and in the Umbrian Appennines (Central Italy, various sources and own data), vs. ∂13C isotopes and coastal onlap. For the CTBE only data from the Eastbourne section (see Figure 4) are plotted; compilation of isotope data for the early Cretaceous from Renard [1986] and Weissert [1989], coastal onlap according to Haq et al. [1987, 1988]. Major sequence boundaries and major condensed sections are indicated by thick lines. See text for discussion. Radiolarian-rich intervals are marked. All anoxic spikes in the pelagic realm and the global anoxic events are accompanied by radiolarian-rich layers. Correlation between foraminiferal zonation, coastal onlap and oceanic anoxic events is still influenced by the different timescales used by the authors. Regional anoxic events are dated and correlated with the foraminiferal zonation plotted. The global coastal onlap curve correlates exactly to the Selli-Event and the CTBE and the cycles inbetween are fit in.

ubiquitous. This figure also indicates the Galicia margin (a characteristic example for sub-CCD CTBE sites), the likely position of the Northwest African marginal basin setting and the Cape Verde Basin (Site 367) deep setting. Figure 4B shows the "normal" open ocean situation with an expanded and intensified OMZ during the CTBE, comparable to the present day situation in the northern Arabian Sea for example [von Stackelberg, 1972]. OM-rich sediments were deposited where the OMZ impinged on the continental slope and outer shelf. The latter situation accounts for all open-marine continental margin depositional environments, at least in lower latitudes. Before and after the CTBE, the OMZ was weaker and less expanded. Hence green-grey claystones to marlstones were deposited, while Mn-rich sedimentation was continuous below the OMZ. Therefore, high productivity in the shallow environments was the major driving force for the continuous expansion and intensification of the oceanic OMZ, which reached

abyssal environments as deep as 3500 m [Thurow et al., 1988] in the Atlantic and Tethyan oceans during the CTBE.

If one plots the distribution of the most important mid-Cretaceous events studied so far in detail (Figure 7) on Barron's [1990] paleo-upwelling map (Figure 8), it is evident that all CTBE occurrences caused by anoxic conditions in the water column are, in addition to those in coastal basins or epicontinental seas, confined to small oceanic basins with restricted circulation, juvenile, less dense crust and therefore a shallower basin floor. Records of the CTBE belonging to the "normal type" (i.e. Northwest Australia, Cyprus) occur along continental margins bordering the open ocean or are confined to the slopes of oceanic plateaus.

Modelling the global distribution of upwelling cells with varying pCO_2 (Figure 8) does not change the pattern significantly. Thurow et al. [1982] explained the CTBE on the basis of global upwelling. However, there are too many CTBE deposits outside of potential upwelling areas for

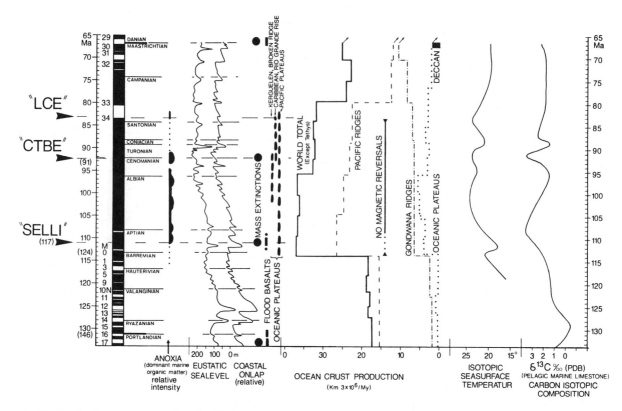

Fig. 10. Middle Cretaceous anomalies. Compilation of magnetic reversals, anoxia, eustatic sea level changes, change of relative coastal onlap, flood basalts/volcanic oceanic plateau formation, production of oceanic crust, sea surface temperature, and carbon isotopic composition of the Cretaceous marine record. Compiled from Thurow and Kuhnt [1986b], Arthur et al. [1987], Vogt [1989], Larson [1991].

upwelling and related nutrient transport to be responsible for the observed black shales and radiolarian sands.

There was a significant change in the facies of North Atlantic/Western Tethys sediments during the Cenomanian/Turonian: from black/green-grey (Hatteras-Formation) to reddish- brown/multicolored (Plantagenet-Formation), which marks the change from frequently O_2-depleted bottom water to well-oxygenated bottom water in the Atlantic Ocean and its marginal seas. The final change in paleoceanographic conditions is almost coeval with the onset of the CTBE in the Atlantic coastal basins. Therefore the CTBE seems to be correlated with the major change from Mesozoic sluggish oceanic circulation without pronounced temperature gradients [Bralower and Thierstein, 1984] to the modern current patterns and climate zones. The CTBE itself would mark a last reversal towards previous paleoceanographic conditions. The first evidence for a reorganization of oceanic circulation patterns (replacement of warm/ saline deep oceanic waters by cold/oxygenated waters) can be seen in the lower Aptian. It is strongly intensified in the late Albian with the first spikes of oxic facies, i.e. intercalations of red/brown-multicolored facies in the deep basins of the Tethys/ Atlantic-transition [Thurow and Kuhnt, 1986a,b; Thurow, 1987]. After the CTBE, red sediments were distributed globally and no further global black shale events are recorded. The earlier Cretaceous may have been cooler, with evidence of

sporadic high-latitude glacial regimes persisting into the early Albian [Frakes and Francis, 1988]. It is possible that the widespread Aptian-Albian volcanic episode caused the onset of overall warmer and equable climate since the Aptian, especially at higher latitudes [i.e. Barron and Washington, 1984; Lasaga et al., 1985; Arthur et al., 1985, 1988].

Considering the results presented so far, most of the peculiar features related to the CTBE and displayed in the sedimentary record can be explained, that is, (a) distribution of black shale facies, (b) faunal patterns, (c) geochemical patterns, and (d) isotopic patterns. The observed features are in accordance with the paleoceanographic model. Although this model is somewhat different from those presented so far, it does not explain why this event happened. In order to proceed with this explanation, several intimate links between mid-Cretaceous events, especially the CTBE, and geological processes (Figures 9 and 10) must be considered. The CTBE

1. corresponds to the major transgressive trend in the Phanerozoic

2. corresponds to a major warming trend in Earth's climate

3. is accompanied by excessive volcanism, especially in the Pacific

4. displays faunal features which can be explained by excess nutrient availability

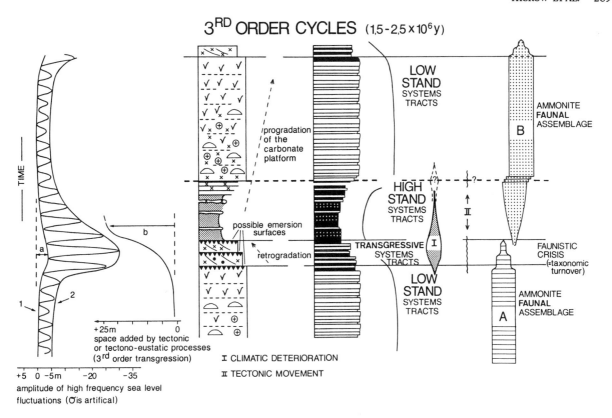

3RD ORDER CYCLES (1,5 - 2,5 × 10⁶ y)

Fig. 11. Sedimentary and biotic expression of high-frequency sea level fluctuations in a carbonate platform environment and in a hemipelagic/pelagic depositional environment. Modified from Ferry [1991].

5. is represented by sediments which are net sinks for carbon

6. accounts for oxygen depletion in large oceanic areas

These features must be discussed to see if and how they are related, and finally to discuss a scenario which describes the oceanic processes related to the CTBE. In order to understand this event it is important to distinguish between the general trends mentioned above and the short-term fluctuations which are contrary to the general trend.

There are certainly many causes for the CTBE along the Northwest Australian margin, as for the formation of the bulk of the mid-Cretaceous black shales. However, if there was a single initial mechanism, its moving force has to be in the Earth's interior. Its primary expression was the intense mid-Cretaceous oceanic volcanism/ridge production leading to the formation of huge plateaus in the Cretaceous Pacific and sea level rise, accompanied by excess CO_2 and climatic deterioration [Figure 10, Schlanger et al., 1981; Vogt, 1989; Larson, 1991]. This culminated in the peak of the Cenomanian/Turonian transgression which corresponds not only to the CTBE, but also to the maximum highstand of the sea level in the Phanerozoic (Figures 9 and 10). Variations in terrigenous-detrital input, variations in availability of nutrients, degree of oxidation of the water column, and decline or expansion of life habitats were the consequences. Flooding of large coastal areas increased the shelf environment considerably. Starvation from terrestrial input and landward shift of the productive areas/carbonate deposition are evident and may also have been responsible for the strong rise of the CCD in the open ocean. The observed frequent volcanism was responsible for increased CO_2-exhalations which led to (a) general tendency towards global warming since the early Aptian, (b) strong plankton pulse, additionally enhanced by the leaching of large amounts of nutrients from the flooded areas, and (c) subsequent oxygen-depleted anoxic conditions and storage of large amounts of carbon in the sediments (positive $\partial^{13}C$ excursion).

The direct cause for the particular deposits and some of the changes in oceanography is seen in small-scale sea level fluctuations (5th order) superimposed on transgressive systems within 3rd order cycles. Each of these may display a distinct black shale interval (i.e. Selli-Event, CTBE, Figures 9 and 11). The idea of a temporary amplification of high-frequency sea level oscillations as a result of global climate coolings during 3rd order rises in sea level explains how a few, intense sea level drops may occur paradoxically at the top of lowstand systems tracts and last over most of the next transgressive systems tract [Figure 11, Ferry, 1991, Ferry and Rubino, 1989].

Ferry proposes that 3rd order cycles are a modulation of glacio-eustatic high-frequency cycles (orbitally-controlled Milankovitch cycles). This modulation should be primarily tectonic in origin, and it may be linked to changes in global mantle dynamics leading to volcanic surges (hotspot activities, mantle super plume).

These volcanic pulses are believed to cool global climate over the depositional time of a few parasequences during which the usually "low-amplitude, orbitally-controlled high-frequency glacioeustatic oscillation" [Ferry, 1991] may be dramatically shifted. This would result in a short-term amplitude of glacio-eustatic oscillation greater than the amplitude of 3rd order changes in sea level. Cathless and Hallam [1991] propose a different origin for the short-term fluctuations. They argue that the rapid creation of new rifts would change plate density to an extent that a plate could rapidly react to compression produced by the new rifts. Short-term fluctuations (0.6×10^5 yr/50m magnitude) of the global sea level are the result. This explanation is in better accordance with the mid-Cretaceous geological record than the glacial model. However, a convincing test of the Cathless and Hallam [1991] model is still lacking.

Both hypotheses would easily explain why platform floodings coincide with increased transport of nutrients and changes in the nature of sediments supplied to basins; more terrigenous material is due to a generally more humid climate.

Overall warm and humid climate in the mid-Cretaceous enhanced this process by supporting extensive weathering and soil formation, especially when 4rth or 5rth order cyclic variations (Figure 11) account for short-term periodic seaward transport of trace metals and nutrients during relative lowstands. This results in subsequent high-productivity and storage of a great volume of biomass. These 4rth or 5rth order variations may also account for the frequently observed radiolarian sand/black shale rhythms in the pelagic environment, or for the limestone/black shale rhythms, with a higher proportion of terrestrial OM, in the epicontinental environment during the CTBE.

High-frequency oscillations of the sea level around the Cenomanian/Turonian boundary are recorded in sedimentary deposits of many areas. These fluctuations together with the overall transgressive trend, with the maximum of the Phanerozoic sea level recorded in the highstand deposits above the CTBE black shale, gave rise to rhythmic nutrient and SiO_2 flux to the oceanic reservoir. Such fluctuations can also explain the cyclic nature of many of the CTBE deposits (i.e. black shale/radiolarite).

Because the lowland areas were increasingly flooded, the nutrient supply never ceased at this time. An enormous amount of nutrients could be progressively leached from land and transported to accumulate in the oceans. Resulting high plankton productivity led to strong intensification and finally expansion of the OMZ both horizontally and vertically in marginal seas, but only to weak OMZ intensification along margins bordering the open ocean. The record of Northwest Australia is the best example of the latter process known so far.

Trace metal concentrations augment biological evidence, such as lack of benthic burrowing, to demonstrate the existence of anoxia in marginal seas and along continental margins. Mechanisms known so far to produce such short-term sea level fluctuations are either glacial, or rapid changes in intraplate stress. Although polar ice caps are not proven for the Cretaceous period, there may have been small ice-covered areas sufficient for small sea level fluctuations.

All these features are part of a mid-Cretaceous scenario displaying a general transgressive trend and a general climatic trend towards global warming, i.e. installation of "greenhouse" conditions. These trends are related to strong magmatic activity in the mantle after the early Aptian ("mantle superplume") which finally resulted in rejuvenation of oceanic crust, subsequently shallower oceanic sea floor, reduced water depth, and flooding of lowlands. While mantle degassing and increased weathering raised the level of atmospheric CO_2, polar ice caps disappeared.

The attainment of the modern deep-water circulation system (north-south instead of east-west, horizontal instead of vertical transport) with its highly oxygenated deep-water masses, together with the end of continuous nutrient flux to the basins, terminated global black shale sedimentation to the present.

Acknowledgments. J. Thurow thanks the Ocean Drilling Program (ODP) for the invitations to participate on ODP Legs 103 (Galicia margin) and 123 (Northwest Australia) and P. Meyers is grateful for the opportunity to participate in ODP Leg 122 (Northwest Australia). The Deutsche Forschungsgemeinschaft (DFG) provided financial support for most of the analytical work. This contribution benefited considerably from discussions and reviews by K. Emeis, R. Stein, and U. von Rad.

REFERENCES

Arthur, M. A., North Atlantic Cretaceous Black Shales: The record at Site 398 and a brief comparison with other occurrences, in *Init. Repts DSDP, 47*, pp. 719-738, U.S. Government Printing Office, Washington, D.C., 1979.

Arthur, M. A., and Premoli-Silva, I., Development of widespread organic carbon-rich strata in the Mediterranean Tethys, in *Nature and Origin of Cretaceous Organic Carbon-Rich Facies*, edited by S. O. Schlanger, and M. B. Cita, 7-54, Academic Press, London, UK, 1982.

Arthur, M. A., and Schlanger, S. O., Cretaceous "Oceanic Anoxic Events" as causal factors in development of reef-reservoired giant oil fields, *Am. Assoc. Petrol. Geol. Bull., 63*, 870-885, Tulsa, Oklahoma, 1979.

Arthur, M. A., Dean, W. E., and Claypool, G. E., Anomalous ^{13}C enrichment in modern marine organic carbon, *Nature, 315*, 216-218, 1985.

Arthur, M. A., Dean, W. E., and Pratt, L. M., Geochemical and climatic effects of increased marine organic burial at the Cenomanian/Turonian boundary, *Nature, 335*, 714-717, 1988.

Arthur, M. A., Schlanger, S. O., and Jenkyns, H. C., The Cenomanian-Turonian oceanic anoxic event, II, Paleoceanographic controls on organic-matter production and preservation, in *Marine Petroleum Source Rocks, Geol. Soc. London, Spec. Publ., 26*, edited by J. Brooks, and A. J. Fleet, 401-420, Blackwell, Oxford, 1987.

Arthur, M. A., Jenkyns, H. C., Brumsack, H.-J., and Schlanger, S. O., Stratigraphy, geochemistry, and paleoceanography of organic carbon-rich Cretaceous sequences, in *Cretaceous Resources, Events and Rhythms, NATO ASI Series, Series C, 304,* edited by R. N. Ginsburg, and B. Beaudoin, 75-119, 1987.

Barber, P. M., The Exmouth Plateau deep water frontier: A case of history, in *The North West Shelf Australia,* edited by P. G. Purcell, and R. R. Purcell, 173-187, Perth, W.A., 1988.

Barron, E.J., and Washington, W.M., The role of geographic variables in explaining paleoclimates: Results from Cretaceous climate model sensitivity studies, *J. Geophys. Res., 89,* 1267-1279, Washington, D.C., 1984.

Betzer, P. R., Showers, W. J., Laws, E. A., Winn, C. D., Ditullo, G. R., and Kroopnick, P. M., Primary productivity and particle fluxes on a transect of the equator at 153°W in the Pacific Ocean, *Deep-sea Res., 31,* 1-11, 1984.

Bishop, J. K. B., The barite-opal-organic carbon association in oceanic particulate matter, *Nature, 233,* 241-243, 1988.

Bralower, T. J., and Thierstein, H. R., Low productivity and slow deep-water circulation in mid- Cretaceous oceans, *Geology, 12,* 614-618, Boulder, Colorado, 1984.

Bralower, T. J., and Thierstein, H. R., Organic-carbon and metal accumulation rates in Holocene and mid-Cretaceous sediments: palaeoceanographic significance, in *Marine Petroleum Source Rocks, Geol. Soc. London, Spec. Publ., 26,* edited by J. Brooks, and A. J. Fleet, 345-369, Blackwell, Oxford, 1987.

Brass, G. W., Southam, J. R., and Peterson, W. H., Warm saline bottom water in the ancient ocean, *Nature, 296,* 620-623, 1982.

Bréhéret, J. G., Episodes de sédimentation riche en matière organique dans les marnes bleues d'âge Aptien et Albien de la partie pélagique du bassin vocontien, *Bull. Soc. Géol. France, 4,* 349-356, 1988.

Bruland, K. W., Oceanographic distribution of cadmium, zinc, nickel, and copper in the North Pacific, *Earth Planet. Sci. Lett., 47,* 176-198, 1980.

Brumsack, H.-J., Geochemistry of Cretaceous black shales from the Atlantic Ocean (DSDP Legs 11, 14, 36 and 41), *Chem. Geol., 31,* 1-25, 1980.

Brumsack, H.-J., and Thurow, J., The geochemical facies of black shales from the Cenomanian/Turonian Boundary Event, in *Biogeochemistry of Black Shales, Mitt. geol.-paläont. Inst. Univ. Hamburg, 60,* edited by E. T. Degens, P. A. Meyers, and S. C. Brassell, 247-265, 1986.

Brumsack, H.-J., Inorganic geochemistry of the German `Posidonia Shale´: palaeoenvironmental consequences, in *Modern and Ancient Continental Shelf Anoxia, Geol. Soc. London, Spec. Publ., 58,* edited by R. V. Tyson, and T. H. Pearson, 353-362, 1991.

Cathless, L. M., and Hallam, A., Stress-induced changes in plate density, Vail sequences, epeirogeny, and short-lived global sea level fluctuations, *Tectonics, 10,* 659-671, 1991.

Chester, R., *Marine Geochemistry,* Unwin Hyman, London, 698pp, 1990.

De Carlo, E. H., McMurtry, G. M., and Kim, K. H., Geochemistry of ferro-manganese crusts from the Hawaiian Archipelago, I, Northern survey areas, *Deep Sea Res., 33,* 441-467, 1987.

de Graciansky, P. C., Deroo, G., Herbin, J. P., Montadert, L., Müller, C., Schaaf, A., and Sigal, J., Ocean wide stagnation episode in the Late Cretaceous, *Nature, 308,* 346-349, 1984.

Dean, W. E., and Arthur, M. A., Inorganic and organic geochemistry of Eocene to Cretaceous strata recovered from the Lower Continental Rise, North American Basin, Site 603, Deep Sea Drilling Project Leg 93, *Init. Repts. DSDP, 93,* U.S. Government Printing Office, Washington, D.C., 1093-1137, 1987.

Degens, E. T., and Ross, D. A. (eds.), *The Black Sea - Geology, Chemistry, and Biology,* Am. Assoc. Petrol. Geol., Mem., 20, 1974.

Demaison, G. J., and Moore, G. T., Anoxic environments and oil source bed genesis, *Am. Assoc. Petrol. Geol., Bull., 64,* 1179-1209, 1980.

Dymond, J., Suess, E., and Lyle, M., Barium in deep-sea sediment: a geochemical proxy for paleoproductivity, *Paleoceanography, 7,* 163-181, 1992.

Einsele, G., and Wiedmann, J., Turonian black shales in the Moroccan Coastal Basins: first upwelling in the Atlantic Ocean?, in *Geology of the Northwest African Continental Margin,* edited by U. von Rad, K. Hinz, M. Sarnthein, and E. Seibold, 396-414, Springer, Berlin, 1982.

Emerson, S., and Hedges, J. I., Processes controlling the organic carbon content of open ocean sediments, *Paleoceanography, 3,* 621-634, 1988.

Ferry, S., Une alternative au modèle de stratigraphie séquentielle d'Exxon: la modulation tectono-climatique des cycles orbitaux, *Géol. Alpine, Mém. H.S., 18,* 47-99, 1991.

Ferry, S., and Rubino J. L., Mesozoic eustacy record on Western Tethyan Margins, *Publ. Assoc. Séd. Fr., 12,* 1-141, 1989.

Fischer, A. G., Long-term oscillations recorded in stratigraphy, in *Climate in Earth History,* edited by W. Berger, and J. C. Crowell, National Acad. Press, Washington, D.C., 97-104, 1982.

Force, E. R. et al., Influences of oceanic anoxic events on manganese deposition and ophiolite-hosted sulfide preservation, in *Paleoclimate and Mineral Deposits, U.S. Geol. Surv. Circ. 822,* edited by R. M. Cronin et al., 26-29, 1983.

Frakes, L. A., and Francis, J. E., A guide to Phanerozoic cold polar climates from high-latitude ice-rafting in the Cretaceous, *Nature, 333,* 547-549, 1988.

Gradstein, F. M., Gibling, M., Jansa, L. F., Kaminski, M., Ogg, J. G., Sarti, M., Thurow, J., Westermann, G. E. G., and von Rad, U., Mesozoic Stratigraphy of Thakkhola, Central Nepal, *Centre Mar. Geol. Spec. Rept, 1,* Dalhousie Univ., Halifax Canada, 1989.

Gradstein, F. M., Ludden, J. N., Adamson, A. et al., *Proc. ODP, Init. Repts., 123,* Ocean Drilling Program, College Station, TX , 1990.

Gradstein, F. M., Ludden, J. N. et al., *Proc. ODP, Sci. Results, 123,* Ocean Drilling Program, College Station, TX, 1992.

Haq, B. U., Boyd, R. L., Exon, N. F., and von Rad, U., Evolution of the central Exmouth Plateau: a post-drilling synthesis, in *Proc. ODP, Sci. Results, 122,* 801-816, Ocean Drilling Program, College Station, TX, 1992.

Haq, B. U., Hardenbol, J., and Vail, P. R., Chronology of fluctuating sea levels since the Triassic (250 million years ago to present), *Science, 235,* 1156-1167, 1987.

Haq, B. U., Hardenbol, J., and Vail, P. R., Mesozoic and Cenozoic chronostratigraphy and cycles of sea- level change, in *Sea level Changes: An Integrated Approach, Soc. Econ. Paleont. Min., Spec. Publ., 42,* edited by C. K. Wilgus, B. S. Hastings, H. W. Posamentier et al., 73-108, Tulsa, Oklahoma, 1988.

Haq, B. U., von Rad, U., O`Connell, S. et al., *Proc. ODP, Init. Repts., 122,* Ocean Drilling Program, College Station, TX, 1990.

Heirtzler, J. R., Cameron, P. J., Cook, P. J., Powell, T. G., Roeser, H. A., Suhardi, S., and Veevers, J. J., The Argo Abyssal Plain, *Earth Planet. Sci. Lett., 41,* 21-31, 1978.

Herbert, T. D., and Sarmiento, J. L., Ocean nutrient distribution and oxygenation: Limits on the formation of warm saline bottom water over the past 91 m.y., *Geology, 19,* 702-705, 1991.

Herbin, J. P., Magniez, F., Müller, C., and De Graciansky, P. C., Mesozoic organic-rich sediments in the South Atlantic: Distribution in time and space, in *Biogeochemistry of Black Shales, Mitt. Geol.-Paläont. Inst. Univ. Hamburg, 60,* edited by E. T. Degens, P. A. Meyers, and S. C. Brassell, 71-97, 1986b.

Herbin, J. P., Montadert, L., Müller, C., Gomez, R., Thurow, J., and Wiedmann, J., Organic-rich sedimentation at the Cenomanian-Turonian boundary in oceanic and coastal basins in the North Atlantic and Tethys, in *North Atlantic Palaeoceanography, Geol. Soc. London, Spec. Publ., 21,* edited by C. P. Summerhayes, and N. J. Shackleton, 389-422, Blackwell, London, UK, 1986a.

Jenkyns, H. C., Cretaceous anoxic events - from continents to oceans, *J. Geol. Soc. London, 137*, 171-188, 1980.

Jenkyns, H. C., Impact of Cretaceous Sea Level Rise and Anoxic Events on the Mesozoic Carbonate Platform of Yugoslavia, *Am. Assoc. Petrol. Geol., Bull., 75*, 1007-1017, 1991.

Jenkyns, H. C., Géczy, B., and Marshall, J. D., Jurassic manganese carbonates of Central Europe and the Early Toarcian anoxic event, *J. Geol., 99*, 137-149, 1991.

Kempe, S., Alkalinity: the link between anaerobic basins and shallow water carbonates, *Naturwissenschaften, 77*, 426-427, 1990.

Klinkhammer, G. P., and Bender, M. L., The distribution of manganese in the Pacific Ocean, *Earth Planet. Sci. Lett., 46*, 361-384, 1980.

Kuhnt, W., Herbin, J. P., Thurow, J., and Wiedmann, J., Distribution of Cenomanian-Turonian organic facies in the Western Mediterranean and along the adjacent Atlantic margin, in *Deposition of Organic Facies, Am. Assoc. Petrol. Geol., Stud. Geol. Ser., 30*, edited by A. Huc, 133-160, 1990.

Larson, R. L., The latest pulse of the earth: Evidence for a mid-Cretaceous superplume, *Geology, 19*, 547-550, 1991.

Lasaga, A. C., Berner, R. A., and Garrels, R. M., An improved geochemical model of atmospheric CO_2 fluctuations over the past 100 million years, in *The Carbon Cycle and Atmospheric CO_2: Natural Variations Archean to Present, Geophys. Monogr., 32*, edited by E. T. Sundquist, and W. S. Broecker, 397-411, 1985.

Martin, J. H., and Knauer, VERTEX: manganese transport through the oxygen minima, *Earth Planet. Sci. Lett., 67*, 35-47, 1984.

Meyers, P. A., Dunham, K., and Dunham, P. L., Organic geochemistry of Cretaceous organic-carbon-rich shales and limestones from the Western North Atlantic Ocean, in *North Atlantic Palaeoceanography, Geol. Soc. London, Spec. Publ., 21*, edited by C. P. Summerhayes, and N. J. Shackleton, 333-345, 1986.

Meyers, P. A., Leenheer, M. J., Kawka, O. E., and Trull, T. W., Enhanced preservation of marine-derived organic matter in Cenomanian black shales from the Southern Angola Basin, *Nature, 312*, 356-359, 1984.

Müller, P. J., and Suess, E., Productivity, sedimentation rate, and sedimentary organic matter in the oceans - I. Organic carbon preservation, *Deep-Sea Research, 26A*, 1347-1362, 1979.

Pedersen, T. F., and Calvert, S. E., Anoxia vs. Productivity: What controls the formation of organic-rich sediments and sedimentary rocks?, *Am. Assoc. Petrol. Geol. Bull., 74*, 454-466, 1990.

Pratt, L. M., and Threlkeld, C. N., Stratigraphic significance of $^{13}C/^{12}C$ ratios in Mid-Cretaceous rocks of the Western Interior, U.S.A, in *The Mesozoic of Middle North America, Can. Soc. Petrol. Geol. Mem., 9*, edited by D. F. Stott, and D. J. Glass, 305-312, 1984.

Renard, M., Pelagic carbonate chemostratigraphy (Sr, Mg, ^{18}O, ^{13}C), *Mar. Micropal., 10*, 117-164, 1986.

Robertson, A. H. F., and Hudson, J. D., Cyprus umbers: chemical precipitates on a tethyan ocean ridge, *Earth Planet. Sci. Lett., 18*, 93-101, 1973.

Rullkötter, J., Littke, R., Radke, M., Horsfield, B., and Thurow, J., Petrography and geochemistry of organic matter in Triassic and Cretaceous deep-sea sediments from the Wombat and Exmouth plateaus and nearby abyssal plains off Northwest Australia, in *Proc. ODP, Sci. Res., 122*, 317-333, Ocean Drilling Program, College Station, TX, 1992.

Rullkötter, J., Mukhopadhyay, P. K., and Welte, D. H., Geochemistry and petrography of organic matter from Deep Sea Drilling Project Site 603, lower continental rise off Cape Hatteras, *Init. Repts DSDP, 93*, 1163-1176, (U.S. Govt. Printing Office), Washington, D.C., 1987.

Sarmiento, J. L., Herbert, T. D., and Toggweiler, J. R., Causes of anoxia in the world ocean, *Global Biogeochemical Cycles, 2*, 115-128, 1988.

Schlanger, S. O., High frequency sea level fluctuations in Cretaceous time: an emerging geophysical problem, in *Mesozoic and Cenozoic Oceans, Am. Geophys. Union, Geodyn. Ser., 15*, edited by K. J. Hsü, 61-74, Washington, D.C., 1986.

Schlanger, S. O., and Cita, M. B., Introduction, in *Nature and Origin of Cretaceous Organic Carbon-Rich Facies*, edited by S. O. Schlanger, and M. B. Cita, 1-6, Academic Press, London, UK, 1982.

Schlanger, S. O., and Jenkyns, H. C., Cretaceous anoxic events: causes and consequences, *Geol. Mijnbouw, 55*, 179-184, 1976.

Schlanger, S. O., Arthur, M. A., Jenkyns, H. C., and Scholle, P. A., The Cenomanian-Turonian oceanic anoxic Event. I. Stratigraphy and distribution of organic carbon-rich beds and the marine $\partial^{13}C$ excursion, in *Marine Petroleum Source Rocks, Geol. Soc. London, Spec. Publ., 26*, edited by J. Brooks, and A. J. Fleet, 371-399, Blackwell, Oxford, 1987.

Schlanger, S. O., Jenkyns, H. C., and Premoli-Silva, I., Volcanism and vertical tectonics in the Pacific Basin related to global Cretaceous transgressions, *Earth Planet. Sci. Lett., 52*, 435-449, 1981.

Schmitz, B., Barium, equatorial high productivity, and the northward wandering of the Indian continent, *Paleoceanography, 2*, 63-78, 1987.

Scholle, P. A., and Arthur, M. A., Carbon isotope fluctuations in Cretaceous pelagic limestones: Potential stratigraphic and petroleum exploration tool, *Am. Assoc. Petrol. Geol. Bull., 64*, 67-87, 1980.

Shipboard Scientific Party Leg 123, Sedimentology of the Argo and Gascoyne Abyssal Plains, NW Australia: Report on Ocean Drilling Program Leg 123, *Carbonates and Evaporites, 3*, 201-212, 1988.

Stein, R., Surface-water paleo-productivity as inferred from sediments deposited in oxic and anoxic deep-water environments of the Mesozoic Atlantic Ocean, in *Biogeochemistry of Black Shales Mitt. Geol.-Paläont. Inst. Univ. Hamburg, 60*, edited by E. T. Degens, P. A. Meyers, and S. C. Brassell, 55-70, 1986.

Stein, R., Accumulation of organic carbon in marine sediments, *Lecture Notes Earth Sci., 34*, 1-217, 1991.

Stein, R., and Rullkötter, J., Littke, R., Schaefer, R. G., and Welte, D. H., Organofacies reconstruction and lipid geochemistry of sediments from the Galicia margin, Northeast Atlantic (ODP Leg 103), *Proc. ODP, Sci. Res., 103*, 567-585, Ocean Drilling Program, College Station, TX, 1988.

Stein, R., Rullkötter, J., and Welte, D. H., Accumulation of organic-carbon-rich sediments in the late Jurassic and Cretaceous Atlantic Ocean - A synthesis, *Chem. Geol., 56*, Amsterdam, 1-32, 1986.

Thurow, J., Die kretazischen Turbiditserien im Gibraltarbogen: Bindeglied zwischen atlantischer und tethyaler Entwicklung, Ph. Diss. Univ. Tübingen, 496pp, 1987.

Thurow, J., Cretaceous radiolarians of the North Atlantic Ocean: ODP Leg 103 (Sites 638, 640, and 641) and DSDP Legs 93 (Site 603) and 47B (Site 398), *Proc. ODP, Sci. Res., 103*, 379-418, Ocean Drilling Program, College Station, TX, 1988a.

Thurow, J., Diagenetic history of Cretaceous radiolarians, North Atlantic Ocean (ODP Leg 103 and DSDP Holes 398D and 603B), *Proc. ODP, Sci. Res., 103*, 531-555, Ocean Drilling Program, College Station, TX, 1988b.

Thurow, J., Modelling late Mesozoic paleoceanographic events: The example of the Cenomanian/ Turonian Boundary Event (CTBE), Habilitation Thesis, Univ. Tübingen, 77pp., 1992.

Thurow, J., and Gibling, M., Hydrocarbon potential, organic matter diagenesis, sedimentology, and paleoenvironment of upper Jurassic dark shales, Northern Himalaya and Argo Abyssal Plain, *Abstracts 28, Int. Geol. Congr.*, Washington, D.C., 3/238, 1989.

Thurow, J., and Kuhnt, W. The Cretaceous seaway between Northern Atlantic and Western Tethys - the paleostrait of Gibraltar, *Abstr. 2 Int. Conf. Paleoceanogr.*, Woods Hole, MA., 1986b.

Thurow, J., and Kuhnt, W., Mid-Cretaceous of the Gibraltar Arch Area, in *North Atlantic Palaeoceanography, Geol. Soc. London, Spec. Publ., 21*, edited by C. P. Summerhayes, and N. J. Shackleton, 423-445, Blackwell, Oxford, UK, 1986a.

Thurow, J., Kuhnt, W., and Wiedmann, J., Zeitlicher und paläogeographischer Rahmen der Phthanit- und Black Shale-Sedimentation in Marokko, *N. Jb. Geol. Paläont., Abh., 165*, 147-176, Stuttgart, 1982.

Thurow, J., Moullade, M., Brumsack, H.-J., Masure, E., Taugourdeau, J., and Dunham, K., The Cenomanian/Turonian Boundary event (CTBE) at Hole 641A, ODP Leg 103 (compared with the CTBE interval at Site 398), *Proc. ODP, Sci. Res., 103*, 587-634, Ocean Drilling Program, College Station, TX, 1988.

Veevers, J. J., Heirtzler, J. R., Bolli, H. M., Carter, A. N., Cook, P. J., Krasheninnikov, V. A., McKnight, B. K., Proto-Decima, F., Renz, G. W., Robinson, P. T., Rocker, K., and Thayer, P. A., *Init. Repts. DSDP, 27*, 1974.

Vogt, P. R., Volcanogenic upwelling of anoxic, nutrient-rich water: A possible factor in carbonate-bank/reef demise and benthic faunal extinction, *Geol. Soc. Am. Bull., 101*, 1225-1245, 1989.

von Rad, U., Haq, B. U. et al., *Proc. ODP, Sci. Results, 122*, Ocean Drilling Program, College Station, TX, 1992.

von Rad, U., Thurow, J., Haq, B. U., Gradstein, F. M., Ludden, J. and Leg 122/123 Shipboard Scientific Parties, Triassic to Cenozoic evolution of the NW-Australian continental margin and the birth of the Indian ocean (preliminary results of ODP Legs 122 and 123), *Geol. Rdsch., 78*, Stuttgart, 1189-1209, 1989.

von Stackelberg, U., Faziesverteilung in Sedimenten des indisch-pakistanischen Kontinentalrandes (Arabisches Meer),*"Meteor"-Forsch. Ergebnisse, C, 9*, 1-73, 1972.

von Stackelberg, U., Pumice and buried manganese nodules from the equatorial North Pacific Ocean, *Geol. Jb., D87*, 229-285, 1987.

Waples, D. W., Reappraisal of anoxia and organic richness, with emphasis on Cretaceous of North Atlantic, *Am. Assoc. Petrol. Geol. Bull., 67*, 963-978, 1983.

Wedepohl, K. H., Environmental influences on the chemical composition of shales and clays, in *Physics and Chemistry of the Earth, 8*, edited by L. H. Ahrens et al., 305-333, Pergamon Press, Oxford and New York, 1970.

Weissert, H., C-Isotope stratigraphy, a monitor of paleoenvironmental change: A case study from the early Cretaceous, *Surveys in Geophysics, 10*, 1-61, 1989.

Wiedmann, J., Butt, A., and Einsele, G., Vergleich von marokkanischen Kreide-Küstenaufschlüssen und Tiefseebohrungen (DSDP): Stratigraphie, Paläoenvironment und Subsidenz an einem passiven Kontinentalrand, *Geol. Rdsch., 67*, 454-508, Stuttgart, 1978.

Wiedmann, J., Butt, A., and Einsele, G., Cretaceous stratigraphy, environment, and subsidence history at the Moroccan continental margin, in *Geology of the Northwest African Continental Margin*, edited by U. von Rad, K. Hinz, M. Sarnthein, and E. Seibold, Springer, Berlin, 366-395, 1982.

Zimmerman, H. B., Boersma, A., and McCoy, F. W., Carbonaceous sediments and palaeoenvironment of the Cretaceous South Atlantic Ocean, in *Marine Petroleum Source Rocks, Geol. Soc. London, Spec. Publ., 26*, edited by J. Brooks, and A. J. Fleet, 271-286, Blackwell, Oxford, 1987.

The Cretaceous/Tertiary Boundary in the Southern Indian Ocean: Results from the Coring Operations of the Ocean Drilling Program

JAMES J. POSPICHAL

Department of Geology, Florida State University, Tallahassee, FL, 32306, USA

BRIAN T. HUBER

Department of Paleobiology, NHB-121, National Museum of Natural History, Smithsonian Institution, Washington, D.C., 20560, USA

Recent coring in the Indian Ocean during Ocean Drilling Program (ODP) Legs 119-122 has resulted in the recovery of several continuous or near continuous Cretaceous/Tertiary (K/T) boundary sections. Paleontological, sedimentological, and geochemical data have been compiled from these sequences and are summarized in this report with emphasis on calcareous nannofossil and foraminiferal distributions.

Leg 119 recovered a complete K/T section at Site 738 on the southern tip of the Kerguelen Plateau at a latitude of 63°S. This high-latitude sequence was complemented by the Leg 120 Site 750 recovery of a nearly complete section at 55°S, also on the Kerguelen Plateau. These two sections are important in that they provide data from the high latitudes, where few sites have been drilled.

Leg 121 recovered an expanded K/T section at Site 752 on Broken Ridge in the eastern Indian Ocean. Broken Ridge comprised the northern portion of the Kerguelen Plateau and was located at about 50-55°S latitude during K/T time, before the onset of rifting during the middle Eocene caused northward movement along the Southeast Indian Ridge to its present location at 30°S. This site received a continuous supply of volcanic ash throughout K/T time, which resulted in a highly expanded Danian sequence comparable in thickness to such well-studied sections as El Kef.

Leg 122 recovered an essentially complete K/T boundary at Site 761 off the northwest Australian margin. This site, with a paleolatitude of approximately 30°S, provides the northern anchor of the Indian Ocean transect.

The sections mentioned above, with the exception of Site 750, contain iridium anomalies at or close to the paleontologically defined K/T boundaries. Magnetostratigraphy is available for Sites 738, 752 and 761 where the boundaries fall within zones of reversed polarity assigned to Subchron C29R. Biostratigraphic analyses indicate that the K/T sections of Sites 738 and 752 are most complete, while Sites 750 and 761 are missing portions of the lowest Danian biozones. The topmost Maastrichtian may be absent at Site 761. Preservation of calcareous microfossils is generally moderate or poor at each site, but intervals with good preservation are present above and below the boundary at Sites 738 and 750.

INTRODUCTION

The biotic crisis at the Cretaceous/Tertiary (K/T) boundary has received considerable attention from scientists especially in the last decade. This effort has been fueled by the controversial proposal by Alvarez et al. [1980] that the mass extinctions in both the marine and the terrestrial realms were caused by the impact of an asteroid. Evidence for this theory came in the form of anomalous abundances of iridium found in the K/T boundary clay at Gubbio, Italy. Conversely, the excess iridium has also been attributed to extensive volcanism at K/T time [Officer et al., 1987]. The emplacement of the Deccan Trap flood basalts during this time has been suggested to have created the environmental perturbations necessary for mass extinctions [Caldiera and Rampino, 1990].

Deep sea drilling (DSDP and ODP) has contributed important data on this problem with the recovery of

Synthesis of Results from Scientific
Drilling in the Indian Ocean
Geophysical Monograph 70
Copyright 1992 American Geophysical Union

the subject of intense multidisciplinary studies over the past two decades [for summaries, see Thierstein, 1981; Perch-Nielsen et al., 1982; Smit and Romein, 1985]. Deep-sea cores in many cases can provide a relatively undisturbed record of the sedimentological and biological processes which occurred across the K/T boundary. It is well known that the terminal Cretaceous crisis in the marine realm is best illustrated by the abrupt extinctions of the calcareous microplankton, which include planktonic foraminifers and photosynthetic nannoplankton. Fossils of these organisms are usually abundant in pelagic sediments, which facilitates detailed study of the biotic response to environmental changes across the K/T boundary.

The recent coring effort in the Indian Ocean by ODP (Figures 1 and 2) has produced several sequences which allow the database to be extended to this ocean for detailed comparisons. The following is a summary of the results from studies of the K/T cores recovered during Legs 119-122. The boundaries discussed here are considered to be the most complete recovered in the Indian Ocean based on a combination of magneto-, chemo-, and biostratigraphic

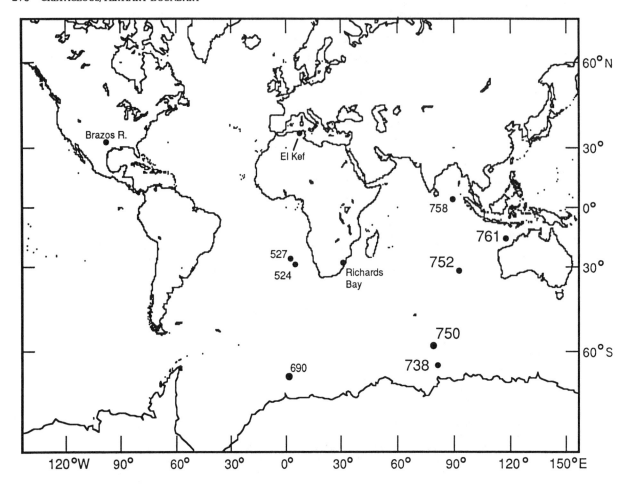

Fig. 1. Location of Indian Ocean ODP Sites 738, 750, 752, 761, and several other DSDP and outcrop K/T sites referred to in this report.

criteria. Figure 3 provides a comparison of the core recovery, lithology, and magnetobiostratigraphy for the K/T boundaries of Sites 738, 750, 752, and 761. Emphasis is placed on the comparison of calcareous nannofossil and foraminiferal assemblages from each boundary sequence.

Other Indian Ocean boundary sections that have been cored by ODP/DSDP, but are not reviewed in this report include northern Ninetyeast Ridge Sites 216 and 217 [Proto-Decima, 1977], and recently cored Site 758 [Peirce, Weissel et al., 1989]. Each of these sections have been determined to be biostratigraphically incomplete.

SITE 738-SOUTHERN KERGUELEN PLATEAU
Lithology and Clay Mineralogy

During ODP Leg 119, Hole 738C was cored at a water depth of 2252.5 m at 62.7°S, 82.8°E on the southern Kerguelen Plateau (Figure 1). Cretaceous/Tertiary boundary sediments were recovered by the rotary coring process in Section 119-738C-20R-5 at 377.16 meters below sea floor (mbsf). The section includes uppermost Maastrichtian sediments comprised of well-indurated white chalk with chert layers and overlying less indurated Paleocene chalks and oozes (Figures 3,4) [Barron, Larsen et al., 1989].

The boundary in Section 119-738C-20R-5 occurs within a carbonate and clay-rich thinly laminated interval approximately 15 cm thick (Figures 3,4). It is placed at the level of a distinct 2 mm thick gray clay layer at 96.0-96.2 cm, about 2 cm above the base of the laminated interval. Black to dark gray chert is present below the boundary clay at the base of Section 119-738C-20R-5 from 103 to 117 cm.

As detailed in Ehrmann [1991] and Thierstein et. al. [1991], clay analysis of the K/T boundary interval of Site 738 indicates an overall dominance of smectite (88%-97.5%) over illite (2%-12%), chlorite (<4%), and kaolinite (<2%). No significant difference was noted in the clay mineral composition between the 2mm gray clay layer and the sediments above and below. This led Thierstein et al. [1991] to suggest that the boundary gray clay was not derived from impact or impact-related processes.

Biostratigraphy

Calcareous nannofossils. Calcareous nannofossils are generally poorly to moderately preserved in K/T boundary sediments at Site 738 [Thierstein et al., 1991]. Specimens are often fragmented, etched or overgrown. Nannofossils

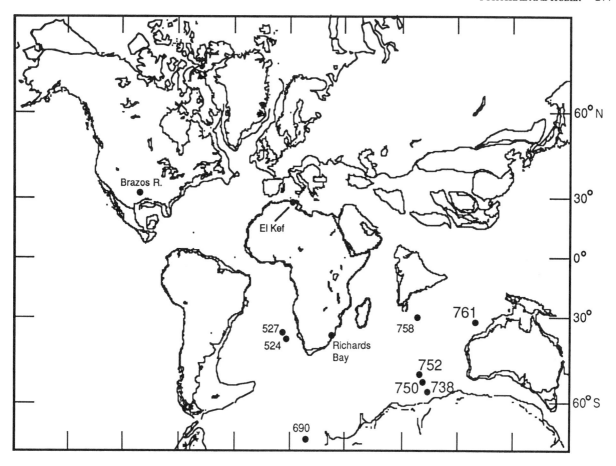

Fig. 2. Tectonic reconstruction of the Indian Ocean for 66 Ma (Scotese and Denham, Terra Mobilis^{tm}) with approximate locations of ODP, DSDP and outcrop K/T sections referred to in this paper.

are less well preserved in the 2 mm boundary clay than in the surrounding chalks.

The biostratigraphy and magnetostratigraphy for this site, as well as those discussed in the following sections, are summarized in Figure 3. Wei and Thierstein [1991] assign the uppermost Maastrichtian of Site 738 to Zone NC23 based on the first occurrence (FO) of *Nephrolithus frequens* near the top of Core 119-738C-24R. The commonly used low- to mid-latitude markers for the uppermost Maastrichtian, *Micula murus* and *M. prinsii* were not recorded at this site.

The uppermost Maastrichtian of Site 738 appears to be intact based on the austral high-latitude calcareous nannofossil zonation scheme of Pospichal and Wise [1990a] (Figure 3). Sediments just below the boundary can be assigned to the *Cribrosphaerella daniae* Subzone based on the presence of that species down to the top of Core 119-738C-24R and the absence of the "middle" to lower Maastrichtian species *Nephrolithus corystus*, *Biscutum magnum*, and *Reinhardtites levis*. When the more detailed high austral latitude scheme of Watkins et al. [1991] is applied, the uppermost portion of the *C. daniae* Subzone can be further subdivided by the presence of an acme of *Prediscosphaera stoveri*. This species was noted in high

abundances in samples just below the boundary at Site 738 [Wei and Thierstein, 1991; J. J. Pospichal, unpublished data, 1992]. Additional nannofossil taxa present in the uppermost Maastrichtian include the common high-latitude species, *Kamptnerius magnificus*, *Arkhangelskiella cymbiformis*, *Prediscosphaera cretacea*, and *Acuturris scotus*. The assemblage is similar to that reported from austral high latitude Sites 689 and 690 cored on Maud Rise in the Weddell Sea [Pospichal and Wise, 1990a, 1990b].

The laminated interval above the boundary clay (96.1 cm) in Section 119-738C-20R-5 can be assigned to lowermost Danian Zone NP1 (CP1a). The marker for the top of this zone, *Cruciplacolithus tenuis*, was reported by Thierstein et al. [1991] to first occur at 70.5 cm, although Wei and Pospichal [1991] note its FO above this level between 35 cm and 2 cm in Section 119-738C-20R-5. With a closer sampling interval, the FO was found at 16 cm [J. J. Pospichal, unpublished data, 1992]. Regardless of these discrepancies, the presence of an interval above the boundary, which lacks *C. tenuis* and its predecessor, *C. primus*, is generally indicative of a biostratigraphically complete section. This interval can be further subdivided by the FO of *Hornibrookina* at 86 cm. This event denotes the top of high austral latitude Zone NA1 of Wei and

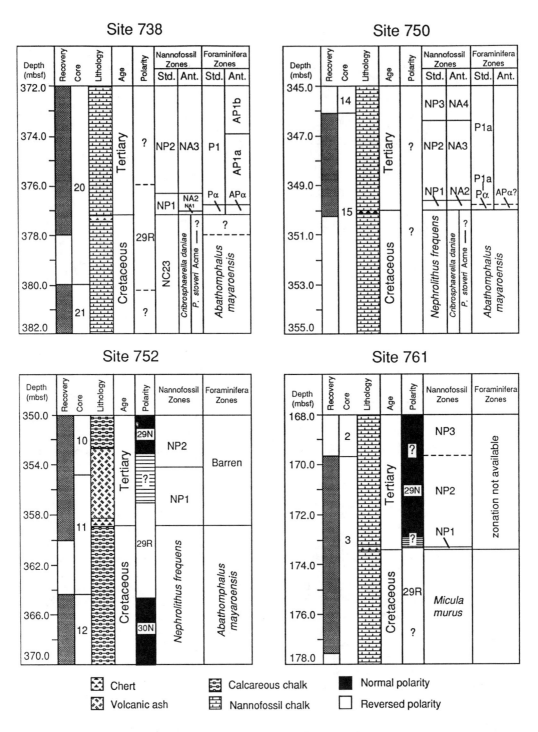

Fig. 3. Summary of core recovery, litho-, magneto-, and biostratigraphy for K/T boundary cores discussed in this report. For Sites 738 and 750, Std. = standard zonation schemes; Ant. = high austral latitude zonation schemes.

Pospichal [1991] (Figure 3).

Zone NP1 is characterized by the common presence of species which are considered to have crossed the boundary after having comprised a very small portion of the Cretaceous assemblage. These species, referred to as survivors or persistent species, sometimes exhibit abundance blooms as reported from a number of K/T sections [see Thierstein, 1981]. Commonly a successional dominance occurs above the boundary, most notably with species of *Thoracosphaera*, a calcareous dinoflagellate, and *Braarudosphaera* [e.g. Jiang and Gartner, 1986]. Cretaceous species are also common in Zone NP1 and gradually

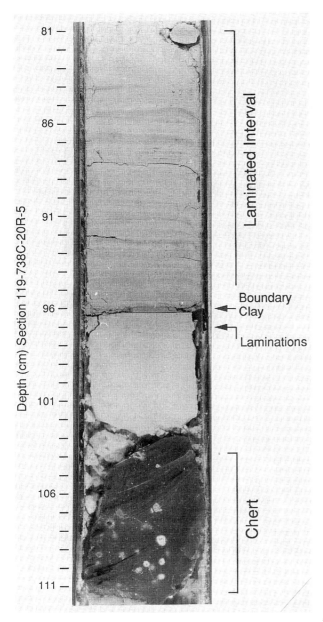

Fig. 4. The K/T boundary of Site 738 (ODP Section 119-738C-20R-5, 81-111 cm). The 2 mm thick boundary clay is indicated by an arrow. Dark material below 103 cm is chert.

decrease in abundance upsection. These species have generally been considered reworked although this has been challenged [Perch-Nielsen et al., 1982; Thierstein et al., 1991, see discussion].

At Site 738, the persistent or survivor assemblage consists of common to abundant *Placozygus sigmoides* with few *Markalius inversus, Biscutum castrorum* (*B. constans* or oval morphotype of *M. inversus* of some authors), *Neocrepidolithus* spp., and very rare *Cyclagelosphaera reinhardtii*. The assemblage is most similar to that reported from Site 690 in the Weddell Sea [Pospichal and Wise, 1991b] and those of the northern high latitudes [Perch-Nielsen et al., 1982]. *Thoracosphaera*

is also present in Site 738 sediments, but is not as abundant as at sites such as El Kef and Brazos River where it is considered to have bloomed [Perch-Nielsen et al., 1982; Jiang and Gartner, 1986]; there it comprises over 50% of the assemblage immediately above the boundary. *Braarudosphaera*, also noted for abundance blooms, is not present in sediments above the boundary at Site 738.

The first calcareous nannofossil species to appear (in an evolutionary sense) above the boundary at Site 738 are rare *Biantholithus sparsus* and few *Hornibrookina* at 10 cm above the boundary. *Hornibrookina*, assigned to *H. teuriensis* in Wei and Pospichal [1991], becomes abundant at about 35 cm above the boundary. The next to appear, *Cruciplacolithus primus* occurs 60 cm above the boundary and is followed by *C. tenuis* and *Prinsius dimorphosus* about 20 centimeters upsection. *Prinsius dimorphosus* is very abundant and completely dominates the assemblage of NP2.

The earliest Danian assemblage of Site 738 lacks several species noted at many low- and mid-latitude sections. These include, *Neobiscutum romeinii, N. parvulum*, and *Futyania petalosa*, which have been noted in abundance especially in the Tethyan sequences [see summary of Perch-Nielsen et al., 1982].

Foraminifers. Foraminiferal preservation is quite poor and very few identifiable specimens were isolated in the well-indurated chalk immediately below the K/T boundary at Site 738 [Huber, 1991]. Identified planktonic species include *Heterohelix globulosa, Globigerinelloides multispina* and *G. subcarinatus*, all of which are typical components of Maastrichtian assemblages of the Austral Realm. The upper Maastrichtian marker, *Abathomphalus mayaroensis* is not present immediately below the boundary, but it occurs in the core-catcher of Core 119-738C-20R and down through Core 119-738C-23R (see discussion below). Planktonic foraminifers from immediately above the boundary clay are diagnostic of the lowermost Danian *Eoglobigerina fringa* Zone (APα Zone) of Stott and Kennett [1990] (Figure 3). The nominate taxon of the *Parvularugoglobigerina eugubina* Zone (Pα Zone) used in tropical zonal schemes [Berggren and Miller, 1988] is notably absent from the Site 738 core. *Parvularugoglobigerina eugubina* and *Woodringina* spp. have not been reported at any K/T site in the circum-Antarctic region. Also absent is the *Guembelitria cretacea* Zone (P0 Zone), which is recognized in more expanded nearshore sections such as El Kef [Smit, 1982].

Other planktonic species identified in the APα Zone at Site 738 include *Eoglobigerina fringa, E. eobulloides*, and *Chiloguembelina crinita*. The base of the overlying *Globoconusa daubjergensis* Subzone (AP1a) of Huber [1991] occurs within 39 cm above the boundary. It is characterized by the presence of the nominate taxon along with *Subbotina pseudobulloides* and *E. simplicissima*. Foraminiferal preservation dramatically improves within this subzone and is good for most of the overlying Danian sequence. Reworked Cretaceous specimens occur throughout most of the Paleocene sequence at Site 738.

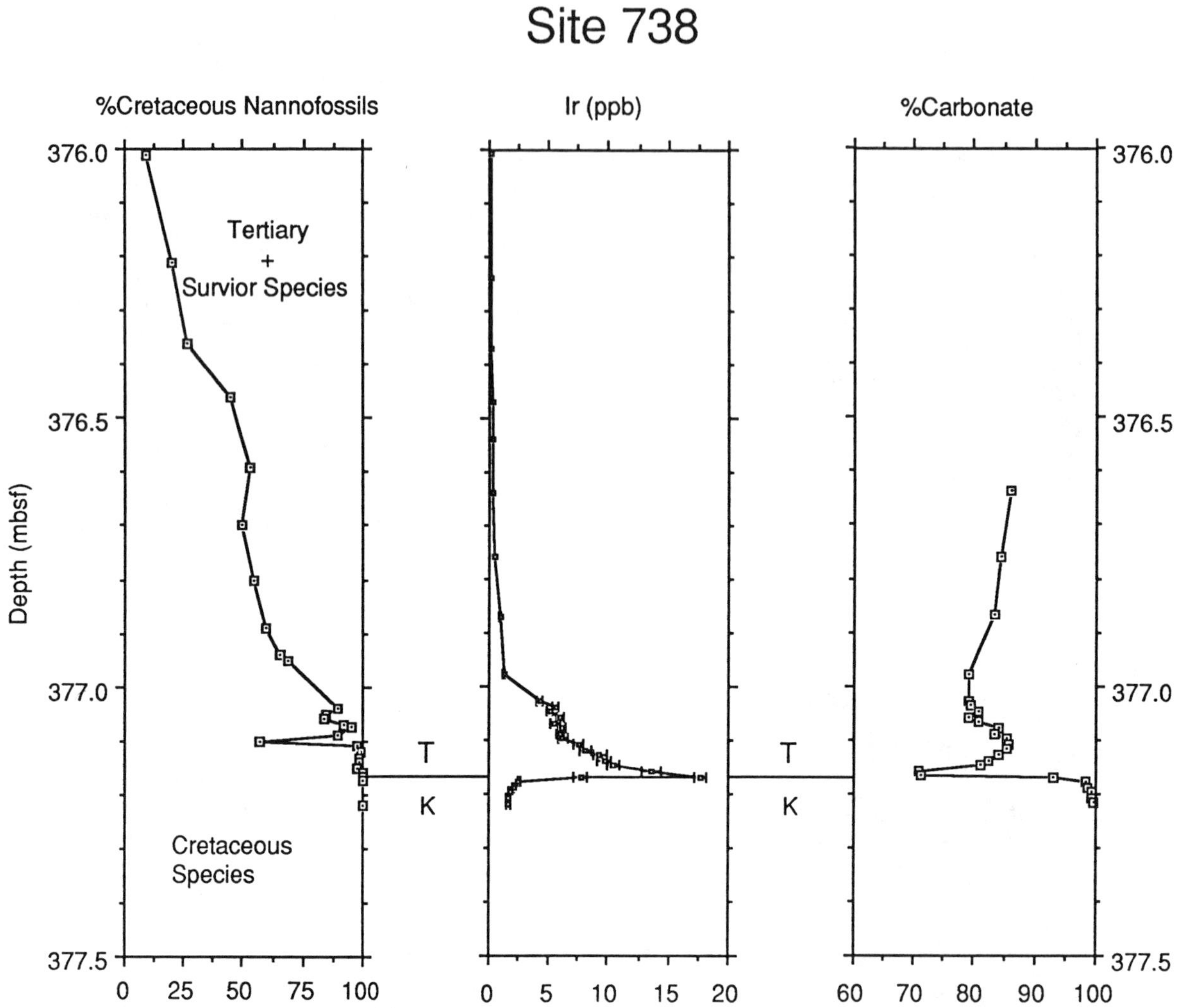

Site 738

Fig. 5. Iridium abundance in parts per billion (ppb), percent carbonate, and calcareous nannofossil turnover across the K/T boundary of Site 738. Nannofossil data are from J.J. Pospichal [unpub. data, 1992]. Iridium and carbonate data replotted from Schmitz et al. [1991].

Although some observations on the distribution of benthic foraminifers above and below the K/T boundary were included in the report of Huber [1991], a detailed biostratigraphic study of the benthic turnover is still needed. Assemblage changes that were noted include: (1) disappearance within 3 m below the boundary clay of the large and ornate species *Neoflabellina praereticulata*, *Frondicularia* spp., and *Bolivinoides draco*, (2) the test size of *Gavelinella beccariiformis* is reduced and benthic species diversity decreases within 0.5 m below the boundary clay, and (3) the benthic assemblages immediately above the boundary are not significantly different from those just below. Benthic foraminifers comprise from 35% to 81% of assemblages within the lower Danian portion of the laminated interval. Species that dominate the uppermost Maastrichtian and lowermost Danian assemblages include *G. beccariiformis*, *Nuttalides truempyi*, and *Gyroidinoides globulosus*. Assemblages in both intervals indicate deposition in a middle to upper bathyal paleodepth.

Discussion. The calcareous nannofossil and planktonic foraminiferal assemblage turnover which occurs within the 15 cm laminated interval is viewed by Thierstein et al. [1991] as a gradual transition. Figure 5 illustrates the nannoplankton transition from an assemblage dominated by Cretaceous species to one primarily composed of Tertiary taxa including persistent or survivor species. The near absence of bioturbation in this boundary may be consistent with the interpretation that there was a gradual reduction of Maastrichtian taxa over a period of time. Hence, the Maastrichtian specimens present above the K/T boundary may be considered *in situ* and not present as the result of upward reworking by bioturbation as has been classically viewed. However, redeposition by current activity cannot be ruled out. In addition, evidence from the combination of iridium and nannofossil data of Site 752 (see appropriate section below) may also suggest that Cretaceous species present above the boundary are the result of redeposition.

Huber [1991] cited evidence indicating that environmental changes may have begun just prior to the deposition of the boundary clay. These include: (1) disappearance of several benthic foraminifer species within 3 m below the K/T boundary, (2) occurrence of several specimens that resemble the Tertiary genus *Chiloguembelina* within 5 cm below the boundary clay, (3) absence of keeled planktonic foraminifers from upper Maastrichtian sediments between the core-catcher sample and the boundary clay in Section 119-738C-20R-5, (4) appearance of laminae and hence, loss of benthic bioturbation 2 cm below the boundary clay, and (5) presence of diminutive forms of planktonic and benthic foraminifers. However, only the presence of laminae below the boundary clay can be considered as unequivocal evidence of pre-boundary change, as the strongly lithified nature of the K/T sediments biases against recovery of larger specimens, including the keeled planktonic and larger benthic taxa, and downhole contamination of the chiloguembelinid morphotypes cannot be ruled out. Detailed thin-section study of this K/T interval would provide a much more accurate determination of foraminiferal size and relative abundance changes that occur in this K/T sequence.

Magnetostratigraphy

The Cretaceous/Tertiary boundary at Site 738 lies within an interval of reversed magnetic polarity [Sakai and Keating, 1991], which most likely corresponds to Subchron C29R (Figure 3). Consistent magnetic data could not be attained from cores above and below K/T boundary in Core 119-738C-20R.

Geochemistry

The distribution of rare-earth elements was analyzed by Schmitz et al. [1991] in a 1-m-thick interval across the K/T boundary (Figure 5). The highest concentration of iridium was noted within the distinct 2 mm gray clay layer, which lies 2 cm above the base the laminated interval. The iridium enrichment of 320 ng/cm² (18 ppb Ir, whole-rock samples) measured in the gray boundary clay is one of the highest of all known K/T boundary sections.

Bulk carbonate content analysis provided by Thierstein et al. [1991] indicates a drop in the percentage of carbonate from 96% in the uppermost Maastrichtian to 91% in the 2 cm interval below the boundary clay layer (Figure 5). The carbonate content in the boundary clay of 69% was the lowest value measured in the interval investigated.

SITE 750-KERGUELEN PLATEAU

Lithology

During Leg 120, Hole 750A was rotary cored at a water depth of 2030.5 m at 57°35.54'S and 81°14.42'E on the Kerguelen Plateau (Figure 1). The K/T boundary was recovered in Section 120-750A-15R-3 at 91.5 cm (349.50 mbsf). The uppermost Maastrichtian sediments consist of semi-indurated white nannofossil chalk, which is overlain by more indurated lower Danian grayish green marl grading upward to white chalk [Schlich, Wise et al., 1989] (Figure 6). The lithology is similar to that at Site 738, but no

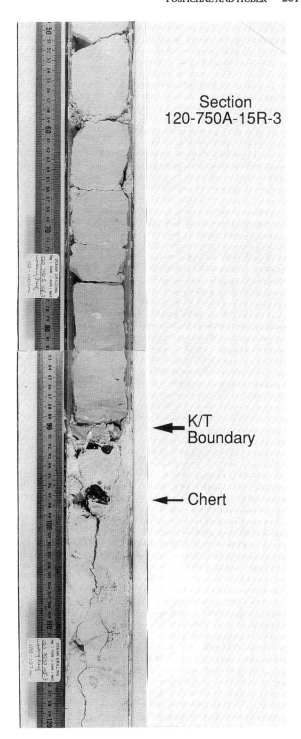

Section
120-750A-15R-3

← K/T Boundary

← Chert

Fig. 6. The K/T boundary from Site 750 (Section 120-750A-15R-3, 50-120 cm). The black, cm size, subangular clasts are chert fragments. Darker sediments above the boundary are calcareous marls. The white sediment below is nannofossil chalk.

laminations or distinct clay layer are present. The sharp contact between Maastrichtian and Paleocene sediments is fractured and disturbed by the drilling process. Fragments of chert that appear to have been emplaced as the result of drilling disturbance are also present in the boundary interval [Schlich, Wise et al., 1989]. No iridium anomaly was noted at this site which may be due to this drilling disturbance, poor recovery, and/or the presence of a disconformity at the boundary as discussed in the following section.

Biostratigraphy

Calcareous nannofossils. Calcareous nannofossils from the K/T boundary of Site 750 were studied by Ehrendorfer and Aubry [1992], who noted assemblages similar to those of Site 738 and Site 690 (65°S, Weddell Sea). Nannofossils in the boundary interval are abundant and moderately well preserved.

The uppermost Maastrichtian from the base of Core 120-750A-19R to the K/T boundary at Sample 120-750A-15R-3, 91.5 cm is assigned to the *Nephrolithus frequens* Zone (Figure 3). Nannofossil assemblages of this interval appear typical of the high austral latitudes and include the characteristic high abundance of *Prediscosphaera stoveri* [Ehrendorfer and Aubry, 1992; Watkins, 1992]. Other species present include *Nephrolithus frequens, Kamptnerius magnificus, Prediscosphaera cretacea, Cribrosphaerella daniae,* and *Lucianorhabdus cayeuxii.* The topmost Maastrichtian may not be represented at this site. The sharp irregular contact at the K/T boundary may indicate the presence of a disconformity, or it may be an artifact of drilling disturbance. In either case, it is likely that stratigraphic recovery across this boundary interval is incomplete.

Only 5 cm of sediment separate the K/T contact and the FO of *Cruciplacolithus* above, which may further indicate that some the lowermost Danian sediment is missing. The first *Cruciplacolithus,* assigned to *C. primus* by Ehrendorfer and Aubry [1992], was noted in Sample 15R-3, 86 cm. This species was not considered to be *C. tenuis* because it lacked "feet" on the ends of its crossbars, which span the central area of the coccolith. Rare specimens with "feet" were noted in the sample at 51 cm where Ehrendorfer and Aubry [1992] place the NP1/NP2 boundary. It should be noted that this is not the same criterion used by Wei and Pospichal [1991] at Site 738. At that site, as well as at Sites 752 and 761, the NP1/NP2 boundary was determined by the FO of *Cruciplacolithus* greater than or equal to 7μm. At Site 750, these forms are present 5 cm above the K/T boundary. Nonetheless, the thickness of NP1 in Figure 3 is shown as reported by Ehrendorfer and Aubry [1992]. Furthermore, the FO of *Hornibrookina* (assigned to *H.* sp. cf. *H. teuriensis*) at Site 750 is noted right at the boundary (91.5 cm), which indicates that the austral high latitude Zone NA1 of Wei and Pospichal [1991] is not present.

The persistent or survivor assemblage immediately above the boundary in Zone NP1 is similar to that of Site 738 and consists of *Placozygus sigmoides, Thoracosphaera* spp., *Markalius inversus, Biscutum castrorum,* and *Neocrepidolithus* spp. This assemblage is replaced several centimeters above the K/T boundary by one comprised of *Cruciplacolithus, Hornibrookina,* and small *Prinsius.* As at Site 738 and the other Indian Ocean sites, *Prinsius* spp.

completely dominate the assemblage of Zone NP2.

The earliest Danian species of *Neobiscutum,* well known from the Tethyan K/T sequences, were also not noted at Site 750. However, several specimens of *Futyania petalosa* were observed in Zone NP2 sediments [J. J. Pospichal, unpublished data, 1992]. This extends the geographic range of this species to at least 57°S in the Indian Ocean and reflects its wider distribution when compared to *Neobiscutum* as noted by Perch-Nielsen et al. [1982].

Foraminifers. The presence of *Pseudotextularia elegans* in the upper part of the *Abathomphalus mayaroensis* Zone at Site 750 [Quilty, 1992] enables correlation with the *P. elegans* Subzone, and indicates that the upper Maastrichtian is biostratigraphically complete [Huber, 1992]. Preservation is quite good throughout most of the Maastrichtian interval. Although the oldest Danian sediments contain planktonic foraminifers that resemble assemblages from the APα Zone, their preservation is too poor for diagnostic taxonomic identification [Berggren, 1992]. *Globoconusa daubjergensis* and *Subbotina pseudobulloides* first appear 20-25 cm above the boundary, which is indicative of Zone P1a. The interval between this level and the K/T boundary is assigned to undifferentiated Zones Pα-P1a [Berggren, 1992]. Reworking of Cretaceous planktonic foraminifers occurs throughout the Danian section.

Benthic foraminifers are quite rare in the Maastrichtian and Danian sequence at Site 750, comprising less than 5% of the assemblage. A detailed study of the samples bounding the K/T boundary at Site 750 has not been published, so only general comments can be drawn from the shipboard report of Schlich, Wise et al. [1989]. *Bolivinoides draco, Gavelinella beccariiformis, Nuttalides truempyi,* and *N. florealis* are mentioned as important components of the uppermost Maastrichtian assemblages, while earliest Danian assemblages are predominantly composed of *G. beccariiformis* and *N. truempyi,* with less abundant occurrences of *Bolivinoides delicatulus, Anomalina praeacuta, Pullenia coryelli, Bulimina trinitatensis,* and *Neoepinides hillebrandti.* The faunal composition of the benthic assemblages suggests that the paleobathymetry increased from upper bathyal during the late Maastrichtian to middle to lower bathyal during the early Danian.

Magnetostratigraphy

Because of poor core recovery, polarity reversal boundaries could not be constrained with confidence. Thus, a complete magnetostratigraphy of the K/T boundary interval of Site 750 could not be achieved [Zachos et al., 1992].

Geochemistry

Stable isotopes and carbonate content were analyzed by Zachos et al. [1992] (Figure 7). Isotopic values measured from planktonic and benthic foraminifera tests reflect a characteristic absence of $\delta^{13}C$ vertical gradients during earliest Danian Zone NP1, which has been attributed to the

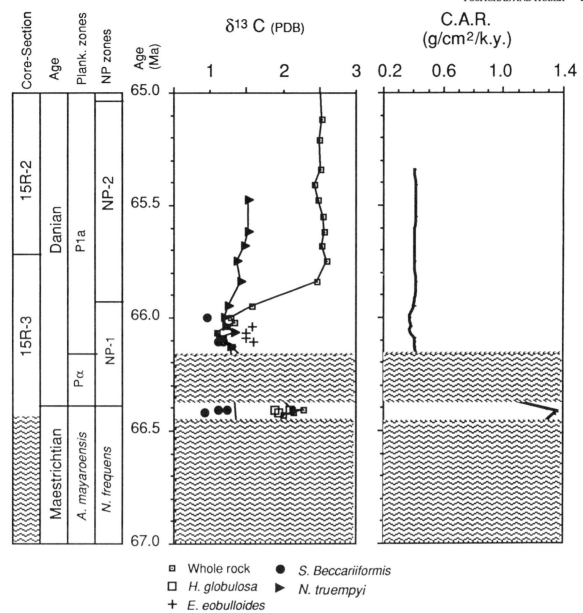

Fig. 7. Whole rock, fine fraction (<63 µm), planktonic and benthic foraminifera stable carbon isotope values and accumulation rates across the K/T boundary in Hole 750A [modified from Zachos et al., 1992]. Absolute ages are based on Berggren et al. [1985] for the Danian and Pospichal and Wise [1990a] for the Maestrichtian. Shaded region represents missing sediment.

significant drop in surface productivity at K/T time [Hsü et al., 1982; Zachos and Arthur, 1986]. Concurrent with the lack of surface to deep water $\delta^{13}C$ gradients at Site 750 is a drop in biogenic carbonate accumulation rates. Vertical isotopic gradients and carbonate accumulation rates were reestablished beginning at about the NP1/NP2 boundary commensurate with an increase in abundance of Tertiary nannoplankton such as *Prinsius* spp. and *Cruciplacolithus* [Ehrendorfer and Aubry, 1992]. The estimated time for recovery at Site 750, considering the probable stratigraphic gap, is about 0.5 m.y. This is consistent with previous calculations from low- and mid-latitude K/T sections elsewhere [e.g. Zachos et al., 1989].

SITE 752-BROKEN RIDGE

Lithology

The Cretaceous/Tertiary boundary of Leg 121 Site 752 was cored at a water depth of 1086.3 m on Broken Ridge (30°53.483'S, 93°34.652'E) (Figure 1). At K/T time, Broken Ridge comprised the northern portion of the Kerguelen Plateau (Figure 2) before rifting and subsequent northward movement began in the middle Eocene. The estimated paleolatitude for Site 752 is 50-55°S [Peirce, Weissel et al., 1989]. The K/T boundary is placed at Sample 121-752B-11R-3, 94-95 cm (358.75 mbsf) within a sequence of ash, chert, and chalk which immediately underlies a 6-m to 6.5-m ash layer (Figure 8). The lithology

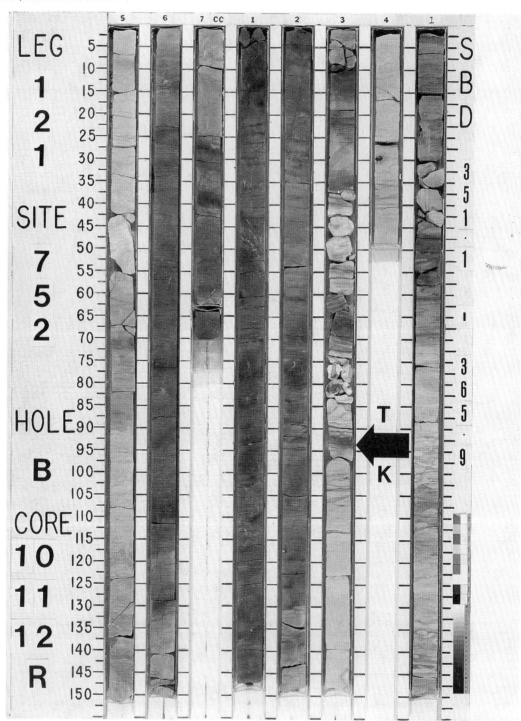

Fig. 8. ODP Leg 121 Cores 121-752B-10R, -11R, and -12R. The K/T boundary is located in Section 121-752B-11R-3 at 94-95 cm. The prominent thick, dark sediment above the boundary is volcanic ash. The light-colored sediment is chalk. The nannofossil NP1/NP2 boundary is in Section 121-752B-10R-6 at 135-136 cm [from Pospichal, 1991].

of the boundary interval (Figure 9) is described in detail by Peirce, Weissel et al. [1989]. The ash layer present between 90 and 95 cm overlies light gray, mottled, and well-indurated uppermost Maastrichtian chalk. The section from 85 to 90 cm, is a light-colored chalk with soft-sediment deformation structures. The overlying interval between 75 and 85 cm is composed of gray chert and porcellanite, which underlies chalk and grades upward into ash at about 60 cm. This is overlain by 25 cm of chalk, which lies just below an approximately 6-m-thick ash layer.

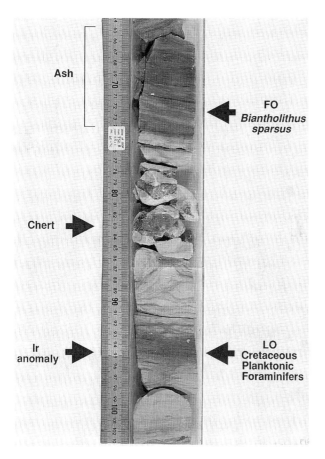

Fig. 9. Enlarged view of the K/T boundary (Section 121-752B-11R-3, 64-103 cm). The first occurrence (FO) of *Biantholithus sparsus* and the last occurrence (LO) of Cretaceous foraminifers are indicated, along with the location of the iridium anomaly, chert, and ash layers.

This K/T boundary interval is composed of numerous drilling "biscuits" produced by twisting and breaking during the rotary coring process (Figure 9). Detailed interpretations of this interval must be made with caution as much sediment can be missing from between the "biscuits." Recovery in Core 121-752B-11R was only 52% of the 9.6 m cored. However, logging data [Peirce, Weissel et al., 1989] suggest that most of the overlying 6 m of ash was recovered, and recovery in overlying Core 121-752B-10R was 100%.

Ash and CaCO3-Site 752

Mass accumulation rates on Broken Ridge for volcanic ash and calcium carbonate (Figure 10) are given in Peirce, Weissel et al. [1989] and Rea et al. [1990]. Ash flux from a local source was high during the Turonian through Santonian and gradually decreased through the Late Cretaceous. Ash is the major sedimentary component immediately overlying the K/T boundary as indicated by the presence of a 6 m to 6.5 m ash layer. Magnetic susceptibility data suggest that this ash unit is the result of multiple ash falls and not a single event [Peirce, Weissel et al., 1989]. This ash unit does not represent a sudden influx

of ash, but is the result of normal ash accumulation in the absence of calcium carbonate sedimentation.

Thus, the significant change in the sedimentary component across the K/T boundary is in the accumulation of the biogenic component. Ash flux increases only slightly. This sharp reduction in carbonate flux represents a major drop in productivity during the early Tertiary. The recovery of calcareous microplankton productivity is manifested in the sediments by an increase in carbonate content, which coincides with the transition from the thick basal Tertiary ash to chalk (Figure 8). This transition occurs almost 2 m above the NP1/NP2 boundary at Site 752. At nearby Site 750, the reestablishment of vertical carbon isotope gradients and high carbonate accumulation is shown by Zachos et al. [1992] to occur at the NP1/NP2 boundary. This discrepancy may be incidental in light of the strong ash component at Site 752, or it may reflect local variations in the pattern of productivity recovery in early Tertiary seas. However, it should be pointed out that, as mentioned previously, the method for denoting the NP1/NP2 boundary at Site 750 differs from that used for Site 752. If the method used for the latter site (the FO of *Cruciplacolithus* > 7μm) is employed at Site 750, then the NP1/NP2 boundary there would be placed about 35 cm below the level where carbonate accumulation values abruptly increase. This would produce a pattern similar to that at Site 752 where the NP1/NP2 boundary is below the level at which carbonate flux values begin to increase. Nonetheless, the estimated duration of the low-productivity interval at both Sites 750 and 752 is consistent with that noted by Zachos and Arthur (1986) and Zachos et al. [1989] for deep-sea Site 524 (South Atlantic) and Site 577 (North Pacific), respectively.

Biostratigraphy
Calcareous nannofossils. Nannofossil assemblages in K/T boundary sediments of Site 752 are similar to those found at Sites 738 and 750 with a few exceptions possibly due to the more northerly location and perhaps preservational differences [Pospichal, 1991]. Calcareous nannofossils are poorly preserved in the well-indurated chalks immediately below the boundary at 121-752B-11R-3, 94-95 cm. Preservation is equally poor and nannofossil abundance low in the porcellanitic sediments above the boundary. In ash layers above and below the boundary, nannofossil preservation is better as reflected by the presence of more diverse assemblages.

Figure 11 illustrates the drop in the percent abundance of Cretaceous calcareous nannofossils across the K/T boundary. The assemblage turnover is similar to other sites with the rapid replacement of Cretaceous species by survivor or persistent taxa, which are in turn replaced by an assemblage dominated by newly evolved Tertiary species. An interval about 80 cm above the boundary is characterized by a large increase in the percentage of Cretaceous species, which correlates well with an additional peak in iridium abundance (see discussion below).

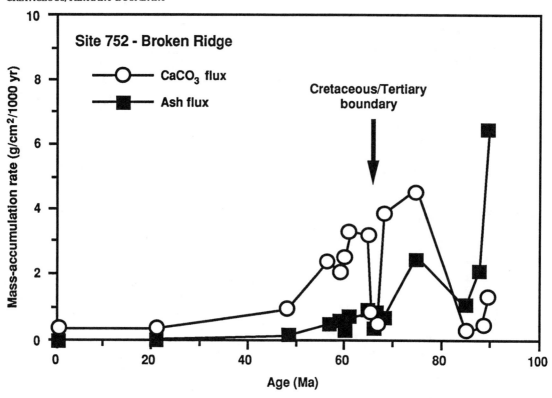

Fig. 10. Mass-accumulation rates of ash and calcium carbonate on Broken Ridge from the mid-Cretaceous to present [from Peirce, Weissel et al., 1989]. Note the major decline in CaCO₃ at the K/T boundary.

The uppermost Maastrichtian sediments of Site 752 are assigned to the *Nephrolithus frequens* Zone, which extends down to Section 121-752B-17R-1 [Resiwati, 1991] (Figure 3). As previously mentioned, the top of this zone in high austral latitudes can be delineated by the presence of an acme of *Prediscosphaera stoveri* [Pospichal and Wise, 1990a; Watkins et al., 1991]. However, at this site, poor preservation of nannofossils immediately below the boundary precludes a confident identification of such an interval [Pospichal, 1991]. Although higher abundances of *P. stoveri* were noted in the ash-rich layers where preservation was improved, abundances were still lower than at Sites 738 and 750. The lower abundance may also be attributed to the more northerly location of Site 752 or alternatively, the presence of a disconformity.

Additional constituents of the uppermost Maastrichtian nannofossil assemblage include *N. frequens, Kamptnerius magnificus, Arkhangelskiella cymbiformis,* and *Cribrosphaerella daniae.* The dissolution resistant species, *Micula decussata,* is very abundant in sediments below the boundary where preservation is poorest. Like Sites 738 and 750, the assemblage is characteristic of higher latitudes and lacks the low- to mid-latitude upper Maastrichtian components such as *Micula murus, M. prinsii,* and abundant *Watznaueria barnesae.*

The sediments immediately above the K/T boundary and up to the FO of *Cruciplacolithus tenuis* near the base of Section 121-752B-10R-6 are placed in Zone NP1

[Pospichal, 1991] (Figure 3). The expanded nature of this boundary is well illustrated when the thickness of this zone is compared to the approximately 20 to 80 cm at Site 738 and the 40 cm thickness at Site 750 (Figure 3). Moreover, unlike at Sites 738 and 750, small delicate forms of *Cruciplacolithus primus* appear just 73 cm above the boundary, well below the first *C. tenuis* at Site 752. Another difference is the near absence of *Hornibrookina* in Site 752 sediments. This taxon is a major component of the assemblage in the upper part of NP1 and in NP2 at Sites 738 and 750, but it is very rare throughout the section at Site 752 and only first appears near the top of NP2.

Similar to Sites 738 and 750, the assemblage of Zone NP1 at Site 752 is dominated by *Placozygus sigmoides.* Other taxa present include common *Thoracosphaera, Markalius inversus,* and *Biscutum castrorum. Biantholithus sparsus* first appears about 20 cm above the boundary and is rare (Figure 9). *Cruciplacolithus tenuis* and abundant to very abundant *Prinsius dimorphosus* are the dominant constituents of Zone NP2.

Site 752 was proximal to Site 750 on the northern portion of the Kerguelen Plateau (Figure 2) until rifting in the middle Eocene, and thus, similar nannofossil assemblages may be expected. However, as mentioned above, *Hornibrookina* is not nearly as abundant in the Danian assemblages of Site 752 as at Sites 738 or 750. A disconformity may account for this, but the expanded thickness of the section plus the observation that this

Site 752

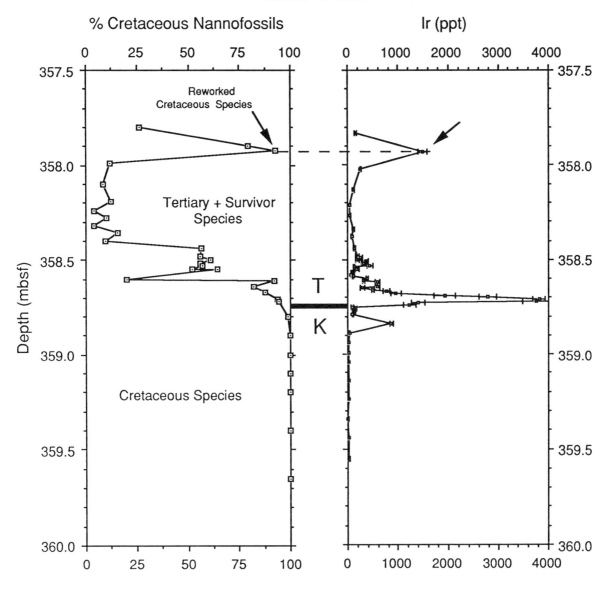

Fig. 11. Iridium abundance in parts per trillion (ppt) and percent abundance of Cretaceous calcareous nannofossils across the K/T boundary of Site 752. The iridium peak above the boundary at 357.90 mbsf corresponds to an interval with abundant reworked Cretaceous nannofossils. The iridium is replotted from data of Michel et al. [1991] and the nannofossil data are from Pospichal [1991].

species has a rare and spotty occurrence throughout the entire Danian may suggest otherwise. Preservation also does not appear to be the problem as *Hornibrookina* is a distinct and robust form. Pospichal [1991] concluded that in the Indian Ocean, the northern limit of the biogeographic province represented by abundant *Hornibrookina* lies somewhere between Site 750 and the pre-rift location of Site 752. This was also suggested to explain the differences between these sites in the abundance of *Prediscosphaera stoveri* in uppermost Maastrichtian sediments, although this is probably related

more to preservational differences.

As at Sites 738 and 750, Site 752 sediments lack the low-latitude species *Neobiscutum romeinii* and *N. parvulum*. One specimen of *Futyania petalosa* was noted near the base of Zone NP2 [Pospichal, 1991].

Foraminifers. Planktonic foraminifers from Site 752 were studied by van Eijden and Smit [1991] who used the LO of Cretaceous species to delimit the K/T boundary in Sample 121-752B-11R-3, 96-97 cm (Figure 9). Indurated chalks below the boundary are assigned to the uppermost

Maastrichtian *Abathomphalus mayaroensis* Zone, which extends down to near the top of Core 121-752B-17R. Foraminifer preservation in this zone is poor. Besides rare *A. mayaroensis*, more common taxa in this zone include *Heterohelix planata*, *Rugoglobigerina rugosa*, and near the top, *Planoglobulina acervulinoides*.

Strong induration of lower Danian sediments at Site 752 has inhibited detailed foraminiferal biostratigraphy above the K/T sequence. Samples from Sections 121-752B-10R-7, 11R-1, -11R-2 were barren. The lowest datable samples in Sections 121-752B-9R-1 to -10R-6 are placed in Zone P1a based on the presence of rare and poorly preserved *Eoglobigerina fringa* and *E. edita*.

There is little change among the benthic foraminifer assemblages across the K/T boundary at Site 752 as only seven species disappeared [Nomura, 1991]. Assemblages comprised of *Gavelinella beccariiformis*, *Cibicidoides velascoensis*, and *Gyroidinoides globosus* indicate middle to lower bathyal paleodepths for the Maastrichtian and Paleocene at this site.

Magnetostratigraphy

The Cretaceous/Tertiary boundary of Site 752 lies well within Subchron C29R [Gee et al., 1991]. Unfortunately, correlation of important early Danian biostratigraphic events such as the NP1/NP2 boundary with the magnetostratigraphy could not be achieved because the upper limit of the C29N/C29R boundary could not be precisely determined [Pospichal et al., 1991]. The C29R/C30N boundary below was noted near the top of Core 121-752B-12R.

Geochemistry

Michel et al. [1991] measured iridium abundances across the boundary of Site 752 and report anomalous values at two horizons (Figure 11). An iridium peak of 3800 parts per trillion (ppt) was noted at 91-93 cm in Section 121-752B-11R-3 near the extinction of Cretaceous planktonic foraminifers and the calcareous nannofossil K/T boundary. One additional peak (1500 ppt) was noted above in the same section at 13-14 cm. Schuraytz et al. [1991] also noted two anomalous peaks in iridium abundance at approximately the same levels as Michel et al. [1991]. In addition, Schuraytz et al. [1991] noted that no allogenic quartz grain >10 μm could be found and that the presence of shocked quartz grains in K/T boundary sediments of Site 752 could not be confirmed.

Iridium and Nannofossils

Figure 11 illustrates the remarkable correlation between iridium abundance and Cretaceous nannofossils at and above the K/T boundary. A similar relationship between nannofossils and iridium *below* the K/T boundary was demonstrated previously at Site 690 except that additional minor peaks of iridium were shown to coincide with abundant Tertiary and survivor species reworked downward by bioturbation [Pospichal et al., 1990]. These reworked nannofossils corresponded to large dark-colored burrows

easily distinguished within the lighter-hued Maastrichtian chalk. At Site 690, the iridium and the younger nannofossils were transported as much as 1.3 m below the boundary by burrowing organisms. Subsidiary iridium peaks were also noted below the boundary at Pacific DSDP Site 577 [Michel et al., 1985; Wright et al., 1985], whereas enrichments were noted to extend 2 m above and below the boundary at Gubbio, Italy [Crocket et al., 1988]. This iridium distribution pattern was cited by the former authors as possible evidence for multiple impacts as suggested by the models of Davis et al. [1984] and Whitmire and Jackson [1984]. The pattern was also used in support of extended volcanism across the K/T boundary [Officer et al., 1987; Crocket et al., 1988].

The nannofossil-iridium-burrow relationship in the K/T section from Site 690 was used to suggest that multiple iridium peaks might not be the result of either multiple impacts or extended periods of volcanism, but merely due to the redistribution of iridium up- and down-section via bioturbation [Pospichal et al., 1990]. In the Site 752 section, the iridium and the percent abundance of Cretaceous nannofossils correspond extremely well just above the boundary and especially in the sample at about 80 cm above the boundary. The section just above the boundary is bioturbated with some soft sediment deformation, which can account for the extended enrichment of iridium and Cretaceous species in that interval (Figure 11). On the other hand, there are no apparent burrow structures in the ash-rich sediment 80 cm above the boundary where the secondary peak in iridium abundance and Cretaceous nannofossils coincide. Thus, it is difficult to ascribe this to burrowing activity. It instead appears here that the relationship between iridium and Cretaceous nannofossils at Site 752 may be attributed to the nearby erosion of the iridium rich boundary layer plus some subjacent Cretaceous nannofossils with subsequent transportation and redeposition. The redistribution of iridium and Cretaceous nannofossils at other localities, which cannot be attributed unequivocally to borrowing may also have occurred in this manner.

Some workers have suggested that the drop in the percentage of Cretaceous nannofossil species across the boundary reflects a gradual or step-wise reduction of these forms through time [e. g., Thierstein et al., 1991], while others consider the extinction to be abrupt with all specimens above the boundary reworked. Likewise, for example, a stepwise reduction of planktonic foraminifer across the K/T boundary was noted by Keller [1988] at El Kef, whereas Smit [1982] noted simultaneous extinctions of the foraminifers in the very same section. Perch-Nielsen et al. [1982] measured the stable isotopic signal of bulk sediments just above and below the boundary at Site 524 and noted two distinct signals. They surmised that since the sediments measured just above the boundary contain predominantly Cretaceous nannofossil specimens, these specimens with an apparent Tertiary isotopic signal must have lived in Tertiary seas. Similarly, Barrera and Keller [1990] presented isotopic evidence for planktonic foraminifera survivorship in the earliest Danian.

The strikingly close relationship between iridium and Cretaceous nannofossil species at Site 752 located 80 cm above the boundary supports the classic view of reworking. We see no other mechanism by which Cretaceous species could nearly disappear from the section for an extended interval only to reappear coincident with anomalously high concentrations of iridium. Thus we conclude that the minor iridium peaks above and below the K/T boundary at several sites mentioned are the result of bioturbation and/or redeposition.

SITE 761-NORTHWEST AUSTRALIAN MARGIN

Lithology

The complete core descriptions for Site 761 are given in Haq, von Rad, O'Connell et al., [1990]. Hole 761C, located at 16°44.23'S, 115°32.10'E, was rotary cored at a water depth of 2167.9 m (Figure 1). The Cretaceous/Tertiary boundary occurs in Section 122-761C-3R-3 between 66 and 75 cm (173.40 mbsf) (Figure 12). The lower Paleocene consists of massive nannofossil chalk with foraminifers and is variably structureless or bioturbated and mottled. The lowest Paleocene is comprised of mottled and bioturbated clay-rich nannofossil chalk with foraminifers and changes from dark greenish gray to light greenish gray 16 cm above the boundary. A sharp contact caused by drilling disturbance is present at 67 cm (Figure 12). Below this level is a heavily burrowed pale brown chalk about 8 cm thick. A sharp color change occurs at the base of this interval. The underlying Maastrichtian consists of white nannofossil chalk with foraminifers and with grayish brown chert nodules at the top (Figure 12).

Between 66 cm and 68 cm a major change occurs in the abundance of micrite particles and clay. At 66 cm, the percentage of clay and silt-sized disseminated carbonate debris increases sharply while the bulk percentage of nannofossils decreases from about 40% to 10% of the total sediment.

A distinct "boundary clay" as noted at Site 738 and many other K/T boundary localities is not present in this section, although the amount of clay does increase coincident with the color change at the boundary level. An unconformity may account for the absence of a "boundary clay" or this layer may have been smeared out by bioturbation as suggested by the abundance of burrows in the boundary interval. The contact at 76 cm in Core 122-761C-3R-3 is sharp and uneven (Figure 12), which probably indicates the presence of an unconformity at least at this level.

Biostratigraphy

Calcareous nannofossils The Cretaceous/Tertiary boundary at Site 761, placed by nannofossils at 67 cm in Section 122-761C-3R-3, is based on a pronounced change in the nannofossil assemblage [Pospichal and Bralower, 1992]. As mentioned above, the contact at 67 cm appears unconformable either naturally or by drilling disturbance and hence, some of the section may be missing. As at the other Indian Ocean sites, the nannofossil assemblage change across the K/T boundary is characterized by an

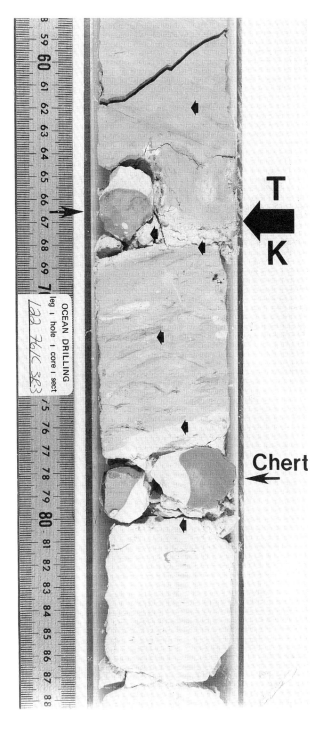

Fig. 12. Enlargement of Cretaceous/Tertiary boundary interval of Site 761 (Section 122-761C-3R-3, 58-88 cm). The nannofossil K/T boundary, Ir- enriched zone, and chert nodules are shown with arrows. Sediment above 68 cm is Danian, dark greenish gray nannofossil chalk. Interval 68 cm to 76 cm is bioturbated, light brownish gray chalk overlying Maastrichtian white chalk. Small arrows superimposed on core photo denote nannofossil sample locations. Note: Because of sediment movement within the core liner, sample locations in this photo differ from original sample designations by approximately 1 cm.

Site 761

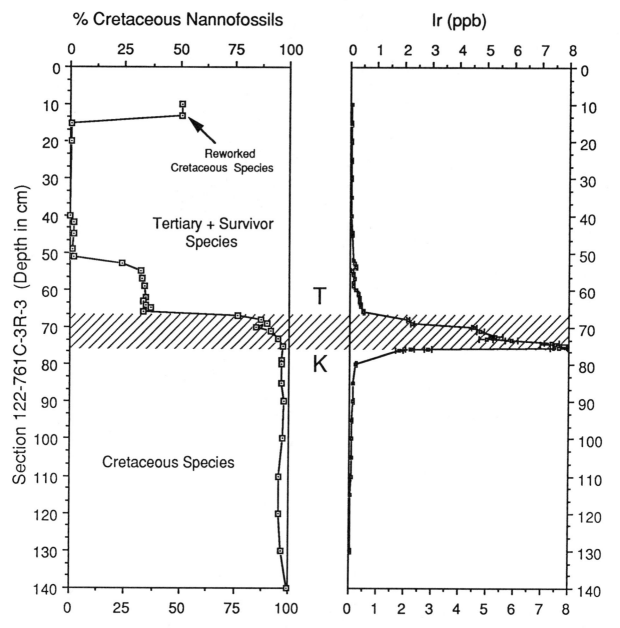

Fig. 13. Site 761 iridium abundance in parts per billion (ppb) and percent abundance of Cretaceous nannofossils across the K/T boundary in Section 122-761C-3R-3. Note reworking of Cretaceous nannofossils at the top of the section. Shaded region corresponds to the 8 cm bioturbated interval between the peak Ir abundance and the abrupt drop in the percentage of Cretaceous nannofossils. Ir profile is replotted from data of Rocchia et al. [1992]. Nannofossil data are from Pospichal and Bralower [1992].

increase in abundance of survivor or persistent species at the expense of Cretaceous species (Figure 13). Similar to Site 752, an interval of abundant reworked Cretaceous species is also present about 60 cm above the boundary at this site. However, unlike in the Site 752 section, an additional iridium anomaly was not noted coincident with this interval of reworking.

Sediments immediately below the K/T boundary at Site

761 contain a distinctly mid latitude upper Maastrichtian assemblage assigned to the *Micula murus* Zone (Figure 3). The moderately preserved assemblage differs greatly from the other Indian Ocean sites by the presence of more abundant *Watznaueria barnesae*, *Prediscosphaera majungae*, and *Lithraphidites quadratus*, in addition to few *Micula murus*. Only rare to few higher latitude species such as *Cribrosphaerella daniae*, *Kamptnerius magnificus*, and *N.*

frequens are present. *Prediscosphaera stoveri* is also present but much less abundant than at the higher latitude Indian Ocean sites.

The uppermost Maastrichtian marker, *Micula prinsii*, is not present at this site. This could be due to: 1) preservational bias, 2) environmental exclusion due to a more mid-latitude position of Site 761 at K/T time, or 3) the absence of the topmost Maastrichtian [Pospichal and Bralower, 1992]. The latter is more likely as *M. prinsii* has been noted in higher latitude Danish sections and at South Atlantic Site 524 [Perch-Nielsen et al., 1982] and Site 527 [Manivit, 1984], which are of similar paloeolatitudes (30°S) to Site 761 (Figure 2). In addition, *M. prinsii* was noted in a core from Richards Bay in South Africa, which would have been at an equal or probably higher latitude in the southwest Indian Ocean [Perch-Nielsen *in* Tredoux et al., 1988] (Figure 2). The absence of an acme of *P. stoveri* just below the boundary may also be due to missing section, although the lower latitude position of Site 761 could equally account for this.

Sediments in a 16-cm interval immediately above the boundary belong to Zone NP1 [Pospichal and Bralower, 1992] (Figure 3). The nannofossil assemblage of this zone is dominated by the survivor or persistent species, *Cyclagelosphaera reinhardtii*, which is very rare at the other Indian Ocean K/T sites. In addition to this species, the assemblage includes, *Placozygus sigmoides*, *Markalius inversus*, *Biscutum castrorum*, *Neocrepidolithus* spp., and *Thoracosphaera*. Reworked Cretaceous species are also common in NP1. The Tertiary species, *Biantholithus sparsus*, is rare at the other Indian Ocean sites but is common just below the NP1/NP2 boundary at Site 761.

The NP1/NP2 boundary is marked by an erosional contact and color change between 51 and 53 cm in Section 122-761C-3R-3. A hiatus is indicated by an abrupt appearance of common to abundant *Prinsius* spp., *Cruciplacolithus tenuis,* and *C. primus*. The latter is found to first occur below *C. tenuis* in the most complete K/T sections such as at Site 752. No specimens of *Hornibrookina* were noted at this site, which may further emphasize the preference of these taxa for higher latitudes.

Similar to Sites 738, 750, and 752, an abundance bloom of *Braarudosphaera* or *Thoracosphaera* was not recorded at Site 761 as has been reported in a number of low latitude and shelfal K/T sections [Thierstein, 1981; Perch-Nielsen et al., 1982]. The absence of these blooms may be due to unconformities at and above the boundary or to environmental factors. In the case of *Braarudosphaera*, specimens are only very rare throughout the section, which may suggest they never achieved any great abundance.

Additionally, the low latitude earliest Danian species, *Neobiscutum romeinii* and *N. parvulum* were not observed at this site. Although rare, *Futyania petalosa* was noted at higher latitude Sites 750 and 752, and thus its presence may be expected at Site 761, but none were observed. Furthermore, its presence was also noted at Richards Bay, South Africa [Perch-Nielsen *in* Tredoux et al., 1988]. The absence of these taxa, especially *F. petalosa*, may be due to

missing section, or, on the other hand, it may also illustrate the provinciality or narrow range of ecologic tolerance of these species.

Foraminifers. No information on K/T boundary planktonic or benthic foraminifers from Hole 761C of Site 761 is given in the shipboard reports [Haq, von Rad, O'Connell et al., 1990]. The information presently available is provided by Wonders [1992], who reported on the planktonic foraminiferal biostratigraphy of Upper Cretaceous sediments from Leg 122 Holes 761B, 762C, and 763B. Assemblages of the uppermost Maastrichtian *Abathomphalus mayaroensis* Zone in Holes 761B and 762C are dominated by *Heterohelix* spp. and *Rugoglobigerina* spp., along with consistent occurrences of *A. mayaroensis, Globotruncana arca, Pseudotextularia elegans, Racemiguembelina fructicosa,* and *Gublerina cuvillieri. Contusotruncana contusa* is common at the top of the zone.

Magnetostratigraphy

The Cretaceous/Tertiary boundary of Hole 761C lies within an interval of reversed polarity assigned to Subchron C29R [Rocchia et al., 1992] (Figure 3). The C29N/C29R boundary lies between samples at 22 and 53 cm in Section 122-761C-3R-3. The lower boundary of C29R could not be determined.

Geochemistry

Rocchia et al. [1992] report a peak iridium abundance of 7.680 ± 0.310 parts per billion (ppb) (80 ng/cm^2) in Sample 122-761C-3R-3, 75-76 cm (Figure 13). This peak coincides with a change in sediment color from pale white to light brown (Figure 12) and is 8 cm below the most distinct shift in nannofossil assemblages at 67 cm (Figure 13). If these two events are considered to have been synchronous then the 8 cm interval, as previously mentioned, may be the result of the intense bioturbation as evidenced by the abundant burrows in this short interval (Figure 12).

SUMMARY

Detailed bio-, magneto-, and chemostratigraphic analyses indicate that K/T sections from Sites 738 and 752 are the most complete while Site 750 and 761 are missing portions of the lowest Danian biozones. Site 761 may also be missing the topmost Maastrichtian. Anomalous concentrations of iridium in boundary sediments are present at Sites 738, 752, and 761. Magnetic polarity measurements reveal that the K/T boundary of Sites 738, 752, and 761 are located in intervals of reversed polarity assigned to Subchron C29R, while the polarity of Site 750 sediments can not be constrained.

The recovery of complete or near complete K/T boundaries during Legs 119-122 in the Indian Ocean provides a good opportunity for comparison of foraminifer and nannofossil assemblages across latitude in a single ocean basin. Furthermore, the laminated sequence of Site 738 offers excellent opportunity for study of the evolution of microplankton groups across a K/T boundary that is uncomplicated by the problems of bioturbation. Likewise, the highly expanded section of Site 752, which is comparable in thickness to the well-studied Tethyan sequences such as El Kef, provides similar opportunities.

With regards to biogeography, planktonic foraminifer and calcareous nannofossil K/T assemblages of Sites 738 and 750 compare well with those from other high-latitude South Atlantic sites such as Weddell Sea Sites 689 and 690. Within the Indian Ocean, a distinct biogeographic gradient across the Kerguelen Plateau can be illustrated by the distribution of the calcareous nannofossil, *Hornibrookina*. This taxon is abundant at southern and middle Kerguelen Sites 738 and 750 but very rare at Site 752, which was situated on the plateau north of Site 750 at K/T time. In addition, K/T boundary calcareous nannofossil and planktonic foraminifer assemblages of northwest Australian margin Site 761 differ significantly from the Kerguelen Plateau sites and provide a mid- to low-latitude end member for biogeographic comparison.

To conclude, the recent drilling by ODP Legs 119-122 has extended the K/T boundary data base to the Indian Ocean for which very little had been previously compiled. Notably, studies of the K/T boundaries from the high latitude Kerguelen Plateau sites complement those of the Weddell Sea region and greatly improve our knowledge of this critical event in the Southern Ocean.

Acknowledgments. This study was supported by Leg 121 United States Science Advisory Committee (USSAC) funds, a Smithsonian Fellowship, Geological Society of America Grant #4772-91 to JJP, and NSF grant DPP 91-18480. Reviews and critical comments by Katharina von Salis Perch-Nielsen, Tony Hallam, S. W. Wise, Frank Wind, and Gerta Keller greatly improved this manuscript. Special thanks also go to Janalisa Soltis and Jennifer Pattison-Hall at ODP for help in compiling information for this paper and to Kelly Dowler for assistance with preparation of the manuscript.

REFERENCES

Alvarez, L. W., Alvarez, W., Asaro, F., and Michel, H., Extraterrestrial cause for the Cretaceous-Tertiary extinction, *Science, 208*, 1095-1108, 1980.

Barrera, E., and Keller, G., Stable isotope evidence for gradual environmental changes and species survivorship across the Cretaceous/Tertiary boundary, *Paleoceanography, 5*, 867-890, 1990.

Barron, B. Larsen, B., et al., *Proc. ODP, Init. Repts., 119*, Ocean Drilling Program, College Station, TX, 1989.

Berggren, W. A., Kent, D. V., Flynn, J. J., and van Couvering, J. A., Cenozoic geochronology, *Geol. Soc. Amer. Bull., 96*, 1407-1418, 1985.

Berggren, W. A., and Miller, K. G., Paleogene tropical planktonic foraminiferal biostratigraphy and magnetobio-chronology, *Micropaleontology, 34(4)*, 362-380, 1988.

Berggren, W. A., Paleogene planktonic foraminiferal magnetobiostratigraphy of the Southern Kerguelen Plateau (Sites 747-749). in *Proc. ODP, Sci. Results, 120*, edited by S.W. Wise, Jr., R. Schlich et al., pp. 551-568, Ocean Drilling Program, College Station, TX, 1992.

Caldiera, K., and Rampino, M. R., Deccan volcanism, greenhouse warming, and the Cretaceous/Tertiary boundary, in *Proceedings of the Conference on Global Catastrophes in Earth History: An Interdisciplinary Conference on Impacts, Volcanism, and Mass Mortality, 247*, edited by V. Sharpton and P. Ward, pp. 497-508, Spec. Pap. Geol. Soc. Amer., 1990.

Crockett, J. H., Officer, C. B., Wezel, F. C., and Johnson, G. D., Distribution of noble metals across the Cretaceous/Tertiary boundary at Gubbio, Italy: iridium variation as a constraint on the duration and nature of Cretaceous/Tertiary boundary events, *Geology, 16*, 77-80, 1988.

Davis, M., Hut, P., and Muller, R. A., Extinctions of species by periodic comet showers, *Nature, 308*, 715-717, 1984.

Ehrendorfer, T., and Aubry, M. -P., Calcareous nannoplankton changes across the Cretaceous/Paleocene boundary in the Southern Indian Ocean (Site 750), in *Proc. ODP, Sci. Results, 120*, edited by S. W. Wise, Jr., R. Schlich et al., pp. 451-470, Ocean Drilling Program, College Station, TX, 1992.

Ehrmann, W. U., Implications of sediment composition on the southern Kerguelen Plateau for paleoclimate and depositional environment, in *Proc. ODP, Sci. Results, 119*, edited by J. Barron, B. Larsen et al., pp. 185-210, Ocean Drilling Program, College Station, TX, 1991.

Gee, J., Klootwijk, C. T., and Smith, G. M., Magnetostratigraphy of Paleogene and Upper Cretaceous sediments from Broken Ridge, eastern Indian Ocean, in *Proc. ODP, Sci. Results, 121*, edited by J. Weissel, J. Peirce, E. Taylor, J. Alt et al., pp. 359-376, Ocean Drilling Program, College Station, TX, 1991.

Haq, B. U., von Rad, U., O'Connell, S. et al., *Proc. ODP, Init. Repts., 122*, Ocean Drilling Program, College Station, TX, 1990.

Hsü, K. J., Mckenzie, J. A., He, Q. X., Terminal Cretaceous environmental and evolutionary changes, in *Geological implications of impact of large asteroids and comets on earth, 190*, edited by L. T. Silver, and P. H. Schultz, pp. 317-328, Spec. Pap. Geol. Soc. Amer., 1982.

Huber, B. T., Maestrichtian planktonic foraminifer biostratigraphy and the Cretaceous/Tertiary boundary at ODP Hole 738C (Kerguelen Plateau, Southern Indian Ocean), in *Proc. ODP, Sci. Results, 119*, edited by J. Barron, B. Larsen et al., pp. 451-466, Ocean Drilling Program, College Station, TX, 1991.

Huber, B. T., Upper Cretaceous planktonic foraminiferal zonation for the Austral Realm, *Mar. Micropaleontology*, 1992.

Jiang, M. J., and Gartner, S., Calcareous nannofossil succession across the Cretaceous/Tertiary boundary in east-central Texas, *Micropaleontology, 32(3)*, 232-255, 1986.

Keller, G., Extinction, survivorship and evolution of planktonic foraminifers across the Cretaceous/Tertiary boundary at El Kef, Tunisia, *Marine Micropaleontology, 13*, 239-263, 1988.

Manivit, H., Paleogene and Upper Cretaceous calcareous nannofossils from Deep Sea Drilling Project Leg 74, in *Init. Repts. DSDP, 74*, edited by T. C. Moore, Jr., P. D. Rabinowitz et al., 475-499, U.S. Government Printing Office, Washington D.C., 1984.

Michel, H. V., Asaro, F., Alvarez, W., and Alvarez, L. W., Elemental profile of iridium and other elements near the Cretaceous/Tertiary boundary in Hole 577B, in *Init. Repts. DSDP, 86*, edited by G. R. Heath, L. H. Burkle et al., pp. 533-538, U.S. Government Printing Office, Washington, D.C., 1985.

Michel, H. V., Asaro, F., and Alvarez, W., Geochemical study of the Cretaceous-Tertiary boundary region in ODP 121-752B, in *Proc. ODP, Sci. Results, 121*, edited by J. Weissel, J. Peirce, E. Taylor, J. Alt et al., pp. 415-422, Ocean Drilling Program, College Station, TX, 1991.

Nomura, R., Paleoceanography of upper Maestrichtian to Eocene benthic foraminiferal assemblages at Sites 752, 753, and 754, eastern Indian Ocean, in *Proc. ODP, Sci. Results, 121*, edited by J. Weissel, J. Peirce, E. Taylor, J. Alt et al., pp. 267-355, Ocean Drilling Program, College Station, TX, 1991.

Officer, C. B., Hallam, A., Drake, C. L., and Devine, J. D., Late

Cretaceous and paroxysmal Cretaceous/Tertiary extinctions, *Nature, 326,* 143-149, 1987.

Peirce, J., Weissel, J. et al., *Proc. ODP, Init. Repts., 121,* Ocean Drilling Program, College Station, TX, 1989.

Perch-Nielsen, K., McKenzie, J. A., and He, Q., Bio- and isotope-stratigraphy and the 'catastrophic' extinction of calcareous nannoplankton at the Cretaceous/Tertiary boundary, in Geological Implications of Impact of Large Asteroids and Comets on Earth., 190, edited by L. T. Silver, and P. H. Schultz, 353-371, Spec. Pap. Geol. Soc. Amer., 1982.

Pospichal, J. J., Calcareous nannofossils across the Cretaceous/Tertiary boundary at Site 752, Eastern Indian Ocean, in *Proc. ODP, Sci. Results, 121,* edited by J. Weissel, J. Peirce, E. Taylor, J. Alt et al., 395-414, Ocean Drilling Program, College Station, TX, 1991.

Pospichal, J. J., and Wise, S. W., Jr., Maestrichtian calcareous nannofossil biostratigraphy of Maud Rise ODP Leg 113 Sites 689 and 690, Weddell Sea, in *Proc. ODP, Sci. Results, 113,* P. F. Barker, J. P. Kennett et al., 465-487, Ocean Drilling Program, College Station, TX, 1990a.

Pospichal, J. J., and Wise, S. W., Jr., Calcareous nannofossils across the K/T boundary, ODP Hole 690C, Maud Rise, Weddell Sea, in *Proc. ODP, Sci. Results, 113,* edited by P. F. Barker, J. P. Kennett et al., pp. 515-532, Ocean Drilling Program, College Station, TX, 1990b.

Pospichal, J. J., and Bralower, T. J., Calcareous nannofossils across the Cretaceous/Tertiary boundary, Site 761, Northwest Australian Margin, in *Proc. ODP, Sci. Results, 122,* U. von Rad, B. U., Haq et al., pp. 735-752, Ocean Drilling Program, College Station, TX, 1992.

Pospichal, J. J., Wise, S. W., Asaro, F., and Hamilton, N., The effects of bioturbation across a biostratigraphically complete, high southern latitude K/T boundary, in Proceedings of the Conference on Global Catastrophes in Earth History: An Interdisciplinary Conference on Impacts, Volcanism, and Mass Mortality, 247, edited by V. Sharpton and P. Ward, pp. 497-508, Spec. Pap. Geol. Soc. Amer., 1990.

Pospichal, J. J., et al., Cretaceous-Paleogene biomagneto-stratigraphy of ODP Sites 752-755, Broken Ridge: a synthesis, in *Proc. ODP, Sci. Results, 121,* edited by J. Weissel, J. Peirce, E. Taylor, J. Alt et al., pp. 721-742, Ocean Drilling Program, College Station, TX, 1991.

Proto Decima, F., Paleocene to Eocene calcareous nannoplankton of the Indian Ocean, in *Indian Ocean Geology and Biostratigraphy,* edited by J. R. Heirtzler et al., pp. 353-370, Amer. Geophys. Union, Washington, 1977.

Quilty, P. G., Upper Cretaceous planktonic foraminifers and biostratigraphy, ODP Leg 120, Southern Kerguelen Plateau, in *Proc. ODP, Sci. Results, 120,* edited by S. W. Wise, Jr., R. Schlich et al., pp. 371-392, Ocean Drillng Program, College Station, TX, 1992.

Rea, D. K., Dehn, J., Driscoll, N. W., Farrell, J. W., Janecek, T. R., Owen, R. M., Pospichal, J. J., Resiwati, P., and ODP Leg 121 Scientific Party, Paleoceanography of the eastern Indian Ocean from ODP Leg 121 drilling on Broken Ridge, *Geol. Soc. Amer. Bull., 102,* 679-690, 1990.

Resiwati, P., Upper Cretaceous calcareous nannofossils from Broken Ridge and Ninetyeast Ridge, Indian Ocean, in *Proc. ODP, Sci. Results, 121,* edited by J. Weissel, J. Peirce, E. Taylor, J. Alt et al., pp. 141-170, Ocean Drilling Program, College Station, TX, 1991.

Rocchia, R., Boclet, D., Bonte, P., Froget, L., Galbrun, B., Jehanno, C., and Robin, E., Iridium and other element distributions, mineralogy and magnetostratigraphy near the Cretaceous-Tertiary boundary in Hole 761C, in *Proc. ODP, Sci. Results, 122,* edited by U. von Rad, B. U. Haq et al., pp. 753-762, Ocean Drilling Program, College Station, TX, 1992.

Sakai, H., and Keating, B., Paleomagnetism of Leg 119-Holes 737A, 738C, 742A, 745B, and 746A, in *Proc. ODP, Sci. Results, 119,* edited by J. Barron, B. Larsen et al., pp. 751-770, Ocean Drilling Program, College Station, TX, 1991.

Schlich, R., Wise, S. W., Jr., et al., *Proc. ODP, Init. Repts., 120,*

edited by S.W. Wise,Jr., R. Schlich et al., Ocean Drilling Program, College Station, TX, 1989.

Schmitz, B., Asaro, F., Michel, H.V., Thierstein, H.R., and Huber, B.T., Element stratigraphy across the Cretaceous/Tertiary boundary in ODP Hole 738C, in *Proc. ODP, Sci. Results, 119,* edited by J. Barron, B. Larsen et al., pp. 719-730, Ocean Drilling Program, College Station, TX, 1991.

Schuraytz, B. C., O'Connell, S., and Sharpton, V. L., Iridium and trace element measurements from the Cretaceous-Tertiary boundary, ODP Site 752, Broken Ridge, Indian Ocean, in *Proc. ODP, Sci. Results, 121,* edited by J. Weissel, J. Peirce, E. Taylor, J. Alt et al., pp. 913-920, Ocean Drilling Program, College Station, TX, 1991.

Smit, J., Extinction and evolution of planktonic foraminifera after a major impact at the Cretaceous/Tertiary boundary, in *Geological Implication of Impact of Large Asteroids and Comets on Earth, 190,* edited by L. T. Silver, and P. H. Schultz, pp. 329-352, Spec. Pap. Geol. Soc. Amer., 1982.

Smit, J., and Romein, A. J., T., A sequence of events across the Cretaceous-Tertiary boundary, *Earth Planet. Sci. Lett., 74,* pp. 155-170, 1985.

Stott, L. D,. and Kennett, J. P., Antarctic Paleogene planktonic foraminifer biostratigraphy: ODP Leg 113, Sites 689 and 690, in *Proc. ODP, Sci. Results, 113,* edited by P. F. Barker, J. P. Kennett et al., pp. 549-569, Ocean Drilling Program, College Station, TX, 1990.

Thierstein, H. R., Late Cretaceous nannoplankton and the change at the Cretaceous/Tertiary boundary, in *The Deep Sea Drilling Project: A Decade of Progress., 32,* edited by J. E. Warme, R. G. Douglas, and E. L. Winterer, pp. 355-394, Spec. Publ. Soc. Econ. Paleontol. Mineral., 1981.

Thierstein, H. R., Asaro, F., Ehrmann, W. U., Huber, B., Michel, H., Sakai, H. Schmitz, B., The Cretaceous-Tertiary boundary at Site 738, south Kerguelen Plateau, in *Proc. ODP, Sci. Results, 119,* edited by J. Barron, B. Larsen et al., pp. 849-868, Ocean Drilling Program, College Station, TX, 1991.

Tredoux, M., Verhagen, B. Th., Hart, R. J., De Wit, M. J., Smith, C. B., Perch-Nielsen, K., Sellschop, J. P. F., Geochemical comparison of K-T boundaries from the Northern and Southern Hemispheres, Conference on Global Catastrophies: Impacts, Volcanism and Mass Mortality. Snowbird, Utah, *Lunar Planetary Institute* Contr. No. 673 (abstract), 198-199, 1988.

van Eijden, A. J. M., and Smit, J., Eastern Indian Ocean Cretaceous and Paleogene quantitative biostratigraphy, in *Proc. ODP, Sci. Results, 121,* edited by J. Weissel, J. Peirce, E. Taylor, J. Alt et al., pp. 77-124, Ocean Drilling Program, College Station, TX, 1991.

Watkins, D. K., Crux, J. A., Pospichal, J. J., and Wise, S. W., Jr., Upper Cretaceous nannofossil biostratigraphy of the Southern Ocean and its paleobiogeographic implications, *Int. Nannoplank. Assoc. Newsl., 13, 2,* (abstracts), 72, 1991.

Watkins, D. K., Upper Cretaceous nannofossils from ODP Leg 120, Kerguelen Plateau, Southern Ocean, in *Proc. ODP, Sci. Results, 120,* edited by S. W. Wise, Jr., R. Schlich et al., pp. 343-370, Ocean Drilling Program, College Station, TX, 1992.

Wei, W. and Pospichal, J.J., Danian calcareous nannofossil succession at ODP Site 738 in the southern Indian Ocean, in *Proc. ODP, Sci. Results, 119,* edited by J. Barron, B. Larsen et al., pp. 495-512, Ocean Drilling Program, College Station, TX, 1991.

Wei, W. and Thierstein, H. R., Upper Cretaceous and Cenozoic calcareous nannofossils of the Kerguelen Plateau (southern Indian Ocean) and Prydz Bay (East Antarctica), in *Proc. ODP, Sci. Results, 119,* edited by J. Barron, B. Larsen et al., pp. 467-494, Ocean Drilling Program, College Station, TX, 1991.

Whitmire, D. P., and Jackson, A. A., IV, Are periodic mass extinctions driven by a distant solar companion? *Nature, 308,* 713-715, 1984.

Wonders, A. A. H., Cretaceous planktonic foraminiferal biostratigraphy, Leg 122, Exmouth Plateau, Australia, in *Proc. ODP, Sci. Results, 122,* edited by U. von Rad, B. U. Haq et al., pp. 587-600, Ocean Drilling Program, College Station, TX, 1992.

Wright, A., Bleil, U., et al., Summary of Cretaceous/Tertiary boundary studies, Deep Sea Drilling Project Site 577, Shatsky Rise, in *Init. Repts. DSDP*, *86*, edited by G. R. Heath, L. H. Burkle et al., pp. 799-804, U.S. Government Printing Office, Washington, D.C., 1985.

Zachos, J. C., and Arthur, M. A., Paleoceanography of the Cretaceous/Tertiary boundary event: Inferences from stable isotopic and other data, *Paleoceanography*, *1*, 5-26, 1986.

Zachos, J. C., Arthur, M. A., and Dean, W. E., Geochemical evidence for suppression of pelagic marine productivity at the Cretaceous-Tertiary boundary, *Nature*, *337*, 61-64, 1989.

Zachos, J. C., Aubry, M.-P., Berggren, W. A., Ehrendorfer, T., Heider, F., and Lohmann, K. C., Chemobiostratigraphy of the Cretaceous/Paleocene boundary at ODP Site 750, Southern Kerguelen Plateau, in *Proc. ODP, Sci. Results, 120*, edited by S. W. Wise, Jr., R. Schlich et al., pp. 961-977, Ocean Drilling Program, College Station, TX, 1992.

Accumulations of Organic Matter in Sediments of the Indian Ocean: A Synthesis of Results from Scientific Deep Sea Drilling

PHILIP A. MEYERS AND GERALD R. DICKENS

Department of Geological Sciences, The University of Michigan, Ann Arbor, MI, 48109-1063, USA

Scientific ocean drilling has documented a number of episodes of organic-carbon-rich-sediment accumulation which occurred during the history of the Indian Ocean. These accumulations are noteworthy because organic-rich marine sediments are uncommon, and their existence indicates special paleodepositional conditions. Two general types of organic-carbon-rich sediments appear in the drilled record of Indian Ocean sedimentation - those that contain marine organic matter and those that have land-derived organic matter.

Sediments rich in marine organic matter accumulated during times of enhanced preservation of organic matter and of elevated marine productivity. Two thin layers of black shales with up to 26% marine-derived organic carbon were deposited at the Cenomanian/Turonian boundary on the Exmouth Plateau, evidently as a result of an intensified midwater oxygen minimum. Temporary uplift of Broken Ridge during the Eocene caused an interval of more rapid burial and consequently enhanced accumulation of marine organic matter. The dramatic uplift of the Himalayas initiated the high productivity associated with monsoonal upwelling in the middle Miocene and led to the continued accumulation of sediments rich in organic matter off Somalia, Oman, and Pakistan.

The episodes of accumulation of land-derived organic matter in Tethys and the Indian Ocean represent mostly periods of enhanced continental erosion. During the Late Triassic and the Late Jurassic, the Tethyan margins off northwest Australia and northern India accumulated thick layers of silty claystones enriched in continental organic matter which were deposited as parts of prograding fluvio-deltaic and submarine fan systems. Similar deposits formed again on these margins and also on the southeast African margin in Berriasian and Valanginian times. Claystones containing several percent continental organic carbon were deposited in submarine fans off southeast Africa and northwestern Australia in the Aptian-Albian. Uplift of the Himalayas and the Tibetan Plateau in the Miocene led to the thick and extensive Indus and Ganges-Brahmaputra submarine fan deposits which contain one to two percent continental organic matter. A change in sediment organic matter type, from continental to marine, records the uplift of the Owen Ridge from an earlier location within the Indus Fan during the middle Miocene. Upper Pleistocene turbidites in the Red Sea contain up to 1.4% organic carbon, much of which is from land-derived organic matter.

Some accumulations of land-derived organic matter record the existence of formerly elevated areas and times of milder climate. Freshwater coals were laid down on the margin of East Antarctica during temperate Albian climates. These coals are the probable source of continental organic matter present in Eocene diamictites in Prydz Bay. Silty claystones and sandstones containing up to 1% coaly organic matter accumulated on the Kerguelen Plateau in the late Albian and suggest the former existence of freshwater swamps on this now submerged plateau. The presence of coal in Paleogene sediments on Ninetyeast Ridge indicates that this submarine ridge was formerly above sealevel and supported continental vegetation.

INTRODUCTION

The expected fate of most organic matter in the oceans is oxidation and destruction, rather than accumulation on the seafloor. For example, Müller and Suess [1979] estimate that less than 1% of the organic matter produced in the photic zone reaches the sea bottom. Furthermore, results obtained by Cobler and Dymond [1980] from near-bottom sediment traps indicate that only 10% of the organic matter reaching surficial sediments escapes benthic reprocessing and becomes incorporated in deeper sediments.

Sediments of the deep ocean typically contain little organic matter. Degens and Mopper [1976] have estimated an average concentration of 0.2% organic carbon for modern deep-sea sediments. This value compares well with

Synthesis of Results from Scientific
Drilling in the Indian Ocean
Geophysical Monograph 70

the average organic carbon concentration of 0.3% calculated for Deep Sea Drilling Project Legs 1 through 31 by McIver [1975]. Many pelagic biogenic oozes, however, have concentrations that are below the limits of detection of modern carbon analyzers (<0.05%). Sediments of this type are common in the modern Indian Ocean [e.g., Snowdon and Meyers, 1992].

For most of their history, Indian Ocean sediments have received little organic matter, either from marine production or from continental erosion. Some regions of the Indian Ocean, however, contain sediments having elevated amounts of organic matter, recording interesting intervals in the geological evolution of this ocean. These unusual accumulations required special depositional conditions, ranging from enhanced delivery of organic matter to the seabottom to its improved preservation. In this review of the occurrences of such organic-carbon-rich sediments encountered during scientific ocean drilling, we

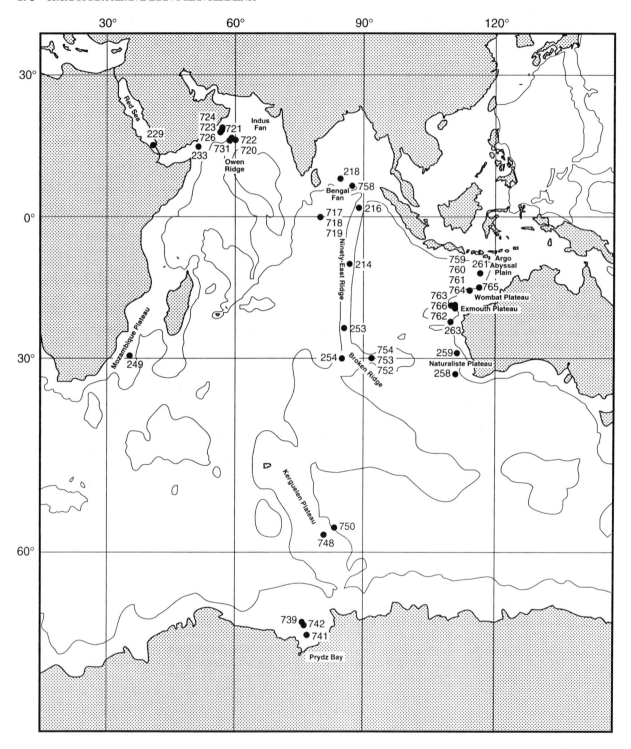

Fig. 1. DSDP and ODP sites in the Indian Ocean at which sediments and rocks having elevated concentrations of organic carbon have been found. Important bathymetric features mentioned in the text are labeled. The 4 km isobath is approximated.

discuss their paleoceanographic significance to the history of the Indian Ocean and surrounding continental margins.

ENHANCED ACCUMULATIONS OF ORGANIC MATTER

Sediments ranging in age from Late Triassic to modern have been recovered from the Indian Ocean during scientific ocean drilling done by the Deep Sea Drilling Project (DSDP) and the Ocean Drilling Program (ODP). Some of the drill sites have yielded sediments that contain organic matter concentrations which are elevated above the values typical of most oceanic areas (Figure 1). Nine episodes of regionally enhanced accumulation of organic matter have been documented during the 14 drilling legs in the Indian Ocean (Table 1). Some of these oceanic accumulations have counterparts which have also been described from continental areas. •

Organic matter types are routinely determined on sediments recovered by ocean drilling using Rock-Eval pyrolysis. This procedure consists of controlled heating of sediment samples to 600°C and measurement of the amounts of pyrolysis products that are generated from the organic matter contained in these samples. Two parameters are particularly useful in inferring the origin of organic matter. The Hydrogen Index is the quantity of hydrocarbons produced from the organic matter, expressed as mg hydrocarbons per g total organic carbon (TOC). Marine organic matter typically has high Hydrogen Index values [Espitalie et al., 1977]. The Oxygen Index is the quantity of CO_2 generated from the organic matter and is also expressed as mg per g TOC. Cellulose-containing land plants produce organic matter having high Oxygen Indices and low Hydrogen Indices [Espitalie et al., 1977]. These two parameters are commonly plotted against each other in a van Krevelen-type diagram. These diagrams show the diagenetic trends of Type I waxy organic matter, Type II marine organic matter, and Type III land-plant organic matter.

The age range of drilled sediments recovered from the Indian Ocean is greater than that obtained from any other ocean. We describe the nine episodes of enhanced organic matter accumulation in their order of deposition, from oldest to most recent. The age scale we employ is that of the Decade of North American Geology [Palmer, 1983].

Upper Triassic Fluvio-Deltaic Sediments on the Wombat Plateau

A series of shallow-water, lagoonal and reefal sedimentary sequences were deposited on the northwestern Australian continental margin prior to the rifting in the late Valanginian (ca 132 my BP) which separated Australia and India [Exon et al., 1982; von Rad et al., 1992]. This series was investigated by drilling at ODP Sites 759, 760, 761, and 764 on the Wombat Plateau [Haq, von Rad et al., 1990; von Rad et al., 1992]. The results from coring at these sites have been combined to give a composite summary of sediment accumulation on this passive margin during Late Triassic (230-212 Ma) time [Exon et al., 1989; Williamson et al., 1989]. At this time and until the Valanginian, the

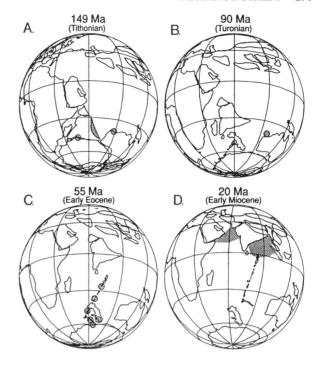

Fig. 2. Plate tectonic reconstructions of the Indian Ocean: A. In the Late Jurassic (149 Ma) prior to rifting of Gondwanaland; B. Near the time of the Cenomanian/Turonian Boundary Event (90 Ma); C. Near the time of the Paleocene/Eocene boundary (55 Ma); D. In the middle early Miocene (20 Ma). General areas which contain elevated amounts of organic matter near the times of these reconstructions are indicated by hatching.

northern margin of Australia and India were united and formed part of the southern coast of Tethys (Figure 2A).

Snowdon and Meyers [1992] summarize total organic carbon concentrations found in the various sedimentary stratigraphic units present in three Wombat Plateau sites (Figure 3). The composite section consists of Norian-to-Carnian (230-218 Ma) fluvio-deltaic clastic sediments which typically contain 1 to 3% total organic carbon (TOC), with occasional samples exceeding 10% TOC. The fluvio-deltaic sediments are overlain by Rhaetian (218-212 Ma) lagoonal sediments which contain 0.1 to 1% TOC. An attenuated thickness of Cretaceous and Tertiary chalks and calcareous oozes covers the Rhaetian sediments. The carbonates generally contain less than 0.2% TOC (Figure 3).

The results of Rock-Eval pyrolysis indicate that the organic matter of the Upper Triassic Wombat Plateau sediments is dominated by debris from cellulosic land plants. Most of the samples from these units yield low Hydrogen Indices (<100 mg/g) characteristic of Type III kerogen (Figure 4). A continental origin of the organic matter is consistent with the mid-latitude deltaic and lagoonal Late Triassic paleoenvironments postulated for this locale by e.g., Ogg et al., this volume [1992].

The Upper Triassic sections encountered in Sites 759, 760, and 761 are equivalent to the Mungeroo Formation

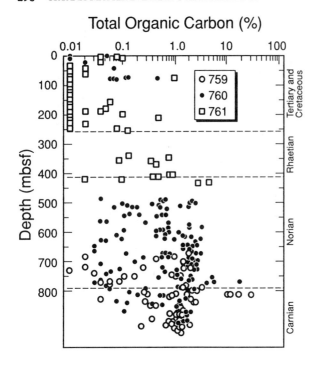

Fig. 3. Relationship of organic carbon content to the composite stratigraphic section from Sites 759, 760, and 761 on the Wombat Plateau, offshore of northwestern Australia. Carnian-Norian (230-218 Ma) fluvio-deltaic siltstones are overlain by Rhaetian (218-212 Ma) lagoonal sediments. Cretaceous and Tertiary carbonates comprise the upper section. An equivalent Triassic sequence is found throughout most of the southern Himalayas.

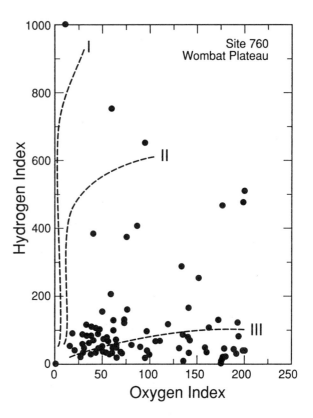

Fig. 4. Rock-Eval diagram of organic matter from Carnian-Norian (230-218 Ma) sediments at Site 760, Wombat Plateau, northwestern Australia. Type III land-derived organic matter dominates in sediments having >0.1% TOC.

which underlies much of the northwestern Australian continental margin. Cook et al. [1985] report the results of organic petrographic examinations of cuttings from sections of the Mungeroo Formation in offshore wells. They find abundant proportions of vitrinite, which is a type of disseminated organic matter derived from woody particles, as well as thin coaly layers. The vitrinite is thermally immature in the upper parts of the formation, showing that the Triassic sediments were never deeply buried. Both the vitrinite and coaly deposits confirm the dominance of vascular land-plant organic matter in these sediments.

The Upper Triassic shallow-water sedimentary sequence on the Wombat Plateau is similar to the Norian shallow-water siltstones and Rhaetian subtidal carbonate-rich sediments found in the Zanskar Range of northern India [Gaetani and Garzanti, 1991] and fluvio-deltaic sediments of the Thini Formation in central Nepal [Gradstein et al., 1989]. The Thini Formation includes coals [U. von Rad, personal communication, 1992]. Little additional information about the organic matter contents of these Himalayan rocks is provided by these authors, yet the lithologic similarity and their paleogeographic proximity to the Wombat Plateau during Triassic times (Figure 2A) suggest that the organic matter type and amounts may also correspond. It appears

that the organic-carbon-rich Upper Triassic sequence on the Wombat Plateau belongs to a Tethyan continental margin that extended across northwestern Australia and the northern part of greater India and consequently was the site of an important coastal marine accumulation of land-derived organic matter.

Other Tethyan margin accumulations of continental organic matter exist as the Kimmeridgian (156-152 Ma) Spiti Shale of northern India [e.g. Gaetani and Garzanti, 1991], the Oxfordian-Tithonian (163-144 Ma) Nupra Shale exposures of Nepal [e.g. Gradstein et al, 1989], and the Oxfordian-Tithonian Upper Dingo Claystone of northwestern Australia [e.g. Barber, 1982]. These three accumulations appear to be parts of the worldwide marine organic matter accumulated in ocean margin sediments at many locations [e.g., Hallam, 1987]. Scientific ocean drilling has not investigated these sequences, although commercial exploratory wells have penetrated the Dingo Claystone in the Kangeroo Trough between the Exmouth Plateau and the Australian mainland [e.g. Barber, 1982]. Organic carbon concentrations in exposures of the Nupra Shale range between 0.7 to 2.1% [Gradstein et al., 1989]. Organic matter in the Nupra Shale is thermally overmature with respect to oil generation. Vitrinite reflectance values of samples from Nepal are between 1.5 and 2% [Gradstein

TABLE 1. Episodes of Organic Matter Accumulation in Sediments of the Indian Ocean

Sediment Accumulation	Location	Organic Matter Type	Organic Carbon Concentration	Relative Size
Carnian-Rhaetian siltstones	Northwest Australian margin and Himalayas	continental	1-2%	thick and extensive
Oxfordian-Tithonian siltstones and claystones	Northwest Australian margin and Himalayas	continental	1-2%	thick and extensive
Berriasian-Valanginian siltstones	Northwest Australian margin and Himalayas	continental	1-2%	thick and extensive
Aptian-Albian deposits	East Antarctica			
coals	and southern India	continental	up to 20%	regional
coaly siltstones	Southern Kerguelen Plateau	continental	up to 10%	local
black claystones	Exmouth Plateau	continental	1-2%	regional
Cenomanian/Turonian claystones	Exmouth Plateau	marine	up to 26%	thin and localized
Paleocene coaly layers	Ninety East Ridge	continental	unknown	limited
Eocene diamictites	Prydz Bay	continental	0.4-1.6%	regional
Eocene biogenic oozes	Broken Ridge	marine (?)	up to 0.5%	limited
Miocene-Holocene deposits				
submarine fans	Bengal and Indus fans	continental	up to 2%	thick and extensive
biogenic oozes	Oman Margin-Owen Ridge	marine	1-6%	regionally thick
Pleistocene-Holocene turbidites	Red Sea	continental	0.7-1.4%	regionally thick

et al., 1989]. This maturity must arise either from deep burial of the Nupra Shale within its deltaic sequence or from regional heating which accompanied the Himalayan orogeny.

Neocomian Submarine Fan Deposits on the Exmouth Plateau and Mozambique Ridge

A sequence of fluvio-deltaic silty claystones and siltstones reaching 10 km in thickness accumulated during the Berriasian-Valanginian (144-131 Ma) on the Exmouth Plateau off northwest Australia [Exon et al., 1982; Haq et al., 1992]. Drilling at ODP Sites 762 and 763 sampled the upper parts of this sequence, which is known onshore as the Barrow Formation. Deeper sections of the Barrow Formation have generated oil and gas and are the source of commercial production in northwestern Australia. Offshore production is located on the continental shelf and on Barrow Island offshore of northwestern Australia [cf. Campbell et al., 1984].

The silty claystones and siltstones typically contain about 1% TOC, with occasional samples having up to 1.5% at Site 762 and up to 2% TOC at Site 763 [Snowdon and Meyers, 1992]. The results of Rock-Eval pyrolysis indicate that the organic matter in these distal fluvio-deltaic sediments is derived mostly from land plants (Figure 5). Organic carbon isotopic ratios (average ca -27.5‰), biomarker hydrocarbon distributions (dominance of C_{29} steranes and plant wax n-alkanes), and organic petrography support this source identification [Barber, 1982; Campbell

et al., 1984; Cook et al., 1985; Meyers and Snowdon, 1992; Snowdon and Meyers, 1992]. The level of thermal maturity as indicated by vitrinite reflectance values of 0.2 to 0.4% is immature to marginally mature with respect to the onset of petroleum generation [Snowdon and Meyers, 1992].

Basal sediments at DSDP Site 249 on the Mozambique Ridge in the western Indian Ocean comprise Valanginian-Hauterivian (138-124 Ma) black silty claystones averaging 1.5% organic carbon [Girdley, 1974; Girdley et al., 1974]. These sediments are postulated by Girdley et al. [1974] to be the precursors of the south-to-north transgressive Aptian-Cenomanian Domo Formation described by Flores [1973] from wells in eastern Mozambique. The sources of the organic matter in the Site 249 sediments have not been identified, but the clastic lithology suggests predominance of a continental origin. These sediments were probably deposited in swampy lowlands created during the early stages of separation of Madagascar from Africa.

Aptian-Albian Coaly Sediments and Black Shales

Two types of organic-carbon-rich sediments were deposited in the Indian Ocean during Aptian-Albian times (119 to 97 Ma). Accumulations rich in coaly, land-derived organic matter have been recovered from several locations that were coastal swamps or deltas on the northern borders of rifting Gondwana. Other sediments containing mostly marine-derived organic matter have been found in deep-water paleoenvironments. These "black shales" are generally contemporaneous with similar deposits known from the Atlantic and Pacific Oceans and appear to be part of a global pattern of episodically enhanced organic matter accumulation during the Aptian-Albian [e.g., Weissert, 1981].

The upper 100 m of a sequence of coaly siltstones and sandstones believed to be as much as 3 km thick was recovered at Site 741 at the mouth of Prydz Bay, East Antarctica. Organic carbon concentrations are highly variable, typically ranging between 0.2 to 7% but reaching as high as 62% in coaly fragments [Barron, Larson et al., 1989]. Palynological examination shows the particulate organic matter to consist wholly of nonmarine debris, principally carbonized particles, spores, and pollen. A middle Albian (ca 105 Ma) age is assigned to the spores and pollen [Truswell, 1991]. Rock-Eval pyrolysis indicates that the organic matter in samples from this siliciclastic sequence is altered Type III continental material [McDonald et al., 1991], consistent with the evident coalification. Analyses of extractable hydrocarbons shows good retention of their original land-plant biomarker signatures, indicating low thermal maturity despite their great age [Kvenvolden et al., 1991; McDonald et al., 1991]. Vitrinite reflectance values range consistently between 0.30 and 0.32% [Turner and Padley, 1991], confirming that the samples are thermally immature. Kerogen carbon isotope ratios are -23.5‰ [McDonald et al., 1991], a value which is somewhat positive for organic matter derived from typical C3 land plants [e.g., Emerson and Hedges, 1988; Jasper and Gagosian, 1989] and may indicate contributions from C4 grasses in coastal marshes. Turner and Padley [1991] interpret the sequence as representing a gradually subsiding alluvial plain crossed by meandering streams and having numerous shallow lakes. They note that a similar coal-bearing siliciclastic sedimentary sequence occurs in the Mahanadi Basin of southern India, which was adjacent to East Antarctica until the end of the Early Cretaceous (Figure 2A).

A number of locations in the eastern Indian Ocean contain intervals of dark-colored Aptian-Albian claystones that were probably deposited in deep water [Kidd and Davies, 1977, Figure 1]. Concentrations of organic carbon in sediments from DSDP Site 258 on the Naturaliste Plateau, Site 259 on the Perth Abyssal Plain, and Site 261 on the Argo Abyssal Plain are not particularly elevated, never exceeding 0.6% [Bode, 1974], and indicate that organic matter was consumed in depleting porewater oxygen so that sulfide minerals could form. Organic carbon ranges between 0.7 and 2.1%, however, in the black claystones from DSDP Site 263 on the eastern edge of the Cuvier Abyssal Plain [Bode, 1974]. Although initially interpreted as being a subsided shallow-water deposit [Heirtzler, Veevers et al., 1974], the Site 263 black shales are probably a turbiditic deep-water deposit similar to those described on basin margins of the Atlantic Ocean by Dean et al. [1984]. A single organic carbon isotopic measurement of an Albian sediment sample from Site 263 gave a δ13C value of -24.6‰ [Erdman et al., 1974]. This sample, containing

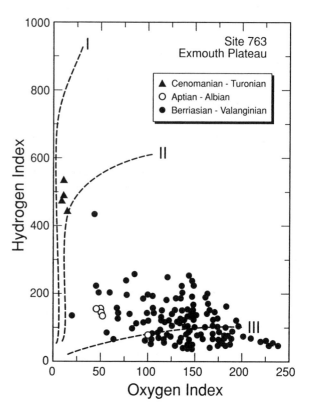

Fig. 5. Rock-Eval diagram of organic matter from Lower Cretaceous (144-97 Ma) and Cenomanian/Turonian boundary (91 Ma) sediments at Site 763, Exmouth Plateau, northwestern Australia. Type III land-derived organic matter is dominant, although Type II marine organic matter contributes to the Cenomanian/Turonian Boundary Event black shales.

0.7% organic carbon and having an isotopic value intermediate between continental and marine organic carbon [e.g., Emerson and Hedges, 1988; Jasper and Gagosian, 1989], probably contains a mixture of terrigenous and marine organic matter, which is consistent with a turbiditic origin.

Drilling at Sites 762 and 763 on the Exmouth Plateau, offshore of northwestern Australia, recovered an early Aptian dark-colored calcareous claystone equivalent to the onshore Muderong Shale. Organic carbon concentrations reach 2% in this claystone. Rock-Eval pyrolysis (Figure 5) and a kerogen carbon isotopic measurement of -28.4‰ indicate that the organic matter is predominantly continental in origin [Snowdon and Meyers, 1992; Meyers and Snowdon, 1992]. A significant marine contribution, however, is evident in the distribution of extractable straight-chain hydrocarbons and from the abundance of C_{27} biomarker steranes [Meyers and Snowdon, 1992]. This subtle change to a larger contribution of marine organic matter is consistent with the lithologic change to a deeper water depositional setting as this passive margin continued to subside and became more isolated from continental influences.

Lower Cretaceous silty claystones similar to the Muderong Shale of northwestern Australia exist across the suture zone between the Indian plate and the Asian plate. Drilling in the onshore portion of the Ganges-Brahmaputra Delta has recovered the Barremian-Aptian (124-113 Ma) Bolpur Formation, which is described by Lindsay et al. [1991] as a wedge of clastic fluvio-deltaic sediment deposited in a restricted marine environment.

Aptian-Albian sediments at DSDP Site 249 on the Mozambique Ridge are laminated silty claystones containing between 1 and 1.7% organic carbon [Girdley, 1974]. Girdley et al. [1974] note that these sediments are correlative with the Aptian-Cenomanian Domo Formation described in eastern Mozambique by Flores [1973]. They were deposited during a northward transgression of a postulated rift graben structure, which probably accompanied the early separation of Madagascar from Mozambique. Although the type of organic matter in the Site 249 Aptian-Albian sediments has not been identified, the clastic lithology and the postulated depositional setting imply continental sources.

Albian silty claystones containing up to 10.5% TOC have been found in ODP Sites 748 and 750 on the southern Kerguelen Plateau [Schlich, Wise et al., 1989a, 1989b]. Most of the organic matter in these accumulations consists of coaly particles that are thermally mature and were probably derived from coal layers on land. Rock-Eval pyrolysis shows the organic matter in Site 750 Albian sediments to resemble humic coals [Schlich, Wise et al., 1989b]. The Kerguelen Plateau is bathymetrically separated from East Antarctica where Aptian-Albian coals occur, making the Antarctic an unlikely source for the Site 748 sediment components. It is more reasonable that coal swamps existed on the southern portions of the Plateau in the mid-Cretaceous and produced coal deposits which subsequently were eroded. These source areas have subsided and are now submarine. The former existence of coal swamps implies a moist, temperate climate, and erosion of coal layers suggests sufficient land elevation so that rivers could erode and then transport organic-carbon-rich particles to the sea.

The Cenomanian/Turonian Boundary Event

The latest Cenomanian-earliest Turonian (ca 91 Ma) was a time during which accumulations of sediments especially rich in organic carbon occurred in marine sediments. The occurrence of organic-carbon-rich sediments from this period has been documented over much of the world [e.g., Arthur et al., 1987; Schlanger et al., 1987; Kuhnt et al., 1990]. No examples were known from the Indian Ocean, however, until ODP Leg 122 drilled on the Exmouth Plateau off northwestern Australia. Two thin layers (4 cm and 12 cm) of organic-carbon-rich black claystone were encountered at the Cenomanian/Turonian boundary at Site 763 (Figure 2B). Organic carbon content measures as high as 26% in the thinner of these layers and 9% in the thicker one [Rullkötter et al., 1992]. Rock-Eval Hydrogen Index values in the range of 350 to 500 mg hydrocarbon/g TOC

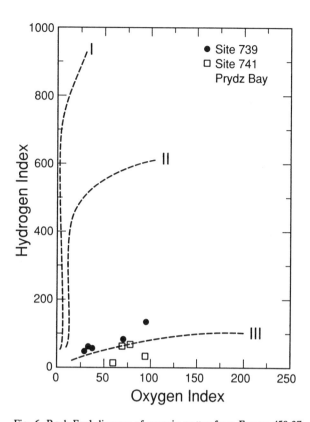

Fig. 6. Rock-Eval diagram of organic matter from Eocene (58-37 Ma) glacial sediments from Site 739 and Albian (113-97 Ma) coaly sediments from Site 741 in Prydz Bay, Antarctica. Close similarity implies a common source from the Albian coals.

(Figure 5) indicate that these layers contain Type II marine organic matter [Haq, von Rad et al., 1990; Rullkötter et al., 1992]. Visual maceral analyses confirm the marine origin [Rullkötter et al., 1992]. The existence of two organic-carbon-rich layers present in the Site 763 sequence reveals that multiple events of enhanced production and preservation of marine organic matter occurred at the time of the Cenomanian/Turonian boundary. Evidence of multiple episodes is not found at other localities in the Indian Ocean and at few locations elsewhere. A single, thin layer of Cenomanian/Turonian boundary claystone was recovered at nearby Site 762, but contained only ~0.01% TOC [Haq, von Rad et al., 1990]. Other boundary claystones were found at Site 765 on the Argo Abyssal Plain and Site 766 on the Gascoyne Abyssal Plain, where they contain 0.05 to 0.12% TOC [Rullkötter et al., 1992]. Finally, Schlanger et al. [1987] describe an organic-carbon-lean Cenomanian/Turonian carbonate sequence from DSDP Site 258 on the Naturaliste Plateau off southwestern Australia.

The existence of contrasting types of Cenomanian/Turonian boundary sequences in relatively close proximity (Figure 1) in the northeastern Indian Ocean is interesting. The presence of high concentrations of marine organic matter in sediments from Site 763 suggests some combination of enhanced productivity and improved preservation of organic matter was involved in their deposition. The absence of organic-carbon-rich sediments at Sites 258, 762, 765, and 766 shows, however, that not all parts of the Indian Ocean experienced the combination of high productivity and slow circulation that led to accumulation of organic matter in the seabottom. Rullkötter et al. [1992] postulate that the Cenomanian/Turonian Boundary Event represents a time of worldwide oxygen depletion in bottom waters which improved preservation of the organic matter produced in overlying waters. Differences in surface paleoproductivity were responsible for site-to-site contrasts. Alternatively, e.g., Thurow et al., this volume, conclude that the oxygen depletion occurred at midwater depths and that deeper waters were oxygenated, resulting in organic-carbon-rich sediments accumulating only in the oxygen minimum zone. The intensity of the oxygen minimum zone, coupled with water depths, controlled organic matter accumulation. Both scenarios imply that local paleoenvironmental conditions, particularly those influencing marine productivity and near-bottom oxygen availability, apparently overprinted global factors important to the Cenomanian/Turonian Boundary Event in the Indian Ocean.

Maestrichtian-Paleocene Coaly Deposits on Broken Ridge and Ninetyeast Ridge

The Maestrichtian-Paleocene (74-58 Ma) was an interval of low accumulation of organic matter in most parts of the Indian Ocean. Sediments containing ca 0.5% organic carbon accumulated at Sites 752 and 754 on Broken Ridge [Peirce, Weissel et al., 1989; Littke et al., 1991]. The depositional setting for these sites during the Cretaceous/Tertiary boundary was probably open-ocean mid-slope on this ridge (Figure 2C). Rock-Eval pyrolysis yielded Hydrogen Indices below 100 and Oxygen Indices in the 100-200 range [Peirce, Weissel et al., 1989], indicating that the organic matter is Type III continental material. A land-derived origin of the organic matter is supported by a kerogen carbon isotope value of -29.6‰ measured on a solitary sample of Paleocene sediment from DSDP Site 254 on Broken Ridge [Erdman et al., 1974].

The source of the land-derived organic matter found in the Maestrichtian-Paleocene sections on Broken Ridge may be from Ninetyeast Ridge (Figure 1). This ridge is believed to have contained a series of islands in the Late Cretaceous and Early Tertiary, which subsided during the Eocene [von der Borch, Sclater et al., 1974]. Shallow-water sediments containing wave-rounded cobbles and shell hash are found in Paleocene sections of Sites 214 and 216 on Ninetyeast Ridge [von der Borch, Sclater et al., 1974]. Furthermore, brown coal and a postulated soil horizon were encountered in Paleocene sections at 425 meters below seafloor (mbsf) at Site 214 [Cook, 1974]. It is possible that the southern sections of this ridge accumulated subaerial deposits of organic matter and that fine-sized particles of this material were carried by ocean currents to Broken Ridge. An alternative scenario is proposed by Littke et al. [1991],

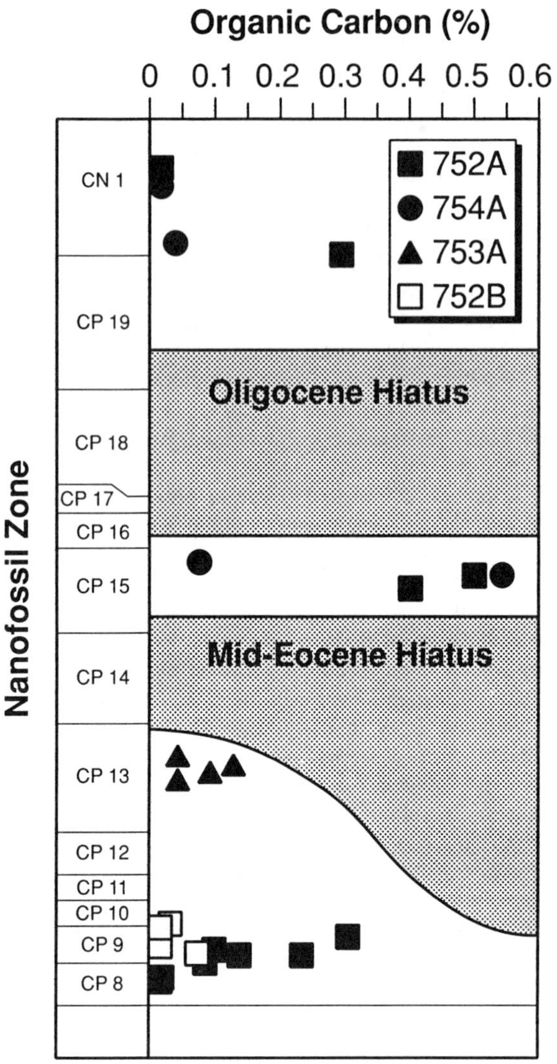

Fig. 7. Concentrations of organic carbon in Tertiary sediments from three ODP sites on Broken Ridge. Nannofossil zones are from Pospichal et al. [1991]. Enhanced concentrations coincide with temporary uplift of Broken Ridge to 300 m water depth in the Eocene.

who find that the size of land-derived particles of sedimentary organic matter decreases from east to west across Broken Ridge. They postulate that this pattern may indicate transport of organic matter by surface ocean currents from land sources on Australia.

Eocene Diamictites in Prydz Bay, East Antarctica

Most Paleogene (67-24 Ma) sediments in the Indian Ocean contain very little organic matter. Eocene (58-37 Ma) diamictites from Sites 739 and 742 in Prydz Bay, however, contain between 0.4 to 1.6% organic carbon [Barron, Larson et al., 1989; McDonald et al., 1991]. Distributions of extractable hydrocarbons in samples from Sites 739, 741, and 742 are dominated by continental plant biomarkers [Kvenvolden et al., 1991; McDonald et al., 1991]. Rock-Eval pyrolysis (Figure 6) shows that the organic matter from Site 739 closely resembles the Type III coaly organic matter present in middle Albian sediments from Site 741, which is closer to the Antarctic mainland (Figure 1). The probable sources of the enhanced amounts of organic matter in the submarine tills at Sites 739 and 742 are the Albian freshwater coals found at Site 741. The organic-carbon-rich diamictites seaward of Site 741 record times of more extensive Antarctic glaciation during the Eocene.

Eocene Nannofossil Oozes on Broken Ridge

Compilation of organic carbon concentrations in Tertiary sediments from Sites 752, 753, and 754 on Broken Ridge (Figure 7) shows that enrichments to over 0.5% TOC occurred in the mid-Eocene [Peirce, Weissel et al., 1989; P. A. Meyers and G. R. Dickens, unpublished data, 1992]. The elevated organic carbon values appears to be related to the tectonic history of this ridge. Broken Ridge uplifted in the Middle Eocene from a former depth of 1200 m to a paleodepth of 300 m before subsiding to its present depth of 1100 m [Pospichal et al., 1991]. It is during the interval of shallow paleodepth (nannofossil zone CP15, Figure 7) that enrichment of organic matter in sediments occurred. Three processes may have contributed to this enrichment: (1) diminished oxidation of organic matter made possible by sinking through a shallower water column, (2) retarded oxidation of organic matter because of deposition within the oxygen minimum zone, and (3) enhanced marine production of organic matter as a result of bathymetric upwelling. Available data do not permit us to discriminate among these three possible scenarios, and we hope that this observation encourages future studies.

Effects of Miocene Uplift and Erosion of the Himalayas

The rapid uplift of the Tibetan Plateau and the Himalayan Mountains appears to have started in the early Miocene (ca 21 Ma) and to have achieved its present elevation by about 8 Ma [Harrison et al., 1992]. Accumulation of organic matter in the sediments of the Indian Ocean has been enhanced as a consequence of the uplift and erosion of the Himalayan Mountains (Figure 2). Organic matter has been transported from land by the Ganges-Brahmaputra river system to accumulate in the Bay of Bengal and by the Indus River to the Arabian Sea. Furthermore, production of marine organic matter has been enhanced since the middle Miocene as a result of the establishment of monsoonal upwelling in the Arabian Sea.

The Bengal submarine fan. ODP Sites 717, 718, and 719 are located on the southern distal edge of the Bengal Fan (Figure 1) and record contributions of material eroded from the southeastern Himalayas by the Ganges and Brahmaputra Rivers. Organic carbon concentrations typically oscillate between 0.2 and 2% in sediments deposited since 6 Ma at these three sites. One sample from 165 mbsf at Site 717 reaches the exceptional value of 7.4% (Figure 8). The

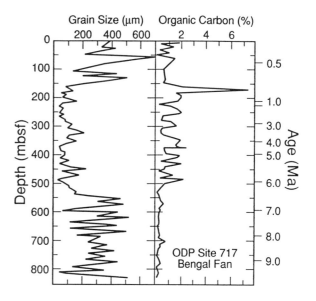

Fig. 8. Downhole variations in sediment grain size and organic carbon content at Site 717, distal Bengal Fan. Approximate ages are from Cochran, Stow et al. [1989a].

variation evidently reflects episodes of turbiditic downslope redeposition of fluvial sediments originally deposited in regions closer to the river delta.

Suspended particles are also carried to the northern Indian Ocean by the Ganges and Brahmaputra Rivers, and these particles can settle in areas bathymetrically higher than the Bengal Fan. Littke et al. [1991] report organic carbon concentrations as high as 0.9% in Late Neogene-Quaternary sediments at Site 758 on the northern Ninetyeast Ridge (Figure 1). This organic matter is associated with clays and is composed of the oxidized debris of vascular land plants. It begins to appear at Site 758 ca 12 Ma, evidently as a result of increased fluvial input of sediments to the Bay of Bengal as the Himalayas approached their present elevation.

Organic carbon concentrations of sediments deposited in the Bengal Fan prior to 6 Ma are less variable than those of subsequent sediments. The percentages are lower in the older deposits, typically about 0.5%. This record of this pattern extends to 9.5 Ma at Site 717 (Figure 8) and to 17 Ma at Site 718. Rock-Eval pyrolysis (e.g., Site 717, Figure 9) indicates that most of the sedimentary organic matter deposited at this distal location on the Bengal Fan since the middle Miocene has consisted of Type III continental material [Cochran, Stow et al, 1989b]. Study of extractable lipids from several samples from Site 717 suggests that the organic constituents of sediments at these distal locations originated from the upper slope of the western Bay of Bengal and from the Ganges-Brahmaputra delta [Poynter and Eglinton, 1990]. A significant land-plant contribution is evident in the molecular lipid compositions of these sediments. Microscopic examination of kerogen, the non-soluble fraction of organic matter, isolated from a middle Miocene sample from more northerly DSDP Site 218 (Figure 1) showed it to consist of woody, herbaceous organic matter [Hunt, 1974a]. Isotopic study of several samples from Site 218 reveals variability in the proportion of marine-to-continental organic matter, with kerogen carbon isotope values varying between -23.8 and -18.5‰ [Erdman et al., 1974]. The variability is consistent with turbiditic deposition, in which sediments can originate from different upslope sources. The appearance of major proportions of land-derived organic matter in the sediments of the Bengal Fan evidently extends at least to the middle Miocene, although the rate of supply seems to have accelerated in the late Miocene, concordant with the history of Himalayan uplift [Harrison et al., 1992].

The Indus submarine fan. ODP Site 720 is on the western distal edge of the Indus Fan, where it records the more recent part of the history of Indus River sediment transport from the western Himalayas. The oldest sediments recovered from Site 720 date from ca 1 Ma, and organic carbon concentrations are generally between 0.1 and 1% at this site [Prell, Niitsuma et al., 1989]. Concentrations reach values as high as 3.7% in some intervals, however, and reflect the variability of this turbiditic sequence. Rock-Eval pyrolysis (Figure 9) and organic petrologic examinations reveal that the organic matter in these sediments comprises predominantly reworked and detrital land-derived material which evidently originated from the Indus River erosional system [Bertrand et al., 1991].

Upwelling in the Indian Ocean. Upwelling has the potential of increasing marine production and thereby enhancing the accumulation of organic matter in underlying sediments. The dominant upwelling system in the Indian Ocean is the monsoon-driven system along the Somali-Omani coast. Equatorial upwelling is not significant in the Indian Ocean, and the western coast of Australia is unique in being the world's only Eastern Boundary Current region lacking upwelling-enhanced high productivity.

The uplift of the Himalayas and the associated elevation of the Tibetan Plateau led to establishment of the monsoonal wind system of the northern Indian Ocean. Variations in the percentage of organic carbon in the lithologic units present in sediments from Site 726 at 331m water depth on the Oman Margin (Figure 1) suggest that enhancements in both production and preservation of organic matter have occurred since the middle Miocene. Peak concentrations of organic carbon (1 to 5%) are found between sub-bottom depths of about 45m to 125m in Pleistocene/upper Miocene partially laminated oozes and chalks. Shallower and deeper sediments display organic matter that is less well preserved than in this lithologic unit, along with evidence of bioturbation. Prell, Niitsuma et al., [1989] postulate that the organic-carbon-rich unit represents a period of shallower or expanded oxygen minimum zone and

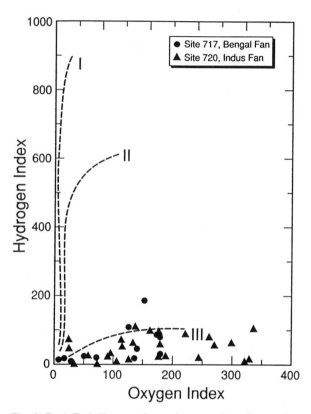

Fig. 9. Rock-Eval diagram of organic matter in sediments from Site 717, distal Bengal Fan, and Site 720, distal Indus Fan. Type III land-derived organic matter dominates in sediments from both locations.

consequent enhanced preservation of organic matter.

Studies of the factors important to accumulation of marine organic matter in sediments, however, suggest that a low level of bottom-water oxygenation is of minor importance [e.g., Emerson and Hedges, 1988; Pederson et al., 1992]. The principal factor is a high settling flux, which typically accompanies high surface productivity. The Pleistocene-late Miocene sub-bottom TOC maximum may represent an interval of particularly enhanced productivity at Site 726, which would contribute to diminished bottom water oxygen content and lead to laminated sediments.

All eight drill sites on the Oman margin occupied by ODP Leg 117 recovered biogenic oozes rich in organic matter. Organic carbon concentrations reach as high as 18% in upper Pliocene (~2 Ma) sediments at Site 724 [Muzuka et al., 1991], although most values are typically in the range of 1 to 6%. Rock-Eval pyrolysis (Figure 10) indicates that marine production is the dominant source of the organic matter in these oozes [Prell, Niitsuma et al., 1989]. The marine origin is supported by the results of kerogen petrographic analyses [Betrand et al., 1991] and carbon stable isotope analyses [Muzuka et al., 1991]. Molecular and isotopic analyses of organic matter in organic-carbon-rich sediments from DSDP Leg 24 Sites 232 and 233 in the nearby Gulf of Aden similarly indicate a predominantly marine origin [Erdman et al., 1974; Hunt, 1974b].

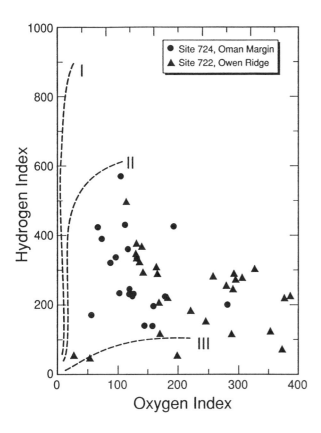

Fig. 10. Rock-Eval diagram of organic matter in sediments from Site 724, Oman Margin, and Site 722, Owen Ridge. Oxidized marine organic matter dominates in sediments on the Oman Margin, whereas upwelling-derived marine organic matter overlies middle Miocene detrital continental organic matter in sediments from the Owen Ridge.

Atomic C/N ratios of organic matter from Oman Margin Sites 723 and 724 are high for marine organic matter. They range between 15 and 35, with a few values reaching as high as 52 [Seifert and Michaelis, 1991]. Although values in this range are more typical of land-plant organic matter [e.g., Jasper and Gagosian, 1989], similarly elevated C/N ratios are reported for marine organic matter deposited under the Benguela Current upwelling system in the eastern South Atlantic Ocean [Meyers et al., 1984]. Meyers [1992] postulates that upwelling-enhanced marine productivity may lead to nitrogen depletion in the photic zone, thereby forcing organisms to synthesize organic matter with high C/N ratios. Alternatively, phytoplankton may synthesize organic matter that is lipid-rich and consequently has high C/N ratios under conditions of high productivity.

Estimates of surface seawater paleotemperatures using the U^K_{37} index [Brassell et al., 1986] yields temperature ranges of 23.0 to 26.5 C at Site 723 on the Oman Margin and 25.0 to 28.5 C at Site 721 on the Owen Ridge [Prell, Niitsuma et al., 1989]. The cooler marginal temperatures are consistent with the presence of stronger upwelling along the coast of Oman. Upper Eocene sediments from

Site 726 on the Oman Margin are relatively poor (0.24%) in organic carbon [Prell, Niitsuma et al., 1989] and evidently were deposited prior to the onset of upwelling in the western Arabian Sea.

Organic matter in sediments on the Owen Ridge.
Sediments recovered by drilling on the Owen Ridge contain organic matter which records both erosion of the Himalayas and the onset of monsoonal upwelling. Lower Miocene sediments from Sites 722 and 731 contain up to 3% organic carbon. Rock-Eval pyrolysis shows the organic matter to be Type III continental material [ten Haven and Rullkötter, 1991]. A coal layer containing ca 50% organic carbon was recovered at Site 722 and was interpreted to be a turbidite deposit [Prell, Niitsuma et al., 1989]. The part of the seafloor that is now on the Owen Ridge was evidently formerly lower and within the turbiditic depositional regime of the Indus Fan [Prell, Niitsuma et al., 1989]. The lower Miocene sediments enriched in continental organic matter therefore indicate that the rise of the Himalayas and consequent Indus River erosion of the Indian subcontinent date from at least this time.

Middle Miocene to Holocene sediments from Sites 721, 722, and 731 on the Owen Ridge contain highly variable concentrations of organic carbon (0.1 to 3.7%). The variability results from fluctuations in the amounts of marine-derived organic matter superimposed on a low background of detrital land-derived organic matter deposited subsequent to uplift of the Owen Ridge above the Indus Fan. High-resolution comparisons of percent CaCO3, percent organic matter, and Rock-Eval Hydrogen Index values were done in lower Pliocene Core 117-722A-16X and upper Miocene Core 117-722A-29X [Prell, Niitsuma et al., 1989]. Both showed that periods of higher accumulation of organic carbon typically had higher proportions of hydrogen-rich, marine organic matter and lower concentrations of CaCO3, presumably because of opal dilution (Figure 11). The opal dilution precedes enhanced accumulation of marine organic matter in the sediments, possibly indicating that the initial pulse of organic matter is consumed in improving conditions for subsequent preservation of organic matter in the seafloor.

The cycles in organic matter concentrations on Owen Ridge presumably reflect fluctuations in marine productivity and subsequent preservation. Their origin, however, remains unknown. Monsoonal productivity is less during glacial times, but both the cycles illustrated in Figure 11 are in Neogene sediments deposited prior to the onset of northern hemisphere glacial periods. Similar productivity cycles have been described in middle-to-upper Miocene sediments from DSDP Sites 362 and 532 on the Walvis Ridge in the South Atlantic [Diester-Haass et al., 1990]. For these, variations in the size of the Antarctic polar ice volume and concomitant sealevel fluctuations have been postulated to be responsible. It is possible that Neogene cycles in Antarctic glaciation caused cycles in Arabian Sea upwelling, via either changes in sealevel or strengths of winds.

Rock-Eval pyrolysis of samples from Site 722 on the Owen Ridge shows that both Type II marine organic matter and Type III continental organic matter (Figure 10) exist in these sediments. This site is between the center of upwelling on the northwest margin and the Indus Fan in the northeast Arabian Sea (Figure 1). Its organic matter appears in the van Krevelen plot to be a blend of oxidized marine and detrital continental material. This combination actually results from accumulation of calcareous oozes containing marine organic matter on top of lower Miocene sediments containing detrital continental organic matter. At Site 731 near the crest of the Owen Ridge, accumulation of upwelling-produced organic-carbon-rich oozes does not occur until ca 10 Ma in the early late Miocene (Figure 12). A hiatus in the Pliocene abbreviates the record of high productivity at this location [D. W. Murray, personal communication, 1991], but shows more readily the lack of enhanced marine production in the middle and early Miocene.

Sedimentary Organic Matter in the Red Sea

Calcareous oozes recovered from Site 229 in the axial trough of the southern Red Sea (Figure 1) contain organic carbon concentrations in the range of 0.7 to 1.4% [Hunt, 1974b; McIver, 1974]. These upper Pleistocene sediments exhibit abundant slump and turbidite features and have an exceptionally high sedimentation rate [Whitmarsh, Weser et al., 1974, 580 m/m.y.]. Microscopic examination of the kerogen shows that it comprises a mixture of amorphous, herbaceous, and woody particles [Hunt, 1974b; McIver, 1974]. The organic matter evidently is a blend of marine and land-derived components. The land-derived organic matter is probably redeposited material which was originally deposited in shallow coastal areas. Its presence in the axial trough is consistent with the evidence for downslope transport of sediments at this site. The detrital material augments the production by marine organisms and contributes to the elevated concentration of organic matter. These Red Sea deposits are modern analogs of organic-carbon-rich sediments laid down during the early stages of rifting of ancient passive ocean margins, such as those of Neo-Tethys and the Mesozoic Indian Ocean.

SUMMARY

Episodes of accumulation of sediments having elevated concentrations of organic carbon have occurred in a number of places and at a variety of times in the history of the evolution of the Indian Ocean and the preceding Neo-Tethys. These various accumulations of organic-carbon-rich sediments record times of enhanced marine productivity, improved organic matter preservation, and increased continental erosion during the closing of Tethys and the opening of the Indian Ocean. Most occurrences have been accumulations of continental organic matter in passive margin, shallow-water, coastal settings. These accumulations are common worldwide in the early stages of continental rifting and ocean basin evolution. Particularly impressive amounts of continental organic matter,

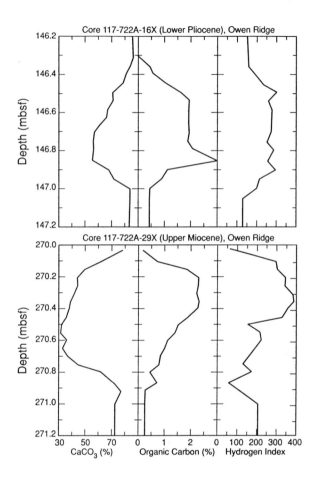

Fig. 11. Comparison of concentrations of $CaCO_3$ and organic carbon and Rock-Eval Hydrogen Index (HI) values of closely spaced samples in Core 117-722A-16X (lower Pliocene) and Core 117-722A-29X (upper Miocene) from Owen Ridge. Correspondence in increases in marine character (higher HI values), higher organic carbon concentrations, and greater opal dilution of $CaCO_3$ indicate additions of marine organic matter to background levels of continental organic matter. From Prell, Niitsuma et al. [1989].

however, have been deposited in the thick and extensive fan deposits peculiar to the northern Indian Ocean. In addition, sediments rich in marine organic matter have accumulated under the monsoonal upwelling system offshore of western India-Pakistan, Arabia, and Somalia. This monsoonal system is an oceanographic feature that is unique to the Indian Ocean.

The episodes of accumulation of land-derived organic matter in Tethys and the Indian Ocean represent mostly periods of enhanced continental erosion. During the Late Triassic and the Late Jurassic, the Tethyan margins off northwest Australia and northern India accumulated thick layers of silty claystones enriched in continental organic matter which were deposited as parts of prograding fluvio-deltaic and submarine fan systems. The existence of these sequences implies that young mountain ranges were being actively eroded on nearby continental regions during these times. Similar deposits formed again on these margins and also on the southeast African margin in Berriasian and Valanginian times. Claystones containing several percent continental organic carbon were deposited in submarine fans at Site 249 off southeast Africa and at Sites 762, 763, and 766 off northwestern Australia in the Aptian-Albian. These relatively thin, distal fan sequences imply diminished erosion rates on nearby continents. Uplift of the Himalayas and the Tibetan Plateau in the Miocene led to the thick and extensive Indus and Ganges-Brahmaputra submarine fan silty-claystone deposits which contain one to two percent continental organic matter. The uplift of the Himalayas and their consequent erosion may be an analog for the conditions which led to the thick and extensive upper Triassic, upper Jurassic, and lower Cretaceous fan deposits now present in Nepal and offshore of northwestern Australia.

Some accumulations of land-derived organic matter in the Indian Ocean record the former existence of elevated areas and times of milder paleoclimate. Freshwater coals present at Site 741 on the margin of East Antarctica indicate temperate Albian climates on Gondwana. Coaly organic matter is also found at Sites 748 and 750 on the Kerguelen Plateau in upper Albian silty claystones and sandstones. These sediments suggest the former existence of freshwater swamps on this now submerged plateau. The presence of coal in Paleogene sediments at Site 214 on Ninetyeast Ridge indicates that this ridge was formerly above sealevel and was heavily vegetated.

Contributions of land-derived organic matter significantly increase the concentrations of organic carbon in the ocean sediments in two ways. First, the land-derived material augments the supply of marine-derived organic matter. Second, land plants produce organic matter which contains large proportions of slowly degraded woody tissues, making land-derived organic matter more likely to be preserved in deep-sea sediments than marine organic matter [e.g. Emerson and Hedges, 1988].

Sediments rich in marine organic matter accumulated in the Indian Ocean during times of enhanced preservation of organic matter and of elevated marine productivity. Two thin layers of black shales containing as much as 26% marine-derived organic carbon were deposited at the Cenomanian/Turonian boundary at Site 763 on the Exmouth Plateau. Absence of similar organic carbon enrichments in Cenomanian/Turonian boundary sequences at nearby Sites 258, 762, 765, and 766 suggests that organic matter preservation was enhanced as a result of deposition within an intensified midwater oxygen minimum. Temporary uplift of Broken Ridge during the Eocene caused an interval of more rapid burial and consequently enhanced accumulation of marine organic matter at Sites 752 and 754. The dramatic uplift of the Himalayas initiated the high productivity associated with monsoonal upwelling in the middle Miocene and led to the continued accumulation of sediments rich in organic matter off Somalia, Oman, and Pakistan. Sediments deposited at

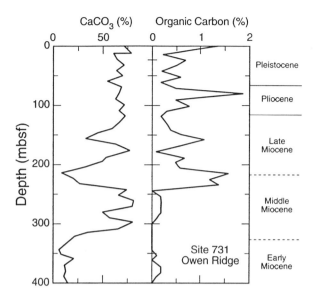

Fig. 12. Changes in concentrations of $CaCO_3$ and organic carbon in sediments from Site 731 on the Owen Ridge. Low $CaCO_3$ and organic carbon percentages in the early Miocene record accumulation of Indus Fan detrital sediments until uplift of the Owen Ridge in the early middle Miocene. Enhanced concentrations of organic carbon beginning at ca 10 Ma in the early late Miocene record the onset of monsoonal upwelling. From Prell, Niitsuma et al. [1989].

Site 731 on the Owen Ridge record the onset of monsoonal upwelling as the sudden appearance of organic carbon concentrations of 1 to 1.4% at ca 10 Ma.

Increases in the production of marine organic matter have a profound effect on the accumulation of organic carbon in marine sediments. The availability of organic matter that can become buried in the seafloor is obviously increased when marine production is enhanced. Equally important to accumulating organic carbon in sediments is the improved preservation that typically accompanies higher production rates. Marine organic matter is more susceptible to oxidative remineralization during sinking and early sedimentation than is continental organic matter. Flux rates to the seafloor typically increase, thereby diminishing destruction of organic matter in the water column during times of high marine production. Furthermore, the dissolved oxygen supply in bottom seawater can become exhausted and eliminate benthic degradation of organic matter. In areas like the coasts of western India-Pakistan, Arabia, and Somalia, an expanded and intensified oxygen minumum zone can be created under the upwelling area, thus improving the preservation and accumulation of the organic matter that reaches the seafloor and would otherwise be metabolized.

Acknowledgments. Comments from K. A. Kvenvolden, B. R. T. Simoneit, C. P. Summerhayes, and U. von Rad did much to improve this paper. We thank F. M. Gradstein and D. W. Murray for sharing their information and insights with us, J. C. Zachos for providing the Indian Ocean paleoreconstructions, and D. E. Austin for creating the illustrations. PAM strangely enjoyed the experience of spending 9 weeks at sea during ODP Leg 122.

REFERENCES

Arthur, M. A., Schlanger, S. O., and Jenkyns, H. C., The Cenomanian-Turonian Oceanic Anoxic Event, II. Palaeoceanographic controls on organic-matter production and preservation, in *Marine Petroleum Source Rocks*, edited by J. Brooks and A. J. Fleet, pp. 401-420, Blackwell Scientific Publishers, Oxford, 1987.

Barber, P. M., Palaeotectonic evolution and hydrocarbon genesis of the central Exmouth Plateau, *Aust. Pet. Explor. Assoc. J.*, 22, 131-144, 1982.

Barron, J., Larson, B., and Shipboard Scientific Party, Site 741, *Proc. ODP, Init. Repts.*, 119, 377-395, 1989.

Bertrand, P., Lallier-Verges, E. and Gral, H., Organic petrology of Neogene sediments from north Indian Ocean (Leg 117), amount, type, and preservation of organic matter, *Proc. ODP, Sci. Results*, 117, 587-594, 1991.

Bode, G. W., Carbon and carbonate analyses, Leg 27, *Init. Repts. DSDP*, 27, 499-502, 1974.

Brassell, S. C., Eglinton, G., Marlowe, I. T., Sarnthein, M., and Pflaumann, U., Molecular stratigraphy: a new tool for climatic assessment, *Nature*, 320, 129-133, 1986.

Campbell, I. R., Tait, A.M., and Reiser, R.F., Barrow Island oilfield, revisited, *Aust. Pet. Explor. Assoc. J.*, 24, 289-298, 1984.

Cobler, R., and Dymond, J., Sediment trap experiment on the Galapagos Spreading Center, equatorial Pacific, *Science*, 209, 801-803, 1980.

Cochran, J. R., Stow, D. A. V., and Shipboard Party, Site 717, *Proc. ODP, Init. Repts.*, 116, 45-89, 1989a.

Cochran, J. R., Stow, D. A. V., and Shipboard Party, Site 718: Bengal Fan, *Proc. ODP, Init. Repts.*, 116, 91-154, 1989b.

Cook, A. C., Report on the petrography of a Paleocene brown coal sample from the Ninetyeast Ridge, Indian Ocean, *Init. Repts. DSDP*, 22, 485-488, 1974.

Cook, A. C., Smyth, M., and Vos, R. G., Source potential of Upper Triassic fluvio-deltaic systems of the Exmouth Plateau, *Aust. Pet. Expl. Assoc. J.*, 25, 204-215, 1985.

Dean, W. E., Arthur, M. A., and Stow, D. A. V., Origin and geochemistry of Cretaceous deep-sea black shales and multicolored claystones, with emphasis on Deep Sea Drilling Project Site 530, southern Angola Basin, *Init. Repts. DSDP*, 75, 819-844, 1984.

Degens, E. T., and Mopper, K., Factors controlling the distribution and early diagenesis of organic material in marine sediments, in *Chemical Oceanography*, 5, edited by J. P. Riley and R. Chester, pp. 59-113, Academic Press, London, 1976.

Diester-Haass, L., Meyers, P. A., and Rothe, R., Miocene history of the Benguela Current and Antarctic ice volumes: evidence from rhythmic sedimentation and current growth across the Walvis Ridge (Deep Sea Drilling Project Sites 362 and 532), *Paleoceanogr.*, 5, 685-707, 1990.

Emerson, S., and Hedges, J. I., Processes controlling the organic carbon content of open ocean sediments, *Paleoceanogr.*, 3, 621-634, 1988.

Erdman, J. G., Schorno, K. S., and Scanlan, R. S., Geochemistry of carbon: DSDP Legs 22, 24, 26, 27, and 28, *Init. Repts. DSDP*, 24, 1169-1176, 1974.

Espitalié, J., Laporte, J. L., Madec, M., Marquis, F., Leplat, P., Paulet, J., and Boutefeu, A., Méthode rapide de caractérisation des roches mères, de leur potential pétrolier et de leur degré d'évolution, *Revue de l'Institut Francais du Pétrole*, 32, 23-42, 1977.

Exon, N. F., von Rad, U., and von Stackelberg, U., The geological development of the passive margins of the Exmouth Plateau off northwest Australia, *Mar. Geol.*, 47, 131-152, 1982.

Exon, N. F., Williamson, P. E., von Rad, U., Haq, B. U., and O'Connell, S., Ocean drilling finds Triassic, *Oil & Gas J.*, October 30, 46-52, 1989.

Flores, G., The Cretaceous and Tertiary sedimentary basins of Mozambique and Zululand, in *Sedimentary Basins of the African Coasts, Part II, South and East Coasts*, edited by G. Blant, pp. 81-112, Assoc. African Geol. Surv., Paris, 1973.

Gaetani, M., and Garzanti, E., Multicyclic history of the northern India continental margin (northwestern Himalaya), *AAPG Bull.*, 75, 1427-1446, 1991.

Girdley, W. A., Appendix III, Carbon and carbonate carbon, *Init. Repts. DSDP*, 25, 841, 1974.

Girdley, W. A., Leclaire, L., Moore, C., Vallier, T. L., and White, S. M., Lithologic summary, Leg 25, Deep Sea Drilling Project, *Init. Repts. DSDP*, 25, 725-741, 1974.

Gradstein, F.M., Gibling, M. R., Jansa, L. F., Kaminski, M.A., Ogg, J. G., Sarti, M., Thurow, J. W., von Rad, U., and Westermann, G. E. G., *Mesozoic Stratigraphy of Thakkhola, Central Nepal*, Spec. Report No. 1, Dalhousie University, Halifax, 1989.

Hallam, A., Mesozoic marine organic-rich shales, in *Marine Petroleum Source Rocks*, edited by J. Brooks and A. J. Fleet, pp. 251-261, Blackwell Scientific Publishers, Oxford, 1987.

Haq, B. U., von Rad, U., and Shipboard Party, *Proc. ODP, Init. Repts.*, 122, Ocean Drilling Program, College Station, TX, 1990.

Haq, B. U., Boyd, R., Exon, N. and von Rad, U., Evolution of the central Exmouth Plateau: a post-drilling perspective, *Proc. ODP, Sci. Results*, 122, 801-816, 1992.

Harrison, T. M., Copeland, P., Kidd, W. S. F., and Yin, A., Raising Tibet, *Science*, 255, 1663-1670, 1992.

Heirtzler, J. R., Veevers, J. J., and Shipboard Scientific Party, Site 263, *Init. Repts. DSDP*, 27, 279-335, 1974.

Hunt, J. M., Hydrocarbon and kerogen studies, *Init. Repts. DSDP*, 22, 673-675, 1974a.

Hunt, J. M., Hydrocarbon and kerogen studies on Red Sea and Gulf of Aden cores, *Init. Repts. DSDP*, 24, 1165-1167, 1974b.

Jasper, J. P., and Gagosian, R. B., Glacial-interglacial climatically forced $\delta^{13}C$ variations in sedimentary organic matter, *Nature*, 343, 60-62, 1989.

Kidd, R. B., and Davies, T. A., Indian Ocean sediment distribution since the Late Jurassic, *Mar. Geol.*, 26, 49-70, 1977.

Kuhnt, W., Herbin, J. P., Thurow, J. and Wiedmann, J., Distribution of Cenomanian-Turonian organic facies in the western Mediterranean and along adjacent Atlantic margin, in *Deposition of Organic Facies*, edited by A. Y. Huc, pp. 133-160, AAPG Studies in Geology #30, Am. Assoc. Petrol. Geol., Tulsa, 1990.

Kvenvolden, K. A., Hostettler, F. R., Rapp, J. B., and Frank, T. J., Aliphatic hydrocarbons in sediments from Prydz Bay, Antarctica, *Proc. ODP, Sci. Results*, 119, 417-424, 1991.

Lindsay, J. F., Holliday, D. W., and Hulbert, A. G., Sequence stratigraphy and the evolution of the Ganges-Brahmaputra Delta complex, *AAPG Bull.*, 75, 1233-1254, 1991.

Littke, R., Rullkötter, J. and Schaefer, R. G., Organic and carbonate carbon accumulation on Broken Ridge and Ninetyeast Ridge, central Indian Ocean, *Proc. ODP, Sci. Results*, 121, 467-487, 1991.

McDonald, T. J., Kennicutt, M.C., Rafalska, J. K., and Fox, R. G., Source and maturity of organic matter in glacial and Cretaceous sediments from Prydz Bay, Antarctica, ODP Holes 739C and 741A, *Proc. ODP, Sci. Results*, 119, 407-416, 1991.

McIver, R., Residual gas contents of organic-rich canned sediment samples from Leg 23, *Init. Repts. DSDP*, 23, 971-973, 1974.

McIver, R., Hydrocarbon occurrences from JOIDES Deep Sea Drilling Project, *Proc. Ninth World Pet. Congr.*, 269-280, 1975.

Meyers, P. A., Organic matter variations in sediments from DSDP Sites 362 and 532: evidence of changes in the Benguela Current upwelling, in *Evolution of Upwelling Systems since the Early Miocene*, in press, edited by C. P. Summerhayes, W. L. Prell, and K.-C. Emeis, Geological Society, London, 1992.

Meyers, P.A., and Snowdon, L. R., Extractable hydrocarbon and carbon isotope geochemistry of Early Cretaceous sediments from Sites 762 and 763 on the Exmouth Plateau, northwest Australian margin, *Proc. ODP, Sci. Results, 122,* 855-860, 1992.

Meyers, P.A., Brassell, S. C., and Huc, A. Y., Geochemistry of organic carbon in South Atlantic sediments from Deep Sea Drilling Project Leg 75, *Init. Repts. DSDP, 75,* 967-981, 1984.

Müller, P. J., and Suess, E., Productivity, sedimentation rate, and sedimentary organic matter in the oceans. I. Organic matter preservation, *Deep-Sea Res., 26,* 1347-1362, 1979.

Muzuka, A. N. N., Macko, S. A., and Peresen, T. F., Stable carbon and nitrogen isotope compositions of organic matter from Sites 724 and 725, Oman Margin, *Proc. ODP, Sci. Results, 117,* 571-586, 1991.

Palmer, A. R., The Decade of North American Geology 1983 Geologic Time Scale, *Geology, 11,* 503-504, 1983.

Pederson, T. F., Shimmield, G. B., and Price, N. B., Lack of enhanced preservation of organic matter in sediments under the oxygen minimum on the Oman Margin, *Geochim. Cosmochim. Acta, 56,* 545-551, 1992.

Peirce, J., Weissel, J., and Shipboard Scientific Party, Broken Ridge Summary, *Proc. ODP, Init. Repts., 121,* 457-516, 1989.

Pospichal, J. J., Dehn, J., Driscoll, N. W., van Eijden, A. J. M., Farrell, J. W., Fourtanier, E., Gamson, P., Gee, J., Janacek, T. R., Jenkins, D. G., Klootwijk, C., Nomura, R., Owen, R. M. Rea, D. K., Resiwati, P., Smit, J., and Smith, G., Cretaceous-Paleogene biomagnetostratigraphy of Sites 752-755, Broken Ridge: a synthesis, *Proc. ODP, Sci. Results, 121,* 721-741, 1991.

Poynter, J., and Eglinton, G., Molecular composition of three sediments from Hole 717C: the Bengal Fan, *Proc. ODP, Sci. Results, 116,* 155-161, 1990.

Prell, W. L., Niitsuma, N., and Shipboard Party, *Proc. ODP, Init. Repts., 117,* Ocean Drilling Program, College Station, TX, 1989.

Rullkötter, J., Littke, R., Radke, M., Disko, U., Horsfield, B., and Thurow, J., Petrography and geochemistry of organic matter in Triassic and Cretaceous deep-sea sediments from the Wombat and Exmouth Plateaus and nearby abyssal plains off northwest Australia, *Proc. ODP, Sci. Results, 122,* 317-335, 1992.

Schlanger, S. O., Arthur, M. A., Jenkyns, H. C., and Scholle, P. A., The Cenomanian-Turonian Oceanic Anoxic Event, I, Stratigraphy and the distribution of organic carbon-rich beds and the marine $\delta^{13}C$ excursion, in *Marine Petroleum Source Rocks,* edited by J. Brooks and A. J. Fleet, pp. 371-399, Blackwell Scientific Publishers, Oxford, 1987.

Schlich, R., Wise, S. W., Jr., and Shipboard Scientific Party, Site 748, *Proc. ODP, Init. Repts., 120,* 157-235, 1989a.

Schlich, R., Wise, S. W., Jr., and Shipboard Scientific Party, Site 750, *Proc. ODP, Init. Repts., 120,* 277-337, 1989b.

Seifert, R., and Michaelis, W., Organic compounds in sediments and pore waters of Sites 723 and 724, *Proc. ODP, Sci. Results, 117,* 529-545, 1991.

Snowdon, L. R., and Meyers, P. A., Source and maturity of organic matter in sediments and rocks from Sites 759, 760, 761, and 764 (Wombat Plateau) and Sites 762 and 763 (Exmouth Plateau), *Proc. ODP, Sci. Results, 122,* 309-315, 1992.

ten Haven, L., and Rullkötter, J., Preliminary lipid analyses of sediments recovered during Leg 117, *Proc. ODP, Sci. Results, 117,* 561-569, 1991.

Truswell, E. M., Data report: Palynology of sediments from Leg 119 drill sites in Prydz Bay, East Antarctica, *Proc. ODP, Sci. Results, 119,* 941-945, 1991.

Turner, B. R., and Padley, D., Lower Cretaceous coal-bearing sediments from Prydz Bay, East Antarctica, *Proc. ODP, Sci. Results, 119,* 57-60, 1991.

von der Borch, C. C., Scalter, J. G., and Shipboard Scientific Party, Site 214, *Init. Repts. DSDP, 22,* 119-191, 1974.

von Rad, U., Exon, N., and Haq, B. U., Rift-to-drift history of the Wombat Plateau, northwest Australia: Triassic to Tertiary Leg 122 results, *Proc. ODP, Sci. Results, 122,* 756-800, 1992.

Weissert, H., The environment of deposition of black shales in the Early Cretaceous: an ongoing controversy, in *The Deep Sea Drilling Project: A Decade of Progress,* edited by J. C. Warme, R. G Douglas, and E. L. Winterer, pp. 547-560, SEPM, Tulsa, 1981.

Whitmarsh, R. B., Weser, O. E., and Shipboard Scientific Party, Site 229, *Init. Repts. DSDP, 23,* 753-807, 1974.

Williamson, P. E., Exon, N. F., Haq, B., von Rad, U., and Leg 122 Shipboard Scientific Party, A North West Shelf Triassic reef play: results from ODP Leg 122, *Aust. Pet. Explor. Assoc. J., 29,* 328-344, 1989.

Cenozoic Carbonate Accumulation and Compensation Depth Changes in the Indian Ocean

L. C. PETERSON

Rosenstiel School of Marine and Atmospheric Science, University of Miami, Miami, FL, 33149, USA

D. W. MURRAY

Department of Geological Sciences, Brown University, Providence, RI, 02912, USA

W. U. EHRMANN

Alfred-Wegener-Institut für Polar- und Meeresforschung, D-2850 Bremerhaven, Germany

P. HEMPEL

GEOMAR - Research Center for Marine Geosciences, Wischhofstrasse 1-3, D-2300 Kiel, Germany

Quantitative estimates of the flux of biogenic carbonate to the sea floor, and of the rates and patterns of its subsequent accumulation, are key components to understanding past ocean history. As one of its principal objectives, Leg 115 of the Ocean Drilling Program drilled a depth transect of sites in the western equatorial Indian Ocean for detailed studies of the carbonate system. Here, we review the major results of the Leg 115 program and compare them with data on carbonate accumulation rates derived from eighteen additional sites drilled during the remaining legs of the Indian Ocean campaign. Regional accumulation patterns are found that are variably affected by the changing balance between productivity and dissolution, and by physical processes such as erosion. In the Paleogene, tectonic activity and changing bottom circulation resulted in a patchy record of carbonate sedimentation across the basin. In the early and middle Miocene, greatly reduced carbonate accumulation in the equatorial region was the result of a shallow CCD and a probable drop in surface productivity. Winnowing was also common at shallow Indian Ocean sites during this interval, suggesting a more vigorous intermediate water circulation. At about 8-9 Ma, carbonate accumulation rose abruptly in shallow low latitude sites and a strong dissolution gradient was first established in the water column. Arabian Sea sites at this time show accumulation patterns similar to Indo-Pacific sites far removed from the direct influence of monsoonal upwelling, suggesting that carbonate deposition in this region is responding more to global-scale processes than to the localized influence of the monsoon.

A comparison of the Neogene depth history of the CCD to the record of eustasy supports the argument that sea level exerts a first-order control on the deposition of carbonate through shelf-basin fractionation. Regional variations in carbonate accumulation above the CCD are probably more directly affected by carbonate production rates in the overlying surface waters.

INTRODUCTION

The spatial and temporal accumulation patterns of calcium carbonate in the marine stratigraphic record represent a primary source of data about the chemistry and circulation of past oceans, and hence the global geochemical cycle of CO_2. The construction of well-constrained models of the deep sea carbonate budget and of its changes through time and space are of critical importance to understanding the history of both ocean circulation and global climate. Such models can be developed only by the study and quantification of those processes that control the calcium carbonate system.

Synthesis of Results from Scientific
Drilling in the Indian Ocean
Geophysical Monograph 70
Copyright 1992 American Geophysical Union

Deep-sea drilling has long since established that carbonate dissolution levels in the different ocean basins have fluctuated markedly during the Cenozoic [e.g., Berger and Winterer, 1974; van Andel, 1975; Heath et al., 1977; Hsü et al, 1984; Rea and Leinen, 1985]. In general, these studies have shown that long-term changes in the patterns of carbonate distribution have been similar between the major ocean basins, suggesting a global forcing mechanism perhaps linked to changing sea level or to a changing supply of carbonate to the oceans [Kennett, 1982]. Over the shorter term, distinct interocean differences in this general pattern have been found to exist that are likely related to basin-to-basin fractionation of carbonate [Berger, 1970] through changes in deep circulation, surface productivity, and/or interbasinal hypsometry.

Prior to the 1987-88 phase of drilling by the Ocean

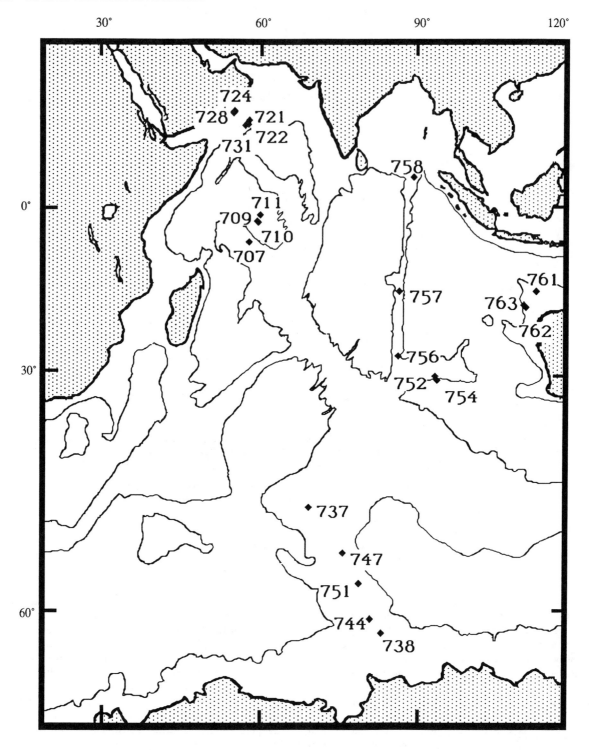

Fig. 1. Index map illustrating locations of Ocean Drilling Program sites utilized in this study. Water depths of individual sites are listed in Figures 3 through 7. Regional bathymetry is defined by an approximation of the 4000 m isobath.

Drilling Program (ODP), efforts to reconstruct the Cenozoic history of carbonate sedimentation in the Indian Ocean were hampered by the lack of suitable stratigraphic sequences available for study. Of the more than 50 locations originally drilled in 1972 by the Deep Sea Drilling Project (DSDP), only a small subset of sites reached basement and were continuously cored, and all sites were drilled in the rotary mode. DSDP drilling was also largely restricted to the low latitudes with an uneven distribution of sites relative to water depth. Despite these shortcomings, studies by van Andel [1975], Sclater et al. [1977], Davies and Kidd [1977], and Kidd and Davies [1978] provided an important first-order picture of changing carbonate sedimentation patterns in the Indian Ocean basin through time. Because of data limitations, however, the focus of these earlier studies was generally restricted to mapping sediment distributions for various time slices and to evaluating the history of the carbonate compensation depth (CCD) from the stratigraphic and paleodepth distribution of carbonate/clay transitions.

Leg 115, the first leg of the nine leg ODP venture into the Indian Ocean, had as its major paleoceanographic objective the drilling of a depth transect of sites in the western equatorial Indian Ocean specifically for studies of the deep-sea carbonate system. Combining data on sediment carbonate concentrations with a well resolved bio- and magnetostratigraphy, Peterson and Backman [1990] produced the first detailed compilation of Cenozoic carbonate mass accumulation rates (MARs) for the tropical Indian Ocean. These data, when combined with subsidence histories modeled for the individual transect sites, were also used to produce a better constrained record of CCD motion for the Neogene.

In this paper, our intent is to review the principal results of the Leg 115 "carbonate transect" program, and to compare and contrast these results with data on carbonate accumulation rates and patterns derived from selected sites drilled during the eight other Indian Ocean ODP legs (Figure 1). By focusing on the rates at which the biogenic carbonate components have accumulated, rather than on the gross distribution of carbonate-rich facies, our immediate goal is to produce a more rigorously quantified and regionally cohesive portrait of carbonate sedimentation in the Indian Ocean than previously available. The longer term challenge will be to integrate these data with similar results from a growing global array of drillsites to better understand the temporal evolution of the carbonate and ocean-climate systems.

STRATEGIES, MATERIALS, AND METHODS

Carbonate accumulation in open-ocean environments is largely a balance between the production rate of planktonic foraminifers and calcareous nannoplankton in surface waters and their flux to and subsequent dissolution on the sea floor. Productivity at the sea surface is primarily determined by the availability of nutrients, whereas dissolution on the sea floor is largely a function of the calcium carbonate saturation state of seawater at the sediment/water interface. Both processes are themselves intimately linked to changes in surface- and deep-water circulation that control the rates of vertical mixing and the interbasinal distribution of nutrients, alkalinity, oxygen, and carbon [e.g., Broecker and Peng, 1982].

Any attempt to model the carbonate budget of the oceans and its variability in the past requires knowledge of the rates at which the carbonate components have been produced and have accumulated or dissolved through time. Such information is best collected within the framework of a depth transect of sites since the dissolution of carbonate is a depth-dependent process. The dissolution gradient in the ocean can ideally be thought of as extending from the lysocline, the depth level at which carbonate dissolution becomes clearly apparent [Berger, 1968, 1975], to the CCD, the still deeper level separating carbonate-bearing sediments above from carbonate-free sediments below [Bramlette, 1961]. In general, the working premise of transect studies has been that the carbonate accumulation records at sites shallow enough to have always remained above the lysocline provide a baseline measure of the surface production of carbonate, while the depth-dependent differences in accumulation between deeper sites reflect progressive loss of carbonate from the sediments to dissolution. The success of this strategy in studies of late Quaternary sediments [e.g., Curry and Lohmann, 1985, 1986, 1990; Peterson and Prell, 1985a,b; Farrell and Prell, 1989, 1991; Wu et al., 1991] has motivated its application to the longer time scales accessible to scientific drilling.

Five closely spaced sites (Sites 707 through 711) were drilled in the carbonate depth transect of Leg 115. These sites, ranging in water depth from 1541 to 4428 m, are located in the western equatorial region on the northern Mascarene Plateau, Madingley Rise, and on the adjacent basin floor (Figure 2). The depth interval sampled by the five sites encompasses the present depth range of intermediate-, deep-, and bottom-water masses in the area. The latter two in the Indian Ocean have their origins in the North Atlantic and Antarctic, respectively, as neither deep nor bottom-water masses are presently formed in the Indian Basin [Tchernia, 1980; Warren, 1981]. In the intermediate-depth range, a high-salinity water mass, largely of Red Sea and Persian Gulf origins, fills much of the mid-depths of the northwestern Indian Ocean [Wyrtki, 1973]. This northern Indian Ocean intermediate water can be traced today to depths of 1500 m and as far south as the Mascarene Plateau, where it mixes with northward-flowing Antarctic Intermediate Water. Sites 707 through 711 currently lie to the north of the sharp hydrochemical front at 10°S that today separates surface waters of the unique seasonally reversing monsoon gyre to the north from those of the subtropical gyre to the south [Wyrtki, 1973]. Surface waters to the north of this front have high nutrient levels and correspondingly high productivity [Koblentz-Mishke et al., 1970], whereas to the south they are characterized by low levels of both. The productivity gradient in surface waters today is clearly reflected in the distribution of biogenic opal and barite in the underlying sediments

[Leinen et al., 1986; Schmitz, 1987].

Although Sites 707 through 711 were drilled specifically for studies of the carbonate system, they only provide insight into the dynamics of carbonate accumulation in the tropical Indian Ocean. In this review, we have attempted to extend the regional coverage by assimilating data from a selected total of 18 additional sites drilled during the ODP Indian Ocean campaign (Figure 1). Leg 117 sites in the Arabian Sea, though their records are limited in time, provide important data on carbonate accumulation in the region of the Indian Ocean strongly influenced by monsoonal upwelling. Sites drilled on the Kerguelen Plateau during Legs 119 and 120 add a critical perspective from the high latitudes, while sites from Broken Ridge and Ninetyeast Ridge (Leg 121) and from off northwest Australia (Leg 122) yield the opportunity to fill in mid-latitude gaps. We have not included data from sites drilled during Legs 116, 118, and 123 since their leg objectives largely precluded recovery of materials suitable for this study. We have also not attempted to directly integrate data from earlier Indian Ocean DSDP sites since their stratigraphic and chronologic control is, in general, greatly inferior to that of sites drilled by the ODP. To develop the appropriate correlations and chronostratigraphies for these older sites is beyond the scope of this effort.

The underlying rationale for converting data on sediment carbonate content (in wt. %) into estimates of carbonate MARs has been discussed extensively by van Andel et al. [1975], Moore et al. [1984], Pisias and Prell [1985], and Broecker and Peng [1987] among others. Carbonate MAR estimates are based on simple computations involving the sedimentation rate (cm/kyrs), the dry-bulk density of the sediment (g/cm^3), and the measured weight percent calcium carbonate. Of the parameters used to calculate the carbonate MAR records, the estimates of sedimentation rate are most typically what drive the overall calculations.

For the Leg 115 transect sites, the age control used to calculate sedimentation rates comes from a combination of biostratigraphic and magnetostratigraphic data, all placed within the framework of the geomagnetic polarity time scale of Berggren et al. [1985a,b]. An exception to this occurs in the Miocene where the revised age estimates of Backman et al. [1990] have been used for certain nannofossil events. An extensive discussion of the age models used for the Leg 115 sites can be found in Appendix B of Peterson and Backman [1990]. For all of the other sites utilized here, we have used stratigraphic data and age models from the *Initial Reports* volume for each leg, updated where possible by additional data or modifications presented in those *Scientific Results* volumes published to date. If not done so already, the published stratigraphic data were calibrated to the Berggren et al. [1985a,b] time scale for consistency of comparison, and include the modifications suggested by Backman et al. [1990]. A compilation of the stratigraphic and carbonate MAR data used in this paper can be obtained from the senior author by direct request.

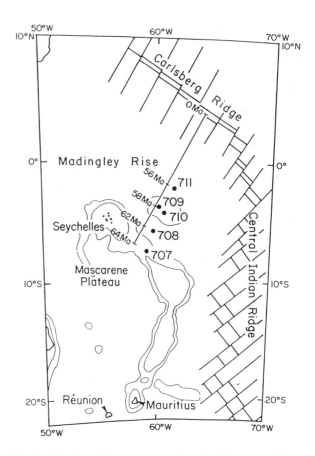

Fig. 2. Physiographic map centered on the Mascarene Plateau showing location of drillsites that comprise the Leg 115 "carbonate depth transect". Basement ages are estimates that were used by Simmons [1990] to model subsidence histories for each site.

In the Leg 115 transect sites, samples were analyzed for carbonate content at the rate of three to four samples per 1.5-m core section. A total of more than 2200 measurements are reported in Peterson and Backman [1990], along with their associated estimates of carbonate MAR. Although bulk sedimentation rates vary over different intervals and between sites, an average sampling interval of 75-120 k.y. was achieved for much of the transect sequence. For the additional ODP sites summarized here, data on carbonate contents and sediment dry-bulk densities were taken directly from the *Initial Reports* for each leg. In general, carbonate contents were measured much less frequently at most of these sites, with measurements being made on the same samples for which physical properties data (including dry-bulk density) were generated. These data are typically available at the rate of two to three measurements per 9.5-m core, but still provide sufficient resolution to examine the long-term Cenozoic trends in carbonate accumulation. Where additional post-cruise data have been incorporated from reports in the *Scientific Results* volumes, their source is acknowledged in the following discussion.

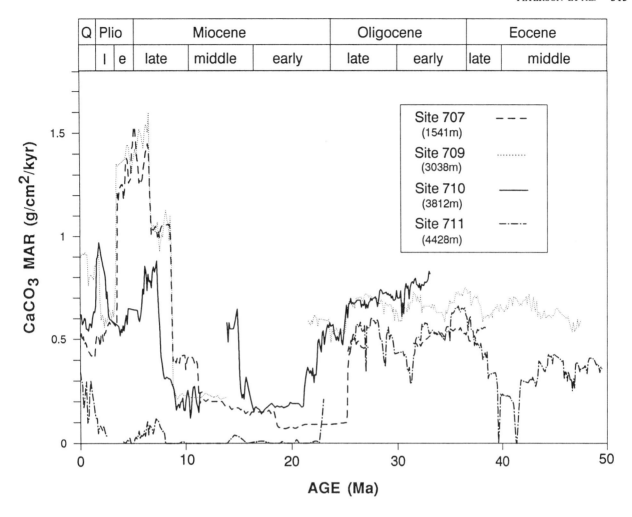

Fig. 3. Combined carbonate accumulation histories (in g/cm²/kyr) for Mascarene Plateau Sites 707, 709, 710, and 711. The data used in this figure have been smoothed by means of a five-point running average.

CARBONATE ACCUMULATION HISTORIES FOR LEG 115 SITES

The Eocene to Holocene carbonate MAR data for all Leg 115 transect sites but 708 (this site was riddled with turbidites and was of substantially poorer quality than the others) are plotted together in Figure 3. To simplify their comparison to less densely sampled sites, the data here have been smoothed slightly using a five-point running mean.

The cumulative MAR data set for the Leg 115 sites reveals three distinct periods of past carbonate accumulation for this region: 1) the Paleogene, a time of intermediate carbonate accumulation rates (0.4-0.7 g/cm²/kyrs) and generally reduced between-site accumulation differences; 2) the early and middle Miocene, a period characterized by greatly reduced carbonate MARs (typically <0.2 g/cm²/kyrs) at all sites; and 3) the late Miocene to Holocene, a span of time marked by the highest accumulation rates observed (up to 1.5 g/cm²/kyrs), and by the first sustained appearance of substantial contrasts in

carbonate accumulation as a function of the water depth of the drill sites.

Within the Eocene of the western tropical Indian Ocean, our knowledge of carbonate MARs comes mostly from the two long records of Sites 709 and 711. The Site 709 record indicates relatively uniform carbonate accumulation (~0.6 g/cm²/kyrs) through this interval. In contrast, carbonate MARs at the deeper Site 711 drop to near zero over two short intervals centered near 40 Ma, and are significantly reduced relative to the comparatively stable rates in Site 709 over a broad surrounding interval. Eocene sediments at both sites consist primarily of nannofossil oozes, though intense carbonate dissolution at Site 711 in the two short intervals near 40 Ma has concentrated the biogenic opal fraction and produced nearly pure radiolarian oozes. The steady increase in carbonate MARs at Site 711 thereafter (i.e., between about 39 and 35 Ma) is probably related to the well-documented deepening of the CCD across the Eocene/Oligocene boundary [Heath, 1969; Berger, 1973;

van Andel, 1975; Hay, 1988].

Throughout most of the Oligocene, carbonate MAR changes can be traced reasonably well across the four sites shown in Figure 3, and generally average about the same or slightly less (0.5-0.6 g/cm²/kyrs) than the Eocene rates in Site 709. These rates suggest moderate levels of surface productivity, an observation consistent with the relatively common occurrence of biogenic opal in the sediments [e.g., Mikkelsen, 1990; Johnson, 1990; Baldauf et al., this volume]. On average, carbonate MARs in the Oligocene vary by only about 0.2-0.3 g/cm²/kyrs among the sites, an observation which suggests that the accumulation gradient between sites was considerably reduced relative to the gradient around 40 Ma, or to what is observed for the late Neogene. If all of the sites initially received the same pelagic input, a standard working assumption of the transect strategy, we would tend to interpret this pattern as evidence for a reduced dissolution gradient over the depth interval between sites. Given the shallower paleodepths of these sites in the Oligocene (see Discussion) such a reduced dissolution gradient might be anticipated, though to date systematic comparisons of calcareous microfossil preservation between sites have not been completed. On the other hand, the fact that MAR data from the deeper Site 710 exceed those from all other sites between about 28 and 33 Ma suggests either an age model problem in this interval or that some reworking and downslope transport must have occurred. Premoli-Silva and Spezzaferri [1990] noted that reworking of foraminiferal assemblages is a persistent feature of much of the Paleogene section at this site. Okada [1990], however, reported that the calcareous nannofossils were relatively unaffected, with all of the zones and subzones capable of being recognized in their correct order. Such apparently inconsistent observations might be accounted for by a rather continual downslope drift of fine coccolith carbonate from shallower depths, as opposed to more sporadic transport events that move the coarser materials downslope. Size fraction data from DSDP Site 237 [Thiede, 1974], a shallow site (1623 m) located adjacent to Site 707, show coarse fractions (> 63 microns) of only 10-20% in the Oligocene section, values that would seem to argue against extensive winnowing of the fines from the top of the plateau.

The second major sedimentary regime apparent in the Leg 115 accumulation data encompasses most of the early and middle Miocene and is characterized by greatly reduced carbonate MARs at all of the transect sites (Figure 3). The transition from higher Oligocene MARs to the low rates of this interval apparently began in the latest Oligocene (~24-25 Ma) and concluded somewhere between 22.5 Ma (at Site 711) and 21.2 Ma (at Site 710). The lack of carbonate accumulation at Site 711 over much of the early and middle Miocene interval indicates a CCD shallower than the paleodepth of this site, a shoaling trend consistent with the observations of most other CCD reconstructions [e.g., van Andel, 1975; Heath et al., 1977; Hsü et al., 1984; Rea and Leinen, 1985].

As in the Oligocene, little contrast in carbonate accumulation between shallow and deep Leg 115 sites is observed in the early and middle Miocene. The rather steady tracking of carbonate MARs through their drop across the Oligocene-Miocene transition and the surprisingly similar between-site gradients before and after would seem to imply some simple forcing mechanism that affects all sites in an equal manner. Data available at the moment, however, suggest that the combined influence of several interrelated processes is probably at work here, with the relative contributions of carbonate dissolution, low productivity, and winnowing most likely varying as a function of site depth. As would be expected with a shallow CCD, increased dissolution of the calcareous components during the early and middle Miocene is observed at most sites in this interval [Rio et al., 1990; Boersma, 1990; Vincent and Toumarkine, 1990]. Lower surface productivity would seem to be generally indicated by the disappearance of biogenic opal at all sites in the late Oligocene and its absence in the stratigraphic record until the late Miocene [Mikkelsen, 1990; Johnson, 1990; Hempel and Bohrmann, 1990]. Finally, winnowing and downslope redistribution of carbonate has clearly affected the transect sites as well. At Site 707, the shallowest site, carbonate MARs are lower than those calculated for the deeper Site 710 through much of the early and middle Miocene. Although deposition at Site 707 appears to have been continuous, winnowing of the fines has produced a clear coarsening of the sediments, as it has at nearby DSDP Site 237. Coarse fractions in the latter climb from low Oligocene values and average some 50-60% through much of the Miocene [Thiede, 1974]. At deeper sites, the presence of reworked materials and some real and/or probable short hiatuses in the early-middle Miocene interval [Backman, Duncan et al., 1988] also argue that some downslope redeposition has occurred. In Site 709, for example, we chose to discard data between about 14 and 21 Ma (Figure 3) because of sediment disturbance and numerous stratigraphic inversions in the section.

Unlike the gradual transition that separates the accumulation regime of the Paleogene from that of the early and middle Miocene, the transition to the third major regime of high carbonate MARs in the late Miocene seems to have been relatively abrupt. In the deeper Sites 710 and 711, the greater part of the rise appears to occur about 1 m.y. later (~8 Ma) than recorded at shallow Sites 707 and 709 (~9 Ma). Peterson and Backman [1990] considered this event to be synchronous with the discrepancy the result of an age model imperfection. Given that age control in the pertinent intervals is based on magnetostratigraphy in Sites 710 and 711 and biostratigraphy in Sites 707 and 709, we tend to expect the deeper sites to provide a more accurate chronology for the late Miocene MAR increase.

From the time of this MAR rise (~8-9 Ma) to the present in the western equatorial Indian Ocean, shallow Leg 115 sites record the highest carbonate MARs (up to 1.5 g/cm²/kyrs) of the last 50 m.y., biogenic silica reappears in the sediments, and the first substantial depth gradients

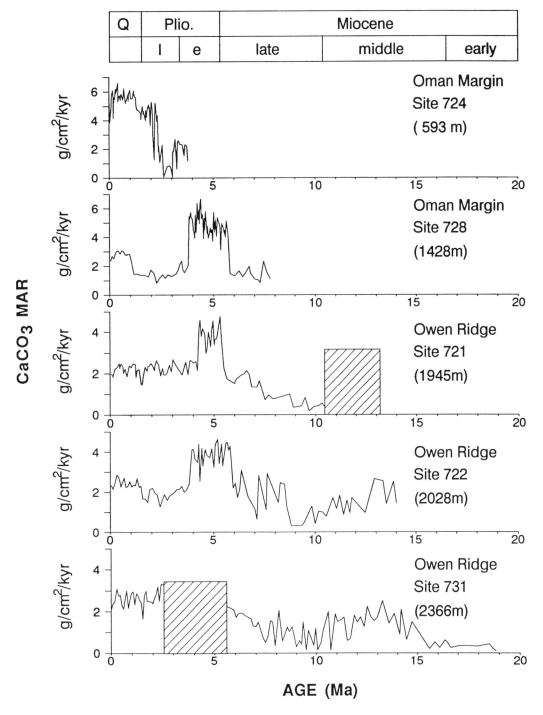

Fig. 4. Calculated carbonate mass accumulation rates (g/cm²/kyr) for Leg 117 sites from the Arabian Sea. Hachured areas indicate hiatuses in the section.

in carbonate accumulation are found. These data, when taken together, indicate that oceanographic conditions in the tropical Indian Ocean were beginning to approximate those found today. The evidence for higher surface productivity in this interval is unmistakable. Although it precedes the abrupt rise in carbonate MARs, biogenic silica

makes its reappearance at these sites in the form of radiolarians and sponge spicules at about 10 Ma, with diatoms later becoming the dominant opal component at about 8 Ma when the carbonate MARs rise [Johnson, 1990; Mikkelsen, 1990; Baldauf et al., this volume]. The diatom assemblages themselves contain high relative abundances of the *Thalassionema* group, taxa generally characteristic of productive upwelling conditions [Mikkelsen, 1990]. At around 3 Ma, a marked decrease in *Thalassionema* abundance and a general decline in silica accumulation is consistent with the scenario of reduced surface productivity implied by the pronounced drop in carbonate MARs at that time.

CARBONATE ACCUMULATION IN THE ARABIAN SEA - LEG 117 RESULTS

During Leg 117, a total of eleven sites were drilled on the Oman margin and Owen Ridge in the northern Arabian Sea [Prell, Niitsuma et al., 1989]. Carbonate MAR data from five of those sites are shown in Figure 4. Together, the five sites are representative of the three major depositional environments cored during Leg 117: the shallow Oman margin, the deep margin, and the crest of Owen Ridge.

Sites from the shallow Oman margin (Sites 723, 724, 725, 726, 727, and 730) were found to contain significant detrital calcite in addition to pelagic biogenic calcite, rendering them of little use to this study. Of the six shallow margin sites, Site 724 (593 m) contained the longest section of continuous deposition (0-3.75 Ma) and is the one illustrated here. Between 3.1 and 3.75 Ma, carbonate sediments accumulated at Site 724 with an average MAR of about 2 g/cm^2/kyrs and an average carbonate content of about 44%. In the interval between ~2.1 and 3.1 Ma, a mixture of carbonate-rich and siliceous-rich facies were found, with carbonate MARs dropping to <1 g/cm^2/kyrs. Within this interval, the carbonate content averages 31% and both foraminifers and coccoliths are poorly preserved. Laminated sequences were observed in this interval of poor preservation, suggesting that high dissolution is related to the conditions that led to anoxic bottom waters at this time [Shipboard Scientific Party, 1989a]. At about 2.1-2.3 Ma the mean carbonate MAR at Site 724 increased sharply, remaining at high levels (~5 g/cm^2/kyrs) for the past 2 m.y.. Although carbonate contents average near 54% in this interval, as much as one half of the material may be detrital [Shipboard Scientific Party, 1989a].

The deep margin Sites 728 and 729 contain carbonate-rich sediments that are largely of pelagic origin. Site 728 spans the last 8 m.y. and contains the longest margin section having continuous deposition (Figure 4). The mean carbonate MARs at Site 728 are similar to those found offshore on the Owen Ridge both in magnitude and temporal variability. Between 8 and 6 Ma, average carbonate accumulation rates were relatively low (~1 g/cm^2/kyrs) at this deep margin location. This corresponds to an interval of more abundant siliceous microfossils and enhanced dissolution of carbonate microfossils [Shipboard Scientific Party, 1989b]. Carbonate MARs increased by a factor of five near 6 Ma, and remained high until about 4 Ma. The mean rates were subsequently low during the late Pliocene and early Pleistocene (1 to 3.5 Ma), but have increased by a factor of two since that time.

Owen Ridge Sites 721, 722, and 731 were cored near the crest of the ridge in water depths ranging from 1945 to 2366 m. Sites 721 and 731 contain hiatuses, but Site 722 records continuous pelagic deposition over the last 14 m.y. and is representative of the other ridge crest sites in the interval of overlap. Prior to 14 Ma, the Owen Ridge crest was still within the depth zone affected by turbidite deposition on the adjacent Indus Fan [Shipboard Scientific Party, 1989c]. At or about 14 Ma, the crest of the Owen Ridge was tectonically elevated above the region of turbidity current influence, and pelagic deposition commenced. Between 14 and about 10.5 Ma, carbonate-rich sediments (mean CaCO$_3$ content = 65%) containing nannofossils characteristic of relatively warm waters were deposited on the ridge crest. Both bulk sedimentation rates (~20 m/m.y.) and carbonate MARs (mean = ~1.6 g/cm^2/kyrs) are relatively low over this section. Above this unit, sediments with abundant siliceous microfossils were deposited between 10.5 and 6.4 Ma (411 to 343 mbsf at Site 722). Calcareous microfossil preservation, as at Site 728 on the margin, is poor in this silica-rich interval, with carbonate MARs in Sites 721, 722, and 731 averaging about 1.5 g/cm^2/kyrs. After this time, carbonate accumulation increased to maximum rates of 3.5-4.5 g/cm^2/kyr between 6 and 4 Ma, with excellent carbonate preservation throughout [Murray and Prell, 1991]. Carbonate MARs decreased abruptly at about 4 Ma at the Owen Ridge sites, but have increased over the last 1 m.y. to their present mean rate of ~2.5 g/cm^2/kyrs. This rate is less than half of the present rate for the hemipelagic sediments of the shallow Oman margin, but is comparable to modern carbonate MARs at the deep Site 728. The present carbonate MARs on Owen Ridge are of similar magnitude to rates found today in the carbonate-rich pelagic sediments that underlie the productive waters of the eastern equatorial Pacific [e.g., Theyer et al., 1985; Leg 138 Shipboard Scientific Party, 1991].

CARBONATE ACCUMULATION ON NINETYEAST AND BROKEN RIDGES - LEG 121 RESULTS

During Leg 121 a total of seven sites were drilled on Broken Ridge and Ninetyeast Ridge in the eastern Indian Ocean. Carbonate MAR data for five of these sites are summarized in Figure 5.

Prior to about 42 Ma, Broken Ridge formed the northern part of the broad Kerguelen-Broken Ridge Plateau. During the middle Eocene, this large mega-plateau was split by the newly forming Southeast Indian Ridge [Peirce, Weissel, et al., 1989; Rea et al., 1990]; the uplift and subsequent erosion of Broken Ridge resulted in a prominent Eocene angular unconformity across its top that removed virtually all of the Paleogene section. Of the four sites drilled on Broken Ridge (Sites 752 to 755), only Site 752 contains a significant record of Paleogene deposition. Chalks of this age at Site 752 indicate open-ocean, high-latitude sedimentation, while paleodepth markers indicate a gradual subsidence of the northern margin of the Kerguelen-Broken Ridge Plateau from upper slope depths to probably as deep as 1000-1500 m [Rea et al., 1990]. Carbonate MARs in the early Eocene and through much of the Paleocene were

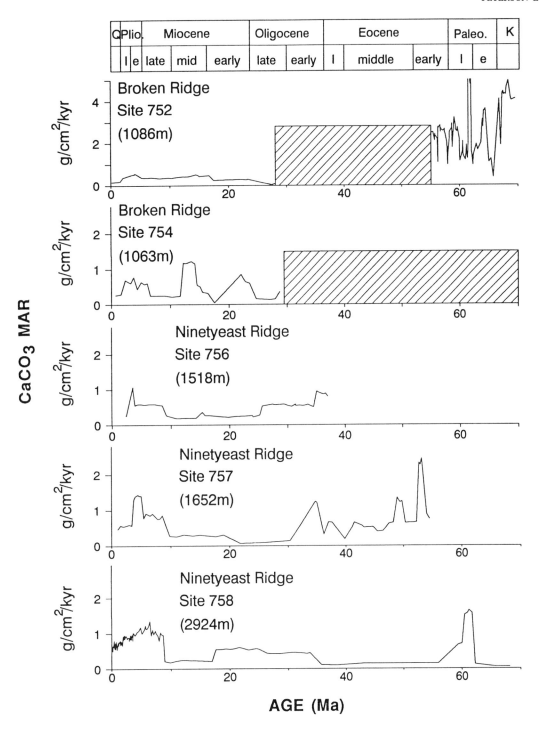

Fig. 5. Calculated carbonate mass accumulation rates (g/cm²/kyr) for Leg 121 Sites 752 and 754 (Broken Ridge) and Sites 756-758 (Ninetyeast Ridge). Hachured areas indicate hiatuses in the section.

relatively high for an open-ocean setting, varying between about 1.0 and 4.0 g/cm²/kyrs (Figure 5). As noted by Rea et al. [1990], MAR values were low just at and above the Cretaceous/Tertiary boundary, dropping from rates of >4.0 g/cm²/kyrs in the underlying Maestrichtian chalks.

The middle Eocene rifting of the Kerguelen-Broken Ridge

Plateau is thought to have produced an uplift of about 2500 m at the southern margin of Broken Ridge [Rea et al., 1990]. The subsequent history of the now separate Broken Ridge is one of post-rifting subsidence and northward drift to its present location near 31°S. The Neogene sediments that form the pelagic cap on Broken Ridge consist

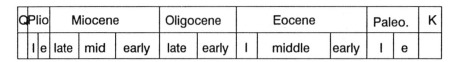

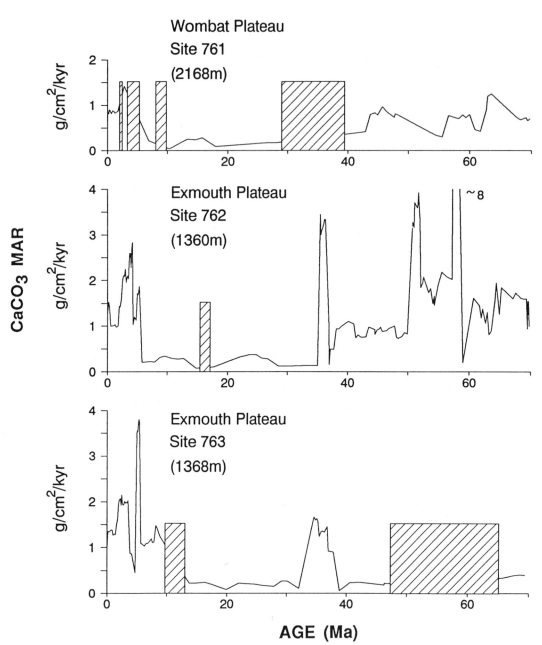

Fig. 6. Calculated carbonate mass accumulation rates (g/cm²/kyr) for Leg 122 Sites from the Wombat and Exmouth Plateaus. Hachured areas indicate hiatuses in the section.

primarily of nannofossil oozes that accumulated at much lower rates than in the underlying Paleogene sequence. Carbonate concentrations in the Neogene sections of Sites 752 and 754 typically exceed 95%, but carbonate MARS only average between 0.35 to 0.5 g/cm²/kyrs. Both the DSDP Leg 26 scientists [Davies, Luyendyk et al., 1974]

and ODP Leg 121 scientists [Peirce, Weissel et al., 1989; Rea et al., 1990] conclude that the low carbonate MARs, combined with the relatively shallow water depth and high foraminifer content, indicate substantial winnowing of the section.

Cenozoic sediments drilled at the three latitudinally-distributed sites along the Ninetyeast Ridge (Sites 756-758; see Figure 1) also consist of relatively pure carbonate oozes. The exception occurs in the late Miocene-Pleistocene section of Site 758, the northernmost site near 5°N, where dilution by terrigenous clay is important [Shipboard Scientific Party, 1989d]. In the early and middle Cenozoic, the record of carbonate deposition varies somewhat between the Ninetyeast Ridge sites (Figure 5). At Site 758, most of the Eocene section is condensed or missing, as reflected in the very low carbonate MARs between about 35 and 55 Ma. At Site 757, Eocene MARs average about 0.5 to 0.6 g/cm²/kyrs and range as high as >2.0 g/cm²/kyrs, but much of the Oligocene is greatly condensed. Site 756 appears to have had continuous deposition since the Oligocene, but has recorded low carbonate MARs (0.3-0.5 g/cm²/kyrs) throughout. As noted by the Shipboard Scientific Party [1989d], the relatively low rates of carbonate accumulation on Ninetyeast Ridge during the Eocene through middle Miocene occurred over a broad latitudinal range beneath the Southern Hemisphere subtropical gyre, covering paleopositions of roughly 10°S to 40°S over the interval between about 9 and 58 Ma.

At about 9 Ma, all three of the Ninetyeast Ridge sites experienced a sudden, several-fold increase in carbonate MAR. The recorded rise in accumulation is greater in Sites 757/758 than in Site 756, with carbonate MAR values in the former two averaging about 1.0 g/cm²/kyrs after 9 Ma. Though a slight rise in carbonate MAR is also observed in the latest Miocene (~6 Ma) at Broken Ridge Site 754, grain size evidence for enhanced winnowing between about 13 and 6 Ma [Rea et al., 1990] suggests that this MAR increase is more likely an artifact of decreased circulation intensity than it is the result of increased carbonate flux from the surface. As already noted, terrigenous accumulation at Site 758 began to increase after about 8 Ma as this site drifted north beneath the equator and came under the increasing influence of Bengal Fan deposition [Shipboard Scientific Party, 1989d].

CARBONATE ACCUMULATION OFF NW AUSTRALIA - LEG 122
RESULTS

ODP Legs 122 and 123 together constituted a program to drill one of the oldest passive continental margins in the world. Because scientific objectives in this region centered on sampling the thick Mesozoic section, drillsites on these legs were generally targeted where the Cenozoic sediment cover was relatively thin and perhaps not fully representative of depositional patterns over that interval. Nevertheless, three Leg 122 sites (761-763) provide general details of Cenozoic carbonate accumulation (Figure 6).

Site 761 was drilled at a water depth of 2168 m on the Wombat Plateau. The Cenozoic section here consists of a total of 176 m of foraminifer-nannofossil oozes and chalks that exhibit excellent microfossil preservation. Carbonate MARs in the Paleocene and early and middle Eocene range between about 0.5 and 1.0 g/cm²/kyrs. Above a late

Eocene-Oligocene hiatus spanning some 12 m.y., carbonate MARs drop to values typically less than 0.3 g/cm²/kyrs that then persist until the late Miocene. Although three short hiatuses in the late Miocene and Pliocene section obscure the timing, a clear rise in carbonate MARs is evident at this site after about 7-8 Ma.

Site 762 on the Exmouth Plateau recovered a much thicker and relatively complete Cenozoic sequence, with the K/T boundary identified at ~555 mbsf. Despite the thicker section, the nannofossil assemblages throughout generally indicate a low-latitude open-ocean environment during the Cenozoic [Haq, von Rad, O'Connell et al., 1990]. Paleocene and Eocene accumulation rates are higher here than at Site 761, ranging between 1.0 and about 2.0 g/cm²/kyrs except for short intervals where calculated MARs reach as high as ~8 g/cm²/kyrs (these latter are presumed to be largely age model artifacts). Low carbonate MAR values (0.2-0.4 g/cm²/kyrs), and overall low bulk sedimentation rates (~2 m/m.y.), in the Oligocene to late Miocene section were attributed by Haq, von Rad, O'Connell et al. [1990] to winnowing and the activity of erosive bottom currents. As at Site 761, carbonate MARs rise abruptly in the late Miocene to values exceeding 1.0 g/cm²/kyrs.

Exmouth Plateau Site 763 encountered an unexpected hiatus that removed much of the Paleocene and early Eocene. With the exception of a relatively short interval of high MARs (>1.0 g/cm²/kyrs) in the late Eocene, most of the pre-middle Miocene section in Site 763 is characterized by very slow bulk sedimentation (1-3 m/m.y.) and low (<0.3 g/cm²/kyrs) carbonate MARs. Pelagic sedimentation rates and carbonate accumulation picked up considerably after the middle Miocene hiatus found at this site, though the increase here seems to predate the rise observed at Sites 761 and 762. It is not clear at this time whether this offset is real or an artifact of the age models used.

CARBONATE ACCUMULATION IN THE SOUTHERN OCEAN - KERGUELEN PLATEAU DATA FROM LEGS 119 AND 120

A total of eleven sites were drilled on the Kerguelen Plateau by Legs 119 and 120, five of which had records suitable for summary here (Figure 7). For Sites 738 and 744, carbonate data supplementing those reported by Barron, Larsen et al. [1989] were obtained from Ehrmann [1991], while additional carbonate data for Site 737 were taken from Bohrmann and Ehrmann [1991]. Carbonate MARs for Site 751 were calculated using the larger carbonate data set in Mackensen et al. [1992].

The carbonate MAR records from the Kerguelen Plateau are arranged in latitudinal order from north to south in Figure 7. Site 737, the northernmost site in the array (~50°S), is located close to the present Polar Front, while Site 738 lies some 1300 km to the south(~63°S). Sedimentation on the plateau throughout most of the Cenozoic has been biogenic, with a major change from carbonate to biosiliceous sediment accumulation having occurred in the latest Miocene. Before this time, sediments deposited consist primarily of nannofossil oozes and chalks, with

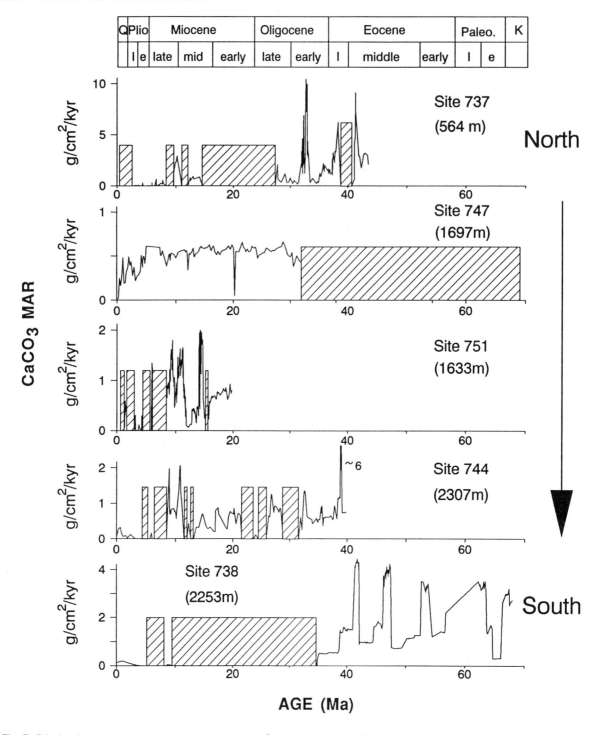

Fig. 7. Calculated carbonate mass accumulation rates (g/cm2/kyr) for Kerguelen Plateau sites drilled during Legs 119 and 120. Sites are arranged in descending order from north to south. Hachured areas indicate hiatuses in the section.

minor occurrences of chert and volcanic sands. After this time, sedimentation shifted across the Plateau to dominantly that of diatom ooze. On the southernmost Kerguelen Plateau, at Site 738, this transition is dated at >6.1 Ma, while at Site 744 it occurred at about 5.8 Ma [Ehrmann, 1991]. At Site 751, on the central Plateau,

nearly pure diatom ooze began to accumulate above a hiatus spanning the time interval 5.9-4.7 Ma [Mackensen et al., 1992]. The somewhat later onset of biosiliceous sedimentation here suggests a gradual northward migration of the facies boundary, probably related to northward movement of the Polar Front. Similar facies changes have been observed in uppermost Miocene sediments recovered on Maud Rise in the Atlantic sector of the Southern Ocean during Leg 113 [Barker, Kennett et al., 1988].

Hiatuses attributable to both tectonic activity and current erosion punctuate the sedimentary sections on Kerguelen Plateau (Figure 7), hindering the reconstruction of continuous carbonate MAR records. Nevertheless, general observations can be made. For the early Paleogene, carbonate accumulation in this region is best estimated from Site 738. Carbonate MARs for this interval at Site 738 vary between about 1 and 4 $g/cm^2/kyrs$, though these variations are primarily a function of the age model since carbonate content is quite uniform throughout at about 90% and bulk density changes are minor. If one assumes instead a long-term average sedimentation rate of 11.8 m/m.y., a mean carbonate MAR of 1.3-1.4 $g/cm^2/kyr$ is suggested for the pre-Oligocene section of this site.

As noted earlier, the Kerguelen Plateau and Broken Ridge were separated in the Eocene (~42 Ma) by spreading at the Southeast Indian Ridge. At Site 747, the Paleocene-early Oligocene section is so highly condensed that carbonate accumulation patterns could not be resolved in this interval (Figure 7). Within this condensed section the most conspicuous individual hiatus, spanning much of the Eocene, probably corresponds to the "discordance A" of Munschy and Schlich [1987] that seems to record uplift/erosion and the separation of pre-rift and post-rift sedimentation. At northern Site 737, the most prominent unconformity occurs later, between about 15 and 28 Ma. Between this hiatus and a much shorter one centered on about 40 Ma, the sediments consist of a thick sequence of calcareous claystones that form an almost continuous hemipelagic section of middle Eocene to late Oligocene in age.

At Site 738, on the southern plateau, there is no evidence that tectonic processes are responsible for the condensed Oligocene and Neogene section found here. Instead, deposition of this abbreviated sequence is thought to have been the result of strong bottom currents and erosion and/or non-deposition [Barron, Larsen et al., 1989]. Increased bottom current activity is probably responsible for many of the short Neogene hiatuses found at Sites 737, 751, and 744. At Site 747, a surprisingly complete late Oligocene and Neogene sequence was recovered. Here, carbonate MARs tend to average 0.5-0.6 $g/cm^2/kyrs$ until the late Miocene onset of biosiliceous sedimentation. Although carbonate MARs progressively drop through the Pliocene and into the Quaternary, bulk sedimentation rates actually rise in response to the overall increase in surface productivity that accompanied the northward migration of the Polar Front.

DISCUSSION

The Indian Ocean is the youngest of the three major ocean basins and is tectonically the most complex. Its formation began with the breakup of Gondwanaland in the Mesozoic and continues today through the interaction of the African, Antarctic, and Indo-Australian tectonic plates.

Paleogeographic and tectonic reconstructions of the Indian Ocean are numerous and include, among others, those of McKenzie and Sclater [1971], Sclater and Fisher [1974], Pimm et al. [1974], Norton and Sclater [1979], and Royer and Sandwell [1989].

Davies and Kidd [1977] and Kidd and Davies [1978] have previously reviewed the sedimentation history of the Indian Ocean based on the drilling results of the DSDP. Taking into account the changing paleobathymetry and tectonic setting through time, they constructed a series of facies maps showing major sediment distribution patterns for a selected number of important time slices. From their work it is clear that carbonate accumulation in the Indian Ocean has been strongly affected by the progressive physical evolution of the basin as a whole. Carbonate deposition through much of the Mesozoic, for example, was more widespread because individual basins were small and a higher percentage of the ocean floor was relatively shallow and above the CCD. With increasing maturity of the Indian Ocean basin into the Cenozoic, and a higher proportion of ocean floor at greater depths, carbonate deposition became more areally restricted and more subject to control by changing production and/or dissolution.

With the exception of the sites drilled during the Leg 115 program, almost all of the ODP sites summarized here were drilled in present water depths of less than 2500 m. Though individual sites show evidence of both subsidence and uplift, each of the non-Leg 115 sites has been shallow enough to have always remained above the CCD over the length of its record. Because of their lack of major facies changes and their limited vertical and geographic distribution, we have not tried to utilize the newer ODP sites for mapping purposes. Instead, in the discussion that follows, we have tried to relate in a general way our carbonate MAR data to the earlier observations of Davies/Kidd.

The Paleogene

By the beginning of the Cenozoic, the development of the Indian Ocean was such that open-ocean pelagic sedimentation began to prevail across much of the basin. Davies and Kidd [1977] showed carbonate sediments to be common on the tops of present-day ridges and plateaus, with pelagic clays in the deep basins and coarse terrigenous sediments collecting off east Africa and west of India. By the early Oligocene, the Indian Ocean had begun to more closely resemble its present-day configuration with three distinct regions: an almost totally enclosed northwestern region bounded to the south by the Carlsberg Ridge and with limited access to Tethys; a central region split by the inverted "Y" of the spreading ridges; and an eastern region, dominated by the deep Wharton Basin, and still in free communication with the western Pacific at low- to mid-latitudes [Davies and Kidd, 1977; Kidd and Davies, 1978]. Additional throughflow to the Pacific would have increasingly occurred to the southeast with the progressive separation of Antarctica from Australia [e.g., Kennett, 1982]. Much of the Paleogene record in the Indian Ocean

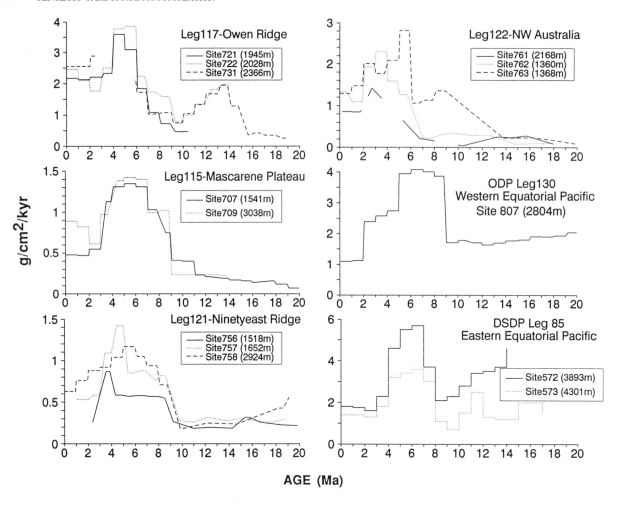

AGE (Ma)

Fig. 8. Expanded view of the last 20 m.y. of carbonate accumulation at Indian Ocean ODP sites drilled on Legs 115, 117, 121, and 122. Also shown for comparison are carbonate mass accumulation rates over the same interval for equatorial Pacific sites cored during ODP Leg 130 [data from Kroenke, Berger, Janecek et al., 1991] and DSDP Leg 85 [data from Theyer et al., 1985]. Calculated carbonate MARs in this figure have been expressed as average accumulations over 1 m.y. intervals to account for differing sample densities. With the exception of the Leg 85 sites, all of the sites included here are from relatively shallow water depths and hence are more likely to record variations in biogenic carbonate production through time. All of the sites exhibit a marked increase in carbonate MARs near 9-10 Ma, and a subsequent decrease near 3.5 Ma. More recent data from Leg 138 sites in the eastern equatorial Pacific show similar temporal trends in carbonate accumulation over the past 10 m.y. [Leg 138 Shipboard Scientific Party, 1991]. The overall similarity of the Leg 117 records to those from locations far removed from the region of strong monsoonal upwelling implies that carbonate deposition in the Arabian Sea is responding more to global-scale processes than to the localized influence of the monsoon.

was found by DSDP drilling to be patchy and discontinuous as a result of tectonic activity and bottom-water erosion. Similar conclusions can be drawn based on the results of the ODP campaign.

Paleogene sediments from the Leg 115 sites on and around the Mascarene Plateau consist predominantly of nannofossil oozes and chalks of middle Eocene and Oligocene age, with carbonate MARs that generally average about 0.5-0.6 g/cm2/kyrs. These rates are slightly higher than those typically found in oligotrophic regions [0.3-0.5 g/cm2/kyrs; e.g., Thiede and Rea, 1981; Rea and Leinen, 1986], but are perhaps not as high as might be expected for a near- equatorial location. Tectonically, all of

the Leg 115 sites are attached to the African Plate; plate reconstructions [e.g., McKenzie and Sclater, 1971] suggest that their positions have not changed substantially with respect to the equator over the spans of time they record.

Leg 121 sites from the Ninetyeast Ridge that recovered Paleogene sediments record variable carbonate MARs that range from <0.5 to upwards of 2.0 g/cm2/kyrs, but generally average <1.0 g/cm2/kyrs. Uplift and erosion accompanying the middle Eocene rifting and separation of Kerguelen Plateau from Broken Ridge has removed most of the Paleogene section from both, with only southern Kerguelen Site 738 preserving a record of deposition. Carbonate MARs here are again variable and average, over the long-term, about 1.3-1.4 g/cm^2/kyrs. Sites off western Australia (Leg 122) indicate Paleogene carbonate deposition at rates typically between 0.5 and 2.0 g/cm^2/kyrs.

One of the most dramatic climatic steps of the Cenozoic occurred near the Eocene/Oligocene boundary [see Corliss et al., 1984, and Corliss and Keigwin, 1986, for reviews]. At this time, distinct changes in the stable isotope record of foraminifers and in biotic patterns occurred that have been variously interpreted as evidence for global cooling, build-up of continental ice sheets, and the development of psychrospheric conditions in the deep sea [e.g., Shackleton and Kennett, 1975; Benson, 1975; Keigwin, 1980; Haq, 1981; Shackleton, 1986; Miller et al., 1987]. Coincident with this event, a pronounced drop in the depth of the CCD in all of the oceans [Heath, 1969; Berger, 1975; van Andel, 1975; Hay, 1988] has been attributed to the resulting changes in the character and turnover of oceanic bottom waters.

As noted earlier, none of the shallow non-Leg 115 sites examined here show evidence of ever having been below the CCD. Of the Leg 115 sites, only the deep Site 711 seems to record motion of the CCD near the Eocene/Oligocene boundary. Here, the steady increase in carbonate MARs between about 39 and 35 Ma (Figure 3) is presumably related to this well-documented deepening of the CCD. Although the MAR rise in Site 711 is consistent with a fall of the CCD, it is interesting to note that the accumulation record of the shallower Site 709 gives no clear indication of related changes in surface carbonate productivity. In a previous study of DSDP Sites 214 (Ninetyeast Ridge) and 219 (eastern Arabian Sea), Thunell and Corliss [1986] report an increase in carbonate accumulation rates across the Eocene/Oligocene boundary, a rise they attribute to higher carbonate input from the surface. At sites off northwestern Australia (Figure 6) carbonate accumulation appears to increase at about this time, though the period of higher rates is shortlived. In sites from the Ninetyeast Ridge and Kerguelen Plateau (Figures 5 and 7), the record is equivocal at best. These observations seem to differ significantly from those made on the opal system. Baldauf et al. [this volume] report that low latitude opal records from the Indian Ocean show a marked decline in Oligocene biosiliceous sedimentation from higher levels of deposition in the Eocene. Southern Ocean sediments seem to show the opposite trend, with increases in diatom abundance and opal content observed in the Oligocene sections. Baldauf et al. [this volume] attribute these patterns to the progressive restriction of global equatorial (Tethyan) surface currents and the expansion of Southern Ocean opal production as a result of polar cooling and the beginnings of circum-Antarctic flow. Whether the apparently more complicated patterns of carbonate accumulation are somehow artifacts of the quality of sites and/or age models used, or represent real intrabasinal variations in biogenic carbonate production requires further evaluation.

The Early to Middle Miocene

During the early and middle Miocene, carbonate MARs are greatly reduced at all of the Leg 115 transect sites (Figure 3). As noted earlier, the combined influence of several interrelated processes seems to be operating to produce such low accumulation rates over this interval. At each of the sites, the effects of increased dissolution on foraminiferal and nannofossil preservation are clearly observed in sediments of early and middle Miocene age. Carbonate lost from the sediments as a result of this dissolution must certainly account for a large part of the explanation for why carbonate MARs are so reduced, particularly at the deeper transect sites. In Site 711, the complete absence of carbonate over much of the early and middle Miocene interval clearly indicates a shoaling of the CCD to a depth above the paleodepth of this site. Evidence for a shallower CCD in the Miocene has existed since DSDP Leg 3 recovered red clays of that age in the South Atlantic [Maxwell et al., 1970]. In the Indian Ocean, Davies and Kidd [1977] note that the distribution of deep sea clays reached their maximum extent in the early Miocene.

Winnowing is evident not only here in the shallow sites of the Mascarene Plateau region, but is apparent in the Neogene sections of intermediate depth sites across the Indian Ocean. As this review indicates, coarse sediment textures and the presence of condensed sections and/or hiatuses are also reported at sites from the tops of the Ninetyeast Ridge, Kerguelen Plateau, and the Wombat and Exmouth plateaus off Australia. The shallow winnowing and downslope redistribution of fine carbonates would be expected to reduce between-site carbonate MAR contrasts. Bulk grain size records, such as produced for sediments from Broken Ridge by Rea et al. [1990], would seem to have great potential for producing better quantified histories of intermediate-water current velocities.

As a final factor, low surface productivity in the early and middle Miocene may have additionally contributed to the decreased carbonate MARs observed at the Leg 115 sites. This general inference is based on the already noted disappearance of biogenic silica in late Oligocene sediments and its absence from Leg 115 sites until the late Miocene. Hempel and Bohrmann [1990] report decreased concentrations of barite in the same sections. A scenario calling for lower surface productivity, and a reduced flux of carbonate to the ocean floor, is at least locally consistent with evidence for a shallower CCD during this period. This results from the fact that the depth of the CCD is effectively determined by mass-balance considerations (input vs. loss). Since carbonate preservation is at least partly a function of burial rate, lower bulk sedimentation rates could also contribute to enhanced dissolution and even chemical backstripping [e.g., Berger, 1978].

During the late Oligocene and throughout the Miocene, the gradual closure of the Indonesian and Tethyan seaways and the northward migration of the Indian subcontinent out from the equatorial region [e.g., McKenzie and Sclater, 1971; Norton and Sclater, 1979] are likely to have had dramatic effects on sedimentation in the tropical Indian

Ocean. Coincident with changes in the paleogeography of the tropical Indian Ocean, tectonic events elsewhere, such as the opening of the Drake Passage near the Oligocene/Miocene boundary [Barker and Burrell, 1977], began to configure the ocean basins into an increasingly modern approximation. In the middle Miocene, increased Antarctic ice volume and deep water cooling are indicated by the well-known enrichment of foraminiferal $\delta^{18}O$ that occurred at about 14 Ma [e.g., Savin et al., 1975; Shackleton and Kennett, 1975; Vincent et al., 1985; Miller et al., 1987].

Although our survey has shown carbonate MARs to be low at a majority of ODP sites during the early and middle Miocene, it is interesting to note that only the Leg 115 sites (Figure 3), and perhaps Site 758 on the northern Ninetyeast Ridge (Figure 5), show the pronounced drop in MARs from higher late Oligocene values that occurs near the Oligocene/Miocene boundary. A nearly identical decrease in both bulk and carbonate accumulation at the beginning of the Neogene was more recently observed from Leg 130 drilling on the Ontong-Java Plateau in the western equatorial Pacific [Leg 130 Shipboard Scientific Party, 1990; Kroenke, Berger, Janecek et al., 1991]. Kroenke, Berger, Janecek et al. [1991] speculate that the apparent "carbonate drought" recorded in the early Neogene sediments of the Ontong-Java Plateau was related to global causes, and suggest high sea levels [Haq et al., 1987] and the sequestering of carbonate on continental shelves as one possible mechanism.

Based on a compilation of Miocene benthic foraminifer stable isotope and faunal abundance data, Woodruff and Savin [1989] suggest that Miocene thermohaline circulation patterns were distinctly different from those of today. In particular, their examination of carbon isotope gradients between the major basins led them to speculate that the early Miocene ocean may have been strongly influenced by an influx of warm saline water from the Tethyan region into intermediate depths of the northern Indian Ocean. This influx was suggested to diminish or terminate at about 14 Ma as the Tethyan-Indian connection became permanently closed [Rogl and Steininger, 1983] and as Antarctic cooling enhanced the production of southern source deep waters. Zachos et al. [this volume] support the interpretation of a northern (Tethyan) source of intermediate waters before about 12 Ma, but indicate that a southern source must have existed as well.

Such a scenario is at least partially consistent with observations made here and elsewhere. On the basis of benthic foraminifers from the Leg 115 sites, Boersma [1990] concludes that well-oxygenated conditions over the depth range of intermediate- and deep-waters were prevalent in the equatorial zone through most of the early and middle Miocene. The increased winnowing recorded at shallow sites over much of the Indian Ocean would also seem to be compatible with a more vigorous circulation of intermediate water of local origins. On the other hand, the called for proximity to a region of downwelling and source of well-oxygenated intermediate-waters seems inconsistent

with the visual evidence for enhanced carbonate dissolution in Leg 115 sediments during this interval. It is also interesting to note that the MAR records at most Indian Ocean sites surveyed here do not indicate a significant response in carbonate accumulation to the prominent cooling and ice volume increase that occurred at 14 Ma. Instead, the major step-like increase in carbonate accumulation seems to substantially postdate this mid-Miocene glacial event. Further work is needed to unravel the relationship between carbonate accumulation and global climate in the early Neogene.

The Late Miocene to Recent

The abrupt rise in carbonate MARs that occurs in the late Miocene of the shallower Leg 115 sites (Figure 3) results in the highest accumulation rates of the last 50 m.y. on the Mascarene Plateau. From ~8-9 Ma through approximately 3 Ma, the carbonate MARs at Sites 707 and 709 range between 1.0 and 1.6 $g/cm^2/kyrs$ and are virtually identical to each other, implying rapid deposition and very little loss of carbonate to dissolution over the depth range that separates the two sites (~1500 m). As noted earlier, biogenic silica reappeared in these records in the form of radiolarians and sponge spicules at about 10 Ma, with diatom taxa characteristic of upwelling conditions becoming the dominant constituent around 8 Ma and persisting until 3 Ma. In the equivalent stratigraphic intervals in Sites 709, 710, and 711, sediment colors were found to change from the yellowish-white and buff colors of sediments above and below to various shades of greenish-gray, apparently reflecting a change to reducing conditions and probably related to some combination of high initial organic carbon input or enhanced organic carbon preservation [Backman, Duncan et al., 1988]. Anomalously low whole-core magnetic susceptibility values in this "reduced" interval were also attributed by Robinson [1990] to the effects of suboxic diagenesis on the NRM-carrying iron oxides and oxyhydroxides in the sediment. All of these lines of evidence point to increased surface productivity as the major factor in producing the high carbonate MARs of this interval.

Between ~8-9 and 3 Ma, the contrast between carbonate accumulation at shallow Leg 115 sites and deep Leg 115 sites reached a maximum (Figure 3). Because these closely-spaced sites should have experienced a common subsidence history, the development of a large depth-dependent carbonate MAR contrast in the late Miocene must indicate the first appearance of a strong vertical dissolution gradient in the water column. If we interpret the "lysocline" as the depth level where carbonate loss from the sediments begins in earnest, then this level must clearly have fallen between the paleodepths of Sites 709 and 710 during the late Miocene and early Pliocene. The clear thickening of the sublysocline zone (i.e., the vertical zone separating the top of the lysocline from the CCD) in this interval is also consistent with a higher overall level of surface productivity since modeling results directly relate the thickness of this zone to the downward flux of carbonate

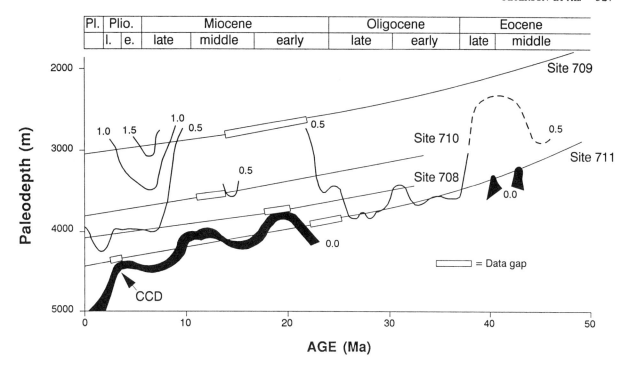

Fig. 9. Bathymetric variations in carbonate mass accumulation (g $CaCO_3/cm^2/kyr$) in the western equatorial Indian Ocean [from Peterson and Backman, 1990]. Carbonate MAR data from Leg 115 Sites 709, 710, 708, and 711 were plotted as a function of estimated age and water depth before contouring. Details on the subsidence models used for these sites can be found in Simmons [1990]. The zero $g/cm^2/kyr$ isopleth records the time-dependent history of the CCD.

from the surface [e.g., Broecker and Peng, 1982].

In the Mascarene Plateau region, the record of Site 709 is probably most indicative of shallow accumulation patterns in the late Pliocene-Recent interval since Site 707 shows evidence of increased winnowing after about 3.5 Ma. Carbonate MARs at Site 709 drop markedly above the early/late Pliocene boundary, only to rise again to peak values of 0.8-0.9 $g/cm^2/kyrs$ after about 2.5 Ma. The latter rates are comparable to those found in Sites 757 and 758 from the northern Ninetyeast Ridge (Figure 5), but are lower than Quaternary carbonate MARs reported from the equatorial Pacific [~1-3 $g/cm^2/kyrs$; Chuey et al., 1987; Lyle et al., 1988]. The observed MAR differences tend to faithfully parallel the modern differences in primary productivity between the two oceans [Berger et al., 1987].

In the Arabian Sea to the north, the first significant accumulation of opal-rich sediments occurred in the late middle Miocene. Shipboard scientists from DSDP Leg 23 [Whitmarsh et al., 1974] initially suggested that the onset of opal deposition at about 11 Ma on Owen Ridge marked the beginning of a strong monsoon circulation associated with orographic changes in Asia [e.g., Ruddiman and Kutzbach, 1991]. More recent studies of Leg 117 materials, however, indicate that the foraminiferal and radiolarian faunas uniquely associated with the upwelling of cold, nutrient-rich waters do not appear in the area until about 8 Ma [Nigrini, 1991; Kroon et al., 1991]. Carbonate MAR

data from the Leg 117 sites (Figure 4) also show a general increase in carbonate accumulation after about 8-9 Ma, reaching peak levels between about 6 and 3.5 Ma.

The general similarity of Leg 117 depositional patterns to those observed on and around the Mascarene Plateau would seem to indicate a common Neogene history for surface productivity in the western equatorial Indian Ocean and Arabian Sea. Whether these results are part of a signal that is unique to this region (i.e., the monsoon), or one that is more regional or global in scope, is unclear. The initial development and subsequent evolution of the Indian Ocean monsoon has been linked to uplift of the Himalayas and Tibetan Plateau resulting from the collision of India and Asia [Ruddiman and Kutzbach, 1991]. The onset of monsoon circulation in the Arabian Sea has been cited as the cause of regional vegetation changes in Africa and Asia from forests to grasslands, and of faunal turnovers from browsers to grazers [e.g., Flynn and Jacobs, 1982; Barry et al., 1985; Quade et al., 1989]. Molnar and England [1990], however, have questioned the data supporting a late Miocene phase of plateau uplift and have proposed instead that global climatic cooling, independent of uplift, is responsible for the observed regional changes.

Because the monsoon has such a pronounced effect on the oceanographic characteristics of the Arabian Sea, changes in monsoon intensity through time should produce a clear signal in the underlying sediments. As already noted, the initiation of opal accumulation at about 11 Ma in the northwestern Arabian Sea [Whitmarsh et al., 1974] was at first cited as marking the beginning of strong monsoon-driven upwelling along the coasts of Somalia and Oman. However, opal deposition at the tropical Indian Ocean sites cored by Leg 115 also commenced near 10 Ma in an area not as directly influenced by monsoon circulation [Mikkelsen, 1990; Johnson, 1990]. In addition, the carbonate accumulation records at shallow Leg 117 sites atop Owen Ridge show similarities not only to Leg 115 sites, but to other shallow sites in the tropical Indian (Legs 121, 122) and Pacific Oceans (Figure 8). Some of the differences between these records can perhaps be attributed to the local imprint of monsoonal upwelling on the record from the Arabian Sea. The overall similarity of the MAR records from these widely scattered sites (Figure 8), however, would seem to imply that the first-order changes in carbonate accumulation observed in the Arabian Sea and western tropical Indian Ocean are not uniquely related to localized monsoonal conditions. Instead, carbonate production in this region, and its subsequent accumulation at shallow Leg 117 and 115 sites, is probably responding to more regional and/or global-scale processes such as eustatic sea level [Haq et al., 1987], continental weathering [Davies and Worsley, 1981; Delaney and Boyle, 1988; Raymo et al., 1988], deep water flow [Woodruff and Savin, 1989], and/or trade wind variations [Rea and Bloomstine, 1986].

In the high southern latitudes (Figure 7), the major change during the late Miocene-Recent interval is the dramatic switchover from carbonate to biosiliceous sedimentation that took place about 5-6 Ma. The increase in opal deposition on Kerguelen Plateau coincided with a substantial decline in opal accumulation rates in the equatorial Pacific [Leinen, 1979; Brewster, 1980] and marked the beginning of the Antarctic's role as a major silica sink [Barron and Baldauf, 1989]. A stabilizing of the West Antarctic ice cap at this time and northward expansion of the Polar Front [Kennett, 1978; Barker et al., 1987] are most likely related to this event.

Although records from throughout the Indian Ocean and elsewhere show differences in the response of local sedimentation over the late Neogene, the large increase in tropical carbonate accumulation at about 8-9 Ma, the corresponding evidence for enhanced dissolution gradients, and the qualitative assessment of opal deposition patterns together seem to indicate the onset of near-modern circulation patterns in the surface and deep ocean.

History of Carbonate Compensation Changes

On a global basis, it is the ratio of carbonate production in the surface waters to the riverine input of Ca^{+2} that determines the saturation chemistry of the deep ocean. Since production rates in the surface ocean today greatly exceed the rate of supply of Ca^{+2} [Broecker and Peng, 1982], "compensation" occurs through the dissolution of carbonate on the sea floor. As the relative importance of production and supply vary through time, the amount of

dissolution in the deep ocean changes and the CCD adjusts its depth in order to keep Ca^{+2} levels in balance.

Van Andel [1975] attempted the first analysis of Indian Ocean CCD behavior on the basis of drilling results from DSDP Legs 22-27. His reconstruction for the Cenozoic showed a generally shallow (<4000 m) CCD until the late Eocene, followed by a steady drop to around 4400 m in the late Oligocene and a steep rise to above 4000 m again in the middle Miocene. This was followed by an equally steep drop to its present-day level. Sclater et al. [1977], with an expanded suite of sites, arrived at a second CCD reconstruction for the Indian Ocean that differed in several important ways. This latter analysis shows the CCD to have remained at or about 4000 m for most of the Cenozoic, dropping sharply to it present level only after the Miocene.

Using the Leg 115 depth transect sites, Peterson and Backman [1990] combined carbonate accumulation data with back-tracked estimates of site-specific paleobathymetry to produce an age-paleodepth reconstruction of the CCD and of carbonate deposition above it (Figure 9). The zero isopleth in Figure 9, the level at which no net accumulation occurs, by definition records the depth of the CCD. For most of the Neogene, the range and timing of CCD motion can be reasonably well constrained with the Leg 115 data. From its initial crossing of Site 711 near the Oligocene/Miocene boundary on its rise to an early Miocene high, the CCD appears to have deepened and shoaled twice more before beginning the final drop to its present-day level at about 5000 m. Although the timing of these excursions is largely based on the record at Site 711, the motions are consistent with the changes in carbonate MARs observed at the shallower sites. The total range of motion described for the Neogene CCD is on the order of a kilometer or more.

Data from the Leg 115 sites cannot be used to place direct constraints on the depth of the CCD in the Paleogene. The moderately high carbonate MARs, however, and their uniformity among the sites would seem to support the deeper depth estimates of van Andel [1975] for at least the Oligocene. It is clear that all of the sites must also have been above the CCD when its well known drop occurred across the Eocene/Oligocene boundary. Nevertheless, the absence of carbonate over several short intervals near the middle/late Eocene boundary at Site 711 suggests that the regional CCD reached its shallowest position of the past 50 m.y. at about this time.

The CCD reconstruction for the tropical Indian Ocean shown in Figure 9, and particularly the early and middle Miocene shoaling, is generally compatible with first-order reconstructions from the other major ocean basins [see Hay, 1988, for a compilation]. Berger [1970] and Berger and Winterer [1974] were among the first investigators to suggest that carbonate partitioning between shallow and deep seas, modulated by sea level changes, was the causal mechanism responsible for the common element of long-term CCD variations. As noted by Hay [1988], general support for this hypothesis has come with the subsequent

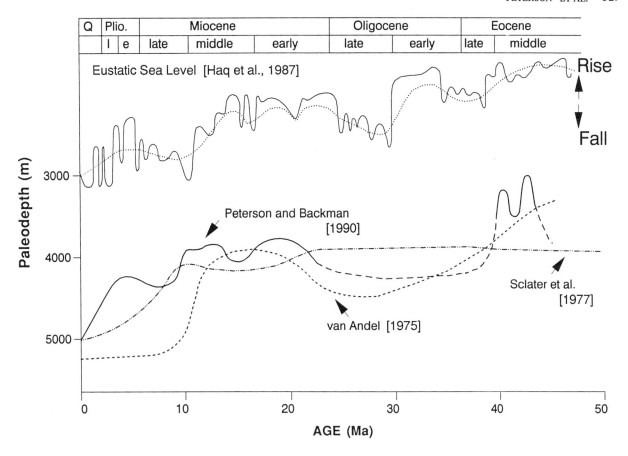

Fig. 10. Comparison of the late Cenozoic CCD reconstructions of van Andel [1975], Sclater et al. [1977], and Peterson and Backman [1990] to the eustatic sea level record of Haq et al. [1987]. The dotted line superimposed on the Haq et al. curve is a visual attempt to filter out the longer-term component of variability. The increased resolution of the Peterson and Backman [1990] CCD reconstruction for the Neogene compares favorably with the somewhat smoothed record of eustasy over the same interval. This observation supports models that invoke sea level changes and the partitioning of carbonate between the shelf and deep sea as having a first-order control on the depth of the CCD.

publication of eustatic sea level records by Vail et al. [1977] and Haq et al. [1987]. In Figure 10, we compare the CCD analyses of van Andel [1975], Sclater et al. [1977], and Peterson and Backman [1990] to the eustatic sea level record of Haq et al. [1987]. As can be seen, the increased resolution afforded by the higher quality of ODP drillsites does nothing if not strengthen the overall comparison between CCD and sea level trends. Data compiled here also suggest, however, that carbonate MAR variations at sites above the CCD are not necessarily linked in a simple manner to whatever mechanism controls the depth of the CCD. For example, comparison of the carbonate MAR records from shallow Sites 707 and 709 (Figure 8) with the CCD history shown in Figure 9 shows the CCD to deepen during the abrupt carbonate MAR rise of 8-9 Ma, and then to deepen further when carbonate MARs decrease after about 3 Ma. The lack of direct correspondence between MAR variations and motion in the CCD implies that the respective controlling processes in this interval are uncoupled, with shallow carbonate accumulation more likely related to changes in regional surface fertility, and

the CCD itself responding to a forcing that is more global in nature (e.g., sea level or mean Ca^{+2} concentrations). Continued refinement of CCD reconstructions and of our knowledge of carbonate accumulation patterns in all the major basins will lead to a better understanding of how the carbonate system has responded to the Cenozoic evolution of the global ocean.

CONCLUSIONS

Our review of results from the Leg 115 "carbonate depth transect" and from a suite of other Indian Ocean ODP sites shows regional carbonate accumulation patterns that are variably affected by the changing balance between productivity and dissolution, and by physical processes such as erosion. In the Paleogene, both the character and accumulation rates of carbonate sediments indicate relatively open ocean conditions, though tectonic activity and changing bottom circulation has resulted in a patchy and discontinuous record across the basin. In the early and middle Miocene, greatly reduced carbonate accumulation in the Mascarene Plateau region can be attributed to increased dissolution and a shallow CCD, enhanced winnowing, and a probable drop in surface productivity. Winnowing is common at shallow sites throughout the Indian Ocean at this time, suggesting a more vigorous intermediate water circulation. At about 8-9 Ma, carbonate MARs rise abruptly in most low-latitude sites and the first substantial contrasts in accumulation as a function of water depth appear in the Leg 115 sites. These observations indicate the onset of near-modern depositional conditions, with increased levels of surface productivity and the development of a strong dissolution gradient in the water column. The similarities between carbonate MAR and opal records from the Arabian Sea and from Indo-Pacific locations not directly influenced by monsoon circulation imply that the bulk of biogenic sediment accumulation in the northwestern Indian Ocean is controlled more by global-scale processes than by conditions specific to monsoonal upwelling.

A comparison between the depth history of the tropical Indian Ocean CCD and the record of eustatic sea level is favorable, strengthening the perception that enhanced shallow deposition of carbonate during sea level highstands comes at the expense of deposition in the deep sea. Carbonate accumulation patterns above the CCD are more complex in origin and probably reflect the greater influence of regional variations in carbonate production at the ocean's surface.

Acknowledgments. The authors thank the convenors for their efforts in organizing the JOIDES Indian Ocean Workshop that stimulated this review. We also thank USSAC and DFG for providing financial support for travel expenses. Thoughtful reviews by W. B. Curry and R. B. Dunbar were most helpful and improved this manuscript. Graphics assistance from H.-L. Lin and C. Moss is greatly appreciated, as was the help of M. Waldorf in compiling data from Legs 117 and 121. Thanks also go to W. Busch for providing corrected bulk density data for Leg 117 sediments. This is a contribution from the Rosenstiel School of Marine and Atmospheric Science, University of Miami.

REFERENCES

Backman, J., Duncan, R. A. et al., *Proc. ODP, Init. Repts., 115,* Ocean Drilling Program, College Station, TX, 1988.

Backman, J., Schneider, D. A., Rio, D., and Okada, H., Neogene low-latitude magnetostratigraphy from Site 710 and revised age estimates of Miocene nannofossil datum events., in *Proc. ODP., Sci. Results, 115,* edited by R. A. Duncan, J. Backman, L. C. Peterson et al., 271-276, Ocean Drilling Program, College Station, TX, 1990.

Barker, P. F., and Burrell, J., The opening of Drake Passage, *Mar. Geol., 25,* 15-34, 1977.

Barker, P. F., Kennett, J. P., and the ODP Leg 113 shipboard scientific party, Glacial history of Antarctica, *Nature, 328,* 115-116, 1987.

Barker, P. F., Kennett, J. P. et al., *Proc. ODP, Init. Repts., 113,* Ocean Drilling Program, College Station, TX, 1988.

Barron, J. A., and Baldauf, J. G., Tertiary cooling steps and paleoproductivity as reflected by diatoms and biosiliceous sediments, in *Productivity of the Ocean: Past and Present,* edited by W. H. Berger, V. S. Smetacek, and G. Wefer, 341-354, John Wiley and Sons, 1989.

Barron, J., Larsen, B. et al., *Proc. ODP., Init. Repts., 119,* Ocean Drilling Program, College Station, TX, 1989.

Barry, J. C., Johnson, N. M., Raza, S. M., and Jacobs, L. L., Neogene mammalian faunal change in southern Asia: correlations with climatic, tectonic, and eustatic events, *Geology,* 637-640, 1985.

Benson, R. H., The origin of the psychrosphere as recorded in changes of deep-sea ostracode assemblages, *Lethaia, 8,* 69-83, 1975.

Berger, W. H., Planktonic foraminifera: selective solution and paleoclimatic interpretation, *Deep-Sea Res., 15,* 31-43, 1968.

Berger, W. H., Biogenous deep-sea sediments: fractionation by deep-sea circulation, *Geol. Soc. Am. Bull., 81,* 1385-1402, 1970.

Berger, W. H., Cenozoic sedimentation in the eastern tropical Pacific, *Geol. Soc. Am. Bull., 84,* 1941-1954, 1973.

Berger, W. H., Deep-sea carbonates: dissolution profiles from foraminiferal preservation, in *Dissolution of Deep-Sea Carbonate,* Cushman Foundation Foraminiferal Res. Spec. Publ., 13, edited by W. V. Sliter, A. W. H. Be, and W. H. Berger, 82-86, 1975.

Berger, W. H., Sedimentation of deep-sea carbonate: maps and models of variations and fluctuations, *J. Foraminiferal Res., 8,* 286-302, 1978.

Berger, W. H., Fischer, K., Lai, C., and Wu, G., Ocean productivity and organic carbon flux. Part I. Overview and maps of primary production and export production, Univ. of California, San Diego, SIO Reference 87-30, 1987.

Berger, W. H., and Winterer, E. L., Plate stratigraphy and the fluctuating carbonate line, in *Pelagic Sediments on Land and Under the Sea,* edited by K. J. Hsü, and H. C. Jenkyns, 11-48, Blackwell Sci. Publ., London, 1974.

Berggren, W. A., Kent, D. V., and Van Couvering, J. A., The Neogene: Part 2. Neogene geochronology and chronostratigraphy, in*The Chronology of the Geologic Record, 10,* edited by N. J. Snelling, 211-260, Geol. Soc. Mem., London, 1985a.

Berggren, W. A., Kent, D. V., and Flynn, J. J., Jurassic to Paleogene: Part 2. Paleogene geochronology and chronostratigraphy, in *The Chronology of the Geologic Record, 10,* edited by N. J. Snelling, 141-195, Geol. Soc. Mem., London, 1985b.

Boersma, A., Late Oligocene to late Pliocene benthic foraminifers from depth traverses in the central Indian Ocean, in *Proc. ODP., Sci. Results, 115,* edited by R. A. Duncan, J. Backman, L. C. Peterson et al., 315-380, Ocean Drilling Program, College Station, TX, 1990.

Bohrmann, G., and Ehrmann, W. U., Analysis of sedimentary facies using bulk mineralogic characteristics in Cretaceous to Quaternary sediments from the Kerguelen Plateau: Sites 737, 738, and 744, in *Proc. ODP, Sci. Results, 119,* edited by J. Barron, B. Larsen et al., 211-224, Ocean Drilling Program, College Station, TX, 1991.

Bramlette, M. N., Pelagic sediments, in *Oceanography, 67,* edited by M. Sears, 345-366, Am. Assoc. Adv. Sci. Publ., 1961.

Brewster, N. A., Cenozoic biogenic silica sedimentation in the Antarctic Ocean, *Geol. Soc. Amer. Bull., 91,* 337-347, 1980.

Broecker, W. S., and Peng, T.-H., *Tracers in the Sea,* Eldigio Press, Palisades, NY, 1982.

Broecker, W. S., and Peng, T.-H., The role of $CaCO_3$ compensation in the glacial to interglacial atmospheric CO_2 change, *Global Biogeochem. Cycles, 1,* 15-29, 1987.

Chuey, J. M., Rea, D. K., and Pisias, N. G., Late Pleistocene paleoclimatology of the central equatorial Pacific: a quantitative record of eolian and carbonate deposition, *Quat. Res., 28,* 323-339, 1987.

Corliss, B. H., Aubry, M.-P., Berggren, W. A., Fenner, J. M., Keigwin, L. D., and Keller, G., The Eocene/Oligocene boundary event in the deep sea, *Science, 226,* 806-810, 1984.

Corliss, B. H., and Keigwin, L. D., Eocene/Oligocene paleoceanography, in *Mesozoic and Cenozoic Oceans,* edited by K. J. Hsü, 101-118, AGU, Geodynamics Series, 15,

Washington, 1986.

Curry, W. B., and Lohmann, G. P., Carbon deposition rates and deep water residence time in the equatorial Atlantic Ocean throughout the last 160,000 years, in *The Carbon Cycle and Atmospheric CO_2, Natural Variations Archean to Present*, edited by E. T. Sundquist, and W. S. Broecker, 285-301, AGU, Geophysical Monograph 32, Washington, D.C., 1985.

Curry, W. B., and Lohmann, G. P., 1986. Late Quaternary carbonate sedimentation at the Sierra Leone Rise (eastern equatorial Atlantic Ocean), *Mar. Geol., 70*, 223-250.

Curry, W. B., and Lohmann, G. P., Reconstructing past particle fluxes in the tropical Atlantic Ocean, *Paleoceanography, 5*, 487-505, 1990.

Davies, T. A., and Kidd, R. B., Sedimentation in the Indian Ocean through time, in *Indian Ocean Geology and Biostratigraphy*, edited by J. R. Heirtzler, H. M. Bolli, T. A. Davies, J. B. Saunders, and J. G. Sclater, 61-86, AGU, Washington, 1977.

Davies, T. A., Luyendyk, B. P. et al., *Init. Repts. DSDP, 26*, U.S. Government Printing Office, Washington, D.C., 1974.

Davies, T. A., and Worsley, T. R., Paleoenvironmental implications of oceanic carbonate sedimentation rates, *Soc. Econ. Pal. Mineral., Spec. Publ. No. 32*, 169-179, 1981.

Delaney, M. L., and Boyle, E. A., Tertiary paleoceanic chemical variability: unintended consequences of simple geochemical models, *Paleoceanography, 3*, 137-156, 1988.

Ehrmann, W. U., Implications of sediment composition on the southern Kerguelen Plateau for paleoclimate and depositional environment, in *Proc. ODP, Sci. Results, 119*, edited by J. Barron, B. Larsen et al., 185-210, Ocean Drilling Program, College Station, TX, 1991.

Farrell, J. W., and Prell, W. L., Climatic change and $CaCO_3$ preservation: an 800,000 year bathymetric reconstruction from the central equatorial Pacific Ocean, *Paleoceanography, 4*, 447-466, 1989.

Farrell, J. W., and Prell, W. L., Pacific $CaCO_3$ preservation and $\delta^{18}O$ since 4 Ma: paleoceanographic and paleoclimatic implications, *Paleoceanography, 6*, 485-498, 1991.

Flynn, L. J., and Jacobs, L. L., Effects of changing environments on Siwalik rodent faunas of northern Pakistan, *Palaeogeogr., Palaeoclimatol., Palaeoecol., 38*, 129-138, 1982.

Haq, B. U., Paleogene paleoceanography: early Cenozoic oceans revisited, *Oceanol. Acta*, Proceedings 26th International Geological Congress, Geology of Oceans Symposium, Paris, July 7-17, 71-82, 1981.

Haq, B. U., Hardenbol, J., and Vail, P. R., Chronology of fluctuating sea levels since the Triassic, *Science, 235*, 156-1167, 1987.

Haq, B. U., von Rad, U., O'Connell, S. et al., *Proc. ODP, Init. Repts., 122*, Ocean Drilling Program, College Station, TX, 1990.

Hay, W. W., Paleoceanography: a review for the GSA Centennial, *Geol. Soc. Am. Bull., 100*, 1934-1956, 1988.

Heath, G. R., Carbonate sedimentation in the abyssal equatorial Pacific during the past 50 million years, *Geol. Soc. Am. Bull., 80*, 689-694, 1969.

Heath, G. R., Moore, T. C., Jr., and van Andel, T. H., Carbonate accumulation and dissolution in the equatorial Pacific during the past 45 million years, in *The Fate of Fossil Fuel CO_2 in the Oceans*, edited by N. R. Andersen, and A. Malahoff, 627-639, Plenum Press, New York, 1977.

Hempel, P., and Bohrmann, G., Carbonate-free sediment components and aspects of silica diagenesis at Sites 707, 709, and 711 (Leg 115, western Indian Ocean), in *Proc. ODP., Sci. Results, 115*, edited by R. A. Duncan, J. Backman, L. C. Peterson et al., 677-698, Ocean Drilling Program, College Station, TX, 1990.

Hsü, K. J., McKenzie, J. A., Oberhansli, H., Weissert, H., and Wright, R. C., South Atlantic Cenozoic paleoceanography, in *Init. Repts. DSDP, 73*, edited by K. J. Hsü, J. L. LaBrecque, et al., 771-785, U.S. Government Printing Office, Washington, D.C., 1984.

Johnson, D. A., Radiolarian biostratigraphy in the central Indian

Ocean, Leg 115, in *Proc. ODP., Sci. Results, 115*, edited by R. A. Duncan, J. Backman, L. C. Peterson et al., 395-409, Ocean Drilling Program, College Station, TX, 1990.

Keigwin, L. D., Paleoceanographic change in the Pacific at the Eocene-Oligocene boundary, *Nature, 287*, 722-725, 2980.

Kennett, J. P., The development of planktonic biogeography in the Southern Ocean during the Cenozoic, *Mar. Micropaleontology, 3*, 301-345, 1978.

Kennett, J. P., *Marine Geology*, Prentice-Hall, Englewood Cliffs, NJ, 1982.

Kidd, R. B., and Davies, T. A., Indian Ocean sediment distribution since the late Jurassic, *Mar. Geol., 26*, 49-70, 1978.

Koblentz-Mischke, O. J., Volkovinsky, V. V., and Kabanova, J. G., Plankton primary production of the world ocean, in *Scientific Exploration of the South Pacific*, edited by W. S. Wooster, 183-193, National Academy of Sciences, Washington, D.C., 1970.

Kroenke, L. W., Berger, W. H., Janecek, T. R. et al., *Proc. ODP, Init. Repts., 130*, Ocean Drilling Program, College Station, TX, 1990.

Kroon, D., Steens, T. N. F., and Troelstra, S. R., Onset of monsoonal related upwelling in the western Arabian Sea, in *Proc. ODP, Sci. Results, 117*, edited by W. L. Prell, N. Niitsuma et al., 257-264, Ocean Drilling Program, College Station, TX, 1991.

Leg 130 Shipboard Scientific Party, Reading the ocean's diary, *Nature, 346*, 111-112, 1990.

Leg 138 Shipboard Scientific Party, Ancient-ocean climate links, *Nature, 353*, 304-305, 1991.

Leinen, M., Biogenic silica accumulation in the central equatorial Pacific and its implications for Cenozoic paleoceanography, *Geol. Soc. Am. Bull., Part II, 90*, 1310-1376, 1979.

Leinen, M., Cwienk, D., Heath, G. R., Biscaye, P. E., Kolla, V., Thiede, J., and Dauphin, J. P., Distribution of biogenic silica and quartz in recent deep-sea sediments, *Geology, 14*, 199-203, 1986.

Lyle, M., Murray, D. W., Finney, B. P., Dymond, J., Robbins, J. M., and Brooksforce, K., The record of late Pleistocene biogenic sedimentation in the eastern tropical Pacific Ocean, *Paleoceanography, 3*, 39-59, 1988.

Mackensen, A., Barrera, E., and Hubberten, H. W., Neogene circulation in the southern Indian Ocean: evidence from benthic foraminifers, carbonate data, and stable isotope analyses (Site 751), in *Proc. ODP, Sci. Results, 120*, edited by S. W. Wise Jr., R. Schlich et al., 867-878, Ocean Drilling Program, College Station, TX, 1992.

Maxwell, A. E., Von Herzen, R. P. et al., *Init. Repts. DSDP, 3*, U.S. Government Printing Office, Washington, D.C., 1970.

McKenzie, D. P., and Sclater, J. G., The evolution of the Indian Ocean since the late Cretaceous, *Geophys. J. Roy. Astron. Soc., 25*, 437-528, 1971.

Mikkelsen, N., Cenozoic diatom biostratigraphy and paleoceanography of the western equatorial Indian Ocean, in *Proc. ODP., Sci. Results, 115*, edited by R. A. Duncan, J. Backman, L. C. Peterson et al., 411-432, Ocean Drilling Program, College Station, TX, 1990.

Miller, K. G., Fairbanks, R. G., and Mountain, G. S., Tertiary oxygen isotope synthesis, sea level history, and continental margin erosion, *Paleoceanography, 2*, 1-19, 1987.

Molnar, P., and England, P., Late Cenozoic uplift of mountain ranges and global climate change: Chicken or egg? *Nature, 346*, 29-34, 1990.

Moore, T. C., Jr., Rabinowitz, P. D. et al., *Init. Repts. DSDP, 74*, U.S. Government Printing Office, Washington, 1984.

Munschy, M., and Schlich, R., Stucture and evolution of the Kerguelen-Heard Plateau (Indian Ocean) deduced from seismic stratigraphy studies, *Mar. Geol., 76*, 131-152, 1987.

Murray, D.W., and Prell, W.L., Pliocene to Pleistocene variations in calcium carbonate, organic carbon, and opal on the Owen Ridge, northern Arabian Sea, in *Proc. ODP, Sci. Results, 117*, edited by W. L. Prell, N. Niitsuma et al., 343-363, Ocean Drilling Program, College Station, TX, 1991.

Nigrini, C., Composition and biostratigraphy of radiolarian assemblages from an area of upwelling (northwestern Arabian Sea, Leg 117), in *Proc. ODP, Sci. Results, 117*, edited by W. L. Prell, N. Niitsuma et al., Ocean Drilling Program, 89-126, College Station, TX, 1991.

Norton, I. O., and Sclater, J. G., A model for the evolution of the Indian Ocean and the breakup of Gondwanaland, *J. Geophys. Res., 84*, 6803-6830, 1979.

Okada, H., Quaternary and Paleogene calcareous nannofossils, Leg 115. In: Duncan, R.A., Backman, J., Peterson, L.C., et al., *Proc. ODP., Sci. Results, 115*, 129-174, Ocean Drilling Program, College Station, TX, 1990.

Peirce, J., Weissel, J. et al., *Proc. ODP, Init. Repts, 121*, Ocean Drilling Program, College Station, TX, 1989.

Peterson, L. C., and Backman, J., Late Cenozoic calcium carbonate accumulation and the history of the carbonate compensation depth in the western equatorial Indian Ocean, in *Proc. ODP., Sci. Results, 115*, edited by R. A. Duncan, J. Backman, L. C. Peterson et al., 467-507, Ocean Drilling Program, College Station, TX, 1990.

Peterson, L. C., and Prell, W. L., Carbonate dissolution in Recent sediments of the eastern equatorial Indian Ocean: preservation patterns and carbonate loss above the lysocline, *Mar. Geol., 64*, 259-290, 1985a.

Peterson, L. C., and Prell, W. L., Carbonate preservation and rates of climate change: an 800 kyr record from the Indian Ocean, in *The Carbon Cycle and Atmospheric CO$_2$: Natural Variations Archean to Present*, edited by E. T. Sundquist, and W. S. Broecker, 251-270, AGU Geophysical Monograph 32, Washington, D.C., 1985a.

Pimm, A. C., McGowran, B., and Gartner, S., Early sinking history of the Ninetyeast Ridge, northwestern Indian Ocean, *Geol. Soc. Amer. Bull., 85*, 1219-1224, 1974.

Pisias, N. G., and Prell, W. L., High resolution carbonate records from the hydraulic piston cored section of Site 572, in *Init. Repts. DSDP, 85*, edited by L. Mayer, F. Theyer et al., 711-722, U.S. Government Printing Office, Washington, D.C., 1985.

Prell, W. L., Niitsuma, N. et al., *Proc. ODP, Init. Repts., 117*, Ocean Drilling Program, College Station, TX, 1989.

Premoli-Silva, I., and Spezzaferri, S., Paleogene planktonic foraminifer biostratigraphy and paleoenvironmental remarks on Paleogene sediments from Indian Ocean sites, Leg 115, in *Proc. ODP., Sci. Results, 115*, edited by R. A. Duncan, J. Backman, L. C. Peterson et al., 277-314, Ocean Drilling Program, College Station, TX, 1990.

Quade, J., Cerling, T. E., and Bowman, J. R., Development of the Asian monsoon revealed by marked ecological shift during the latest Miocene in northern Pakistan, *Nature, 342*, 163-166, 1989.

Raymo, M. E., Ruddiman, W. F., and Froelich, P. N., Influence of the late Cenozoic mountain building on ocean geochemical cycles, *Geology, 16*, 649-653, 1988.

Rea, D. K., and Bloomstine, M. K., Neogene history of the South Pacific tradewinds: evidence for hemispherical asymmetry of atmospheric circulation, *Palaeogeogr., Palaeoclimatol., Palaeoecol., 55*, 55-64, 1986.

Rea, D. K., Dehn, J., Driscoll, N. W., Farrell, J., Janecek, T. R., Owen, R. M., Pospichal, J. J., Resiwati, P., and the ODP Leg 121 Scientific Party, Paleoceanography of the eastern Indian Ocean from ODP Leg 121 drilling on Broken Ridge, *Geol. Soc. Am. Bull., 102*, 679-690, 1990.

Rea, D. K., and Leinen, M., Neogene history of the calcite compensation depth and lysocline in the South Pacific Ocean, *Nature, 316*, 805-807, 1985.

Rea, D. K., and Leinen, M., Neogene controls on hydrothermal activity and paleoceanography of the southeast Pacific, in *Init. Repts. DSDP, 92*, edited by M. Leinen, D. K. Rea et al., 597-617, U.S. Government Printing Office, Washington, D.C., 1986.

Rio, D., Fornaciari, E., and Raffi, I., Late Oligocene through early Pleistocene calcareous nannofossils from western equatorial Indian Ocean (Leg 115), in *Proc. ODP., Sci. Results, 115*, edited by R. A. Duncan, J. Backman, L. C. Peterson et al., 175-236, Ocean Drilling Program, College Station, TX, 1990.

Robinson, S. G., Applications for whole-core magnetic susceptibility measurements of deep-sea sediments: Leg 115 results, in *Proc. ODP., Sci. Results, 115*, edited by R. A. Duncan, J. Backman, L. C. Peterson et al., 737-772, Ocean Drilling Program, College Station, TX, 1990.

Rogl, F., and Steininger, F. F., Vom zerfall der Tethys und Paratethys, die Neogene palaeographie und palinspastik des zirkum-Mediterranean Raumes, *Ann. Naturhist. Mus. Wien, 85/A*, 135-163, 1983.

Royer, J.-Y., and Sandwell, D. T., Evolution of the eastern Indian Ocean since the late Cretaceous: constraints from Geosat altimetry, *J. Geophys. Res., 94*, 13,755-13,782, 1989.

Ruddiman, W. F., and Kutzbach, J. E., Plateau uplift and climate change, *Sci. Amer., 264*, 66-75, 1991.

Savin, S. M, Douglas, R. G., and Stehli, F. G., Tertiary marine paleotemperatures, *Geol. Soc. Amer. Bull., 86*, 1499-1510, 1975.

Schmitz, B., Barium, equatorial high productivity, and the northward wandering of the Indian continent, *Paleoceanography, 2*, 63-77, 1987.

Sclater, J. G., Abbott, D., and Thiede, J., Paleobathymetry and sediments of the Indian Ocean, in *Indian Ocean Geology and Biostratigraphy*, edited by J. R. Heirtzler, H. M. Bolli, T. A. Davies, J. B. Saunders, and J. G. Sclater, 25-60, AGU, Washington, D.C., 1977.

Sclater, J. G., and Fisher, R. L., Evolution of the east central Indian Ocean, with emphasis on the tectonic setting of of the Ninetyeast Ridge, *Geol. Soc. Amer. Bull., 85*, 683-702, 1974.

Shackleton, N. J., Paleogene stable isotope events, *Palaeogeogr., Palaeoclimatol., Palaeoecol., 57*, 91-102, 1986.

Shackleton, N. J., and Kennett, J. P., Paleotemperature history of the Cenozoic and the initiation of Antarctic glaciation: oxygen and carbon isotope analyses in DSDP Sites 277, 279, and 281, in *Init. Repts. DSDP, 29*, edited by J. P. Kennett, R. E. Houtz et al., 743-755, U.S. Government Printing Office, Washington, D.C., 1975.

Shipboard Scientific Party, Site 724. In: Prell, W.L., Niitsuma, N. et al., *Proc. ODP, Init. Repts., 117*, edited by J. P. Kennett, R. E. Houtz et al., College Station, TX, 1989a.

Shipboard Scientific Party, Site 728, in *Proc. ODP, Init. Repts., 117*, edited by W. L. Prell, N. Niitsuma et al., 495-545, Ocean Drilling Program, College Station, TX, 1989b.

Shipboard Scientific Party, Site 722, in *Proc. ODP, Init. Repts., 117*, edited by W. L. Prell, N. Niitsuma et al., 255-317, Ocean Drilling Program, College Station, TX, 1989c.

Shipboard Scientific Party, 1989d. Ninetyeast Ridge Summary, in *Proc. ODP, Init. Repts., 121*, edited by J. Peirce, J. Weissel et al., 517-537, Ocean Drilling Program, College Station, TX, 1989d.

Simmons, G. R., Subsidence history of basement sites and sites along a carbonate dissolution profile, Leg 115, in *Proc. ODP., Sci. Results, 115*, edited by R. A. Duncan, J. Backman, L. C. Peterson et al., 123-126, Ocean Drilling Program, College Station, TX, 1990 .

Tchernia, P., *Descriptive Regional Oceanography*, Pergamon Press, New York, 1980.

Theyer, F., Mayer, L. A., Barron, J. A., and Thomas, E., The equatorial Pacific high-productivity belt: elements for a synthesis of Deep Sea Drilling Project Leg 85 results, in *Init. Repts. DSDP, 85*, edited by L. Mayer, F. Theyer et al., 971-985, U.S. Government Printing Office, Washington, D.C., 1985.

Thiede, J., Sediment coarse fractions from the western Indian Ocean and the Gulf of Aden (Deep Sea Drilling Project Leg 24), in *Init. Repts. DSDP, 24*, edited by R. L. Fisher, E. T. Bunce et al., 651-765, U.S. Government Printing Office, Washington, D.C., 1974.

Thiede, J., and Rea, D. K., Mass accumulation rates of Barremian to Recent biogenic sediments from the Mid-Pacific Mountains

(Deep Sea Drilling Project Site 463) and Hess Rise (Sites 464, 465, 466) central North Pacific Ocean, in *Init. Repts. DSDP, 62*, edited by J. Thiede, T. L. Vallier et al., 637-651, U.S. Government Printing Office, Washington, D.C., 1981.

Thunell, R. C., and Corliss, B. H., Late Eocene-Early Oligocene carbonate sedimentation in the deep sea, in *Terminal Eocene Events*, edited by C. Pomerol, and I. Premoli Silva, 63-380, Elsevier, Amsterdam, 1986.

Vail, P. R., Mitchum, R. M., Jr., and Thompson, S., III, Global cycles of relative changes of sea level, in *Seismic Stratigraphy - Applications to Hydrocarbon Exploration*, edited by C. E. Payton, AAPG Mem. 26, 83-97, 1977.

van Andel, T. H., Mesozoic-Cenozoic calcite compensation depth and the global distribution of calcareous sediments, *Earth Planet. Sci. Lett., 26*, 187-194, 1975.

van Andel, T. H., Heath, G. R., and Moore, T. C., Jr., Cenozoic history and paleoceanography of the central equatorial Pacific, *Geol. Soc. Am. Mem., No. 143*, 1975.

Vincent, E., Killingley, J. S., and Berger, W. H., Miocene oxygen and carbon isotope stratigraphy of the tropical Indian Ocean, in *The Miocene Ocean: Paleoceanography and Biogeography,*

Mem. Geol. Soc. Amer., 163, edited by J. P. Kennett, 103-130, 1985.

Vincent, E., and Toumarkine, M., Neogene planktonic foraminifers from the western tropical Indian Ocean, Leg 115, in *Proc. ODP., Sci. Results, 115*, edited by R. A. Duncan, J. Backman, L. C. Peterson et al., 795-836, Ocean Drilling Program, College Station, TX, 1990.

Warren, B. A., Transindian hydrographic section at Lat. 18° S: property distributions and circulation in the South Indian Ocean, *Deep-Sea Res., 28A*, 759-788, 1981.

Whitmarsh, R. B., Weser, O. E., Ross, D. A. et al., *Init. Repts., DSDP, 23*, U.S. Government Printing Office, Washington, D.C., 1974.

Woodruff, F., and Savin, S. M., Miocene deepwater oceanography, *Paleoceanography, 4*, 87-140, 1989.

Wu, G., Yasuda, M. K., and Berger, W. H., Late Pleistocene carbonate stratigraphy on Ontong-Java Plateau in the western equatorial Pacific, *Mar. Geol., 99*, 135-150, 1991.

Wyrtki, K., Physical oceanography of the Indian Ocean, in *The Biology of the Indian Ocean*, edited by B. Zeitzschel, and S. A. Gerlach, 18-36, Springer-Verlag, Berlin, 1973.

Biosiliceous Sedimentation Patterns for the Indian Ocean During the Last 45 Million Years

J. G. BALDAUF

Department of Oceanography and Ocean Drilling Program, Texas A&M University, College Station, TX, 77845, USA

J. A. BARRON

U.S. Geological Survey, MS915, Menlo Park, CA , 49025, USA

W. U. EHRMANN

Alfred-Wegner-Institut fur Polar-und Meeresforschung, D-2850 Bremerhaven, FRG

P. HEMPEL

GEOMAR, Forschungszentrum fur marine Geowissenschaften and der Christian Albrechts Universität zu Kiel, Wischhofstrasse 1-3, D-2300 Kiel FRG

D. MURRAY

Department of Geological Sciences, Brown University, RI, 02912-1846, USA

A spatial and temporal framework of the history of biosiliceous sedimentation patterns in the Indian Ocean for the last 45 million years is identified based on the occurrence of diatom and percent opal composition of the sedimentary sequences recovered from scientific drilling in the Indian Ocean by the Deep Sea Drilling Project and the Ocean Drilling Program. The northwest Indian Ocean exhibits moderate biosiliceous sedimentation during the Eocene with a decline in such sedimentation during the Oligocene. The early and middle Miocene is characterized by low biosiliceous sedimentation. During the late middle Miocene (~11.0 Ma) both diatom abundance and percent opal values increase, most likely reflecting increased surface productivity resulting from the establishment of the monsoonal gyral circulation. Biosiliceous sediments are sparse in the northeast Indian Ocean limiting paleoceanographic interpretations.

Unlike the northwest Indian Ocean, the southern Indian Ocean exhibits low diatom abundance and opal values during the late Eocene, followed by an abrupt increase in both the diatom abundance and in the percent opal composition at about 35.8 Ma, most likely reflecting increased productivity in the southern ocean resulting from destratification of the water column and increased polar cooling. Opal values remain moderate and fairly constant during the Oligocene through the middle Miocene and increase to present-day values during the latest Miocene.

INTRODUCTION

Modern surface circulation of the Indian Ocean consists of three distinct domains that include (i) the monsoon gyre, (ii) the southern hemisphere subtropical gyre and (iii) the Indian Ocean sector of the southern ocean influenced by Antarctic waters and the Circumpolar current [Wyrtki, 1973]. Unique to the Indian Ocean is the monsoonal circulation which is seasonal and which greatly influences upwelling along the coast of Arabia, the Somali Current, and to a lesser extent some parts of the east coast of India. The monsoonal gyre is separated from the southern subtropical gyre by a pronounced hydrochemical front at

Synthesis of Results from Scientific
Drilling in the Indian Ocean
Geophysical Monograph 70
Copyright 1992 American Geophysical Union

about 10°S (Figure 1). These surface domains are reflected by the distribution of the surface chemical and biological properties, especially by the seasonal sea surface temperatures, phosphate distribution, and primary productivity.

Regions of moderate to high phosphate (PO_4) contents are associated with the monsoon region and the area of the southern divergence. The monsoon area has PO_4 values between 1.0-2.0 ug-atl-1 with concentrations dropping to less than 0.4 ug as one goes from the monsoon region across the front at 10°S into the region of the subtropical gyre. The region associated with the Antarctic divergence is rich in PO_4 attaining values up to 2.2 ug [Wyrtki, 1973]. Clearly this pattern indicates that regions of high productivity characterize the monsoon and Antarctic regions whereas low productivity characterizes the

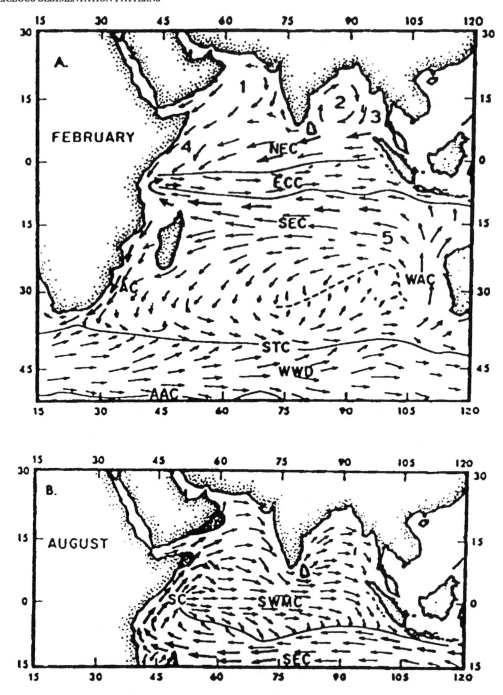

Fig. 1. Surface circulation and surface watermasses of the Indian Ocean during February (A) and August (B), modified from Prell et al., [1980]. AAC=Antarctic Convergence, WWD=West Wind Drift, STC=Subtropical Convergence, AC=Agulhas Current, WAC=West Australian Current, SEC=South Equatorial Current, ECC=Equatorial Countercurrent, NEC=North Equatorial Current, SC=Somali Current, SWMC=Southwest Monsoon Current, 1=Arabian Sea, 2= Bay of Bengal, 3=Andaman Sea, 4=Somali Basin, 5=Wharton Basin.

subtropical waters.

Biological productivity studies for surface waters of the Indian Ocean parallel PO_4 values with productivity values ranging from 0.01 to 249 mg $C/m^3/hr$ [Qasim, 1982].

Figure 2 from Lisitzin [1972] illustrates the total annual production of silica in surface waters of the world's oceans in $g/m^2/yr$. With respect to the Indian Ocean, the areas of high production are associated with the Antarctic

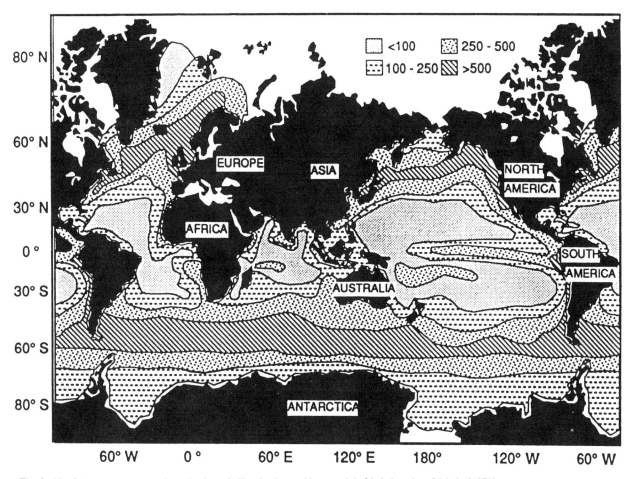

Fig. 2. Absolute masses or annual production of silica in the world ocean (g/m2/yr), based on Lisitzin [1972].

divergence, and to a lesser extent the coastal regions of Oman and India and waters adjacent to Sri Lanka, Burma, and Java.

The present-day oceanic and climatic conditions in the Indian Ocean are in part reflected in the distribution of modern sediments. Biosiliceous sediments accumulate today beneath regions of high surface water productivity associated with regions of oceanic upwelling [Berger, 1970; Calvert, 1974; Leinen, 1979]. This present-day pattern is dynamic and has evolved through geological time in response to global and regional changes in chemical budgets, tectonics, climate, sea-level and changes in the chemical and sediment cycles. [see Baldauf and Barron, 1990 for detailed discussion].

Although the spatial and temporal framework for the distribution of biogenic sediments is established for the Atlantic, Pacific and to a lesser degree the Southern Ocean [see Ehrmann and Thiede, 1985; Barron and Baldauf, 1989; Baldauf and Barron, 1990], our understanding of biosiliceous sedimentation in the Indian Ocean is less refined. Previous studies on the distribution of biosiliceous sedimentation in the Indian Ocean can be summarized by the works of Distanov et al. [1971], Lisitzin [1972],

Leclaire [1974, 1978], Caulet [1977, 1978], Udintsev [1975], Burckle [1989], Baldauf and Barron [1990], and Mikkelsen and Swart [1992]. Figure 3 shows the distribution of diatoms in the surface sediments from the Indian Ocean based on the studies of Udintsev [1975], Leclaire [1978] and Lisitzin [1972]. In general sediments enriched with diatoms are restricted to regions associated with the southern divergence or to the low-latitude region south of the equator. Diatoms are less abundant along the coast of Arabia. In the southern region associated with the polar front diatoms attain values greater than 100×10^6 valves per gram of sediment [Lisitzin, 1972; Udintsev, 1975]. In the equatorial region diatoms attain values of $1-5 \times 10^6$ valves per gram sediment, while in the surface sediments offshore Arabia diatom concentrations of $<1 \times 10^6$ valves per gram occur [Lisitzin, 1972].

SPATIAL AND TEMPORAL FRAMEWORK

Scientific drilling during Deep Sea Drilling Project Legs 22 through 28 and Ocean Drilling Program Legs 115-122 has provided a database from over 100 sites in the Indian Ocean. Analysis of the sediment from these sites allows determination of the spatial and temporal distribution of

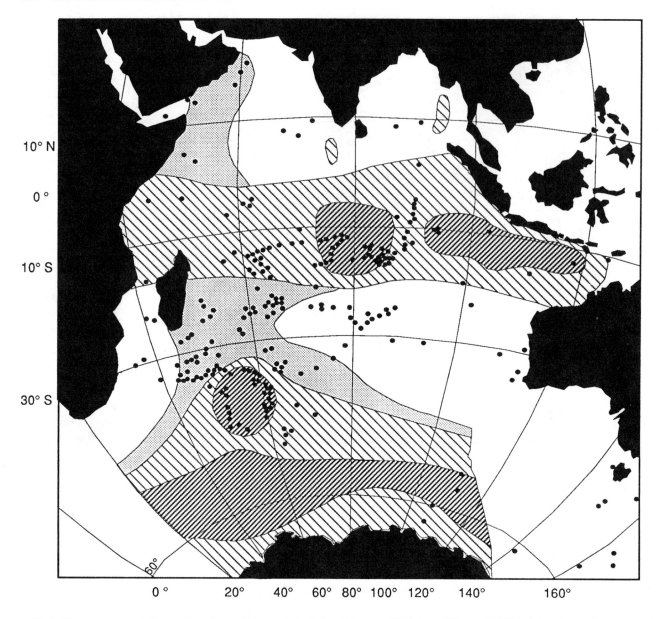

Fig. 3. The occurrence of diatoms in surface sediments of the Indian Ocean, modified from Udintsev [1975] and Leclaire [1978]. Dark hactured area represents zones of enriched diatom abundances, stippled area represents areas of low diatom abundances, clear regions represents area where diatoms are absent.

biosiliceous sediments in the Indian Ocean during the last 45 million years. For the purposes of this study the Indian Ocean has been divided into 3 quadrants. Quadrant 1, representing the northwest Indian Ocean, consists of the region from 90°E to 30°E and north of 40°S; Quadrant II, representing the northeast Indian Ocean consists of the interval from 90°E to 150°E and north of 40°S. Quadrant III represents the Indian Ocean south of 40°S. DSDP and ODP sites within these quadrants are shown in Figure 4. Quadrant I contains over 60 sites most of which are located in the area influenced by seasonal monsoons, along the Ninetyeast, Central Indian, or Chagos-Laccadive ridges, or positioned in the Madagascar region. Quadrant II contains

30 sites concentrated in two areas including the western and northwestern margin of Australia and Broken Ridge. Quadrant III in the southern Indian Ocean contains 23 sites of which all but 7 are located on a north-south transect from the Kerguelen Plateau to Prydz Bay, east Antarctica.

The somewhat patchy spatial arrangement of the sites in the Indian Ocean allows limited insight into the spatial and temporal distribution of biosiliceous sediments and the response of productivity and sedimentation to changing oceanographic, tectonic and climatic conditions. Figures 5-7 show the spatial and temporal distribution of middle Eocene-Quaternary sediments recovered from DSDP and ODP sites located in each of these quadrants.

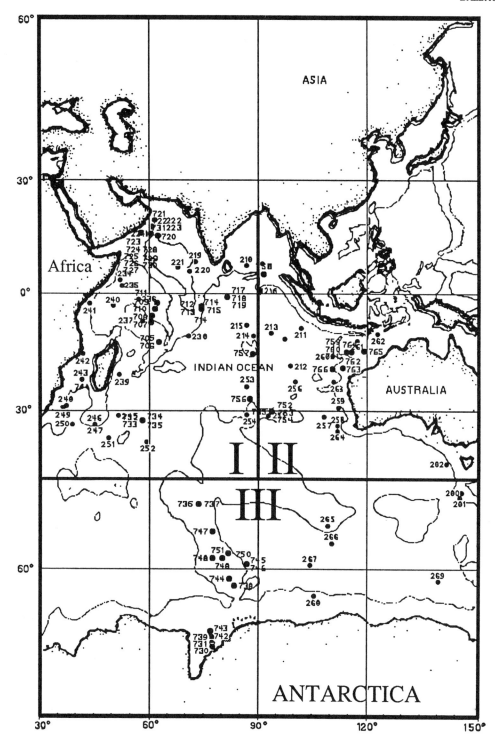

Fig. 4. Location of DSDP and ODP sites in the Indian Ocean. For the purpose of this study the Indian Ocean has been divided into three quadrants: the northwest Indian Ocean (Quadrant I); the northeast Indian Ocean (Quadrant II), and the southern Indian Ocean (Quadrant III).

Chronostratigraphic control for these sequences is based on numerous studies including those of Barron, Larsen, Baldauf et al., [1989], Baldauf and Barron [1990], Fenner and Mikkelsen [1990], Mikkelsen [1990], Baldauf and Barron [1991], Barron et al. [1991], Spaulding et al. [1991], and Harwood and Maruyama [1992]. The chronology of Berggren et al. [1985] is used throughout this study. Diatom occurrence incorporates smear and strewn slide data [Fisher, Bunce et al., 1974; Schrader, 1974; Simpson, Schlich et al., 1974; von der Boch, Sclater et al., 1974; Whitmarsh, Weser, Ross et al., 1974; Backman, Duncan et al., 1989; Prell, Niitsuma et al., 1989; Fenner and Mikkelsen, 1990; Mikkelsen, 1990; Baldauf and Barron, 1991; Fourtanier, 1991; Harwood and Maruyama, 1992; Baldauf, 1992; Baldauf, unpublished data, 1992].

In the northwest Indian Ocean (Quadrant I, Figure 5) few sites represent a near continuous stratigraphic record. This in part reflects spot coring procedures used during early DSDP cruises, the occurrence of stratigraphic breaks, or unsuccessful recovery of the sedimentary sequence. Stratigraphic sequences containing consistent diatoms occur in the Oman Margin-Arabian Sea-Owen Ridge areas (Sites 222-224, 721, 722, and 729-731), the Mascarene Plateau region (Sites 237, 707-711) and the Central Indian and Chagos-Laccadives ridges (Sites 238, 712-715).

From a temporal perspective these stratigraphic intervals enriched with diatoms are generally restricted to two time intervals, the Paleogene (~30-42 Ma) and the late Neogene-Quaternary (~0-11 Ma). Paleogene sediments containing diatoms occur at Sites 219, 220, 236, 237, and 706-711, 713. Upper Neogene-Quaternary sediments containing diatoms are numerous (see Figure 5).

The dataset for the northeastern Indian Ocean (Quadrant II, Figure 6) consists of 8 sites (Sites 211, 213, 216, 259, 262, and 758) having a documented occurrence of diatoms for the last 45 million years. These are located in proximity to the Wharton Basin and Java Trench, on the northern end of the Ninetyeast Ridge, northwest Australia and the Java Sea. With the exception of poorly preserved diatoms in the Eocene-Oligocene interval at DSDP Sites 216 and 217 [Fenner, 1984] and the upper Oligocene-middle Miocene interval at ODP Site 758 [Fourtanier, 1991], diatoms are restricted to sediments of late Neogene-Quaternary age (~0-8 Ma).

Unlike the northern regions, the southern Indian Ocean (Quadrant III) generally contains diatoms (Figure 7) from all sites occupied throughout the region. The one exception occurs from sites positioned on the continental margin of Prydz Bay, east Antarctica, which either contain sporadic diatoms or are barren of diatoms [Baldauf and Barron, 1991]. Within quadrant III diatoms are generally prevalent in sediments of Oligocene or younger age.

PERCENT OPAL DATA

Quantitative analysis of percent biogenic opal have been completed for only a few selected sites in the Indian Ocean. Included in these studies are Sites 707, 709, and 711 on the Mascarene Plateau [Hempel and Bohrmann, 1990] and Site 722 from the Oman Ridge [Murray, unpublished data, 1991] all located in Quadrant I; Sites 213-217 on the Ninety-East Ridge [Quadrant II; Schmitz, 1987]; and Sites 737, 738, and 744 on the Kerguelen Plateau [Bohrmann and Ehrmann, 1991] located in Quadrant III (Figure 8).

Middle Miocene to Quaternary opal data have been collected by Murray [unpublished data, 1991] from Site 722

located on Owen Ridge. Quaternary data also have been generated by Murray and Prell (1991) for Core RC27-61 on Owen Ridge. Both datasets are based on a modified version of the extraction techniques of Mortlock and Froelich [1989]. The dataset indicates that opal values are low (<10 percent) in the early middle Miocene, increase to values of 40-60 percent during the early late Miocene and decrease to values generally less then 20 percent for the latest Miocene to Pliocene interval. Opal values in the Quaternary fluctuate between 15 and 60 percent [Murray and Prell, 1991].

Opal data for the Leg 115 sites (Sites 711, 709, and 707) are based on the percent opal in carbonate-free sediment using the method of Eisma and Van der Gaast [1971; see Hempel and Bohrmann, 1990]. Opal data are available for the middle Eocene through Miocene of Site 711. At this site opal values are 80-100 percent in the middle and late Eocene interval, gradually declining from about 80 percent in the latest Eocene to generally less than 20 percent in the early Oligocene. Values remain low and are generally less than 5 percent for the remainder of the Oligocene and Miocene portion of the sequence.

Opal data from Hempel and Bohrmann [1990] for Site 709 indicate values of 60-100 percent opal for the middle and late Eocene. Values gradually decline in the latest Eocene and Oligocene from values of 50 percent to values less than 20 percent in the late Oligocene. Opal values for the early and middle Miocene are generally less than 5 percent. In the earliest late Miocene, opal values increase attaining up to 20 percent composition of the carbonate-free sediment.

The stratigraphic sequence at Site 707 consists of a near continuous record from the late Paleocene to the Quaternary. One hiatus occurs and separates the lower Eocene from the middle Eocene. Opal data from Hempel and Bohrmann [1990] show values from 0-100 percent for the Paleocene to late Miocene interval. The late Paleocene and early Eocene interval is characterized by opal-CT with the transition from opal-A to opal-CT coinciding with the early Eocene/middle Eocene unconformity. Two intervals of high opal values occur in the middle Eocene to late Miocene sequence. The first interval occurs in the middle Eocene to late Oligocene (attaining values of ~90 percent) and the second interval occurs in the late Miocene (attaining values of 20-60 percent). These two intervals are separated by a late Oligocene to middle Miocene interval in which opal values are generally less than 20 percent. No data are available from the Pliocene or Quaternary.

Opal and barium contents have been collected by Schmitz [1987] for a few selected samples from DSDP Sites 213-217 following the methods of Bostrom et al. [1972]. Sites 213-217 are located in Quadrant II north of and within the modern equatorial high-productivity zone. According to Schmitz [1987], the opal and Ba contents of these sediments confirm the northward passage of the Indian plate beneath a persistent equatorial high-productivity zone during the Cenozoic. Schmitz [1987] assumed near-constant 8° width for the equatorial high-productivity zone and predicted changes in the northward plate motion at about 53, 40, 17, and 6 Ma to account for temporal changes

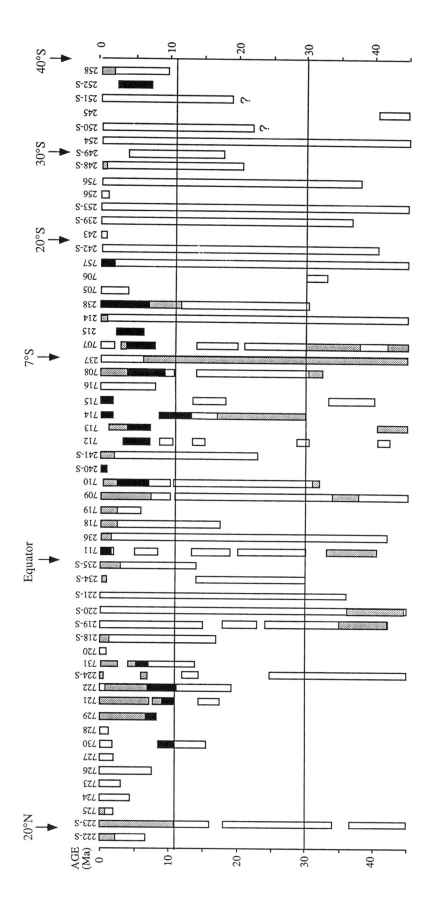

Fig. 5. Spatial and temporal framework of selected DSDP and ODP sites from the northwest Indian Ocean (Quadrant I). Sites are arranged along a north-south transect (5A, sites from about 20°N to 7°S; 5B, sites from about 7°S to about 40°S). Diatom abundance is based on smear and strewn slide data (see text). Black intervals represent abundant diatoms; stippled intervals represent rare to few diatoms; clear intervals represent intervals were diatoms are absent. Figure shows two intervals of diatom abundance. The first in the Eocene and the second in the late middle Miocene through Quaternary.

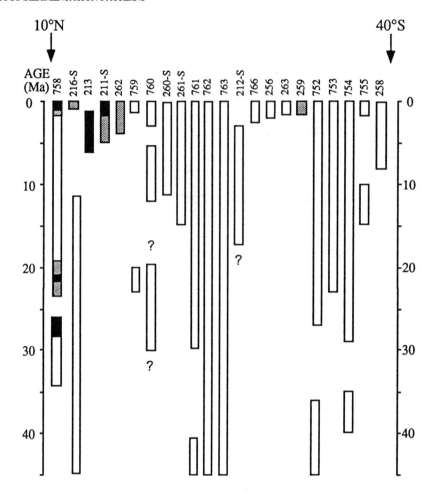

Fig. 6. Spatial and temporal framework of selected DSDP and ODP sites from the northeast Indian Ocean (Quadrant II). Sites are arranged along a north-south transect from about 10°N to 40°S. Diatom abundance is based on smear and strewn slide data (see text). Black intervals represent abundant diatoms; stippled intervals represent rare to few diatoms; clear intervals represent intervals where diatoms are absent. Diatoms are generally absent from this quadrant.

in the opal and Ba content at the five sites. Schmitz [1987] suggests that the high opal contents in the Eocene sediments imply that the equatorial productivity zone was 4° wider during the Eocene. If the northward motion of the Indian Plate has remained at about 3.7 cm/yr for the last 45 Ma, as predicted by Duncan [personal communication, 1991], Schmitz's [1987] opal data also suggest a late Miocene expansion of the equatorial high-productivity zone.

Opal data from quadrant III consists of data from Bohrmann and Ehrmann [1991] for Sites 737 and 744 and from Ehrmann and Grobe [1991] for Sites 745 and 746 based on the methods of Eisma and Van der Gaast [1971]. Opal data are also available for Sites 738 and 744 in Ehrmann [1991]. Data exists from Site 737 for the middle Miocene-early Pliocene interval. Values within this interval generally average greater than 40 percent with maximum values greater than 80 percent occurring in the late Miocene and early Pliocene. No data are available for

the Eocene to earliest middle Miocene portion of Site 737 as it was not possible to estimate the opal-A content because of the large quantity of volcanic glass present in this interval [Bohrmann and Ehrmann, 1991].

The opal data for Sites 745 and 746 represent a near continuous composite record for the last 10 million years. Opal values are variable, but average about 40 percent throughout these sequences. A minimum value of less than 5 percent occurs in the late Miocene at Site 746 and maximum values of about 80 percent occur in the latest Pliocene of Site 745 and the Quaternary of Site 746.

The Eocene through Quaternary sequence at Site 744 contains both biosiliceous and biocalcareous sediments. Seven unconformities occur in the sequence (32.0-28.5, 26.3-24.5, 24.0-21.3, 13.4-12.5, 12.2-11.2, 8.9-6.0, 5.6-4.2 Ma). Opal-A values are low (<20 percent) in the Eocene, but show an abrupt increase (from 0 to 100 percent) at about the Eocene/Oligocene boundary. Opal values remain high (80-100 percent) in the early Oligocene

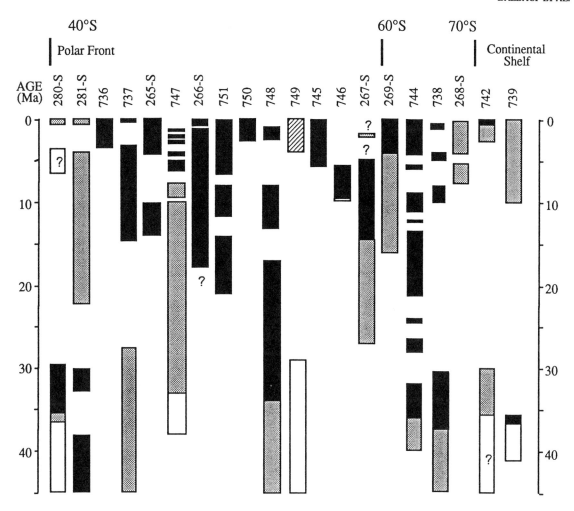

Fig. 7. Spatial and temporal framework of selected DSDP and ODP sites from the southern Indian Ocean (Quadrant III). Sites are arranged along a north-south transect from 40°S to the continental shelf of eastern Antarctica. Diatom abundance is based on smear and strewn slide data (see text). Black intervals represent abundant diatoms; stippled intervals represent rare to few diatoms; clear intervals represent intervals where diatoms are absent. Figure shows that diatoms generally increase in abundance at about the Eocene/Oligocene boundary (36.7 million years) and are generally present throughout the remaining portion of the Oligocene-Pleistocene.

and decrease to values less than generally 40 percent from the late early Oligocene to the middle Miocene. Values once again increase (attaining values of 80 percent) in the middle and late Miocene.

Opal data for Site 738 consist of data for the Late Cretaceous through the early Oligocene and for the Pliocene and Quaternary [see Ehrmann, 1991]. Opal present in the Cretaceous through middle Eocene interval consists of opal-CT. Opal is nearly absent from the late Eocene and present in the early Oligocene and Pliocene-Quaternary intervals where it attains values of about 30 percent.

The above qualitative and quantitative datasets indicate that two distinct intervals of high biosiliceous sedimentation occurred in the Indian Ocean during the last 45 m.y. The first approximates the Eocene/ Oligocene boundary and is most pronounced in the southern Indian Ocean (Quadrant III). The second approximates the late middle Miocene to late Miocene transition and is most pronounced in the northwestern Indian Ocean (Quadrant I). The intervening late Oligocene to middle Miocene interval is generally a period of low biosiliceous deposition.

EOCENE/OLIGOCENE TRANSITION

Figure 8 compares the late Eocene to early Oligocene percent opal data from Sites 707, 709 and 711 from the northwest Indian Ocean and that from Sites 738 and 744 from the Kerguelen Plateau. These records indicate that opal deposition increased from 0 to almost 100 percent during the middle Eocene at Site 707. However, it is not possible to determine the exact timing of this increase because it is associated with both an unconformity and a transition from opal-CT below to opal-A above. The middle Eocene to earliest Oligocene of Site 707 is characterized by high opal values (attaining values close to 100 percent). These high opal values are also characteristic of Sites 709 and 711, which attain opal values greater than 80 percent for the

Opal-A (%)

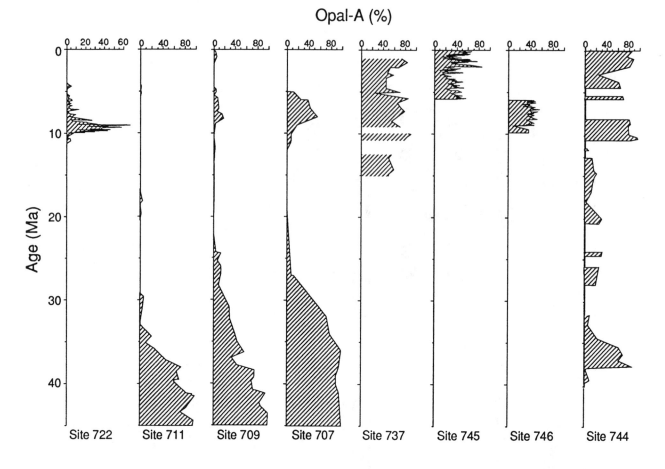

Fig. 8. Percent opal data for carbonate-free sediments from DSDP and ODP sites in the Indian Ocean Data is based on results of Hempel and Bohrmann [1990] from Sites 707, 709, and 711 on the Mascarene Plateau; Murray [unpublished data, 1992] from Site 707; Bohrmann and Ehrmann [1991] from Sites 737 and 744; and Ehrmann et al. [1991] for Sites 745 and 746.

middle and late Eocene interval. A latitudinal pattern in opal distribution is observed in the Eocene and Oligocene with higher values present at the southernmost site (Site 707) and lower values present at the northern most site (Site 711)[see Hempel and Bohrmann 1991]. The Eocene diatom assemblages at these sites are dominated by planktonic diatoms representing a low to middle latitude assemblage which is expected based on plate reconstructions indicating an estimated paleo latitude of 20°S to 5°N for these sites [Fenner and Mikkelsen, 1990; McKenzie and Sclater, 1971].

Commencing in the early Oligocene opal values decrease at all three sites, generally attaining values of less than 20 percent in the early Oligocene and reaching values close to zero by the Oligocene/Miocene boundary. This general decline in opal values corresponds approximately to a general increase in carbonate which changes from average values of 80 percent in the Eocene to values of 90 percent in the Oligocene [see Hempel and Bohrmann, 1991]. The Oligocene diatom assemblage shows a transition from a cosmopolitan assemblage to a more restricted assemblage at these sites [see Fenner and Mikkelsen, 1990]. This

transition corresponds to a similar change in floral composition observed elsewhere in the world's oceans [see Fenner, 1982 and Baldauf, 1992]. It should be noted that at several sites (specifically Sites 707, 709 and 714), the diatom assemblage contains freshwater elements which Fenner and Mikkelsen [1990] interpret as windblown since neither benthic or brackish waters occur.

Comparing these results with those from Site 744 shows a slight difference between opal patterns in the northwest Indian Ocean with that of the southern Indian Ocean. At Site 744 opal values in the late Eocene are much less than that in the northwestern Indian Ocean (<20 percent at Site 744 compared with >90 percent in the northwestern Indian Ocean) and increase to approximately 80 percent at 35.8 Ma. Although the transition from low to high opal values is abrupt at Site 744, it corresponds to the period of gradual decline in opal values observed in the northwestern Indian Ocean.

Comparing the opal values at Site 744 to the oxygen isotope record of Barrera and Huber [1991; Figure 9] indicates that this abrupt transition in opal values at Site 744 approximates an increase of about 1.2 ppm in benthic

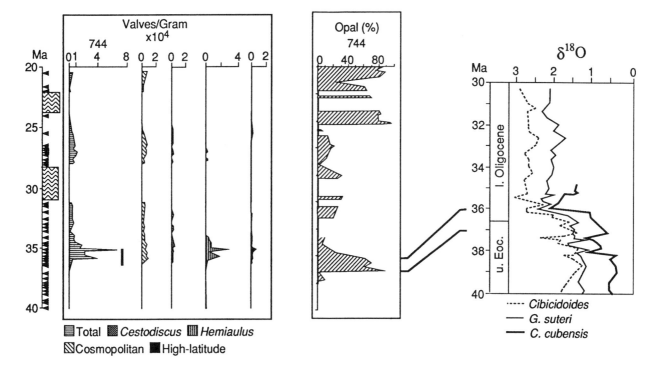

Fig. 9. Percent opal data for carbonate-free sediments from ODP Site 744 versus age in million years. Percent opal is compared with the oxygen isotope data of Barrera and Huber [1991] and diatom floral data of Baldauf (in press). Arrows indicate sample locations for floral analysis.

and planktonic foraminiferal oxygen isotope values which occurs at about 35.9 Ma. This isotopic shift has been interpreted by numerous workers to represent the increased ice volume resulting from the development of continental glaciation on east Antarctica [Ehrmann et al., this volume and Zachos et al., 1992], and/or the result of cooler water temperatures without a significant increase in ice volume [Wei, 1991].

The shift in isotope values of about equal magnitudes in both benthic and planktonic signals has been interpreted by Barrera and Huber [1991], and Baldauf [1992], to represent destratification of the water column, increased mixing and upwelling resulting in increased productivity. This destratification results in part by ocean reorganization and ocean response to climatic change [see Baldauf and Barron, 1990]. Such reorganization is also supported by the diatom flora which exhibits an abrupt increase in abundance at an interval equivalent to that of the isotope shift. Diatoms are virtually absent in Eocene sediments at Site 744 and increase in abundance in the earliest Oligocene from values of less than 1×10^{14} valves per gram to greater than $4 \text{-} 8 \times 10^{14}$ valves per gram [Baldauf, 1992].

Additional support for oceanic response to climate change is derived from (i) the occurrence of ice rafting detritus present at Site 744 for the interval from 35.8-36 Ma [Ehrmann, 1991; Breza and Wise, 1992]; (ii) decrease in carbonate values from about 95 percent in the Eocene to values between 65 and 90 percent in the Oligocene [Barron, Larsen et al., 1991], and (iii) the floral composition which

consists of both species endemic to the high southern latitudes as well as cosmopolitan species [Figure 9; Baldauf and Barron, 1991; Baldauf, 1992].

The above discussion suggests the relatively high opal accumulation rates at low latitudes during the Eocene most likely resulted from uninterrupted global equatorial (Tethyan) surface currents [Kennett, 1982; Schmitz, 1987, Baldauf and Barron, 1990]. During the Oligocene this global equatorial circulation became more and more restricted in the Indian Ocean as the Australian land mass moved northward resulting in the decline of low-latitude opal accumulation rates. Meanwhile, during the Oligocene the expansion of Southern Ocean biosiliceous sedimentation occurred associated with polar cooling and the beginning of the circum-Antarctic Current.

LATE MIDDLE MIOCENE TO LATE MIOCENE CHANGES IN OPAL SEDIMENTATION

Figure 10 compares the middle Miocene to early Pliocene (4-14 Ma) percent opal data and sediment accumulation rates in a south to north transect of the western Indian Ocean with the composite benthic foraminiferal oxygen isotope record from southwest Pacific DSDP Sites 588 and 590 [Kennett, 1986; Barton and Bloemendal, 1986 for Leg 90 stratigraphy]. The transect extends from Site 737, near the Polar Front on the Kerguelen Plateau to Site 722 on Owen Ridge in the northwestern Indian Ocean (Figure 4).

In comparing these records of opal sedimentation, one must consider the influence of the northward motion of the

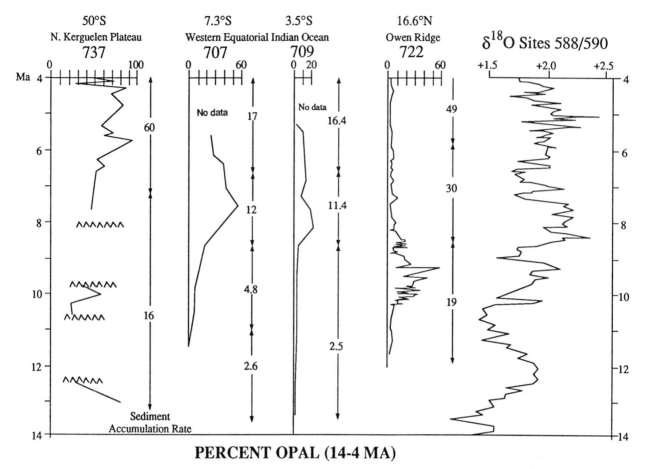

PERCENT OPAL (14-4 MA)

Fig. 10. Comparison of percent biogenic opal (%) and sedimentation accumulation rates (m/m.y.). From ODP Sites 737 (ODP Leg 119, Kerguelen Plateau); 707 and 709 (ODP Leg 115, western equatorial Indian Ocean), and 722 (ODP Leg 117, Owen Ridge) with a composite isotopic record from DSDP Sites 588 and 590 (DSDP Leg 90, western Pacific) for the interval from 14 - 4 Ma. Percent biogenic opal data is from Site 737 [Bohrmann and Ehrmann, 1991], Sites 707 and 709 (Hempel and Bohrmann, 1991) and Site 722 [Murray, unpublished data, 1992]. Isotope data is from Kennett [1986]. Sedimentation accumulation rates are based on interpretations of this study using data from ODP Legs 119 [Barron et al., 1991], Leg 117 [Spaulding et al., 1991]; and Leg 115 [Backman, Duncan et al., 1990]. Wavy lines represents an unconformity

Indian Plate on the dataset from Sites 707, 709 and 722. However, a rate of 3.7 cm/yr northward movement of the Indian Plate for the period after 45 Ma [R. Duncan, personal communication, 1991], suggests at most 520 km (or less than 5° of latitude) of northward displacement in the last 14 m.y.

Figure 10 shows that biosiliceous sedimentation was most likely continuous during the 14 to 4 Ma interval at Polar Front Site 737 with no major changes until the latest Miocene. The Site 737 record, however, is interrupted by hiatuses between 12.5 and 11.0 Ma and between 9.9 and 8.2 Ma [Barron et al., 1991].

Late Neogene opal sedimentation apparently began at about 11 Ma at tropical Site 707 and Owen Ridge Site 722 (Figure 10). Nigrini [1991] places the first appearance of radiolarians at 11.9 Ma in ODP Leg 117 sites in the northwest Indian Ocean, while Johnson (1990), argues for a 10 to 11 Ma onset of late Neogene biosiliceous sedimentation in the tropical Indian Ocean. According to Johnson [1990], silica is absent or sparsely present in

sediments of early and middle Miocene age at virtually all ODP Leg 115 sites, whereas silica is common and radiolarians are well preserved in upper Miocene sediments after 10 Ma.

Rather than relate this event to the onset of monsoonal upwelling, Johnson [1990] points out that a reversed record of biogenic silica preservation is present in the tropical and temperate Atlantic, and he suggests that the 10 to 11 Ma event reflects a change in deep water circulation between the Atlantic and the Indian oceans. Johnson [1990] cites Woodruff and Savin's [1989] isotope arguments, which call for the intensification of North Atlantic Deep Water (NADW) formation at 11 Ma and argue that a decline in biosiliceous sedimentation in the North Atlantic and an increase in the North Pacific and Indian surface water circulation to explain the sudden increase inoceans at 11 Ma is evidence of a change in deep-water circulation.

Burckle [1989], on the other hand, calls for a change in biosiliceous sediments in the tropical Indian Ocean between 10 and 11 Ma. In the modern northern Indian Ocean, Burckle [1989] notes that a hydrochemical boundary at 10-12°S separates nutrient-depleted waters to the south from highly productive waters to the north. He observes that diatom abundance is high north of this boundary, whereas south of the boundary diatom productivity is low and diatoms are depleted (or dissolved) in surface sediments. Burckle [1989] argues that since this hydrochemical boundary is maintained by the reversing monsoon system, the onset of the monsoon system should be recorded by the initiation of diatom occurrence in cores that were north of this hydrochemical boundary during the middle and late Miocene. Burckle [1989] cites the record of biosiliceous sedimentation at DSDP Site 238 to infer that the seasonally-changing monsoon gyre was initiated in the late middle Miocene at 10-11 Ma. Our studies place the onset of diatom deposition at Site 238 at about 11 Ma.

Percent opal increases at northwest Site 722 between about 10.4 and 9.3 Ma, corresponding to a period of high latitude cooling revealed by a positive oxygen isotope shift of about +0.5 per mil between 10.3 and 9.2 Ma (Figure 10). Presumably, upwelling increased in the northwest Indian Ocean region of Site 722 during this earliest late Miocene polar cooling event, but no changes are apparent at this time in the rather scantly opal records of western Indian Ocean Sites 707 and 709. A hiatus at Polar Front Site 737 removes most of this portion of the record.

A second positive oxygen isotope shift of about +0.4 per mil between 9.0 and 8.3 Ma coincides with increased opal percentages in Sites 707 and 709, suggesting an intensification of upwelling; however, opal percentages at more northern Site 722 decline at this time (Figure 10). Again, the presence of a hiatus between 9.9 and 8.2 Ma obscures the opal record at Site 737. The declining percent opal data at Site 722 seem to suggest that upwelling was decreasing in the northwestern Indian Ocean after 9 Ma; however, Kroon et al. [1991] argue that increased abundances of the planktonic foraminifer *Globigerina bulloides* beginning at 8.6 Ma signal either the beginning of the monsoonal upwelling system or strong intensification of monsoonal winds.

The sediment accumulation rate at Site 722 increases at about 8.6 Ma [Spaulding et al., 1991]; however, this increase is apparently restricted to the carbonate component [Murray, unpublished data, 1992]. The intensification of gyral circulation at this time may have caused an shoreward contraction of the region of high opal sedimentation in the northwest Indian Ocean to a pattern resembling that of the present day (Figure 2). Studies in the North Pacific indicate the intensification of gyral circulation there at about 8.6 Ma [Barron and Baldauf, 1990], corresponding perhaps to a period of major Antarctic glaciation [Ciesielski and Weaver, 1983].

Both percent opal and the sediment accumulation rates increase dramatically between 7 and 6 Ma at Site 737 (Figure 10). A major increase in diatom sedimentation at about 6 Ma is recorded throughout the Southern Ocean (see

ODP volumes 113, 114, 119, and 120) and presumably marks a northward shift in the Antarctic Polar Front in response to the latest Miocene polar cooling. Although the percent opal plots for the Sites 707 and 709 show decreasing values after 8 Ma (Figure 10), Mikkelsen [1990] records an abrupt increase in diatom abundance at Sites 707-710 and 712 between 7 and 6 Ma, and Schrader [1974] reports an increase in diatom abundance in Core 22 of Site 238, which corresponds to an age of about 6.7 Ma. The percent opal data for site 722 show little change between 7 and 6 Ma; however the sediment accumulation rate at this site increases from 30 to 49 m./m.y. at about 6.7 Ma, caused by an increase in the carbonate accumulation rate [Murray, unpublished data, 1992].

Upwelling and biogenic opal sedimentation increased dramatically in the eastern equatorial Pacific between 7 and 6 Ma [Shipboard Scientific Party, Leg 138, in press; Baldauf, unpublished data, 1992], so it is not unreasonable to expect an increase in upwelling in the equatorial Indian Ocean at the same time. Evidence for a slightly older onset of the monsoonal wind system comes from the isotopic record of paleosol carbonate in the northern Pakistan which Quade et al. [1989] interpret to show a major change in vegetation cover from forest to grassland at 7.4 Ma.

SUMMARY

Examination of the sediment (diatom abundance and percent opal) recovered from DSDP and ODP sites in the Indian Ocean provides a history of biosiliceous sedimentation in the Indian Ocean for the last 45 million years. These qualitative and quantitative datasets indicate the two distinct intervals (Eocene-Oligocene and late middle Miocene-late Miocene) of high biosiliceous sedimentation occur in the Indian Ocean. The intervening late Oligocene to middle Miocene interval is generally a period of low biosiliceous deposition.

The Eocene/Oligocene transition is most pronounced in the southern Indian Ocean where diatom abundances and opal values are low during the late Eocene. An abrupt increase in both the diatom abundance and in the percent opal composition at about 35.8 Ma most likely reflects increased productivity in the Southern Ocean resulting from destratification of the water column and increased polar cooling. Opal values remain moderate and fairly constant during the Oligocene through middle Miocene and increase to present-day values during the latest Miocene-early Pliocene in the Southern Ocean.

The second transition approximates the late middle Miocene to late Miocene interval (~11 Ma) and is most pronounced in the northwestern Indian Ocean. The northwest Indian Ocean exhibits moderate biosiliceous sedimentation during the Eocene with a gradual decline in such sedimentation during the Oligocene through early middle Miocene. During the late middle Miocene (~11.0 Ma) both diatom abundance and percent opal values increase, most likely reflecting increased surface productivity resulting from the establishment of the seasonal monsoonal system.

Acknowledgments. The authors thank the organizors of the Cardiff Workshop for a well structured and effective meeting. The authors acknowledge financial support from DFG, ODP, and USSAC to assist in travel expenses. P. Hempel appreciates fruitful discussion with J. Bohrmann. Recognition is also due to Doris Cooley for typing the manuscript. The manuscript was reviewed by Ted Moore and Jean-Pierre Caulet.

REFERENCES

Backman, J., Duncan, R., et al., *Proc. ODP, Init. Repts., 115* , Ocean Drilling Program, College Station, TX, 1989.

Baldauf, J. G., Middle Eocene through early Miocene diatom evolution, in *Eocene-Oligocene Climatic and Biotic Evolution*, edited by W. Berggren and D. Prothero, 310-326, Princeton University Press, 1992.

Baldauf, J. G., and Barron, J. A., Diatom biostratigraphy: Kerguelen Plateau and Prydz Bay regions of the Southern Ocean, in *Proc. ODP, Sci. Results, 119*, edited by J. A. Barron, B. Larsen et al., 547-598, Ocean Drilling Program, College Station, TX, 1991.

Baldauf, J. G., and Barron, J. A., Evolution of biosiliceous sedimentation patterns-Eocene through Quaternary: Paleoceanographic response to polar cooling, in *Geological History of the Polar Oceans: Arctic versus Antarctic*, edited by U. Bleil, J. Thiede, 575-607, Kluwer Academic Publishers, 1990.

Barron, J. A., and Baldauf, J. G., Tertiary cooling steps and paleoproductivity as reflected by diatoms and biosiliceous sediments, in *Productivity of the Ocean: Present and Past, Dahlem Workshop Repts.*, edited by W. Berger, V. Smetack, and G. Wefer, 341-354, Wiley, New York, 1989.

Barron, J. A., and Baldauf, J. G., Development of biosiliceous sedimentation in the North Pacific during the Miocene and early Pliocene, in *Pacific Neogene Events-Their Timing, Nature and Interrelationship*, edited by R. Tsuchi, 43-63, University of Tokyo Press, Tokyo, 1990.

Barron, J. A., Baldauf, J. G., Barrera, E., Caulet, J. P., Huber, B. T., Keating, B. H., Lazarus, D., Sakai, H., Thierstein, H. R., and Wei, W., Biochronologic and magnetochronologic synthesis of Leg 119 sediments from the Kerguelen Plateau and Prydz Bay Antarctica, in *Proc. ODP, Sci. Results, 119*, edited by J. A. Barron, B. Larsen et al., 813-849, Ocean Drilling Program, College Station, TX, 1991.

Barron, J. A., Larsen, B., Baldauf, J. G. et al., *Proc. ODP. Init. Results, 119*, 942p., Ocean Drilling Program, College Station, TX, 1989.

Barrera, E., and Huber, B., Paleogene and early Neogene oceanography of the southern Indian Ocean: Leg 119 foraminifer stable isotope results, in *Proc. ODP, Sci. Results, 119*, edited by J. A. Barron, B. Larsen et al., 693-718, Ocean Drilling Program, College Station, TX, 1991.

Barton, C. E. and Bloemendal, J., Paleomagnetism of sediments collected during Leg 90, southwest Pacific, in *Init. Repts., DSDP, 90 (2)*, edited by J. P. Kennett, C. C. von der Borch et al., 1273-1316, U.S. Government Printing Office, Washington, D. C., 1986.

Berger, W. H., Biogenous deep-sea sediments: Fractionation by deep sea circulation, *Geol. Soc. Am. Bull. 81*, 1385-1402, 1970.

Berggren, W. A., Kent, D. U., Flynn, J. J., and Van Couvering, J. A., Cenozoic geochronology, *Geol. Soc. Am. Bull., 96*, 1407-1418, 1985.

Bohrmann, G., and Ehrmann, W.U., Analysis of sedimentary facies using bulk mineralogic characteristics in Cretaceous to Quaternary sediments from the Kerguelen Plateau: Sites 737, 738 and 744, in *Proc. ODP, Sci. Results, 119*, edited by J. A. Barron, B. Larsen et al., 211-223, Ocean Drilling Program, College Station, TX, 1991.

Bostrom, K., Joensuu, O., Valde, S., and Riera, M., Geochemical history of south Atlantic Ocean sediments since Late Cretaceous, *Mar. Geol., 12*, 85-121, 1972.

Breza, J. R., and Wise, S. W., Jr., Lower Oligocene ice-rafted debris on the Kerguelen Plateau: evidence for East Antarctic continental glaciation, in *Proc. ODP, Sci. Results, 120*, edited by S. W. Wise Jr., R. Schlich et al., 151-178, Ocean Drilling Program, College Station TX, 1992.

Burckle, L. H., Distribution of diatoms in sediments of the northern Indian Ocean: relationship to physical oceanography, *Mar. Micropaleontol. 15(1/2)*, 53-65, 1989.

Calvert, S. E., Deposition and diagenesis of silica in marine sediments, *Spec. Publ., Inter. Assoc., Sediment.*, 273-299, 1974.

Caulet, J. P., La silice biogene dans les sediments Neogenes et Quaternaires de l'ocean Indien austral., *Bull. Soc. Geol. France 7*, 1019-1030, 1977.

Caulet, J. P., Sedimentation biosiliceuse Neogene et Quaternaire dans l'ocean Indien, *Bull Soc., Geol., Fr., 4*, 577-583, 1978.

Ciesielski, P. F., and Weaver, F. M., Neogene and Quaternary paleoenvironmental history of Deep Sea Drilling Project Leg 71 sediments, southwest Atlantic Ocean, in *Init. Repts., DSDP, 71*, edited by W. J. Ludwig, V. A. Krasheninnikov et al., 461-477, U.S. Government Printing Office, Washington, D.C., 1983.

Distanov, U. G., Kopejkin, V. A., Kuznetsova, T. A., and Silantiev, V. N., Partcularites de l'accumulation de sliice dans les bassins marins au cours du Mesozoique et du Cenozoique, *Dokl. Academic Nauk., 201 (3)*, 668-671, 1971.

Ehrmann, W.U., Implications of sediment composition on the southern Kerguelen Plateau for paleoclimate and depositional environment, in *Proc. ODP, Sci. Results, 119*, edited by J. A. Barron, B. Larsen et al., 185-210, Ocean Drilling Program, College Station, TX , 1991.

Ehrmann, W. U., and Grobe, H., Cyclic sedimentation at Sites 745 and 746, in *Proc ODP, Sci. Results, 119*, edited by J. A. Barron, B. Larsen et al., 225-237, Ocean Drilling Program, College Station, TX, 1991.

Ehrmann, W. U., and Thiede, J., History of Mesozoic and Cenozoic sediment fluxes to the North Atlantic Ocean., in *Contributions to Sedimentology, 15*, edited by Fuchtbauer, A. P. Lisitzyn, J. D. Milliman, and E. Seibold, 1-109, Schweizerbart, Stutgartt, 1985.

Ehrmann, W. U., Grobe, H., and Futterer, D., Late Miocene to Holocene glacial history of east Antarctica by sediments from Sites 745 and 746, in *Proc. ODP, Sci., Results, 119*, edited by J. A. Barron, B. Larsen et al., 239-260, Ocean Drilling Program, College Station, TX, 1991.

Eisma, D., and Van der Gaast, S. J., Determination of opal in marine sediments by X-Ray diffraction, *Neth. J. Sea Res., 5*, 382-389, 1971.

Fenner, J., Diatoms in Eocene and Oligocene sediments off NW Africa, their stratigraphic and paleoceanographic occurrences (PhD thesis), Univ. Kiel, FRG, 1982.

Fenner, J., Eocene-Oligocene planktic diatom stratigraphy in the low latitudes and the high southern latitudes, *Micropaleontol., 30*, 319-342, 1984.

Fenner, J., and Mikkelsen, N., Eocene-Oligocene diatoms in the western Indian Ocean: Taxonomy, stratigraphy, and paleoceanography, in *Proc. ODP, Sci., Results, 115*, edited by R. Duncan, J. Backman et al., 433-463, Ocean Drilling Program, College Station, TX, 1990.

Fisher, R. L., Bunce, E. T. et al., *Init. Repts., DSDP, 24*, U.S. Government Printing Office, Washington, D.C., 1183p, 1974.

Fourtainer, E., Diatom biostratigraphy of equatorial Indian Ocean Site 758, in *Proc. ODP, Sci. Results, 121*, edited by J. Weissel, J. Peirce, E. Taylor J. Alt et al., 189-210, Ocean Drilling Program, College Station, TX, 1991.

Harwood, D., and Maruyama, T., Middle Eocene to Pleistocene diatom biostratigraphy of Southern Ocean sediments from the Kerguelen Plateau, Leg 120, in *Proc. ODP, Sci., Results, 120*, edited by S. W. Wise Jr., R. Schlich et al., 683-733, Ocean Drilling Program, College Station, TX, 1992 .

Hempel, P., and Bohrmann, G., Carbonate free sediment components and aspects of silica diagenesis at Sites 707, 709 and 711, in *Proc. ODP, Sci., Results, 115*, edited by R. A. Duncan, J. Backman, L. Peterson et al., 677-698, Ocean Drilling Program, College Station, TX, 1990.

Johnson, D.A., Radiolarian biostratigraphy in the central Indian Ocean, Leg 115, in *Proc. ODP, Sci. Results, 115*, edited by R. A. Duncan, J. Backman, L. Peterson et al., 395-409, Ocean Drilling Program, College Station, TX, 1990.

Kennett, J. P., *Marine Geology*, 813p, Prentice Hall, Inc, Englewood Cliffs, N. J., 1982.

Kennett, J. P., Miocene to early Pliocene oxygen and carbon isotope stratigraphy of the southwest Pacific, in *Init. Repts, DSDP, 90*, edited by J. P. Kennett, C. C. von der Borch et al., 1383-1411, U.S. Government Printing Office, Washington, D.C., 1986.

Kroon, D., Steens, T., and Troelstra, S. R., Onset of monsoonal related upwelling in the western Arabian Sea as revealed by planktonic foraminifers, in *Proc. ODP Sci. Results 117*, edited by W. Prell, N. Niitsuma et al., 257-263, Ocean Drilling Program, College Station, TX, 1991.

Leclaire, L., Late Cretaceous and Cenozoic pelagic deposits. Paleoenvironmental and paleoceanography of the central western Indian Ocean, in *Init. Repts. DSDP, 25*, edited by E. Simpson, R. Schlich et al., 481-512, U.S. Government Printing Office, Washington, D.C., 1974.

Leclaire, L., Les depots siliceous neogenes des oceans actuels: produits de l'evollution climatique terrestre. *Bull. Soc. Geol. France, XX(4)*, 559-568, 1978.

Leinen, M., Biogenic silica accumulation in the central equatorial Pacific and its implication for Cenozoic paleoceanography, *Geol. Soc. Am. Bull. 90*, 1310-1376, 1979.

Lisitzin, A. P., Sedimentation in the World Ocean, *Soc. Ec. Paleontol. Mineral., Special Publ.., n 17*, 218p, Tulsa, 1972.

McKenzie, D. P., and Sclater, J. G., The evolution of the Indian Ocean since the late Cretaceous, *Geophys. J. R., Astron. Soc. 24*, 437-528, 1971.

Mikkelsen, N., Cenozoic diatom biostratigraphy and paleoceanography of the western equatorial Indian Ocean, in *Proc. ODP, Sci., Results, 115*, edited by R. A. Duncan, J. Backman, L. Peterson et al., 411-432, Ocean Drilling Program, College Station, TX, 1990.

Mikkelsen, N., and Swart, P., Preservation patterns of Cenozoic biogenic silica in the tropical Indian Ocean, *Terra Nova*, in press, 1992.

Murray, D., and Prell, W., Pliocene to Pleistocene variations in calcium carbonate, organic carbon, and opal on the Owen Ridge, northern Arabian Sea, in *Proc. ODP Sci. Results, 117*, edited by W. Prell, N. Niitsuma et al., 343-364, Ocean Drilling Program, College Station TX, 1991.

Mortlock, R.A., and Froelich, P.N., A simple method for the rapid determination of biogenic opal in pelagic marine sediments, *Deep Sea Res. Part A, 36*, 1415-1426, 1989.

Schmitz, B., Barium, equatorial high productivity and the

northward wandering of the Indian Continent, *Paleoceanography 2(1)*, 63-77, 1987.

Schrader, H. J., Cenozoic marine planktonic diatom stratigraphy of the tropical Indian Ocean, *Init. Repts. DSDP, 24*, 887-967, U.S. Government Printing Office, Washington, D.C., 974.

Simpson, E. S. W., Schlich, R. et al., *Init. Repts. DSDP, 25* U.S. Government Printing Office, Washington, D.C., 1974.

Spaulding, S. A., Bloemendal, J., Hayashida, A., Hermelin, J. O. R., Kameo, K., Kroon, D., Nigrini, C. A., Sato, T., Steens, T. N. F., Takayama, T., and Troelstra, S. R., Magnetostratigraphic and biostratigraphic synthesis, Leg 117, Arabian Sea, in *Proc. ODP Sci. Results 117*, edited by W. Prell, N. Niitsuma et al., 127-145, Ocean Drilling Program, College Station, TX, 1991.

Udintsev, G. B., Geological-Geophysical atlas of the Indian Ocean, *Publ. Acad. Sc. U.S.S.R., 1*, 151p, Peragon Press, 1975.

von der Boch, C. C., Sclater, J. G., *Init. Repts. DSDP, 22*, 890p. U.S. Government Printing Office, Washington, D.C., 1974.

Wei, W., Calcareous nannofossil stratigraphy and reassesment of the Eocene glacial record in subantarctic piston cores of the southeast Pacific, in *Proc. ODP., Sci. Results, 120*, edited by S. W. Wise Jr., E. Schlich et al., Ocean Drilling Program, College Station TX, 1991.

Whitmarsh, R. B., Weser, O. E., Ross, D. A. et al., *Init Repts, DSDP, 23*, 1177p., U.S. Government Printing Office, Washington, D.C., 1974.

Woodruff, F., and Savin, S., Miocene deepwater oceanography, *Paleoceanography 4(1)*, 87-140, 1989.

Wyrtki, K., Physical oceanography of the Indian Ocean, in *The Biology of the Indian Ocean*, edited by B. Zeitzschel, and S. A. Gerlasch, 18-326, Springer-Verlag, New York, 1973.

Zachos, J. C., Berggren, W. A., Aubry, M. P., and Mackensen A., Isotope and trace element geochemistry of Eocene and Oligocene foraminifers from Site 748, Kerguelen Plateau, in *Proc. ODP, Sci., Results, 120*, edited by S. W. Wise Jr., R. Schlich et al., Ocean Drilling Program, College Station, TX, 1992.

Nigrini, C., Composition and biostratigraphy of radiolarian assemblages from an area of upwelling (northwestern Arabian Sea, Leg 117), in *Proc. ODP Sci. Results 117*, edited by W. Prell, N. Niitsuma et al., 89-126, Ocean Drilling Program, College Station, TX, 1991.

Prell, W., Niitsuma, N., et al., *Proc. ODP Sci. Results 117*, Ocean Drilling Program, College Station, TX, 1989.

Prell, W. L., Hutson, W. H., Williams, D. F., Be, A. W. H., Geitzenauer, K., and Molfino, B., Surface circulation of the Indian Ocean during the last glacial maximum, approximately 18,000 yr. B.P., *Quat. Res., 14*, 309-336, 1980.

Quade, J., Cerling, T. E., and Bowman, J. R., Development of Asian Monsoon revealed by marked ecological shift during the latest Miocene in northern Pakistan, *Oceanol. Acta, 4*, 91-91, 1989.

Quasim, S. Z., Biological productivity of the Indian Ocean, *Indian J. Mar. Sci. 6*, 122-137, 1982.

Paleogene and Early Neogene Deep Water Paleoceanography of the Indian Ocean as Determined from Benthic Foraminifer Stable Carbon and Oxygen Isotope Records

JAMES C. ZACHOS[1] AND DAVID K. REA

Department of Geological Sciences, University of Michigan, Ann Arbor, MI, 48109-1063, USA

KOJI SETO

Institute of Geology and Mineralogy, Hiroshima University, Hiroshima, 730 Japan

RITSUO NOMURA

Department of Earth Sciences, Shimane University, Matsue, 690 Japan

NOBUAKI NIITSUMA

Institute of Geosciences, Shizuoka University, Shizuoka, 422 Japan

[1]*Present address: Dept. of Earth Sciences, University of California, Santa Cruz, CA , 95064, USA*

We have compiled Paleogene and early Neogene benthic foraminifer stable isotope records from 20 Indian Ocean DSDP and ODP sites. Subsidence histories and age models were developed for each site using nannofossil biochronology as the primary means of age control. With this information a continuous early Paleocene to late Miocene isotopic time series was assembled for the Indian Ocean. In addition, two-dimensional plots were constructed showing vertical and latitudinal oceanic isotopic distributions for five time-slices in the Paleogene and early Neogene.

This set of reconstructions provides a comprehensive record of Paleogene and early Neogene deep-water paleoceanography of the Indian Ocean. Important changes include gradual warming of deep waters from the late Paleocene (60 Ma) to early Eocene (55 Ma) followed by long-term, step wise cooling through the Eocene. During the warm early Eocene, the deep Indian Ocean was essentially isothermal. However, as cooling progressed during the middle Eocene the basin became thermally segregated as cool deep-waters entered from the south and warm shallow intermediate-waters entered from the north, possibly Warm Saline Tethyan Waters. The flux of cooler bottom waters, presumably from the marginal seas of Antarctica, intensified in early Oligocene time and then again in middle Miocene time. Warm saline waters were present at shallow intermediate depths in the northern half of this basin until at least mid-Miocene time. Carbon isotope records indicate that a component of North Atlantic deep water had established a presence in the Indian Ocean by late Miocene time.

INTRODUCTION

Our understanding of pre-historic climate change has rapidly improved since the inception of the Deep Sea Drilling Project (DSDP) over two decades ago. Deep-sea sediments recovered from the ocean floor have yielded a relatively detailed and nearly complete account of changes in the ocean environment over the last 100 m.y. Although evidence for reconstructing paleo-environments is derived from a variety of biogenic and non-biogenic sedimentary constituents, much is obtained from one relatively minor component of deep sediments, fossil benthic foraminifers. The chemical makeup and relative species abundances of benthic foraminifers provide a faithful record of ambient environmental conditions. As a result, by determining faunal and isotope and trace element compositions for benthic foraminera from various regions of the ocean, it has been possible to resolve the physical and chemical properties of ancient deep waters on a global scale. These records have in turn been extremely useful for reconstructing both large-scale patterns of deep-water circulation and long-term changes in ocean chemistry. Moreover, foraminifer stable carbon and oxygen isotope records have proven to be extremely useful for global correlations. As such, benthic foraminifers have become indispensable in the quest to define the climate history of Earth.

Although the more recent paleoceanographic record has been reconstructed from study of globally distributed sedimentary sequences, the record of pre-Neogene climates has been based largely on information derived from

Synthesis of Results from Scientific
Drilling in the Indian Ocean
Geophysical Monograph 70

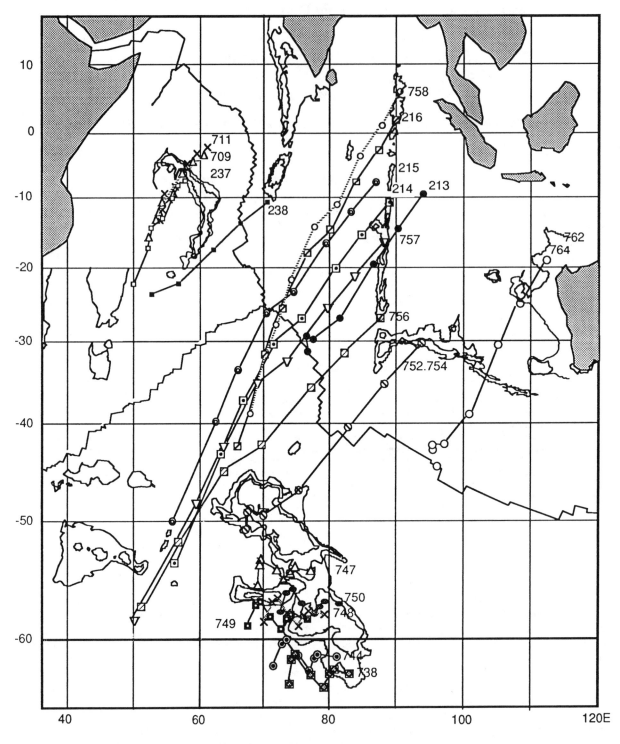

Fig. 1. Present day and backtracked paleocoordinates of Indian Ocean DSDP and ODP sites. Traces show migration path of each site. Symbols represent 0, 10, 20 , 35, 45, 55, 60, and 66 Ma [Royer and Sandwell, 1989; Royer and Chang, 1991; Royer and Coffin, 1992].

Atlantic deep-sea sequences [e.g. Miller et al., 1987a]; only limited numbers of complete Paleogene sequences were recovered from the Pacific and Indian oceans during the DSDP. And those few sections from the Indian Ocean typically suffered from poor recovery and hiatuses.

Consequently, the prevalent global patterns and mode(s) of pre-Neogene deep water circulation are still poorly understood.

One goal of the Ocean Drilling Program (ODP) has been to improve the spatial resolution of pelagic sequences for paleoclimatic reconstructions, particularly in climatically "sensitive" regions. To this end, drilling was conducted in several critical regions of the Indian Ocean including previously unexplored high latitudes regions, Prydz Bay and Kerguelen Plateau (Legs 119 and 120). Drilling was also conducted along the Broken and Ninetyeast ridges (Leg 121), and in the high productivity area of the western equatorial Indian Ocean (Leg 115). Although hiatuses and poor recovery were still a problem, a sufficient number of continuous, undisturbed sections were recovered during these ODP legs to significantly improve the spatial and temporal distribution of Paleogene sedimentary sequences in the Indian Ocean. During the last three years a number of isotopic studies were undertaken on sedimentary sections recovered during these ODP legs [Woodruff et al., 1990; Barrera and Huber, 1991; Rea et al., 1991; Seto et al., 1991; Mackensen et al., 1992; Zachos et al., 1992a,b; unpublished manuscript, 1992]. These studies essentially quadrupled the amount of Paleogene and early Neogene stable isotopic data available and have provided new insight into the paleoceanography of the Indian Ocean.

In this report we have compiled published Paleogene and early Neogene benthic foraminifer isotopic records generated for twelve ODP and eight DSDP sites ranging from just north of the equator to 65°S latitude (Figure 1). Although present-day depths range from 1090 to 5600 m, the deepest paleodepth when backtracked to the early Paleogene is 3800 mbsf. Thus, our discussion of paleotemperature and paleocirculation patterns is limited in most intervals to ocean shallower than 3500 m. The age models for many of the sites are based on calcareous nannofossil datums placed within the framework of the geomagnetic polarity time scale of Berggren et al. [1985c]. At several sites Sr isotope stratigraphy provides additional chronostratigraphic control of Oligocene sequences. In a few cases magnetostratigraphy served as an independent check of our age assignments.

BACKGROUND AND PREVIOUS WORK

Earlier reconstructions of benthic foraminifer isotopic distributions have demonstrated that a number of important changes occurred in the physical and chemical characteristics of the deep ocean during the Paleogene and early Neogene [e.g., Shackleton and Kennett, 1975; Savin, 1977; Shackleton et al., 1984; Oberhänsli and Toumarkine, 1985; Keigwin and Corliss, 1986; Miller et al., 1987a; 1987b; 1988; Woodruff and Savin, 1989; 1991; Kennett and Stott, 1990]. Of these changes the most notable was the long-term cooling in deep sea temperatures. Oxygen isotope records show that the early Paleogene ocean was relatively warm, with bottom water temperatures ranging from 10 to 11°C in the late Paleocene to as high as 14 to 16°C by early Eocene time. In contrast, the Oligocene and Miocene ocean was cooler, with deep water temperatures of less than 6 to 8°C. The overall transition from warm to cool took place over 10 m.y. proceeding in steps with the largest step occurring in the

earliest Oligocene time. This event marked the initial appearance of large ice-sheets on Antarctica [Miller et al., 1987a]. Another major step toward cooler temperatures and greater ice-volume occurred in middle Miocene time [Savin et al., 1981; Shackleton, 1984; Vincent et al., 1985; Woodruff and Savin, 1989; Wright et al., 1991].

This record of Paleogene deep water temperature change has generated considerable speculation on the nature of Paleogene and early Neogene global climate conditions and deep water circulation, especially during the warm intervals. Based on the existence of high(er) deep water temperatures, it has been suggested that for much of the Paleogene the rate of deep water formation at high latitudes might have been less than that of today. Instead, deep water might have formed in low-latitude, high salinity marginal seas [e.g. Matthews and Poore, 1980; Shackleton and Boersma, 1981; Brass et al., 1982; Woodruff and Savin, 1989; Kennett and Stott, 1991]. The exact locations of these hypothetical low latitude source regions are still unknown, although it has been suggested that conditions in the western Tethys may have been suitable for deep water formation [Berggren and Hollister, 1977; Woodruff and Savin, 1989; Barron and Peterson, 1990].

Such speculation on the nature of early Cenozoic deep water formation has, in part, been supported by isotopic data. Recent reconstructions of Cenozoic intermediate and deep water carbon isotope distributions demonstrate that the modern pattern of thermohaline circulation, with deep water formation in the Antarctic and North Atlantic, evolved only after the early late Miocene, and that prior to this time North Atlantic deep water production was either nonexistent or intermittent [Woodruff and Savin, 1989; Wright et al., 1991; 1992; Nomura et al., 1991b]. The reconstructions further suggest that during the early Miocene Tethyan/Indian saline waters flowed at intermediate depths southward to Antarctica where they cooled and became a major component of deep waters [Woodruff and Savin, 1989]. Other evidence of unusual early Cenozoic deep water circulation patterns includes a recent isotopic study of two vertically offset sites at Maud Rise in the south Atlantic which showed a slight thermal inversion of deep water masses during much of the Eocene and Oligocene reflecting the possible presence of warmer, more saline bottom waters [Kennett and Stott, 1990; 1991].

If warm, saline waters were indeed a major component of intermediate and deep waters at any time during the Paleogene, as suggested for the early Miocene, it is very probable that these waters flowed from somewhere in the sub-tropical Indian Ocean or Tethys [e.g. Tethyan Indian Saline Water (TISW), Woodruff and Savin, 1989]. If so, evidence for such a flux would exist in sediments of the Indian Ocean. Prior to recent drilling, however, only a few low-resolution Paleogene isotopic records existed for the Indian Ocean [Oberhansli, 1986] making it virtually impossible to create meaningful reconstructions of isotopic distributions in this basin.

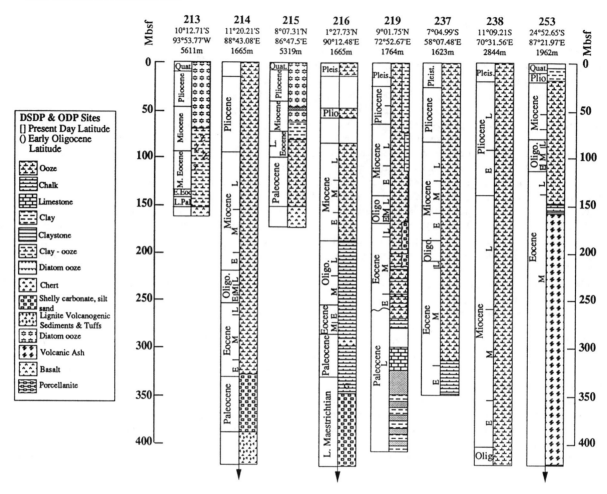

Fig. 2. Coordinates, water depths, chrono- and lithostratigraphies of each Indian Ocean DSDP and ODP site considered in this study.

INDIAN OCEAN SITE LOCATIONS, LITHOSTRATIGRAPHY AND
STABLE ISOTOPES

Isotopic records from ten ODP (Legs 115, 119, 120 and 121) and eight DSDP sites (Legs 22, 23, 24, 28) were compiled for this study. Present day coordinates and water depths, as well as complete litho- and chronostratigraphies for each site are provided in Figure 2 and Table 1. Here we furnish additional details concerning regional settings, lithostratigraphy, depositional history, and the nature and primary sources of the stable isotope records compiled herein.

Northwestern Indian Ocean and Arabian Sea

Holes 709A and 709B were drilled during Leg 115 in the western equatorial Indian Ocean near the summit of Madingley Rise at a water depth of 3041 m. The recovered sequences consist mainly of middle Eocene to Pleistocene deep water nannofossil chalk and ooze (Figure 2). Although Eocene and Oligocene sediments show evidence of extensive reworking [Premoli-Silva and Spezzaferri, 1990], Miocene sediments appear to be undisturbed and relatively complete with the exception of a brief hiatus in

the middle Miocene (16 to 17 Ma) [Rio et al., 1990]. Miocene sedimentation rates varied from 2 to 10 m/m.y. A detailed benthic foraminifer stable isotope record was generated for the 125 m thick Miocene portion of this sequence by Woodruff et al. [1990] (Figure 3). The most conspicuous feature of the two-hole, composite record, based mainly on analyses of *Cibicidoides* spp., is a 1.0‰ increase in $\delta^{18}O$ in the middle Miocene.

Site 219, drilled during Leg 23, is located on the crest of the Chagos-Laccadive Ridge to the southwest of India at a water depth of 1764 m. This site contains a 100 m sequence of middle Eocene to upper Oligocene chalk and ooze [Whitemarsh, Weser, Ross et al., 1974]. Major unconformities were recognized at the middle and upper Eocene boundaries, and a minor unconformity at the Eocene/Oligocene (E/O) boundary. A middle Eocene to late Oligocene stable isotope record was constructed for this site [Keigwin and Corliss, 1986]. Two species of benthic foraminifera, *Oridorsalis tener* and *Cibicidoides ungerianus*, were analyzed. Unlike other locations, the benthic foraminifer do not show a distinct ^{18}O enrichment in the early Oligocene, supporting the possibility of a

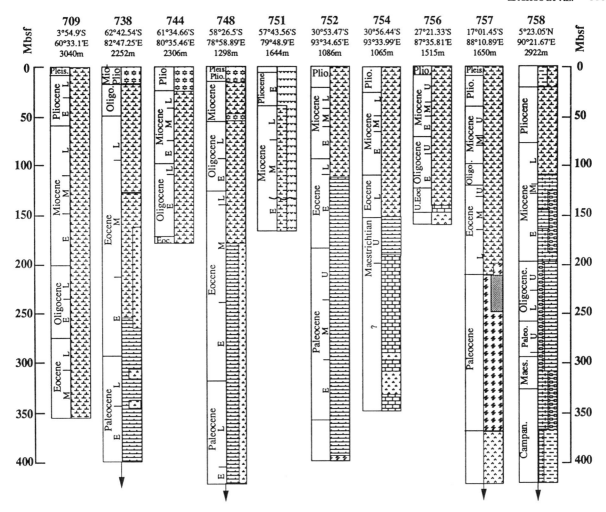

short hiatus spanning the E/O boundary [Keigwin and Corliss, 1986].

Sediments of Miocene age were also recovered at two DSDP sites from the northwest Indian Ocean: Site 237, located on the Mascarene Plateau at a water depth of 1623 m, and Site 238, located near the southern end of the Chagos-Laccadive Ridge in the Argo Fracture Zone at a water depth of 2844 m [Fisher, Bunce et al., 1974]. The Miocene sequence at both sites consists of nannofossil ooze, approximately 100 m thick at Site 237 and 260 m at 238. Sediment accumulation, with the exception of the early Miocene at Site 237, was rapid and essentially uninterrupted at both sites. High resolution Miocene benthic foraminifer stable isotope records were developed for both sites [Vincent et al., 1985]. Isotopic analyses were mostly of *Oridorsalis umbonatus*. As at Site 709, a prominent ~1.0‰ $\delta^{18}O$ enrichment was recorded near the middle Miocene at both sites.

Southern Indian Ocean; Kerguelen Plateau

Kerguelen Plateau is comprised of two separate elevations, a southern and northern sector. The southern and much of the northern sector were formed in the late Albian as India

separated from Antarctica [Schlich, Wise et al., 1989]. Further enlargement of the northern sector might have occurred in the middle Eocene during rifting of Broken Ridge [Royer and Coffin, 1992]. During Legs 119 and 120, seven sites were drilled on the southern plateau [Barron, Larsen et al., 1989; Schlich, Wise et al., 1989]. The two highest latitude sites, 738 and 744, were drilled during Leg 119 on the southernmost flank of the Kerguelen Plateau at water depths of 2252 and 2306 m, respectively. A 360 m thick sequence of Paleogene chalk and nannofossil ooze was recovered at Site 738, and a 160 m sequence of late Eocene to early Pliocene nannofossil ooze was recovered at Site 744. Recovery rates were relatively high above the middle Eocene at both sites. Late Paleocene to early Oligocene sediment accumulation at Site 738 was essentially continuous, followed by a period of non-deposition to latest Miocene time. Sedimentation rates were high at this location, in excess of 11 m/m.y. for most of the Paleogene. As a hiatus formed at Site 738 during the Oligocene, deposition at Site 744 was continuous with average sedimentation rates of about 4.6 m/m.y. Traces of ice-rafted debris were recognized in upper Eocene and lower Oligocene sediments at Site 744 [Ehrmann, 1991; Hambrey

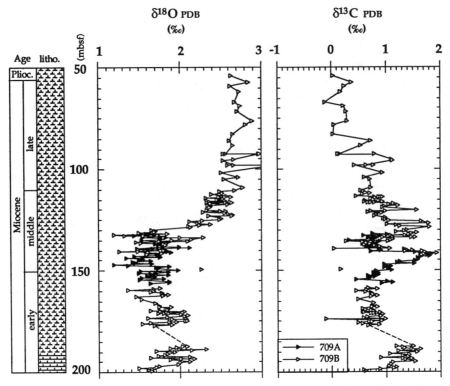

Fig. 3. Benthic foraminifer isotope record for Holes 709A and 709B plotted versus depth [Woodruff et al., 1990]. All values are from measurement of *Cibicidoides* spp. relative to PDB.

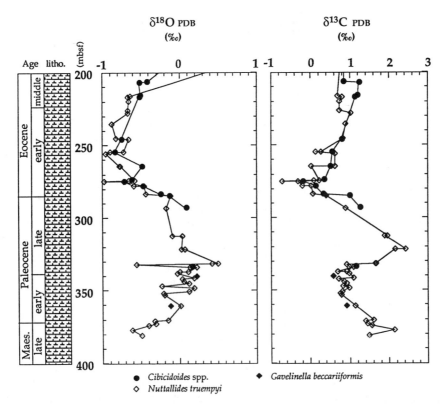

Fig. 4. Paleocene and Eocene benthic foraminifer stable isotope record of Hole 738C [Barrera and Huber, 1991]. Measured values relative to PDB.

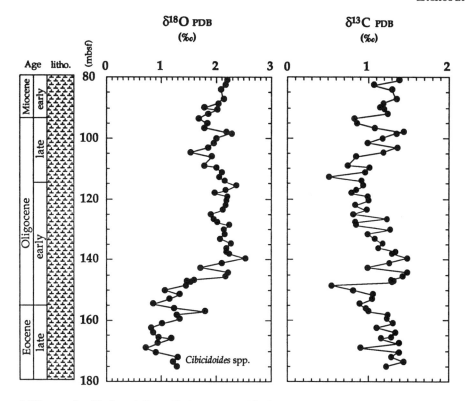

Fig. 5. Eocene and Oligocene benthic foraminifer stable isotope record for Hole 744A [Barrera and Huber, 1991].

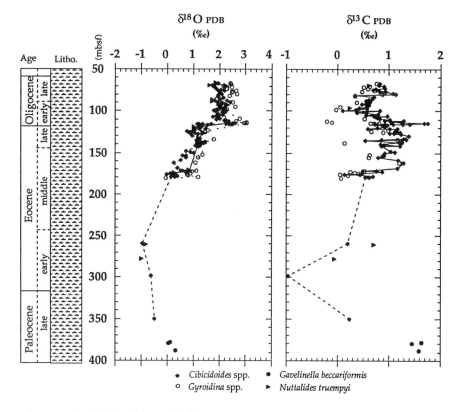

Fig. 6. Paleocene to Oligocene benthic foraminifer stable isotope record of Holes 748B and 748C [Zachos et al., 1992a].

et al., 1991]. Detailed benthic foraminifer isotopic records were produced for both sites, providing a nearly continuous record of Paleogene and early Neogene environmental change for the southern Indian Ocean (Figures 4 and 5) [Barrera and Huber, 1991].

Site 748 is located just north of the Leg 119 sites in the Raggatt Basin of the southern plateau at a water depth of 1298 m [Schlich, Wise et al., 1989]. Drilling at this location yielded a 380 m sequence of upper Paleocene to middle Miocene chalk and nannofossil ooze. Recovery was excellent in the upper part of the section but decreased below the middle Eocene where cherts were encountered. A thin horizon of ice-rafted debris was discovered in a lower Oligocene core [Breza and Wise, 1992]. Sedimentation rates ranged from 15 to 20 m/m.y in the late Paleocene and early Eocene to 5 m/m.y. in the late Eocene to middle Miocene. A high resolution benthic foraminifer stable isotopic record was constructed for the lower Eocene to lower Miocene portion of the sequence (Figures 6 and 7) [Zachos et al., 1992a; 1992b]. Foraminifera preservation was excellent over the entire interval. The most noteworthy feature of this isotopic record is an abrupt 1.6‰ increase in $\delta^{18}O$ within the lower Oligocene ice-rafted debris bearing interval. This increase at Site 748 was accompanied by a +0.8‰ excursion in $\delta^{13}C$.

Site 751 is located just a few hundred kilometers to the northwest of Site 748 at a water depth of 1644 m [Schlich, Wise, et al., 1989]. Drilling at this location produced an stratigraphically expanded 125 m Miocene sequence of nannofossil ooze with diatoms. Sedimentation rates at this location were relatively high during the Miocene, averaging about 15 to 20 m/m.y. Three hiatuses were recognized at Site 751, from 13.4 to 14.8 Ma, 12 to 12.8 Ma, and 15.8 to 16.2 Ma [Harwood et al., 1992]. A high resolution benthic foraminifer (Cibicidoides spp.) stable isotope record was built for the Miocene (Figure 8) [Mackensen et al., 1992]. Sampling resolution of this record was high, exceeding 1 sample/50 k.y. in most intervals. The most conspicuous feature of this record is a 1.2‰ $\delta^{18}O$ increase in the middle Miocene.

Broken Ridge

Sites 752 and 754 are located on Broken Ridge, at water depths of 1086 and 1065 m, respectively [Peirce, Weissel, et al., 1989]. The sedimentary sequence at Site 752 includes 220 m of Paleocene and lower Eocene nannofossil chalks and oozes unconformably overlain by 65 m of Miocene ooze. Paleocene sedimentation rates were very high, exceeding 25 m/m.y. while Neogene rates were low, less than 3.6 m/m.y. The Neogene at Site 754 is represented by a nearly continuous 100 m sequence of upper Oligocene, Miocene, and Pliocene ooze. Miocene sedimentation rates averaged 4.7 m/m.y. Benthic foraminiferal stable isotope records were constructed for both sites [Seto et al., 1991; Rea et al., 1991; Seto et al., unpublished data, 1992]. Several benthic taxa were analyzed including Stensionia beccarlliformis and Anomalinoides spp. through the Paleocene and lower

Eocene at Site 752 (Figure 9)(Seto et al., 1991), and Oridorsalis and Cibicidoides spp. through the Neogene at both sites (Figures 9 and 10)[Rea et al., 1991; Seto et al., unpublished data, 1992]. The Paleocene/Eocene carbon isotopic excursion and benthic foraminifer extinction event (latest Paleocene (CP8)) have been recognized at Site 752 [Nomura, 1991a]. Prominent oxygen and carbon isotope excursions are recorded in the middle Miocene intervals (CN5a-b) of both sites.

Ninetyeast Ridge

Ninetyeast Ridge is a major north-south lineament which extends nearly 5000 km from 34°S to 10°N latitude. Basement ages and burial depths increase northward indicating that the ridge formed near its southern end as a trace of the Kerguelen/Ninetyeast hotspot on the Indian Plate [von der Borch, Sclater et al., 1974]. A total of seven sites were drilled on the ridge during DSDP Legs 23, 26, and ODP Leg 121. The sedimentary records of these sites together with those drilled on Kerguelen Plateau and Broken Ridge provide a high resolution north to south transect of the Indian Ocean.

Three sites, 756, 757, and 758, were drilled on the Ninetyeast Ridge during ODP Leg 121 [Peirce, Weissel et al., 1989]. Two sites were drilled on the southern half of the ridge, Site 756 near the crest of the southernmost tip of the ridge at water depth of 1515 m, and Site 757 located approximately 11° further north at a water depth of 1650 m. Recovery rates were generally high at both sites; a 130 m thick sequence of upper Eocene to Pleistocene ooze was recovered in two holes at Site 756, and a 200 m thick sequence of lower Eocene to Pleistocene ooze was recovered from two holes at Site 757. Sedimentation at Site 756, albeit slow (<4 m/m.y.), appears to have proceeded uninterrupted from late Eocene to Pleistocene. At Site 757 sedimentation rates were moderate through the Eocene, averaging better than 5 m/m.y. Rates slowed to less than 2 m/m.y. during the Oligocene and early Miocene, but increased to 6.5 m/m.y. by late middle Miocene time .

Eocene to late Miocene benthic foraminifer stable isotope records were assembled for both sites (Figures 11 and 12)[Rea et al., 1991; Nomura et al., in press; Seto et al., unpublished data, 1992]. A number of benthic taxa were analyzed, including Cibicidoides, Gyroidina, Uvigerina, Oridorsalis, and Stilostomella. In general, preservation of foraminifers deteriorated from younger to older intervals. Some specimens from the lower Eocene at Site 757 were infilled and slightly overgrown with secondary calcite.

Site 758, the most northern of sites on Broken Ridge, was drilled at a water depth of 2922 m. Basement age at this site is estimated to be 80 Ma [Duncan, 1978]. The recovered sedimentary section consists of a relatively condensed and hiatus-ridden sequence of lower Paleogene chalks, unconformably overlain by a thick and continuous sequence of Neogene ooze. Sedimentation rates in the Miocene averaged 7 m/m.y. Despite the fact that much of the lower Paleocene and nearly the entire Eocene is absent at Site 758, a somewhat condensed (sedimentation rates

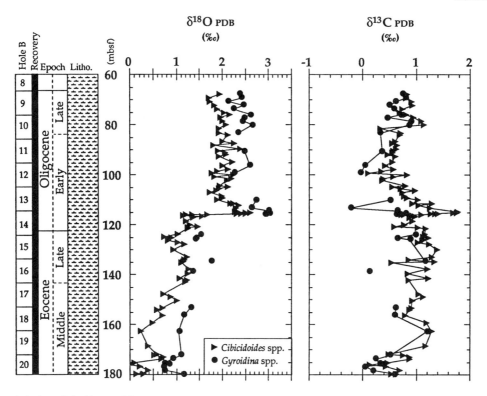

Fig. 7. Expanded view of the Eocene/Oligocene benthic foraminifer stable isotope record of Hole 748B [Zachos et al., 1992a; 1992b].

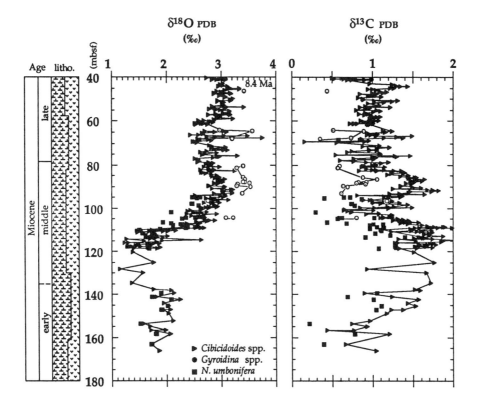

Fig. 8. Benthic foraminifer stable isotope record of Hole 751A [Mackensen et al., 1992].

TABLE 1. Neogene Nannofossil Zonations for Indian Ocean DSDP and ODP Sites

Neogene Datum	Zone (base) Okada and Bukry [1980]	Martini [1971]	Age (Ma)	Ref	DSDP and ODP Sites — Depth (mbsf) 214*	216*	237*	238*	709**	752***	754***	756***	757***	758***
FO Discoaster tamalis	CN11b		3.80	1							17.7-19.2	8.00-8.50	26.6-28.1	
C3N-2 top			4.10	1										56.20
C3R-2 top			4.24	1										58.70
C3N-3 top			4.40	1										61.50
C3R-3 top			4.47	1										63.05
C3N-4 top			4.57	1										64.00
FO Ceratolithus rugosus	CN10c	NN13	4.66	3					54.8-53.3					
C3R-4 top			4.77	1										66.70
FO Ceratolithus acutus	CN10b		4.85	4			82.5	138.3	57.8-56.3				38.2-39.7	
LO Discoaster quinqueramus	CN10a	NN12	5.26	2	92.9			144.4	63.0-60.8	19.1	22.7-25.3		42.8-44.3	
C3AN-1 top			5.35	1										76.40
C3AR-1 top			5.53	1										80.50
C3AN-2 top			5.68	1										82.70
LO Discoaster berggrenii			5.80	2					107.6-104.6					
C3AR-2 top			5.89	1										84.30
C3AN-3 top			6.37	1										90.20
C3AR-3 top			6.50	1										92.10
FO Amaurolithus primus	CN9b		6.70	4	119.4		100.0	186.3		25.1	31.8-34.9		52.5-54.0	83.1-84.6
FO Discoaster quinqueramus			7.46	4									62.2-63.6	102.4-103.9
FO Discoaster berggrenii	CN9a	NN11	8.00	3	133.3	111.0	111.1	214.6						
FO Discoaster neorectus	CN8b		8.50	1	133.3		120.6	218.3						
LO Discoaster hamatus	CN8a	NN10	8.67	4	152.7	119.0	124.8	268.4	111.2-109.1			34.70-35.67	71.2-71.8	116.6-118.1
FO Discoaster hamatus	CN7	NN9	10.50	4	162.0		130.0	280.6	114.2-112.7			37.30-38.80		
FO Catinaster coalitus	CN6	NN8	11.10	4	170.8	124.5	130.0	290.5	115.7-114.2					
FO Discoaster pentaradiatus			12.00	1							46.5-48.5			
LO Coccolithus floridanus	CN5b		13.10	1, 4	181.1	139.5	135.5				73.5-77.0		80.9-81.5	
LO Sphenolithus heteromorphus	CN5a	NN6	13.60	4	190.8		149.0			58.1		45.40-46.90	83.0-84.5	
FO Discoaster exilis			15.40	1							80.0-83.2		89.0-90.5	
LO Helicosphaera ampliaperta	CN4	NN5	16.00	1, 4		149.0	158.5	362.1						
FO Sphenolithus heteromorphus	CN3		18.40	4	201.0	158.3		386.3		73.1	86.7-89.7	54.70-56.20	95.7-97.2	
LO Sphenolithus belemnos		NN4	18.80	4										144.0-145.5
FO Sphenolithus belemnos	CN2		20.00	4	208.2	167.2	177.5	390.3			89.7-92.9		98.7-100.2	
FO Discoaster druggii	CN1c	NN2	23.60	4			187.0	395.6					100.2-100.8	
LO Dictyococcites bisectus	CN1a	NN1	23.70	1	221.2	195.4	193.0	453.9	201.1-198.1	91.7	107.1-108.6		100.8-102.3	195.3-196.8
LO Sphenolithus ciperoensis			23.70	1								73.70-74.70		
LO Chiasmolithus altus			28.20	1							116.8-113.8			
FO Sphenolithus ciperoensis	CP19a	NP24	30.20	1						93.4		94.20-96.10	105.3-106.8	218.3-219.8

Note: FO = first occurrence, LO = last occurrence. The references refer to the age column and represent (1) Berggren et al. [1985a,b,c]; (2) Gartner [1990]; (3) Rio et al. [1990]; (4) Backman et al. [1990].
* Vincent et al. [1985]; ** Rio et al. [1990]; *** Peirce, J., Weissel, J. et al. [1989].

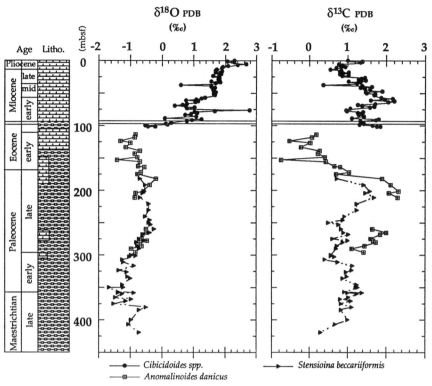

Fig. 9. Benthic foraminifer stable isotope record of Hole 752A [Rea et al., 1991; Seto et al., 1991].

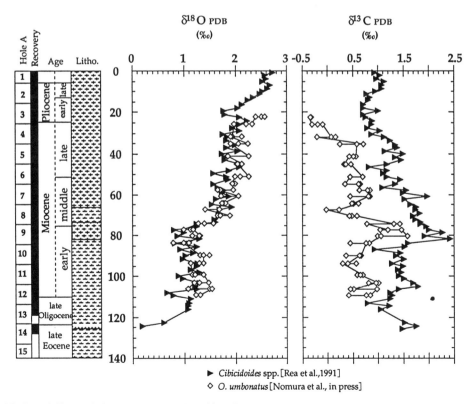

Fig. 10. Benthic foraminifer stable isotope record of Hole 754A [Rea et al., 1991; Nomura et al., in press].

<4 m/m.y.), but complete upper Paleocene interval was recovered here. The Neogene portion of the benthic foraminifer stable isotope record of Site 758 was constructed from *Cibicidoides* and *Oridorsalis* data (Figure 13)[Hovan and Rea, unpublished manuscript, 1992; Seto et al., unpublished data, 1992] and the middle and late Paleocene portion from *Nuttallides truempyi* (Zachos et al., unpublished manuscript, 1992]. Preservation of both Miocene and Paleocene foraminifers is excellent, while preservation of Oligocene foraminifers is generally poor.

Sites 214 and 216 were drilled during DSDP Leg 22 and are located on the Ninetyeast Ridge between Sites 757 and 758 at water depths of 1665 and 2922 m, respectively. Thick sections of Paleogene and Neogene oozes and chalks were penetrated at each site although core recovery was generally poor below the Oligocene. With the exception of a few minor hiatuses, the Oligocene and Miocene sequences appear to be complete. A high resolution Miocene isotopic record was established from analyses of *O. umbonatus* at both sites [Vincent et al., 1985]. As at other Miocene sections, an abrupt, greater than 1.0‰ $\delta^{18}O$ increase was found in the middle Miocene.

Site 253 is located on the southern portion of Ninetyeast Ridge just north of Site 756 at a water depth of 1962 m [Davies, Luyendyk et al., 1974]. It consists of a sequence of upper Eocene and Oligocene nannofossil ooze. A slight hiatus exists at the E/O boundary. A minor early Oligocene oxygen isotope increase was recorded in *Cibicidoides* [Oberhänsli, 1986].

Sites 213 and 215 were also drilled during Leg 22, in the Wharton and Argo basins to the east and west of the Ninetyeast Ridge at water depths of 5611 and 5319 m respectively. These sites were the two deepest considered in this study and, as such, are the only sites capable of providing a true deep water record. During most of the Cenozoic, however, both sites were positioned either below the lysocline or, on occasion, just slightly above [von der Borch, Sclater et al., 1974]. As a result, calcite sediment accumulation was minimal. The only carbonate-bearing intervals present at either site are short intervals of late Paleocene and early Eocene nannofossil ooze, bounded by red clays above and below (Figure 2). The nannofossil stratigraphies indicate that carbonate deposition at both sites was essentially continuous from the late Paleocene to mid-early Eocene [Gartner, 1974]. The late Paleocene benthic foraminifer extinction event was recognized at each site providing a precise time line for correlation [McGowran, 1974; L.D. Stott, personal communication, 1992]. Isotopic records were established for both sites based on measurement of *N. truempyi* [Hovan and Rea, 1992; Zachos et al., unpublished manuscript, 1992]. The Site 215 record, which extends further back into the Paleocene than that of Site 213, displays a nearly 3.0‰ decrease in $\delta^{13}C$ from the late Paleocene to early Eocene.

CHRONOSTRATIGRAPHY

A critical component of any global reconstruction is the chronostratigraphy. The age models developed for this synthesis are based mainly on published calcareous nannofossil datums (Tables 1 and 2)[Gartner, 1974; Roth, 1974; Thierstein, 1974; Rio et al., 1990; Wei, 1991; Wei and Thierstein, 1991; Pospichal et al., 1991a; Aubry, 1992; Wei et al., 1992], and, where available, magnetostratigraphies [Keating and Sakai, 1991; Inokuchi and Heider, 1992]. In addition, published radiogenic strontium (Sr) isotope stratigraphies were used to improve correlations of Oligocene sequences at several sites [Barrera et al., 1991; Pospichal et al., 1991b; Zachos et al., 1992a]. We attempted to be consistent in applying our age model by strictly adhering to the nannofossil/isotopic and benthic foraminifer extinction chronology wherever possible. In a few rare cases where nannofossil stratigraphy was not available or was considered unreliable, we employed other biozonations in constructing age models.

For each of the sites involved in this study, we obtained the initial shipboard generated nannofossil zonations as well as any subsequent revisions made by shore-based investigations (Tables 1 and 2). Standard Cenozoic nannofossil zonations were employed for most sites [Martini, 1971; Okada and Bukry, 1980; Berggren et al., 1985a,b,c]. Because many of the marker species of the standard zonations are absent in high latitude sequences, for southern ocean sites we applied new polar calcareous nannofossil zonations developed by Wei and Wise [1989, 1990, Wei and Thierstein [1991], and Wei et al. [1992]. We did not attempt to "fine tune" any of the age models using stable isotope records. An *a priori* assumption in applying this method is that most isotope excursions recorded in deep sea sequences are ubiquitous. In reality, this can only be assumed for some of the more prominent events which are long in duration relative to the mixing time of the ocean e.g., the early Oligocene and middle Miocene oxygen isotope events related to ice-volume changes. Short-term events can be used for fine tuning only if (1) it has unequivocally been demonstrated that such events are isochronous, and (2) if the sampling resolution of every site that is fine tuned is high enough to resolve that particular event. Failure to meet either criterion could lead to severe biases. Because many of the Paleogene records considered here lack adequate sampling resolution for fine tuning, we chose not to apply this method. Admittedly, application of stable isotope stratigraphy in some intervals would help improve chronostratigraphic correlations.

Nannofossil datums provide the primary age control for the Paleocene and Eocene since magnetostratigraphy over this interval was uninterpretable or unreliable at most sites. In addition, the late Paleocene benthic foraminifer extinction event [Tjalsma and Lohmann, 1983; Thomas, 1990; Nomura, 1991a] provides an additional time horizon. For the early Paleocene to middle Eocene interval, it was possible to use most standard biomarker species in constructing the age models of both low and high latitude sections. We found that the relative order of occurrences remained remarkably consistent from location

TABLE 2. Paleogene Nannofossil Zonations for Indian Ocean DSDP and ODP Sites

Paleogene Datum	Zone (base)		Age (Ma)	Ref.	DSDP and ODP Sites Depth (mbsf)											
	Okada & Bukry [1980]	Martini [1971]			213*	214*	215*	219**	253***	738†	744†	748††	752§	756§	757§	758§
LO Reticulofenestra bisecta	CP19b	NP25	24.0	1,3							100.2	67.0				196.0
LO Sphenolithus ciperoensis			25.2	3		220.5			86.7					74.2	101.3	
LO Chiasmolithus altus			25.5	1,3							106.9	73.5				
LO Reticulofenestra umbilica	CP17	NP23	33.0	3		247.5		165.8	101.0		127.1	104.2		118.4	116.0	
LO Isthmolithus recurvus			34.3	1					113.5		130.7	109.5				
LO Reticulofenestra oamauruensis			36.0	3						24.4	147.9	116.0				
LO Discoaster saipanensis	CP16a	NP21	36.7	1		257.0		175.3	116.1		161.6			135.6	124.0	
FO Reticulofenestra oamauruensis			38.0	3						35.4	167.0	125.7				
FO Isthmolithus recurvus	CP15b	NP19/20	38.7	3					130.3	39.9	169.6	127.2			129.5	
LO Reticulofenestra reticulata			38.7	3								128.8				
FO Chiasmolithus oamauruensis	CP15a	NP18	41.0	3					132.5	70.8		146.4				
LO Chiasmolithus solitus	CP14b	NP17	41.2	1,3		290.0			155.3	70.8		150.0				
FO Reticulofenestra reticulata			42.1	3						97.4		170.9				
FO Reticulofenestra umbilica	CP14a	NP16	44.6	3				246.0	151.5	119.2						
FO Nannotetrina fulgens	CP13	NP15	49.8	3						204.9						
FO Discoaster sublodoensis	CP12	NP14	52.6	1		314.0				227.0						
LO Tribrachiatus orthostylus	CP11	NP13	53.7	1												
FO Discoaster lodoensis	CP10	NP12/13	55.4	1	137.3	328.0	83.5			264.4			114.2			
LO Tribrachiatus contortus	CP9a	NP11	~56.3	1,4												
FO Discoaster diastypus	CP9	NP10	~56.8	1						283.9						
LO Fasciculithus spp.		NP10	57.6	4												
LO Benthic foram. extinction event			57.8	5	141.8		101.8									
FO Tribrachiatus bramlettei		NP10	57.6	1,4												
FO Discoaster multiradiatus	CP8	NP9	59.1	1,2	151.5					302.7			171.1			
FO Discoaster nobilis/H. riedelii	CP7	NP8	59.8	1,2			114.5									265.0
FO Discoaster mohleri	CP6	NP7	60.4	1,2			130.0			326.8			251.4			
CO Heliolithus kleinpellii	CP5	NP6	61.6	1,2			150.0			334.9						271.6
FO Fasciculithus tympaniformis	CP4	NP5	62.0	1									297.7			291.0
FO Prinsius martinii	CP3	NP4	63.7	1						360.7						
FO Chiasmolithus danicus	CP2	NP3	64.8	1						364.6						
FO Chiasmolithus tenuis	CP1b	NP2	65.9	1						376.4						
K/T boundary	CP1a	NP1	66.4	1												

The references pertain to the age column and represent: (1) Berggren et al. [1985a,b,c]; (2) Wei and Wise [1989]; (3) Wei and Wise [1990]; (4) Aubry et al. [18]; (5) Thomas [1990].
* Gartner [1974]; ** Boudreaux [1974]; *** Thierstein [1974];† Wei and Thierstein [1991];†† Wei and Wise [1991]; § Peirce, Weissel et al. [1989].

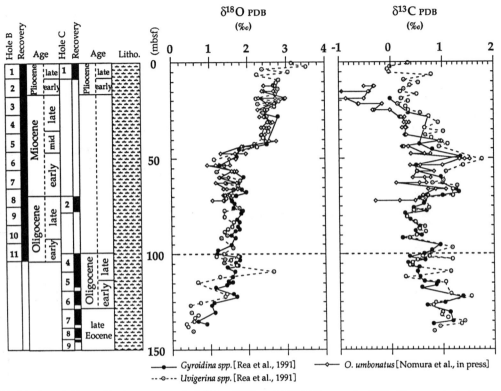

Fig. 11. Benthic foraminifer stable isotope record of Holes 756B and 756C [Rea et al., 1991; Nomura et al., in press]. Note the stratigraphic overlap between holes.

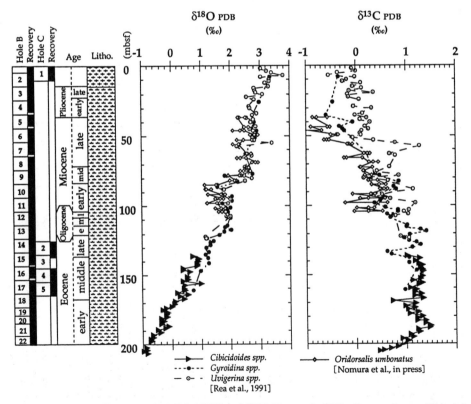

Fig. 12. Benthic foraminifer stable isotope record of Holes 757B and 757C [Rea et al., 1991; Seto et al., unpublished data, 1992].

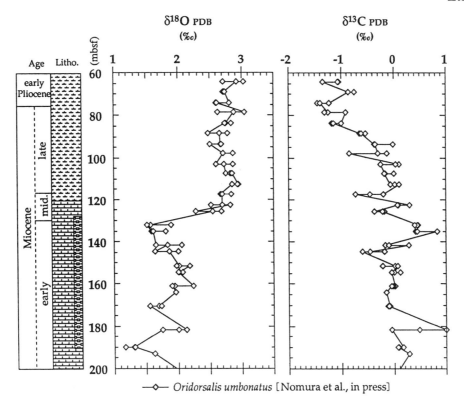

$\delta^{18}O$ PDB
(‰)

$\delta^{13}C$ PDB
(‰)

—◇— *Oridorsalis umbonatus* [Nomura et al., in press]

Fig. 13. Benthic foraminifer stable isotope record of Hole 758A [Seto et al., unpublished data, 1992].

to location.   However, in the middle Eocene to early Oligocene portion of the sections, because of increased provinciality and decreasing diversity of the high latitude floras, it was necessary to rely on alternative species for correlation [Wei, 1991; Wei and Thierstein, 1991].   The first and last occurrences (FO and LO) of these alternative species have been time calibrated at a few southern ocean sites with reliable magnetostratigraphy [Wei and Wise, 1989, 1990; Wei, 1991].

In order to enhance correlations of Oligocene sections, Sr-isotope age datums were considered in the age assignments of several sites.   Sr isotope stratigraphies were available for the four most complete Oligocene Indian Ocean sequences from sites 744, 748, 756 and 757 [Barrera et al., 1991; Pospichal et al., 1991b; Zachos et al., 1992a].   A numeric age was assigned to each Sr isotope data point using the Sr isotope age equation of Miller et al. [1988],

$$Ma = 20392.79 - 28758.84 \, (^{87}Sr/^{86}Sr) \qquad (1)$$

This equation was originally derived through calibration of the Site 522 Sr isotope and magnetic stratigraphies to the geomagnetic polarity time scale of Berggren et al. [1985c].   As such, Sr isotopes should provide a chronostratigraphy that is consistent with that obtained from magnetically time-calibrated nannofossil zonations.

Excluding a few anomalous data points, comparisons of the Sr isotope and calcareous nannofossil determined ages at the four sites show good overall agreement between the

two approaches [Pospichal et al., 1991b].   For example, at Sites 744 and 748 the age of the Sr isotope Eocene/Oligocene boundary is close to the last occurrence (LO) of *Reticulofenestra oamaruensis*, an alternative marker species for the boundary at high latitudes.   At lower latitude Site 756, the Sr isotope E/O boundary coincides with the LO of *Discoaster saipanensis*, the traditional marker species for the E/O boundary, indicating that the LO of these two species provide a reliable time horizon for the boundary.   The Sr isotope records from Holes 756B and 756C also revealed the presence of 6 m.y. of overlap in the Oligocene between the two holes, despite the apparent lack of overlap in coring depths [Pospichal et al., 1991b].

The Miocene age models are based primarily on the nannofossil occurrences listed in Table 1.   Although magnetostratigraphic records were comparatively more reliable in Miocene than in older sequences, in order to be consistent in the application of the chronostratigraphy, magnetostratigraphy was only used to improve correlations in a few special cases where age diagnostic nannofossils either were not present (e.g. Site 758) or yielded conflicting results.   Sr isotope stratigraphies might also help improve Miocene correlations, however, at the time of this writing few had been constructed for Indian Ocean sites.   For the Miocene intervals of Southern Ocean Sites 744 and 751 we relied on published age assignments [Barrera and Huber, 1991; Mackensen et al., 1992; Harwood et al., 1992].   These sites, however, were not included in the continuous time series comparisons since

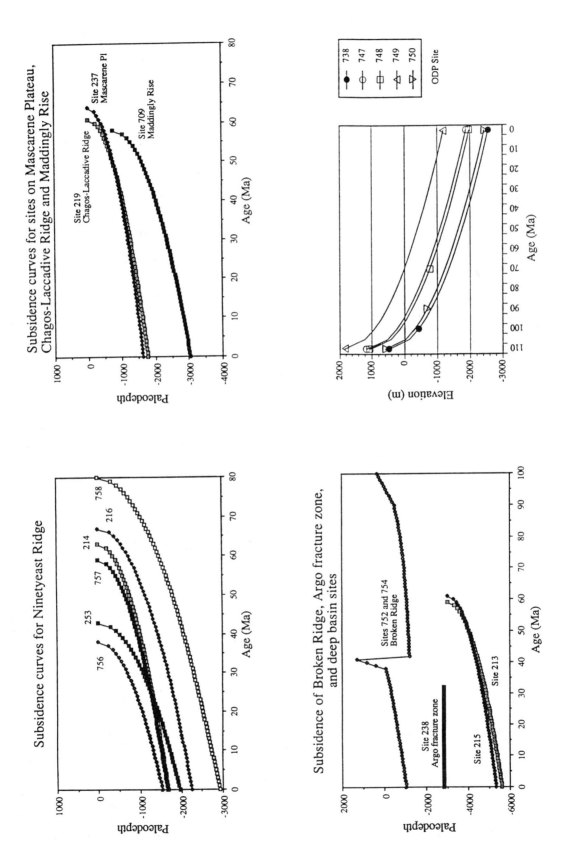

Fig. 14. Indian Ocean paleodepths calculated for sites located on Ninetyeast Ridge, Mascarene Plateau, Chagos-Laccadive Ridge, Maddingley Rise, Broken Ridge, and Argo fracture zone. The lower, right-hand graph represents the subsidence curves for Kerguelen Plateau sites as calculated by Coffin [1992].

application of calcareous nannofossil biostratigraphy to high latitudes still remains problematic for this particular interval of time.

SUBSIDENCE CALCULATIONS FOR INDIAN OCEAN DRILLSITES

Subsidence history of the drillsites were determined by using the relationship of increasing depth to the square root of age:

$$\Delta Z = kt^{1/2} \tag{2}$$

thus Z_t (depth at age t) $= Z_0$ (depth at age zero) $+ kt^{1/2}$, where Z is in kilometers, t is in millions of years, and k is the subsidence parameter (Figure 14)[Sclater et al., 1971; Davis and Lister, 1974; Rea and Leinen, 1986]. The accuracy of these sorts of calculations, when ages and initial depths are well constrained, is on the order of ±100 meters. The more poorly constrained locations are mentioned below; the age-depth history for Site 238 is the least understood. Van Andel [1975] may have been the first to utilize subsidence curves in deciphering the sedimentary history of the Indian Ocean.

For most sites the first task is to determine the subsidence parameter which differs slightly for each site [cf: Rea and Leinen, 1986] but for the "normal" Indian Ocean sites always fall into the range of 0.27 to 0.34. Next, the depth at time of crustal formation is estimated. The age of crustal formation is taken from the lowest biostratigraphic zone as given on the timescale of Berggren et al. [1985c]. For the two ocean floor sites on either side of the Ninetyeast Ridge, Sites 213 and 215, we used an axial paleodepth of 3000 meters, a rather deep value but one in accordance with present depths along the Southeast Indian Ocean Ridge and very similar to those used by van Andel [1975] for these sites. Ninetyeast ridge sites (214, 216, 253, 756, 757 and 758) all bottomed in very shallow to subaerial deposits and were assigned an initial depth of 0 m. Site 219 on the Chagos-Laccadive Ridge was also assigned an initial depth of 0 m, but appears to have subsided more slowly than the more normal sites. The subsidence history of Site 709 on the crest of Madingley Rise had been determined by Simmons [1990] and we have adopted his estimates. Coffin [1992] determined paleodepths for Sites 738, 747, 748, 749, and 750 on Kerguelen Plateau. Unlike most subsiding oceanic regions, Kerguelen Plateau has an initial elevation well above sealevel, with perhaps two kilometers total relief. This conclusion is supported by evidence of extensive subaerial weathering of basement basalts recovered from the plateau [Coffin, 1992].

Two drilling regions present more than normal difficulties in paleodepth determination. Site 238 drilled into 506 meters of sediment perched on a tectonic sliver within the Argo fracture zone. The water depth there is 2844 meters, so basement depth is 3338 meters. The Argo transform offsets the Central Indian Ridge, which had rift-valley floor depths in the general range of 3500 meters [Tapscott et al., 1980]. Sclater et al. [1985] imply that Site 238 has subsided with a normal $kt^{1/2}$ history, but since the present

depth is less than that of the axial valley and since the mesoscale tectonics of the Argo transform are unknown, we present a more conservative estimate of no depth change for this site rather than construct any ad-hoc and untestable model.

The Broken Ridge drillsites 752 - 755 have a subsidence history complicated by the Eocene separation of Kerguelen Plateau and Broken Ridge and the subsequent uplift of Broken Ridge. Careful plotting of subsidence curves, allowing for sediment loading and taking into account the information based on benthic foraminifer assemblages, permits a reasonable estimation for Broken Ridge paleodepths [Rea et al., 1990]. Broken Ridge itself was uplifted about 2500 m in the middle Eocene in response to the intra-plateau rifting event. It is unlikely that the Broken Ridge sites were ever deeper than about 1200 m in the early Eocene.

STABLE ISOTOPE RECONSTRUCTIONS: APPROACH AND PRINCIPAL RESULTS

Using the individual benthic foraminifer isotope records, a number of isotopic reconstructions were assembled showing the temporal evolution as well as spatial distribution of isotopic values in the Indian Ocean during the Paleogene and early Neogene. First, using the above age models, benthic foraminifer carbon and oxygen isotope data from all sites were plotted as a composite times series for the Paleogene and early Neogene (5 - 65 Ma). For convenience of presentation and discussion, this isotopic record was sub-divided into three 20 m.y. intervals, 5-25, 25-45, and 45-65 Ma. Second, using results derived from the continuous time series, paleodepths calculated from the subsidence curves, and back-tracked paleolatitude estimates of Royer and Sandwell [1989], Royer and Chang [1991], and Royer and Coffin [1992] (Figure 1; Appendix 1), we reconstructed oceanic isotopic distributions for five specific time slices, the late Paleocene (59-60 Ma), the early Eocene (55-56 Ma), the early Oligocene (34-35 Ma), early Miocene (21-23 Ma), and the middle Miocene (12-13 Ma). Data from some sites not shown in the continuous time comparisons (i.e., Sites 744 and 751) were included in the time slices to improve resolution. The isotopic values displayed for each site in the time slice reconstructions represent a mean for the values from that interval. Finally, employing the subsidence curves we developed time-continuous carbon and oxygen isotopic contour plots showing the evolution of oceanic isotope distributions with depth for the Miocene Indian Ocean. We limited this form of comparison to Miocene time since the depth distribution of Indian Ocean sites for older periods was inadequate for developing meaningful isotopic contour reconstructions.

In all above reconstructions, carbon and oxygen isotopic compositions of benthic foraminifers were adjusted to correct for genus specific non-equilibrium isotope fractionation effects. The isotopic adjustments, listed in Table 3, were taken from previous investigations in which intra-species isotope differences were firmly established

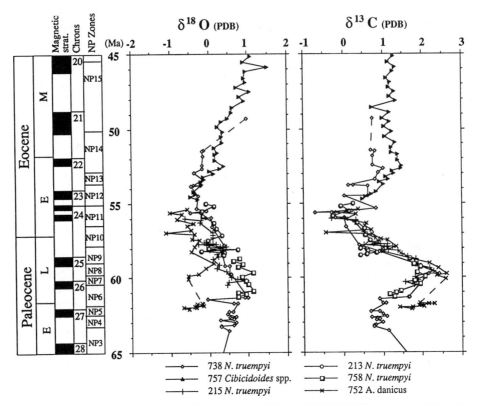

Fig. 15. Benthic foraminifer stable isotope comparison for the Paleocene/Eocene (45-65 Ma) [Barrera et al., 1991; Rea et al., 1991; Seto et al., 1991; Hovan and Rea, 1992; Zachos et al., unpublished manuscript, 1992]. Age calculations are based on stratigraphic data provided in Table 2. Values have been adjusted using correction factors in Table 3. Diagenesis has overprinted the oxygen isotope values from Site 752.

through paired analyses [Shackleton et al., 1984; Keigwin and Corliss, 1986; Woodruff et al., 1990; Barrera and Huber, 1991; Zachos et al., 1992a]. Because they are genetically controlled, the vital effect differences are assumed to be invariant through time.

TABLE 3. Isotopic Correction Factors for Individual Benthic Foraminifer Genera.[5]

Correction Factors Taxa	$\partial^{18}O$	$\partial^{13}C$	Ref.
Oridorsalis spp.	0.00	0.90	1,2,3
Nuttallides spp.	0.50	0.00	1
Cibicidoides spp.	0.50	0.00	1,2,3,4
Anomalinoides spp.	0.30	0.30	1
Gyroidina spp.	0.00	0.00	1,4
Uvigerina spp.	0.00	0.90	1,2

1 - Shackleton et al. [1984];
2 - Keigwin and Corliss [1986];
3 - Woodruff et al. [1991];
4 - Zachos et al. [1992]a
5 - These values were added to measured values (PDB) to correct for disequilibrium isotopic effects.

Paleocene/Eocene (45-65 Ma)

The continuous time series for the Paleocene and early Eocene is comprised of isotopic records from six sites: 213, 215, 738, 752, 757, and 758 (Figure 15). The time

segment represented at each site is quite variable; no record is complete over the entire 20 m.y. period and no interval of time is represented at all seven sites. Benthic values from Site 748 were included in the 55 to 56 time slice. Paleodepths of the seven sites are between 600 and 3500 m. Sites 752 and 757 were most shallow with paleodepths of less than 1000 m, while Sites 213 and 215 were deepest with paleodepths in excess of 3500 m. Despite the wide range of paleodepths represented, in those intervals of stratigraphic overlap precise fits of first and second order isotopic trends were obtained between sites.

Several distinct carbon isotopic trends emerge from this comparison. A general 1.5‰ increase is recorded at all sites in the late Paleocene beginning near 61 Ma and peaking at about 60 Ma. This is followed by a gradual 2.5‰ decrease from 60 to 56 Ma, and a 1.0‰ increase from 56 to 52.5 Ma. For the period 52.5 to 45 Ma, which is represented at only two sites, 738 and 757, there is little change in $\delta^{13}C$.

The late Paleocene and early Eocene time slice comparisons (Figures 16 and 17) indicate that the overall intrabasin range of $\delta^{13}C$ values was small. For the interval 59-60 Ma the entire range of $\delta^{13}C$ values observed across the basin is less than 0.5‰, with the heaviest values at Site 752, ~2.4‰, and the lowest values at Site 758, ~1.9‰ (Figure 16). The $\delta^{13}C$ values at Site 738 and the deepest

Late Paleocene Indian Ocean
(59-60 Ma)

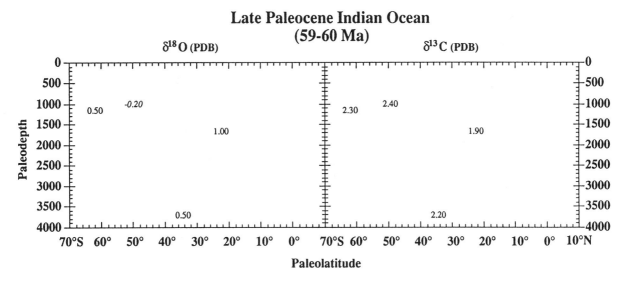

Fig. 16. Mean stable carbon and oxygen isotope values of late Paleocene (59-60 Ma) benthic foraminifers from the Indian Ocean plotted at the appropriate paleolatitude and paleodepth of each site. Values have been adjusted with the correction factors shown in Table 3. Values in italics (Site 752) represent diagenetically altered values.

Early Eocene Indian Ocean
(55-56 Ma)

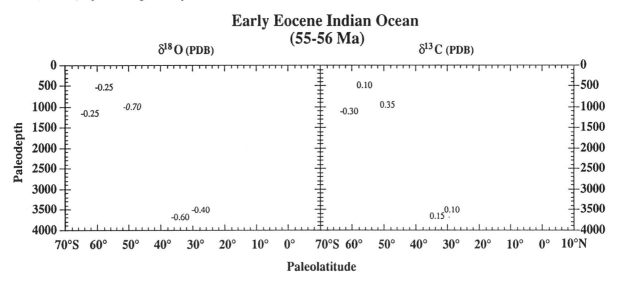

Fig. 17. Mean stable carbon and oxygen isotope values of early Eocene (55-56 Ma) benthic foraminifers from the Indian Ocean plotted at the appropriate paleolatitude and paleodepth of each site.   Values have been adjusted with the correction factors shown in Table 3. Values in italics (Site 752) represent diagenetically altered values.

site, 215, are similar to or just slightly lower than those of Site 752. The lack of intrabasin isotopic heterogeneities was apparently maintained through the early Eocene, even as mean values for the basin decreased by 3.0‰ (Figure 17).

The oxygen isotope record also shows a distinct long-term trend despite high(er) frequency noise in individual records. A general 1.0‰ decline is recorded from 61 to 56 Ma followed by a gradual 1.2‰ increase from 54 to 45 Ma (Figure 15). Only the record from the shallowest site, 752, appreciably departs from this long-term pattern. As with carbon, the overall range of $\delta^{18}O$ values observed for the late Paleocene and early Eocene is also minimal, less than

0.5‰, with no obvious spatial patterns in either the late Paleocene or early Eocene time slices (Figures 16 and 17).

Eocene-Oligocene (25-45 Ma)

The Eocene-Oligocene reconstruction contains benthic foraminifer isotope records from six sites: 738, 744, 748, 756, 757, and 253.   All isotopic values are from *Cibicidoides* spp. or *Gyroidina* spp.   The temporally highest resolution records available for this interval are those of Sites 744 and 748 which were plotted together in Figure 18.   The carbon and oxygen isotope records of these two Kerguelen Plateau sites show a remarkably close correlation over most of the late Eocene and Oligocene.

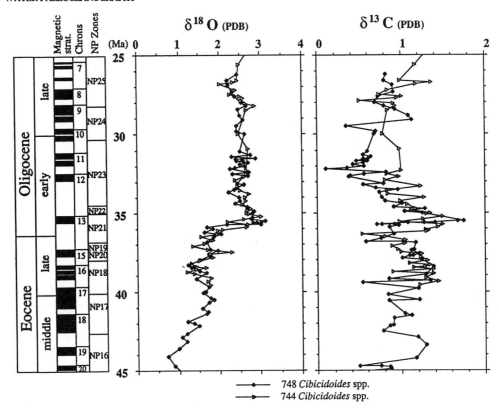

Fig. 18. Eocene/Oligocene benthic foraminifer stable isotope time series for Kerguelen Plateau Sites 744 and 748 [Barrera et al., 1991; Zachos et al., 1992a; 1992b]. Age assignments are based on datums provided in Table 2. Values have been adjusted using correction factors in Table 3.

Records from Broken and Ninetyeast Ridge sites are compared with the Site 748 record in Figure 19.

For this interval the most prominent changes are recorded in the composite benthic foraminifer δ18O record. An overall 2.0 to 2.5‰ decrease is recorded from the lower to upper Eocene. Although only a few of the individual records are continuous over the entire length of this interval, it is obvious that the long-term pattern differs slightly between some sites. From about 45 to 42 Ma the decline in absolute values at Sites 748 and 757 is nearly equivalent, not entirely unexpected since the sites were at nearly identical depths. However, at about (the) 42 Ma (level), a clear divergence occurs between the two records. The δ18O values at Site 748 rapidly increase by 0.5‰ while the values at Site 757 remain constant. For most of the next 5 m.y. the two records show little change but remain separated by >0.5‰. The records briefly converge once at the 38 Ma level and again at about 36 Ma (see arrows on Figure 19). The δ18O record of Site 756, which was at a paleodepth (~400 m) placing it within the surface water zone, parallels that of 757 for much of the late Eocene and Oligocene, although separated by -0.2 to -0.4‰.

The largest step in the Eocene/Oligocene δ18O record is a 0.8 to 1.6‰ increase at 35.5 to 36 Ma. The increase is greatest at Sites 744 and 748 (~1.6‰), smallest at Sites 756 and 757 (~0.8‰), and nearly absent at Site 253, where a hiatus may exist. This increase is followed immediately

by a slight 0.3 to 0.6‰ decrease at all sites. Above this level the δ18O differences between the Indian Ocean sites are larger and more permanent than for any preceding time of the Cenozoic. With the exception of a slight 0.6‰ decrease from 28 to 26.5 Ma, there appears to be little additional change in benthic δ18O before Miocene time. The difference in the magnitude of the early Oligocene shift between sites appears to be an artifact of sampling resolution. At Sites 744 and 748 the sample resolution is 2 to 4 times higher than that of any other site across this interval. Sample resolution at Sites 253, 756, and 757 appears to be inadequate for resolving the full expression of this event.

The benthic carbon isotope comparison reveals several short-term synchronous excursions over the Eocene and Oligocene. The first is a slight 0.4‰ carbon isotope decrease at about 38 to 39 Ma. This excursion is most distinct at Sites 744 and 748, but can also be detected at Sites 756 and 757. The second and more prominent event is an abrupt 0.5 to 0.8‰ increase from 36 to 35 Ma. This increase is recorded at all sites, including those with relatively low sample resolution, and coincides with the large increase in δ18O values described above. Similar carbon isotope excursions have been recognized in Atlantic sites [Miller and Thomas, 1985; Miller et al., 1988]. As with the oxygen isotope increase, not all sites in the Indian Ocean show the same magnitude δ13C

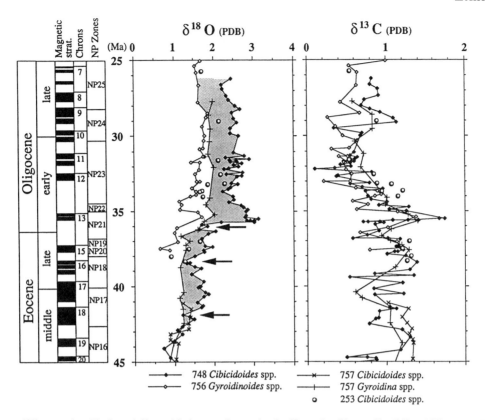

Fig. 19. Eocene/Oligocene benthic foraminifer stable isotope time series for Kerguelen Plateau Site 748 and Ninetyeast Ridge Sites 253, 756, 757 [Rea et al., 1991; Oberhansli; 1986; Zachos et al., 1992a, 1992b]. Age calculations are based on stratigraphic data provided in Table 2. Values have been adjusted using correction factors in Table 3. Arrows represent intervals of shallow intermediate water warming in high latitudes.

increase; sites sampled at low resolution (Sites 756 and 757) show somewhat smaller increases while the excursions recorded at Sites 744 and 748 are larger and appear more abrupt. Following the increase at 35 Ma, $\delta^{13}C$ values at every site gradually decrease by 0.75 to 1.0‰, eventually stabilizing by 32 Ma. Values remain constant until the 29 Ma level where a slight 0.2 to 0.3‰ positive excursion occurs.

Carbon isotope ratios were fairly homogeneous in the Indian Ocean during much of the Oligocene. The overall range of carbon isotope values for any given level of the Oligocene was typically less than 0.3‰. Close examination of the 33 to 34 Ma time slice reconstruction reveals the presence of a slight south to north gradient, with the most ^{13}C enriched values recorded at southernmost Sites 744 and 748 and the most depleted values at northernmost Sites 758 and 219 (Figure 20).

Miocene (5 - 25 Ma)

The Miocene time series reconstruction is comprised of benthic foraminifer isotope records from ten sites (Figure 21). For sites where both *Cibicidoides* and *Oridorsalis* data were available, we plotted only *Oridorsalis* since it was the most commonly analyzed taxa over the Miocene. To correct for vital effects, adjustments were applied to the

carbon isotope values of *Oridorsalis* spp. (+0.95‰) and, as in previous plots, the oxygen isotope values of *Cibicidoides* spp (+0.5‰). Also, time slice comparisons were assembled for 21 to 23 and 12 to 13 Ma (Figures 22 and 23), and continuous depth-age-isotopic contour plots were constructed for the entire Miocene (Figures 24 and 25). Isotope data from Kerguelen Sites 744 [Barrera and Huber, 1991] and 751 [Mackensen et al., 1992] were only included in the time slice reconstructions.

The overall correlation of oxygen isotope records is very tight for most of the Miocene. The most prominent first order feature is a greater than 1.0‰ $\delta^{18}O$ increase in the middle Miocene at 13 to 15 Ma (Figure 21). This middle Miocene increase has been recognized at nearly every Miocene sequence studied to date [e.g. Miller et al., 1987a; Woodruff and Savin, 1989; 1991]. Slight offsets (~0.1 - 0.5 m.y.) in the timing of this increase between sites may not be real but artifacts of errors in the age models. The magnitude of increase, ~1.0 to 1.2‰, is nearly identical at every site with the exception of the shallower sites where the increase is slightly smaller. This large mid-Miocene increase was preceded by a less distinct 0.3‰ $\delta^{18}O$ increase near 24 Ma, and followed by a 0.5‰ positive shift at about 9 Ma.

As observed in the Oligocene, distinct patterns are evident

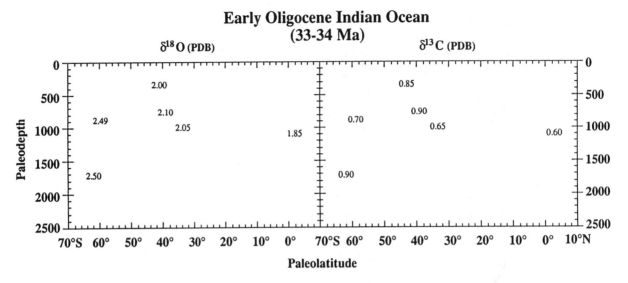

Fig. 20. Mean stable carbon and oxygen isotope values of early Oligocene (33-34 Ma) benthic foraminifers from the Indian Ocean plotted at the appropriate paleolatitude and paleodepth of each site. Values have been adjusted with the correction factors shown in Table 3.

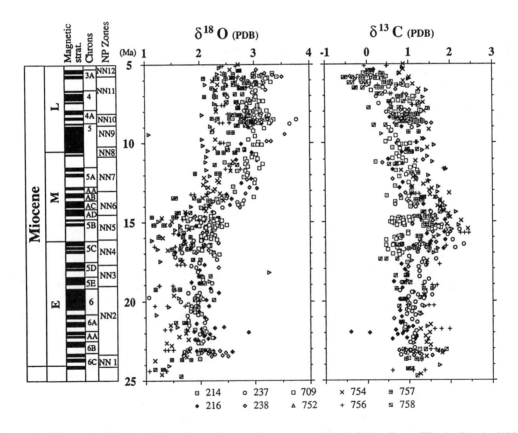

Fig. 21. Composite Miocene benthic foraminifer carbon and oxygen isotope record for the Indian Ocean [Woodruff et al., 1990; Vincent et al., 1985; Rea et al., 1991; Nomura et al., in press]. Age calculations are based on stratigraphic data provided in Table 2. *Cibicidoides* spp. or *Oridorsalis umbonatus* values were adjusted for species specific vital effects using the correction factors in Table 3.

in the Miocene $\delta^{18}O$ distributions across the basin. Over much of the lower Miocene the most ^{18}O depleted values (~1.2‰) are observed at the shallowest sites, 754 and 756, while the most enriched values (2.5‰) are found at the highest latitude site, 751 (Figure 22). This spatial pattern persists following the middle Miocene $\delta^{18}O$ increase, but the range of values expands slightly due to a slightly larger increases at the deeper sites. At 12 Ma values of greater than 3.0‰ are recorded only at Site 751, but by 8 Ma similar values are recorded at nearly all sites below 1700 m (Figures 23 and 24).

Several synchronous shifts or excursions can be recognized in the Miocene benthic foraminifer carbon isotope record (Figure 21). The most prominent is at 15-16 Ma when $\delta^{13}C$ values abruptly increase by as much as 0.8‰ with the largest increases at sites with paleodepths in the range of 750 to 1500 m. Another basin wide event occurs at 7 Ma when values at most sites decrease by about 0.7‰. The relative distribution of carbon isotopes within the basin during the Miocene can be discerned from the time-slice and age-depth-isotope contour plots (Figures 22, 23 and 25). In the early Miocene reconstruction (Figure 22), the overall range of $\delta^{13}C$ between sites is less than 0.75‰. The heaviest benthic foraminifer $\delta^{13}C$ values over this and subsequent intervals occur at the shallowest sites 752 and 754 on Broken Ridge. It appears that for a brief 1 to 2 m.y. interval centered at 22 Ma, masses of comparatively ^{13}C enriched waters were present at two distinct paleodepths in the Indian Ocean, one between 750 and 1250 m and a second between 2400 and 2750m.

DEEP SEA PALEOTEMPERATURES AND GLOBAL ICE VOLUME

Secular variations in the oxygen isotopic composition of benthic foraminifers reflect either changes in ambient water temperature and/or isotopic composition. Changes in the latter are primarily driven by global ice volume. In sediments older than Oligocene, temperature was probably the primary factor determining $\delta^{18}O$, since there is little physical evidence of large scale early Cenozoic glacial activity (i.e. ice-rafted debris)[e.g. Kennett and Shackleton, 1976; Kennett and Barker, 1990].

The long- and short-term oxygen isotopic variations recorded during the Paleogene and early Neogene by benthic foraminifers within the Indian Ocean are in most respects remarkably consistent with those recorded elsewhere. Beginning with the early Cenozoic all sites with the exception of Site 752 show a gradual 1.0‰ decrease from the late Paleocene to early Eocene, reaching a minimum at about 55 Ma. This decrease, which reflects approximately a 4°C increase in temperature (assuming no ice-volume change) is relatively uniform at all depths in the basin. This trend has previously been recognized in sites from the Atlantic and Pacific and is attributed to gradual warming of deep waters synchronous with polar and sub-polar warming during the early Eocene [Shackleton and Kennett, 1975; Savin, 1977; Shackleton and Boersma, 1981; Miller et al., 1987a; Katz and Miller, 1991]. Site 752 benthic foraminifers are considerably more depleted in

^{18}O than those of the other Indian Ocean sites and may be overprinted by diagenesis. The preservation of Paleocene foraminifers is extremely poor at this site; many specimens are overgrown with secondary calcite, and planktonic foraminifers are scarce suggesting heavy dissolution. An unusual, but viable, explanation for alteration is slight recrystallization in a fresh water lens. This would have occurred during the early Eocene uplift of Broken Ridge which briefly (1-2 m.y.) exposed the ridge to sub-aerial weathering [Peirce, Weissel, et al., 1989; Rea et al., 1990]. Excluding Site 752, our reconstructions show the range of $\delta^{18}O$ values between Indian Ocean sites over the late Paleocene and early Eocene to have been small and within the range of values observed in the Pacific and Atlantic [Miller et al., 1987a,b], indicating that bottom water temperatures on a global scale were fairly homogeneous during most of this time.

The early Eocene thermal maximum was followed by cooling and increased thermal segregation of deep water masses during the middle and late Eocene in the Indian Ocean. As previously recognized, the rate of cooling was not uniform but occurred in a step-like fashion, with several distinct "events". Cooling began between 53 and 51 Ma and proceeded uniformly to 45 Ma with an overall decrease of 5°C, approximately equivalent to the decrease in surface temperatures recorded near the margin of Antarctica for the same interval [Stott et al., 1990].

An unusual event occurred in the Indian Ocean around 43 to 42 Ma; $\delta^{18}O$ records from several sites which had overlapped previously, diverged. At high latitude Site 748, and later at 744 (Figure 18), benthic $\delta^{18}O$ values continued to increase, while $\delta^{18}O$ values of lower latitude Site 757 stabilized (Figure 19). This divergence occurred despite the fact that Site 757 was at a similar paleodepth. This change in $\delta^{18}O$ patterns at 42 Ma most likely reflects increased thermal segregation of intermediate water masses between these locations, and possibly within the basin in general. For the period prior to 42 Ma the thermal characteristics of intermediate waters at Kerguelen and Broken Ridge were essentially identical, indicating that Indian Ocean intermediate waters were derived from a single location. Barrera and Huber [1991] suggested that this source might have been located at low latitudes. After 42 Ma, at least two thermally distinct water masses existed at intermediate depths in the Indian Ocean, a cool high latitude derived water mass and a warmer lower latitude water mass. The initial separation of benthic foraminifer oxygen isotopes in the Indian Ocean at 42 Ma occurred shortly after the seaway between Antarctica and Australia had begun to widen [McGowran, 1990]. The wider seaway may have accelerated cooling of polar surface waters at this time; by 42 Ma temperatures of high southern latitude surface waters had declined to less than 10°C (Stott et al., 1990). It is possible that during the Eocene, this temperature represented a threshold temperature below which surface waters were cold enough to sink in sufficient volumes and flow northward as an intermediate water mass to middle latitudes (~45°S) of the Indian Ocean. Although it is

Early Miocene Indian Ocean
(21-23 Ma)

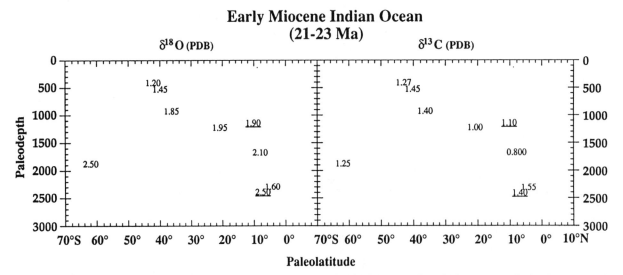

Fig. 22. Mean stable carbon and oxygen isotope values of early Miocene (21-23 Ma) benthic foraminifers from the Indian Ocean plotted at the appropriate paleolatitude and paleodepth of each site. Values have been adjusted with the correction factors in Table 3. The lone high latitude value is from Site 744 [Barrera and Huber, 1991]. Western Indian Ocean sites are underlined.

Middle Miocene Indian Ocean
(12-13 Ma)

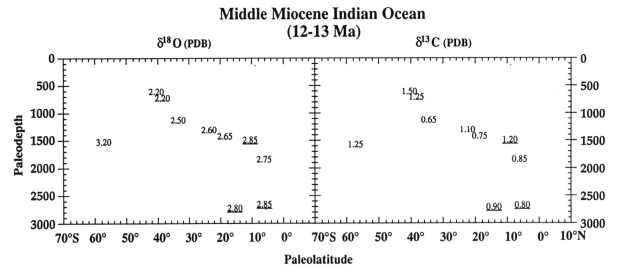

Fig. 23. Mean stable carbon and oxygen isotope values of middle Miocene (12-13 Ma) benthic foraminifers from the Indian Ocean plotted at the appropriate paleolatitude and paleodepth of each site. Values have been adjusted with the correction factors shown in Table 3. The lone high latitude value is from Site 751 [Mackensen et al., 1992]. Western Indian Ocean sites are underlined.

uncertain if these colder intermediate waters were generated on or near the Antarctic margin, we note that after 42 Ma, intermediate water temperatures at Kerguelen appear to have co-varied with those of surface waters of the Southern Ocean as monitored at Sites 689 and 690 [Stott et al., 1990]. Further to the north in the eastern portion of the basin, warmer intermediate waters show little change after 42 Ma indicating that they were derived from a low latitude region of constant temperature. Based on the permanent $\delta^{18}O$ offsets between the sites on Kerguelen and on Ninetyeast Ridge, this arrangement appears to have persisted until at least late Oligocene time. Another positive excursion reflecting short-term cooling occurred

at 38 to 37 Ma. In total, from 50 to 36 Ma deep waters in high latitudes had cooled by nearly 9°C, while low latitude intermediate waters of the Indian Ocean had cooled by about 7°C.

Isotopic evidence for deep water cooling and increased thermal segregation of water masses at 42 Ma is consistent with paleontological evidence for increased cooling and thermal stratification. For example, planktonic foraminiferal assemblages from all basins exhibit changes in both latitudinal and vertical distribution which suggest sudden cooling close to this time [Keller, 1983; Boersma et al., 1987]. Also, warm water calcareous nannofossils, such as *Coccolithus formosus*, disappeared from high latitude

$$\delta^{18}O$$

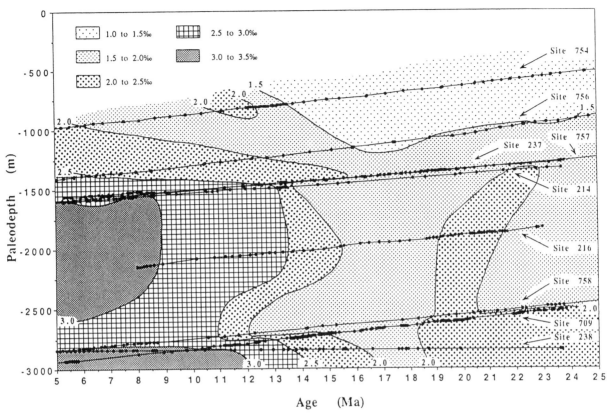

Fig. 24. Miocene (2 to 25 Ma) oxygen isotope contour plot of benthic foraminifer $\delta^{18}O$ values plotted along an age - paleodepth line for each Indian Ocean site. Data are from Figure 21.

surface waters by about 44 Ma [Pospichal and Wise, 1990; Wei and Wise, 1990]. Benthic foraminifer diversity also decreases at this time indicating cooling [Thomas, 1990]. These biotic and isotopic responses highlight the importance of this step in the long-term cooling of Antarctica.

The most prominent event in the long-term cooling of the Indian Ocean occurred at 35-36 Ma when a greater than 1.0‰ increase in $\delta^{18}O$ values was recorded by benthic foraminifers. The timing of this increase as determined from bio- and Sr isotope stratigraphies clearly demonstrates that it is the same event recorded in Atlantic and Pacific ocean deep sea sites [Shackleton and Kennett, 1975; Keigwin, 1980; Poore and Matthews, 1984; Wise et al., 1985; Keigwin and Corliss, 1986; Miller et al., 1987a]. However, the overall character of the increase as recorded in the Indian Ocean is unusual in two important respects. First, at the two Southern Ocean sites, 744 and 748, the oxygen isotope increase coincided with deposition of terrigenous debris [Barrera and Huber, 1991; Zachos et al., 1992a,b]. The texture and size of debris indicate that it was ice rafted in origin [Ehrmann, 1991; Wise et al., 1991; Breza and Wise, 1992; Zachos et al., 1992b] and the presence of heavy mineral components in the debris suggests that it could only have originated from

Antarctica, the nearest metamorphic rock bearing terrain. This would require the production of large icebergs capable of traveling 1000 km north to Kerguelen. The fact that the ice-rafted debris was deposited precisely at the time of the oxygen isotope increase lends further support to arguments in favor of ice volume change as the main cause of the early Oligocene $\delta^{18}O$ increase [e.g. Keigwin, 1980; Matthews and Poore, 1980; Poore and Matthews, 1984; Keigwin and Corliss, 1986; Miller et al., 1987a; Zachos et al., 1992b].

Second, at Sites 744 and 748 the magnitude of increase and the peak absolute values were several tenths of a per mil higher than previously recorded elsewhere. Adjusted $\delta^{18}O$ values of *Cibicidoides* spp. were for a brief interval, in excess of 3.1‰ at both sites. Values this high, assuming close to average salinities and bottom water temperatures no colder than present day would require a large ice-sheet, possibly as massive as the present day Antarctic ice-sheet [Zachos et al., 1992b]. The existence of the $\delta^{18}O$ increase, presumably related to ice volume expansion, should be worldwide, but has not been recognized everywhere because of low sampling resolution. The magnitude of increase as recorded from site to site in the Indian Ocean appears to vary primarily as a function of sampling resolution. At Sites 748 and 744 where the sampling resolution was high, the $\delta^{18}O$ increase is very

$$\delta^{13}C$$

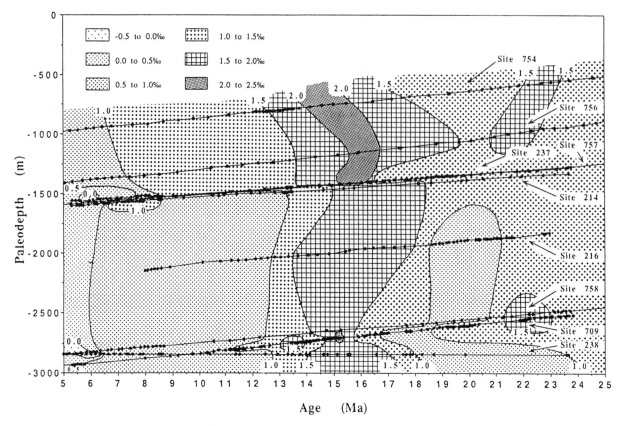

Fig. 25. Miocene (2 to 25 Ma) carbon isotope contour plot of benthic foraminifer $\delta^{13}C$ values plotted along an age - paleodepth line for each Indian Ocean Site. Data are from Figure 21.

large, about 1.6 to 1.8‰, whereas at sites with lower sample resolution, 756 and 757, the recorded increase was noticeably smaller, no more than 1.2‰. We suspect that at these sites the low sample resolution was inadequate to characterize the full magnitude and abrupt nature of the $\delta^{18}O$ event. If Sites 744 and 748 were sampled at the same resolution as most other sites, about 1 sample/section, the 30 to 50 cm interval of peak $\delta^{18}O$ values at each site might have been bypassed and as a result the overall increase underestimated by almost 0.3‰.

Shortly after the early Oligocene oxygen isotope increase, values at all locations decreased by 0.4 to 0.5‰. Miller et al. [1987a] interpreted a similar pattern of an abrupt short-term increase in Atlantic E/O boundary sequences to represent a rapid but temporary increase in ice volume. This scenario is consistent with the brief appearance of ice-rafted debris on Kerguelen Plateau [Breza and Wise, 1992; Zachos et al., 1992b].

Absent from the Indian Ocean records is a distinct $\delta^{18}O$ increase at 30 Ma which has been observed in Atlantic DSDP and ODP sites and attributed to a short-term ice-volume increase [Miller and Thomas, 1985; Miller et al., 1987a; Kennett and Stott, 1990]. One obvious explanation for the discrepancy is the presence of a hiatus

at Sites 744, 748 and 756 in the Indian Ocean.

By early Miocene time the distinct thermal segregation of water masses that had developed during the Oligocene had dissipated slightly as intermediate and deep waters warmed. Benthic foraminifer $\delta^{18}O$ values of high latitude Site 744 were at least 0.5‰ higher than those recorded at lower latitude sites, indicating a temperature difference of about 2 to 3°C. Cooler water bathing Kerguelen at that time may represent a proto circum-polar water. Cool waters were also present at the deeper sites in the western basin. In middle Miocene time a basin wide oxygen isotope increase was recorded. Benthic $\delta^{18}O$ values at every site show a greater than 1.0‰ increase. As observed in other ocean basins, the ^{18}O enrichment initiated near 15 Ma and peaked around 12 Ma. The increase was not entirely uniform but appears to have been slightly greater in the high polar latitudes; adjusted benthic $\delta^{18}O$ values at Site 751 (not included in the Miocene contour plot) were as high as 3.3 to 3.4‰ by late middle Miocene time (Figure 8)[Mackensen et al., 1992]. These values are comparable to values obtained for some Holocene benthic foraminifers [e.g. Zahn and Mix, 1991], and are just 0.2‰ higher than the values obtained for benthic foraminifer at the climax of the early Oligocene increase.

Identical increases in benthic $\delta^{18}O$ have been reported for the middle and late Miocene in other basins [Miller et al., 1987a; 1988; Woodruff and Savin, 1989; 1991]. Although most investigators agree that the increase represents some combination of increased ice volume and lowered temperatures, the relative contributions of each have been debated [e.g. Vincent et al., 1985; Miller et al., 1987a; Woodruff and Savin, 1991]. Typically, globally uniform changes in foraminifer $\delta^{18}O$ are considered to represent ice-volume changes. However, the magnitude and rate of $\delta^{18}O$ change recorded across the middle Miocene interval has been found to differ slightly from site to site. For example, at some Indian and Pacific Ocean locations changes in planktonic foraminifer values either lag or do not parallel the benthic record [Barrera et al., 1985; Vincent et al., 1985; Savin et al., 1985; Woodruff and Savin, 1991]. Moreover, the magnitude of increase recorded by benthic foraminifers appears to have been higher at sites proximal to Antarctica [Woodruff and Savin, 1991].

In the Indian Ocean the magnitude of increase was nearly identical (±0.1‰) at all sites regardless of depth; this implies mainly an ice-volume effect because one would expect temperature change to be more variable spatially. The slightly greater $\delta^{18}O$ increase (>1.2‰) recorded at Site 751 indicates that this was accompanied by cooling in the higher latitudes (~1-2°C)[Mackensen et al., 1992]. Also, the temporal relationship of the $\delta^{18}O$ increase to benthic faunal changes varied considerably from location to location. At Site 756 quantitative analysis revealed that the major change in benthic foraminifer assemblages occurred at 17.1 Ma whereas at the very shallow mid-latitude Site 754 the major change in diversity occurred near the termination of the oxygen isotope increase at about 13.8 Ma [Nomura, 1991b; in press]. Moreover, at Site 751 where oxygen isotope values increased by >1.2‰, benthic faunal assemblages showed little if any variation in the middle Miocene [Mackensen et al., 1992]. We recognize that in terms of sorting temperature from ice volume effects it may not be appropriate to consider changes in benthic foraminifer assemblages since they may be more sensitive to changes in other physical and chemical aspects of the deep water environment (i.e., oxygen content, pH, sediment organic carbon content, corrosiveness, etc.)[Lohmann, 1978; Corliss, 1979]. In fact, faunal composition changes in Miocene benthic foraminifera have been linked mainly to the changes in the rates of deep water formation and ocean turnover on productivity and deep water chemistry [Woodruff, 1985; Thomas and Vincent, 1987; Miller and Katz, 1987; Nomura, 1991b]. However, the uniform nature of the middle Miocene $\delta^{18}O$ increase as well as the lack of a more direct response in benthic foraminifers in the Indian Ocean favors ice volume expansion as the principal cause of this oxygen isotope increase.

DEEP AND INTERMEDIATE WATER PALEOCIRCULATION

Present day sub-surface waters of the Indian Ocean consist of water derived from several sources outside of and within the basin. Waters just below the surface layer south of 10°S are essentially dominated by cold Antarctic Intermediate Water (AAIW) while shallow intermediate waters to the north consist of a saline water mass representing a mixture of Red Sea, Persian Gulf and Arabian Sea waters [Wyrtki, 1973; Warren, 1981]. Red Sea outflow can be traced at depths of 600 m to 25°S along the western side of the basin. Waters below 2000 m are primarily composed of Antarctic Bottom Water (AABW) with a small component of North Atlantic Deep Water (NADW) derived from Circumpolar Deep Water (CDW)[Warren, 1981; Mantyla and Reid, 1983; Metzl et al., 1990]. Abyssal waters of the eastern basin enter through an opening in the Southeast Indian Ocean Ridge south of Australia, while waters in the western basin enter through though the Crozet Basin from the Atlantic-Indian Basin [Kolla et al., 1976]. The southern portions of the Indian Ocean including the Kerguelen Plateau reside within CDW.

One objective of this study was to determine the nature of intermediate and deep water circulation in the Indian Ocean during the Paleogene. Several workers have suggested that during this time, because of reduced planetary temperature gradients, formation of deep waters may have occurred in low latitude regions of high salinity, in a manner similar to the present day contribution of Arabian and Red Sea water to intermediate waters of the Indian Ocean, but on a much larger scale [Brass et al., 1982; Shackleton and Boersma, 1981]. At present very little isotopic evidence exists in support of such a circulation pattern in the Paleogene. Kennett and Stott [1990] recently reported the presence of a $\delta^{18}O$ inversion in Paleogene benthic foraminifer records from two vertically offset sites on Maud Rise in the Atlantic sector of the southern ocean; they argued that this inversion represented the presence of warmer, more saline waters underlying cool waters. They further suggested that these waters originated from the Tethys Sea, which contained extensive shallow water platforms suitable for the production of high salinity waters.

If production of saline waters was occurring in Tethys during the Paleogene, these waters may have passed through the northern Indian Ocean. The oxygen isotope evidence described above clearly indicates the presence of a warmer, lower latitude intermediate water mass for much of the Eocene and Oligocene that prior to 42 Ma might have extended as far south as Kerguelen Plateau and possibly Antarctica. We assume based on the $\delta^{18}O$ differences that a component of this water mass was formed in low latitudes.

A more direct method of determining the direction of bottom water flow is through reconstruction of carbon isotope distributions. As a deep water mass migrates through a basin its carbon isotopic composition becomes progressively lower due to the accumulation of ^{12}C-enriched CO_2 released from decomposition of particulate organic matter [Kroopnick, 1985]. Some benthic foraminifer accurately record the $\delta^{13}C$ of ambient bottom water dissolved inorganic carbon (DIC); as a result it has been possible to retrace the path or source of ancient water masses by reconstructing carbon isotope distributions from fossil foraminifers [e.g. Curry and Lohmann, 1982; Duplessy et al., 1984; Miller and Fairbanks, 1985; Woodruff and Savin, 1989]. In applying this method to reconstructing paleocirculation patterns, however, one must be aware of factors that limit its effectiveness. For example, the average rate of change in $\delta^{13}C$ depends mainly on the rain rate of organic carbon from the surface to deep ocean which in turn is partly a function of primary productivity. The amount of $\delta^{13}C$ change within a given deep water mass as it migrates between two points might diminish during periods of reduced marine productivity (oligotrophic oceans). Also, as a water mass migrates through the oceans it mixes with water masses of different isotopic composition. In such situations the $\delta^{18}C$ of the water mass may not reflect its true "age". These effects tend to reduce the sensitivity of carbon isotopes as a water mass tracer and as a result make it difficult to distinguish water mass distribution and/or migration paths on regional scales. Nonetheless, this method can provide reasonable constraints on interpretations of global-scale patterns of sub-surface circulation (i.e. basin to basin fractionation), and in a few isolated cases where enough data exist, allow us to differentiate more small-scale patterns. In the case of the Paleogene and early Neogene Indian Ocean, we seek patterns in carbon isotopes which may allow us to (1) recognize the existence or lack of basin to basin fractionation and (2) reconstruct vertical distribution of water masses within the basin.

As observed in other basins, the mean carbon composition of the deep Indian Ocean decreased by nearly 3.0‰ from late Paleocene to early Eocene. This was accompanied by substantial warming of polar regions and the deep sea. The exact cause(s) of the carbon isotopic and climatic transition is still unknown. Nonetheless, because of the importance of cooling and deep convection in high latitudes to bottom water formation, it is at this time when changes in the character of deep water circulation are most likely to have occurred.

From the few sites available it is difficult to determine if carbon isotope distributions changed noticeably in the Indian Ocean during the late Paleocene-early Eocene transition. At 59-60 Ma, there appears to have been a slight north to south gradient, indicating a possible northward direction of flow for intermediate waters coming from a high southern latitude source. The absence of a more northern location in the 55-56 Ma reconstruction prevents us from determining if that pattern persisted beyond the Paleocene. One subtle and unusual change in the Paleogene $\delta^{13}C$ distributions of the Indian Ocean occurred in the interval just below the benthic foraminifer extinction horizon. Comparison of Sites 213 and 215, east and west of the Ninetyeast Ridge, reveals that prior to the extinction event at ~57.8 Ma bottom water $\delta^{13}C$ values at Site 213 were noticeably lighter (0.3-0.4‰) than those at 215 for the same period, implying the presence of two distinct water masses. This is unusual in that these sites were just several hundred kilometers apart, and at nearly identical paleodepths (~3200 m). The difference is not an artifact of species vital effects since *N. truempyi* was analyzed at both sites, nor of correlation errors, since the conspicuous benthic foraminifer extinction event serves as the upper datum for this interval (Zone CP8b) at both sites. The offset in $\delta^{13}C$ values most likely reflects differences in deep water ages and sources which resulted from the influence of sub-marine barriers on deep water circulation. It appears that the Ninetyeast Ridge along with Kerguelen Plateau to the south and the Indian Continent to the north was an effective barrier to deep water communication (>2000 m) between the western and eastern Indian Oceans during the late Paleocene. The more negative values for the eastern basin imply the presence of aged deep water, possibly from the Pacific. Following the benthic foraminifer extinction event, the $\delta^{13}C$ compositions of bottom waters at these two locations became homogeneous. This indicates that either Site 213 was no longer receiving aged water from the Pacific and/or that deep waters at these two sites became chemically more homogeneous. The former explanation is somewhat consistent with observed changes in the CCD depth at that time. The CCD was unusually deep in the Indian Ocean during the late Paleocene, deep enough for carbonate accumulation at Sites 213 and 215 below 3200 m [van Andel, 1975]. However, the initiation of carbonate accumulation was delayed by nearly 2 m.y. at Site 213 despite the fact that it was at nearly the same paleodepth as Site 215 (Figure 14). This would indicate that during the late Paleocene bottom waters at Site 213 were more corrosive and, thus, older than at Site 215.

To determine if deep waters were being supplied to the Indian Ocean from the Antarctic, we plotted the benthic foraminifer carbon isotope records of Hole 690B, South Atlantic sector of the Southern Ocean [Kennett and Stott, 1990], and Site 577, northwest Pacific [Miller et al., 1987b] together with those of the Indian Ocean sites. The age model for Site 690B was revised using the interpretation of that site's magnetostratigraphy provided by Spieß [1990] and early Paleocene nannofossil datums of Pospichal and Wise [1990] (see Appendix 2 for the datums). The revised age model produced ages for the lower Eocene that were up to 2 m.y. older than those calculated using the magnetostratigraphic interpretation of Stott and Kennett [1990] [our interpretation is similar to that of Thomas et al., 1990]. During the late Paleocene the mean value of the Indian Ocean was 2.15‰, approximately 0.3‰ lower than the mean value recorded at Hole 690B over the same interval (Figure 26). Values of Pacific Site 577 were similar or slightly lower than those recorded in the Indian Ocean. These patterns persisted up through the early Eocene indicating that during late Paleocene and early Eocene, Indian Ocean waters were older than those of the Antarctic, but similar to or younger than those of the central Pacific. The carbon isotopes therefore suggest that the bulk of deep water entering the Indian Ocean during the late Paleocene and early Eocene was derived from the Southern Ocean surrounding Antarctica in agreement with

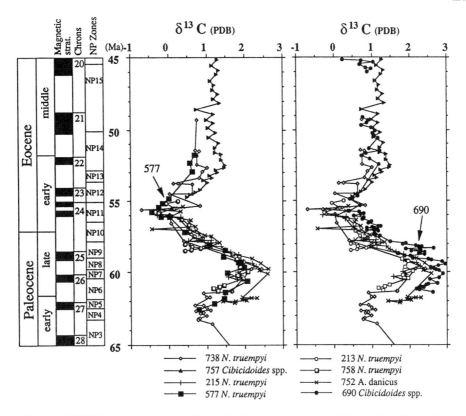

Fig. 26. Paleocene/Eocene (45-65 Ma) benthic foraminifer stable isotope comparison of Indian, Pacific and Southern Ocean records. Included are data from Site 577 in the northwest Pacific [Miller et al., 1987b], and Hole 690B in the south Atlantic sector of the Southern Ocean [Kennett and Stott, 1990]. Values have been adjusted for disequilibrium vital effects using correction factors in Table 3. Age assignments for the Indian Ocean sites are based on datums provided in Table 2. The age model for Site 577 is that of Miller et al. [1987b]. The original age model of Hole 690B [Kennett and Stott, 1990] was revised using the interpretation of that sites magnetostratigraphy provided by Spieß [1990], nannofossil biostratigraphy of Pospichal and Wise [1991], and depth-age relationships shown in Appendix 2.

Katz and Miller [1991].

By early Eocene time, as $\delta^{13}C$ values reached a minimum, oxygen isotope values were low and relatively uniform within the basin indicating warm temperatures (~12°C) and little thermal variability. According to several investigators it was during this time that warm saline bottom water production may have been most significant [Brass et al., 1982; Shackleton and Boersma, 1981; Kennett and Stott, 1990]. Although it is possible that warmer, saline bottom waters were a major component of deep waters, we note that early Eocene Indian Ocean benthic foraminifer $\delta^{13}O$ values were very similar to values obtained for high latitude planktonic foraminifers [Stott et al., 1990; Barrera and Huber, 1991]. Such a pattern indicates that a large component of deep and intermediate water was still formed in association with the cooler surface waters surrounding Antarctica. Thus, both the carbon and oxygen isotopic records support a predominantly high southern latitude source for deep waters in the Indian Ocean during the Paleocene and early Eocene.

The oxygen isotopic reconstructions show that a major change in intermediate water distribution took place in the Indian Ocean during the early middle Eocene. From 45 to 42 Ma, and we assume for some time prior to this, it appears that $\delta^{18}O$ values were relatively homogeneous between Sites 748 and 757 indicating the presence of a single water mass at shallow intermediate depths between 35 and 60°S latitude. At 42 Ma $\delta^{18}O$ values diverge, indicating that at least two distinct water masses were then present at shallow intermediate depths in the Indian Ocean: a cool, high latitude water mass confined to latitudes south of 35°S and a warm, low latitude water mass north of 35°S. Today portions of the southern Kerguelen Plateau are bathed by cold CDW. However, because it is unlikely that a similar type of water mass existed any time prior to the deep opening of the Drake Passage in the Neogene, we refer to the cool water mass of southern origin as Antarctic Intermediate Water (AAIW). The northern, warmer water mass was probably very similar in character to present day intermediate waters from the Arabian and Red Seas. During the Eocene, however, such warm saline water most likely originated from the Tethyan or northern Indian oceans, thus we refer to this as Tethyan-Indian Saline Water (TISW)[e.g. Woodruff and Savin, 1989].

This paleoceanographic event at 42 Ma may reflect the initial effects of geographic thermal isolation on the Antarctic continent which culminated 7 m.y. later with the initial appearance of ice sheet(s). From 42 Ma to 36 Ma,

AAIW was present in the southern Indian Ocean with the exception of two brief intervals centered at 38 Ma and 36 Ma, during which TISW may have extended to the Southern Ocean. At 35 Ma in association with the early Oligocene ice-volume event, AAIW and TISW appear to have undergone additional cooling, after which they were permanently offset for the remainder of the Oligocene. With the exception of a few brief intervals, carbon isotope values for most of the Eocene and Oligocene were fairly homogeneous over most of the basin. In fact, Indian Ocean $\delta^{13}C$ values were similar to those recorded in the other basins [Zachos et al., 1992a]. One exception to this occurred at the peak of the ice volume event in the early Oligocene when ^{13}C-enriched values are recorded in the high southern latitude sites on Kerguelen as seen in the 33 - 34 Ma time slice. These higher values may reflect a brief but substantial increase in the rate of nutrient-depleted AAIW production. Alternatively, the higher values could be an artifact of sampling resolution differences over an interval where the carbon isotope records at all sites show rapid but brief increases.

In early Miocene time from about 23 to 21 Ma, masses of comparatively ^{13}C-enriched waters were present at two distinct levels in the Indian Ocean, one between 750 and 1250 m and a second centered at 2750 m. These water masses might represent young, nutrient-depleted waters coming from either the south or north. We ignore the values for Site 238 whose bottom depth was kept constant as a result of uncertainties in calculating its subsidence history. Two of the deep sites with high $\delta^{13}C$ values, Sites 709 and 758, were located in the northern part of the basin indicating the possible input of nutrient-poor waters from the north [e.g. Woodruff and Savin, 1989; 1991]. Nutrient-depleted waters were present at the shallower Sites 754 and 756, both located in the mid-high latitudes. As noted by Woodruff and Savin [1989], $\delta^{13}C$ values for waters above 3000 m throughout the basin are as high as any values observed in the other basins. Only south Atlantic and south Pacific intermediate depth waters show comparable values. Woodruff and Savin [1989] suggested that these high values resulted from the flux of young waters to the Indian Ocean from a northern source, the Tethyan Seaway. Based on our compilation it appears that a major southern source existed as well.

The most prominent climatic event of the Miocene was the middle Miocene increase in ice volume and cooling of the deep sea. This event was preceded by a relatively rapid 0.6‰ positive excursion in the carbon isotopic composition of the ocean (Figure 21). The sequential order of these events has led to some speculation on a potential cause and effect relationship. For example, it has been suggested that the $\delta^{13}C$ increase might have been caused by large scale sequestering of organic carbon from the ocean into continental shelf sedimentary reservoirs [e.g. Monterey Hypothesis; Vincent et al., 1985; Vincent and Berger, 1985]. This in turn would have reduced CO_2 levels in the ocean and atmosphere system paving the way for middle Miocene cooling and glaciation some 2 m.y. later.

As a result of middle Miocene glaciation and high latitude cooling one might expect a change in ocean circulation patterns. The carbon isotope reconstructions reveal that during the initial phases of the carbon isotope event the distribution of carbon isotopes within the Indian Ocean remained essentially unchanged (Figure 25) as absolute values increased equally at all locations. However, as the mid-Miocene climatic changes peaked at 13 to 12 Ma, the distribution of carbon isotopes within the basin showed a distinct change. By 12 Ma the high $\delta^{13}C$ values characteristic of the deep northern sites during the early Miocene disappeared (Figure 23 and 24). This reversal in $\delta^{13}C$ patterns suggests that the flux of water from Tethyan Sea sources to the Indian Ocean had become substantially reduced [e.g. Woodruff and Savin, 1989]. Although comparison with other records show that $\delta^{13}C$ values on average for the whole basin were still as high as those recorded in other basins, it is clear in the long-term record that circulation patterns were beginning to shift toward a modern configuration.

Toward the end of the Miocene (7-6 Ma), deep waters had cooled to near modern values resulting in increased thermal stratification within the basin (Figure 24). Also, carbon isotope gradients within the basin diminished as the mean $\delta^{13}C$ value for Indian Ocean TDC decreased to about 0.3‰, noticeably less than values in the North Atlantic for the same period, indicating that the Indian Ocean was for the first time receiving some component of North Atlantic Deep Water [e.g. Woodruff and Savin, 1989; 1991; Wright et al., 1991; 1992]. This change in circulation patterns was also concordant with the origination of modern benthic faunas in the Indian Ocean [Nomura et al., in press].

SUMMARY

Early Paleocene to late Miocene benthic foraminifer stable isotope records from over 18 Indian Ocean DSDP and ODP sites have been compiled and placed into age equivalent time scales. Although many of the individual age models were based primarily on calcareous nannofossil biochronology, remarkably good correlations of first and second order isotopic inflections were obtained. These isotopic comparisons were used to reconstruct the Paleogene and early Neogene paleoceanography of the Indian Ocean. The major findings are summarized as follows.

1) Between 60 to 56 Ma intermediate and deep waters warmed uniformly by 4°C, and remained warm until at least 53 Ma. During much of the Paleocene, little or no deep water communication occurred between the western and eastern portions of the basin due to the presence of Ninetyeast Ridge, Kerguelen Plateau, and India, which together acted as an effective barrier to east-west deep water migration. Deep waters of the eastern basin may have been derived from the Pacific.

2) From 53 Ma to 43 Ma intermediate waters cooled uniformly by 5°C. At 42 Ma, temperatures of intermediate waters diverged latitudinally and possibly vertically as a

Appendix 1. Paleolatitudes.

Site/Age	0 Ma	10 MA	15 Ma	20 Ma	35 Ma	45 Ma	55 Ma	60 Ma	66 Ma
213	-10.21	-15.21	-17.61	--20.00	-27.34	-30.00	-29.55	-29.94	-31.73
214	-11.34	-15.95	-18.24	-20.53	-27.36	-30.35	-37.63	-43.56	-54.00
215	-8.12	-12.59	-14.84	-17.10	-23.80	-26.73	-33.97	-39.88	-50.27
216	1.46	-3.30	-5.65	-7.99	-15.17	-18.38	-25.87	-31.92	-42.85
237	-7.08	-8.18	-9.12	-10.06	-13.55	-14.80	-15.99	-17.33	-22.43
238	-11.15	-14.13	-15.91	-17.69	-22.43	(-23.87)			
709	-3.91	-4.82	-5.65	-6.49	-9.67	-10.74	-11.72	-12.96	-15.70
711	-2.74	-3.65	-4.48	-5.31	-8.48	-9.56	-10.54	-11.78	-14.52
738	-62.71	-62.48	-62.65	-62.83	-63.54	-62.66	-61.44	-61.70	-63.34
744	-61.58	-61.32	-61.51	-61.69	-62.41	-61.50	-60.26	-60.53	-62.18
747	-54.81	-54.52	-54.71	-54.91	-55.64	-54.60	-53.92	-54.32	-56.07
748	-58.44	-58.17	-58.36	-58.55	-59.27	-58.34	-57.09	-57.37	-59.02
749	-58.72	-58.42	-58.62	-58.82	-59.55	-58.58	-57.31	-57.61	-59.26
750	-57.59	-57.34	-57.53	-57.71	-58.48	-57.52	-56.29	-56.56	-58.20
752	-30.89	-35.80	-38.15	-40.50	-47.43	-49.77	-49.06	-49.44	-51.18
754	-30.94	-35.85	-38.20	-40.54	-47.47	-49.82	-49.11	-49.48	-51.22
756	-27.35	-31.81	-34.03	-36.25	-42.61	-45.24	(-52.07)	(-57.61)	(-66.44)
757	-17.02	-21.57	-23.83	-26.10	-32.77	-35.64	-42.80	-48.63	(-58.67)
758	5.38	0.60	-1.76	-4.11	-11.36	-14.61	-22.13	-28.20	-39.21
764	-19.89	-25.90	-28.48	-31.08	-39.55	-42.83	-42.98	-43.37	-45.11

Sources: Royer and Sandwell [1989]; Royer and Chang [1991]; Royer and Coffin [1992].

result of increased thermal stratification within the basin. Southern sites were under the influence of a proto-AAIW or CPDW while sites to the north were bathed by a warmer (+2 to 3°C), more saline water mass possibly derived from the Tethyan Sea. These observations imply the formation of a proto-polar front (sub-polar convergence), possibly the delayed effects of the thermal isolation of Antarctica that began several million years earlier. From 42 to 36 Ma bottom waters cooled by an additional 4°C. Carbon isotope distributions indicate that bottom waters of the Indian Ocean were young and nutrient-depleted relative to other basins throughout middle to late Eocene time.

3) During early Oligocene time, southern Indian Ocean sites record in detail a major expansion of Antarctic ice sheets. During this event, bottom waters cooled by an additional 2 to 3°C as the flux of cold, nutrient-depleted water from the Antarctic increased. This cooling reinforced north-south thermal segregation of shallow intermediate waters.

4) Early Miocene bottom waters were slightly warmer than Oligocene. Carbon isotope evidence indicates the occurrence of young, nutrient-depleted waters from both the Tethyan region and the southern high-latitude source. In the middle Miocene, high latitude deep waters cooled by a few degrees as global ice volume increased. This cooling was preceded by a basin-wide positive excursion in carbon isotope values. Near the end of Miocene time (~7 to 8 Ma) the carbon isotope composition of the Indian Ocean decreased to values below that of the North Atlantic indicating a major reconfiguration of intermediate and deep water circulation patterns.

Appendix 2. Hole 690B Age Model Parameters.

Datum	Depth	Age
B C20R	118.70	46.23
T C21N	119.01	49.20
B C21R	130.48	51.95
B CP12	131.40	52.40
B C22R	132.84	52.80
T C23N	132.85	54.65
B C23N	133.18	53.70
FO D. lodoenis	134.41	55.20
B C24R	185.50	58.60
FO H. kleinpelli	213.40	61.60
FO F. tympaniformis	229.40	62.00

Sources: Spieß [1990]; Pospichal and Wise [1990].

Acknowledgments. We thank Lowell Stott, Lisa Sloan, Ellen Thomas, and K. C Lohmann for discussion and comments, and Wolf Berger and Richard Corfield for their reviews. We also thank Linda Albertzart, Paul Stebleton, and James Burdett for technical assistance and Jean-Yves Royer for providing the paleocoordinates of each Indian Ocean DSDP and ODP site. Portions of this investigation were supported by JOI/USSAC awards to Zachos and to Rea, and by NSF grants OCE-8811299 to Rea and OCE-9012389 to Zachos. Tables of the data presented in this paper can be obtained from the principal author upon written request.

REFERENCES

Aubry, M. P., Paleogene calcareous nannofossils from the Kerguelen Plateau, in *Proc. ODP, Sci. Results, 120*, edited by R. Schlich, S. W. Wise Jr. et al., 471-492, 1992.
Aubry, M.-P., Berggren, W. A., Kent, D. V., Flynn, J. J., Klitgord, K. D., Obradovich, J. D., and Prothero, R., Paleogene geochronology: An integrated approach,

Paleoceanography, 3, 707-742, 1988.

Backman, J., Schneider, D. A., Rio, D., and Okada, H., Neogene low-latitude magnetostratigraphy from Site 710 and revised age estimates of Miocene nannofossil datum events, in *Proc. ODP, Sci. Results, 115*, edited by J. Backman, R. A. Duncan, L. C. Peterson et al., 271-276, 1990.

Barrera, E. and Huber, B. T., Paleogene and early Neogene oceanography of the southern Indian Ocean: Leg 119 foraminifer stable isotope results, in *Proc. ODP, Sci. Results, 119*, edited by J. Barron, B. Larsen et al., 693-718, 1991.

Barron, E.J., and Peterson, W.H., The Cenozoic ocean circulation based on ocean general circulation model results, *Paleogeogr., Palaeclimatol., Palaeoecol., 83*, 1-28, 1991.

Barrera, E., Barron J., and Halliday, A., Strontium isotope stratigraphy of the Oligocene-lower Miocene section at Site 744, southern Indian Ocean, in *Proc. ODP, Sci. Results, 119*, edited by J. Barron, B. Larsen et al., 731-738, 1991.

Barrera, E., Keller, G., and Savin, S. M., Evolution of the Miocene ocean in the eastern North Pacific as inferred from oxygen and carbon isotopic ratios of foraminifera, in *The Miocene Ocean, GSA Memoir 163*, edited by J. P. Kennett, 83-102, 1985.

Barron, J., Larsen, B., et al., *Proc. ODP, Init. Repts., 119*, 1989.

Berggren, W. A., Kent, D. V., and Flynn, J. J., Jurassic to Paleogene: Part 2, Paleogene geochronology and chronostratigraphy, in *The Chronology of the Geological Record, Geol. Soc. London Mem. 10*, edited by N. J. Snelling, 141-195, 1985a.

Berggren, W. A., Kent, D. V., and Van Couvering, J. A., The Neogene: Part 2, Neogene geochronology and chronostratigraphy, in *The Chronology of the Geological Record, Geol. Soc. London Mem. 10*, edited by N. J. Snelling, 211-260, 1985b.

Berggren, W. A., Kent, D. V., Flynn, J. J. and Van Couvering, J. A., Cenozoic geochronology, *Geol. Soc. Am. Bull., 96*, 1407-1418, 1985c.

Berggren, W.A., and Hollister, Plate Tectonics and paleocirculation - commotion in the ocean, *Tectonophysics, 38*, 11-48, 1977.

Boersma, A., Premoli-Silva, I., and Shackleton, N.J., Atlantic Eocene planktonic foraminiferal paleohydrographic indicators and stable isotope paleoceanography, *Paleoceanography, 2*, 287-331, 1987.

Boudreaux, J. E., Calcareous nannoplankton ranges, Deep Sea Drilling Project Leg 23, *Init. Repts. DSDP, 23*, 1073-1090, 1974.

Brass, G.W., Southam, J.R., & Peterson, W.H., Warm saline bottom water in the ancient ocean, *Nature, 296*, 620-623, 1982.

Breza, J. and Wise, S. W., Jr., Lower Oligocene Ice-Rafted Debris on the Kerguelen Plateau: evidence for East Antarctic Continental Glaciation, in *Proc. ODP, Sci. Results, 120*, edited by R. Schlich, S. W. Wise Jr. et al., 161-178, 1992.

Coffin, M. F., Subsidence of the Kerguelen Plateau: The Atlantis concept, in *Proc. ODP, Sci. Results, 120*, edited by R. Schlich, S. W. Wise Jr. et al., 945-950, 1992.

Corliss, B. H., Recent deep-sea benthonic foraminiferal distributions in the southeast Indian Ocean: Inferred bottom-water routes and ecological implications, *Mar. Geol., 31*, 115-138, 1979.

Curry, W. B., and Lohmann, G. P., Carbon isotopic changes in benthic foraminifera from the western South Atlantic: Reconstruction of glacial abyssal circulation patterns, *Quat. Res., 18*, 218-235, 1982.

Davies, G. A., Luyendyk, B. P. et al., *Init. Repts. DSDP, 26*, 1974.

Davis, E. E., and Lister, C. R. B., Fundamentals of ridge crest topography, *Earth Planet. Sci. Letts., 21*, 405-413, 1974.

Duncan, R. A., Geochronology of basalts from the Ninetyeast Ridge and continental dispersion in the eastern Indian Ocean, *J. Volcanol. Geotherm. Res., 4*, 283-305, 1978.

Duplessy, J.-C., Shackleton, N. J., Matthews, R. K., Prell, W.,

Ruddiman, W. F., Caralp, M., and Hendy, C. H., C[13] record of benthic foraminifera in the last interglacial ocean: implication for the carbon cycle and the global deep water circulation, *Quat. Res., 21*, 225-243, 1984.

Ehrmann, W. U., Implications of sediment composition on the southern Kerguelen Plateau for paleoclimate and depositional environment, in *Proc. ODP, Sci. Results, 119*, edited by J. Barron, B. Larsen et al., 185-210, 1991.

Fisher, R. L., Bunce, E. T. et al., *Init. Repts. DSDP, 24*, 1183 pp, 1974.

Gartner, S., Neogene calcareous nannofossil biostratigraphy, Leg 116 (central Indian Ocean). in Cochran, J. R., Stow, D. V. et al., *Proc. ODP, Sci. Results, 116*, 165-187, 1990.

Gartner, S., Jr., Nannofossil biostratigraphy, Leg 22, Deep Sea Drill. Proj., in *Init. Repts. DSDP, 22*, edited by C. C. von der Borch, J. G. Sclater et al., 577-600, 1974.

Hambrey, M. J., Ehrmann, W. U., and Larsen, B., Cenozoic glacial record of the Prydz Bay Continental Shelf, East Antarctica, in *Proc. ODP, Sci. Results, 119*, edited by J. Barron, B. Larsen et al., 77-132, 1991.

Harwood, D. M., Lazarus, D. B., Abelman, A., Aubry, M. P., Berggren, W. A., Heider, F., Inokuchi, H., Maruyama, T., McCartney, K., Wei, W., and Wise, S. W., Jr., Neogene intergrated magnetobiostratigraphy of the central Kerguelen Plateau, Leg 120, in *Proc. ODP, Sci. Results, 120*, edited by R. Schlich, S. W. Wise Jr. et al., 1031-1052, 1992.

Hovan, S. A., and Rea, D. K., The Cenozoic record of continental mineral deposition on Broken and Ninetyeast Ridges, Indian Ocean: southern African aridity and sediment discharge from the Himalayas, *Paleoceanography*, in press.

Hovan, S. A., and Rea, D. K., Paleocene-Eocene boundary changes in atmospheric and oceanic circulation: a southern hemisphere record, *Geology, 20*, 15-18, 1992.

Inokuchi, H., and Heider, F., Magnetostratigraphy of sediments from Site 748 and 750, ODP Leg 120, in *Proc. ODP, Sci. Results, 120*, edited by R. Schlich, S. W. Wise Jr. et al., 247-254, 1992.

Katz, M. E. and Miller, K. G., Early Paleogene benthic foraminiferal assemblage and stable isotope composition in the southern ocean, Ocean Drilling Program Leg 114, in *Proc. ODP, Init. Repts. 114 pt. B*, 481-516, 1991.

Keating, B. H., and Sakai, H., Magnetostratigraphic studies of sediment from Site 744, southern Kerguelen Plateau, in *Proc. ODP, Sci. Results, 119*, edited by J. Barron, B. Larsen et al., 771-794, 1991.

Keigwin, L. D. and Corliss, B. H., Stable isotopes in late middle Eocene to Oligocene foraminifera, *Geol. Soc.Am. Bull., 97*, 335-345, 1986.

Keigwin, L. D., Paleoceanographic change in the Pacific at the Eocene-Oligocene boundary, *Nature, 287*, 722-725, 1980.

Keller, G., Biochronology and paleoclimatic implications of middle Eocene to Oligocene planktonic foraminiferal faunas, *Mar. Micropaleontol., 7*, 463-468, 1983.

Kennett, J. P. and Shackleton, N. J., Oxygen isotopic evidence for the development of the psychrosphere 38 m.y. ago, *Nature, 260*, 513-515, 1976.

Kennett, J. P. and Stott, L. D., Proteus and Proto-Oceanus, Paleogene Oceans as revealed from Antarctic stable isotopic results; ODP Leg 113, in *Proc. ODP, Init. Repts., 113*, edited by J. P. Kennett, and P. F. Barker et al., 865-880, 1990.

Kennett, J. P. and Stott, L. D., Abrupt deep-sea warming, paleoceanographic changes and benthic extinctions at the end of the Paleocene, *Nature, 353*, 225-229, 1991.

Kennett, J. P. and Barker, P. F., Climatic and Oceanographic Developments in the Weddell Sea, Antarctica, since the latest Cretaceous, an ocean-drilling perspective, in *Proc. ODP, Init. Repts., 113*, edited by J. P. Kennett, and P. F. Barker et al., 937-962, 1990.

Kolla, V., Sullivan, L., Streeter, S. S., and Langseth, M. G., Spreading of Antarctic bottom water and its effects on the floor of the Indian Ocean inferred from bottom water potential temperature, turbidity, and sea-floor photography,

Marine Geology, 21, 171-189, 1976.

Kroopnick, P., The distribution of ^{13}C of TCO_2 in the world oceans, *Deep Sea Res., 32*, 57-84, 1985.

Lohmann, G. P., Abyssal benthonic foraminifera as hydrographic indicators in the western South Atlantic Ocean, *J. Foraminiferal Res., 8*, 6-34, 1978.

Mackensen, A., Barrera, E., and Hubberton, H.-W., Neogene circulation in the Southern Ocean: Evidence from benthic foraminifers, carbonate data, and stable isotope analyses (Site 751), in *Proc. ODP, Sci. Results, 120*, edited by R. Schlich, S. W. Wise Jr. et al., 867-880, 1992.

Mantyla, A. W., and Reid, J. L., Abyssal characteristics of the World Ocean waters, *Deep Sea Res., 30*, 805-833, 1983.

Martini, E., Standard Tertiary and Quaternary calcareous nannoplankton zonation, in *Proceedings of the Second international Conference on Planktonic Microfossils, 2*, edited by A. Farinacci, 739-785, Roma, Rome (Tecnoscienza), 1971.

Matthews, R. K. and Poore, R. Z., Tertiary $\delta^{18}O$ record and glacio-eustatic sea-level fluctuations, *Geology, 8*, 501-504, 1980.

McGowran, B., Foraminifera, in *Init. Repts. DSDP, 22*, edited by C. C. von der Borch, J. G. Sclater et al., 601-608, 1974.

McGowran, B., Fifty million years ago, *Am. Scientist, 78*, 31-39, 1990.

Metzl, N., Moore, B., and Poisson, A., Resolving the intermediate and deep advective flows in the Indian Ocean by using temperature, salinity, oxygen and phosphate data: the interplay of biogeochemical and geophysical tracers, *Palaeogeog., Palaeoclimat., Palaeoecol., 89*, 81-111, 1990.

Miller, K. G., and Fairbanks, R. G., Oigocene to Miocene carbon isotope cycles and abyssal circualtion changes, in The Carbon Cycle and Atmospheric CO2: Natural Variations Archean to present, *Geophysical Monograph, 32*, 469-486, 1985.

Miller, K. G., and Katz, M. E., Oligocene to Miocene benthic foraminifera and abyssal circulation changes in the North Atlantic, *Micropaleontol., 33*, 97-149, 1987.

Miller, K. G., and Thomas, E., Late Eocene to Oligocene benthic foraminifera isotopic record, Site 574, equatorial Pacific, *Init. Repts. DSDP, 85*, 771-777, 1985.

Miller, K. G., Fairbanks, R. G., and Mountain, G. S., Tertiary oxygen isotope synthesis, sea-level history, and continental margin erosion, *Paleoceanography, 2*, 1-19, 1987a.

Miller, K. G., Janecek, T. R., Katz, M. E., and Keil, D. J., Abyssal circulation and benthic foraminiferal changes near the Paleocene/Eocene boundary, *Paleoceanography, 2*, 741-761, 1987b.

Miller, K. G., Feigenson, M. D., Kent, D., and Olsson, R. K., Upper Eocene to Oligocene isotope (^{87}Sr\^{86}Sr), $\delta^{18}O$, $\delta^{13}C$) standard section, Deep-sea Drill. Proj. Site 522, *Paleoceanography, 3*, 223-233, 1988.

Nomura, R., Paleoceanography of upper Maestrichtian to Eocene benthic foraminiferal assemblages at Site 752, 753, and 754, eastern Indian Ocean, in *Proc. ODP, Sci. Results, 121*, edited by J. Weissel, J. Peirce, E. Taylor, J. Alt et al., 3-29, 1991a.

Nomura, R., Oligocene to Pleistocene benthic foraminifer assemblages at Sites 754 and 756, eastern Indian Ocean, in *Proc. ODP, Sci. Results, 121*, edited by J. Weissel, J. Peirce, E. Taylor, J. Alt et al., 31-75, 1991b.

Nomura, R., Seto, K., and Niitsuma, N., Late Cenozoic deep-sea benthic foraminiferal changes and isotopic records in the eastern Indian Ocean, in *Fourth international Symposium on Benthic Foraminifera*, edited by T. Saito, (Sendai, September, 1990), in press.

Oberhänsli, H., Latest Cretaceous-Early Neogene oxygen and carbon isotopic record at DSDP sites in the Indian Ocean, *Mar. Micropaleontol., 10*, 91-115, 1986.

Oberhänsli, H., and Toumarkine, M., The Paleogene oxygen and carbon isotope history of Sites 522, 523, and 524 from the central South Atlantic, in *South Atlantic Paleoceanography*, edited by K. Hsü, H. J. Weissert, 125-148, Cambridge University Press, Cambridge, England, 1985.

Okada, H., and Bukry, D., Supplementary modification and introduction of code number to the low-latitude coccolith biostratigraphic zonation (Bukry, 1973; 1975), *Mar. Micropaleontol., 5*, 321-325, 1980.

Peirce, J., Weissel, J. et al., *Proc. ODP, Init. Repts., 121*, Ocean Drilling Program, College Station, TX, 1989.

Poore, R. Z. and Matthews, R. K., Oxygen isotope ranking of late Eocene and Oligocene planktonic foraminifers, Implications for Oligocene sea-surface temperatures and global ice-volume, *Mar. Micropaleontol. 9*, 111-134, 1984.

Pospichal, J. J., and Wise, S. W., Jr., Paleocene to Eocene calcareous nannofossils of ODP Sites 689 and 690, in *Proc. ODP, Sci. Results, 113*, edited by J. P. Kennett, and P. F. Barker et al., 613-638, 1990.

Pospichal, J. W. et al., Cretaceous-Paleogene biomagnetostratigraphy of Sites 752-755, Broken Ridge: a synthesis, in *Proc. ODP, Sci. Results, 121*, edited by J. Weissel, J. Peirce, E. Taylor, J. Alt et al., 721-742, 1991a.

Pospichal, J. W., Wei, W., and Wise, Jr., S. W., Data report, Indian Ocean Oligocene strontium isotope stratigraphy, Legs 119-121, Kerguelen Plateau and Ninetyeast Ridge, in *Proc.ODP, Sci. Results, 121*, edited by J. Peirce, J. Weissel et al., 921-932, 1991b.

Premoli-Silva, I. and Spezzaferri, S., Paleogene planktonic foraminifer biostratigraphy and paleoenvironmental remarks on Paleogene sediments from Indian Ocean sites, in *Proc. ODP, Sci. Results, 115*, edited by R. A. Duncan, J. Backman, L. C. Peterson et al., 277-314, 1990.

Rea, D. K., Lohmann, K. C., MacLeod, N. D., House, M. A., Hovan, S. A., and Martin, G. D., Oxygen and carbon isotopic records the oozes of ODP Sites 752, 754, 756, and 757, eastern Indian Ocean, in *Proc. ODP, Sci. Results, 121*, edited by J. Weissel, J. Peirce, E. Taylor, J. Alt et al., 229-240, 1991.

Rea, D. K., and Leinen, M., Crustal subsidence and calcite deposition in the South Pacific Ocean, in *Init. Repts. DSDP, 92*, edited by M. Leinen, D. K. Rea et al., 299-303, 1986.

Rea, D. K., Dehn, J., Driscoll, N.W., Farrell, J., Janecek, T. R., Owen, R. M., Pospichal, J. J., Resiwati, P., and the ODP Leg 121 Scientific Party, Paleoceanography of the eastern Indian Ocean from ODP Leg 121 drilling on Broken Ridge, *Geol. Soc. Am. Bull., 102*, 679-690, 1990.

Rio, D., Fornaciari, F., and Raffi, I., Late Oligocene through early Pleistocene calcareous nannofossils from western equatorial Indian Ocean (Leg 115), in *Proc. ODP Sci. Results, 115*, edited by J. Backman, R. A. Duncan, L. C. Peterson et al., 175-235, 1990.

Roth, P. H., Calcareous nannofossils from the Northwestern Indian Ocean, Leg 24, Deep Sea Drill. Proj., in *Init. Repts. DSDP, 24*, edited by R. L. Fisher, E. T. Bunce et al., 969-994, U.S. Government Printing Office, Washington, D.C., 1974.

Royer J.-Y. and Chang, T., Evidence for relative motions between the Indian and Australian plates during the last 20 Ma from plate tectonic reconstructions: implication for the deformation of the Indo-Australian plate, *J. Geophys. Res.*, 1991.

Royer, J. -Y. and Coffin, M. F., Jurassic to Eocene plate tectonic reconstructions in the Kerguelen Plateau Region, in *Proc. ODP, Sci. Results, 120*, edited by R. Schlich, S. W. Wise Jr. et al., 917-928, 1992.

Royer, J.-Y. and Sandwell, D. T., Evolution of the eastern Indian Ocean since the Late Cretaceous: constraints from GEOSAT altimetry, *J. Geophys. Res., 94*, 13755-13782, 1989.

Savin, S. M., The history of the Earth's Surface temperature during the past 100 million years, *Ann. Re Earth Planet. Sci., 3*, 319-355, 1977.

Savin, S. M., Abel, L., Barrera, E., Hodell, D., Keller, G., Kennett, J. P., Killingley, J., Murphy, M., and Vincent, E., The evolution of Miocene surface and near-surface marine temperatures: Oxygen isotopic evidence, in *The Miocene Ocean, GSA Memoir 163*, edited by J. P. Kennett, 49-82,

1985.

Savin, S. M., Douglas, R. G., Keller, G., Killingley, J. S., Shaughnessey, Sommer, M. A., Vincent, E., and Woodruff, F., Miocene benthic foraminiferal isotope record: a synthesis. *Mar. Micropaleontol., 6,* 423-450, 1981.

Schlich, R., Wise, S. W., Jr. et al., *Proc. ODP, Init. Repts., 120,* Ocean Drilling Program, College Station, TX, 1989.

Sclater, J. G., Anderson, R. N., and Bell, M. L., The elevation of ridges and the evolution of the central Pacific, *J. Geophys. Res., 76,* 7888-7915, 1971.

Sclater, J. G., Meinke, L., Bennett, A., and Murphy, C., The depth of the ocean through the Neogene, in*The Miocene Ocean: Paleoceanography and Biogeography, Geol. Soc. of Am. Memoir 163,* edited by J. P. Kennett, 1-19, Boulder, CO, 1985.

Seto, K., Nomura, R., and Niitsuma, N., Data Report: oxygen and carbon isotopic records of the upper Maestrichtian to lower Eocene benthic foraminifers at Site 752 in the eastern Indian Ocean, in *Proc. ODP, Sci. Results, 121,* edited by J. Weissel, J. Peirce, E. Taylor, J. Alt et al., 885-889, 1991.

Shackleton, N. J., Oxygen isotope evidence for climatic change, in *Fossils and climate,* edited by P. Brenchley, 27-34, John Wiley and Sons, New York,1984.

Shackleton, N. J., Oceanic carbon isotope constraints on oxygen and carbon dioxide in the Cenozoic atmosphere, in *Natural Variations in Carbon Dioxide and the Carbon Cycles,* edited by E. T. Sundquist and W. S. Broecker, Am. Geophys. Union Monograph, 32, 412-417, 1985.

Shackleton, N. J., and Kennett, J. P., Paleotemperature history of the Cenozoic and the initiation of Antarctic glaciation: oxygen and carbon isotope analyses in DSDP sites 277, 279 and 281, in *Init. Repts. DSDP, 29,* 743-755, 1975.

Shackleton, N. J., Hall, M. A., and Boersma, A., Oxygen and carbon isotope data from Leg 74 foraminifers, in *Init. Repts. DSDP, 74,* edited by T. C. Moore Jr., P. D. Rabinowitz et al., 599-612, 1984.

Shackleton, N. J. and Boersma, A., The climate of the Eocene ocean, *J. Geol. Soc., 138,* 153-157, London, 1981.

Simmons, G. R., Subsidence history of basement sites along a carbonate dissolution profile, Leg 115, in *Proc. ODP, Sci. Results, 115,* edited by R. A. Duncan, J. Backman, L. C. Peterson et al., 123-126, 1990.

Spieß, V., Cenozoic magnetostratigraphy of Leg 113 drill sites, Maud Rise, Weddell Sea, Antarctica, in *Proc. ODP, Sci. Results, 113,* edited by J. P. Kennett, and P. F. Barker et al., 261-318, 1990.

Stott, L. D., Kennett, J. P., Shackleton, N. J. and Corfield, R. M., The evolution of Antarctic surface waters during the Paleogene, inferences from the stable isotopic composition of planktonic foraminifera, ODP Leg 113, in *Proc. ODP, Sci. Results, 113,* edited by J. P. Kennett, and P. F. Barker et al., 849-864, 1990.

Tapscott, C. R., Patriat, P., Fisher, R. L., Sclater, J. G., Hoskins, H., and Parsons, B., The Indian Ocean triple junction, *J. Geophys. Res., 85,* 4723-4739, 1980.

Thierstein, H. R., Calcareous Nannoplankton - Leg 24 Deep Sea Drill. Proj., in *Init. Repts. DSDP, 26,* edited by G. A. Davies, B. P. Luyendyk et al., 619-668, 1974.

Thomas, E., Late Eocene to recent deep-sea benthic foraminifers from the central equatorial Pacific Ocean, in *Init. Repts. DSDP, 85,* edited by L. Mayer, F. Theyer et al., 655-694, 1985.

Thomas, E., Changes in composition of Neogene benthic foraminiferal faunas in equatorial Pacific and north Atlantic, *Palaeogeogr., Palaeoclimatol., Palaeoecol., 53,* 47-16, 1986.

Thomas, E., Late Cretaceous through Neogene deep-sea benthic foraminifers (Maud Rise, Weddell Sea, Antarctica), in *Proc. ODP, Sci. Results, 113,* edited by J. P. Kennett, and P. F. Barker et al., 571-594, 1990.

Thomas, E., and Vincent, E., Equatorial Pacific and deep-sea benthic foraminifera: faunal changes before the middle Miocene polar cooling, *Geology, 15,* 1035-1039, 1987.

Tjalsma, R. C. and Lohmann, G. P., Paleocene-Eocene bathyal and

abyssal benthic foraminifera from the Atlantic Ocean, *Micropaleontology Spec. Publ., 4,* 1-90, 1983.

van Andel, T. H., Mesozoic/Cenozoic calcite compensation depth and the global distribution of calcareous sediments, *Earth Planet. Sci. Letts., 26,* 187-194, 1975.

Vincent, E., and Berger, W. H., Carbon dioxide and polar cooling in the Miocene: The Monterey hypothesis, in *Natural Variations in Carbon Dioxide and the Carbon Cycles,* edited by E. T. Sundquist and W. S. Broecker, Am. Geophys. Union Monograph, 32, 455-468, 1985.

Vincent, E., Killingley, J. S., and Berger, W. H., Miocene oxygen and carbon isotope stratigraphy of the tropical Indian Ocean. in *The Miocene Ocean: Paleoceanography and Biogeography, Mem. Geol. Soc. Am., 163,* edited by J. P. Kennett, 103-130, 1985.

von der Borch, C. C., Sclater, J. G., et al., *Init. Rep.DSDP, 22,* (U.S. Govt. Printing Office), Washington, 1974.

Warren, B. A., Deep circulation of the world ocean, in *Evolution of Physical Oceanography,* edited by B. A. Warren and C. Wunsch, 6-41, MIT Press, 1981.

Wei, W., Middle Eocene-lower Miocene calcareous nannofossil magnetobiochronology of ODP Holes 699A and 703A in the subantarctic South Atlantic, *Mar. Micropaleontol., 18,* 143-165, 1991.

Wei, W., and Thierstein, H. R., Upper Cretaceous and Cenozoic calcareous nannofossils of the Kerguelen Plateau (southern Indian Ocean) and Prydz Bay (East Antarctica), in *Proc.ODP, Sci. Results, 119,* Barron, J., Larsen, B., et al., 467-494, 1991.

Wei, W., and Wise, S. W., Jr., Paleogene calcareous nannofossil magnetobiochronology: results from South Atlantic DSDP Site 516, *Mar. Micropaleontol., 14,* 119-152, 1989.

Wei, W., and Wise, S. W., Jr., Middle Eocene to Pleistocene calcareous nannofossils recovered by the Ocean Drill. Progr. Leg 113 in the Weddell Sea, in *Proc. ODP, Sci. Results, 113,* edited by J. P. Kennett, and P. F. Barker et al., 639-666, 1990.

Wei, W., Villa, G., and Wise, S. W., Jr., Paleoceanographic implications of Eocene-Oligocene calcareous nannofossils from ODP Sites 711 and 748 in the Indian Ocean, in *Proc.ODP, Sci. Results, 120,* edited by R. Schlich, S. W. Wise Jr. et al., 979-1001, 1992.

Whitemarsh, R. B., Weser, O. E., Ross, D. A. et al., *Init. Repts. DSDP, 23,* pp. 1180, 1974.

Wise, S. W., Breza, J. R., Harwood, D. M., Wei, W., Paleogene glacial history of Antarctica, in *Controversies in Modern Geology,* 133-171, Academic Press, 1991.

Wise, S. W., Gombos, A. M., and Muza, J. P., Cenozoic evolution of polar water masses, southwest Atlantic Ocean, in *South Atlantic Paleoceanography,* edited by K. Hsü, H. J. Weissert, 283-324, Cambridge University Press, Cambridge, England, 1985.

Woodruff, F., Changes in Miocene deep-sea benthic foraminiferal distribution in the Pacific Ocean: Relationship to paleoceanography, in *The Miocene Ocean: Paleoceanography and Biogeography. Geol. Soc. Am. Mem., 163,* edited by J. P. Kennett, 131-176, 1985.

Woodruff, F., and Savin, S. M., Miocene deep-water oceanography, *Paleoceanography, 4,* 87-140, 1989.

Woodruff, F., Savin, S. M., and Abel, L., Miocene benthic foraminifer oxygen and carbon isotopes, Site 709, Indian Ocean, in *Proc. ODP, Sci. Results, 115,* edited by R. A. Duncan, J. Backman, L. C. Peterson et al., 519-528, 1990.

Woodruff, F., and Savin, S. M., Mid-Miocene isotope stratigraphy in the deep sea: high resolution correlations, paleoclimatic cycles, and sediment preservation, *Paleoceanography, 6,* 755-805, 1991.

Wright, J. D., Miller, K. G., and Fairbanks, R. G., Evolution of modern deepwater circulation: Evidence from the late Miocene Southern Ocean, *Paleoceanography, 6,* 275-290, 1991.

Wright, J. D., Miller, K. G., and Fairbanks, R. G., Early and middle Miocene stable isotopes: implications for deepwater circulation, climate, and tectonics, *Paleoceanography, 7,* 357-390, 1992.

Wyrtki, K. , Physical oceanography of the Indian Ocean, in *The Biology of the Indian Ocean,* 18-36, Springer-Verlag, 1973.

Zachos, J. C., Berggren, W. A., Aubry, M.-P., and Mackensen, A., Isotope and trace element geochemistry of Eocene and Oligocene foraminifers from Site 748, Kerguelen Plateau, in *Proc. ODP, Sci. Results, 120,* edited by R. Schlich, S. W.

Wise Jr. et al., 839-854, 1992a.

Zachos, J. C., Breza, J., and Wise, S. W., Jr., Early Oligocene ice-sheet expansion on Antarctica: sedimentological and Isotopic evidence from Kerguelen Plateau, *Geology, 20,* 569-573, 1992b.

Zahn, R., and Mix, A. C., Benthic foraminiferal $\delta^{18}O$ in the Ocean's temperature-salinity-density field: constraints on Ice Age thermohaline circulation, *Paleoceanography, 6,* 1-20, 1991.

Delivery of Himalayan Sediment to the Northern Indian Ocean and its Relation to Global Climate, Sea Level, Uplift, and Seawater Strontium

DAVID K. REA

Department of Geological Sciences, The University of Michigan, Ann Arbor, MI, 48109-1063, USA

The mass accumulation rate of the terrigenous component of deep sea sediments has been quantified for eleven DSDP and ODP sites in the northern Indian Ocean. Depositional patterns at these sites all show very low input prior to 12 Ma, a five-fold increase in terrigenous flux starting 12 Ma, and distinct flux maxima in the late Miocene and the middle Pliocene. Combining the individual site records into one stacked, averaged and normalized record greatly improves the temporal definition of the overall depositional record. Peaks in sediment delivery clearly do not correspond either with Neogene sea-level changes nor with times of global climate change, implicating Himalayan uplift as the process that controls deposition in the northern Indian Ocean. The late Cenozoic uplift may have occurred in two stages, one in the late Miocene, 9 to 6 Ma, and a later mid-Pliocene phase from 4 to 2 Ma. Comparison of the sediment flux record to the seawater $^{87}Sr/^{86}Sr$ curve shows a distinct lack of correspondence, such that the timing and direction of changes in the slope of the $^{87}Sr/^{86}Sr$ curve do not correspond with the record of clastic input. Other processes influencing the Sr composition of seawater during the past 40 m.y. must be more important than previously considered: variation in sea-floor hydrothermal activity and/or important changes in the $^{87}Sr/^{86}Sr$ ratio of dissolved riverine strontium.

INTRODUCTION

The delivery of clastic sediments to the northern Indian Ocean from south-central Asia is, as for all sub-continental regions, controlled by some combination of uplift, changes in sea level, and the changes in global climates. If it is possible to sort out the effects of climate and sea level on this process, then a record of Himalayan uplift may be achievable. The uplift of the Himalayan ranges and Tibetan Plateau is of itself an important question in tectonics. Further, this uplift may have consequences that extend far beyond the mountainous regions of south-central Asia, consequences that may include ocean geochemistry [Raymo et al., 1988; Raymo, 1991; Richter et al., 1992], the initiation of Asian monsoonal atmospheric and oceanic circulation in the earlier part of the late Miocene [Prell and Kutzbach, 1991; Kroon et al., 1991], and even the mid-Pliocene onset of Northern Hemisphere glaciation [Ruddiman and Kutzbach, 1989]. If this tectono-climatic association is valid, then scenarios which call for modest middle Cenozoic uplift and very rapid late Cenozoic uplift are appropriate. Recent work has demonstrated that the rapid northward motion of the Indian subcontinent slowed markedly at about 55 Ma, presumably upon contact with Asia [Klootwijk et al., 1991, 1992], and that turbidite deposition began in the northern Indian Ocean in early Miocene [Cochran, 1990] or perhaps Oligocene [Kolla and Coumes, 1987] time. Many authors suggest relatively rapid uplift since about the middle Miocene [Gansser, 1964, 1981; Zeitler, 1985; Molnar et al., 1987; Searle et al., 1987; Copeland and Harrison, 1990], but there is not yet a consensus on the temporal details of the mid-to late-Cenozoic uplift of the Himalayan Mountains [see above references and Ruddiman et al., 1989; Cochran, Stow et al., 1990; Molnar and England, 1990; Richter et al., 1991; Harrison et al., 1992]. Quantification of the oceanic depositional record permits an overview of the sediment delivery to the northern Indian Ocean that integrates all the controlling influences of uplift, climate, and sea-level.

Determination of mountain uplift history by investigation of the clastics shed therefrom is a time-honored method in geological sciences. The enormous sedimentary deposits of the Indus and Bengal fans are the direct result of Himalayan erosion and the construction of these fans should contain an unambiguous record of sediment supply from the Indian subcontinent [Curray and Moore, 1971; Kolla and Coumes, 1987]. Here I present the geologic history of the flux of clastic sediments to the northern Indian Ocean. The record of sediment delivery thus obtained can be compared to well known records of sea-level change and of climate change in order to discern any apparent relationships between them. The data presented below are appropriate to answering questions on timescales of hundreds of thousands of years and longer. Thus processes like the Plio-Pleistocene sea-level variations associated with orbital-timescale changes in ice volume will not be individually discernable in our results. On a longer timescale, however, such periods of rapid and large sea level changes should result in enhanced sediment supply to the deep sea.

DETERMINATION OF SEDIMENT FLUXES

The traditional value for sediment deposition rate is the linear sedimentation rate (LSR), commonly given in centimeters per thousand years (cm/ky), or meters per

Synthesis of Results from Scientific
Drilling in the Indian Ocean
Geophysical Monograph 70
Copyright 1992 American Geophysical Union

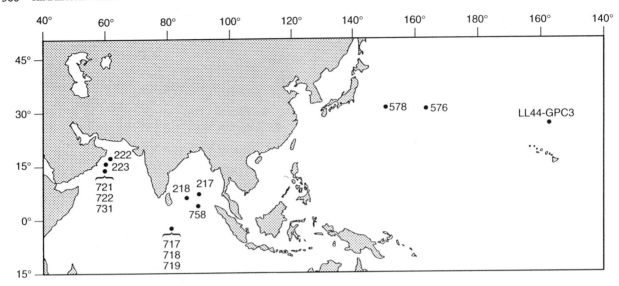

Fig. 1. Index map of the northern Indian Ocean showing locations of drill sites examined for sediment flux data.

million years (m/my). These values, however, have severe limitations and are useful only as a gross descriptor of sedimentation. A much more useful value to describe sedimentation is the mass accumulation rate (MAR). This value is a quantification of the true flux of sediment to the ocean floor measured in mass per unit area and unit time, commonly $g(cm^2_* ky)^{-1}$. The MAR is the product of the linear sedimentation rate and dry bulk density (DBD):

$$MAR(g(cm^2_* ky)^{-1}) = LSR(cm/ky) \times DBD(g/cm^3) \qquad (1)$$

These values have the properties of accounting for downcore compaction via the bulk density values, being comparable from site to site and ocean to ocean, and being useful for any quantitative mass balance calculation. Furthermore, since the relative abundance of any minor sediment component fluctuates in response to and antithetically from the abundance of any major sedimentary component, the only way to examine the true variability of all the sedimentary components in a core is to determine the MAR of each one:

$$MAR_{component} = MAR_{total} \times wt\%_{component} \qquad (2)$$

The Deep Sea Drilling Project and its successor the Ocean Drilling Program have recovered cores from eleven drill sites in the northern Indian Ocean (Figure 1; Table 1) which permit quantification of the sediment input into that ocean. MAR values for the terrigenous sediment component in each of these drill sites were determined using the general methodology established by Rea and Thiede [1981]. For each sediment core recovered, nominally 9.5 meters long, the percentage of terrigenous material was taken from the original lithologic, smear-slide descriptions [von der Borch, Sclater et al., 1974; Whitmarsh, Weser, Ross et al., 1974; Cochran, Stow et al., 1989; Prell, Niitsuma et al., 1989]. Wet-bulk density data were taken from the

shipboard GRAPE (Gamma Ray Attenuation Porosity Evaluator) measurements [ibid.] and converted to DBD values [Rea and Thiede, 1981]. Linear sedimentation rates were determined from comparison of the nannofossil biostratigraphy from the Leg 22 [von der Borch, Sclater et al., 1974] and 23 [Whitmarsh, Weser, Ross et al., 1974] site chapters with the timescale of Berggren et al. [1985]. For the Leg 116 sites I employed the nannofossil biostratigraphy presented by Gartner [1990] and for the Leg 117 sites I used the nannofossil stratigraphy of Sato et al. (1991). The data for Site 758, Leg 121, are from Hovan and Rea [unpublished manuscript, 1992]. Sedimentation rates are determined on a zone-by-zone basis and are well constrained for the Leg 116, 117 and 121 sites but the discontinuous coring practices of the earlier cruises entails deterioration of this type of data. For the Leg 22 and 23 sites a downcore depth-biostratigraphic zone plot was constructed and used to derive a conservative LSR model that had the fewest slope (rate) changes. Every sediment core was assigned an age based upon linear interpolation within the appropriate nannofossil zone. One terrigenous MAR value was calculated for each core from the average values of DBD, LSR and percent terrigenous component for that 9.5-meter interval. The accuracy estimated for these types of calculations is approximately ±25% with much of that error stemming from the smear-slide abundance estimates. This degree of accuracy is not a problem in interpretation as all sites show many-fold to order of magnitude changes in terrigenous MAR (Figures 2 and 3).

The deeper sites investigated here, those situated directly on the Indus (222, 223) and Bengal (218, 717, 718, 719) fans, are dominated by turbidites and are characterized by silts and sands. The sites on Owen Ridge (721, 722, 731) and Ninetyeast Ridge (217, 758) contain pelagic carbonate with a minor silt/clay component. This minor terrigenous component is the result of Himalayan-derived hemipelagic deposition at the northern Ninetyeast Ridge and of a

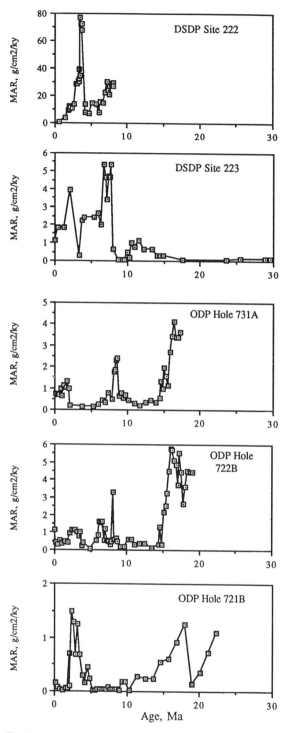

Fig. 2. Mass accumulation rates (MAR) of terrigenous sediments in drill sites on the Indus Fan. Note variable vertical scales.

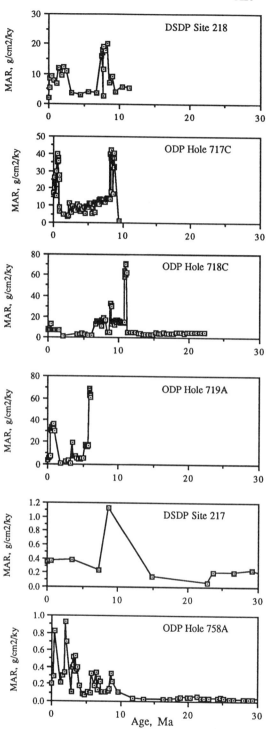

Fig. 3. Mass accumulation rates (MAR) of terrigenous sediments in drill sites on the Bengal Fan. Note variable vertical scales.

combination of hemipelagic and eolian material on Owen Ridge [Peirce, Weissel et al., 1989; Prell, Niitsuma et al., 1989].

The question of a single site on a deep-sea fan, or anywhere, giving a reliable overview of the construction of the entire deposit is relevant. In approaching this question

one can make two observations and one additional calculation. The observations are that the three sites drilled within 10 km of each other on the Bengal Fan (717, 718, 719) show similar sedimentation patterns (Figure 3) indicating this problem to be of far smaller magnitude than assumed previously [Cochran, 1990]. Secondly, the

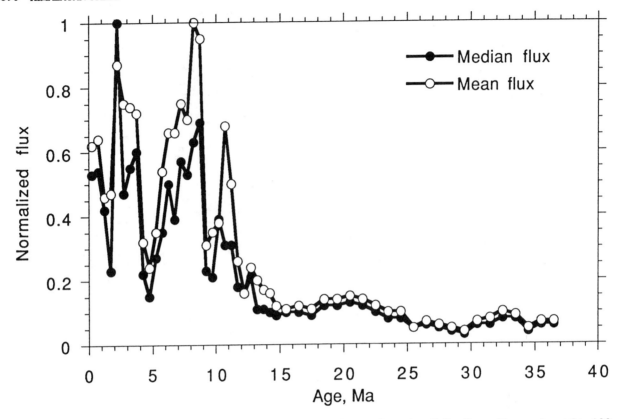

Fig. 4. Normalized, stacked and averaged sediment delivery from the Himalayas to the northern Indian Ocean. Note maxima at 2 to 4 Ma and 6 to 9 Ma.

general pattern of the individual flux records (Figures 2 and 3) are all reasonably similar suggesting similar depositional histories at all sites.

To further elucidate the true temporal pattern of sediment flux to the northern Indian Ocean, I have normalized, stacked and averaged the flux data from all sites. The time interval chosen is 0.5 m.y. for sediments less than 15 m.y. old, and 1.0 m.y. for older sediment. For every site, the flux data for each time interval were averaged and then normalized such that the highest flux at any site has a relative value of 1.0 and all other values are less. The normalized values from all eleven sites are summed for each time interval and divided by the number of sites contributing to the sum. This procedure serves to reduce the noise, often from biostratigraphic inadequacies, and enhance the signal in any data set by a factor equal to the square root of the number of data points included in the average. Commonly 8 or 9 values from the 11 sites were included in every 0.5 m.y. average, enhancing the signal contained therein by a factor of 2.8 to 3.0. Only four sites are useful for times older than 15 Ma. Inherent in this procedure is the assumption that the hemipelagic sites provide equally valid temporal information as the turbidite sites. The final stacked and averaged values were again normalized to a maximum value of 1.0 (Table 2). It is possible to do the same calculations using the median value of each time period. Using the median values has the advantage of excluding extremes that may be included in a

mean value, but the result will display somewhat more statistical variability [Dean and Dixon, 1951]. Figure 4 shows the results of these determinations. The normalized median flux values display a single-value peak at 2.25 Ma that is 30% greater than the next highest point and thus serves to reduce all other relative values. The normalized mean fluxes have several higher values; these normalized mean fluxes will be the basis for the temporal interpretations that follow.

TERRIGENOUS SEDIMENT FLUX

The flux data (Figures 2 and 3) indicate two ocean-wide periods of enhanced terrigenous sediment flux to the region of the Indus and Bengal fans. During the lower portion of late Miocene time clastic fluxes increased 2.5- (Site 223, Figure 2) to 13-fold (Site 719, Figure 3) over the average values at those sites. This late Miocene pulse occurs in ten of the eleven sites examined. It is missing only from Hole 721B (Figure 2) which is at a depth of 1945 meters on the Owen Ridge and is the shallowest of the sites studied. At Site 718, one of three sites on the distal Bengal Fan, this pulse appears to begin a bit earlier than at any of the other sites, with increased fluxes beginning at about 11 Ma. All eleven sites exhibit a Plio-Pleistocene pulse of terrigenous input, commonly of somewhat lesser magnitude than the late Miocene pulse, with generally a two- to five-fold increase in the MAR of terrigenous sediment (although at Sites 222 and 721 [Figure 2] the increase is more than an

TABLE 1. Drillsites Examined for Terrigenous Flux Information.

Leg	Hole	Latitude	E. Long	Depth (m)	Fan	Record length
117	721B	16°07.8'N	59°51.9'	1945	Indus (h)	22.3 Ma, 424 m
117	722B	16°37.3'N	59°47.8'	2028	Indus (h)	19.0 Ma, 566
117	731A	16°28.2'N	59°42.2'	2366	Indus (h)	17.4 Ma, 409 m
23	222	20°05.5'N	61°30.6'	3546	Indus (t)	8.2 Ma, 1300 m
23	223	18°45.0'N	60°07.8'	3633	Indus (t)	58 Ma, 665 m
22	217	08°55.6'N	90°32.3'	3030	Bengal (h)	62 Ma, 402 m
22	218	08°00.4'N	86°17.0'	3737	Bengal (t)	11.5 Ma, 773 m
116	717C	00°55.8'S	81°23.4'	4735	Bengal (t)	9.4 Ma, 828 m
116	718C	01°01.3'S	81°24.1'	4730	Bengal (t)	22.3 Ma, 940 m
116	719A	00°57.7'S	81°24.0"	4737	Bengal (t)	6.0 Ma, 460 m
121	758A	05°23.4'N	90°21.7'	2924	Bengal (h)	80 Ma, 677 m

Hemipelagic sites designated "h", turbidite sites designated "t".

order of magnitude). Sediments deposited during these two times of enhanced flux contain a mineral suite that corresponds to the eroding rocks of the Higher Himalayas [Brass and Raman, 1990; Yokoyama et al., 1990; Amano and Taira, 1992].

There are several other inferences concerning the depositional processes of this region that can be drawn from the flux information. These data support the conclusion of Cochran [1990] that, when studied on this tectonic timescale, local variability is only a minor concern. The spike at Site 718 to MAR values of about 70 g(cm^2$_*$ky)$^{-1}$ at 11 Ma (Figure 3) is all in one nannofossil zone and may be the consequence of poorly constrained biostratigraphy (and thus an extreme LSR), but the general elevation above background level that occurs about 11 Ma is real for that site. There is not quite enough information at nearby Hole 717C to discount firmly this observation at Site 718. Further, because there are limited data available for this particular time interval, this one high value is the entire reason for the relative maxima at 10.5 to 11.5 Ma on the normalized mean flux curve (Figure 4).

The three sites drilled during Leg 117 on Owen Ridge, Sites 721, 722, and 731, all exhibit much higher clastic fluxes prior to 14 Ma (Figure 2). The Leg 117 Shipboard Scientific Party [1989] recognized that the lowermost sedimentary unit in each of these sites consisted of turbidites of the Indus Fan, and reasoned that the middle Miocene reduction in turbidite input here was the result of tectonic uplift of the Owen Ridge raising the sea floor above the level of turbidite deposition and not an indication of any change in clastic flux to the sea floor. It is important to determine whether these lower and middle Miocene higher flux values represent normal sedimentation on the Indus Fan or whether they may reflect a true pulse of sediment from the Indus drainage basin. The clay mineralogy of this lowermost unit is distinctly different from that of the overlying Owen Ridge units and similar to that of the Indus Fan sites [Debrabant et al., 1991]. Flux values of the Owen Ridge turbidites, a few g(cm^2$_*$ky)$^{-1}$, are similar to the average background flux values at Indus Fan Sites 222 and 223 (see Figure 2) and thus are

sedimentologically unremarkable in the fan environment. Further, neither nearby Indus Fan Site 223 nor any of the Bengal Fan sites show any indication of a depositional pulse during the lower or middle Miocene. Thus the interpretation of the Leg 117 Shipboard Scientific Party [1989] regarding the terrigenous sediments at Owen Ridge seems valid, and only the upper hemipelagic portions of those three sites are included in the normalization and averaging procedure. Various Leg 117 authors have suggested an Arabian-eolian source for the clays in the pelagic carbonates above the Owen Ridge turbidites [Clemens and Prell, 1991; deMenocal et al., 1991]. This may be largely correct, but the terrigenous flux maxima in the early portion of the late Miocene and in the Plio-Pleistocene that occur on Owen Ridge are indistinguishable in age from those fan-related maxima seen at all the other sites and probably represent Indus-derived hemipelagic input.

TEMPORAL PATTERN OF SEDIMENT FLUX AND POSSIBLE CONTROLLING FACTORS

The normalized, stacked and averaged MAR values exhibit two distinct sedimentary regimes in the northern Indian Ocean during middle and late Cenozoic time (Figure 4). For the past 12 m.y. the region has been characterized by relatively high and variable mass accumulation rates of terrigenous sediment; the earlier record is one of low clastic input. Since 12 Ma, the flux of sediment to the northern Indian Ocean has, on the average, increased five-fold over earlier values. Two periods of relatively high sediment input dominate the late Cenozoic record, at 2 to 4 Ma and 6 to 9 Ma . Lesser peaks are found at 0 to 1 Ma and 10.5 to 11.5 Ma. The peak at about 11 Ma derives from the single very high flux value at Site 718 and is therefore somewhat less reliable than the others.

Other information from both terrestrial and marine settings also point to the main sedimentary pulses occurring in the late Miocene and Plio/Pleistocene, and not earlier. Amano and Taira [1992], in a study of the heavy mineral assemblages of Bengal Fan Sites 717, 718 and 719, associate minerals derived from the erosion of the

TABLE 2. Normalized Median and Mean Sediment Fluxes.

Age interval	Median flux	Mean flux
0.0-0.5	0.532	0.619
0.5-1.0	0.540	0.637
1.0-1.5	0.419	0.462
1.5-2.0	0.226	0.475
2.0-2.5	1.000	0.872
2.5-3.0	0.468	0.747
3.0-3.5	0.548	0.739
3.5-4.0	0.605	0.720
4.0-4.5	0.218	0.323
4.5-5.0	0.145	0.236
5.0-5.5	0.266	0.348
5.5-6.0	0.355	0.543
6.0-6.5	0.500	0.662
6.5-7.0	0.387	0.664
7.0-7.5	0.573	0.749
7.5-8.0	0.532	0.702
8.0-8.5	0.629	1.000
8.5-9.0	0.694	0.949
9.0-9.5	0.226	0.309
9.5-10.0	0.210	0.346
10.0-10.5	0.387	0.380
10.5-11.0	0.306	0.685
11.0-11.5	0.306	0.503
11.5-12.0	0.177	0.262
12.0-12.5	0.161	0.157
12.5-13.0	0.210	0.242
13.0-13.5	0.113	0.198
13.5-14.0	0.113	0.166
14.0-14.5	0.097	0.159
14.5-15.0	0.089	0.122
15.0-16.0	0.097	0.113
16.0-17.0	0.105	0.117
17.0-18.0	0.089	0.107
18.0-19.0	0.121	0.141
19.0-20.0	0.121	0.141
20.0-21.0	0.129	0.147
21.0-22.0	0.121	0.143
22.0-23.0	0.105	0.124
23.0-24.0	0.081	0.096
24.0-25.0	0.081	0.102
25.0-26.0	0.048	0.051
26.0-27.0	0.065	0.068
27.0-28.0	0.048	0.062
28.0-29.0	0.040	0.046
29.0-30.0	0.032	0.040
30.0-31.0	0.056	0.068
31.0-32.0	0.065	0.077
32.0-33.0	0.081	0.102
33.0-34.0	0.081	0.091
34.0-35.0	0.040	0.048
35.0-36.0	0.065	0.068
36.0-37.0	0.056	0.070

Higher Himalayas only with the large late Miocene and Plio/Pleistocene (Figure 3) sedimentary pulses and interpret two stages of uplift. Kolla and Coumes [1987] in a seismic-profile study of the Indus Fan note that turbidite deposition may have been occurring since the Oligocene, but that the two major depositional episodes occurred in late Miocene and Plio/Pleistocene time. In a study of the age and deposition rate of the Siwalik deposits of Pakistan, Johnson et al. [1985] showed continuous deposition since the early Miocene with a pronounced increase in sedimentation rate in the earliest part of the late Miocene. The heavy mineral assemblage of the Siwalik units studied shows a sudden increase in metamorphic-related minerals coincident with the rapid increase in sedimentation rate, suggesting the unroofing of a newly-exposed metamorphic terrane [Johnson et al., 1985]. Together, these studies and the information compiled above and shown on Figure 4 all show a similar depositional history. Terrigenous clastics were being deposited at continental and marine sites by at least late Oligocene time. Major increases or pulses in this process occurred in late Miocene and Plio/Pleistocene time. None of the records of sediment deposition show indications of important depositional pulses at any time earlier than about the middle/late Miocene boundary.

To determine the processes or factors that may play a determining role in causing the flux patterns observed it is appropriate to examine other records of global change. Two such records pertain. The first is the oxygen isotope record of Cenozoic climate change. The $\delta^{18}O$ record shown on Figure 5 is that compiled from the benthic foraminifer *Uvigerina* sp. for ODP Sites 756, 757 and 758 on the Ninetyeast Ridge [Rea et al., 1991; Hovan and Rea, unpublished manuscript, 1992]. This plot shows the increases in $\delta^{18}O$ values associated with important climatic changes at 36, 13.5 and 2.5 Ma. These events are considered to reflect initiation of significant cooling and ice accumulation on Antarctica at the time of the Eocene /Oligocene boundary, an important cooling and/or enhancement of Antarctic ice volume in the middle Miocene, and the onset of Northern Hemisphere glaciation in the late Pliocene [Kennett, 1982].

The Indian Ocean isotope data show the three important changes quite clearly (Figure 5). If these times of important climate change have had an effect on the delivery of sediment from the Himalayas to the Indian Ocean [Molnar and England, 1990] there should be some clear temporal correspondence between plots of sediment delivery and of $\delta^{18}O$. If one compares these patterns for the length of the record (Figure 6) no temporal correspondence is observed. The early Oligocene cooling and increase in Antarctic ice volume appears to have had no effect on the delivery of sediment from the Himalayan region to the ocean. The isotope event at 13.5 Ma precedes the earliest, minor flux event at 11 Ma by more than 2 million years, an awkwardly long time if climate is the determining factor in the erosion of pre-existing highlands. Details of these two records (Figure 7) further emphasize this non-correspondence. The major late Miocene flux peak bears no relation to the

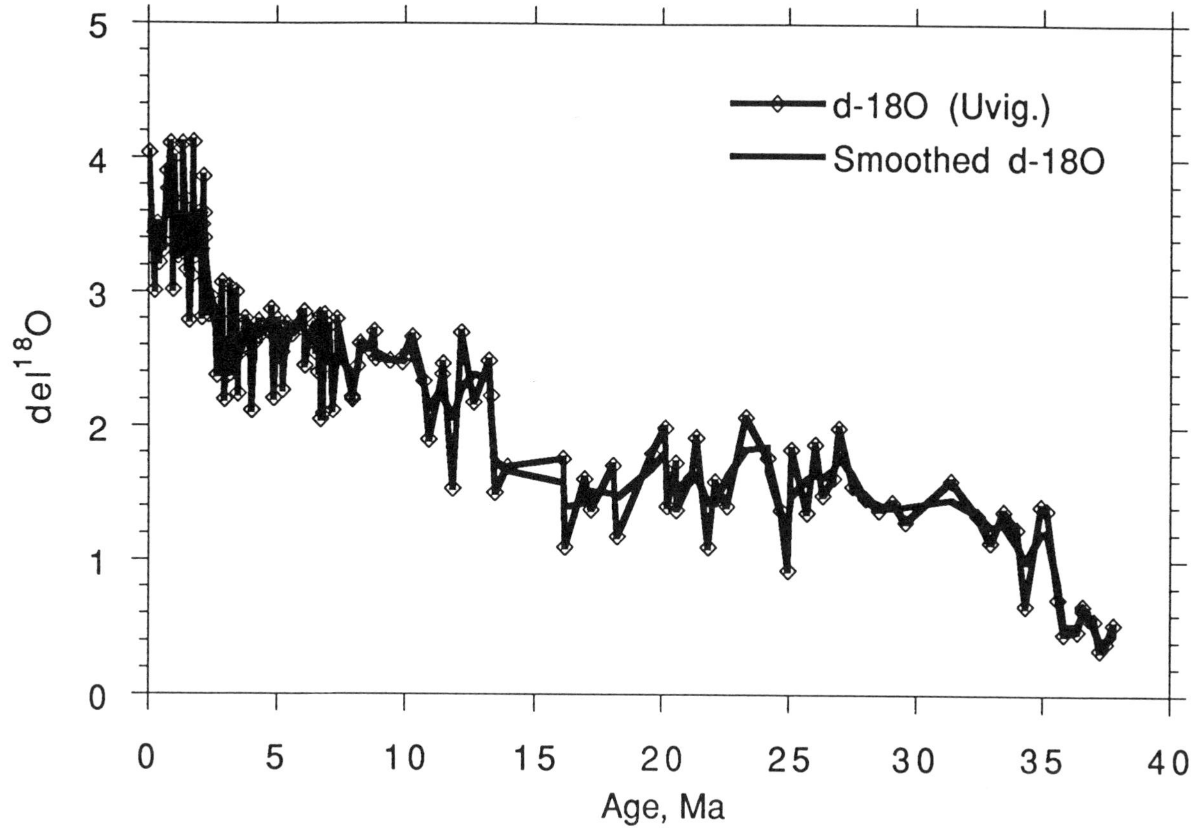

Fig. 5. Oxygen isotopic values of the benthic foraminifer *Uvigerina* sp. from three Indian Ocean drill sites on Ninetyeast Ridge: Sites 756, 757, and 758.

relatively unchanging isotope record at that time and the Pliocene flux maxima occurs before the late Pliocene δ18O shift that signals the onset of Northern Hemisphere glaciation. These comparisons seem to eliminate global climate changes as a determining factor for erosion of the Himalayas.

Changes in global sea level play an important role in the sediment input to the oceans. Times of rising and high sea level generally correspond with times of reduced sediment delivery to the deep sea, and times of falling, rapidly fluctuating, or low sea level generally correspond with times of enhanced input of sediment to the deep sea [Haq et al., 1987; Greenlee and Moore, 1988]. Sea level fluctuations are probably not important to the record of Indian Ocean sedimentation prior to 12 Ma; the geologic history of sea level for the later Neogene, however, may be quite important. Relative sea level records determined by Haq et al. [1987] and by Greenlee and Moore [1988] are shown on Figure 8. These records show higher sea levels in the middle Miocene and a distinct lowstand at 10 Ma. A broad highstand between 9 and 6 Ma is followed by rapidly fluctuating sea levels between 6.5 and 3.5 Ma. The record of the last 2.5 million years is not shown on these sorts of curves but would be represented by dozens of fluctuations of 50 to 100 meters amplitude. As the Greenlee and Moore [1988] record is a bit more detailed, I use it to compare with the normalized sediment flux pattern for the past 15 m.y.

(Figure 9). This comparison shows that the sea-level fall to the 10 Ma lowstand corresponds to the modest flux maximum at about 11 Ma. The highstand at 9 to 6 Ma, however, is exactly coincident with the time of maximum sediment input to the northern Indian Ocean, opposite to the expected relationship. Further, the time of rapidly fluctuating sea levels between 6 and 4 Ma, which should be optimal conditions for sediment delivery to the sea floor, are the time of minimal terrigenous relative fluxes. Relatively high fluxes occur during the following period of reduced sea level change, and the past 2.5 m.y which should be a time of high sediment input is characterized by relative fluxes lower that either of the two major peaks (Figure 9). Since sea level fluctuations and sediment delivery patterns are generally opposite the normal, well-understood, relationship, the conclusion is that sea level does not control the delivery of sediment to the northern Indian Ocean.

Comparison of the sediment flux data to representations of global climate and of sea level indicate that these important aspects of the earth's environmental systems do not control sediment input to the northern Indian Ocean. The dominant pattern of flux maxima at 2 to 4 Ma and 6 to 9 Ma result from another process; the only remaining large scale event that could be responsible for the observed pattern is the late Cenozoic uplift of the Himalayan ranges and Tibetan Plateau. Sediments associated with each of

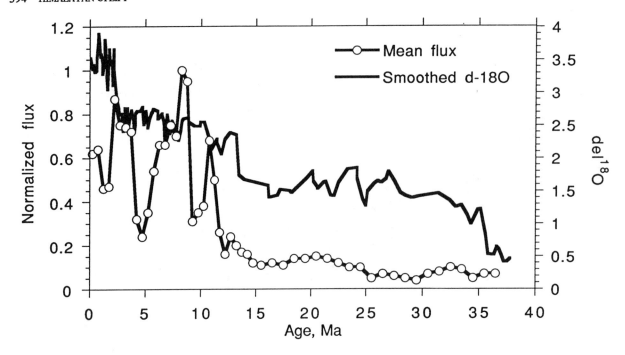

Fig. 6. Comparison of the smoothed oxygen isotope record and the normalized mean sediment flux record in the northern Indian Ocean.

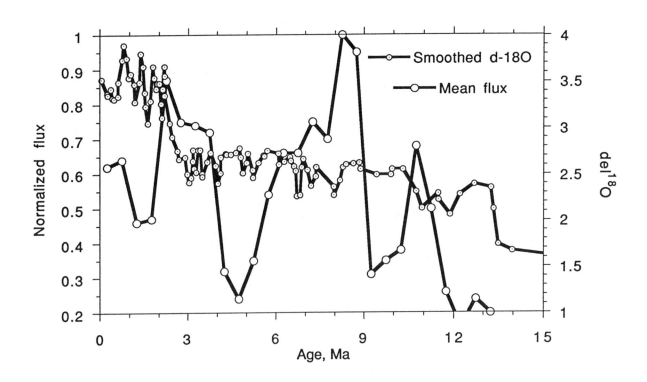

Fig. 7. Details of the Neogene portion of the smoothed oxygen isotope record and the mean sediment flux record. Note that the times of climate change at 13.5 Ma and 2.5 Ma are not associated with any lasting increases in sediment flux.

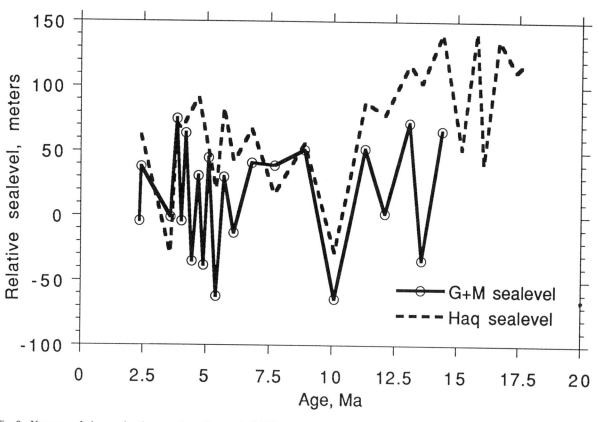

Fig. 8. Neogene relative sea level curves from Haq et al. [1987] and Greenlee and Moore [1988]. Note falling sea level from 11 to 10 Ma and time of stable sea level from 9 to 6 Ma.

these pulses have both clay and heavy mineral assemblages that can be associated with the eroding Himalayas. No evidence of similar sedimentary pulses occur at any time earlier in the Cenozoic.

The mass accumulation rates of terrigenous sediments entering the northern Indian Ocean (Figures 2 and 3) and the overall temporal pattern of their variability (Figure 4) provide strong support for theories of rapid uplift of the Himalayas during the late Cenozoic [Zeitler, 1985; Molnar et al., 1987; Amano and Taira, 1992]. The data do not support suggestions of widespread uplift and erosion beginning in early Miocene time [Copeland et al., 1987; Copeland and Harrison, 1990; Richter et al., 1991, 1992]. The late Cenozoic uplift occurred in two distinctly separate stages, one at about 9 to 6 Ma and a second 4 to 2 Ma. The earlier uplift may have served to enhance land-sea contrast, thus significantly strengthening the intensity of the Asian monsoons in the early part of late Miocene time [Prell and Kutzbach, 1991; Kroon et al., 1991]. The mid-Pliocene uplift began before and ended after the onset of Northern Hemisphere glaciation. Thus a causal relationship [Ruddiman and Kutzbach, 1989] is not clearly demonstrated, although depending on threshold assumptions, such a relationship may not be inconsistent with the information presented.

A SUPPLEMENTARY SECTION ON SEAWATER STRONTIUM

The strontium isotopic composition of seawater has been getting steadily more radiogenic since the middle Eocene (Figure 10). This curve is the integrated result of three fluxes: the input of relatively light strontium into the ocean from hydrothermal and other basalt-related sources, the input of relatively heavy strontium from continental sources, and the input of Sr with near-seawater isotopic values from the dissolution of marine carbonates [DePaolo, 1986; Palmer and Edmond, 1989]. This increasingly radiogenic nature of the seawater Sr curve has been related to enhanced erosion of and runoff from continents during the later portions of the Cenozoic [DePaolo, 1986; Raymo et al., 1988; Hodell et al., 1990; Raymo, 1991]. Hodell et al. [1990, 1991] and Richter et al. [1992], noting that the runoff from the Himalayan region is much more radiogenic than anywhere else [Palmer and Edmond, 1989; Krishnaswami et al., 1992], suggested that for this reason, and since southern Asia is the predominant sediment source to the world ocean, the record of input of strontium to the ocean from the Himalayan region would be of fundamental importance to the Cenozoic $^{87}Sr/^{86}Sr$ history.

The important changes in the seawater strontium curve occur in the middle Eocene, about 42 Ma, when the curve steepens abruptly, and in the middle Miocene, at about 16 Ma, when the slope becomes more gentle (Figure 10). Another steepening of the gradient may begin about 2.5 Ma. The sundry students of strontium have assumed that the

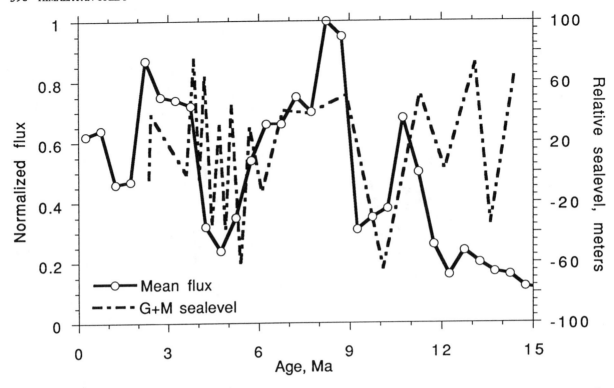

Fig. 9. Comparison of the sea level record of Greenlee and Moore [1988] with the normalized mean sediment flux record in the northern Indian Oce Falling sea level at 10-11 Ma is associated with minor sediment flux peak, but the major flux peak at 6-9 Ma occurs at a time of high sea level and the following flux minima occurs when sea level is lower and highly variable. The last 2.5 Ma is a time of rapid and large variations in sea level but this is not reflected in the sediment delivery curves.

flux of hydrothermal strontium has changed little since the middle Eocene and interpreted the data to represent changing input rates of continentally-derived radiogenic strontium into the ocean. The information gathered and calculated for the study of Himalayan sediment delivery can be compared with the seawater $^{87}Sr/^{86}Sr$ history to observe whether any temporal correspondence occurs (and remembering that the residence time of Sr in the oceans is on the order of 2.5 m.y. [Hodell et al., 1990]).

Comparison of the seawater strontium curve to the record of Himalayan sediment delivery to the Indian Ocean shows a poor correspondence (Figure 11). The most rapid increase in the relative amount of radiogenic strontium in the ocean occurs during times of relatively quite low sediment delivery. The $^{87}Sr/^{86}Sr$ slope decrease at 16 Ma precedes by a few million years the five-fold late Miocene increase in sediment input. Further, this slope change is in the "wrong" sense - towards the less radiogenic end-member at at time when the world-wide continental input of material to the ocean is increasing notably (Raymo et al., 1988). Comparison of the data spanning just the past 15 m.y. (Figure 12) shows that this unanticipated relationship is evident in the details of the data from Miocene and younger sediments. The times of high sediment flux, 9 to 6 and 4 to 2 Ma, are characterized by level or even declining $^{87}Sr/^{86}Sr$ values. Times of relatively reduced sediment delivery are characterized by more steeply rising values (Figure 12). So the expected correlation of the $^{87}Sr/^{86}Sr$ slope with the

oceanic flux of continentally derived material is not achieved either in the long or medium time frame.

The highly radiogenic isotopic values of the dissolved Sr in Himalayan rivers derive from outcrops of Precambrian granites and gneisses in the Ganges drainage basin [Krishnaswami et al., 1992]. If the erosion of these rocks is a controlling factor in the shape of the seawater strontium curve then such erosion has been greatly reduced since 16 Ma, in contrast both to the overall sediment delivery information (Figure 11) and to the data of Amano and Taira [1992] which show erosion of these same sorts of rocks to be enhanced during the late Miocene and Plio/Pleistocene depositional pulses. It seems unlikely that the erosion of these granites and gneisses explains the steep portion of the Cenozoic seawater strontium curve; perhaps it may bear on the more modest variability since 16 Ma (Figure 12).

The $\delta^{18}O$ curve of global climate (Figure 5) shows distinct changes which denote times when climate deterioration presumably has enhanced chemical and physical weathering of the continents, resulting in more sediment delivery to the ocean, although we observed above that this correspondence does not work well for the Himalayan region. Comparison of the $^{87}Sr/^{86}Sr$ curve with the oxygen isotope curve (Figure 13) shows apparently similar problems. The slope change in strontium isotope values begins several million years before the important climate change at the end of Eocene time. I note here, however, that

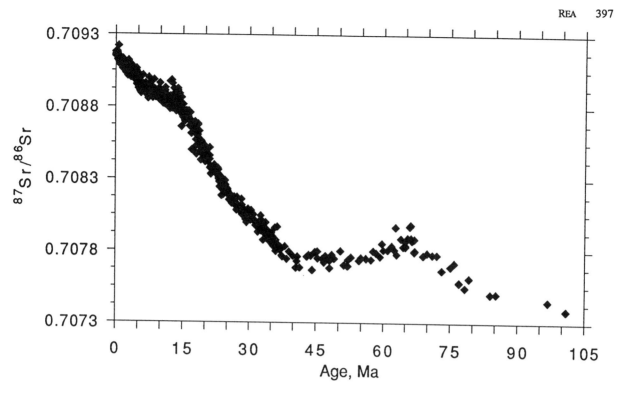

Fig.10. The Late Cretaceous and Cenozoic seawater $^{87}Sr/^{86}Sr$ curve. Data from several sources have been slightly adjusted based on laboratory standards and seawater values reported by each investigator [DePaolo and Ingram, 1985; DePaolo, 1986; Capo and DePaolo, 1990; Hess et al., 1986, 1989; Hodell et al., 1989, 1990, 1991; Miller et al., 1988, 1991]. All data are plotted using the Berggren et al. [1985] timescale.

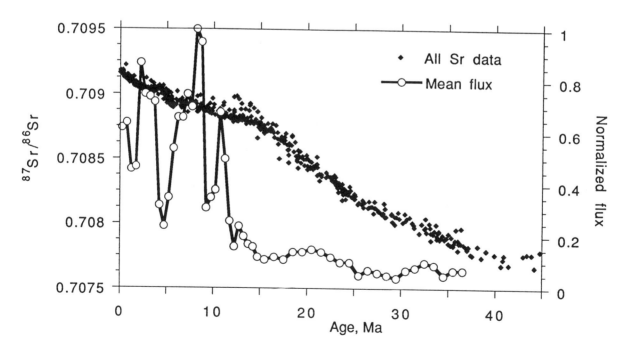

Fig. 11. Comparison of the seawater strontium curve of Figure 10 to the normalized mean sediment flux to the northern Indian Ocean.

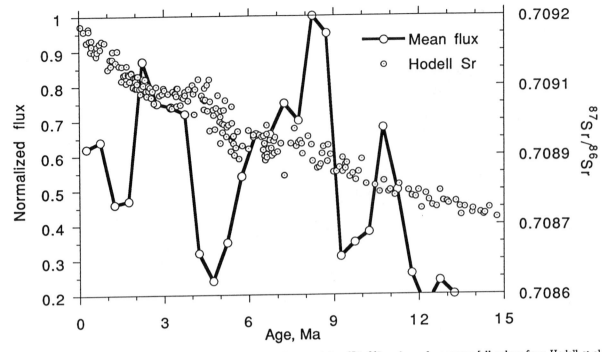

Fig. 12. Details of the past 15 m.y. of Himalayan sediment delivery and the $^{87}Sr/^{86}Sr$ values of seawater [all values from Hodell et al., 1989, 1990, 1991]. Note that the times of reduced sediment flux at 4 to 6 Ma and 0 to 2.5 Ma are the times when the $^{87}Sr/^{86}Sr$ curve is steepest.

recent high-latitude ODP drilling has found evidence of the cooling of and glacial ice upon Antarctica during late Eocene time [Kennett and Barker, 1990; Ehrmann, 1991; Barron et al., 1991]. This earlier cooling of the southern continent might be associated in time with the slope change of the $^{87}Sr/^{86}Sr$ curve [Zachos et al., 1992].

In the middle Miocene, the reduction in slope of the $^{87}Sr/^{86}Sr$ curve precedes the 13.5 Ma change in climate/ isotopes by 2.5 m.y. and is in the sense of suggesting reduced, rather than enhanced, importance of input of radiogenic strontium. The steepening of the Sr-isotope curve in the last 2.5 m.y. does correspond to the mid-Pliocene onset of Northern Hemisphere glaciation (and the associated rapid sea-level fluctuations) as indicated by the $\delta^{18}O$ information.

Sea level control appears poorly constrained. The Cenozoic sea level fluctuations (Figure 8) would seem to call for much more sediment delivery into the ocean during the later portion of the Cenozoic, since the middle Miocene, than before. Comparison of the sea level curves to the $^{87}Sr/^{86}Sr$ curve (Figure 14) for the past 20 m.y. may suggest a correlation between flatter portions of the $^{87}Sr/^{86}Sr$ curve at 8 to 6 Ma and 2.5 to 4 Ma and times of higher sea levels. Periods of rapidly fluctuating sea level at 4 to 6 Ma and 0 to 2.5 Ma, presumably times of higher global sediment delivery to the sea floor, correspond to steeper portions of the curve.

All this information leaves us with this conclusion: the Cenozoic strontium isotopic curve for seawater does not reflect the input of continentally derived clastics to the ocean. Therefore the flux of radiogenic Sr, which is in the

dissolved load of rivers, may have little to do with the flux of sediment, although a general decoupling of gross overall dissolved and particulate sediment delivery, especially in regions like the Himalayas undergoing very rapid erosion, would be difficult to support. The isotopic ratio of the dissolved riverine strontium may have changed markedly with time, being higher from 42 to 16 Ma and lower since. The non-correspondence of clastic fluxes and slopes of the $^{87}Sr/^{86}Sr$ curve (Figure 13) even suggests that the $^{87}Sr/^{86}Sr$ ratio of the river waters varied in the opposite sense of the particulate fluxes, being significantly higher during times of lower particulate flux. Clearly the knowledge of exactly what is being eroded in the Himalayas and when is very important [Palmer and Edmond, 1989; Krishnaswami et al., 1992; Amano and Taira, 1992]. That information lies in the details of the mineralogy and geochemistry of the sediments of the Indus and Bengal Fans.

There is a final observation to be made. All discussions about the nature of the $^{87}Sr/^{86}Sr$ curve assume that it is the continental/radiogenic aspect that is determining the slope of the curve. The potential role of the hydrothermal component, assumed to have changed little since there has not been much change in crustal generation rates [Delaney and Boyle, 1988; Richter et al., 1992], is not considered. However, it has been shown that reliance upon the presumed association of hydrothermal activity with spreading rate is inappropriate at these time intervals of less than tens of millions of years [Owen and Rea, 1985; Lyle et al., 1987]. To determine directly the amount of sea-floor hydrothermal activity during the past 40 m.y. one must quantify the mass accumulation rate of the

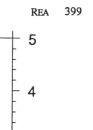

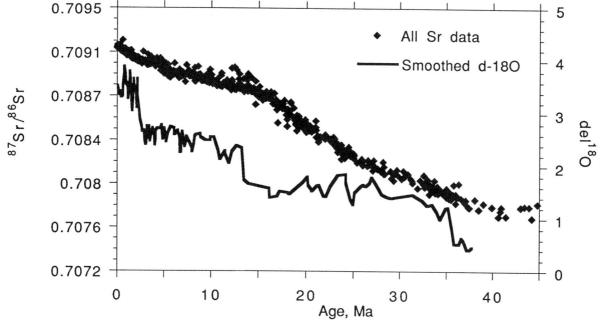

Fig. 13. Comparison of the seawater strontium curve of Figure 10 to the Indian Ocean oxygen isotope record of climate change.

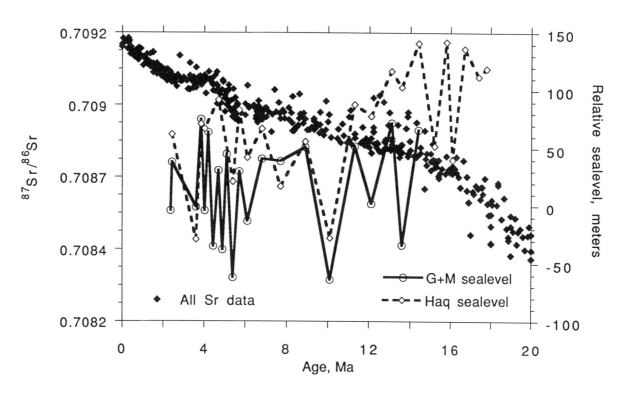

Fig. 14. Details of the past 20 m.y. of sea level changes [from Haq et al., 1987; Greenlee and Moore, 1988] compared to the $^{87}Sr/^{86}Sr$ seawater record.

hydrothermal component in the oceanic sedimentary section. This analysis and calculation has only been done once and in rather rudimentary fashion for the whole Cenozoic but the results may be instructive for considering the strontium data. Leinen [1989] has determined the flux of hydrothermal materials to the central North Pacific pelagic clay core LL44-GPC3 (Figure 1) for the entire Cenozoic. That record shows a five to ten-fold reduction in the flux of hydrothermal materials in sediments younger than late Eocene. Leinen's calculations are in agreement with other indications of Cenozoic hydrothermal activity such as the Indian Ocean hydrothermal flux record [Owen and Zimmerman, 1991] and the age distribution of hydrothermal ore deposits [Olivarez and Owen, 1989].

The rather awkward non-correlation between the increasingly radiogenic strontium composition of seawater and what we presently understand about both climate change and the delivery of continental debris to the oceans suggests that additional processes may be involved in determining the shape of the $^{87}Sr/^{86}Sr$ curve of Figure 10. Based on the work of Leinen [1989], I suggest that one critical additional factor is a marked reduction in the input of the hydrothermal end member, certainly during middle (42 to 16 Ma) and probably during late (16 to 0 Ma) Cenozoic time, thus permitting the seawater strontium curve to trend more rapidly towards a continental-radiogenic signature. Investigations of the Cenozoic fluxes of hydrothermal materials thus hold great potential for increasing our understanding of the Cenozoic record of seawater strontium.

CONCLUSIONS

Temporal patterns of sediment delivery from the Himalayan region to the northern Indian Ocean show higher sediment MAR values since about 12 Ma, with low values for the entire older Cenozoic. Relative flux maxima at 2 to 4 Ma and from 6 to 9 Ma do not correspond to times of either climate change or to times of low or falling sea level. Since the remaining determinant of clastic flux from a continent is the uplift of mountains, the data support hypotheses of rapid late Cenozoic uplift of the Himalayas. The uplift appears to have occurred in two stages in the late Miocene and middle Pliocene and may be ultimately responsible for the origin of Asian monsoons and even of Northern Hemisphere glaciation.

Examination of the seawater strontium isotope curve in light of the record of Himalayan sediment delivery to the ocean shows a distinct non-correspondence between the two. There may also be a question about invoking Cenozoic climate change to explain the increasingly radiogenic nature of seawater strontium. The possibility that other determining factors may have changed in important ways, such as sea-floor hydrothermal activity or the $^{87}Sr/^{86}Sr$ ratio of dissolved Sr in rivers, should be considered more strongly.

Acknowledgments. This investigation arose from discussions at the Indian Ocean Workshop held in Cardiff in July of 1991 and sponsored by JOI/USSAC, NERC, and the University of Wales, College of Cardiff. I would like to thank Charles Jones for his strontium isotope data set which he normalized among the various laboratories and D. A. Hodell and W. F. Ruddiman for thoughtful reviews; earlier versions of this paper were commented upon by B. A. Van der Pluijm, C. Badgley and two anonymous reviewers.

REFERENCES

Amano, K., and Taira, A., Two-phase uplift of Higher Himalayas since 17 Ma, *Geology, 20,* 391-394, 1992.

Barron, J., Larsen, B., and Baldauf, J. G., Evidence for late Eocene to Early Oligocene Antarctic glaciation and observations on late Neogene glacial history of Antarctica: results from Leg 119, in *Proc. ODP, Sci. Results, 119,* edited by J. Barron, G. Larsen et al., pp. 869-891, Ocean Drilling Program, College Station, TX, 1991.

Berggren, W. A., Kent, D. V., Flynn, J. J., and van Couvering, J. A., Cenozoic Geochronology, *Geol. Soc. of Am. Bull., 96,* 1407-1418, 1985.

Brass, G. W., and Raman, C. V., Mineralogy of sediments from the Bengal Fan, in *Proc. ODP, Sci. Results, 116,* edited by J. R. Cochran, D. A. V. Stow et al., pp. 35-41, Ocean Drilling Program, College Station, TX, 1990.

Capo, R. C., and DePaolo, D. J., Seawater strontium isotopic variations from 2.5 million years ago to the present, *Science, 249,* 51-55, 1990.

Clemens, S. C., and Prell, W. L., One million year record of summer monsoon winds and continental aridity from the Owen Ridge (Site 722), northwest Arabian Sea, in *Proc. ODP, Sci. Results, 117,* edited by W. L. Prell, N. Niitsuma et al., pp. 365-388, Ocean Drilling Program, College Station, TX, 1991.

Cochran, J. R., Himalayan uplift, sea level and the record of Bengal Fan sedimentation at the ODP Leg 116 sites, in *Proc. ODP, Sci. Results, 116,* edited by J. R. Cochran, D. A. V. Stow et al., pp. 397-414, Ocean Drilling Program, College Station, TX, 1990.

Cochran, J. R., Stow, D. A. V., et al., *Proc. ODP, Init. Repts., 116,* 388 pp., Ocean Drilling Program, College Station, TX, 1989.

Cochran, J. R., Stow, D. A. V., et al., *Proc. ODP, Sci. Results, 116,* 445 p., Ocean Drilling Program, College Station, TX, 1990.

Copeland, P., and Harrison, T. M., Episodic rapid uplift in the Himalaya revealed by $^{40}Ar/^{39}Ar$ analysis of detrital K-feldspar and muscovite, Bengal fan, *Geology, 18,* 354-357, 1990.

Copeland, P., Harrison, T. M., Kidd, W. S. F., Ronghua, X., and Yuquan, Z., Rapid early Miocene acceleration of uplift in the Gangdese Belt, Xizang (southern Tibet), and its bearing on accommodation mechanisms of the India-Asia collision, *Earth Planet. Sci. Letts., 86,* 240-252, 1987.

Curray, J. R., and Moore, D. G., Growth of the Bengal deep sea fan and denudation in the Himalayas, *Geol. Soc. Am. Bull., 82,* 563-572, 1971.

Dean, R. B., and Dixon, W. J., Simplified statistics for small numbers of observations, *Anal. Chem., 23,* 636-638, 1951.

Debrabant, P., Krissek, L., Bouquillon, A., and Chamley, H., Clay mineralogy of Neogene sediments of the western Arabian Sea: mineral abundances and paleoenvironmental implications, in *Proc. ODP, Sci. Results, 117,* edited by W. L. Prell, N. Niitsuma et al., pp. 183-196, Ocean Drilling Program, College Station, TX, 1991.

Delaney, M. L., and Boyle, E. A., Tertiary paleoceanic chemical variability: unintended consequences of simple geochemical models, *Paleocean., 3,* 137-156, 1988.

deMenocal, P., Bloemendal, J., and King, J., A rock-magnetic record of monsoonal dust deposition to the Arabian Sea: evidence for a shift in the mode of deposition at 2.4 Ma, in:

Proc. ODP, Sci. Results, 117, edited by W. L. Prell, N. Niitsuma, et al., pp. 389-401, Ocean Drilling Program, College Station, TX, 1991.

DePaolo, D. J., Detailed record of the Neogene Sr isotope evolution of seawater from DSDP Hole 590B, *Geology, 14*, 103-106, 1986.

DePaolo, D. J., and Ingram, B. L., High-resolution stratigraphy with strontium isotopes, *Science, 227*, 938-941, 1985.

Ehrmann, W. U., Implications of sediment composition on the southern Kerguelen Plateau for Paleoclimate and depositional environment, in *Proc. ODP, Sci. Results, 119*, edited by J. Barron, G. Larsen et al., pp. 185-210, Ocean Drilling Program, College Station, TX, 1991.

Gansser, A., *Geology of the Himalayas*, Wiley-Interscience, London, 289 p., 1964.

Gansser, A., The geodynamic history of the Himalaya, in *Zagros - Hindu Kush - Himalaya Geodynamic Evolution*, edited by H. K. Gupta and F. M. Delany, American Geophysical Union Geodynamics Series, Volume 3, pp. 111-121, Washington, D.C., 1981.

Gartner, S., Neogene calcareous nannofossil biostratigraphy, Leg 116 (Central Indian Ocean), in *Proc. ODP, Sci. Results, 116*, edited by J. R. Cochran, D. A. V. Stow et al., pp. 165-187, Ocean Drilling Program, College Station, TX, 1990.

Greenlee, S. M., and Moore, T. C., Jr., Recognition and interpretation of depositional sequences and calculation of sea-level changes from stratigraphic data - offshore New Jersey and Alabama Tertiary, in *Sea-Level Changes - An Integrated Approach*, edited by C. K. Wilgus, B. S. Hastings, C. G. St. C. Kendall, H. W. Posamentier, C. A. Ross, and J. C. Van Wagoner, pp. 329-353, Society of Economic Paleontologists and Mineralogists Special Publication No. 42, Tulsa, OK, 1988.

Haq, B. U., Hardenbol, J., and Vail, P. R., Chronology of fluctuating sea levels since the Triassic, *Science, 235*, 1136-1167, 1987.

Harrison, T. M., Copeland, P., Kidd, W. S. F., and Yin, A., Raising Tibet, *Science, 255*, 1663-1670, 1992.

Hess, J., Bender, M. L., and Schilling, J.-G., Evolution of the ratio of strontium-87 to strontium-86 in seawater from Cretaceous to Present, *Science, 231*, 979-984, 1986.

Hess, J., Stott, L. D., Bender, M. L., Kennett, J. P., and Schilling, J.-G., The Oligocene marine microfossil record: age assessments using strontium isotopes, *Paleocean., 4*, 655-680, 1989.

Hodell, D. A., Mead G. A., and Mueller, P. A., Variation in the strontium isotopic composition of seawater (8 Ma to present): implications for chemical weathering rates and dissolved fluxes to the oceans, *Chemical Geology (Isotope Geology Section), 80*, 291-307, 1990.

Hodell, D. A., Mueller, P. A., and Garrido, J. R., Variations in the strontium isotopic composition of seawater during the Neogene, *Geology, 19*, 24-27, 1991.

Hodell, D. A., Mueller, P. A., McKenzie, J. A., and Mead, G. A., Strontium isotope stratigraphy and geochemistry of the late Neogene ocean, *Earth Planet. Sci. Letts., 92*, 165-178, 1989.

Johnson, N. M., Stix, J., Tauxe, L., Cerveny, P. F., and Tahirkheli, R. A. K., Paleomagnetic chronology, fluvial processes and tectonic implications of the Siwalik deposits near Chinji Village, Pakistan, *J . Geol., 93*, 27-40, 1985.

Kennett, J. P., *Marine Geology*, Prentice-Hall, Englewood Cliffs, NJ, 813 p., 1982.

Kennett, J. P., and Barker, P. F., Latest Cretaceous to Cenozoic and oceanographic developments in the Weddell Sea, Antarctica: an ocean-drilling perspective, in *Proc. ODP, Sci. Results, 113*, edited by P. F. Barker, J. P. Kennett et al., pp. 937-960, Ocean Drilling Program, College Station, TX, 1990.

Klootwijk, C. J., Gee, J. S., Peirce, J. W., and Smith, G. M., Constraints on the India-Asia convergence: paleomagnetic results from Ninetyeast Ridge, in *Proc. ODP, Sci. Results, 121*, edited by J. W. Peirce, J. K. Weissel et al., pp. 777-882, Ocean Drilling Program, College Station, TX, 1991.

Klootwijk, C. T., Gee, J. S., Peirce, J. W., Smith, G. M., and McFadden, P. L., An early India-Asia contact: paleomagnetic constraints from Ninetyeast Ridge, ODP Leg 121, *Geology, 20*, 395-398, 1992.

Kolla, V., and Coumes, F.,Morphology, internal structure, seismic stratigraphy, an sedimentation of the Indus Fan, *Am. Assoc. Pet. Geol. Bull., 71*, 650-677, 1987.

Krishnaswami, S., Trivedi, J. R., Sarin, M. M., Ramesh, R., and Sharma, K. K., Strontium isotopes and rubidium in the Ganga-Brahmaputra river system: weathering in the Himalaya, fluxes to the Bay of Bengal and contributions to the evolution of oceanic $^{87}Sr/^{86}Sr$, *Earth Planet. Sci. Letts., 109*, 243-253, 1992.

Kroon, D., Steens, T., And Troelstra, S.R., Onset of monsoonal related upwelling in the western Arabian Sea as revealed by planktonic foraminifers, in *Proc. ODP, Sci. Results, 117*, edited by W. L. Prell, N. Niitsuma et al., pp. 257-263, Ocean Drilling Program, College Station, TX, 1991.

Leg 117 Shipboard Scientific Party, Background and summary of drilling results - Owen Ridge, in *Proc. ODP, Init. Repts., 117*, edited by W. L. Prell, N. Niitsuma et al., pp. 35-42, Ocean Drilling Program, College Station, TX, 1989.

Leinen, M., The pelagic clay province of the North Pacific Ocean, in *The Eastern Pacific Ocean and Hawaii, The Geology of North America Volume N*, edited by E. L. Winterer, D. M. Hussong, and R. W. Decker, pp. 323-335, Geological Society of America, Boulder, CO, 1989.

Lyle, M., Leinen, M., Owen, M. and Rea, D. K., Late Tertiary history of hydrothermal deposition at the East Pacific Rise: correlation to volcano-tectonic events, *Geophys. Res. Letters, 14*, 595-598, 1987.

Miller, K. G., Feigenson, M. D., Kent, D. V., and Olsson, R. K., Upper Eocene to Oligocene isotope ($^{87}Sr/^{86}Sr$, $\delta^{18}O$, $\delta^{13}C$) standard section, Deep Sea Drilling Project Site 522, *Paleocean., 3*, 223-233, 1988.

Miller, K. G., Feigenson, M. D., Wright, J. D., and Clement, B. M., Miocene isotope reference section, Deep Sea Drilling Project Site 608: an evaluation of isotopic and biostratigraphic resolution, *Paleocean., 6*, 33-52, 1991.

Molnar, P., and England, P., Late Cenozoic uplift of mountain ranges and global climate change: chicken or egg, *Nature, 346*, 29-34, 1990.

Molnar, P., Burchfiel, B. C., Ziyun, Z., K'uangyi, L., Shuji, W., and Minmin, H., Geologic evolution of northern Tibet: results of an expedition to Ulugh Muztagh, *Science, 235*, 299-305, 1987.

Olivarez, A. M., and Owen, R. M., Plate tectonic reorganizations: implications regarding the formation of hydrothermal ore deposits, *Mar. Mining, 8*, 123-138, 1989.

Owen, R. M., and Rea, D. K., Sea floor hydrothermal activity links climate to tectonics: the Eocene CO_2 greenhouse, *Science, 227*, 166-169, 1985.

Owen, R. M., and Zimmerman, A. R. B., The geochemistry of Broken Ridge sediments, in *Proc. ODP, Sci. Results, 121*, edited by J. W. Peirce, J. K. Weissel et al., pp. 437-445, Ocean Drilling Program, College Station, TX, 1991.

Palmer, M. R., and Edmond, J. M., The strontium isotope budget of the modern ocean, *Earth Planet. Sci. Letts, 92*, 11-26, 1989.

Peirce, J. W., Weissel J. K. et al., *Proc. ODP, Init. Repts, 121*, 1000 p., Ocean Drilling Program, College Station, TX, 1989.

Prell, W. L., and Kutzbach, J. E., Sensitivity of the Indian monsoon to changes in tectonic (orographic), orbital (solar variation) and glacial boundary conditions: model-data comparison, *Trans. Am. Geophys. Union (EOS), AGU 1991 Fall Meeting Program and Abstracts*, 257-258, 1991.

Prell, W. L., Niitsuma, N. et al., *Proc. ODP, Init. Repts, 117*, 1236 p., Ocean Drilling Program, College Station, TX, 1989.

Raymo, M. E., Geochemical evidence supporting T.C. Chamberlin's theory of glaciation, *Geology, 19*, 344-347, 1991.

Raymo, M. E., Ruddiman, W. F., and Froelich, P. N., Influence of late Cenozoic mountain building on ocean geochemical cycles,

Geology, 16, 649-653, 1988.

Rea, D. K., and Thiede, J., Mesozoic and Cenozoic mass accumulation rates of the major sediment components in the Nauru Basin, western equatorial Pacific, in *Init. Repts. DSDP, 61,* edited by R. L. Larson, S. O. Schlanger et al., pp. 549-555, U.S. Government Printing Office, Washington D.C., 1981.

Rea, D. K., Lohmann, K. C., MacLeod, N. D., House, M. A., Hovan, S. A., and Martin, G. D., Oxygen and carbon isotope records from the oozes of Sites 752, 754, 756, and 757, eastern Indian Ocean, in *Proc. ODP, Sci. Results, 121,* edited by J. W. Peirce, J. K. Weissel et al., pp. 229-240, Ocean Drilling Program, College Station, TX, 1991.

Richter, F. M., Lovera, O. M., Harrison, T. M., and Copeland, P., Tibetan tectonics from $^{40}Ar/^{39}Ar$ analysis of a single K-feldspar sample, *Earth Planet. Sci. Letts, 105,* 266-278, 1991.

Richter, F. M., Rowley, D. B., and DePaolo, D. J., Sr isotope evolution of seawater: the role of tectonics, *Earth Planet. Sci. Letts, 109,* 11-23, 1992.

Ruddiman, W. F., and Kutzbach, J. E., Forcing of late Cenozoic northern hemisphere climate by plateau uplift in southern Asia and the American west, *J. Geophys. Res., 94,* 18,409-18,427, 1989.

Ruddiman, W.F., Prell, W.L., and Raymo, M.E., Late Cenozoic uplift in southern Asia and the American west: rationale for general circulation modeling experiments, *J. Geophys. Res., 94,* 18,379-18,391, 1989.

Sato, T., Kameo, K., and Takayama, T., Coccolith biostratigraphy of the Arabian Sea, in *Proc. ODP, Init. Repts, 117,* edited by W. L. Prell, N. Niitsuma et al., pp. 37-54, Ocean Drilling Program, College Station, TX, 1991.

Searle, M. P., Windley, B. F., Coward, M. P., Cooper, D. J. W., Rex, A. J., Rex, D., Li, T., Xiao, X., Jan, M. Q., Thakur, V. C., and Kumar, S., The closing of the Tethys and the tectonics of the Himalaya, *Geol. Soc. Am. Bull., 98,* 678-701, 1987.

von der Borch, C. C., Sclater, J. G. et al., *Init. Repts. DSDP, 22,* 890 p., U.S. Government Printing Office, Washington, D.C., 1974.

Whitmarsh, R. B., Weser, O. E., Ross, D. A. et al., *Init. Repts. DSDP, 23,* U.S. Goernment .Printing Office, Washington, D.C., 1180 p., 1974.

Yokoyama, K., Amano, K., Tiara, A., and Saito, Y., Mineralogy of silts from the Bengal Fan, in *Proc. ODP, Sci. Results, 116,* edited by J. R. Cochran, D. A. V. Stow et al., pp. 59-73, Ocean Drilling Program, College Station, TX, 1990.

Zachos, J. C., Berggren, W. A., Aubry, M.-P., and Mackensen, A., Isotope and trace element geochemistry of Eocene and Oligocene foraminifer from Site 748, Kerguelen plateau, in *Proc. ODP, Sci. Results, 120,* edited by R. Schlich, S. W. Wise, Jr. et al., Ocean Drilling Program, College Station, TX, 1992.

Zeitler, P. K., Cooling history of the NW Himalaya, Pakistan, *Tectonics, 4,* 127-151, 1985.

An Indian Ocean Framework for Paleoceanographic Synthesis Based on DSDP and ODP Results

Robert B. Kidd, Anthony T. S. Ramsay and Timothy J. S. Sykes

Department of Geology, University of Wales College of Cardiff, Cardiff, UK

Jack G. Baldauf

Department of Oceanography and Ocean Drilling Program, Texas A and M University Research Park, College Station, TX, 77845, USA

Thomas A. Davies

Institute for Geophysics, The University of Texas at Austin, Austin, TX, 78759, USA

D. Graham Jenkins

National Museum of Wales, Cathays Park, Cardiff, UK

Sherwood W. Wise, Jr.

Department of Geology, Florida State University, Tallahassee, FL, 32306, USA

The stratigraphic phase of the paleoceanographic Indian Ocean synthesis project (PALIOS) has established a rigorous framework for paleoceanographic synthesis of both DSDP and ODP results. Most importantly we have completed an integrated stratigraphic framework that allows correlation between the low and high latitudes and takes account of the existence of diachronous fossil datums. A simple spreadsheet routine has been developed to calculate the depth/age curve for each site which in turn provides the basis for assigning an age to each sample/measurement point. This is an essential step in grouping sediment properties by age for the plotting phase of the PALIOS project and has become the basis for a number of target projects that will culminate in time series palinspastic maps showing accumulation rates and paleobathymetry. Preliminary results of two of the target projects suggest that hiatuses linked to bottom water circulation could be more regionalised than previously thought and whilst volcanogenic sediment input appears most closely linked to the evolution of "hotspots" the resulting distributions are controlled by dispersal mechanisms that include paleowinds and surface water mass circulation.

Introduction

The Paleoceanographic Indian Ocean Synthesis (PALIOS) project is a study being conducted by a team of British and American marine geologists with a span of expertise covering Indian Ocean stratigraphy, sedimentology and geophysics. PALIOS is a synthesis effort that contrasts with, and complements, the other contributions in this volume in its attempt to standardize both the Ocean Drilling Program (ODP) and Deep Sea Drilling Project (DSDP) stratigraphies and sedimentologic databases and to deal with paleoceanographic parameters in a truly oceanwide, rather than basin-wide or regional, sense. Having reassessed the DSDP results, it represents the next phase to which the post-ODP synthesis studies must

Synthesis of Results from Scientific
Drilling in the Indian Ocean
Geophysical Monograph 70

progress and begins to provide indications for future Indian Ocean drilling.

This contribution deals with the procedures and pitfalls encountered in the initial phase of PALIOS, during which we have developed an integrated stratigraphic synthesis framework that allows for correlation between high and low latitude regions, and an automated technique for depth-to-age conversion which other workers might wish now to consider for their own more advanced studies of Indian Ocean DSDP and ODP samples.

We present some preliminary results of so-called "target projects" that run parallel with the overall synthesis effort. These include initial findings that are emerging from studies of hiatus occurrence and the distribution of volcanogenic sediments.

The remaining phase of PALIOS will develop new maps showing distributions through time of a wide range of paleoceanographic parameters. We plan to prepare accumulation rate and mass balance data for the Indian

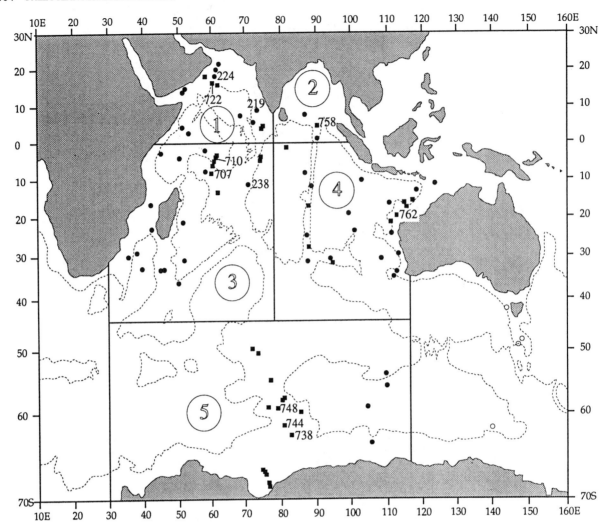

Fig 1. Map showing DSDP (circles) and ODP (squares) sites in the Indian Ocean. Sector boundaries used in PALIOS are shown as straight lines. Key and secondary reference sites are numbered.

Ocean basins and to document quantitatively their sedimentary and biotic responses to tectonic, climatic and oceanographic changes.

BACKGROUND

Synthesis studies based on the sequences drilled during Legs 22 through 27 of the DSDP provided an excellent generalized record of changing sediment distributions during the evolution of the Indian Ocean [see contributions in Heirtzler et al., 1977, Von der Borch, 1978, and texts by Kennett, 1982 and Emiliani, 1981]. Comparisons were possible between empirical subsidence curves and sediment sequences at individual sites, and between reconstructions of paleobathymetry and sediment distributions, which convincingly demonstrated the interrelationship of sedimentary and tectonic developments within this major ocean [Davies and Kidd, 1977; Heirtzler et al., 1977; Kidd and Davies, 1978]. Even at this generalized level,

however, a major gap in the synthesis effort resulted from the absence of drilling data from the high-latitude southern Indian Ocean. This proved to be a serious handicap to any attempt to understand the history of circulation in this ocean. The ODP Kerguelen Plateau and Prydz Bay drilling essentially redressed this problem; together with the DSDP and ODP Ninetyeast Ridge sites there now exists in the Indian Ocean a latitudinal spread of drillsites spanning 75 degrees (Figure 1).

The DSDP Indian Ocean campaign was characterized by a higher incidence of continuous coring than previously had been the case in the DSDP but there remained a large number of 'spot-cored' sites and it was rare, at that time, for more than one hole to be drilled at a site. Similarly, Indian Ocean biostratigraphy was in its infancy in the early 1970's [Vincent, 1977] and only Leg 23 made any attempt at paleomagnetic measurement on sediments [Whitmarsh et al., 1974]. In the intervening fifteen years, advances in

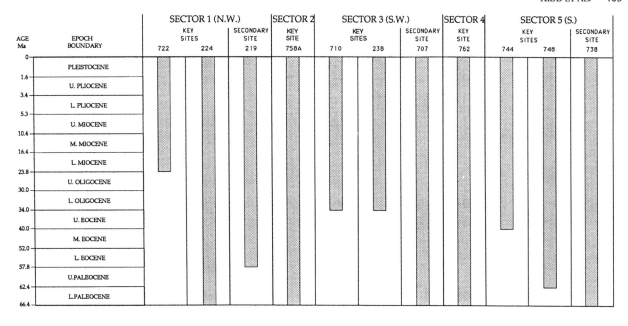

Fig. 2. Age ranges spanned by the PALIOS key and secondary sites.

integrated bio- and magneto-stratigraphic techniques have radically improved stratigraphic resolution and the integrated approach has been extended to stratotypes on land [e.g. Premoli-Silva et al., 1988]. The ODP Indian Ocean sites, in almost all cases, were drilled with a view to complete core recovery, employing double, even triple, coring strategies. The potential for synthesis studies is thus an order of magnitude better from the ODP sites, although the DSDP sites remain essential for both spacial and, as will be seen in the following text, temporal coverage.

Advances on a similar scale have taken place in data availability and techniques for tectonic reconstruction. For previous paleoceanographic synthesis studies we relied on a small number of reconstructions of the ocean based on limited magnetic anomaly compilations and age data largely from drilling [Sclater et al., 1977]. Current paleogeographic reconstructions based on satellite altimetry data and on new syntheses of marine geophysical data have resulted in refined maps of the locations of drillsites that can be displayed, if necessary, at million year intervals; see Royer et al. [this volume]. For the next phase of PALIOS we expect to have reconstructions of paleobathymetry available on which to map sediment distributions and other parameters.

Following the close of the DSDP phase, the JOIDES organization ensured that its database holdings have become integrated into a standardized and more available format. All of the DSDP sedimentary and biostratigraphic data are available in list form on compact disc from which relevant Indian Ocean data can be abstracted in order to plot paleoceanographic parameters against our developing stratigraphic and tectonic frameworks. More recently a similar ODP compact disc has come available.

METHODOLOGY AND APPROACH

General approach

During the initial phase of PALIOS our principal tasks were:

(1) to establish a stratigraphic framework that would allow for correlation between high and low latitudes and between DSDP and ODP sites;

(2) to evaluate the DSDP/ODP sediment data bases and to develop the necessary software to manipulate these; and

(3) to obtain and analyze samples from the repository collections to resolve key biostratigraphic and lithologic uncertainties and to extend the data available for specific target projects.

In order to proceed, the Indian Ocean was divided into five regions or sectors (Figure 1) and a reference stratigraphic column was developed for each sector. Frequently the reference columns have been made up through correlations between two or more "key" sites and often are supplemented by a "secondary" site. Key and secondary sites were carefully selected to maximize their potential for stratigraphic completeness, in terms of age ranges and core recovery (Figure 2). As an example, in Sector 1 two key sites, ODP Site 722 and DSDP Site 219, make up a composite reference column spanning the middle Eocene to present but a secondary site, DSDP Site 224, with only spot coring is required to extend Site 722's temporal range beyond the middle Miocene. We have made the general decision to limit our stratigraphic refinement to the Cenozoic but for some target projects, for example, the studies of volcanic input, we expect to extend analyses beyond the Cretaceous/Tertiary Boundary. (We note that our largely artificial geographical division of the ocean into sectors as in Figure 1 has required refinement for later studies in PALIOS as regional links between parameters have become better refined.)

Stratigraphic Methods

One of the inherited difficulties in using the DSDP/ODP databases for paleoceanographic reconstructions is the inconsistency of the chronological and biostratigraphic frameworks. The framework developed for each cruise results from the current state of scientific knowledge and the individual experience of participants at the time of that specific cruise. Therefore, time-scales, calibration of events, zonal concepts and/or species concepts often differ and are not directly comparable from one cruise to another.

Direct comparison of data sets from different DSDP and ODP cruises requires a standardized and internally consistent stratigraphic framework. The approach applied in the PALIOS Project provides:

(1) standardized stratigraphy that brings together the various time-scales and numerous biostratigraphic zonal schemes used during the DSDP and ODP expeditions in the Indian Ocean;

(2) ocean-wide extent that allows partial correlation between the low and high latitudes; and

(3) integrated stratigraphic events, incorporating both biosiliceous (diatoms) and biocalcareous (calcareous nannofossils and foraminifers).

The resulting stratigraphic framework was developed by evaluating the quality and consistency of both shipboard and shorebased DSDP/ODP paleontological (specifically calcareous nannofossils, foraminifers, diatoms) and magnetostratigraphic databases for all Indian Ocean sites and re-examining key stratigraphical intervals. This was accomplished in the following steps:

(1) Zonal concepts and placement of zonal boundaries were evaluated for each group by reviewing the official databases, the data presented in the DSDP/ODP *Initial Reports* and *Scientific Results* volumes and articles in other journals. Problems occurring within a given group, such as incomplete (or lack of) definition of terms or concepts, inconsistencies in the species concept, or differences in the stratigraphic framework at a site or between sites, were flagged for resampling of the specific interval.

(2) Zonal and datum lists were generated providing downhole depth constraints for each stratigraphic event. The various stratigraphies at each site were compared and

evaluated. Inconsistencies were flagged to determine if sampling was required.

(3) A standardized chronological framework was agreed upon as in Figure 3 [Baldauf et al., unpublished manuscript, 1992].

(4) Datum lists of key events were developed for each group based on previous work. Specific events were evaluated as to their reliability and isochroneity for low to high latitude correlations. This determination was based on currently available scientific literature and recent results from the authors.

(5) About 500 samples were obtained and re-examined to provide a consistent stratigraphy and to further investigate hiatus occurrence.

Biostratigraphy

The primary stratigraphic effort in PALIOS undoubtedly has been the integration of chronological frameworks between the low and high latitudes. A spread of sites covering a wide range of latitude is required, from which can be obtained magneto- and/or isotope-stratigraphies allied to both calcareous and siliceous microfossil zonations. The currently available DSDP/ODP dataset allows only partial integration in that (1) very few sites meet these criteria; (2) the number of sites meeting the requirements is reduced as one continues further back into the geological record (Figure 2); and (3) the spatial distribution of current sites still results in a sampling void spanning over 10 degrees of latitude, from 37-50 degrees S (Figure 1).

Using this available dataset and keeping the above constraints in mind, we have integrated the low-and high latitude chronological and biostratigraphic frameworks in Figure 3 [Baldauf et al., unpublished manuscript, 1992, present a detailed discussion]. The resulting integrated framework relies on calcareous nannofossil, planktonic foraminifers, and diatom biostratigraphies correlated, where possible, with magnetostratigraphy or isotope stratigraphies. Although other stratigraphies, such as radiolarian biostratigraphy, will provide additional information, they are not incorporated into the present framework. The key and secondary sites employed as the basis of our sediment distribution studies reflect the best available stratigraphic coverage (Figure 2).

Fig. 3. (opposite, and following two pages). The Tertiary stratigraphic framework as used in the PALIOS Project; A. (0-25 Ma), B. (25-50 Ma), and C. (50-66.4 Ma). Except where noted, the magnetic polarity timescale follows Berggren et al. [1985a,b]. The hatched area at the Miocene/Pliocene and Eocene/Oligocene boundaries represents differences between placement of these boundaries under discussion by the International Stratigraphic Commission [see Baldauf et al., unpublished manuscript, 1992, for discussion]. Calcareous nannofossil zonations of Martini [1971] and Okada and Bukry [1980] are used for the low-latitudes and that of Wei and Wise [1990, and subsequent papers] for the high latitudes. Calcareous nannofossil biochronology follows Backman et al. [1990] and Berggren et al. [1985a, b] for the low latitudes and Wei and his colleagues for high latitudes. Planktonic foraminiferal zonations include the low-latitude zonation of Blow [1969] with modifications by Kennett and Srinivasan [1983], and Paleogene by Bolli [1957, 1966], Bolli and Premoli-Silva [1973], and Stainforth et al. [1975], the middle-latitude zonation of Jenkins [1971, 1985], and the high-latitude zonations of Berggren [1992], Stott and Kennett [1990], and Huber [1991a,b]. Foraminiferal biochronology follows Berggren et al. [1985a,b], Barron [1985a,b], Jenkins [1985], and Stott and Kennett [1990]. Diatom zonations incorporated into the stratigraphic framework include Barron [1985a,b] for the low latitudes and Baldauf and Barron [1991] for the high latitudes. Diatom biochronology follows Barron [1985a] and Harwood et al. [1990]. The hatched area in the zonal columns represents intervals where no formal zonation is recognized. Dashed lines represent uncertainty in chronostratigraphic placement following Wei and Wise [1992a,b]. In Figure 3A, range 1 = last occurrence of *Cyclicargolithus floridanus*; Arrow 2 = first occurrence of *Calcidiscus macintyrei*, and Arrow 3= last occurrence of *Reticulofenestra bisecta*. In Figure 3C, lower right margin figures indicate precise placement and ages of Wei and Pospichal [1991] high-latitude nannofossil zonal boundaries NA1-NA6.

Geologic time-scale correlation chart

Age			Paleomagnetic scale			Calcareous Nannofossils			Planktonic Foraminifera			Diatoms			Ma
Ma			Chron	Berggren et al.[1985]	Anomaly	Martini [1971]	Okada & Bukry [1980]	Wei & Wise [1990]	Blow [1969]	Jenkins [1971, 1985]	Berggren [in press]	Barron [1985a; 1985b]	Baldauf & Barron [1991]		Ma
	Pleist.		Brun.		1	NN21	CN15 b		N23 - N22			P. doliolus	T. lentiginosa	NSOD21	
	Pliocene late		Matuyama			NN20	CN14 a			SN12		Nitzschia reinholdii B / A	A. ingens	NSOD20	
			Gauss		2	NN19	CN13 b / a		N21			Rhizosolenia praebergonii B / A	R. barboi	NSOD19	
						NN17	CN12 d / c / b / a						T. kolbei	NSOD18	
	Pliocene early				2A	NN16				SN11			T. vulnifica	NSOD17	
			Gilbert		3	NN14-15	CN11 b / a		N19	SN10		Nitzschia jouseae	N. interfrig.- C. insignis	NSOD16	
						NN13							N. barronii	NSOD15	
5						NN12	CN10 c / b		N18		NK7	T. convexa C / B	T. inura	NSOD14	5
	Miocene late		C3A		3A				b	SN9			T. oestrupii	NSOD13	
			C4		4A	NN11	CN9 b		N17			N. miocenica B / A	T. torokina	NSOD12	
										SN8		N. porteri A			
			C4A			NN10	CN8		a			T. yabei B / A			
									N16		NK6		A. fryxellae	NSOD11	
10			C5		5	NN9	CN7 b / a				NK5	Actinocyclus moronensis			10
						NN8	CN6		N15						
								1	N14			Crasepedodiscus coscinodiscus	D. dimorpha	NSOD10	
	Miocene middle		C5A		5A	NN7	CN5 b / a		N13	SN7		C. gigas v. diorama	D. praedimorpha	NSOD9	
						NN6			N12			Coscinodiscus lewisianus	N. denticuloides	NSOD8	
			C5AA / C5AB / C5AC / C5AD						N11	SN6	NK4		D. hustedii- N. grossepunct.	NSOD7	
15			C5B		5B	NN5	CN4		N10 / N9			Cestodiscus peplum B / A	A. ingens - D. maccollumii	NSOD6	15
									N8	SN5			D. maccollumii	NSOD5	
			C5C		5C	NN4	CN3		N7		NK3	Cruciden. nicobarica B / A	C. kanayae	NSOD4	
			C5D		5D			2	N6			T. pileus	Thalassiosira fraga	NSOD3	
	Miocene early		C5E		5E	NN3	CN2		N5	SN4		Craspedodiscus elegans			
20			C6						b		NK2	Rossiella paleacea C / B	Thalassiosira spumellaroides	NSOD2	20
			C6A		6A	NN2	CN1		N4	SN3			Thalassiosira spinosa A	NSOD1	
			C6AA							SN2					
			C6B		6B					SN1	NK1	Rocella gelida			
			C6C		6C	NN1		3	a	SP14 G. euapertura			R. gelida		

Age			Paleomagnetic scale			Calcareous Nannofossils			Planktonic Foraminifera			Diatoms	
Ma			Chron	Berggren et al.[1985]	Anomaly	Martini [1971]	Okada & Bukry [1980]	Wei & Wise [1990]	*1	Jenkins [1971, 1985]	Stott & Kennett [1990]	Baldauf & Barron [1991]	Ma
	Oligocene	late	C6C			NN1	CN1						
			C7		7			R. bisecta				Rocella	
			C7A		7A	NP25	CP19b		G. ciperoensis	G. euapertura		gelida	
			C8		8								
			C9		9	NP24	CP19a		b	b	b	Rocella	
30			C10		10			Chiasmolith.	G. olina		G. labiacrassata	vigilans	30
		early	C11		11	NP23	CP18	altus	a	a	a		
					12					S. angiporoides			
			C12					R. daviesii	G. ampliapertura				B
							CP17			G. brevis		Rhizosolenia	
35						NP22		B. spinosus	C. chipolensis	b		oligocenica A	35
			C13		13	NP21	CP16		P. micra	a	G. angiporoides		
	Eocene	late	C15		15	NP20 - NP19	CP15b	Reticulofen. oamaruensis	G. cerrdazwlansis	S. linaperta			
			C16		16	NP18	CP15a	T. recurvus	T. rhori		G. suteri		
					17			Chiasmolith. oamaruensis					
40			C17			NP17	CP14b	D. saipanensis		G. inconspicwa	G. index		40
			C18		18			Reticulofen. reticulata	O. beckmanni				
			C19		19		CP14a	R. umbilica		G. index	G. index		
45		middle			20	NP16 - NP15			M. lehneri		A. collactea		45
			C20				CP13		G. subconglobata	A. primitiva	P. micra		
			C21		21	NP14	CP12		H. aragonensis		A. bulbrooki		

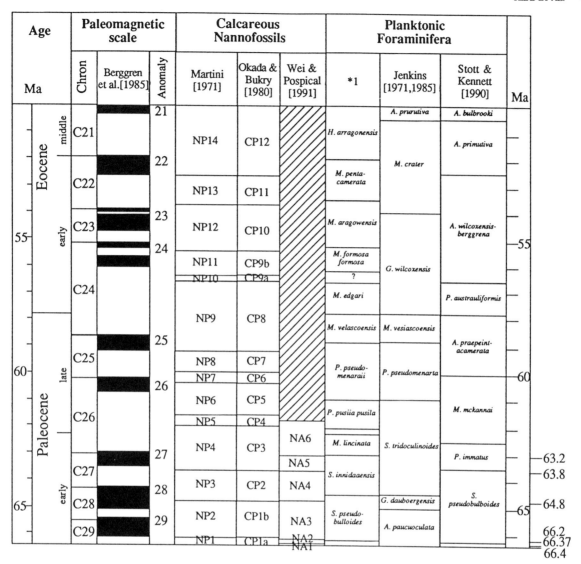

Age		Paleomagnetic scale			Calcareous Nannofossils			Planktonic Foraminifera			
Ma		Chron	Berggren et al.[1985]	Anomaly	Martini [1971]	Okada & Bukry [1980]	Wei & Pospical [1991]	*1	Jenkins [1971,1985]	Stott & Kennett [1990]	Ma
				21					A. prurutiva	A. bulbrooki	
	middle	C21			NP14	CP12		H. arragonensis			
									M. crater	A. primutiva	
		C22		22	NP13	CP11		M. penta-camerata			
	Eocene			23	NP12	CP10		M. aragowensis		A. wilcoxensis-berggrena	
55	early	C23		24	NP11	CP9b		M. formosa formosa	G. wilcoxensis		55
		C24			NP10	CP9a		?			
					NP9	CP8		M. edgari		P. austrauliformis	
				25				M. velascoensis	M. vesiascoensis		
		C25			NP8	CP7				A. praepeint-acamerata	
60	late			26	NP7	CP6		P. pseudo-menaraii	P. pseudomenarta		60
		C26			NP6	CP5				M. mckannai	
	Paleocene				NP5	CP4		P. pusiia pusila			
					NP4	CP3	NA6	M. lincinata	S. tridoculinoides		
		C27		27			NA5	S. innidaaensis		P. immatus	63.2
	early				NP3	CP2	NA4		G. dauboergensis		63.8
65		C28		28	NP2	CP1b	NA3	S. pseudo-bulloides	S. pseudobulloides	64.8
		C29		29					A. paucuoculata		
					NP1	CP1a	NA2 / NA1				66.2 / 66.37 / 66.4

Several zonal schemes are used for calcareous nannofossils. We adhere to the standard Tertiary calcareous nannofossil zonations as proposed by Martini [1971] and Bukry [1973, 1975] and code numbered by Okada and Bukry [1980] for the low latitudes. Because of the reduced abundance of calcareous microfossils in Neogene sediments due to climatic cooling and the corresponding increase in biosiliceous sediments, no calcareous nannofossil zonation is proposed here for the high-latitude Neogene. Instead, several calcareous nannofossil datums are used for biostratigraphic correlation as proposed by Wei and Wise [1992a]. The calcareous nannofossil biostratigraphy for the Paleogene follows that of Martini [1971] and Okada and Bukry [1980] for the low latitudes. Two zonations employed for the high latitudes include the middle Eocene through Oligocene zonation of Wei and Wise [1992a,b] as

modified by Wei et al. [1992] and the Lower Paleocene zonation of Wei and Pospichal [1991].

To reduce difficulties with diachroneity of species of planktonic foraminifers we employ several different planktonic foraminiferal zonations, as detailed in the caption for Figure 3.

Similarly, two diatom zonations are employed in this study, and these together with the nannofossils provide the essential \low to high latitude linkage. The Neogene diatom zonation of Barron [1985a,b] is used for the low-latitude Indian Ocean. This zonation, originally developed for the eastern equatorial Pacific, has been shown to be partially useful in the low-middle latitude Atlantic [Baldauf, 1986; Baldauf and Pockras, 1989] and the low-latitude Indian Ocean [Mikkleson, 1990]. The late Oligocene through Quaternary diatom zonation of Baldauf and Barron

Site 744

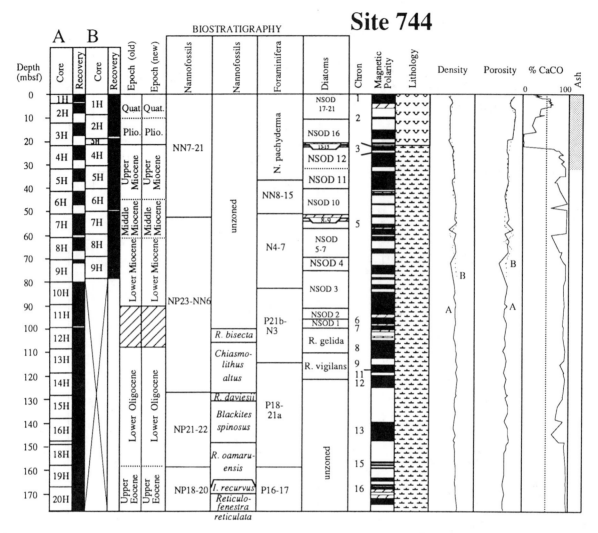

Fig. 4. An example of a PALIOS stratigraphic template. Shown here is the template for Site 744, Holes A and B (ODP Leg 119 positioned on the Kerguelen Plateau, southern Indian Ocean). Depth (meters below seafloor), core, core recovery, Epoch (old) and lithology are from Barron, Larsen et al. [1989]. Two calcareous nannofossil zonal schemes are presented, one showing the low-latitude zonation of Martini [1971] and alongside the new high-latitude zonation of Wei and Wise [1990]. The calcareous nannofossil data incorporate the shipboard work of Wei and Thierstein [1991] and subsequent shorebased studies. The foraminiferal zonation is that of Blow [1969]. The foraminiferal data incorporate the shipboard work of Huber and Schroder and subsequent shorebased studies [Barron, Larsen et al., 1989]. The diatom zonation is the new high-latitude zonation of Baldauf and Barron [1991]. The diatom data incorporate the shipboard work of Baldauf and Barron [1991] and the shorebased work of Baldauf and Barron [1991], Barron et al. [1991], and this PALIOS study. The magnetochronology is from Barron et al. [1991].

[1991] is used for the high latitudes. No Paleogene diatom zonation can yet be adopted for use in either the high or low latitudes.

Chronostratigraphy

Generally we adhere to the Geomagnetic Polarity Time Scale (GPTS) of Berggren et al. [1985a,b] for correlations between the magnetostratigraphy and the chronological scale. This time scale is, however, modified to incorporate the Cobb Seamount and Reunion paleomagnetic events, age assignments for these events following Harland et al. [1990].

Calibration of the biostratigraphy to this time scale follows partly Berggren et al. [1985a,b] for the calcareous

microfossils and Barron et al. [1985] for the siliceous microfossils. In both cases these calibrations were slightly modified to integrate recent drilling results and revised calibrations of specific datums (see Figure 3 caption). Most important are the calcareous nannofossil calibrations which follow Backman et al. [1990] and Berggren et al. [1985a,b] for the low latitudes but Wei and his colleagues for the high latitudes. The latter include a series of papers based on DSDP/ODP drill holes in the South Atlantic, Weddell Sea and Indian Ocean [Wei and Wise, 1989]; [Wei and Thierstein, 1991; Wei and Pospichal, 1991; Wei, 1991; Wei et al., 1992], which are summarized in [Wei and Wise, 1992c]. The studies show

TABLE 1. Age of Epoch Boundaries

Boundary	Age (Ma)	Reference
Pliocene/Pleistocene	1.6[*]	Aguirre and Pasini [1985]; Tauxe et al. [1988]
Miocene/Pliocene	4.9-5.3[*]	Zigderveld et al. [1986]; Berggren et al. [1985a]
Oligocene/Miocene	23.7[*]	Berggren et al. [1985b]
Eocene/Oligocene	34[*]-36.5	Premoli-Silva et al. [1988]; Berggren et al. [1985b
Paleocene/Eocene	57.8[*]	Berggren et al. [1985b]
Cretaceous/Tertiary	66.4[*]	Berggren et al. [1985b]

* Denotes age of boundary used in this study.

that some nannofossil zonal datums are clearly diachronous in the high latitudes, but where time transgressive datums are well constrained by magnetostratigraphy, they can still be used within well defined limits. These newly established calibrations in the high latitudes measurably enhance our ability to correlate these sequences with the low latitudes.

Ages of the Epoch boundaries used in this framework are shown in Table 1. Although the majority of the boundary ages used in the PALIOS study follow Berggren et al. [1985a,b], two boundaries, namely the Eocene/Oligocene boundary and the Miocene/Pliocene boundary, are currently being revised by the International Stratigraphic Commission. The former is particularly important to the overall project because of the paleoceanographic significance of the Eocene to Oligocene period. Detailed discussion of the rationale for the ages assigned for the PALIOS project is contained in Baldauf et al. [unpublished manuscript, 1992].

Files and Templates

The revised stratigraphic framework for each Indian Ocean site was entered into a new PALIOS database. This database consists of Microsoft Excel® and Stratigraphic Template files for each site. The Excel® file provides a list of each datum, the datum calibration (Ma) and the depth constraints. The Stratigraphic Template (see example, Figure 4) graphically presents the revised stratigraphy and other information for each site in a standardized format. The template summarizes information on cores, core recovery, placement of Epoch boundaries (old and revised) and zonal (calcareous nannofossils, foraminifers, diatoms) and magnetic boundaries. The lithology, density, porosity, and carbonate percentages are as discussed in the DSDP and ODP volumes or subsequent publications. This standard presentation facilitates quick comparisons between sites.

Figure 4 is the template for Site 744 positioned on the Kerguelen Plateau, drilled during ODP Leg 119. Site 744 represents one of the two key sites (Sites 744 and 748) for PALIOS Sector 5, the Indian Ocean region of the southern ocean. In addition there is one secondary site, Site 738 (Figure 5). Together these sites provide a composite reference stratigraphy from the present to the Cretaceous. This example template is based on the shipboard studies

[Barron, Larson et al., 1989] and numerous shorebased studies [see Barron et al., 1991]. Unlike the majority of the site data, few revisions of the shipboard data were required for this particular site.

Depth to Age Conversions

Measurement points and sample locations are commonly identified by depth down the drillhole, rather than age of the material being examined. Hence to proceed beyond the summary presentation of the data represented by the templates to a more detailed analysis requires a procedure for assigning ages to individual samples and/or measurements.

In principle, age assignments can be made by plotting the age of known stratigraphic horizons against their depth down the drillhole, and interpolating to determine ages corresponding to the depths of the measurements of interest. In practice the process is complicated by substantial uncertainties in the data. These have been discussed by Moore [1972], Moore and Heath [1978], and Moore and Romine [1981], and include both uncertainties in depths assigned to stratigraphic events and limitations in the stratigraphic data resulting from the presence of hiatuses, diachronism, and problems in the correlation between stratigraphic events and absolute age. Aside from these difficulties, hand plotting of the data is tedious and prone to error, and the resulting depth/age curve is difficult to update since it must be manually replotted as new data become available.

To simplify matters, we have adopted an automated procedure which can be applied consistently to each site to generate a depth/age curve and make age assignments to specified depths. This has several advantages:

(1) The need for meticulous hand plotting is obviated.

(2) Since all the steps in the procedure, and the links between them, are preserved, age assignments can be quickly and easily updated as new information becomes available.

(3) The procedure can readily handle large numbers of points which cannot reasonably be plotted by hand.

The procedure used in PALIOS is based on Microsoft® Excel, Version 2.2, a widely used spreadsheet program. Age

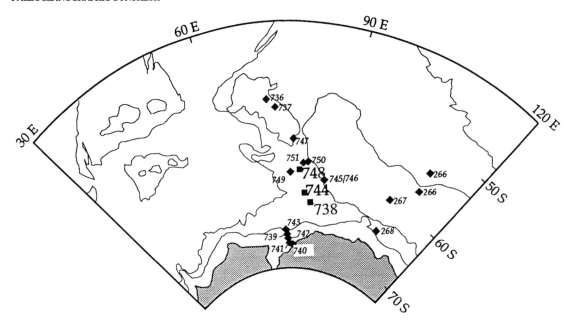

Hole	KEY SITES			SECONDARY SITE
	744A	748A	748B	738
Water depth (m)	2307	1291	1291	2253
Nannofossil zone				
CN15/NN21	X		X	
CN14b/NN20	X		X	
CN13a-b/NN19	X		X	
CN12d/NN18	X		X	
CN12c/NN17	X		X	
CN12a&b/NN16	X		X	
CN11/NN15	X		X	
CN10d/NN14	X		X	
Miocene-Pliocene				
CN10c/NN13	X		X	
CN10a&b/NN12	X		X	
CN9/NN11	X		X	
CN8/NN10	X		X	
CN7/NN9	X		X	
CN6/NN8	X		X	
CN5b/NN7	X		X	
middle Miocene				
CN5a/NN6	X		X	
CN4/NN5	X		X	
CN3/NN4	X		X	
CN1c-2/NN3	X		X	
CN1c/NN2	X		X	
CN1a&b/NN1	X		X	
CP19b/NP25	X		X	
Cp19a/NP24	X		X	
CP18/NP23	X		X	
CP17/NP23	X		X	
Eocene - Oligocene				
CP16c/NP22	X		X	X
CP16a&b/NP21	X		X	X
CP15b/NP19-20	X	X	X	X
CP15a/NP18		X	X	X
CP14		X		X
CP13		X		X
E.M.Eocene				
CP12		X		X
CP11		X		X
CP10		X		X
CP9		X		X
CP8		X		X
CP7		X		X
CP6		X		
CP5		X		X
CP4				X
CP3				X
CP2				X
CP1				X

Fig. 5. Map showing the locations of ODP and DSDP sites in PALIOS sector 5. Key and secondary reference sites are numbered larger. The nannofossil zones recorded at each site are tabulated below the map.

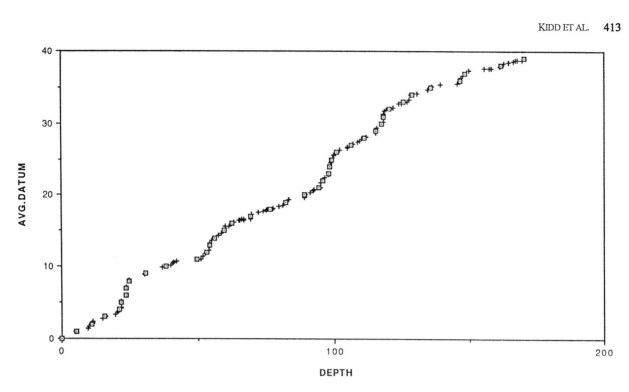

Fig. 6. Depth/age curve for Site 744. Crosses indicate original age datums, plotted at the mid point of the depth range for each datum. Open squares are the depths of 1 m.y. age increments, calculated according to the method outlined in the text.

assignments from the datum list (see above) are smoothed with a 5-point rolling average, and then depths or ages to intermediate sample/measurement points are calculated by linear interpolation (Figure 6). Though not particularly sophisticated, this approach is easy to apply consistently and produces results which are within the resolution of the original data. Davies et al. [1992] describe this method in further detail, and also discuss comparisons between this method and other approaches which might be adopted.

The Excel spreadsheet is used to assign ages to depths corresponding to specific measurements by interpolating between points on the depth/age curve. Once age assignments have been made, it is a simple matter to group measurements according to age intervals and compute parameters such as average carbonate content, accumulation rates, etc.

Accumulation Rates

For calculating accumulation rates, bulk density, porosity and compositional data shown on the templates (as in Figure 4) are grouped in 1 m.y. increments and the following formula [Van Andel et al., 1975; Worsley and Davies, 1979] applied to the averaged values for each increment:

$$R = T.(D - 0.01025.P)/10 \qquad (1)$$

where R = accumulation rate (g/cm^2/Kyr), T = thickness (m), D = bulk density (g/cc), and P = porosity (percent).

Figure 7 shows the results for Site 744. The result of removing the effects of compaction is readily apparent, as

is the fine structure of the data revealed by the use of small, equal time increments, compared to the larger units used in the ODP volumes. On the negative side, it is also apparent that there are significant gaps resulting from lack of data. In some cases gaps in the accumulation rate data reflect hiatuses. But where the depth/age curve indicates sediments are present, bulk density and other parameters can be estimated by interpolation between increments above and below.

Target Projects

The PALIOS project is structured such that its second phase will largely involve plotting sediment parameters as distributions on a series of paleobathymetric maps. In many cases we have found it necessary to supplement the data available in the ODP/DSDP data bases with new data generated in a number of target projects from samples requested from the repositories. The analyses of these new samples have run parallel with the primary tasks in phase one. Preliminary results of two of the target projects (hiatus and volcanogenic sediment distributions) are presented in this contribution.

Throughout the initial evaluation of the DSDP and ODP stratigraphic data and the eventual construction of the templates, careful examination was made of the evidence for hiatuses. Over 50 cases warranted more precise biostratigraphic refinement in the light of our updated stratigraphic framework. Approximately 100 nannofossil samples were selected and examined from above and below each hiatus.

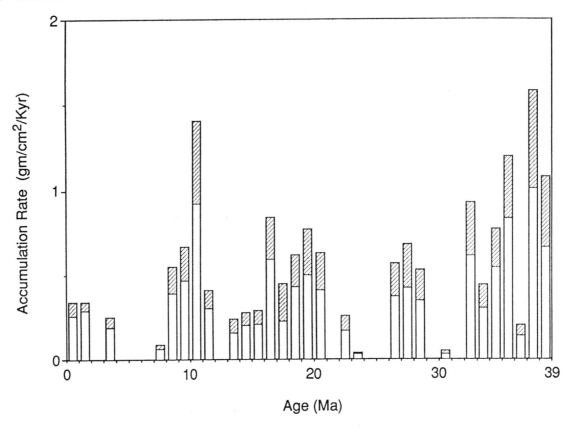

Fig. 7. Total sediment accumulation rates ploted in 1 m.y. increments for Site 744, calculated according to the methods described in the text. Open shading: carbonate fraction; diagonal shading: non-carbonate fraction. Missing columns indicate intervals for which the data are incomplete. Generally these are represented by thin sediments, and likely contain hiatuses.

For the volcanogenic sediments target project our approach has been to supplement the database collected by Vallier and Kidd [1977] with occurrence data from the ODP volumes or database. Initially we looked for occurrences of volcanic glass, zeolites and combinations of certain clay minerals in smear slides or XRD data. We selected intervals in the newly drilled ODP sequences that widened our temporal control or allowed us to concentrate sampling in time ranges that were recognized from the previous study, but were not noted as containing volcanogenic components in the routine shipboard analysis. About 500 samples have been analyzed using standard X-ray diffraction techniques in this study.

PRELIMINARY RESULTS

Through the improvements in stratigraphy and plotting routines developed in the PALIOS project, we can now consider the distributions of parameters against paleogeographic reconstructions developed by the POMP program and its co-workers; see Royer et al. [this volume]. More accurate analysis of spatial distributions will best be served when improved paleobathymetric reconstructions for frequent intervals through geological time become available during this project's second phase. For the moment we present some preliminary results of our target

projects on hiatuses and volcanogenic sediments whose distributions are plotted together on the selected reconstructions shown in Figures 8 through 11.

Hiatuses

An accurate knowledge of the detailed distribution of hiatuses in time and space is clearly fundamental to any subsequent analysis of mass sedimentary budgets on an oceanwide scale. Davies et al. [1975] were first to draw attention to hiatuses of ocean-wide significance in the Indian Ocean. The most widespread hiatuses in the DSDP Indian Ocean sites were generally recognized as Oligocene, early Tertiary and Late Cretaceous in age. The unconformities that were recognized in the carbonate sequences were indeed significant gaps in the stratigraphic record, given the constraints of the DSDP core recovery and biostratigraphy, and were considered as representing erosion or non-deposition of sediment. In deeper sites many of these hiatuses are correlated with undated intervals, which were barren of calcareous microfossils because the site had sunk through its spreading history to levels below the regional calcite compensation depth (CCD) [Kidd and Davies, 1978]. These intervals were considered a "dissolution facies" representing at least very slow or non-deposition, possibly even erosional hiatuses.

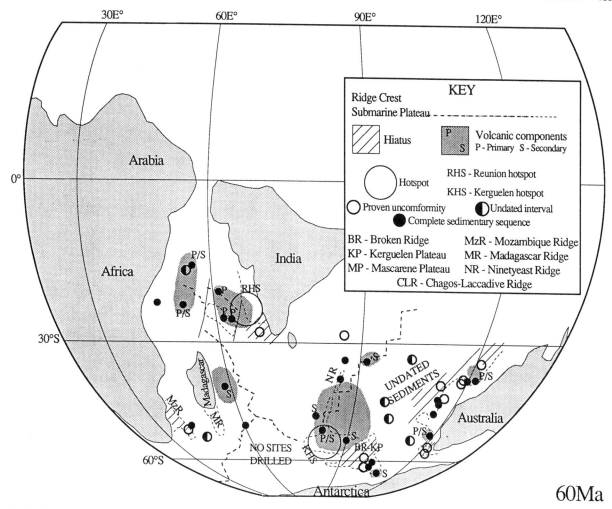

Fig. 8. Paleogeographic reconstruction of the Indian Ocean at 60Ma (Paleocene) showing the distribution of hiatuses and volcanogenic sediments. Hiatus distributions are constrained by proven stratigraphic gaps in sites that penetrate this time period and may extend to undated intervals in pelagic clays. Volcanogenic sediment distributions are probably restricted because of the hiatus development. Sites with volcanic material are classified as containing dominantly primary components, dominantly secondary components or both. See key for location abbreviations.

Davies et al. [1957] consider that the correspondence between documented hiatuses in the carbonate sediments and the dissolution facies intervals was due to periods of incursion of corrosive bottom waters into the deep basins.

Most of the ODP drilling was at relatively shallow carbonate-rich sites so the inventory of stratigraphically well-documented sites is greatly increased and stratigraphic control is much improved. In the hiatuses target project we have conducted a re-sampling program in order to examine closely the timing and likely duration of the hiatuses at both the DSDP and ODP sites. All of the previously identified DSDP hiatuses have been confirmed against the new stratigraphic framework; most importantly none was apparent or spurious. On the other hand, there have been considerable refinements to our interpretation of the time periods represented by those hiatuses.

The addition of the ODP sites and the improved stratigraphy allow a more detailed analysis of the spatial and temporal distribution of hiatuses. The times selected for the figures (60, 50, 35 and 20 Ma), show broad patterns of hiatus distributions in the Indian Ocean.

Paleocene 60Ma (Figure 8) Sediments of this age are absent from topographically elevated areas of the Chagos-Laccadive Ridge, Mozambique Ridge, parts of the southern Kerguelen Plateau and the Australian Margin and the Naturaliste Plateau. Following Davies et al. [1975], we continue to infer the occurrence of basinal hiatuses from poorly-dated dissolution facies in the Wharton Basin.

Eocene 50Ma (Figure 9). Hiatuses are evident on the elevated Mozambique and Madagascar Ridges and on the Ninetyeast Ridge. On the Kerguelen Plateau hiatus distribution is restricted to a single site (Site 747), despite there being continuous deposition at the other plateau sites. (Site 747 is in a col in the topography which may

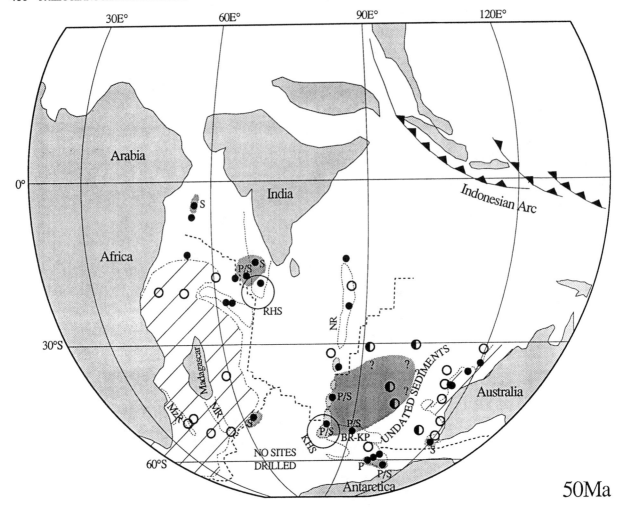

Fig. 9. Paleogeographic reconstruction of the Indian Ocean at 50Ma (Eocene). See explanation of symbols in the Figure 8 caption and key on that figure.

have at this time represented a 'gateway' to bottom water circulation.) Basinal hiatuses are better defined in this interval and are not entirely based on dissolution facies. They are recorded in the Somali, Madagascar and Mozambique basins and along the eastern margin of the Wharton Basin.

Eocene/Oligocene boundary 35Ma (Figure 10). This period was characterized by widespread hiatus development at both elevated and basinal sites. Hiatuses are recorded from the Chagos-Laccadive, Mozambique and Madagascar ridges, part of the southern Kerguelen Plateau, and the Australian margin. The hiatuses are extensively developed in the Somali, Madagascar and Mozambique basins. There are no data for the Crozet Basin and no sites in the southern Central Indian Basin. Hiatuses at sites in the Arabian Sea, and the northern, western and southern margins of the Ninetyeast Ridge suggest that we can confidently extend hiatus distributions throughout the Western and Central basins. Hiatuses in the deeper sites along the Australian

margin might suggest that this distribution could have extended through the undated Wharton Basin and thus might indicate a circulation 'event' that includes this eastern basin also.

Miocene 20Ma (Figure 11). Hiatuses occur predominantly on topographically elevated sites. These include the southern Chagos-Laccadive Ridge, the eastern Mascarene Plateau, the Mozambique Ridge, parts of the Kerguelen Plateau and the Australian margin. The distribution of hiatuses is generally more patchy than at the Eocene/Oligocene boundary.

From the patterns shown in Figures 8 to 11 it is clear that the distribution of hiatuses varied significantly through time. These figures represent "snapshots" in time rather than intervals as have been previously considered, but they do generally confirm the conclusion of Kidd and Davies [1978] that the hiatuses are widespread and occur at both deep and shallow depths. On the other hand, only the Eocene/Oligocene 35Ma pattern suggests a truely

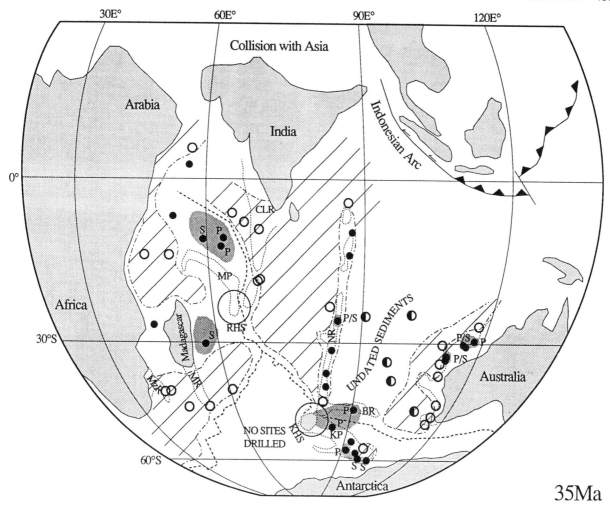

Fig. 10. Paleogeographic reconstructions of the Indian Ocean at 35Ma (Eocene/Oligocene boundary). See explanation of symbols in the Figure 8 caption and key on that figure.

oceanwide hiatus 'event'. The extent of hiatuses within basins is still to some extent obscured by problems in dating dissolution facies. The changing patterns shown by the reconstructions reveal that during the Paleocene hiatuses were mainly restricted to topographically elevated areas. The hiatuses are more extensively developed to include western basinal settings in the Eocene and extend to the eastern basins in the Oligocene. A return to a patchy Paleocene-like distribution occurs in the Miocene. Kidd and Davies [1978] attribute the widespread hiatus distributions in the Oligocene to enhanced circulation of corrosive bottom waters, brought on partly by the onset of Antarctic glaciation and partly by the re-arrangement of the topography of the basins. The glacial model does not account for the post-Oligocene pattern, because continuous sedimentation exists in the Miocene basins. However by this time the modern deep circulation through the basins was well-established through fast spreading south of Australia and the opening of the Drake Passage at ~30Ma [Barker and Burrell, 1977].

We emphasize that these comments are preliminary: we need now to re-examine sections at all of the drill sites to look for previously undetected hiatuses once the new paleobathymetric reconstructions and accumulation rate data become available [Ramsay et al., unpublished manuscript, 1992]; each individual site is still to be backtracked for comparison with subsidence curves in Kidd and Davies [1978] and subsequent studies.

VOLCANOGENIC SEDIMENTS

Volcanogenic components are significant contributors to sediments in the world oceans, both volumetrically and as indicators of ocean basin tectonics. Kennett and Thunnell [1975], from a review of the occurrence of volcanic glass in DSDP drill holes, were able to recognize major explosive events in the Quaternary which they relate to global volcanic activity associated with subduction. Significant explosive activity, however, is apparently also related to tensional tectonic events, particularly within mid-ocean ridge systems. The amount of volcanic detritus in marine

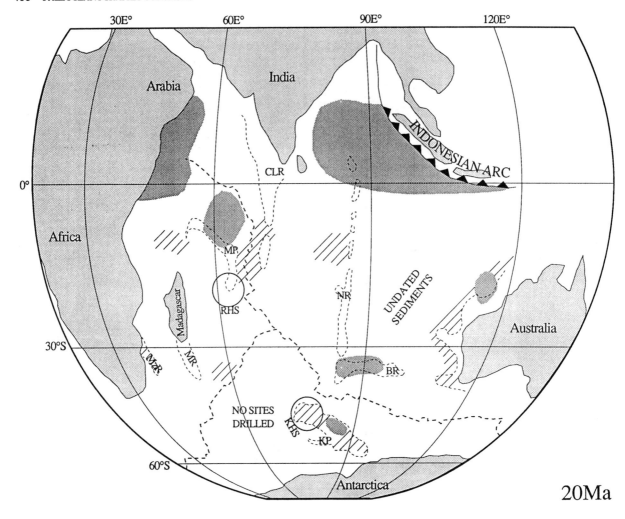

Fig. 11. Paleogeographic reconstruction of the Indian Ocean at 20Ma (Miocene). See explanation of symbols in the Figure 8 caption and key on that figure.

sediments has generally been underestimated because it is difficult to recognize on a cut core surface and it is easily affected by diagenetic processes which destroy primary petrological characteristics.

Vallier and Kidd [1977] used clay mineral ratios to evaluate the relative amounts of volcanic components in DSDP sediment sequences in the Indian Ocean that have undergone diagenesis. By combining these clay mineral ratios with the presence of other components, such as primary volcanic glass and zeolites and other silicates, they were able to determine the relative importance of volcanic material deeper in the sediment columns where diagenesis becomes a major factor. They were able to establish the broad time/space relationships of volcanogenic sediments in the Indian Ocean. The nature of volcanism has changed with time: basaltic volcanism was dominant before the late Cenozoic and silicic volcanism gained in intensity during the late Cenozoic. Vallier and Kidd (1977) suggest that there might be a causal

mechanism between changes in rates or direction of sea-floor spreading and amounts of volcanic input. Pulses of volcanism are recorded in different quadrants of the ocean. The most widespread pulses appear to have been in the Late Cretaceous, the Paleocene-Eocene, the Oligocene, and the Pleistocene but these observations could not be refined further because of the lack of resolution in both core recovery and stratigraphy available from the DSDP sequences. In addition, the importance and cyclic history of the Indian Ocean "hotspots" was not recognized until the ODP campaign and a synthesis of volcanogenic input on an oceanwide scale can now complement the hotspot studies presented by Duncan and Storey [this volume].

At least 18 of the ODP sites had significant intervals of volcanic material reported in them. We targeted an oceanwide core re-examination and sampling program on the ODP drill sites and concentrated initially on the time intervals already identified in the DSDP sites as containing pulses of volcanic input. Intervals lying outside these age ranges were extensively sampled when ODP shipboard analyses indicated significant amounts of the components used by Vallier and Kidd [1977] as volcanogenic indicators. Samples were subjected to routine X-ray diffraction analysis and also to microscope examination. Clay mineral ratios were calculated as in the former study. Many of the DSDP sites were resampled for intercalibration and to refine the timing of input pulses.

Volcanogenic components are here recognized on the basis of primary volcanic material, eg. volcanic glass, erupted phenocrysts, and/or assemblages of clay minerals and zeolites which are their diagenetic derivatives. The occurrence of volcanic components is also revealed by measurements of whole core magnetic susceptibility [Robinson, 1990] and the distribution of natural radioactivity in downhole logs [von Rad and Thurow, 1992]. Primary volcanic components are clearly indicative of derivation from active volcanic centers and show the general evolution of volcanism in the Indian Ocean region. Diagenetic (secondary) derivatives are assumed to have come from erosion of active and extinct volcanic centers and their distributions are used to indicate general transport directions in wind and water mass circulation prevailing at that particular time period. Plotting volcanic distributions on the paleogeographic reconstructions used for the hiatus patterns (Figures 8 to 11) provides indications of some of the changes in volcanic activity but, because of the hiatuses, represents only partial distributions.

Paleocene 60Ma (Figure 8). Primary volcanic components in sediments near the Reunion hotspot indicate active volcanism there at this time. Both primary and secondary components along the East African margin suggest active volcanism in this region possibly related to a leaky transform that was probably the precursor of the Owen Ridge. A mixture of primary and secondary components on Kerguelen Plateau suggests that volcanism was occurring at the Kerguelen hotspot, and secondary components in sediments to the north and east of this may indicate prevailing wind and surface circulation directions not dissimilar to the present day. The area to the east of Madagascar (Site 239) contains primary and diagenetic volcanogenic components. These may have been derived directly from volcanic activity at the spreading center represented by the nearby Central Indian Ridge or, more likely, from the active volcanic center on the Madagascar Ridge identified by DSDP Site 246. Sites 259 and 765 along the Australian margin contain primary and diagenetic components. In the absence of any obvious landward volcanic source components for these sites at this time, they may have been derived from reworked older shelf sediments or transported from the Kerguelen hotspot via ocean currents.

Eocene 50Ma (Figure 9). At this time both primary and secondary volcanic components are present at sites around the Reunion Hotspot and westwards along the Central Indian Ridge. Derived volcanogenic material is also present at the precursor of the Owen Ridge (Site 223). Activity along the Southwest Indian Ridge results in derived components there (Site 245). Mixed primary and secondary assemblages are common in sites on the Kerguelen Plateau and northwards along the Ninetyeast Ridge, indicating continued activity at the Kerguelen Hotspot. Individual sites on the Australian margin and Naturaliste Plateau do not have obvious landward sources and these may be again related to extensive transport in winds and currents from the west. Explosive volcanic activity associated with the Indonesian arc would have begun by this time [Audley-Charles et al., 1988], but there are no indications in the Wharton Basin drill sites which at this time were well to the south.

Eocene/Oligocene Boundary (Figure 10). Despite the widespread hiatuses on this reconstruction, dominantly primary components at the Kerguelen Plateau sites and derived components to the north on the Ninetyeast Ridge attest to the activity of the Kerguelen Hotspot. The hiatuses make the picture indistinct for the Reunion Hotspot, although both derived and primary components are present in the eastern Somali Basin sites and mainly secondary components in the Madagascar Basin. Both of these are deep areas which could both have been drawing sediment from this source. Some of the Somali Basin volcanogenic material is silicic and along with the later presence of volcanic material in the wedge of sediment prograding from the Somali margin, we suggest that land volcanic centers in Ethiopia may have begun to influence sedimentation in this basin at 35 Ma.

Miocene 20Ma (Figure 11). The Kerguelen Hotspot appears to have been active at this time as primary volcanic components dominate in the mid-plateau area (Site 747). Both primary and secondary components on the southern Ninetyeast Ridge and Broken Ridge suggest that the hotspot activity extended to the northern side of the spreading ridge. Derived volcanic material of uncertain origin continued to be deposited on the northeast Australian margin. At Site 239 in the Madagascar Basin no volcanogenic sediments are identified between 24Ma to 17Ma and thus Madagascar sources must have no longer been important. The activity of the Reunion Hotspot may have been reduced. Mixed assemblages of volcanic sediments in the western Somali Basin (Sites 234 and 235) were probably derived from the silicic volcanic centers in northeast Africa and Arabia and from basaltic sources in the Red Sea. The source of eastern Somali Basin volcanogenic sediments (Sites 707-711, 236 and 237) is enigmatic. If the surface circulation pattern in the Indian Ocean at 20Ma was similar to the present, then the Reunion Hotspot might be an unlikely source.

These preliminary results suggest that the distributions of volcanic components are related primarily to the locations of the two main sources of hotspot activity in the Indian Ocean. Other active volcanic centers, probably more short lived, relate to ridge generation, to initiation of fracture zones, and to derivation from landmass volcanism. Volcanic activity and subduction along the Indonesian arc are known to be an important source of derived volcanics during the last 5 m.y. [Dehn et al., 1991]. This source has not been proven from our reconstructions prior to 5Ma.

Since their inception the Kerguelen and Reunion hotspots appear to have been the most important and persistent sources of volcanogenic material within the Indian Ocean. On all of the reconstructions presented, transport away from these hotspot sources can be construed as being consistent with directions of present day prevailing winds and surface circulation, even taking account of the effects of the Oligocene hiatus event. The contribution from ridge generation is difficult to evaluate at this stage and is anyway constrained by the distribution of drillsites.

SUMMARY AND CONCLUSIONS

The PALIOS project represents a longer term synthesis effort than is generally represented in this volume. The stratigraphic phase of the project has established procedures, some of them involving new approaches, that together provide a rigorous framework for paleoceanographic synthesis of both DSDP and ODP results. Most importantly we have been successful in completing an integrated chronological framework that allows correlation between the low and high latitudes. Even at this mid-stage in PALIOS we can recognize constraints that are imposed because of gaps remaining in the distribution of Indian Ocean sites after the ODP campaign and from limited penetration at many of the sites that were drilled. Although reasonably good for the Neogene Indian Ocean the correlations used in our chronological framework can still only be partial. Additional sites are needed from future drilling campaigns to:

(1) eliminate the sampling gap from 37-50 degrees south; and

(2) provide additional sites with the spread of latitudes necessary for calibration of biostratigraphic events to a magnetostratigraphy (most need here is in the low latitudes).

The chronological framework for the Paleogene and older record is much weaker and here the problem is both in the spatial/temporal framework and in the available stratigraphic resolution. Even using composite sites we are left with a stratigraphic record that is essentially incomplete. Numerous additional sites are required for this aspect; they must be drilled deeper with improved coring and logging technology as in the ODP campaign.

Taking as input all available biostratigraphic, isotopic and paleomagnetic datums, a simple spreadsheet routine has been developed for PALIOS to calculate the depth/age curve for each site. This provides the basis for assigning an age to each sample/measurement point and is an essential step in grouping sediment properties by age for the plotting phase of the project. It is the basis for our various target projects and will culminate in a closely-spaced time series of palinspastic maps. Of prime interest

is the calculation of sediment accumulation rates in precise increments for studies of sediment budgets. Detailed analysis of our databases for the PALIOS target projects on hiatuses and volcanogenic sediment distributions, as well as others on faunal and floral distributions, await the paleobathymetric dimension of palinspastic maps that are presently being prepared.

Acknowledgments. This is PALIOS Contribution No. 1, a project funded principally by an ODP Special Topic Grant (R. B. Kidd, Principal Investigator, Marine Geosciences Group, UWC, Cardiff) by the Natural Environment Research Council (NERC Grant No. GST-02434). T. A. Davies and J. G. Baldauf acknowledge support provided under this grant for two NERC Senior Visiting Fellowships at the University of Wales, Cardiff. Additional support was provided by the National Science Foundation under NSF Grants OCE-911893 (T. A. Davies, Principal Investigator), and 91-18480 (S. W. Wise, Jr., Principal Investigator). We thank Dr. W. Wei for kindly providing preprints of his most recent work.

REFERENCES

Aguirre, E., and Pasini, G., The Pliocene/Pleistocene boundary, *Episodes*, 8, 11-120, 1985.

Audley-Charles, M. G., Ballantyne, P. D., and Hall, R., Mesozoic-Cenozoic rift-drift sequence of Asian fragments from Gondwana, in *Mesozoic and Cenozoic Plate Reconstructions*, edited by C. R. Scotese and W. W. Sager, pp. 317-330, Elsevier, Amsterdam-Oxford-New York, 1988.

Backman, J., Schneider, D., Rio, D., and Okada, H., Neogene low-latitude magnetostratigraphy from the Site 710 and revised age estimates of Miocene nannofossils datum events, in *Proc. ODP, Sci. Results, 115*, edited by R. A. Duncan, J. Backman, L. C. Peterson et al., pp. 271-276, Ocean Drilling Program, College Station, TX, 1990.

Baldauf, J. G., Diatom biostratigraphy of the middle- and high-latitude North Atlantic Ocean, Deep Sea Drilling Project, Leg 94, in *Init. Repts. DSDP, 94*, edited by W. F. Ruddiman, R. B. Kidd, E. Thomas et al., pp. 729-762, U.S. Government Printing Office, Washington, D.C., 1986c.

Baldauf, J. G., and Barron, J. A., Diatom biostratigraphy and paleoceanography, in *Proc. ODP, Sci. Results, 119*, edited by J. A. Barron, B. Larsen et al., pp. 547-598, Ocean Drilling Program, College Station, TX, 1991.

Baldauf, J. G., and Pockras, E. M., Diatom biostratigraphy of Leg 108 sediments: Eastern Tropical Atlantic Ocean, in *Proc. ODP, Sci. Results, 108*, edited by W. Ruddiman, M. Sarnthein et al., pp. 23-34, Ocean Drilling Program, College Station, TX, 1989.

Barker, P. F., and Burrell, J., The opening of the Drake Passage, in *Proceedings of the Joint Oceanographic Assembly, 103*, Food and Agriculture Organization of the United Nations, Rome, 1977.

Barron, J. A., Late Eocene to Holocene diatom biostratigraphy of the equatorial Pacific Ocean, Deep Sea Drilling Project Leg 85, in *Init. Repts. DSDP, 85*, edited by L. Mayer, F. Theyer et al., pp. 413-456, (U.S. Govt. Printing Office), Washington, 1985a.

Barron, J. A., Miocene to Holocene planktic diatoms, in *Plankton Stratigraphy*, edited by Cambridge, H. Bolli, J. B. Saunders, and K. Perch-Nielsen, pp. 763-809, Cambridge University Press, 1985b.

Barron, J. A., Baldauf, J. G., Barrera, E., Caulet, J. P., Huber, B. T., Keating, B. H., Lazarus, D., Sakai, H., Thierstein, H. R., and Wei, W., Biochronology and magnetochronologic synthesis of ODP Leg 119 sediments from the Kerguelen Plateau and Prydz Bay, Antarctica, in *Proc. ODP, Sci. Results, 119*, edited by J. A. Barron, B. Larsen et al., pp. 813-847, Ocean Drilling Program, College Station, TX, 1991.

Barron, J. A., Keller, G., and Dunn, D. A., A multiple microfossil biochronology for the Miocene, *Geol. Soc. Am., Memoir 63*, 21-36, 1985.

Barron, J. A., Larsen, L. et al., *Proc. ODP, Init. Repts., 119*, 942 pp., Ocean Drilling Program, College Station, TX, 1989.

Berggren, W. A., Kent, D. V., and Couvering, V., Neogene geochronology and chronostratigraphy, in *The Chronology of the Geological Record, Geological Society Memoir No 10*, edited by N. J. Snelling, pp. 211-260, London, Blackwell Scientific Publications, 1985a.

Berggren, W. A., Neogene planktonic foraminiferal magneto biostratigraphy of the southern Kerguelen Plateau (Sites 747, 748, and 751), in *Proc. ODP, Sci. Results, 120*, edited by R. Schlich, S. W. Wise Jr. et al., pp. 631-647, Ocean Drilling Program, College Station, TX, 1992.

Berggren, W. A., Kent, D. V., and Flynn, J. J., Jurassic to Paleogene: Part 2, Paleogene geochronology and chronostratigraphy, in *The Chronology of the Geological Record, Geological Society Memoir No 10*, edited by N. J. Snelling, pp. 141-195, London, Blackwell Scientific Publications, 1985b.

Blow, W. H., Late middle Eocene to Recent planktonic foraminiferal biostratigraphy, *Proceedings First International Conference on Planktonic Microfossils, Geneva, 1967, 1*, 199-422, 1969.

Bolli, H. M., The genera Globigerina and Globorotalia in the Paleocene-lower Eocene Lizard Springs Formation of Trinidad, BWI, *U.S. Nat. Museum Bull., 215*, 51-81, 1957.

Bolli, H. M., Zonation of Cretaceous to Pliocene marine sediments based on planktonic foraminifera, *Boletin Informativo, Asociacion Venezolana de Geologia, Mineria y Petroleo, 9(1)*, 3-32, 1966.

Bolli, H. M., and Premoli-Silva, I., Oligocene to Recent planktonic foraminifera and stratigraphy of the Leg 15 sites in the Caribbean Sea, in *Init. Repts. DSDP, 15*, edited by N. T. Edgar, J. B. Saunders et al., pp. 475-497, U.S. Government Printing Office, Washington, D.C., 1973.

Bukry, D., Low-latitude coccolith biostratigraphic zonation, in *Init. Repts. DSDP, 15*, edited by N. T. Edgar, J. B. Saunders et al., pp. 487-494, U.S. Government Printing Office, Washington, D.C., 1973.

Bukry, D., Coccolith and silicoflagellate stratigraphy, northwestern Pacific Ocean Deep Sea Drilling Project, Leg 32, in *Init. Repts. DSDP, 32*, edited by R. L. Larson, R. Moberly et al., pp. 677-701, U.S. Government Printing Office, Washington, D.C., 1975.

Davies, T. A., Baldauf, J. G., and Kidd, R. B., A simple spreadsheet routine for calculating depth/age relations, *Computers and Geosciences*, in press, 1992.

Davies, T. A., and Kidd, R. B., Sedimentation in the Indian Ocean through time, in *Indian Ocean Geology and Biostratigraphy*, edited by J. R. Heirtzler, H. M. Bolli, T. A. Davies, J. B. Saunders, and J. G. Sclater, pp. 61-86, American Geophysical Union, Washington, 1977.

Davies, T. A., Weser, O. E., Luyendyk, B., and Kidd, R. B., Unconformities in the sediments of the Indian Ocean, *Nature, 253*, 15-19, 1975.

Dehn, J., Farrel, J. W., and Schmincke, H.-U., Neogene tephrochronology from Site 758 on northern Ninetyeast Ridge: Indonesian arc volcanism of the past 5ma, in *Proc. ODP, Sci. Results, 121*, edited by J. Weissel, J. Peirce, E. Taylor, J. Alt et al., 273-295, Ocean Drilling Program, College Station, TX, 1991.

Emiliani, C. (Ed.), *The Oceanic Lithosphere*, Volume VII of the series: The Sea: Ideas and observations on progress in the study of the sea, 1783 pp., Wiley-Interscience, New York, 1981.

Harland, W. B., Armstrong, R., Cox, A., Craig, L., Smith, A., and Smith, D., *A Geological Time Scale*, 263pp., Cambridge University Press, Cambridge, 1990.

Harwood, G., and Maruyama, T., Middle Eocene to Pleistocene diatom biostratigraphy of Southern Ocean sediments from the Kerguelen Plateau, Leg 120, in *Proc. ODP, Sci. Results, 120*, edited by R. Schlich, S. W. Wise Jr., 1992.

Heirtzler, J. R., Bolli, H. M., Davies, T. A., Saunders, J. B., and Sclater, J. G. (Ed.), *Indian Ocean Geology and Biostratigraphy*, 616 pp., American Geophysical Union, Washington, 1977.

Huber, B. T., Paleogene and early Neogene planktonic foraminiferal biostratigraphy of Sites 738 and 744, Kerguelen Plateau (southern Indian Ocean), in *Proc. ODP, Sci. Results, 119*, edited by J. Barron, B. Larsen et al., pp. 427-449, Ocean Drilling Program, College Station, TX, 1991a.

Huber, B. T., Maestrichtian planktonic foraminiferal biostratigraphy and the Cretaceous Indian Ocean, in *Proc. ODP, Sci. Results, 119*, edited by J. Barron, B. Larsen et al., pp. 451-465, Ocean Drilling Program, College Station, TX, 1991b.

Jenkins, D. G., Cenozoic planktonic foraminifera of New Zealand, *Paleontological Bull. New Zealand Geol. Survey, 42*, 1-278, 1971.

Jenkins, D. G., Southern mid latitude Paleocene to Holocene planktic foraminifera, in *Plankton Stratigraphy*, edited by H. M. Bolli, J. B. Saunders, and K. P. Perch-Nielsen, pp. 263-282, Cambridge University Press, Cambridge, 1985.

Kennett, J. P., *Marine Geology*, 813 pp., Prentice-Hall, Englewood Cliffs, New Jersey, 1982.

Kennett, J. P., and Srinivasan, M. S., *Neogene Planktonic Foraminifera*, 265 pp, Hutchinson Ross Publ. Co., New York, 1983.

Kennett, J. P., and Thunnell, R. C., Global increase in Quaternary explosive volcanism, *Science, 187*, 497-503, 1975.

Kidd, R. B., and Davies, T. A., Indian Ocean Sediment Distribution since the Late Jurassic, *Mar. Geol., 26*, 49-70, 1978.

Martini, E., Standard Tertiary and Quaternary calcareous nannoplankton zonation, in *Proc. II Planktonic Conf. Roma 1970*, 739-785, Ed. Tecnoscienza, Rome, 1971.

Mikkleson, N., Cenozoic diatom biostratigraphy and paleoceanography, in *Proc. ODP, Sci. Results, 115*, edited by R. A. Duncan, J. Backman, L. C. Peterson et al., pp. 411-432, Ocean Drilling Program, College Station,TX, 1990.

Moore, T. C., Jr., DSDP: successes, failures, proposals, *Geotimes, 17*, 27-31, 1972.

Moore, T. C., Jr., and Heath, G. R., Sea-floor sampling techniques, in *Chemical Oceanography, 7*, edited by J. P. Riley and R. Chester, pp. 75-126, Academic Press, London, 1978.

Moore, T. C., Jr., and Romine, K., In search of biostratigraphic resolution, in *The Deep Sea Drilling Project: a decade of progress, Soc. Econ. Paleontol. Miner. Spec. Publ., 32*, edited by J. E. Warme, R. G. Douglas, and E. L. Winterer, pp. 317-334, 1981.

Okada, H., and Bukry, D., Supplementary modification and introduction of code numbers to the low-latitude coccolith biostratigraphic zonation (Bukry, 1973; 1975), *Mar. Micropaleontology, 5*, 321-325, 1980.

Premoli-Silva, I., Coccioni, R., Montonari, A. et al., The Eocene-Oligocene boundary in the Marche-Umbria Basin (Italy), pp. 268, Ancona, Italy, 1988.

Robinson, S. G., Applications for whole-core magnetic susceptibility measurements of deep-sea sediments: Leg 115 results, in *Proc. ODP, Sci. Results, 115*, edited by R. A. Duncan, J. Backman, L. C. Peterson et al., Ocean Drilling Program, College Station, TX, 1990.

Sclater, J. G., Abbott, D., and Thiede, J., Paleobathymetry and sediments of the Indian Ocean, in *Indian Ocean Geology and Biostratigraphy*, edited by J. R. Heirtzler, H. M. Bolli, T. A. Davies, J. B. Saunders, and J. G. Sclater, pp. 25-59, American Geophysical Union, Washington, 1977.

Stainforth, R. M., Lamb, J. L., Luterbacher, H., Bead, J. H., and Jeffords, R. M., Cenozoic planktonic foraminiferal zonation and charateristic index forms, *University of Kansas Paleontological Contributions Article, 62,* 1-425, 1975.

Stott, L. D., and Kennett, J. P., Antarctic Paleogene planktonic foraminiferal biostratigraphy: ODP Leg 113, Sites 689 and 690, in *Proc. ODP, Sci. Results, 113,* edited by P. F. Barker, J. P. Kennett et al., pp. 549-569, Ocean Drilling Program, College Station, TX, 1990.

Tauxe, L., Opdyke, N. D., Pasini, G., and Elui, C., Age of the Plio/Pleistocene boundary in the Vrica Section, southern Italy, *Nature, 304,* 125-129, 1988.

Vallier, T. L., and Kidd, R. B., Volcanogenic sediments in the Indian Ocean, in *Indian Ocean Geology and Biostratigraphy,* edited by J. R. Heirtzler, H. M. Bolli, T. A. Davies, J. B. Saunders, and J. G. Sclater, pp. 87-118, American Geophysical Union, Washington, 1977.

Van Andel, T. H., Heath, G. R., and Moore, T. C., Cenozoic history and paleoceanography of the Central Equatorial Pacific Ocean, *Geol. Soc. Am. Memoir, 143,* 1-134, 1975.

Vincent, E., Indian Ocean Neogene planktonic foraminiferal biostratigraphy and its paleoceanographic implications, in *Indian Ocean Geology and Biostratigraphy,* edited by J. R. Heirtzler, H. M. Bolli, T. A. Davies, J. B. Saunders, and J. G. Sclater, pp. 469-584, American Geophysical Union, Washington, 1977.

Von der Borch, C. (Ed.), Synthesis of Drilling Results in the Indian Ocean, *Elsevier Oceanography Series, 21,* 175 pp., Amsterdam, 1978.

von Rad, U., and Thurow, J., Bentonitic clays as indicators of Early Neocomian post-breakup volcanism off northwest Australia, in *Proc ODP, Sci. Results, 122,* edited by U. von Rad, B. U. Haq et al., pp. 213-232, 1992.

Wei, W., Middle Eocene-lower Miocene calcareous nannofossil magnetobiochronology of ODP Holes 699A and 703A in the subantarctic South Atlantic, *Mar. Micropaleontology, 18,* 143-165, 1991.

Wei, W., and Pospichal, J. J., Danian calcareous nannofossil succession at Site 738 in the Southern Indian Ocean, in *Proc. ODP, Sci. Results, 119,* edited by J. Barron, B. Larsen et al., pp. 495-512, Ocean Drilling Program, College Station, TX, 1991.

Wei, W., and Thierstein, H. R., Upper Cretaceous and Cenozoic calcareous nannofossils of the Kerguelen Plateau (southern Indian Ocean) and Prydz Bay (East Antarctica), in *Proc. ODP, Sci. Results, 119,* edited by J. Barron, B. Larsen et al., pp. 467-493, Ocean Drilling Program, College Station, TX, 1991.

Wei, W., Villa, G., and Wise, S. W., Jr., Paleoceanographic implications of Eocene-Oligocene calcareous nannofossils from ODP Sites 711 and 748 in the Indian Ocean, in *Proc. ODP, Sci. Results, 120,* edited by S. W. Wise Jr., R. Schlich et al., Ocean Drilling Program, College Station, TX, 1992.

Wei, W., and Wise, S. W., Jr., Paleogene calcareous nannofossil magnetobiochronology: results from South Atlantic DSDP Site 516, *Mar. Micropaleontology, 14,* 119-152, 1989.

Wei, W., and Wise, S. W., Jr., Middle Eocene to Pleistocene calcareous nannofossils recovered by Ocean Drilling Program Leg 113 in the Weddell Sea in *Proc. ODP, Sci. Results, 113,* edited by P. F. Barker, J. P. Kennett et al., Ocean Drilling Program, College Station, TX, 1990.

Wei, W., and Wise, S. W., Jr., Biogeographic gradients of middle Eocene-Oligocene calcareous nannoplankton in the South Atlantic Ocean, in press, 1992a.

Wei, W., and Wise, S. W., Jr., Eocene-Oligocene calcareous nannofossil magnetobiochronology of the Southern Ocean, in press, 1992b.

Wei, W., and Wise, S. W., Jr., Selected Neogene calcareous nannofossil index taxa of the Southern Ocean: biochronology, biometrics, and paleoceanography, in *Proc. ODP Sci. Results, 120,* edited by S. W. Wise Jr., R. Schlich et al., Ocean Drilling Program, College Station, TX, in press, 1992c.

Whitmarsh, R. B., Hamilton, N., and Kidd, R. B., Paleomagnetic results from the Indian and Arabian Plates from Arabian Sea cores, in *Init. Repts. DSDP, 23,* pp. 521-525, U.S. Government Printing Office, Washington, D.C., 1974.

Worsley, T. R., and Davies, T. A., Cenozoic sedimentation in the Pacific Ocean: steps toward a quantitative evaluation, *J. Sed. Pet., 49,* 1131-1146, 1979.

Zigderveld, J. D. A., Zachariasse, J.W., Verhallen, P. J. J., and Hilgen, F. J., The age of the Miocene/Pliocene boundary, *News. Stratigir., 6,* 169-181, 1986.

History of Antarctic Glaciation: An Indian Ocean Perspective

W. U. EHRMANN

Alfred-Wegener-Institut für Polar- und Meeresforschung, D-2850 Bremerhaven, FRG

M. J. HAMBREY

Scott Polar Research Institute, Cambridge CB2 1ER, U.K.
now: School of Biological & Earth Sciences, The Liverpool Polytechnic, Liverpool L3 3AF, UK

J. G. BALDAUF

Geological Oceanography and Ocean Drilling Program, Texas A&M University, College Station, TX,
77845-9547, USA

J. BARRON

U.S. Geological Survey, Menlo Park, CA, 94025, USA

B. LARSEN

Geological Survey of Denmark, DK 2400 Copenhagen, Denmark

A. MACKENSEN

Alfred-Wegener-Institut für Polar- und Meeresforschung, D-2850 Bremerhaven, FRG

S. W. WISE, JR.

Department of Geology, Florida State University, Tallahassee, FL, 32306, USA

J. C. ZACHOS

Department of Geology, University of Michigan, Ann Arbor, MI , 48109-1063, USA

Legs 119 and 120 of the Ocean Drilling Program cored 16 sites on a S-N transect from the Antarctic continental shelf of Prydz Bay to the northern Kerguelen Plateau in the Indian Ocean. Thick sequences of glacigenic sediments were recovered in Prydz Bay, whereas the record on Kerguelen Plateau consists mainly of pelagic and, in part, glaciomarine sediments. This paper is a summary of the principle scientific results from the two legs that were concerned with the Cenozoic glacial and climatic history of Antarctica. It integrates a wide range of investigations, such as sedimentological studies including clay sedimentology and ice-rafted debris, studies of the oxygen isotopic composition of planktonic and benthic foraminifers, and paleontological investigations.

The scientific data obtained from these cruises indicate that a long-term cooling trend started at about 52 Ma, after the thermal maximum in early Eocene time. All parameters under review indicate that there has been continental-scale ice in East Antarctica at least since earliest Oligocene time. However, the ice probably was temperate in character, whereas that of the present day is polar with the bulk of ice below the pressure melting point. The question of ice extent, specifically, whether ice had reached the Antarctic coast as early as middle and late Eocene time, is still a matter of dispute. Evidence for that is suggested by the occurrence of isolated middle Eocene sand and gravel grains and by a poorly dated, possibly upper Eocene sequence of thick massive diamictites in Prydz Bay.

From Oligocene to recent time, the ice sheet experienced several major advance and retreat phases, some of them being quite rapid and short-term. However, although we did not find any clear evidence for a disappearance of the ice as postulated from other parts of Antarctica, the fragmentary nature of the stratigraphic record may hide major recessions of the ice sheet from the coast. Major increases of ice volume occurred in both middle Miocene and late Miocene times, the later phase probably associated with the build-up of the West Antarctic ice sheet and the generation of large ice shelves.

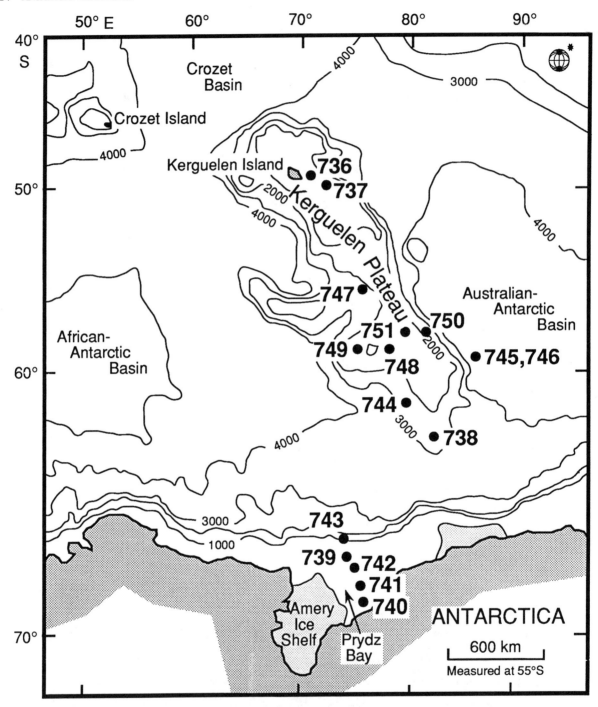

Fig.1. Location map of ODP Legs 119 and 120 drill sites in Prydz Bay and on Kerguelen Plateau. Bathymetry (in meters) is from GEBCO [Hayes and Vogel, 1981; Fisher et al., 1982].

Synthesis of Results from Scientific
Drilling in the Indian Ocean
Geophysical Monograph 70
Copyright 1992 American Geophysical Union

INTRODUCTION

In our quest for understanding the forces and processes that have been responsible for climatic and sea level changes on Earth during the Cenozoic Era, it is essential to determine how the Antarctic ice sheets have evolved during this period. Today, the Antarctic ice sheet exerts significant control on both global atmospheric and

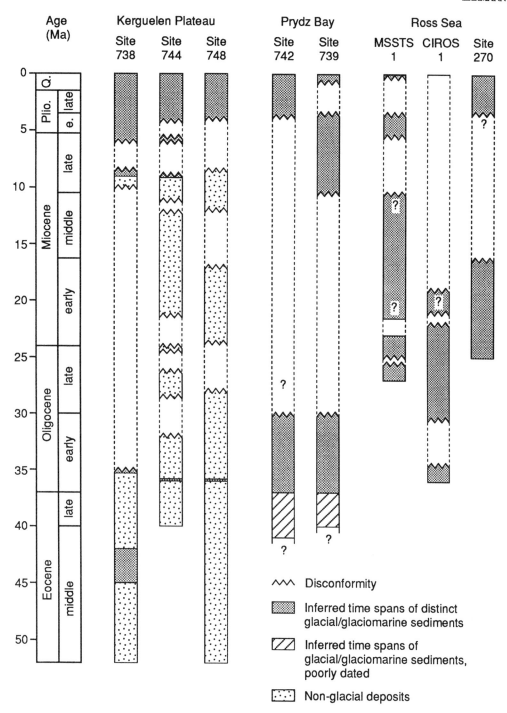

Fig. 2. Occurrence of known or suspected glaciomarine sediments on the Kerguelen Plateau ODP Sites 738, 744 [Ehrmann, 1991; Ehrmann and Mackensen, [1992] and 748 [Breza and Wise, 1992], Prydz Bay ODP Sites 739 and 742 [Barron et al., 1991a; Hambrey et al., 1991], and Ross Sea Sites MSSTS-1 [Barrett, 1986; Harwood et al., 1989), CIROS-1 [Harwood et al., 1989] and DSDP Site 270 [Hayes, Frakes et al., 1975].

oceanic circulation patterns, through the albedo effect of the ice surface, the influence it has on ocean surface temperatures, the production of cold dense bottom waters which penetrate into the northern hemisphere, the effect of the ice margin on the position and width of the subpolar

zone and the associated easterly directed depression systems.

Various means are available for examining the glacial record of Antarctica. On short time-scales, ice cores have yielded a high-resolution climatic record mainly based on

TABLE 1: Location of Sites Drilled in Prydz Bay and on the Kerguelen Plateau During ODP Legs 119 and 120, and Other Important Sites Mentioned in This Review.

Site	Expedition	Location	Latitude, S	Longitude, E	Water Depth, m	Oldest Sediment Cored
736	ODP 119	Kerguelen Plateau	49°24.12′	71°39.61′	629	early Pliocene
737	ODP 119	Kerguelen Plateau	50°13.67′	73°01.95′	564	middle Eocene
738	ODP 119	Kerguelen Plateau	62°42.54′	82°47.25′	2253	early Turonian
739	ODP 119	Prydz Bay	67°16.57′	75°04.91′	412	?Eocene, early Oligocene
740	ODP 119	Prydz Bay	68°41.22′	76°43.25′	807	?Permian-Cretaceous
741	ODP 119	Prydz Bay	68°23.16′	76°23.02′	551	Albian
742	ODP 119	Prydz Bay	67°32.98′	75°24.27′	416	?Eocene, early Oligocene
743	ODP 119	Prydz Bay	66°54.99′	74°41.42′	989	Quaternary
744	ODP 119	Kerguelen Plateau	61°34.66′	80°35.46′	2307	late Eocene
745	ODP 119	Austr.-Antarct. Basin	59°35.71′S	85°51.60′	4093	late Miocene
746	ODP 119	Austr.-Antarct. Basin	59°32.82′	85°51.78′	4070	late Miocene
747	ODP 120	Kerguelen Plateau	54°48.68′	76°47.64′	1697	early Santonian
748	ODP 120	Kerguelen Plateau	58°26.45′	78°58.89′	1287	late Campanian
749	ODP 120	Kerguelen Plateau	58°43.03′	76°24.45′	1070	early Eocene
750	ODP 120	Kerguelen Plateau	57°35.54′	81°14.42′	2031	Albian
751	ODP 120	Kerguelen Plateau	57°43.56′	79°48.89′	1634	early Miocene
CIROS-1		Ross Sea	77°34.55′	164°29.56′	200	early Oligocene
MSSTS-1		Ross Sea	77°33.26′	164°23.13′	195	late Oligocene
270	DSDP 28	Ross Sea	77°26.48′	178°30.19′	633	latest Oligocene

oxygen isotopes and CO_2-contents extending back to about 150,000 years ago [e.g., Lorius et al., 1985], thereby spanning a complete glacial-interglacial cycle and dating back into the previous glacial period. Also, a large number of piston cores and gravity cores have provided us with extensive data on the late Quaternary glacial/interglacial cycles [e.g., Grobe and Mackensen, 1992] and have even allowed us to reconstruct paleoceanography and glacial history as far back as late Pliocene time in great detail [e.g., Abelmann et al., 1990]. However, if climatic model predictions are to be believed, it is possible that projected global warming trends may take us into conditions that were previously evident in the past, prior to the initiation of northern hemisphere ice sheets about 2.5 Ma [Barrett, 1991]. Thus, understanding the evolution of the Antarctic ice sheet in Pliocene and earlier time has gained added importance.

The determination of this earlier glacial history of Antarctica is hindered by the restricted geological record exposed on the continent, owing to the 98% ice cover. However, this record can be obtained from the more accessible sedimentary and paleobiological record that is available beneath the floor of the Southern Ocean. This evidence is both direct, in the form of sediments that have been derived from grounded and floating ice, especially on the continental shelf, and indirect from a variety of sediment parameters and the stable isotope record of the Southern Ocean.

The main problems associated with using the direct glacial record include the large number of hiatuses within the sedimentary sequences, and the relative paucity of the microfossils or sediment suitable for dating. On the other hand the indirect evidence, such as the stable isotopic record, is open to differing interpretations volumes and paleotemperatures [e.g., Shackleton and Kennett, 1975; Shackleton, 1986; Matthews and Poore, concerning global ice 1980]. The main problem in interpreting the indirect sedimentological evidence arises from masking or destruction of the climatic signal by other processes such as sediment redistribution. As a result, the question of Antarctic glaciation has been the subject of vigorous debate since it was recognized by Leg 28 of the Deep Sea Drilling Project [Hayes, Frakes et al., 1975] that the record extended much further back into the Cenozoic Era than was evident prior to ocean drilling.

Major advances have been made within the last five years in elucidating the glacial history of Antarctica through a number of offshore drilling projects, notably Ocean Drilling Program (ODP) Legs 113 [Barker, Kennett et al., 1988, 1990], 114 [Ciesielski, Kristoffersen et al., 1988, 1991], 119 [Barron, Larsen et al. 1989, 1991] and 120 [Schlich, Wise et al., 1989, Wise, Schlich et al., 1992], and New Zealand drilling in the western Ross Sea [Barrett, 1986, 1989]. Of these, we are specifically interested in the results from ODP Legs 119 and 120. The key questions being addressed are:

(1) When did a continental ice sheet first form on Antarctica, and when did it first influence the ocean sedimentary record?

(2) Once formed, was the ice sheet a stable feature, or was it subject to major fluctuations or even disappearances?

South   North

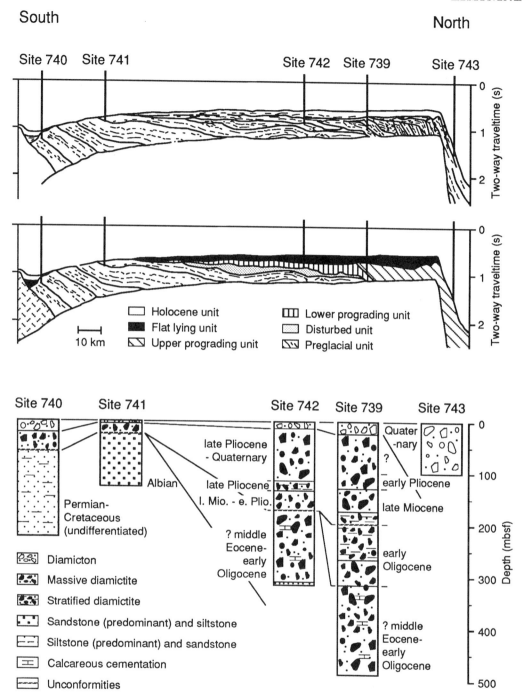

Fig. 3. Geological cross section through ODP Sites 739-743 on the continental shelf of Prydz Bay, based on seismic and drillhole data showing the presence of diamictites in lower Oligocene and possibly older sediments [based on Cooper et al., 1991, and Hambrey et al., 1989].

(3) What has been the thermal character of the ice sheet, i.e. was it comprised of temperate ice and thereby had an association with vegetation in coastal areas, or has it consisted largely of cold ice as it does at the present day?

ODP Legs 119 and 120 drilled 9 sites on the Kerguelen Plateau in the southern Indian Ocean, 2 sites in the Australian-Antarctic Basin and 5 sites on the continental shelf of East Antarctica in Prydz Bay (Figures 1 and 2, Table 1). A range of studies have been undertaken on the drill cores, with a view to elucidating the long-term history of the East Antarctic ice sheet. In particular these cruises

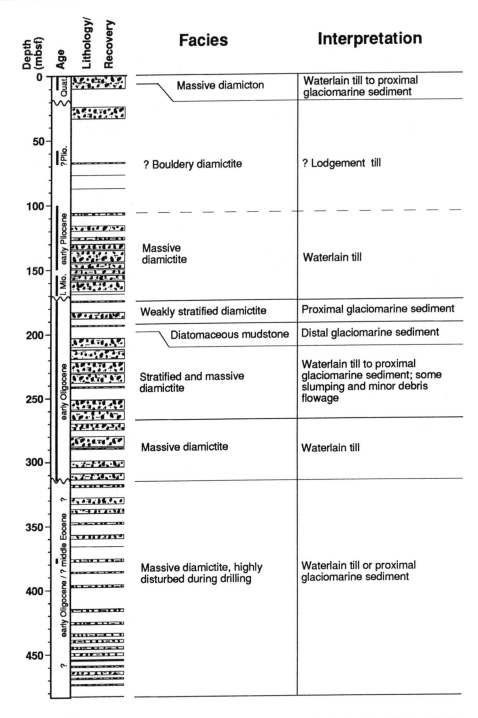

Fig. 4. Stratigraphy, sedimentology and interpretation of the sedimentary sequence at Prydz Bay Site 739 [simplified from Hambrey et al., 1991].

have

(1) addressed the problem of the timing of the onset of glaciation at sea level and the growth and fluctuations of the ice sheet from Oligocene through Neogene and Quaternary time,

(2) investigated the history of glacial erosion of the shelf, which is necessary for determining ice volume changes,

(3) documented other changes in the proximal and distal environments during periods of lesser ice cover, providing indications of climatic change, and

(4) developed a sedimentological model for the development of a high-latitude continental shelf dominated by glacial deposition.

TABLE 2: Characterization and Interpretation of the Main Facies Recovered at the Drill Sites in Prydz Bay [Summarized from Hambrey et al., 1992].

Facies	Description	Interpretation
Massive diamictite (dominant)	Non-stratified sandstone or sandy mudstone with matrix-supported clasts and minor biogenic material.	Lodgement till (preferred clast orientation) or waterlain till (random clast orientation).
Weakly stratified diamictite	As massive diamictite, but with diffuse stratification. Some bioturbation and slumping. Diatomaceous in part; with some shells.	Waterlain till to proximal glaciomarine sediment.
Well stratified diamictite	As massive diamictite, with prominent but generally discontinuous and contorted stratification. Clasts dispersed with occasional dropstone structures. Significant diatom component; with some shells.	Proximal glaciomarine / glaciolacustrine sediment.
Massive sandstone (minor)	Non-stratified, moderately well-sorted to poorly sorted sandstone, often with minor mud and gravel component.	Nearshore to shoreface with minor ice-rafting in distal glaciomarine setting; also gravity flow sediments.
Weakly stratified sandstone	As massive sandstone, but with weak, irregular stratification, and sometimes brecciated.	Nearshore with minor ice-rafting in a distal glaciomarine setting. Some slumping.
Massive mudstone (minor)	Non-stratified, poorly sorted sandy mudstone with dispersed gravel clasts, and sometimes brecciated and bioturbated. Dispersed shells.	Offshore with minor ice-rafting in distal glaciomarine setting. Some slumping.
Well-stratified mudstone (minor)	As massive mudstone, but with discontinuous, well-defined stratification. Syn-sedimentary deformation and bioturbation.	Deeper nearshore with minor ice-rafting in distal glaciomarine setting. Slumping common.
Diatomaceous ooze / diatomite (minor)	Weakly or non-stratified siliceous ooze with >60% diatoms. Minor components include mud, sand and gravel.	Offshore with minor ice-rafting in distal glaciomarine setting.
Diatomaceous mudstone	Massive mud or mudstone with >20% diatoms and minor sand.	Offshore with sedimentation predominantly influenced by ice-rafting and underflows in distal glaciomarine setting.

In the last few years, a number of detailed regional studies on the glacial history of Antarctica have been compiled [e.g., Barrett, 1986, 1989; Kennett and Barker, 1990; Barron et al., 1991a; Wise et al., 1991; Wise et al., 1992]. A comprehensive summary of these studies will not be rendered here. Instead, this review mainly summarizes the scientific results of drilling in the Indian Ocean sector of the Southern Ocean, both on the continental shelf of Prydz Bay and on the Kerguelen Plateau. The basic data and discussions have been already published in several other papers.

THE CONTINENTAL SHELF STRATIGRAPHIC RECORD

The five Prydz Bay sites, 739-743, were drilled on a 180-km-long transect which extended across the continental shelf of Prydz Bay to the continental slope (Figures 1 and 3). The sites are positioned at the mouth of the Lambert Graben, a major tectonic feature dating back to Mesozoic time or earlier, through which about one fifth of the East Antarctic ice sheet currently drains. It was expected that evidence of the earliest coastal ice sheets would be found here. The five sites are described in detail in Barron, Larsen et al. [1989] and Hambrey et al. [1991]. The main seismic stratigraphic sequence across the continental shelf consists of reflectors dipping regionally seaward in the inner and outer parts of the shelf, and reflectors lying in an irregular but nearly flat manner in the middle part of the shelf (Figure 3). In the inner shelf the dipping sequences are uniformly layered, whereas those in the outer shelf comprise more steeply dipping wedges of prograding strata. The stratigraphically lower part of the prograding sequence has less steeply dipping strata. The upper part of the

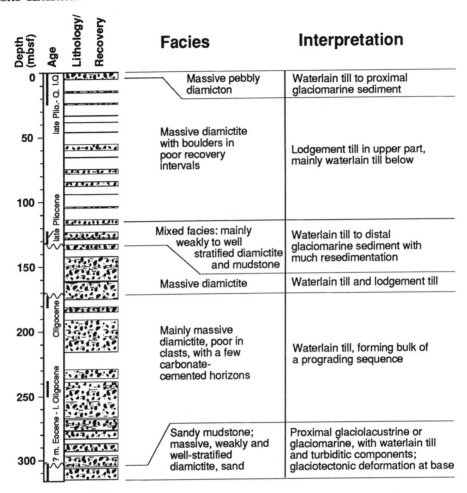

Fig. 5. Stratigraphy, sedimentology and interpretation of the sedimentary sequence at Prydz Bay Site 742 [simplified from Hambrey et al., 1991].

prograding sequence was eroded and truncated and now is overlain by flat-lying reflectors, comprising partly defined topsets within 30-150 m of the seafloor [Cooper et al., 1991].

The sediments recovered from Prydz Bay have been described in terms of a range of lithofacies [Hambrey et al., 1991; Table 2; Figures 4 and 5], the most abundant of which are massive and stratified diamictites[1], a variety of sandstones, and mudstones. Some facies have a biogenic component to a varying degree. Interpretation of the principal lithofacies (Table 2; Figures 4 and 5) is based on lithology, texture, fabric, fossil content, sedimentary structures and relation to adjacent beds. These facies are related to the position of the grounding line of the ice, but their development also depends on whether the grounding line is on the continental shelf or at the shelf break. The Prydz Bay sequence is particularly useful in this respect in that most of the massive diamictites provide evidence of

rain-out of basal glacial debris at the shelf break, with minimal reworking or incorporation of non-glacigenic sediment. The alternative explanation that the massive diamictites could be lodgement tills is discounted on the grounds that seismic data illustrates that it forms part of a prograding wedge. Although the bulk of the prograding succession at Prydz Bay is made up of this facies, open shelf sedimentation has also been important, with diatoms being a significant component of the sediment [Hambrey et al., 1991].

The sedimentary sequence recovered is overwhelmingly of glacial character. It can be considered principally in terms of deposition close to and immediately seaward of the grounding line of an extended Lambert Glacier-Amery Ice Shelf system, probably close to the continental shelf edge, where a break in slope would promote decoupling of the glacier from its bed. Most ice flowing over the Prydz Bay sites probably originated from the eastern margin of the Lambert Trough or from much thinner, less channelized and therefore less dynamic ice flowing off Ingrid Christensen Land [Hambrey, 1991].

---

[1] Diamictite is a non-genetic term for a non-sorted sedimentary rock consisting of a mixture of clay, silt, sand and gravel. Diamicton is the unlithified equivalent and diamict refers to both.

The fresh nature of both the gravel clasts and sand and silt-sized grains throughout the bulk of the sequence points to a source area for the bulk of the sediment that is dominated by physical weathering. The absence of significant angular supraglacial debris layers suggests that little material was derived from rock outcrops projecting above the ice. Probably little or no land was exposed and the bulk of the sediment was derived subglacially and modified by transport at the base of the sliding ice mass, where angular bedrock fragments were reworked into a broader range of shapes.

The glacigenic sequence is remarkable for its apparent uniformity, especially the homogeneity of the hundreds of meters of massive diamict (Figures 4 and 5). This can be explained only if the grounding line was in a relatively stable position or if interbedded lodgement tills and glaciomarine sediments were removed by subsequent glacier advances. The loading history of the sediment indeed suggests a more complex history of glacial advances and retreats than that indicated by the recovered facies [Solheim et al., 1991], as does the downhole logging data, which indicates that sandy horizons (distal glaciomarine or nonglacial) are present, but which were not recovered [Ollier and Mathis, 1991].

In summary, although DSDP and New Zealand drilling on the Ross Sea continental shelf obtained a composite record of glacial sediments dating back to early Oligocene time [Figure 2; Hayes, Frakes et al., 1975; Barrett, 1989], Prydz Bay Sites 739 to 743 have provided for the first time a record of the sequence on a transect from the inner shelf to the continental slope of Antarctica, in an area dominated for a prolonged period by a major glacier complex. The Prydz Bay situation may be typical of other parts of the Antarctic continental margin [e.g., western Antarctic Peninsula; Larter and Barker, 1989], and of some northern high-latitude continental margins, that have been under the influence of major ice drainage systems for millions of years.

THE STRATIGRAPHIC RECORD FROM KERGUELEN PLATEAU

The transect drilled in the continental shelf setting of Prydz Bay can be extended to the much more distal setting of the Kerguelen Plateau, which is situated in the southernmost Indian Ocean (Figure 1). The plateau is about 2500 km long and 500 km wide. It stretches in a northwest-southeasterly direction from about 45°S to 65°S, and rises 2-4 km above the adjacent Australian-Antarctic Basin to the east, the Crozet Basin to the north, and the African-Australian Basin to the west. It is, therefore, well situated for high-latitude paleoceanographic and paleoclimatic studies.

The nine drill sites on Kerguelen Plateau, Sites 736-738, 744 and 747-751, provide a S-N transect along the plateau, from about 62°43′S to 49°40′S. Sites 745 and 746 were drilled just east of the Kerguelen Plateau in the Australian-Antarctic Basin (Figure 1, Table 1). All sites on Kerguelen Plateau recovered pelagic Cenozoic sedimentary sequences [Barron, Larsen et al., 1989; Schlich, Wise et al., 1989],

which generally can be well dated by a combination of calcareous and biosiliceous biostratigraphy and magnetostratigraphy [Barron et al., 1991b; Harwood et al., 1992].

The northern Sites 736 and 737 were drilled close to the present Antarctic Convergence, or Polar Front. This seasonally fluctuating water-mass boundary separates the cold Antarctic Surface Water to the south from the warmer Subantarctic Surface Water to the north. Another important hydrographic feature, the Antarctic Divergence, is situated south of the southernmost Site 738 at about 65°S. The Antarctic Divergence separates the eastward-flowing Antarctic Circumpolar Current [Whitworth, 1988] to the north from the westward-flowing Antarctic Coastal Current to the south. Drilling on Kerguelen Plateau, therefore, should document the development of these oceanographic features, which have a major effect on global climate and surface-water circulation.

PALEOGENE GLACIAL HISTORY

Evidence from the continental shelf

The sparse and imprecise biostratigraphic record precludes detailed assessment of the glacial history of Prydz Bay, but a number of key phases may be inferred (Figure 6). However, age constraints are still limited and biostratigraphic ages for Oligocene time are incompatible with strontium isotope dates from Site 739 [Baldauf and Barron, 1991; Thierstein et al., 1991]. It also has to be borne in mind that hiatuses in the stratigraphic record and unrecovered intervals may hide major fluctuations of the East Antarctic ice sheet. The results from Prydz Bay drilling are therefore compared below with the data from the CIROS-1 core from McMurdo Sound in the Ross Sea, which to date has provided the most complete Oligocene to early Miocene record from the continental shelf [Barrett, 1989].

Thick sequences of preglacial sediments were recovered in the inner part of Prydz Bay, at Sites 741 and 740. Their age increases towards the coast and is probably Albian and undifferentiated Permian-Cretaceous, respectively [Barron, Larsen et al., 1989; Turner, 1991]. The preglacial sediments consist of siltstones, sandstones and conglomerates, occasionally with wood fragments, and are interpreted as fluviatile. These sediments suggest that subaerial, cool- to warm-temperate conditions with extensive vegetation prevailed prior to encroachment by the ice [Turner, 1991; Turner and Padley, 1991; Figure 6A].

Towards the outer shelf, a thicker sequence of glacigenic diamictite overlies the preglacial sediments. The contact between the glacial and preglacial sediments, however, was not recovered. The diamictites are seismically part of an overall prograding sequence. At Site 742, the diamictite is mainly massive and contains only a few fossils, but near the base it is highly disturbed and interbedded with mudstone. At Site 739, the diamictite is especially fossiliferous and more stratified from 173 to 316 mbsf, and there is much evidence of slumping [Hambrey et al., 1991]. The fossil content consists of diatoms, calcareous nannofossils, and dinoflagellates that suggest an earliest

SOUTH NORTH

A. Preglacial Eocene   *(50 Ma)*

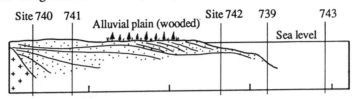

B. Onset of glaciation, Eocene-early Oligocene   *(36-40 Ma)*

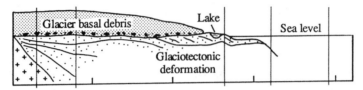

C. Early Oligocene; floating glacier ice at shelf break   *(35 Ma)*

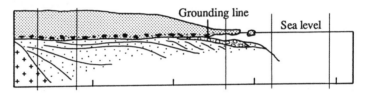

D. Early Olig.; major shelf progradation, ice at shelf break   *(30 Ma)*

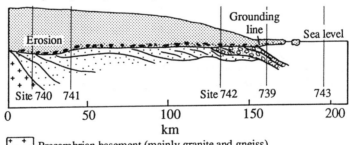

Fig. 6. Summary of the main stages in the history of glaciation and development of the continental shelf, along a transect through the five Prydz Bay drill sites [from Hambrey et al., 1991]. The stages A-I are discussed in the text.

Oligocene age of 36.0 to 34.8 Ma [Barron et al., 1991b].

Strontium isotope studies on molluscan shells from the same interval suggest late Oligocene to earliest Miocene ages ranging from 29.95 to 22.7 Ma [Thierstein et al., 1991]. However, the biostratigraphic data are considered to be more reliable, because: 1) early Oligocene diatoms were encountered from an internal mold of a gastropod isotopically dated at 22.5 Ma; 2) the porewater of Hole 739C is 1.5 to 2 times enriched in dissolved Sr relative to seawater [Thierstein et al., 1991]; and 3) an upsection trend toward younger ages that typifies the lower six Sr ages is countered by relatively random ages obtained in the upper five measured samples.

A thick series of glacial sediments was cored at Sites 739 and 742 beneath the well-dated lowermost Oligocene (36.0 to 34.8 Ma) glacial section. Diatoms, calcareous

E. Late Oligocene-early Miocene *(glacial maximum prior to Quat., 24 Ma)*

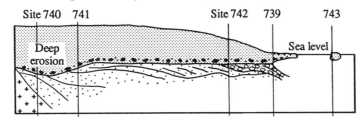

F. Late Miocene; retreat from shelf break *(10 Ma)*

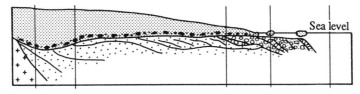

G. Early-late Pliocene transition; retreat phase *(3-3 Ma)*

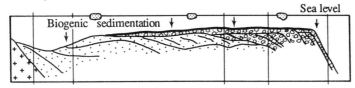

H. Late Pleistocene glacial maximum *(20,000 yr. B.P.)*

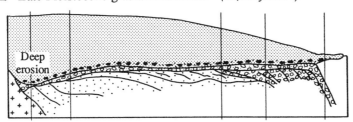

I. Holocene "interglacial" conditions *(10,000 yr. B.P.-present)*

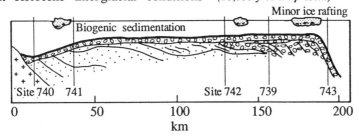

nannofossils, and dinoflagellates indicate only a general middle Eocene to early Oligocene age for this sequence. Magnetostratigraphic data from the lower part of this sequence in Hole 742A (172.5 to 316.0 mbsf) correlate best with magnetic polarity Subchrons C17N-3 through C15N-2 [Sakai and Keating, 1991] and may suggest an age of 40.8 to 37.46 Ma [Barron et al., 1991b]. However, because of the poor core recovery and the lack of information on sedimentation rates, these data seem very uncertain.

The massive diamictites were interpreted largely as the result of deposition of waterlain till close to the grounding line, although such an inference lacks the support of data from analogous modern settings, which are inaccessible. The stratified diamictites have been interpreted as more distal glaciomarine sediment [Hambrey et al., 1991]. The edge of the present-day floating ice shelf is approximately 140 km south of Site 739, the grounding line as much as

410 km. For the grounding line to occur in a position near Sites 739 and 742, a fully established East Antarctic ice sheet has to be inferred. The evidence is both sedimentological and theoretical. In the first place, the absence of angular clasts indicates a lack of subaerial sources, i.e. no exposed mountains or nunataks. Secondly, theoretical ice sheet profiles [Paterson, 1981, Chapter 9] indicate that whichever model is used, ice must have covered all land areas, probably to a greater depth than today - at least in this part of Antarctica. The only circumstance in which ice could have extended so far across the continental shelf, and yet been of substantially less volume than that of today, is if it had surged. A surge possibly could be identified by a sedimentary facies with all components delivered by currents [Grobe, 1986]. However, there is no evidence of surging in the Prydz Bay record.

Following this reasoning, we argue that by early Oligocene time, and possibly as early as middle to late Eocene time, ice advanced across the alluvial plain to within close proximity of Site 742. Seismic data suggest that the advancing ice deformed the underlying sediment to a depth of 100 m [Cooper et al., 1991; Figure 3] and deposited a complex of waterlain tills and proglacial lacustrine or fjord sediments (Figure 6B). Ice then reached the paleocontinental shelf break just landward of Site 742 (Figure 6C) and the phase of early Oligocene waterlain till deposition forming the lower, gently inclined, prograding sequence began [Hambrey et al., 1991]. By later early Oligocene time, grounded ice advanced across Site 742 and continued building up gently dipping, and then more steeply dipping prograding sequences at Site 739 (Figure 6D). Simultaneously, deep erosion took place in the inner Lambert Graben, as a highly active, possibly temperate glacier complex flowed over poorly indurated sediments. No upper Oligocene to middle Miocene sediments were cored in Prydz Bay, because of an erosional unconformity, presumably created by ice erosion. However, sediments of this age are inferred to form the prograding wedge seaward of Site 739 and probably represent continued advance(s) of ice across the expanding shelf, and at the same time eroding it [Hambrey et al., 1991; Figure 6E].

An ice loading event was identified at Site 739 from overcompaction characteristics in early Oligocene sediments between 228 and 186 mbsf [Solheim et al., 1991]. It may have resulted from the glacial advance that produced the early Oligocene to late Miocene hiatus, suggesting that the upper part of the compacted sequence represents reworked sediment associated with the glacial advance. From the CIROS-1 drill hole in the Ross Sea, Barrett et al. [1989] reported major offshore growth of grounded glaciers at about 30.5 Ma and a marked shallowing event which corresponds to a major drop in the eustatic sea level curve [Haq et al., 1987]. Similarly, Bartek et al. [1991] suggested that the first major ice sheet grounding event in the Ross Sea during the late Oligocene was comparable in extent on the shelf with that of the Wisconsin glacial maximum. It is tempting to suggest that

such a late Oligocene period of growth of the East Antarctic ice sheet was responsible for both the erosion of sediments of this age from Site 739 and for the loading event.

The sediments recovered during the CIROS-1 drilling operation in McMurdo Sound of the Ross Sea support the hypothesis of a fully established East Antarctic ice sheet by early Oligocene time [Barrett, 1989]. The CIROS-1 sediments document an early Oligocene phase dominated by the deposition of moderately deep-water, poorly sorted sands under the influence of ice-rafting together with occasional pulses of diamictite sedimentation, forming what is inferred to be waterlain till. A continental glaciation by early Oligocene time (>35 Ma) was also postulated based on glaciomarine sediments from King George Island in the South Shetland Islands [Gazdzicki, 1989]. In CIROS-1, the early Oligocene sequence was terminated by a marked shallowing event, evidence of fluviatile sedimentation, and a hiatus of about 4 m.y. at the early/late Oligocene boundary. The upper Oligocene to lower Miocene sequence is characterized by a series of seven major glacial advances across the site, depositing complexes of lodgement and waterlain tills, and indicating continued ice-rafting between. However, this all took place under a humid, temperate climate that was sufficiently amenable for the growth of a coastal beech forest with podocarps, proteas and other shrubby angiosperms [Hill, 1989; Mildenhall, 1989]. Upper Oligocene continental tillites, with an age of 29.5-25.7 Ma, also have been described from King George Island [Birkenmajer, 1988].

Evidence from Kerguelen Plateau

The sedimentological, paleontological and isotopic data from the Kerguelen Plateau show no sign of glacial conditions at sea level in Paleocene and early Eocene times. The stable oxygen isotopes indicate that, after a thermal maximum in early Eocene time, a long-term cooling trend began at about 52 Ma. This cooling affected both the surface and bottom water [Figure 7; Barrera and Huber, 1991; Zachos et al., 1992; Mackensen and Ehrmann, 1992]. Temperature estimates assuming δ_w=-1.2‰ indicate that the surface water temperature dropped from about 10-14°C at around 52 Ma to about 5-9°C at around 40 Ma. The bottom waters experienced a cooling from about 10-12°C to 3-5°C [Mackensen and Ehrmann, 1992]. These temperatures are in accordance with the assumption that no major ice mass existed in Antarctica prior to the late Eocene, which is in agreement with the findings of Shackleton and Kennett [1975], Shackleton [1986] and Miller et al. [1987, 1991].

The input of terrigenous material reaching the Kerguelen Plateau was low up to early middle Eocene time and no indication of ice-rafting has been detected. The terrigenous material present is mainly clay-sized and implies transport in suspension or by wind. The clay mineral associations are dominated by smectite, indicative for chemical weathering processes under a warm and humid climate [Ehrmann, 1991; Ehrmann and Mackensen, 1992]. The oldest isolated terrigenous sand and gravel grains of

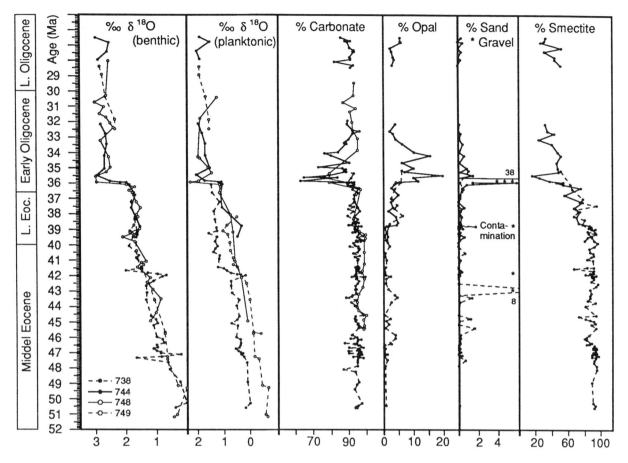

Fig. 7. Oxygen isotope, carbonate, opal, sand, gravel and smectite data of middle Eocene to late Oligocene sediments from drill sites on Kerguelen Plateau, compiled from Ehrmann and Mackensen [1992] and Mackensen and Ehrmann [1992]. The sand content is given as percent of the nonbiogenic sediment fraction, the smectite content is given as percent of the clay fraction of the sediments.

probable ice-rafted origin were detected in middle Eocene sediments at Site 738. Their age is ca. 45-42 Ma [Figure 7; Ehrmann, 1991]. It has been argued recently that those grains might be contaminants [Wise et al., 1992]. However, their *in situ* position has been defended by Ehrmann and Mackensen [1992].

The late middle Eocene (45-40 Ma) sea surface temperatures were ca. 9.5°C over the central Kerguelen Plateau and ca. 8.5°C over the southern Kerguelen Plateau. Deep water temperatures were of the order of 7°C [Mackensen and Ehrmann, 1992]. Assuming a linear decrease in temperature with latitude and extrapolating the cooling trend of surface temperatures to the south, yield a temperature of ca. 6°C at the coast. In our opinion, such a temperature would be low enough to allow some valley glaciers to reach the Antarctic coast, as they do today in southern Chile. Paleotemperature calculations under the assumption of a significant ice sheet, however, result in bottom water temperatures that would be too warm. Therefore, the data are inconsistent with a major glaciation and the formation of sea-ice or ice shelves necessary for the production of cold bottom water.

Glaciological models indicate that towards the end of the

middle Eocene Epoch the equilibrium line dividing the accumulation and ablation zones of ice sheets and glaciers may have fallen to 1000 m above sea level during cold intervals. The whole region >1000 m above sea level would then have been covered by ice. Much of the ice in the lower parts of Antarctica would have been melted, but limited discharge of icebergs calving from valley glaciers into the sea may have been possible [Robin, 1988].

The paleontological record suggests that some environmental change occurred in Paleocene to middle Eocene time. Major benthic foraminiferal faunal changes, as indicated by principal component analysis and first and last appearances of species, occurred at or close to the Paleocene/Eocene boundary [Mackensen and Berggren, 1992]. No middle Eocene change in the assemblages was found on Kerguelen Plateau. In contrast, on Maud Rise a decrease in diversity and a faunal change occurred in ca. 46 Ma sediments [Thomas, 1990]. Although Eocene diatoms are sparse from this region, the assemblage observed consists of cosmopolitan species. On a global scale changes in the diatom assemblages occur at about 43-41 Ma with expansion and provincialism [Baldauf, in press; Baldauf and Barron, 1991].

In summary, it seems possible that ice extended to coastal regions as early as middle Eocene time. In comparison with present-day conditions, and in view of the dominance of smectite over detrital clay minerals, it can be concluded that middle Eocene glacierization was much less extensive than that of today. However, because probable ice-rafted debris is recorded from different localities [Margolis and Kennett, 1970; Birkenmajer, 1988; Ehrmann and Mackensen, 1992; Wei, 1992], there is strong evidence that minor glaciation was occurring in various regions of Antarctica and glaciers reached sea level at several places, while much of the main East Antarctic continent and all of West Antarctica probably remained ice-free. In the ice-free areas, chemical weathering under a humid climate was active, as suggested by the dominance of smectite over detrital clay minerals.

The late Eocene interval was characterized by relatively stable climatic conditions with constant sea surface and bottom water temperatures. Following the middle Eocene cooling, between 40 and 36 Ma, bottom water temperatures over the Kerguelen Plateau varied only slightly around 5°C, while sub-surface water temperatures fluctuated between 6.5 and 8°C, with temperatures increasing from south to north [Mackensen and Ehrmann, 1992]. The near-surface water, however, cooled slightly during the late Eocene Epoch [Zachos et al., 1992]. An intensification of physical weathering on Antarctica is indicated by a slight increase in chlorite and kaolinite influx at 40 Ma, although chemical weathering processes typical of more humid climate were still dominant [Ehrmann and Mackensen, 1992]. Furthermore, a change in the benthic foraminiferal assemblages and decrease in diversity occurred close to the middle/late Eocene boundary at ca. 40 Ma [Schröder-Adams, 1991; Mackensen and Berggren, 1992].

Diatoms and radiolarians are rare and generally poorly preserved in the Eocene sediments. In addition, in these sediments diagenetic alteration products of opal (chert, clinoptilolite) are in evidence [Bohrmann and Ehrmann, 1991; Ehrmann and Mackensen, 1992] and suggest active surface productivity in this region, but at a scale far less than that observed during the Oligocene. At 38.8 Ma the opal content of the sediments clearly increased on the southern Kerguelen Plateau, but the character of the sediments is still dominantly calcareous (Figure 7).

On the Kerguelen Plateau, a dramatic increase in benthic and planktonic foraminiferal $\delta^{18}O$ values of about 1.2 ‰ occurred shortly after the Eocene/Oligocene boundary, at about 35.9 Ma [Figure 7; Barrera and Huber, 1991; Zachos et al., 1992; Mackensen and Ehrmann, 1992]. Because we do not know how much ice was concentrated in Antarctica and what its isotopic composition was, a wide range of explanations of the $\delta^{18}O$ shift is possible. Assuming an ice volume similar in extent and in isotopic composition to that of today, calculation of the early Oligocene bottom water temperatures, compared to the late Eocene values calculated with no ice, reveals that cooling may have been of only minor importance during that time interval. Instead, most of the $\delta^{18}O$ increase may have been caused by an increase in global ice volume, probably representing the onset of continental glaciation of East Antarctica [Zachos et al., 1992; Mackensen and Ehrmann, 1992]. In order to explain the relatively warm deep water temperatures of ca. 5°C over the Kerguelen Plateau, despite the presence of a nearby continental ice sheet, Mackensen and Ehrmann [1992] suggest that Oligocene bottom water production was much less than at present-day. They attribute this to the lack of huge ice shelves in both the Weddell Sea and Ross Sea. Bottom waters were probably only produced close to the ice-covered coasts by katabatic winds and freezing of sea ice, a process which would produce a much smaller volume of bottom water. Thus, the bottom water mass probably was not represented in the samples obtained by drilling on the plateau [Mackensen and Ehrmann, 1992].

The alternative assumption of no or no significant ice sheet on the early Oligocene Earth results in surface and deep water temperatures of about 4°C and 1.5°C, respectively, over the Kerguelen Plateau [Mackensen and Ehrmann, 1992]. This, however, would be too cold without any continental ice on Antarctica and a production of cold and dense bottom water masses at its continental margins. A moderate modification of this hypothesis is discussed by Wei [1991], who, based on calcareous nannofossil evidence, calculated that only a minor part of the early Oligocene $\delta^{18}O$ shift is caused by an increase in global ice volume. Consequently he argues for a >3°C cooling of the ocean surface water masses.

In addition it needs to be mentioned that low-latitude planktonic $\delta^{18}O$ records, although the number is very limited, exhibit a shift of only 0.4 ‰ in early Oligocene $\delta^{18}O$ values [Vergnaud Grazzini and Oberhänsli, 1985; Keigwin and Corliss, 1986]. However, if these values represent a global undisturbed signal rather than a local or regional one [Miller et al., 1991; Mackensen and Ehrmann, 1992], they indicate that the amount of increase attributable to ice volume would have to be about 0.4‰. The remainder would have to be the result of cooling, i.e. over the Kerguelen Plateau ≈3°C.

However, several sediment parameters correspond to this early Oligocene increase in $\delta^{18}O$ values at ca. 36 Ma (Figure 7), and provide additional evidence for major expansion of continental ice at about that time. The clay mineral assemblages began to change at about 36.3 Ma. The smectite concentrations decreased strongly and steadily while detrital illite became more important, indicating that the chemical weathering conditions on East Antarctica were gradually being replaced by physical weathering conditions or mechanical erosion. Minimum smectite concentrations of ca. 15% and maximum illite concentrations were found to occur at ca. 35.5 Ma (Figure 7). This implies a large-scale glacierization rather than local or regional mountain glacierization [Ehrmann, 1991; Ehrmann and Mackensen, 1992].

A very sharp and intense pulse of ice-rafting is recorded at Sites 744 and 748 in sediments that have an age of 36.0-35.8 Ma [Figure 7; Ehrmann, 1991; Breza and Wise, 1992;

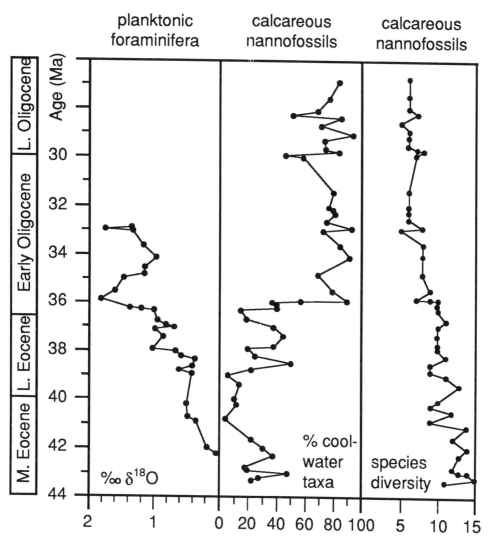

Fig. 8. Oxygen isotope composition of planktonic foraminifers [*Chiloguembelina cubensis*; Zachos et al., 1992], percentage of cool-water taxa and species diversity in the calcareous nannofossil assemblages [Wei et al., 1992] in sediments of Site 748.

Ehrmann and Mackensen, 1992]. Ice-rafted debris is a good indicator of the presence of continental ice reaching the sea. Because fluctuations in the amount of ice-rafted debris may be unrelated to climatic changes or ice sheet fluctuations, ice-rafted debris alone may not be suitable to reconstruct the intensity of glaciation. Therefore, it has to be regarded in combination with other sedimentological, paleontological and isotopic parameters. The early Oligocene peak occurrences of ice-rafted debris at both Site 744 and Site 748 correlate exactly with the oxygen isotope shift and other parameters discussed below. Therefore, the presence of the large amounts of ice-rafted debris as far north as 58°27'S may imply a high frequency of large icebergs calving from tidewater glaciers more typical of ice sheets than local glaciation [Hambrey et al., 1992].

A distinct and sharp drop of carbonate concentrations also occurred at most sites on Kerguelen Plateau in early Oligocene, 36 Ma old sediments [Barron, Larsen et al.,

1989; Schlich, Wise et al., 1989; Zachos et al., 1992; Ehrmann and Mackensen, 1992]. The carbonate content dropped from relatively high and uniform pre-Oligocene values of 90%-95% to values fluctuating between 65% and 95% (Figure 7). The diminished carbonate concentrations in Oligocene strata are due to dissolution, probably caused by enhanced export productivity as a result of the generation of a circum-Antarctic current system with developing oceanic fronts and associated upwelling [Ehrmann and Mackensen, 1992]. This mechanism also serves to explain the simultaneous increase in opal accumulation rates.

Cooling of the surface and intermediate waters over southern Kerguelen Plateau during earliest Oligocene time is revealed by changes in both the calcareous and siliceous biota. For example, several warm to warm-temperate radiolarian species become extinct and are replaced by Antarctic and temperate species [Caulet, 1991]. Increased abundances of the benthic foraminifer *Nuttallides umbonifer* attest to the establishment of a cool, carbonate-corresive water mass over Kerguelen Plateau during early Oligocene time [Schröder-Adams, 1991; Mackensen and Berggren, 1992]. In addition, both the planktonic foraminiferal and calcareous nannofossil assemblages exhibit a continuous decline in diversity through early Oligocene time [Huber, 1991; Wei et al., 1992] with the calcareous nannofossil assemblage also showing an increase in cooler water taxa [Figure 8; Wei, 1991; Wei and Thierstein, 1991]. The diatom assemblage provides further evidence for cooling. This is reflected by an increase in abundance from $<1x10^{14}$ valves/gram prior to the oxygen isotope shift to $4-8x10^{14}$ valves/gram associated with the oxygen isotope shift (Baldauf, in press) and by the floral composition which consists of species restricted to the high latitudes as well as cosmopolitan species. The diatom assemblage is interpreted by Baldauf [in press] and Baldauf and Barron [1991] to reflect an increase in surface productivity and the partial partitioning of surface waters between the middle and high southern latitudes.

Summarizing the above, all the isotopic, sediment-ological and paleontological evidence from Kerguelen Plateau, independently of the record from the continental shelf areas of Prydz Bay and the Ross Sea, points to drop in ocean water temperatures and an expansion of the Antarctic ice cover, probably marking the initiation of a continental-scale East Antarctic ice sheet in earliest Oligocene time.

The oxygen isotope values decreased slightly after the maximum in earliest Oligocene time [Figure 7; Barrera and Huber, 1991; Zachos et al., 1992; Mackensen and Ehrmann, 1992]. Zachos et al. [1992] argue that the ice sheet began to decrease in size, receded from the coast and stabilized as a smaller but still active continental ice sheet inland. However, this idea contradicts the findings from Prydz Bay and the Ross Sea (see above). Mackensen and Ehrmann [1992] calculated mean Oligocene surface and deep water temperatures of ca. 7.8°C and 5°C, respectively, at the central Kerguelen Plateau. At the southernmost Kerguelen Plateau, temperatures seem to have been 1-2°C lower. These temperatures calculated under the assumption of an ice volume similar to the present one (δ_w=-0.28‰) do not contradict a temperate, wet-based ice sheet.

The clay mineral assemblages imply somewhat warmer conditions than in earliest Oligocene time, but colder conditions than in Eocene time. Physical weathering and/or glacial scour was active on the Antarctic continent throughout Oligocene time [Ehrmann, 1991; Ehrmann and Mackensen, 1992]. The amount of ice-rafted debris recovered on southern Kerguelen Plateau shows a strong decrease after its early Oligocene maximum (Figure 7). Angular terrigenous sand grains are rare, but occur in higher amounts than in preglacial Eocene sediments. Terrigenous silt, in contrast, shows a major increase in concentration [Ehrmann, 1991; Ehrmann and Mackensen, 1992]. However, on the central Kerguelen Plateau no terrigenous sand grains were detected in Oligocene sediments [Breza

and Wise, 1992]. In the Atlantic sector of the Southern Ocean, Oligocene ice-rafted debris is found on Maud Rise [Barker, Kennett et al., 1988] and on the continental slope off Kapp Norvegia [Grobe et al., 1990]. Late Oligocene ice-rafted debris occurs also at Site 270 in the Ross Sea [Hayes, Frakes et al., 1975]. Thus, it seems that Oligocene ice rafting was active, but restricted to more proximal settings than Kerguelen Plateau, which supports the hypothesis of a temperate ice sheet and relatively warm surface waters inhibiting long-distance iceberg transport. The decrease in the amount of ice-rafted debris after the early Oligocene maximum may then be explained by the assumption that during the first stage of continental glaciation large quantities of preglacial weathering products were available on the continent and were incorporated into the advancing ice. In later Oligocene time most loose sediment had already been removed from the continent. The glaciers had to erode bed rock mechanically and therefore incorporated less debris.

Several Oligocene hiatuses document a more intense circulation and more erosive behavior of the intermediate water masses than in Eocene time [Barron, Larsen et al., 1989; Schlich, Wise et al., 1989]. The distribution of hiatuses in the deep sea basins around Antarctica further documents that the bottom waters also were more vigorous, probably because of the formation of a bottom water mass at the Antarctic continental margin.

The diatom assemblage also shows a decrease in abundance directly following the initial increase associated with the earliest Oligocene oxygen isotope shift. Abundance values decline from $2-8x10^{14}$ valves/gram associated with the shift to $1-5x10^{14}$ valves/gram for the interval younger than 35.8 Ma. Diatom abundance remains relatively constant throughout the remainder of the Oligocene strata [Baldauf, in press]. The floral composition also shows a decrease in species typical of the high-latitude suggesting a possible decline in surface water partitioning.

In conclusion, the various parameters gained from sediments recovered from the Kerguelen Plateau, together with those from Prydz Bay, imply that no major climatic change took place during Oligocene time, but that glacial conditions continued. The glaciation was probably at least as intense as that of the present-day.

NEOGENE GLACIAL HISTORY
Evidence from the continental shelf

At Prydz Bay Site 739, a phase of deposition following the late Oligocene to middle Miocene hiatus, is represented by horizontally bedded mixed glaciomarine facies of late Miocene age. In contrast, at Site 742 erosion continued, and was also initiated at Sites 740 and 741. This suggests a retreat of the grounding line from the shelf break to a position between Sites 742 and 739 (Figure 6F).

A succession of advances and retreats across the shelf in late Miocene to Quaternary time is recorded, during which lodgement tills and glaciomarine sediments were deposited in alternation, and subjected to loading by overriding ice.

Glacial loading by overriding ice occurs at Site 739 between 154 and 138 mbsf [Solheim et al., 1991], within or just above sediments dated as middle late Miocene (7.4-6.6 Ma) by diatoms [Baldauf and Barron, 1991]. This loading event may coincide with a latest Miocene (Messinian age) ice advance or series of ice advances occurring between about 6.2 and 4.8 Ma. A mid-Miocene to early Pliocene phase of glacial erosion is represented in Prydz Bay by an unconformity at Site 742. Thus, more extensive ice than that of the present day existed, at least for part of this period. Furthermore, drill site MSSTS-1 and DSDP Sites 270, 272, and 273 in the Ross Sea indicate a major regional unconformity representing at least 10 m.y., probably created as a result of erosion by mid-Miocene to early Pliocene glaciers advancing across the shelf and responding in part to the development of the West Antarctic ice sheet [Savage and Ciesielski, 1983].

Solheim et al. [1991] describe a further loading event at both Sites 739 and 742 within sediments dated as early Pliocene by diatoms [about 4.6 to 3.6 Ma; Baldauf and Barron, 1991]. This event probably coincides with the unconformity at 134.4 mbsf at Hole 742A (Figure 5). In late early Pliocene to early late Pliocene time, it is apparent from Site 742 in Prydz Bay that there was a glacial recession, but not total retreat of ice from coastal areas (Figure 6G), since waterlain tills and a 56 cm- thick diatom-rich horizon containing ice-rafted material were deposited. On the ice-free Vestfold Hills, bordering Prydz Bay to the southeast, Pickard et al. [1988] reported marine sediments about 4.5 to 3.5 Ma old (late early Pliocene). These contain diatoms and molluscs of strongly interglacial character and indicate a sea level that was 15 m higher over this site than today. The absence of significant ice-rafted material suggests an even more marked glacial recession than indicated by the Prydz Bay sediments.

There is a variety of evidence from other proximal parts of Antarctica indicating major collapse of the ice sheet at least once during mid-Pliocene time. The main evidence comes from the Sirius Group, which is a terrestrial glacial deposit unconformably covering pre-Tertiary rocks, deposited high up in the Transantarctic Mountains and originating from the Antarctic interior [Webb et al., 1984; McKelvey et al., 1991]. It contains reworked marine microfossils of Late Cretaceous and Tertiary age, the youngest of which are mid-Pliocene [3.1-2.5 Ma; Webb and Harwood, 1991; Barrett et al., submitted]. The Sirius microfossils were inferred to have been emplaced when a latest Pliocene to Quaternary ice sheet expanded and moved over relatively warm marine basins extending across East Antarctica. Different hypotheses on the origin and age of the Sirius Group are discussed by Clapperton and Sugden [1990].

After the early Pliocene to early late Pliocene retreat of the ice, ice reached the continental shelf break in Prydz Bay on an unknown number of occasions, and progradation of the shelf by grounding line sedimentation continued (Figure 6H). Beneath the grounded ice sheet, lodgement tills were deposited. Better age constraints are needed, but it is

evident that this was a major period of erosion of the shelf and a time of expanded ice sheets, during which most sedimentation was concentrated seaward of the continental shelf break.

Finally, following the last Pleistocene ice sheet advance and the onset of interglacial conditions, typified by sedimentation of diatom ooze and terrigenous mud, the area ceased to be directly influenced by glacier ice, although a minor amount of ice rafting continued at these sites (Figure 6I). For the Holocene Epoch, a detailed chronology has been worked out by Domack et al. [1991]. Their results indicate the development of open marine conditions in inner Prydz Bay around 10,000 yrs BP, and a phase of increased ice rafting between 7300 and 3800 yrs BP. The increased ice rafting may have been due to accelerated retreat of the Lambert Ice Shelf.

Evidence from Kerguelen Plateau

Many sedimentary sequences in the southern Indian Ocean are affected by hiatuses at the Oligocene/Miocene boundary [Barron et al., 1991b; Schlich, Wise et al., 1989]. These hiatuses can possibly be interpreted as a result of an intensified circulation caused by a strengthening of glaciation and cooling. The long-term Cenozoic cooling trend was interrupted by a warming episode in late early Miocene time. On Kerguelen Plateau, this warming is best documented by the decrease in $\delta^{18}O$ of benthic and planktonic foraminifers at Sites 747 and 751 between ca. 17.5 and 15 Ma [Figure 9; Mackensen et al., 1992; Wright and Miller, 1992]. However, it is debatable, whether this isotope shift is a pure temperature signal, or whether it also reflects a decrease in East Antarctic ice volume. A late early Miocene warming and exposure of some land in Antarctica may also be indicated by the highest Neogene smectite concentrations [Ehrmann, 1991]. During both early and middle Miocene time, sand-sized ice-rafted sediment components occurred only in very low quantities, indicating that, although glaciers retreated, some of them probably still reached the East Antarctic shoreline [Ehrmann, 1991].

In middle Miocene time, a rapid, drastic increase in $\delta^{18}O$ values is documented worldwide [Savin et al., 1975; Shackleton and Kennett, 1975; Miller et al., 1987, 1991]. At Site 751 on the central Kerguelen Plateau, this $\delta^{18}O$ shift started at ca. 14.9 Ma, but is not totally recorded because of a hiatus spanning the interval from 14.2 to 13.4 Ma (Figure 9). The $\delta^{18}O$ values of benthic foraminifers increased by 1.2 ‰, which is too much to be explained exclusively by an increase of ice. A combination of cooling and increased global ice volume is therefore assumed [Mackensen et al., 1992]. At Site 747, an increase in $\delta^{18}O$ can be observed in two steps at 14.5-13.3 Ma and 12.9-12.1 Ma [Wright and Miller, 1992]. The middle Miocene decrease in smectite content also may suggest a major cooling phase and disappearance of exposed Antarctic land masses [Ehrmann, 1991]. However, there is no evidence from benthic foraminifers for a major change in bottom-water circulation associated with the isotopic

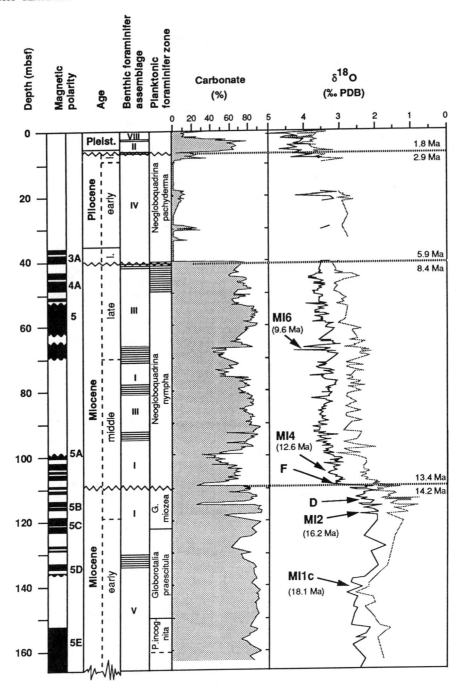

Fig. 9. Site 751 [after Mackensen et al., 1992]. Carbonate contents and oxygen isotope data of benthic (solid line) and planktonic (dotted line) foraminifera plotted vs. depth, magnetic polarities, benthic foraminifer assemblages and planktonic foraminifer zones. Distinct Miocene isotope events are indicated. I: *Nuttallides umbonifer* assemblage; II: *Trifarina angulosa* assemblage; III: *Astrononion pusillum* assemblage; IV: *Epistominella exigua* assemblage; V: *Uvigerina hispidocostata* assemblage; VIII: *Bulimina aculeata* assemblage.

shift [Figure 9; Mackensen, 1992]. The benthic assemblages had probably already become so adapted to cool bottom waters that a further decrease in temperature did not affect them. Possibly the $\delta^{18}O$ shift was mainly caused by an increase in ice volume and only to a minor part by cooling. A drastic increase in the global ice volume is also

indicated by the sea level, which dropped by >200 m in several steps between 14 and 10 Ma [Haq et al., 1987].

A late Miocene shift in the benthic foraminiferal assemblages of Site 751 at ca. 9.6 Ma is interpreted as indicating the injection of a water mass similar to the North Atlantic Deep Water into the Antarctic Circumpolar

Current [Mackensen, 1992]. Ice rafting over the Kerguelen Plateau and in the Australian-Antarctic Basin intensified slightly at ca. 9 Ma and distinctly at ca. 8.5 Ma (Figure 10). This intensification of ice-rafting may be related to an increase in ice discharge (Robin, 1988), the build-up of a West Antarctic ice sheet [Ciesielski et al., 1982; Savage and Ciesielski, 1983], or a cooling of the surface water, allowing further offshore transport of the debris by icebergs. However, the oxygen isotope data from Kerguelen Plateau are relatively constant throughout late Miocene time and do not indicate a major change around 9 Ma [Figure 9; Mackensen et al., 1992; Wright and Miller, 1992].

A sharp latest Miocene decrease in carbonate deposition and the onset of intense diatom ooze sedimentation suggests pronounced intensification of Antarctic glaciation, combined with a northward expansion of cool surface waters and a northward migration of the Polar Front, and a marked shallowing of the Calcite Compensation Depth. The transition from mixed calcareous-biosiliceous ooze to essentially pure biosiliceous ooze occurred at Sites 744 and 751 on the southern and central Kerguelen Plateau, respectively, at 5.8 Ma [Ehrmann, 1991; Mackensen et al., 1992]. At the northern Kerguelen Plateau Site 737, however, the transition occurred somewhat earlier, at ca. 6.7 Ma [Barron et al., 1991a]. Maxima in the influx of ice-rafted debris are recorded at 6.6-6.1 Ma and 5.8-5.3 Ma (Figure 10). They may reflect the cooling of Antarctica and of the Southern Ocean and possibly document the advances and retreats of the ice (see above).

At Site 737 a transition in the diatom assemblage from warm and warm-temperate to temperate species can be observed in sediments 4.8 to 4.2 Ma old. During this period, species characteristic and endemic of the Southern Ocean increase in abundance at the expense of species with geographically widespread distribution. The timing of this transition corresponds in part to a decrease in $\delta^{18}O$ values, suggesting that the increased endemism of the diatom flora corresponds to generally warmer, not cooler, conditions and may relate to stronger development of the Antarctic Polar Front and increased volume deep water of a northern source in the vicinity of Site 737 during the early Pliocene Epoch [Barron et al., 1991a].

A further strong influx of ice-rafted debris has been reported in sediments some 4.5-4.3 Ma old (Figure 10). It can probably be linked to a considerable rise in sea level as a result of a marked early Pliocene deglaciation [Ciesielski et al., 1982; Hodell and Kennett, 1986; Pickard et al., 1988]. As a result of the rise in sea level, ice shelves decayed and grounded ice with sediment incorporated in its base decoupled from the shelf, which led to enhanced calving of icebergs and input of ice-rafted debris. The decay of ice shelves furthermore led to an increase in dirty icebergs from tidewater glaciers. Pickard et al. [1988] argued that the coastline and ice margin of Prydz Bay were as much as 50 km farther inland during the interval from 4.5 to 3.5 Ma. Evidence for surface-water warming at about 4.2 to 4.1 Ma comes from the presence of abundant

calcareous nannofossil assemblages in sediments of this age at both southern Kerguelen Site 744 and northern Kerguelen Site 737 [Wei and Thierstein, 1991].

The next peak in ice-rafted debris accumulation follows the early Pliocene warmer interval. It was identified in 3.2-2.9 Ma old sediments at Site 751 (Figure 10). The ice, expanding after the deglaciation, incorporated large amounts of debris, which had been deposited during the deglaciation in the coastal areas and on the shelf. Similar as in the early Oligocene, calving icebergs therefore may have been rich in debris.

A further step towards present-day conditions took place in late Pliocene time. Both benthic and planktonic $\delta^{18}O$ values increased rapidly. At the same time, a benthic foraminiferal faunal change occurred which indicates strongly that high bottom current activity has prevailed since the late Pliocene Epoch. Current velocities probably increased concurrently with the onset of the major glaciation in the northern hemisphere at about 2.6 Ma [Mackensen, 1992; Mackensen et al., 1992].

CONCLUSIONS

When did the Antarctic glaciation begin?

The onset of glaciation at sea level in Antarctica is still an open question as the base of the glacial sequence cored has been reached neither in Prydz Bay nor in the Ross Sea, nor have the oldest Prydz Bay glacial sediments been well dated. On the Kerguelen Plateau, dating is less of a problem. However, in this distal region the glacial signals are less pronounced. The sediments from the Kerguelen Plateau, as from some other regions, indicate that several glaciers may even have reached the sea as early as middle Eocene time. However, it is unlikely that most of East Antarctica was glacierized at this time, and the coastal climate was probably temperate. There can be no doubt, from both the continental shelf record and the distal marine record as well as the permanent increase in benthic foraminiferal $\delta^{18}O$, that full-scale glaciation over a significant part of East Antarctica was underway by earliest Oligocene time.

We know that the Prydz Bay shelf was under the influence of a fluvial regime with a temperate woodland climate in mid-Cretaceous time. Thus, in order to finally resolve the question of the onset of glaciation at sea level, we need access to a good offshore Upper Cretaceous to Eocene sedimentary record.

The question concerning the timing of the initial development of the East Antarctic ice sheet cannot be determined precisely, as its influence on the marine environment for that period of time during which it failed to reach the sea may not be obvious. Even if warm conditions prevailed at the fringes of Antarctica, it is conceivable that, in view of the high elevation of the continent, ice caps could have existed while the continent was located over the South Polar region. Indeed, some ice may have existed on Antarctica throughout much of the Phanerozoic Eon, even though evidence of full-scale glaciation prior to the Cenozoic Era is limited to the

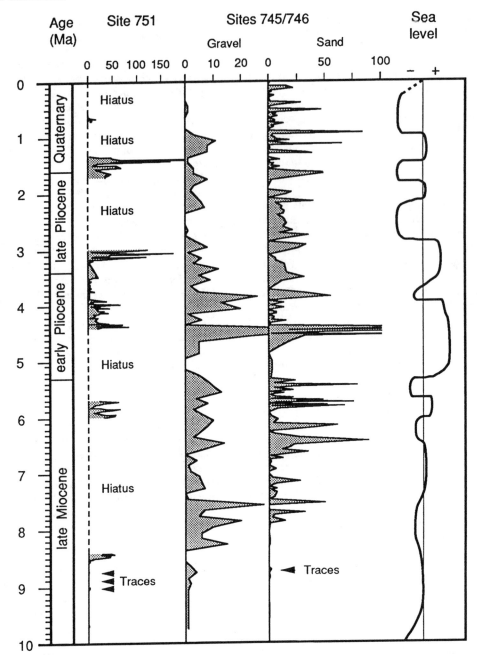

Fig. 10. Late Miocene to Quaternary ice-rafting record in the southern Indian Ocean. The raw data for Site 751 are from Breza [1992] and are plotted *versus* an age model combined from Harwood et al. [1992] and Mackensen et al. [1992]. The accumulation rates are given in grains >250 μm per ka and cm². The raw data for Sites 745 and 746 are from Ehrmann et al. [1991] and are plotted *versus* the age model by Barron et al. [1991b]. The accumulation rates of gravel are given in grains >2 mm per ka visible on the cut surface of the core. The accumulation rates of sand are given in mg sand per ka and cm². The sea level curve is that according to Haq et al. [1987].

Carboniferous and Permian Periods. Thus, it is probable that during Late Cretaceous time the highest parts of Antarctica were glacierized and that valley glaciers descended to <1000 m above sea level [Matthews and Poore, 1980; Robin, 1988; Spicer, 1990].

Was the ice sheet a permanent feature?

Perhaps more important than the timing of the onset of continental Antarctic glaciation is the stability of the ice sheet. There is still considerable disagreement about whether the Cenozoic ice sheet was a stable feature or whether it was subject to major fluctuations or even disappearances. If we can obtain a more comprehensive record of Antarctic ice volume fluctuations through time, we can better interpret the oxygen isotope record of the oceans in terms of paleotemperatures.

From the drilling record in Prydz Bay and in the southern Indian Ocean it seems that the ice never totally disappeared from the Antarctic continent once the ice sheet was established over East Antarctica in earliest Oligocene time. However, several major ice advances and retreats took place during this time interval. Throughout early Oligocene time grounded ice occupied Prydz Bay. No upper Oligocene to middle Miocene sediments were cored in Prydz Bay, because of an unconformity presumably created by ice erosion [Hambrey et al., 1991]. In the Ross Sea, however, four major late Oligocene ice advances separated by partial retreats have been recorded. Each of these advances probably lasted less than one million years [Barrett et al., 1989]. This indicates that major changes in the budget of Antarctic ice volume may occur quite rapidly, and may easily be lost in the drilling record through the development of hiatuses or through poor core recovery. Several further ice advances and retreats are documented by upper Miocene to Quaternary sediments in Prydz Bay [Hambrey et al., 1991], as well as by the amount of ice-rafted debris on the Kerguelen Plateau and in the Antarctic-Australian Basin [Ehrmann, 1991; Ehrmann et al., 1991; Breza, 1992].

Although we have obtained a relatively detailed record of ice advance and retreat, albeit with parts of the stratigraphic record missing or poorly dated, we have not yet acquired quantitative data on Antarctic ice volume. To address this problem, we need further combined effort from glaciologists, sedimentologists and numerical modellers.

Warm or cold ice?

In order to understand the climatic influences on the thermal regime of ice sheets, we need to consider what sedimentary evidence there is for deposition from warm or cold glacier ice. Such considerations have important implications for interpreting the isotopic record because ice volume and paleotemperature estimates may need to be modified if we are dealing with a temperate ice sheet comprising isotopically heavier water.

Evidence of thermal regime comes from vegetation, meltwater sediments, the character of weathering of bedrock, and indications of frozen ground. In the marine environment we need to turn to other indirect indications, such as the composition and texture of the sediments, facies associations, and whether the glaciers were floating or grounded. The latter is important because, if there is evidence of floating glacier ice, then the bulk of the ice is likely to have been below the pressure melting point, because today temperate glaciers are invariably grounded when they enter the sea, whereas it is only cold glaciers that are able to float.

In Prydz Bay there are no direct indications of what the vegetative cover of the land was like at any time during glacial conditions, if indeed it was exposed at all. For example, no contemporaneous pollen has been recovered from the cores nor are any terrestrial sediments preserved. On the other hand, clay mineralogy indicates the presence of chemically weathered bedrock, but this probably

indicates warmer, moister conditions prior to glaciation. The CIROS-1 core in McMurdo Sound has provided stronger evidence of thermal regime than those from Prydz Bay. Muddy sediments, pollen and the presence of a *Nothofagus* leaf all point to a temperate ice mass in Oligocene time.

Nothofagus leaves were also found at various other locations, amongst them Site 693 near South Orkney, King George Island and Seymour Island. They suggest that beech forests were growing in coastal areas during early Oligocene time, and possibly also in protected areas during late Oligocene time [Mildenhall, 1989; Mohr, 1990]. Furthermore, the presence of relatively warm Oligocene surface water masses of ca. 8°C over the Kerguelen Plateau, in spite of the nearby ice sheet, is easily explained by the assumption of a temperate, wet-based nature of the ice sheet rather than a polar, dry-based one [Mackensen and Ehrmann, 1992]. The relatively warm climate would have facilitated the transport of moisture onto the continent and resulted in a higher accumulation of snow and ice. Relatively warm surface water temperatures probably inhibited the long-distance transport of debris by icebergs, and thus explain the relative lack of larger amounts of ice-rafted debris on the Kerguelen Plateau.

In summary, we cannot be certain about the thermal characteristics of the Prydz Bay ice masses. We believe that the Oligocene ice sheet was temperate and wet-based, but we have no clear idea about the characteristics of the Neogene ice sheet. We require more independent lines of evidence of paleoclimate for a reasoned assessment. In reconstructing glacial history, paleoclimate and depositional environment, the transition from a temperate ice sheet to a polar one would be of considerable interest.

How does the ice sheet react on greenhouse warming?

The concern that the greenhouse warming will lead to ice sheet melting and global sea-level rise of catastrophic proportions is a problem that can be addressed by examination of the glacial sedimentary record. The evidence from Prydz Bay and McMurdo Sound of more extensive ice throughout much of the Oligocene Epoch than today, seems to coincide with relatively high temperatures derived from the deep-sea record. This suggests that the build up of ice may take place over East Antarctica when the climate is warmer as a result of greater precipitation of snow, a view already expressed by Robin [1988]. Understanding of the complexities of the interaction between ice, climate and sedimentation is still in an early stage, and future research in this field in Antarctica should be directed toward a better understanding of the processes involved, as well as to obtaining more complete stratigraphic records of glaciation through deep drilling.

Acknowledgments. We thank the crew of JOIDES Resolution, our shipboard and shore-based colleagues, as well as all technical staff involved in the sample preparation and analysis. We further thank P.J. Barrett for carefully reviewing the manuscript. W.U.E. acknowledges the financial support by the DAAD/NATO Research Programme and the Deutsche Forschungsgemeinschaft, M.J.H. the financial assistance from the B.B. Roberts Fund, University of Cambridge, and B.L. the financial support of the Danish Research Council. These three authors together also held a NATO Collaborative Research Grant. This is Contribution No. 554 of the Alfred Wegener Institute for Polar and Marine Research.

REFERENCES

Abelmann, A., Gersonde, R. and Spiess, V., Plio-Pleistocene paleoceanography in the Weddell Sea - siliceous microfossil evidence, in *Geological History of Polar Oceans: Arctic versus Antarctic, NATO/ASI Series C, 308*, edited by U. Bleil, and J. Thiede, 729-759, Kluwer Academic Publishers, Dortrecht, The Netherlands, 1990.

Baldauf, J. G., Middle Eocene through early Miocene diatom floral turnover, in *Eocene-Oligocene Climate and Biotic Evolution*, edited by D. Prothero and W. Berggren, in press, 1992.

Baldauf, J. G. and Barron, J. A., Diatom biostratigraphy: Kerguelen Plateau and Prydz Bay regions of the Southern Ocean, in *Proc. ODP, Sci. Results, 119*, edited by J. Barron, B. Larsen et al., 547-598, Ocean Drilling Program, College Station, TX, 1991.

Barker, P. F., Kennett, J. P. et al., *Proc. ODP, Init. Repts., 113*, 785 pp., Ocean Drilling Program, College Station, TX, 1988.

Barker, P. F., Kennett, J. P. et al., *Proc. ODP, Sci. Results, 113*, 1033 pp., Ocean Drilling Program, College Station, TX, 1990.

Barrera, E. and Huber, B. T., Paleogene and early Neogene oceanography of the southern Indian Ocean: Leg 119 foraminifer stable isotope results, in *Proc. ODP, Sci. Results, 119*, edited by J. Barron, B. Larsen et al., 693-717, Ocean Drilling Program, College Station, TX, 1991.

Barrett, P. J. (Ed.), Antarctic Cenozoic History from the MSSTS-1 Drillhole, McMurdo Sound, *DSIR Bull., 237*, 174 pp, 1986.

Barrett, P. J. (Ed.), Antarctic Cenozoic History from the CIROS-1 Drillhole, McMurdo Sound, *DSIR Bull., 245*, 254 pp, 1989.

Barrett, P. J., Antarctica and global climatic change: a geological perspective, in *Antarctic and Global Climatic Change*, edited by C. M. Harris and B. Stonehouse, 35-50, Belhaven Press, London, 1991.

Barrett, P. J., Hambrey, M. J., Harwood, D. M., Pyne, A. R. and Webb, P.-N., Synthesis, in Antarctic Cenozoic History from the CIROS-1 Drillhole, McMurdo Sound, edited by P. J. Barrett, 241-251, *DSIR Bull., 245*, 1989.

Barrett, P. J., Adams, C. J., Grapes, R. H., McIntosh, W. C., Swisher, C. C. and Wilson, G. S., Radiometric ages support Antarctic deglaciation about three million years ago, Submitted to Nature, 1992.

Barron, J., Larsen, B. et al., *Proc. ODP, Init. Repts., 119*, 942 pp., Ocean Drilling Program, College Station, TX, 1989.

Barron, J., Larsen, B. et al., *Proc. ODP, Sci. Results, 119*, 1003 pp., Ocean Drilling Program, College Station, TX, 1991.

Barron, J., Larsen, B. and Baldauf, J. G., Evidence for late Eocene to early Oligocene Antarctic glaciation and observations on late Neogene glacial history of Antarctica: Results from Leg 119, in *Proc. ODP, Sci. Results, 119*, edited by J. Barron, B. Larsen et al., 869-891, Ocean Drilling Program, College Station, TX, 1991a.

Barron, J. A., Baldauf, J. G., Barrera, E., Caulet, J.-P., Huber, B. T., Keating, B. H., Lazarus, D., Sakai, H., Thierstein, H. R. and Wei, W., Biochronologic and magnetochronologic synthesis of Leg 119 sediments from the Kerguelen Plateau and Prydz Bay, Antarctica, in *Proc. ODP, Sci. Results, 119*, edited by J. Barron, B. Larsen et al., 813-847, Ocean Drilling Program, College Station, TX, 1991b.

Bartek, L. R., Vail, P. R., Anderson, J. B., Emmet, P. A. and Wu, S., Effect of Cenozoic ice sheet fluctuations in Antarctica on the stratigraphic signature of the Neogene, *J. Geophys. Res.,*

96, B4, 6753 6778, 1991.

Birkenmajer, K., Tertiary glacial and interglacial deposits, South Shetland Islands, Antarctica: geochronology versus biostratigraphy (a progress report), *Bull. Pol. Ac. Earth Sci., 36*, 133-144, 1988.

Bohrmann, G. and Ehrmann, W. U., Analysis of sedimentary facies using bulk mineralogical characteristics of Cretaceous to Quaternary sediments from the Kerguelen Plateau: Sites 737, 738, and 744, in *Proc. ODP, Sci. Results, 119*, edited by J. Barron, B. Larsen et al., 211-223, Ocean Drilling Program, College Station, TX, 1991.

Breza, J. R., High-resolution study of Neogene ice-rafted debris, Site 751, southern Kerguelen Plateau, in *Proc. ODP, Sci. Results, 120*, edited by S. W. Wise, Jr., R. Schlich et al., 207-221, Ocean Drilling Program, College Station, TX, 1992.

Breza, J. R. and Wise, S. W., Jr., Lower Oligocene ice-rafted debris on the Kerguelen Plateau: evidence for East Antarctic continental glaciation, in *Proc. ODP, Sci. Results, 120*, edited by S. W. Wise, Jr., R. Schlich et al., 161-178, Ocean Drilling Program, College Station, TX, 1992.

Caulet, J.-P., Radiolarians from the Kerguelen Plateau, Leg 119, in *Proc. ODP, Sci. Results, 119*, edited by J. Barron, B. Larsen et al., 513-546, Ocean Drilling Program, College Station, TX, 1991.

Ciesielski, P. F., Ledbetter, M. T. and Ellwood, B. B., The development of Antarctic glaciation and the Neogene paleoenvironment of the Maurice Ewing Bank, *Mar. Geol., 46*, 1-51, 1982.

Ciesielski, P. F., Kristoffersen, Y. et al., *Proc. ODP, Init. Repts., 114*, 815 pp., Ocean Drilling Program, College Station, TX, 1988.

Ciesielski, P. F., Kristoffersen, Y. et al., *Proc. ODP, Sci. Results, 114*, 826 pp., Ocean Drilling Program, College Station, TX, 1991.

Clapperton, C.M. and Sugden, D.E., Late Cenozoic glacial history of the Ross Sea embayment, Antarctica, *Quat. Res. Rev., 9*, 253-272, 1990.

Cooper, A., Stagg, H. and Geist, E., 1991. Seismic stratigraphy and structure of Prydz Bay, Antarctica: implications from Leg 119 drilling, in *Proc. ODP, Sci. Results, 119*, edited by J. Barron, B. Larsen et al., 5-25, Ocean Drilling Program, College Station, TX, 1991.

Domack, E. W., Jull, A. J. T. and Donahue, D. J., Holocene chronology for the unconsolidated sediments at Hole 740A: Prydz Bay, East Antarctica, in *Proc. ODP, Sci. Results, 119*, edited by J. Barron, B. Larsen et al., 747-750, Ocean Drilling Program, College Station, TX, 1991.

Ehrmann, W. U., Implications of sediment composition on the southern Kerguelen Plateau for paleoclimate and depositional environment, in *Proc. ODP, Sci. Results, 119*, edited by J. Barron, B. Larsen et al., 185-210, Ocean Drilling Program, College Station, TX, 1991.

Ehrmann, W. U. and Mackensen, A., Sedimentological evidence for the formation of an East Antarctic ice sheet in Eocene/Oligocene time, *Palaeogeogr., Palaeoclimatol., Palaeoecol., 93*, 85-112, 1992.

Ehrmann, W. U., Grobe, H. and Fütterer, D. K., Late Miocene to Holocene glacial history of East Antarctica revealed by sediments from Sites 745 and 746, in *Proc. ODP, Sci. Results, 119*, edited by J. Barron, B. Larsen et al., 239-260, Ocean Drilling Program, College Station, TX, 1991.

Fisher, R. L., Jantsch, M. Z. and Comer, R. L., General Bathymetric Chart of the Oceans (GEBCO), Scale 1:10,000.000. 5-9, *Canadian Hydrographic Service*, Ottawa, 1982.

Gazdzicki, A., Planktonic foraminifera from the Oligocene Polonez Cove Formation of King George Island, West Antarctica, *Polish Polar Research, 10*, 47-55, 1989.

Grobe, H., Spätpleistozäne Sedimentationsprozesse am antarktischen Kontinentalhang vor Kapp Norvegia, östliche Weddell See, *Repts. Polar Res., 27*, 128 pp., Alfred Wegener Institute Bremerhaven, 1986.

Grobe, H. and Mackensen, A., Late Quaternary climatic cycles as recorded in sediments from the Antarctic continental margin, in *The Role of the Southern Ocean and Antarctica in Global Change*, edited by J. P. Kennett, Am. Geophys. Union, Antarct. Res. Ser., in press, 1992.

Grobe, H., Fütterer, D. K. and Spieß, V., Oligocene to Quaternary sedimentation processes on the Antarctic continental margin, ODP Leg 113, Site 693, in *Proc. ODP, Sci. Results, 113*, edited by P. F. Barker, J. P. Kennett et al., 121-131, Ocean Drilling Program, College Station, TX, 1990.

Hambrey, M. J., Structure and dynamics of the Lambert Glacier-Amery Ice Shelf system: implications for the origin of Prydz Bay sediments, in *Proc. ODP, Sci. Results, 119*, edited by J. Barron, B. Larsen et al., 61-75, Ocean Drilling Program, College Station, TX, 1991.

Hambrey, M. J., Larsen, B., Ehrmann, W. U. and ODP Leg 119 Shipboard Scientific Party, Forty million years of Antarctic glacial history yielded by Leg 119 of the Ocean Drilling Program, *Polar Record, 25 (153)*, 99-106, 1989.

Hambrey, M. J., Ehrmann, W. U. and Larsen, B., Cenozoic glacial record of the Prydz Bay continental shelf, East Antarctica, in *Proc. ODP, Sci. Results, 119*, edited by J. Barron, B. Larsen et al., 77-132, Ocean Drilling Program, College Station, TX, 1991.

Hambrey, M. J., Barrett, P. J., Ehrmann, W. U. and Larsen, B., Cenozoic sedimentary processes on the Antarctic continental margin and the record from deep drilling, *Z. Geomorph. N.F., Suppl.-Bd. 86*, 73-99, 1992 .

Haq, B. U., Hardenbol, J. and Vail, P. R., Chronology of fluctuating sea levels since the Triassic, *Science, 235*, 1156-1167, 1987.

Harwood, D. M., Barrett, P. J., Edwards, E. R., Rieck, J. J. and Webb, P. N., Biostratigraphy and chronology, in Antarctic Cenozoic History from the CIROS-1 Drillhole, McMurdo Sound, edited by P. J. Barrett, 231-239, *DSIR Bull., 245*, 1989.

Harwood, D. M., Lazarus, D. B., Abelmann, A., Aubry, M.-P., Berggren, W. A., Heider, F., Inokuchi, H., Maruyama, T., McCartney, K., Wei, W. and Wise, S. W., Jr., Neogene integrated magnetobiostratigraphy of the central Kerguelen Plateau, Leg 120, in *Proc. ODP, Sci. Results, 120*, edited by S. W. Wise, Jr., R. Schlich et al., 1031-1052, Ocean Drilling Program, College Station, TX, 1992.

Hayes, D.E., Frakes, L.A. et al., *Init. Repts. DSDP 28*, 1017 pp, U.S. Govt. Printing Office, Washington, D.C., 1975.

Hayes, D. E. and Vogel, M., General Bathymetric Chart of the Oceans (GEBCO), Scale 1:10,000.000. 5-13, *Canadian Hydrographic Service*, Ottawa, 1981.

Hill, R.S., Fossil leaf, in Antarctic Cenozoic History from the CIROS-1 Drillhole, McMurdo Sound, edited by P. J. Barrett, 143-144, *DSIR Bull., 245*, 1989.

Hodell, D. A. and Kennett, J. P., Late Miocene - early Pliocene stratigraphy and paleoceanography of the South Atlantic and Southwest Pacific oceans: a synthesis, *Paleoceanography, 1*, 285-311, 1986.

Huber, B. T., Paleogene and early Neogene planktonic foraminifer biostratigraphy of Sites 738 and 744, Kerguelen Plateau (southern Indian Ocean), in *Proc. ODP, Sci. Results, 119*, edited by J. Barron, B. Larsen et al., 427-449, Ocean Drilling Program, College Station, TX, 1991.

Keigwin, L. D. and Corliss, B. H., Stable isotopes in late middle Eocene to Oligocene foraminifera, *Geol. Soc. Am. Bull., 97*, 335-345, 1986.

Kennett, J. P. and Barker, P. F., Latest Cretaceous to Cenozoic climate and oceanographic developments in the Weddell Sea, Antarctica: An ocean-drilling perspective, in *Proc. ODP, Sci. Results, 113*, edited by P. F. Barker, J. P. Kennett et al., 937-960, Ocean Drilling Program, College Station, TX, 1990.

Larter, R. D. and Barker, P. F., Seismic stratigraphy of the Antarctic Peninsula Pacific margin: a record of Pliocene-Pleistocene ice volume and paleoclimate, *Geology, 17*, 731-734, 1989.

Lorius, C., Jouzel, J., Ritz, C., Merlivat, L., Barkov, N. I., Korotkevich, Y. S. and Kotlyakov, V. M., A 150.000 years cimate record from Antarctic ice, *Nature, 316*, 591-596, 1985.

Mackensen, A., Neogene benthic foraminifers from the southern Indian Ocean (Kerguelen Plateau): biostratigraphy and paleoecology, in *Proc. ODP, Sci. Results, 120*, edited by S. W. Wise, Jr., R. Schlich et al., 649-673, Ocean Drilling Program, College Station, TX, 1992.

Mackensen, A. and Berggren, W. A., Paleogene benthic foraminifers from the southern Indian Ocean (Kerguelen Plateau): biostratigraphy and paleoecology, in *Proc. ODP, Sci. Results, 120*, edited by S. W. Wise, Jr., R. Schlich et al., 603-630, Ocean Drilling Program, College Station, TX, 1992.

Mackensen, A. and Ehrmann, W. U., Middle Eocene through Early Oligocene climate history and paleoceanography in the Southern Ocean: stable oxygen and carbon isotopes from ODP Sites on Maud Rise and Kerguelen Plateau, *Mar. Geol., 108*, in press, 1992.

Mackensen, A., Barrera, E. and Hubberten, H.-W., Neogene circulation in the southern Indian Ocean: evidence from benthic foraminifers, carbonate data, and stable isotope analyses (Site 751), in *Proc. ODP, Sci. Results, 120*, edited by S. W. Wise Jr., R. Schlich et al., 867-880, Ocean Drilling Program, College Station, TX, 1992.

Margolis, S. V. and Kennett, J. P., Antarctic glaciation during the Tertiary recorded in sub-Antarctic deep-sea cores, *Science, 170*, 1085-1087, 1970.

Matthews, R. K. and Poore, R. Z., Tertiary [18]O record and glacioeustatic sea-level fluctuations, *Geology, 8*, 501-504, 1980.

McKelvey, B. C., Webb, P. N., Harwood, D. M. and Mabin, M. C. G., The Dominion Range Sirius Group: a record of late Pliocene - early Pleistocene Beardmore Glacier, in *Geological Evolution of Antarctica*, edited by M. R. A. Thomson, J. A. Crame and J. W. Thomson, 675-682, Cambridge University Press, Cambridge, 991.

Mildenhall, D. C., Terrestrial palynology, in Antarctic Cenozoic History from the CIROS-1 Drillhole, McMurdo Sound, edited by P. J. Barrett, 119-127, *DSIR Bull., 245*, 1989.

Miller, K. G., Fairbanks, R. G. and Mountain, G. S., Tertiary oxygen isotope synthesis, sea level history, and continental margin erosion, *Paleoceanography, 2*, 1-19, 1987.

Miller, K. G., Wright, J. D. and Fairbanks, R. G., Unlocking the ice house: Oligocene-Miocene oxygen isotopes, eustasy, and margin erosion, *J. Geophys. Res., 96, B4*, 6829-6848, 1991.

Mohr, B. A. R., Eocene and Oligocene sporomorphs and dinoflagellate cysts from Leg 113 drill sites, Weddell Sea, Antarctica, in *Proc. ODP, Sci. Results, 113*, edited by P. F. Barker, J. P. Kennett et al., 595-612, Ocean Drilling Program, College Station, TX, 1990.

Ollier, G. and Mathis, B., Lithologic interpretation from geophysical logs in Holes 737B, 738C, 739C, and 742A, in *Proc. ODP, Sci. Results, 119*, edited by J. Barron, B. Larsen et al., 263-289, Ocean Drilling Program, College Station, TX, 1991.

Paterson, W. S. B., *The physics of glaciers*, 380 pp, Pergamon Press, Oxford, 1981.

Pickard, J., Adamson, D. A., Harwood, D. M., Miller, G. H., Quilty, P. G. and Dell, R. K., Early Pliocene marine sediments, coastline, and climate of East Antarctica, *Geology, 16*, 158-161, 1988.

Robin, G. de Q., The Antarctic ice sheet, its history and response to sea level and climatic changes over the past 100 million years, *Palaeogeogr., Palaeoclimatol., Palaeoecol., 67*, 31-50, 1988.

Sakai, H. and Keating, B., Paleomagnetism of Leg 119 - Holes 737A, 738C, 742A, 745B, and 746A, in *Proc. ODP, Sci. Results, 119*, edited by J. Barron, B. Larsen et al., 751-770, Ocean Drilling Program, College Station, TX, 1991.

Savin, S. M., Douglas, R. G. and Stehli, F. G., Tertiary marine paleotemperatures, *Geol. Soc. Am. Bull., 86*, 1499-1510, 1975.

Savage, M. L. and Ciesielski, P. F., A revised history of glacial sedimentation in the Ross Sea, in *Antarctic Earth Science*, edited by R. L. Oliver, P. R. James and J. B. Jago, Cambridge University Press, Cambridge, 555-559, 1983.

Schröder-Adams, C. J., Middle Eocene to Holocene benthic foraminifer assemblages from the Kerguelen Plateau (southern Indian Ocean), in *Proc. ODP, Sci. Results, 119*, edited by J. Barron, B. Larsen et al., 611-630, Ocean Drilling Program, College Station, TX, 1991.

Schlich, R., Wise, S. W., Jr., et al., *Proc. ODP, Init. Repts., 120*, 648 pp., Ocean Drilling Program, College Station, TX, 1989.

Shackleton, N. J., Paleogene stable isotope events, *Palaeogeogr., Palaeoclimatol., Palaeoecol., 57*, 91-101, 1986.

Shackleton, N. J. and Kennett, J. P., Paleotemperature history of the Cenozoic and the initiation of Antarctic glaciation: oxygen and carbon isotope analyses in DSDP Sites 277, 279, and 281, in *Init. Repts. DSDP, 29*, edited by J. P. Kennett, R. E. Houtz et al., 743-755, U.S. Government Printing Office, Washington, D.C., 1975.

Solheim, A., Forsberg, C. F. and Pittenger, A., Stepwise consolidation of glacigenic sediments related to the glacial history of Prydz Bay, East Antarctica, in *Proc. ODP, Sci. Results, 119*, edited by J. Barron, B. Larsen et al., 169-182, Ocean Drilling Program, College Station, TX, 1991.

Spicer, R. A., Reconstructing high-latitude Cretaceous vegetation and climate: Arctic and Antarctic compared, in *Antarctic Paleobiology*, edited by T. N. Taylor, and E. L. Taylor, 27-36, Springer, New York, 1990.

Thierstein, H. R., Macdougall, J. D., Martin, E. E., Larsen, B., Barron, J. and Baldauf, J., Age determinations of Paleogene diamictites from Prydz Bay (Site 739), Antarctica, using Sr isotopes of mollusks and biostratigraphy of microfossils (diatoms and coccoliths), in *Proc. ODP, Sci. Results, 119*, edited by J. Barron, B. Larsen et al., 739-745, Ocean Drilling Program, College Station, TX, 1991.

Thomas, E., Late Cretaceous through Neogene deep-sea benthic foraminifers (Maud Rise, Weddell Sea, Antarctica), in *Proc. ODP, Sci. Results, 113*, edited by P. F. Barker, J. P. Kennett et al., 571-594, Ocean Drilling Program, College Station, TX, 1990.

Turner, B.R., Depositional environment and petrography of preglacial continental sediments from Hole 740A, Prydz Bay, East Antarctica, in *Proc. ODP, Sci. Results, 119*, edited by J. Barron, B. Larsen et al., 45-56, Ocean Drilling Program, College Station, TX, 1991.

Turner, B. R. and Padley, D., Lower Cretaceous coal-bearing sediments from Prydz Bay, East Antarctica, in *Proc. ODP, Sci. Results, 119*, edited by J. Barron, B. Larsen et al., 57-60, Ocean Drilling Program, College Station, TX, 1991.

Vergnaud Grazzini, C. and Oberhänsli, H., Isotopic events at the Eocene/Oligocene transition: a review, in *Terminal Eocene Events*, edited by C. Pomerol, and I. Premoli-Silva, 311-329, Elsevier, 1985.

Webb, P.-N. and Harwood, D.M., Late Cenozoic glacial history of the Ross Embayment, Antarctica, *Quat. Sci. Rev., 10*, 215-223, 1991.

Webb, P.-N., Harwood, D. M., McKelvey, B. C., Mercer, J. H. and Stott, L. D., Cenozoic marine sedimentation and ice volume variation on the East Antarctic craton, *Geology, 12*, 287-291, 1984.

Wei, W., Evidence for an earliest Oligocene abrupt cooling in the surface waters of the Southern Ocean, *Geology, 19*, 780-783, 1991.

Wei, W., Calcareous nannofossil stratigraphy and reassessment of the Eocene glacial record in subantarctic piston cores of the southeast Pacific, in *Proc. ODP, Sci. Results, 120*, edited by S. W. Wise, Jr., R. Schlich et al., 1093-1104., Ocean Drilling Program, College Station, TX, 1992.

Wei, W. and Thierstein, H. R., Upper Cretaceous and Cenozoic calcareous nannofossils of the Kerguelen Plateau (southern Indian Ocean) and Prydz Bay (East Antarctica), in *Proc. ODP, Sci. Results, 119*, edited by J. Barron, B. Larsen et al., 467-493, Ocean Drilling Program, College Station, TX, 1991.

Wei, W. Villa, G. and Wise, S. W., Jr., Paleoceanographic implications of Eocene-Oligocene calcaareous nannofossils from Sites 711 and 748 in the Indian Ocean, in *Proc. ODP, Sci. Results, 120*, edited by S. W. Wise, Jr., R. Schlich et al., 979-999, Ocean Drilling Program, College Station, TX, 1992.

Whitworth, T., III, The Antarctic Circumpolar Current, *Oceanus, 31*, 53-58, 1988.

Wise, S. W., Jr., Schlich, R., et al., *Proc. ODP, Init. Repts., 120*, 1155 pp., Ocean Drilling Program, College Station, TX, 1992.

Wise, S. W., Jr., Breza, J. R., Harwood, D. M. and Wei, W., Paleogene glacial history of Antarctica., in *Controversies in Modern Geology*, edited by J. A. McKenzie, D. W. Müller, and H. Weissert, 133-171, Cambridge University Press, Cambridge, 1991.

Wise, S. W., Jr., Breza, J. R., Harwood, D. M., Wei, W. and Zachos, J. C., Paleogene glacial history of Antarctica in light of Leg 120 drilling results, in *Proc. ODP, Sci. Results, 120*, edited by S. W. Wise, Jr., R. Schlich et al., 1001-1030, Ocean Drilling Program, College Station, TX, 1992.

Wright, J. D. and Miller, K. G., Miocene stable isotope stratigraphy, Site 747, Kerguelen Plateau, in *Proc. ODP, Sci. Results, 120*, edited by S. W. Wis, Jr., R. Schlich et al., 855-866, Ocean Drilling Program, College Station, TX, 1992.

Zachos, J. C., Berggren, W. A., Aubry, M.-P. and Mackensen, A., Isotope and trace element geochemistry of Eocene and Oligocene foraminifers from Site 748, Kerguelen Plateau, in *Proc. ODP, Sci. Results, 120*, edited by S. W. Wise, Jr., R. Schlich et al., 839-854, Ocean Drilling Program, College Station, TX, 1992.

Evolution and Variability of the Indian Ocean Summer Monsoon: Evidence from the Western Arabian Sea Drilling Program

WARREN L. PRELL, DAVID W. MURRAY AND STEVEN C. CLEMENS

Department of Geological Sciences, Brown University, Providence, RI, 02912-1846,USA

DAVID M. ANDERSON

NOAA Paleoclimatology Program, 325 Broadway, Boulder, Colorado, 80303-3328, USA

A number of forcing factors, including the tectonic evolution of Himalaya-Tibet and orbitally-induced changes in seasonal radiation, combine to cause the initiation, evolution, and variability of the Indian Ocean monsoon. Although climate model experiments can be used to estimate the variability attributed to each forcing factor, the only record of past monsoonal variation lies in the sediments of the northern Indian Ocean and the adjacent continents. A major goal of the regional survey cruise (RC27-04) and ODP Leg 117 was to recover the marine geologic record necessary to understand the history of the initiation, evolution and variability of the Indian Ocean summer monsoon and to provide an observational data set for comparison with model simulations of monsoon circulation.

General Circulation Model (GCM) experiments show that orbitally-induced increases in solar radiation significantly strengthen the monsoon winds and precipitation over southern Asia, but that surface boundary conditions (including sea surface temperature, albedo) associated with glacial phases weaken monsoon winds and precipitation. Experiments with full (modern elevations) and reduced plateau-mountain elevations reveal stronger winds and higher precipitation as mountain elevation increases. These results indicate that monsoon strength is equally sensitive to changes in solar radiation (on orbital time scales) and orographic changes (on longer time scales). They also indicate that global cooling cannot intensify the monsoon, so that the onset of the monsoon is most likely related to increased mountain elevation.

Sediments in the northwest Arabian Sea exhibit characteristic fauna (radiolarians and foraminifers) that are endemic to areas of strong upwelling. In the Arabian Sea, intense seasonal upwelling is induced by the southwesterly monsoon winds. Miocene to Recent sediments from the northwest Arabian Sea show distinct geochemical and biological changes which suggest that monsoonal upwelling conditions (abundant nutrients and cold temperatures) were established near 8 Ma. Pelagic sediments deposited before 10.5 Ma contain nannofossils characteristic of warm waters and relatively low surface productivity. Opal-rich sediments, previously thought to reflect the initiation of the strong monsoon circulation, were deposited between 10.5 Ma and 8.0 Ma. However, the fauna in these sediments are not characteristic of the species associated with strong upwelling. Near 8 Ma, the relative abundance of endemic upwelling species increases and is interpreted to reflect the intensification and onset of the strong modern monsoon circulation. Terrestrial climate indicators from adjacent Pakistan are consistent with monsoon intensification at this time.

The comparison of ODP sediment records and climate model simulations of monsoon circulation suggests that the combined effects of strong solar radiation and increased elevations (at least half of the modern orography) forced a strong monsoonal circulation about 8 Ma, which produced intense upwelling in the Arabian Sea and more seasonal climates over southern Asia.

MONSOONAL CLIMATES OF THE WESTERN ARABIAN SEA

Understanding the initiation, evolution, and variability of the Indian Ocean summer monsoon was a major focus of the regional survey cruise RC27-04 and Ocean Drilling Program (ODP) Leg 117. The summer monsoon is the dominant climatic feature of the Indian Ocean tropics and the adjacent continents. The success or failure of the monsoon rains affects almost every phase of life in southern Asia and northern Africa. Monsoonal regions are defined as having highly seasonal climates with wet summers and dry winters [Hastenrath, 1985; Webster, 1987]. In fact, the term "monsoon" is derived from the Arabic word *mausim*, which refers to the seasonal reversal of winds around the Arabian Sea [Warren, 1966]. The climatological region of Earth considered to be truly under the influence of monsoonal circulation includes most of Africa, southern Asia and the Indonesian archipelago (Figure 1).

The winter season of the Asian sector is characterized by low solar radiation, cold temperatures, and northeasterly winds which flow from the cold Asian continent towards the Arabian Sea (Figure 1). These continental winter monsoon winds carry little moisture and have relatively low velocity [Hastenrath and Lamb, 1979]. In contrast, the

Synthesis of Results from Scientific
Drilling in the Indian Ocean
Geophysical Monograph 70

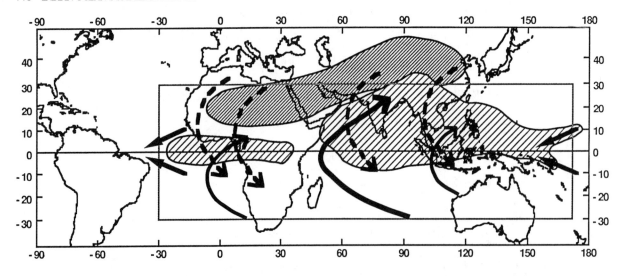

Fig. 1. The region of the tropics that is considered truly monsoonal on the basis of seasonally reversing winds and highly seasonal precipitation with wet summers (area within the solid rectangle). Dense cross-hatched pattern indicates areas of surface heating during the northern summer. Broad cross-hatched patterns outline areas that receive maximum precipitation during northern hemisphere summer. Solid arrows represent the southwesterly winds of the summer monsoon and dashed arrows northeasterly winds of the winter monsoon. Modified from Hastenrath [1985] and Webster [1987].

summer season is characterized by high solar radiation that causes intense sensible and latent heating over northern India and the Tibetan Plateau. This pattern of heating causes ascending air flow and the development of an intense low-pressure cell that is centered over Asia at about 30°N (Figures 1, 2). The atmospheric pressure gradient between the Asian continent (low) and the cooler southern Indian Ocean (the Madagascar high) induces a large-scale meridonal overturning, with the lower circulation limb being the strong low-level southwesterly summer monsoon winds of the western Indian Ocean. The convergence of these marine air masses over the Indian subcontinent and their uplift due to heating and orographic steering causes the seasonal monsoon rains [Hastenrath, 1985; Webster, 1987]. Hence, the monsoonal wind and precipitation patterns are often used as estimates of monsoon intensity because they strongly affect both the modern and past environments of the western Arabian Sea and the climates of tropical Africa and southern Asia, including Arabia and India.

The response of the Arabian Sea to the summer monsoon is driven by the strong southwesterly winds which cause Ekman transport of surface waters away from the Oman coast and develop intense centers of seasonal upwelling [Currie et al., 1973; Wyrtki, 1973; Bruce, 1974; Hastenrath and Lamb, 1979]. However, the Arabian Sea upwelling extends almost 500 km offshore because of the curl-induced upwelling related to the structure of the low-level Findlater Jet [Smith and Bottero, 1977; Luther and O'Brien, 1985; Luther et al., 1990; Anderson et al., 1992]. Conversely, the center of the Arabian Sea is a region of convergence. Experiments that incorporate the wind fields from GCM simulations (increased summer solar radiation at 9 Ka) into the detailed dynamical ocean models of the

Arabian Sea [Luther et al., 1990; Bigg et al., 1992] have shown distinct patterns of upwelling that are consistent with reconstructions of the faunal and SST response at 9 Ka [Prell et al., 1990]. The existence of the coastal and open ocean mechanisms of upwelling also raises the possibility that the changing structure and position of the Findlater Jet may induce differences in the amplitude and phase of the upwelling signal between the proximal Oman margin and the distal Owen Ridge [Anderson, 1990; Anderson et al., 1992].

This coastal and open-ocean upwelling system develops rapidly (usually during June) and dominates the surface ocean properties such as temperature and nutrient content (Figure 3A)[Wyrtki, 1971; Hastenrath and Lamb, 1979], mixed layer thickness [Hastenrath and Greischar, 1989; Rao et al., 1989], and productivity [Brock et al., 1991, 1992]. The upwelling brings cold, nutrient-rich waters from several hundred meters' depth to the surface and triggers high productivity in the euphotic zone [Krey and Babenerd, 1976]. Because these upwelling events are largely driven by Ekman transport [Wyrtki, 1973; Smith and Bottero, 1977; Prell and Streeter, 1982], their duration and intensity reflect the structure and intensity of the monsoonal winds. Thus, the biological, chemical, and sedimentological record of this upwelling system has a direct link to the structure and intensity of the monsoonal winds.

The high productivity and warm water sources (Red Sea and Persian Gulf) of the Arabian Sea lead to another characteristic feature of this area, namely the extensive Oxygen Minimum Zone (OMZ) [Wyrtki, 1973; Slater and Kroopnick, 1984], which has low O_2 concentrations (<0.2 mL/L) between depths of about 200m and 1500m [Spencer et al., 1982; Olson et al., 1992]. Although the OMZ is not

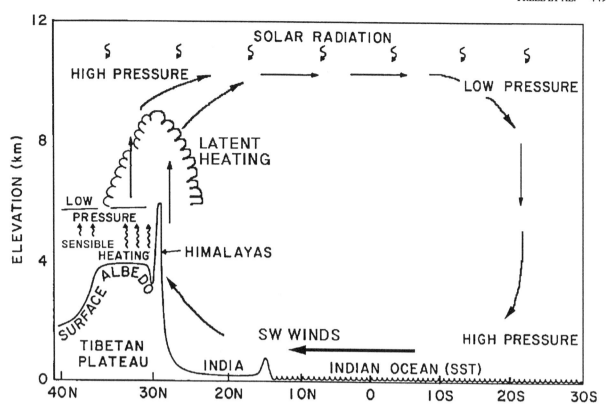

Fig. 2. Schematic diagram of the Indian Ocean monsoon geometry and processes. Summer season solar radiation causes sensible heating of the Tibetan Plateau and subsequent latent heating of the atmosphere above southern Asia. This heating pattern sets up pressure gradients in the upper and lower atmospheres that result in the strong southwesterly low-level winds (the Findlater jet) in the western Indian Ocean. Factors thought to be important in the initiation and maintenance of the monsoon circulation are the elevation of the Tibetan Plateau, the seasonal variation of summer solar radiation, changes in vegetation-albedo, the temperature gradients in the Indian Ocean, and the CO_2 concentration of the atmosphere.

directly forced by the monsoon winds and precipitation, the monsoon upwelling productivity patterns and regional precipitation-evaporation balance are factors that partially determine the extent and intensity of the OMZ.

This brief summary of the monsoon climate and its oceanographic impact clearly demonstrates that the oceanography, productivity, and sediments of the Arabian Sea are strongly impacted by the monsoon. Below we examine the factors that might cause major variations in the strength of the monsoon circulation on geologic time scales. We then examine the geologic record on the monsoon, especially as recorded in marine sediments such as those recovered on ODP Leg 117.

CAUSES OF MONSOON VARIABILITY

Given our current understanding of monsoon circulation, we can identify a variety of factors, not all independent, that could cause changes in the strength of the monsoon on geological time scales. On the basis of GCM experiments, the factors that have the most impact on monsoon strength are the elevation of Himalaya-Tibet and the orbitally-induced variation of northern hemisphere summer radiation [Kutzbach and Guetter, 1986; Prell and Kutzbach, 1987;

Kutzbach et al., 1989, in press; Ruddiman and Kutzbach, 1989]. Additional monsoon variability could be caused by changes in surface boundary conditions, such as the albedo of Africa-Asia, the extent of snow cover over Tibet, the sea surface temperature (SST) of the Indian Ocean, and the concentration of atmospheric CO_2; all factors that are associated with glacial-interglacial changes in climate. The relative impact of several of these factors on monsoon precipitation and winds has recently been estimated by Prell and Kutzbach [in press]. Here, we summarize current thoughts on the sensitivity of the monsoon to major forcing factors, their time-series, and estimates of the time-history of the monsoon.

Orography of Himalaya-Tibet

Experiments with atmospheric general circulation models (GCM) have shown that changes in elevation of Himalaya-Tibet have large effects on the intensity of the monsoon [Hahn and Manabe, 1975; Kutzbach et al., 1989; Ruddiman and Kutzbach, 1989, 1990; Prell and Kutzbach, in press]. Simulations with no-mountains or reduced elevation have less sensible and latent heating of the upper troposphere over southern Asia and thus a significantly weaker (or even

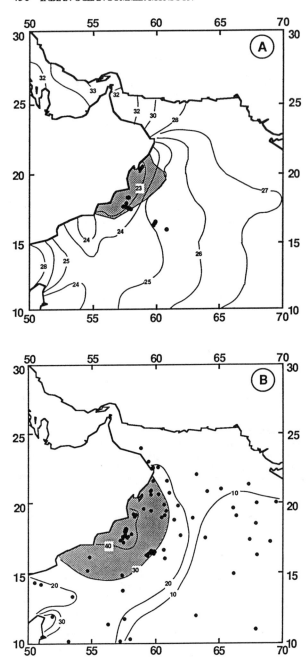

Fig. 3. A. Location of the ODP Leg 117 drilling sites in the Arabian Sea relative to monsoonal upwelling patterns of low sea surface temperature (°C) [from Hastenrath and Lamb, 1987] and high (>1 µm/l) surface phosphorous content [Wyrtki, 1971]. Note the coincidence of low sea surface temperatures and high nutrient content along the margin of Oman and the gradient increasing temperature away from the upwelling zone. The Leg 117 drilling sites are clustered along the margin of Oman (Sites 723 to 730), on the Owen Ridge (Sites 721, 722, and 731), and on the Indus Fan (Site 720).

B. The distribution of *Globigerina bulloides* (relative %) in sediments of the Arabian Sea. Note that high abundances (>30% shaded) occur along the coast of Oman and decrease seaward. This pattern of high percent *G. bulloides* is coincident with low sea surface temperatures and high nutrient content (see A above), both of which are indicative of active upwelling.

(about 20% less precipitation) compared to modern circulation (Figure 4A). These experiments imply that prior to distinct monsoon circulation and monsoonal upwelling, the mountain-plateau elevations were lower than half of present elevation. These implications need to be tested with paleoclimate and sedimentologic data.

Within the late Neogene, surface uplift rates of the Himalaya are controversial but thought to increase dramatically, especially in the Plio-Pleistocene [Mercier et al., 1987; Ruddiman et al., 1989]. Also, a variety of data indicates that the interval 10 Ma to 8 Ma was a time of active deformation and erosion in the Himalayan complex [Raymo and Molnar, 1991; Harrison et al., 1992; Amano and Taira, 1992; Rea, this volume; Weissel et al., this volume]. Initial GCM results indicate that the increase in monsoon intensity is approximately linear with surface uplift. Given this theoretical context, an abrupt appearance of monsoon indicators might be interpreted as evidence for rapid surface uplift or that the actual monsoon response is highly non-linear and dependent on a "critical" elevation.

Solar Radiation

The seasonal distribution of solar radiation due to changes in the Earth's orbit (i.e., the Milankovitch mechanism) is the second most important forcing of the monsoon response [Prell and Kutzbach, 1987; in press]. Both modelling results and paleoclimate data have shown that the Milankovitch mechanism has strong effects on the hydrologic budgets and wind fields of the tropics and, in particular, on the Indian-African summer monsoon [Kutzbach, 1981; Prell, 1984a; b; Kutzbach and Street-Perrott, 1985; Prell and Kutzbach, 1987, Street-Perrott et al., 1990]. The increased summer radiation over Asia, related to the precessional component of the Earth's orbit, results in stronger monsoons due to the heating of the Asian continent and associated dynamical responses (Figures 1,2). Over southern Asia, a full range of orbitally-induced summer radiation (Figure 4A, from 12.5% greater than modern to 5.8% less than modern) produced about 30% greater monsoon precipitation (relative to the model control case) with greater radiation and about 10% less monsoon precipitation with the lower radiation (Figure 4A). This range of monsoon response (about 40%) is comparable with the response (about 50% with modern

no) monsoon circulation. Prell and Kutzbach [in press] estimate that (relative to the full-mountain control case) the no-mountain orography gives a 50% to 60% reduction of monsoon precipitation over southern Asia (Figure 4A) and a 70% decrease in the Arabian Sea winds. They concluded that a prerequisite for strong monsoons (at least as strong as modern) is an orography of Himalaya-Tibet that is at least half the present elevation. With modern solar forcing (see below), half-mountain monsoons are only as strong as glacial-age monsoons, which are weak

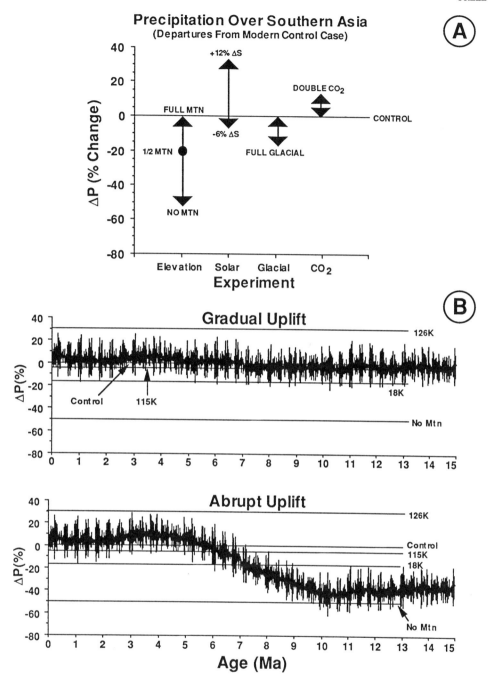

Fig. 4. A. Summary of GCM experiments with monsoon forcing factors. Arrows show the range of monsoonal precipitation over southern Asia (in percent change from the modern control case) for changes in the following monsoon factors: a) elevation (no mountain and half mountain), b) summer solar radiation (+12.5% to -5.8%), c) surface boundary conditions associated with glacial phases, and d) double atmospheric CO_2 concentration [modified from Prell and Kutzbach, 1987, and in press]. Note that the responses due to changes in elevation and solar radiation are twice or more as large as those changes due to changes in glacial surface boundary conditions or CO_2 composition.

B. Estimated paths of monsoon evolution based on the sensitivities of the monsoon to the forcing factors (discussed in text and summarized in 3A) and time-series of the forcing factors [modified from Prell and Kutzbach, in press]. Note that the gradual uplift case continuously supports monsoons stronger than modern (control) for the past 15 m.y., but that the monsoons in the abrupt uplift case first reach modern strength about 7 Ma when the elevations are about half the modern elevation.

radiation forcing) to the no-mountain experiment. Because the radiation changes are orbitally-induced, they have distinct periodicities (precession at 19 k.y. and 23 k.y., tilt at 41 k.y., and eccentricity at 100 k.y. and 414 k.y.). To the degree that each periodicity and its pattern of radiation are important in forcing the monsoon, one should expect the paleomonsoon record to reflect the amplitude and timing of that orbital component and to have a distinct spectral signature.

Surface Boundary Conditions

Changes in the Earth's surface boundary conditions, including the pattern of SST in the central and southern Indian Ocean, have also been correlated with the strength of the monsoon (Figure 4A). During the last glacial maximum, the SST of the west-central Indian Ocean is estimated to be slightly higher than modern SST [CLIMAP, 1981]. One effect of higher SST is to decrease the pressure gradient between Asia and the Indian Ocean and thus to weaken the monsoon circulation. This relationship has been inferred in studies of modern climates [Clemens and Oglesby, in press] and the last glacial maximum climates [Prell and Kutzbach, 1987, in press], but its magnitude is only about half the response due to changes in elevation or solar radiation (Figure 4A). Changes in the extent or duration of snow and ice cover have been suggested as a cause of monsoon variability due to the albedo feedback of snow and ice fields [Shukla, 1987; Barnett et al., 1988]. Increased albedo (i.e., more snow) is thought to delay the seasonal heating cycle over Asia and thus to retard and weaken the monsoon. Several researchers have inferred higher albedo [CLIMAP, 1981] and even the existence of ice caps on the Tibetan Plateau [Khule, 1987] during glacial periods. Although this albedo increase on the plateau is not well documented, it would lead to weaker monsoons. In general, increased "glacial" boundary conditions tend to weaken the monsoon so that "global cooling" trends [Molnar and England, 1990] cannot account for monsoon intensification during the Neogene.

Atmospheric CO$_2$

Higher atmospheric CO$_2$ concentration is another potential forcing for stronger monsoons as a warmer greenhouse climate would have a more active hydrologic budget. Over the past 15 m.y., CO$_2$ concentration is thought to have decreased from double modern values [Arthur et al., 1991] so that late Miocene monsoons should have been strengthened by the greater CO$_2$ concentration. However, model sensitivity experiments [Oglesby and Saltzman, 1990, 1992; Prell and Kutzbach, in press] find that the monsoon is only slightly enhanced by increased CO$_2$ (Figure 4A).

Possible Paths of Monsoon Evolution

Because all these potential forcing factors vary on timescales from 10^3 to 10^7 years and interact in combination, the monsoonal response is complex and partially dependent on the time-history of the important forcing functions. The sensitivity to individual forcing factors may also change with time because of interactions, and this possibility is being evaluated with new model experiments. If both the monsoon sensitivity to the important forcing factors (above, Figure 4A) and the time series of the forcing factors are known, they can be combined to estimate the time-history of monsoonal variation. The most controversial forcing factor is the time-history of the elevation of the Himalaya-Tibet [Molnar and England, 1990; Mercier et al., 1987; Harrison et al., 1992]. To capture the end-member cases, Prell and Kutzbach [in press] used two elevation-surface uplift paths to estimate the pattern of monsoon variability over the past 15 m.y., an interval that contains the onset of strong monsoonal circulation. One path assumes linear surface uplift from near sea level to modern elevation over the past 45 m.y. and the other assigns most of the surface uplift and elevation change to the interval between 10 Ma and 5 Ma. These two elevation paths control the long-term trend of monsoon evolution (Figure 4B), and the summer solar radiation patterns largely determine the "short-term" variability of the monsoon. Because the monsoon response to changes in elevation and solar radiation is approximately equal, "strong" and "weak" monsoons occur at all elevations (Figure 4B). However, the radiation-forced "strong" monsoons do not exceed modern monsoon strength until the circulation is at least half of the full mountain control case (at about 7 Ma in the abrupt uplift case) (Figure 4B). Note that in the gradual uplift case (3B), "strong" monsoons exceed the modern monsoon throughout the past 15 m.y., which is not supported by the paleo data (see below). Also, note that global cooling processes tend to weaken the monsoon and cannot account for its intensification or onset. Although these paths of monsoon evolution are not exact, they provide a physically-based framework for comparison to sediment-based paleo-monsoon time series.

MONSOON RESPONSE AND THE GEOLOGIC RECORD

The monsoonal climate imposes dramatic seasonal changes on the ocean properties and particle fluxes of the western Arabian Sea. In turn, these oceanographic, biologic, and sedimentologic responses give rise to a variety of sediment components that might be expected to reflect monsoonal variability. Here, we examine records of the concentration and flux of biogenic components, such as calcium carbonate, opal, and organic carbon, and the composition of plankton assemblages as potential indicators of monsoonal upwelling intensity. We also focus on the concentration, flux, and grain size of terrigenous material as indicators of eolian activity associated with the monsoon winds. All these components are related, in part, to the monsoon circulation and upwelling, but they also reflect other patterns associated with glacial-interglacial circulation, productivity, sediment flux and preservation. Hence, the inter-comparison of potential monsoon indicators with each other and with other environmental records, such as δ^{18}O (a

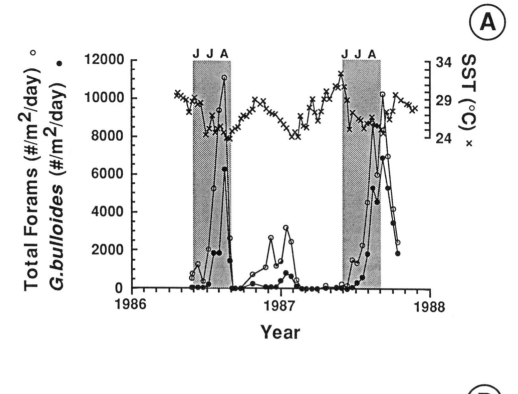

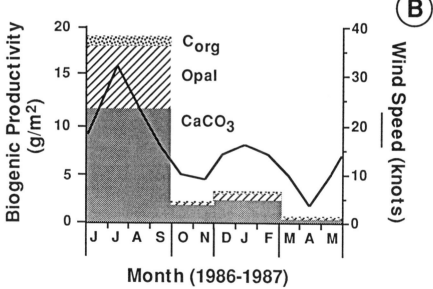

Fig. 5. Sediment trap records of the flux of planktonic foraminifers and biologic components in the western Arabian Sea (1986-1987) and their relation to SST.

A. The flux (#/m2/day) of total foraminifers (>150 μm) and *G. bulloides* during 1986 and 1987 [from Curry et al., 1992] and the SST derived from AVHRR Satellite data [from Curry and Ittekkot, personal communication]. The foraminiferal fluxes in this area are the highest ever recorded in sediment traps and are restricted to the months of the summer monsoon (shaded) and sometimes September (1987) when SST's are lower due to the upwelled waters. Although many species increase their flux during the monsoon season, *G. bulloides* dominates the flux and also exhibits significant interannual variability at this site (62° E) near the Owen Ridge.

B. The flux (g/m2/season) of carbonate, opal, and organic carbon during 1986. Fluxes are highest during the southwest summer monsoon season and thus, on an annual basis, the accumulation of these biogenic sediment components represents monsoon productivity and circulation. [Modified after Nair et al., 1989].

measure of the extent of continental glaciation), is an important strategy to identify the monsoon-related changes as opposed to other global changes.

Sediments of the Western Arabian Sea

Sediments of the western Arabian Sea reflect both the changing productivity associated with monsoon upwelling and the eolian transport associated with the monsoon climates. The sediments on both the Oman margin and the offshore Owen Ridge are predominantly mixtures of biogenic carbonate and terrigenous components with admixtures of opal and organic carbon [see Prell, Niitsuma et al., 1989]. Some detrital carbonate and dolomite is transported by winds and is more abundant on the margin than on the Owen Ridge [Sirocko and Sarnthein, 1989, site chapters in Prell, Niitsuma et al., 1989].

The sediments of the Oman margin are green to olive green, foraminifer-bearing, marly, nannofossil oozes and calcitic, clayey silts. All margin sediments are relatively high in organic carbon (2 to 8%), especially in the shallow slope basin. Although laminated sediments were not found on the margin by the regional survey (RC27-04), laminations of monospecific diatoms interbedded with organic carbon-rich detrital layers were found in the upper basin (300 m to 1000 m) ODP sites in the upper Pliocene and lowermost Pleistocene [see site chapters 723 to 726 in Prell, Niitsuma et al., 1989]. Accumulation rates for these Pliocene and Pleistocene sediments range from 160-260 m/m.y. in the center to 60-130 m/m.y. along the margins of the upper slope basin.

The more offshore, lower basin (~1500 m) sites (728, 730) accumulated less terrigenous material and thus the sediments are mainly foraminiferal oozes, marly calcitic oozes, and marly nannofossil oozes with admixtures of diatomaceous nannofossil chalk and silty clay. The Plio-Pleistocene sediments in this basin accumulated over a range of 20-60 m/m.y. but the middle to upper Miocene sediments accumulated at only 20 to 30 m/m.y. and are primarily nannofossil and marly nannofossil chalks and foraminifer-nannofossil chalks. The composition and accumulation rates of these latter sediments reflect a pelagic depositional environment that is similar to Miocene sediments on the Owen Ridge, about 350 km further offshore. Thus, during the late Neogene, the lithofacies of the Oman margin have evolved from pelagic nannofossil oozes and chalks to more terrigenous marly nannofossil oozes with varying amounts of opal, organic carbon, and laminations in the late Pliocene and early Pleistocene. The Pleistocene sediments are mainly foraminifer-bearing marly nannofossil oozes that are organic-rich but have neither laminations nor significant opal.

The sediments of the Owen Ridge are primarily pelagic in character (Units I, II, III) and are underlain by terrigenous turbidites (Unit IV) [see site chapters 721, 722, and 731 in Prell, Niitsuma et al., 1989]. The base of the section (Unit IV) is a series of coarse-grained sand and silt turbidites and mud turbidites that range from the late Oligocene (?) to the

middle Miocene. The upper 70 to 80 m of this turbidite unit (lower to middle Miocene) has fine-grained upward-fining turbidites that are interbedded with nannofossil chalks. This upsection increase in pelagic versus terrigenous components reflects the uplift of the surface of the Owen Ridge above the influence of turbidite deposition.

Unit III is a low porosity, high density nannofossil chalk of early Miocene age. The carbonate content of the chalk reaches 80 to 90%, silica preservation is poor, and accumulation rates are only about 8 to 15 m/m.y. The sediments of Unit II are foraminifer-nannofossil chalks and oozes and siliceous nannofossil chalks and oozes of late Miocene age. They are distinguished by their abundant opal content and the resultant low bulk and grain densities and high porosity. Accumulation rates range from 20 to 50 m/m.y. Unit I is primarily foraminifer-bearing to foraminifer-nannofossil ooze, and nannofossil ooze of late Pleistocene to late Miocene age. The sediments of Unit I are characterized by sparse opal and exhibit strong cyclicity that is indicated by changes in color (alternating light and dark layers), bulk density, carbonate content, and magnetic susceptibility. Accumulation rates in Unit I range from 20 to 45 m/m.y.

The succession of lithofacies on both the Oman margin and the Owen Ridge reflects both the changing origin and character of sedimentary components and the changing depositional environments of the Oman margin and the Owen Ridge. Below, we examine which aspects of this sediment record can be interpreted to reflect changes in the monsoonal climates as opposed to tectonic or global climate/ocean changes.

Biogenic Sediment Composition

The biogenically-produced calcium carbonate, opal, and organic carbon content of the sediments is widely used to interpret the productivity of overlying waters, but is also related to other factors of deposition and preservation. In the western Indian Ocean, increased flux or concentration of these sediment components has often been attributed to monsoon-induced productivity, a claim that is supported by recent sediment trap results from the western Arabian Sea [Nair et al., 1989]. These data indicate that about 70% of the annual biogenic (carbonate, opal, and organic carbon) flux to the sea floor occurs during the summer monsoon season (Figure 5B). However, on longer time scales (10^3-10^5 yr) the accumulation of these components is complicated by the preservation-related patterns that are associated with the incorporation of organic carbon in the sediments and with changes in the chemistry of the global ocean.

For example, the concentration and flux of $CaCO_3$ decreases due to carbonate dissolution (indicated by increased fragmentation of foraminifers) at the same time that opal and upwelling forams (G. bulloides) increase (Figure 6, 7A). Murray and Prell [1992] interpret this coincidence to be caused by in situ carbonate dissolution associated with increased organic rain into the sediments. Thus $CaCO_3$ concentration and flux are not monsoon

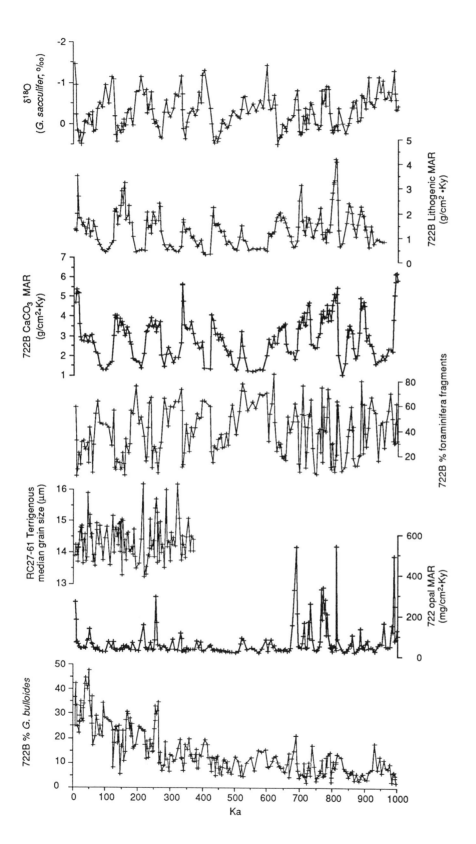

Fig. 6. Late Quaternary climate records from ODP Hole 722B and piston core RC27-61 from the Owen Ridge in the distal upwelling region. The timescales are based on oxygen isotopic stratigraphy [Imbrie et al., 1984]. Calcium carbonate and terrigenous material are the primary sediment components and exhibit significant 100 k.y. variability. The record of lithogenic mass accumulation is strongly correlated with the oxygen isotope (ice-volume) record and is interpreted to indicate maximum flux of eolian material during times of greater aridity, lower sea level and greater deflation of sediments in the source areas. The CaCO$_3$ mass accumulation rate (MAR) and foraminiferal fragmentation index records are driven by glacial-interglacial dissolution cycles. Biologic indicators of monsoonal upwelling include opal flux, percent G. bulloides, and excess barium flux (not shown). Grain size of the lithogenic component is an indicator of the transport capacity of summer-season southwest monsoon and Shamal winds. Although the various monsoon indicators have distinct characteristics, they are strongly correlated over the precessional frequency band.

indicators on the orbital time scale. On longer time scales (10^6 yr), the similarity of Arabian Sea carbonate records to other Indo-Pacific sites indicates that the accumulation of $CaCO_3$ in the Arabian Sea reflects more global-scale processes rather than local monsoon-driven processes [Peterson et al., this volume].

The accumulation of opal in sediments is widely interpreted as an indication of overlying productivity [see Baldauf et al., this volume for discussion and references]. On orbital time scales, the opal accumulation in the western Arabian Sea sediments covaries with other potential monsoon upwelling indicators and is thus interpreted as an indication of monsoon upwelling-induced productivity (Figures 6, 7A). On longer time scales (10^6 years), the high abundance of siliceous microfossils in the middle Miocene sediments of the Arabian Sea has been cited as evidence for the onset of monsoonal upwelling [Whitmarsh, Weser, Ross et al., 1974]. However, other workers [for example Johnson, 1990] have questioned whether the productivity associated with this accumulation of opal was induced by the monsoon upwelling or more global circulation patterns (see Baldauf et al., this volume for a summary of this discussion). Thus, opal accumulation is probably an indicator of productivity but all changes in the productivity of the Arabian Sea are probably not uniquely related to monsoon circulation.

The concentration and flux of organic carbon are also affected by processes other than productivity. In many areas, the abundance of organic carbon covaries with sedimentation rate. This relation is partially true for the western Arabian Sea and thus the concentration of organic carbon is more related to the times of enhanced terrigenous sedimentation than to the times of estimated monsoon upwelling productivity (Figure 7A).

Plankton Composition

The plankton composition of the surface waters and sediments is perhaps the most straightforward indicator of monsoonal upwelling. Numerous studies have documented that certain species of both phyto and zooplankton preferentially inhabit upwelling zones and are often characterized by high productivity during the upwelling season. Although the specific physiological, behavioral or other reasons for these associations are not well known, the empirical correlation of species composition and upwelling is quite strong [see papers in Boje and Tomczak 1978; Suess and Thiede, 1983; Thiede and Suess, 1983; Summerhayes et al., 1992].

In the Arabian Sea, the spatial distribution of several species of planktonic foraminifers (primarily *Globigerina bulloides*, [Figure 3B] *Globigerinita glutinata*, and *Neogloboquadrina dutertrei*) has been correlated with the monsoonal upwelling patterns of temperature, nutrient concentration and primary productivity [Prell et al., 1990; Anderson and Prell, 1991; Brock et al., 1992] of the surface waters (compare Figures 3A and B). Recent sediment trap data [Curry et al., 1992] have confirmed that about 75% of the foraminiferal flux of the western Arabian Sea occurs in the three summer months and that *G. bulloides* comprises about 50% of the flux during the upwelling season (Figure 5A). Previous work on late Quaternary sediments from the Owen Ridge [Prell, 1984a,b; Clemens and Prell, 1990] has shown that the relative abundance of *G. bulloides* varies coherently with the 23 k.y. precessional component of the Earth's orbit, which is thought to be a major radiation component forcing the monsoon (Figures 7A and B).

The radiolarian fauna in the western Arabian Sea sediments are also commonly found in sediments beneath active tropical and sub-tropical upwelling regions [Johnson and Nigrini, 1980; Nigrini, 1991; and Nigrini and Caulet, 1992]. Groups of species categorized as endemic upwelling, displaced temperate, and enhanced tropical are found in sediments beneath the Peru, Oman, and Somali upwelling areas, but are rarely found in open ocean sediments [Nigrini and Caulet, 1992]. Endemic upwelling radiolarians including the *Actinoma* spp. group and *Collosphaera* sp. aff. *C. huxleyi* have relatively long stratigraphic ranges, and may be useful in studying the onset and variation in monsoonal upwelling. Caulet et al. [1992] used the relative abundance of these species to provide an index of upwelling changes for Somali upwelling region over the past 160 k.y. and Nigrini [1991] examined the relative abundance changes of the endemic upwelling group in the Leg 117 sites to infer the onset of intense upwelling.

The isotopic and chemical composition of the plankton also records aspects of the monsoonal environment related to the ocean temperature, CO_2 concentration, productivity and nutrient concentrations. Increased upwelling would be associated with lower SST and more positive $\delta^{18}O$ [Prell and Curry, 1981]. However, the sense of the $\delta^{13}C$ in planktonic foraminifera is not as clear and may vary considerably according to species [Prell and Curry, 1981; Kroon, 1988; Steens et al., 1992].

The concentration of barium in Arabian Sea sediments is interpreted to indicate higher productivity [Shimmield et al., 1990; Shimmield and Mowbray, 1991; Clemens et al., 1991; Shimmield, 1992]. These studies have demonstrated that high barium concentration covaries with other upwelling indicators, especially in the precessional frequency band (1/23 k.y.). However, the barium data have also shown covariance with $\delta^{18}O$ over all major orbital frequency bands. This relation has been interpreted to indicate that some productivity increases are associated with ocean-wide circulation changes or addition of nutrients to the coastal zone through runoff.

Terrigenous Composition

The terrigenous composition of the sediments is useful in interpreting past monsoonal variations because it contains potential information of the strength of the winds (usually on the basis of grain size), the conditions in the source areas (usually inferred from terrigenous flux or pollen data), and the direction of atmospheric circulation (based on the mineralogy of source areas). Sediment trap data also document that 80% of the modern lithogenous fraction is

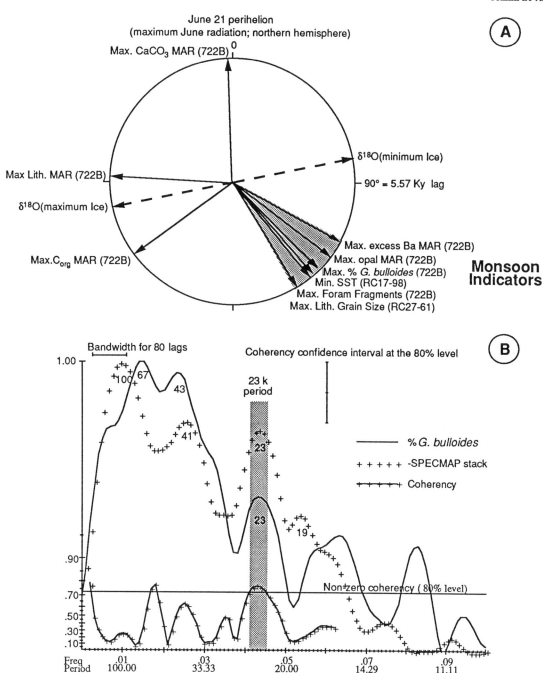

Ⓐ

June 21 perihelion
(maximum June radiation; northern hemisphere)
Max. CaCO₃ MAR (722B)

δ¹⁸O(minimum Ice)

90° = 5.57 Ky lag

Max Lith. MAR (722B)

δ¹⁸O(maximum Ice)

Max.C_org MAR (722B)

Max. excess Ba MAR (722B)
Max. opal MAR (722B)
Max. % G. bulloides (722B)
Min. SST (RC17-98)
Max. Foram Fragments (722B)
Max. Lith. Grain Size (RC27-61)

Monsoon Indicators

Ⓑ

Bandwidth for 80 lags

Coherency confidence interval at the 80% level

23 k period

—— % G. bulloides
+ + + + + -SPECMAP stack
+⊢+⊢+⊢+ Coherency

Non-zero coherency (80% level)

Freq / Period: .01 / 100.00, .03 / 33.33, .05 / 20.00, .07 / 14.29, .09 / 11.11

Cross Spectral Analysis:RC27-61 % G. bulloides vs. SPECMAP stack 2-370 k Δt=2

Fig. 7. Summary of the frequency domain responses of Owen Ridge records over the precessional frequency band (0 - 350 k.y.).

A) Precession phase wheel shows that upwelling indices (opal flux, % G. bulloides, and excess barium flux), foraminiferal fragmentation, and lithogenic grain size) have a consistent response over the past 350 k.y. These monsoon indicators lag the maxima in summer insolation by about 8 k.y. but are in phase with minimum SST in the southern subtropical Indian Ocean. Terrigenous sediment indicators, (lithogenic and CaCO₃ MAR maxima) occur in the opposite hemicycle. See text for discussion of phases.

B) Cross spectrum of percent G. bulloides and the SPECMAP stacked oxygen isotope record, an indicator of continental ice volume. Note the significant coherency over the precession band (shaded bar), the concentration of variance near 41 k.y. (the obliquity cycle), the lack of variance at 100 k.y. (the major ice-volume cycle) and the occurrence of variance at 67 k.y. (a possible non linear combination of the 100 k.y. and 41 k.y. band). Although not shown, G. bulloides is coherent with obliquity at the 41 k.y. band.

transported to the western Arabian Sea during the summer monsoon [Nair et al., 1989]. Another measure of the abundance of the terrigenous component in this region is the magnetic susceptibility of the sediments [Site 721 chapter in Prell, Niitsuma et al., 1989; Bloemendal and deMenocal, 1989; Clemens and Prell, 1991b; deMenocal et al., 1991]. These rapidly-acquired data enable high resolution study of long sections of core that can be compared to the measured terrigenous flux and grain size. Studies of Holocene and late Quaternary sediments have shown that the flux of eolian material to the Indian Ocean is related to the waxing and waning of the northern hemisphere ice sheets through changes in the climate and exposure (through lowered sea level) of potential source areas [Sirocko and Sarnthein, 1989; Clemens, 1990; Clemens and Prell, 1990; Sirocko et al., 1991]. Eolian flux increases during glacial phases of greater aridity in the tropics and exposure of the Persian Gulf due to lowered sea level. However, the median grain size of the lithogenic component is larger during more interglacial phases and covaries with planktonic and sediment indicators of strong monsoonal upwelling (Figure 7A). Thus, the flux of lithogenic material (and therefore magnetic susceptibility) is not a monsoonal indicator, but the grain size seems to be consistently related to other indices of strong monsoonal circulation.

The clay mineral composition of the Arabian Sea sediments had also been used to infer the trajectories of eolian transport and the character of the source areas. Sirocko and Sarnthein [1989], Sirocko and Lange [1991], and Sirocko et al. [1991] have documented the Holocene and last glacial distribution patterns and accumulation rates of various clay minerals. On longer time scales, Krissek and Clemens [1991] and Debrabant et al. [1991] have identified the temporal variations in the western Arabian Sea on Quaternary and Neogene time scales, respectively. These studies concur that the palygorskite content of the sediments is a relatively straightforward indicator of eastward eolian transport by southwesterly and westerly winds associated with the monsoon circulation. The source areas are primarily in the Arabian peninsula and its pattern of variability might be expected to covary with other wind-induced indicators. The terrigenous or clastic flux to the deep sea fans and ridges has also been used to interpret the erosional and uplift history of the adjacent continents, especially the Himalaya-Tibet [see Rea, this volume for a discussion].

This brief summary of potential monsoon indicators demonstrates that the history of monsoon can be reconstructed from the sediments and can be used to test various hypotheses about the initiation and variability of the Indian Ocean summer monsoon.

SCALES OF MONSOON VARIABILITY

The sediments of the northern Indian Ocean record a wide variety of paleoceanographic and paleoclimtologic responses, many of which are related to the southwest monsoon circulation. These sedimentary records exhibit variability on many time scales ranging from annual cycles (recorded by laminations) to long-term trends of millions of years. This variety of time scales should be no surprise as many of the forcing functions (discussed above) for monsoon and other climatic change also exhibit a range of time scales. Here, we summarize some of the records of monsoon-related variability that have resulted from study of the sediments and data obtained on the regional survey cruise (RC27-04) and ODP Leg 117. Although a few data exist, we do not include the decadal to century scale, but rather focus on three somewhat arbitrary time scales: orbital, Plio-Pleistocene, and Neogene. The orbital time scale focuses on variability ranging from 10^3 to 10^5 years and includes the familiar glacial-interglacial changes as well as many other responses that are related to orbital forcing. The Plio-Pleistocene time scale (represented by records 2 to 4 m.y. long) brings in longer term trends but still includes orbital-scale variation. The Neogene time scale (here represented by records about 17 m.y. long) reflects the more slowly changing components of the Earth's climate and tectonic systems in addition to the variations contained in the shorter timescales. The geologic record of the monsoon exhibits variations at all these time scales with shorter time scale variations superimposed on longer time scale trends and variations. Hence, the record is wonderfully complex and reflects multiple forcing, causation, and response.

Orbital Scale Variations

As discussed above, the periodic variations of the Earth's orbit alter the seasonal distribution of solar radiation and provide a major forcing for changes in monsoon strength on time scales of 10^4 to 10^5 k.y. Spectral analyses of time series data generated from the RC27-04 and Leg 117 cores has documented that variability in the major orbital periodicities of 100 k.y., 41 k.y., and 23 k.y. is pervasive in the sediments of the western Arabian Sea. This orbital scale variability occurs in virtually every site and in an astounding number of biological, chemical, and sedimentological records. This is not to say that all the spectra look the same. They differ in their distribution of variance according to the time interval analyzed, the location of the site, and the type of sediment component analyzed. Although all sites exhibited such variations, the periodicities have been documented in the following sites: RC27-61 [Clemens and Prell, 1990, 1991b; Clemens et al., 1991], Leg 117 Site 721 [Bloemendal and deMenocal, 1989; deMenocal et al., 1991], Site 722 [Clemens and Prell, 1991b; Busch, 1991; Murray and Prell, 1991, 1992; Shimmield et al., 1990; Shimmield and Mowbray, 1991; and Weedon and Shimmield, 1991], Site 723 [Anderson, 1990; Anderson and prell, in press], Site 724 [Shimmield et al., 1990, Weedon and Shimmield, 1991], and Site 728 [Busch, 1991; Steens et al., 1991]. Here, we discuss some of the orbital scale records that can be reasonably attributed to monsoon-induced changes in the environment.

Detailed records of monsoon variability for the past 1000 k.y. are available from the Owen Ridge (the upper portions of ODP Hole 722B and companion core RC27-61) and from the Oman margin (ODP Holes 723A, B). Analyses of the various indicators of the monsoonal upwelling have shown that the sedimentary record of the monsoon related to orbital forcing is complex, but that the relationships between different indicators can be used to constrain the interpretation of past monsoonal variation.

In Site 722, the high resolution oxygen isotope chronostratigraphy has been extended to over one million years along with the faunal, biochemical, and terrigenous indicators of upwelling and monsoonal variability (Figure 6). As noted earlier, the sediments are largely a mixture of biogenic carbonate and eolian-derived terrigenous material, with opal and organic carbon comprising <10% by weight. Although carbonate and terrigenous content and accumulation have concentrations of variance at the primary orbital (Milankovitch) periodicities, various studies [Clemens and Prell, 1990, 1991a; Murray and Prell, 1991, 1992; Shimmield et al., 1990; Shimmield and Mowbray, 1991; Shimmield, 1992] show that these major compositonal variations are not directly linked to the monsoon. The accumulation of bulk lithogenic material is closely linked to changes associated with continental glaciation and reflect variations in source area aridity and exposure [Clemens and Prell, 1990; deMenocal et al., 1991]. On the Owen Ridge, terrigenous dilution causes the major cycles in carbonate content (%); whereas, dissolution of calcium carbonate is responsible for the orbital-scale cycles in calcium carbonate accumulation ($g/m^2/k.y.$) observed in cores from the Owen Ridge. Though the sites are located above the lysocline [Cullen and Prell, 1984], the high flux of organic carbon and subsequent decomposition in the sediments enhance changes in mid- to deep-water carbonate ion concentrations that vary on glacial-interglacial cycles [Boyle, 1988a; Murray and Prell, 1992]. This interpretation of the carbonate accumulation is supported by it's strong inverse correlation with foraminifera fragmentation (Figures 6 and 7A).

Modelling experiments [Kutzbach, 1981; Kutzbach and Guetter, 1986; Prell and Kutzbach, 1987] have shown that the summer radiation pattern associated with increased precession of the Earth's axis (23 k.y. periods) induces stronger monsoons. At this frequency band, several indices reflecting upwelling (G. bulloides, opal and excess barium accumulation) and monsoonal wind indicators (maximum lithogenic grain size) are strongly coherent with precession and are in phase with each other (Figure 7A). However, timing of maxima in monsoon response occurs ~8 k.y. after maxima in precessional insolation forcing (Figure 7A). This phase lag is interpreted to reflect both the timing of minima in regional glacial conditions, such as Eurasian snow cover, (minima support stronger monsoons) and the timing of maxima in latent heat flux from the southern subtropical Indian Ocean [Clemens, 1990; Clemens and Prell, 1990, 1991a; Clemens et al., 1991]. On the Oman margin (Site 723), variations of G. bulloides in the 1/23 k.y. frequency band also lag

minimum glacial conditions ($\delta^{18}O$) but somewhat less than the Owen Ridge sites [Anderson, 1990; Anderson and Prell, in press]. The exact phase difference between the margin and the ridge and whether it is significant is still unclear and is the topic of current research.

Organic carbon concentration and accumulation also exhibit variations within the precessional band, but these changes do not match those of the monsoon indices. The record of organic carbon accumulation in the western Arabian Sea is partially driven by change in bulk sedimentation rate and is not considered a simple index of paleoproductivity [Murray and Prell, 1992; Shimmield, 1992].

While the monsoon is commonly thought of as a tropical/subtropical phenomenon, its origins range into the mid-latitudes (i.e., albedo and sensible heating over the Asian Plateau ~35°N and latent heat extraction from the southern hemisphere Indian Ocean ~35°S). Accordingly, some monsoon indicators (G. bulloides, eolian grain size) contain significant concentrations of variance at the 1/41 k.y. frequency band and are coherent and near zero phase with obliquity. Several other monsoon indicators (opal, barium flux) are not coherent but have similar near zero phase [Clemens et al., 1991] Almost all indices associated with terrigenous flux or dilution by terrigenous material show high coherence and near zero phase with the $\delta^{18}O$ proxy for continental glaciation. Thus, the Ba/Al records may reflect the dilution of terrigenous material and thus be related to the increased flux during more glacial intervals. If so, then their pattern of variability does not reflect a simple productivity signal. Two mechanisms have been proposed to account for the variance in monsoon indicators in the obliquity (41 k.y.) band: (1) increased physical upwelling driven by stronger winds [Clemens et al., 1991], and (2) increased nutrient content of the upwelled source waters due to vertical partitioning of the nutrients between deep and intermediate waters [Shimmield et al., 1990; Shimmield and Mowbray, 1991]. This latter process has been proposed for the global ocean by Boyle [1988a, b] and for the Indian Ocean by Kallel et al. [1988] but is not tied directly to the productivity of the overlying waters. Thus, although many the monsoon indicators are consistent over the 41 k.y. obliquity band, the mechanism(s) driving productivity at this period are still not uniquely defined and may well reflect a combination of processes.

In most late Quaternary marine records, such as the oxygen isotope record, the 100 k.y. period is dominant [Imbrie et al., 1984]. Records of G. bulloides from the Oman margin Site 723 have relatively strong 100 k.y. variability that is coherent and in phase with minimum ice volume [Anderson, 1990; Anderson and Prell, in press]. However, most of the records of monsoon strength from the Owen Ridge (~350 km to the southeast) possess relatively little coherent variance at the 100 k.y. band, with the exception of excess Ba flux which leads minimum ice volume by ~30 k.y. The lack of 100 k.y. variability in the G. bulloides record from the ridge is likely due to slight

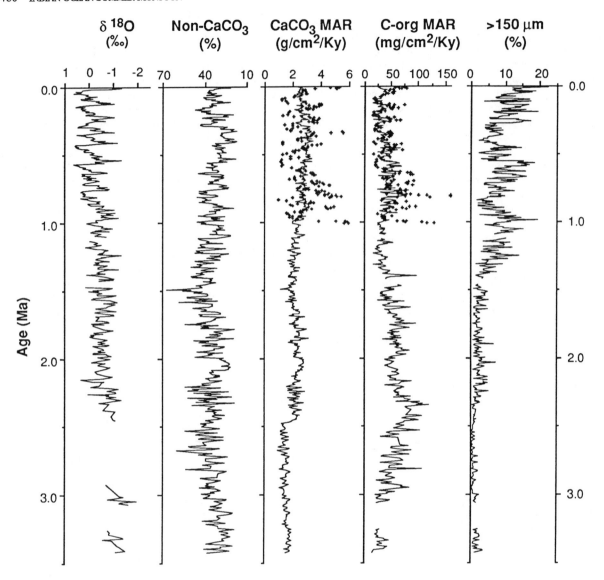

Fig. 8. High-resolution time series of δ18O (*G. sacculifer*), non-carbonate percent, calcium carbonate and organic carbon MAR, and foraminifera content (> 150 μm) in late Neogene sediments of Hole 722B (Owen Ridge). The age model for these time series is based on magnetostratigraphy and is discussed in Murray and Prell [1991]. Note that the more variable CaCO₃ and C-org MAR's (indicated by +) are based on a higher resolution oxygen isotope stratigraphy and chronology. In addition to the Milankovitch scale variability, these records exhibit long term trends that are driven by gradual changes in monsoon circulation superposed on changes in the global ocean. Even though Site 722 is relatively shallow (2028 meters), long term cycles of foraminiferal preservation are apparent in the upper 1.5 m.y. and preservation is particularly poor and carbonate MAR is low in the late Pliocene (3.5-2.5 Ma).

changes in the carbonate preservation related to global carbonate balances as well as to enhanced local dissolution driven by high upwelling productivity. In support of this hypothesis, the associated record of foraminiferal fragments (a measure of dissolution of carbonate tests) does have 100 k.y. variance and is in phase with interglacial climate (Figure 6). High fragmentation also occurs in Oman margin Site 728 (1428 m water depth) where Steens et al. [1991] interpret the increased fragmentation of foraminiferal tests to reflect higher productivity and

stronger upwelling during warmer periods. However, carbonate preservation cannot account for the lack of 100 k.y. cyclicity in the opal and grain size records. Thus, the explanation as to why some ridge and margin indices possess 100 k.y. variability while others do not, and the phase difference between those that do possess 100 k.y. variability, has yet to be fully understood.

The indices of monsoon changes from Site 722 and RC27-61 also exhibit long term trends (e.g., *G. bulloides*) and variability at non-primary (heterodyne) orbital frequencies

Owen Ridge Site 722

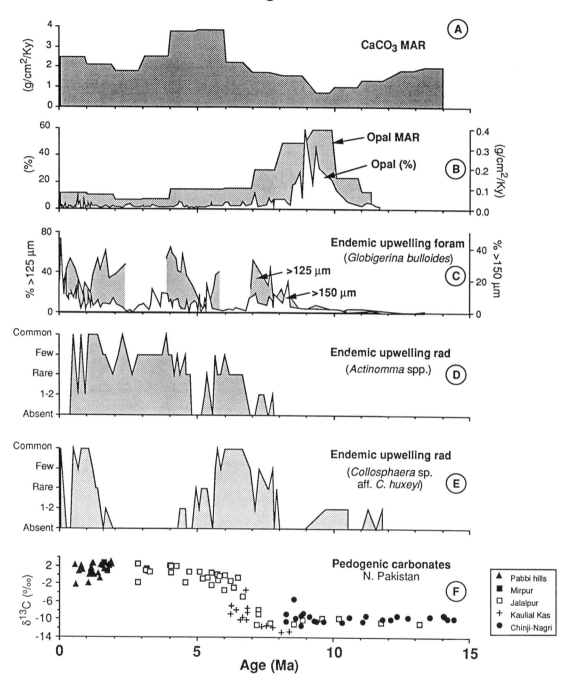

Fig. 9. Summary of biogenic sediment properties (CaCO₃ MAR and opal % and MAR) and endemic upwelling plankton over the past 15 m.y. The MAR for CaCO₃ (A) and opal (B) are averaged over 1 m.y. intervals for the past 15 m.y. at ODP Site 722. Opal concentration (%) is shown at about 100 k.y. resolution. (C) The relative percent of *G. bulloides* is shown both at >150 mm (this study) and >63 μm [from Kroon et al., 1991]. (D,E) The relative abundance of the endemic upwelling radiolarians are from Nigrini [1991]. The foraminifers and radiolarians indicative of cold nutrient-rich waters consistently increase at about 8 Ma. Note that opal % and MAR increased at about 10 Ma and is decreasing when the upwelling-related species become abundant. The bulk carbonate MAR is not easily related to the upwelling indicators and probably represents regional oceanographic changes rather than local monsoon changes. (F) The record of δ¹³C from soil carbonates in Pakistan [Quade et al., 1989] shows a shift towards C4 vegetation near the appearance of endemic upwelling plankton (see text for discussion).

as well as the primary orbital changes. Detailed analysis of the last 350 k.y. shows that the *G. bulloides* index has dominant frequencies at 67 k.y., 43 k.y. and 23 k.y. (Figure 7B). The grain size record contains variance at 29 k.y., 35 k.y., and 54 k.y. periods [Clemens and Prell, 1991b]. The concentration of variance at non-primary orbital periods suggests that non-linear processes are an important component of monsoon variability and require further study.

Plio/Pleistocene Variations

On longer time scales, detailed records of monsoon-related sediment properties from the Owen Ridge (Site 722) exhibit both orbital-scale (Milankovitch) variations and longer-term trends that reflect other types of forcing and processes (Figure 8). For example, records of carbonate accumulation and coarse fraction (>150 μm, mostly foraminifera) from Site 722 show a distinct increase in carbonate accumulation and preservation from 3.5 Ma to the Holocene (Figure 8). Better preservation is indicated by the abundance of whole foraminifera (i.e., higher coarse fraction and less fragmentation). Portions of the late Pliocene (especially from 3.5 Ma to 2.5 Ma) have extremely poor preservation of planktonic foraminifera (the reason for gaps in the $\delta^{18}O$ record) and low rates of carbonate accumulation. During this same interval, carbonate accumulation is also low or rapidly decreasing on the Oman margin (Site 728), on the Mascarene Ridge (Leg 115), and on the Ninetyeast Ridge (Leg 121) [Peterson et al., this volume]. Thus, the low carbonate accumulation of this interval is a basin-wide feature and cannot be attributed only to the monsoon-related processes. During this same interval, organic carbon accumulation is increasing, possibly due to increased productivity (Figure 9). Although carbonate is low or decreasing, organic carbon flux is increasing from 3.5 Ma to 2.5 Ma and its incorporation into the sediments probably contributes to the carbonate dissolution and poor preservation of foraminifera in the western Arabian Sea [Murray and Prell, 1992]. On the Oman margin, laminated sediments, indicating anoxia and a lack of bioturbation, occur during this same time interval. In addition, orbital scale variations of upwelling-related and terrigenous flux-related components are superimposed on these longer term trends, giving a complex record. Hence, the local monsoon-related processes may enhance basin or ocean-wide trends in deep ocean chemistry and the preservation of carbonates. Syntheses of carbonate accumulation rates and lysocline depths, such as in Peterson et al. [this volume] are necessary to differentiate between the global and local contributions to these records.

Another type of long-term trend is the shift in variance from one orbital periodicity to another periodicity. The records of magnetic susceptibility in Owen Ridge Sites 721 [Bloemendal and deMenocal, 1989; deMenocal et al., 1991] and the non-carbonate content in Site 722 [Murray and Prell, 1992] both exhibit a shift from predominantly higher frequencies near 1/23 k.y. to lower frequencies near

1/41 k.y. at about 2.5 Ma. A similar shift is observed in the wet-bulk density and optical density of Site 722, but is less clear in Oman margin Site 728 [Busch, 1991]. This transition from precession-dominated spectra to obliquity and precession-dominated spectra (except for the 100 k.y. periods in the past 800 k.y.) has been interpreted to indicate changes in the aridity and sediment availability in the Arabian-African source areas associated with the rapid expansion of continental ice sheets at about 2.4 Ma [Chapters 3,10 in Prell, Niitsuma et al., 1989; Bloemendal and deMenocal, 1989; deMenocal et al., 1991; Murray and Prell, 1992]. The emergence of the 41 k.y. variability is attributed to the high latitudes becoming more sensitive to the obliquity modulated radiation budget, which has the largest changes poleward of 60°.

In summary, the sediments of the western Arabian sea exhibit a wide variety of variability with periods between 10^4 and 10^5 years. Much of this variability is shown to be coherent with the orbital periods of eccentricity (~100 k.y.), obliquity (~41 k.y.), and precession (~23 k.y.). Sediment components related to the flux of terrigenous material tend to show phasing consistent with glacial-interglacial changes in the climate system, whereas sediment components thought to be associated with the monsoon-induced upwelling and winds exhibit phases consistent with each other but different from the $\delta^{18}O$ "glacial-interglacial" phases. These coherency and phase relationships are important because they provide some constraints on the interpretation of which processes form the paleoceanographic record of climate change.

Neogene Scale Variations

The initiation of strong monsoonal circulation in the northern Indian Ocean can be estimated from longer time-scale (but currently lower resolution) records of the first appearance of various monsoonal indicators. Depending on the choice of monsoon indicators, the age of the onset of strong monsoonal circulation ranges from the late middle Miocene to late Miocene. The sediments from the western Arabian Sea (Site 722, Owen Ridge) provide a continuous 15 m.y.-long record to examine the onset and variability of the various monsoon indicators. Here, we summarize several records that have been linked to the evolution of oceanic environments in the Arabian Sea: the mass accumulation rates (MAR) of calcium carbonate and opal MAR (this study) and monsoonal upwelling indices from planktonic foraminifers [this study; Kroon et al., 1991] and radiolarians [Nigrini, 1991](Figure 9).

The calcium carbonate MAR on the relatively shallow Owen Ridge might be expected to record productivity changes, but its long-term record of mean accumulation is markedly different from the opal and faunal records (Figure 9). All three Owen Ridge sites (721, 722, 731) exhibit low mean accumulation during the interval of maximum opal deposition (~10-8 Ma), with rates subsequently increasing to a maximum near 5 Ma and then decreasing toward the present. This maximum in mean accumulation is about twice the average rate at Site 722 over the past 1 m.y., but equals the mean rate of carbonate accumulation during glacial intervals which are characterized by excellent carbonate preservation [Murray and Prell, 1992]. These records of mean carbonate accumulation from the Arabian Sea are similar to records from other shallow sites in the tropical Indian Ocean and in the equatorial Pacific [Peterson et al., this volume]. Although some of the differences may be attributed to the local imprint of monsoonal upwelling, the overall similarity of the calcite MAR records implies that the major changes in the Arabian Sea and the tropical Indian Ocean do not reflect localized monsoonal upwelling, but are responding to regional and global processes such as eustatic sea level [Haq et al., 1987], continental weathering [Davies and Worsley, 1981; Delaney and Boyle, 1988; Raymo et al., 1988], deep-water flow [Woodruff and Savin, 1989], and/or trade wind strength [Rea and Bloomstine, 1986].

The record of bulk opal accumulation has been previously used to infer the onset of high productivity in the middle Miocene and has been interpreted to mark the beginning of strong monsoon circulation in the Arabian Sea [DSDP Leg 23, Whitmarsh, Weser, Ross et al., 1974; and ODP Leg 117, Prell, Niitsuma et al., 1989]. The initiation of opal accumulation at Site 722 occurred near 11.5 Ma and rapidly increased to a maximum between 10 and 9 Ma (Figure 9). Consistent with these data, Burckle [1989] proposed that the onset of diatom accumulation at DSDP Site 238 (currently 11°09'S, 70°32'E) near 10-11 Ma signaled the initiation of the modern monsoon circulation. Opal deposition also occurred near 10-11 Ma in the tropical Indian Ocean sites cored during ODP Leg 115 [Johnson, 1990; Mikkelsen, 1990; Baldauf et al., this volume]. However, all of these sites are located in the equatorial circulation system and are not under the direct influence of the monsoon circulation. Johnson [1990] has suggested that the renewed opal deposition in the tropical Indian Ocean resulted from the northward movement of the Indian subcontinent, which allowed trans-Indian zonal circulation to be established in the late Miocene. The increase in opal accumulaion has also been related to the onset of NADW production [Woodruff and Savin, 1989; Baldauf et al., this volume]. In addition, the faunal data (see below) suggest that the enhanced productivity did not have the character of intense monsoon-driven upwelling. In summary, the initiation of opal accumulation in the Arabian Sea is certainly related to increased productivity, but the regional data from Leg 115 and Leg 117 indicate that it is also strongly influenced by regional tectonic history and ocean circulation. This result differs somewhat from the orbital scale where opal accumulation covaries with other recognized monsoon indicators. In the orbital-scale case, the opal productivity reflects monsoon upwelling, whereas, on the Neogene time scale, the onset of opal deposition seems to reflect regional or global-scale processes. However, detailed records of opal deposition over the Neogene may still show a direct relationship with other monsoon indicators. The acquisition of such data is in progress.

Preliminary records of the relative abundance of both radiolarians and foraminifers show marked increases in abundance of monsoonal upwelling indices well after the initiation of opal accumulation (Figure 9). The relative abundance of G. bulloides in two size fractions [>150µm, this study; >125 µm, Kroon et al., 1991] is very low until about 8.5 Ma, when they increase to 20% or more of the fauna. Although some gaps and low abundances in the preliminary data are caused by poor preservation, G. bulloides seems to inhabit the western Arabian Sea continuously since about 8.5 Ma. Preliminary data for two groups of radiolarian species (Actinomma spp. and Collosphaera sp.) thought to be endemic to active upwelling zones show similar patterns of appearance in the Arabian Sea [Johnson and Nigrini, 1980; Nigrini, 1991; Nigrini and Caulet, 1992]. Although Collosphaera sp. exhibits some earlier occurrences, both groups appear near 8 Ma and indicate a stronger upwelling circulation at that time. The increase of the foraminiferal and radiolarian upwelling indicators occurs within about 0.5 m.y. of each other at 8.0-8.5 Ma. Higher resolution studies are underway to determine the exact timing of the different monsoonal indicators across this interval. Thus, the abundance of plankton species indicative of strong upwelling is not consistent with the inference that the onset of opal accumulation reflects high productivity associated with the onset of strong monsoonal upwelling during the late middle Miocene.

The terrestrial indicators of monsoonal influence also indicate a late Miocene onset of strong monsoonal circulation. One of the most striking records of terrestrial climate change is the $\delta^{13}C$ and $\delta^{18}O$ of pedogenic carbonates from the Siwalik Group in northern Pakistan [Figures 9,10; Quade et al., 1989]. The $\delta^{13}C$ data are interpreted to show a rapid ecological transition from C3-dominated vegetation (mostly trees and shrubs) to C4-dominated vegetation (mostly grasslands) during the late Miocene (about 7.4-7.0 Ma) [Quade et al., 1989]. The onset and dominance of C4 grasses is thought to represent high summer temperatures, highly seasonal summer rainfall, and the prevalence of fires.

Although most attention has been focused on the dramatic $\delta^{13}C$ data, the $\delta^{18}O$ show an equally impressive shift that slightly preceeds the $\delta^{13}C$ shift [Quade et al., 1989]. The shift from $\delta^{18}O$ values of -10°/oo to about -6 °/oo at about 8 Ma is consistent with the depletion expected for intense monsoonal precipitation [Cerling, 1984]. Over this same interval, the depth of soil leaching decreased from >1 m prior to 7 Ma to < 0.5 m after 7 Ma. This decrease in the depth of the leached zone also implies significant regional climate change and is consistent with a decrease and/or more seasonal precipitation. The combination of the $\delta^{18}O$ and $\delta^{13}C$ data indicates a shift from coastal carbonates to monsoonal carbonates [see Figures 4 and 5 in Cerling, 1984] and is much more persuasive of monsoonal onset than merely the shift from C3 to C4 vegetation. Quade et al. [1989] infer these changes represent the development of strong monsoonal circulation and local climate change.

Quade et al. [1989] also cite major turnovers in several mammal groups at about 7 Ma (related to the change from forests to grasslands) as evidence of major regional climate change at this time.

Recently the $\delta^{13}C$ shift associated with the transition from C3 to C4 vegetation has been identified in Africa and North America [T. E. Cerling, personal communication, 1992]. The global nature of this $\delta^{13}C$ shift is interpreted to reflect the global lowering of atmospheric CO_2, which tends to favor expansion of the C4 vegetation. Although this global shift may be correct, the combined isotopic, soil character, and faunal data from Pakistan still indicate a significant regional change in the environment, most likely related to the onset of strong monsoonal circulation. The potential coincidence between lowered atmospheric CO_2, increased monsoonal circulation, and rapid surface uplift leads to the interesting speculation that the increased surface uplift may act as a positive feedback to enhance erosion rates (rapid uplift and more precipitation) and thereby decrease atmospheric CO_2 by increased weathering reactions [Raymo et al., 1988; Raymo, 1991].

The regional record of terrestrial climate change from the Siwaliks of Pakistan is generally consistent with the marine records from the western Arabian Sea (Figure 9). The transition to C4 vegetation (~7.4 to 7 Ma) occurs somewhat later than the onset of major upwelling plankton (about 8 Ma), but the $\delta^{18}O$ shift, reflecting the shift from coastal to monsoonal precipitation, occurs at about the same time as the plankton indicators of monsoonal upwelling.

The rapid building of the Ganges and Indus deep sea fans is often cited to be an indicator of the uplift-elevation changes in the Himalayas that are associated with the onset of the monsoon [Cochran, 1990; Rea, this volume]. Monsoon precipitation and runoff might also accelerate the transport of sediments to the deep sea. The drilling results of DSDP Leg 23 [Whitmarsh, Weser, Ross et al., 1974] and more recently ODP Leg 116 [Cochran et al., 1990] clearly demonstrate that the Ganges-Indus Fan complex has been accumulating for at least 20 m.y. Even the most distal part of the Ganges Fan exhibits turbidite fan deposition back to at least 17-17.5 m.y. [Cochran, 1990]. Sediment delivery to the fans and their growth patterns are also closely tied to changes in sea level (Figure 10) [Cochran, 1990]. Although the transport of sediments to the fans may well be related to monsoon runoff, existing studies [Cochran, 1990] indicate that uplift-erosion rates and sea level changes are probably the dominant factors in fan growth and that the onset and development of the Ganges-Indus Fan complex represents the interplay of these two factors well before the onset of strong monsoonal circulation. The observation that the fan system was active long before the onset of the monsoon (Figure 10) seems consistent with the modelling results that high elevations (at least half modern) are a prerequisite for strong monsoonal circulation. Hence, much uplift and erosion could occur and the fans could develop before the mountain-plateau complex reached the elevation that would induce a strong

monsoonal response. Recently, Rea [this volume] has attempted to calculate the clastic flux to the northern Indian Ocean. He finds that the flux increases rapidly about 10 Ma and that clastic flux to the deep sea fans was extremely low prior to 10 Ma. Unfortunately, the accumulation rates for the fan sites are poorly constrained, eolian sediments are considered as fluvial, and numerous significant hiatuses are ignored. Hence, the clastic fluxes to the fans, especially prior to 10 Ma, may be quite different than those summarized by Rea [this volume].

In summary, although the timing of all the potential indicators of monsoonal onset does not agree in detail, the ODP Leg 117 results strongly indicate that the onset of strong monsoonal circulation (i.e., strong enough to induce intense upwelling) occurred in the late Miocene at about 8 Ma. The terrestrial response (at least in northern Pakistan) occurred over an interval from 8 to 7 Ma (Figures 9,10) [Quade et al., 1989].

SUMMARY AND CONCLUSIONS

The regional survey cruise (RC27-04) and the ODP Leg 117 drilling and research have greatly expanded our knowledge of the geologic record of the monsoon and its initiation and evolution in the late Neogene. As we have previously stressed, the monsoon is a strong local to regional climate response that is superimposed on other regional and global trends in climate and ocean circulation and chemistry. One of the major challenges in understanding the history of the monsoon system is to unravel its influence on the oceanography and sediments from other synchronous processes that are not directly related to the monsoon system. In spite of this complexity, we have learned a great deal about the western Arabian Sea and the monsoon system. Some of the major findings are:

1. The sediments of the Owen Ridge and the Oman margin are predominantly mixtures of pelagic carbonate and terrigenous components with smaller amounts of opal and organic carbon. Both areas provide relatively continuous sections of the late Neogene and record many aspects of monsoon-related sedimentation. Accumulation rates decrease rapidly offshore with the result that the sediments and accumulation rates of the deeper marginal basins (e.g., Site 728 at 1500 m) are more similar to the Owen Ridge than to the adjacent shallow margin sites.

2. Spectral analyses of detailed time series (from 300 k.y. to 1 m.y. long) of many sedimentary components from both margin and ridge sites have documented that the sedimentary record contains pervasive variability at the orbital (Milankovitch) scale over a wide range of time scales during the late Neogene. Much of this variability is coherent with either orbital variations (eccentricity, obliquity, or precession) or with glacial-interglacial changes, as recorded by $\delta^{18}O$. In general, components related to the terrigenous flux, bulk sedimentation rate, or global ocean chemistry changes are coherent and nearly in

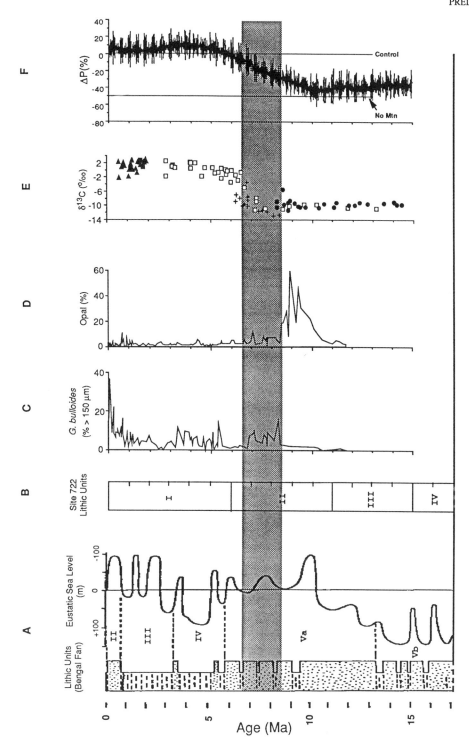

Fig. 10. Summary of various records (to 15 Ma) that are related, at least in part, to the onset and evolution of the monsoon. (A) Lithic units fo the Bengal Fan and their relation to eustatic sea level [from Cochran, 1990], (B) Lithic units from Owen Ridge (Site 722) [from Prell, Niitsuma et al., 1989], (C) The percent of upwelling species, *G. bulloides* in Site 722 (this study), (D) The % opal in Site 722 (this study), (E) The δ13C evidence for a shift from C3 to C4 vegetation in Pakistan [from Quade et al., 1989], and (F) A model estimate of one possible path of monsoon evolution [from Prell and Kutzbach, in press]. Although the Bengal-Indus fans have been building for at least 17 m.y., the coincidence of specific monsoon indicators and modes estimates indicates that either onset or intensification of the monsoon circulation occurred in the interval from about 8.5 Ma to 6.5 Ma (shaded area).

phase with $\delta^{18}O$, especially at periods of 100 k.y. and 41 k.y. A suite of monsoon-related indicators (including *G. bulloides*, opal, eolian grain size, foraminiferal fragmentation index) are all coherent and in phase with each other at the 1/23 k.y. frequency band (precession) but lag orbital precession by about 9 k.y. (Figure 7A). The coincidence of these different indicators of monsoon upwelling lends credence to the variability of the monsoon at this periodicity. However, understanding the productivity changes at the 100 k.y. and 41 k.y. periods still requires a better distinction between the hypotheses of direct wind-induced upwelling and nutrient partitioning between the intermediate and upper waters. In addition, the possibility that changes in wind stress gradients can cause phase differences between the margin and ridge records needs to be considered in the interpretation of the differences between similar upwelling indicators from the proximal and distal regions.

3. Among the longer term trends in the Arabian Sea sediments, one of the more interesting is the shift (at about 2.5 Ma) in variance from frequencies of predominantly 1/23 k.y. to the combination of 1/41 k.y. and 1/23 k.y. The shift is observed in variables that reflect the terrigenous flux to the Arabian Sea (magnetic susceptibility, non-carbonate content, wet-bulk density, and optical density). The emergence of the 41 k.y. periodicity has been interpreted to indicate changes in the aridity and sediment availability associated with the rapid expansion of continental ice sheets at that time. In this interpretation, the 23 k.y. variability would represent the aridity cycles associated with monsoonal variability and should, therefore, be dominant prior to the existence of large ice sheets in the Northern Hemisphere. This assumption infers that the phase of the 23 k.y. variability should change with respect to ice volume and needs to be tested. Hence, the emergence of the 41 k.y. variability represents a new process in the evolution of the tropical climates and in the interpretation of monsoon evolution.

4. On the basis of endemic upwelling species, the onset of intense monsoonal upwelling in the western Arabian Sea occurred in the late Miocene about 8 m.y. ago, distinctly after the onset of high opal accumulation. The appearance of opal in the Arabian Sea sediments, which occurs at about 11 Ma, is attributed to ocean-wide or global changes in circulation rather than directly to the monsoon. Terrestrial indicators of climate change in Pakistan also show large changes consistent with the onset of strong monsoon circulation between 8 Ma and 7 Ma.

5. Active uplift and erosion of Himalaya-Tibet and deposition of the Ganges-Indus submarine fan complexes have been ongoing since at least 17 Ma, and began well before the monsoon circulation was established, and have continued during the monsoon evolution (Figure 10). Thus, mountain chains and plateaus existed during the past 20 m.y. but they were not large (or high) enough to induce

the strong monsoonal circulation. Modelling studies suggest that the mountain-plateau elevations must be at least half the modern elevation and have strong solar forcing to induce monsoons comparable to the modern circulation. Some evidence [this volume] exists for increased clastic flux from the Himalaya-Tibet at about 10 Ma, which may represent the rapid uplift phase necessary to produce the elevations required for initiation of monsoon circulation. However, as Molnar and England [1990] have recently summarized, the Neogene is also a period of significant climate change, and separation of tectonic "uplift" and climate "glacial cooling" effects is often difficult.

The results of Leg 117 research have shown that the geologic and oceanographic record of the monsoon circulation is complex and that not all potential monsoonal indicators respond at the same time and with the same amplitude. No single indicator is uniquely monsoonal, and thus the use of multiple indicators is necessary to separate the other climatic and geologic processes from those associated with the monsoon circulation. The coincidence of monsoonal indicators that individually have differing sources, processes, compositions, and preservational characteristics gives confidence that the monsoonal circulation is responsible for the geologic record. One approach to address these questions is to identify patterns of monsoon circulation and climate change through model experiments that can be attributed to one specific mechanism and to compare these patterns with geologic data sets of monsoonal indicators. This strategy of model-data comparison can identify areas of agreement and disagreement and focus our attention on those those elusive aspects of the monsoon that still evade our understanding.

Acknowledgments. We thank the JOIDES committee structure and the Indian Ocean Panel for their support in planning Leg 117 and the SEDCO and ODP crews for making Leg 117 a reality. This research has been funded by NSF grants OCE-8511571 to W. Prell, OCE-9103353 to W. Prell and S. Clemens, OCE-8911874 to D. Murray, a JOI/USSAC Ocean Drilling Fellowship to S. Clemens and post cruise USSAC grants to Anderson, Murray, and Prell. Support for curating the RC27-04 cores at LDGO is from NSF grant OCE-91-01689 and ONR Grant N00014-90-J-1060. We greatfully acknowledge J. Donnelly, P. Howell, A. Martin, and L. Sheehan for their assistance in compiling and checking this manuscript. Lastly, we thank N. Pisias and T. Hagelberg for their prompt reviews, J. Ortiz for proof-reading, and Bob Duncan for his patience.

REFERENCES

Amano, K., and Taira, A., Two phase uplift of the high Himalayas since 17 MA, *Geology, 20*, 391-394, 1992.

Anderson, D. M., Foraminifer evidence of monsoon upwelling off Oman during the Late Quaternary, Ph.D. thesis, Brown University, Providence, RI, 1990.

Anderson, D. M., and Prell, W. L., A new 300 k.y. record of upwelling off Oman during the late Quaternary: Evidence of the

Asian southwest monsoon, *Paleoceanography*, in press, 1992a.

Anderson, D. M., and Prell, W. L., The structure of the southwest monsoon winds over the Arabian Sea during the late Quaternary: Observations, simulations, and marine geologic evidence, *J. Geophys. Res. Oceans*, in press, 1992b.

Anderson, D. M., and Prell, W. L., Coastal upwelling gradient during the late Pleistocene, in *Proc. ODP, Sci., Results, 117*, edited by W. L. Prell, N. Niitsuma et al., pp. 265-276, Ocean Drilling Program, College Station, 1991.

Anderson, D. M., Brock, J. C., and Prell, W. L., Physical upwelling processes, upper ocean environment, and the sediment record of the southwest monsoon, in *Evolution of Upwelling Systems since the Early Miocene, Geological Society Special Publication No. 62*, edited by C. Summerhayes, W.L. Prell, and K.-C. Emeis, Blackwell Scientific Publications, London, 1992.

Arthur, M. A., Miller, K., and Crowley, T., Global episodes of moderate to extreme warmth, An Earth system history initiative of the global change research program, in *Advisory Panel Report on Earth System History*, edited by G. S. Mountain and M. E. Katz, 51-74, National Science Foundation, Division of Ocean Sciences, Washington, D.C., 1991.

Barnett, T. P., Dumenil, L., Schlese, U., and Roeckner, E., The effect of Eurasian snow cover on global climate, *Science, 239*, 504-507, 1988.

Bigg, G. R., Jiang, D., and Mitchell, J. F. B., A general circulation model of the Indian Ocean at 9000 years B.P., *Paleoceanography, 7*, 119-135, 1992.

Bloemendal, J., and deMenocal, P., Evidence for a shift in the climatic variability of the African and Asian monsoons at 2.5 Ma: An application of whole-core magnetic susceptibility measurements to paleoclimatology, *Nature, 342*, 897-900, 1989.

Boje, R., and Tomczak, M., editors, *Upwelling Ecosystems*, 303 pp., Springer-Verlag, Berlin, Heidelberg, New York, 1978.

Boyle, E. A., The role of vertical chemical fractionation in controlling late Quaternary atmospheric carbon dioxide, *J. Geophys. Res., 93*, 15,701-15,714, 1988a.

Boyle, E. A., Cadmium: Chemical tracer of deep water paleoceanography, *Paleoceanography, 3*, 471-489, 1988b.

Brock, J. C., McClain, C. R., Luther, M. E., and Hay, W. W., The phytoplankton bloom in the northwestern Arabian Sea during the southwest monsoon of 1979, *J. Geophys. Res., 96*, 20,623-20,642, 1991.

Brock, J. C., McClain, C. R., Anderson, D. M., Prell, W. L., and Hay, W. W., Southwest monsoon circulation and environments of Recent planktonic foraminifera in the northwestern Arabian Sea, *Paleoceanography*, in press, 1992.

Brock, J. C., McClain, C. R., and Hay, W. W., A southwest monsoon hydrographic climatology for the northwestern Arabian Sea, *Nature, 97*, 9455-9465, 1992.

Bruce, J. G., Some details of upwelling off the Somali and Arabian coasts, *J. Mar. Res., 32*, 419-423, 1974.

Burckle, L. H., Distribution of diatoms in sediments of the northern Indian Ocean, *Mar. Micropaleontology, 15*, 53-65, 1989.

Busch, W. H., Analysis of wet bulk density and sediment color cycles in Pliocene-Pleistocene sediments of the Owen Ridge (Site 722) and Oman Margin (Site 728), in *Proc. ODP, Sci. Results, 117*, edited by W. L. Prell, N. Niitsuma et al., pp. 239-253, Ocean Drilling Program, College Station, 1991.

Caulet, J. P., Venec-Peyre, M-T., Vernaud-Grazzini, C., and Nigrini, C., Variation of south Somalian upwelling during the last 160 KYR: Radiolarian and foraminifera records in Core MD 85674, in *Evolution of Upwelling Systems since the Early Miocene, Geological Society Special Publication No. 62*, edited by C. Summerhayes, W. L. Prell, and K.-C. Emeis, Blackwell Scientific Publications, London, 1992.

Cerling, T. E., The stable isotopic composition of modern soil carbonate and its relationship to climate, *Earth Planet. Sci. Lett., 71*, 229-240, 1984.

Clemens, S. C., Quaternary variability of Indian Ocean monsoon winds and climate, Ph.D. Thesis, Brown University, Providence, RI, 1990.

Clemens, S. C., and Oglesby, R. J., Role of interhemispheric moisture transport in the Indian Ocean summer monsoon: Data-model and model-model comparison, *Paleoceanography*, in press, 1992.

Clemens, S. C., and Prell, W. L., Late Quaternary forcing of Indian Ocean summer-monsoon winds: A comparison of Fourier model and general circulation model results, *J. Geophys. Res., 96*, 22,683-22,700, 1991a.

Clemens, S. C., and Prell, W. L., One-million year record of summer-monsoon winds and continental aridity from the Owen Ridge (Site 722), northwest Arabian Sea, in *Proc. ODP, Sci. Results, 117*, edited by W.L. Prell, N. Niitsuma et al., pp. 365-388, Ocean Drilling Program, College Station, TX, 1991b.

Clemens, S. C., and Prell, W. L., Late Pleistocene variability of Arabian Sea summer-monsoon winds and dust source-area aridity: Eolian record from the lithogenic component of deep-sea sediments, *Paleoceanography, 5*, 109-145, 1990.

Clemens, S., Prell, W., Murray, D., Shimmield, G., and Weedon, G., Forcing mechanisms of the Indian Ocean monsoon, *Nature, 353*, 720-725, 1991.

CLIMAP Project Members, Seasonal reconstruction of the Earth's surface at the last glacial maximum, *Geological Society of America Map and Chart Series, 36*, 21 p., 1981.

Cochran, J. R., Himalayan uplift, sea level, and the record of Bengal Fan sedimentation at the ODP Leg 116 sites, in *Proc. ODP, Sci. Results, 116*, edited by Cochran, J. R., J. R. Curray, W. W. Sager, and D. A. V. Stow, pp. 397-414, Ocean Drilling Program, College Station, TX, 1990.

Cochran, J. R., J. R. Curray, W. W. Sager, and D. A. V. Stow, editors, *Proc. ODP, Sci. Results, 116*, 445 p., Ocean Drilling Program, College Station, TX, 1990.

Cullen, J. L., and Prell, W. L., Planktonic foraminifera of the Northern Indian Ocean: distribution and preservation in surface sediments, *Mar. Micropaleontology, 9*, 1-52, 1984.

Currie, R. I., Fisher, A. E., and Hargreaves, P. M., Arabian Sea Upwelling, in *The Biology of the Indian Ocean*, edited by B. Zeitzschel, 37-52, Springer, New York, 1973.

Curry, W. B., Ostermann, D. R., Guptha, M. V. S., and Ittekkot, V., Foraminiferal production and monsoonal upwelling in the Arabian Sea: Evidence from sediment traps, in *Evolution of Upwelling Systems since the Early Miocene, Geological Society Special Publication No. 62*, edited by C. Summerhayes, W. L. Prell, and K.-C. Emeis, Geological Society, London, 1992.

Davies, T. A., and Worsley, T. R., Paleoenvironmental implications of oceanic carbonate sedimentation rates, *Society of Economic Paleontologists and Mineralogists Special Publication 32*, 169-179, 1981.

Debrabant, P., Krissek, L., Bouquillon, A., and Chamley, H., Clay mineralogy of Neogene sediments of the western Arabian Sea: mineral abundances and paleoenvironmental implications, in *Proc. ODP, Sci. Results, 117*, edited by W.L. Prell, N. Niitsuma et al., pp. 183-196, Ocean Drilling Program, College Station, TX, 1991.

Delaney, M. L., and Boyle, E. A., Tertiary paleoceanic chemical variability: Unintended consequences of simple geochemical models, *Paleoceanography, 3*, 137-156, 1988.

deMenocal, P., Bloemendal, J., and King, J., A rock-magnetic record of monsoonal dust deposition to the Arabian Sea: Evidence for a shift in the mode of deposition at 2.4 Ma, in *Proc. ODP, Sci. Results, 117*, edited by W. L. Prell, N. Niitsuma et al., pp. 389-408, Ocean Drilling Program, College Station, TX, 1991.

Hahn, D. G., and Manabe, S., The role of mountains in the South Asian monsoon circulation, *J. Atmos. Sci., 32*, 1515-1541, 1975.

Haq, B. U., Hardenbol, J., and Vail, P. R., Chronology of fluctuating sea levels since the Triassic, *Science, 235*, 1156-1166, 1987.

Harrison, T. M., Copeland, P., Kidd, W. S. F., and An Yin, Raising Tibet, *Science, 255*, 1663-1670, 1992.

Hastenrath, S., *Climate and the Circulation of the Tropics*, D. Reidel, Boston, 1985.

Hastenrath, S., and Greischar, L. L., *Climatic Atlas of the Indian Ocean, Part III: Upper-Ocean Structure*, University of Wisconsin Press, Madison, Wisconsin, 1989.

Hastenrath, S. and Lamb, P. J., *Climatic Atlas of the Indian Ocean, Part I: Surface Climate and Atmospheric Circulation*, University of Wisconsin Press, Madison, Wisconsin, 1979.

Imbrie, J., et al., The orbital theory of Pleistocene climate: Support from a revised chronology of the marine delta $\delta^{18}O$ record, in *Milankovitch and Climate, 1*, edited by A. Berger, J. Imbrie, J. Hays, G. Kukla, and B. Salzman, pp. 269-305, Reidel, Dordrecht, 1984.

Johnson, D. A., Radiolarian biostratigraphy in the central Indian Ocean, Leg 115, in *Proc. ODP, Sci. Results, 115*, edited by R. A. Duncan, J. Backman, L. C. Peterson et al., pp. 395-409, Ocean Drilling Program, College Station, TX , 1990.

Johnson, D. A., and Nigrini, C., Radiolarian biogeography in surface sediments of the western Indian Ocean, *Mar. Micropaleontology, 5*, 111-152, 1980.

Kallel, N., Labeyrie, L. D., Juillet-Leclerc, A., and Duplessy, J.-C., A deep hydrological front between intermediate and deep-water masses in the glacial Indian Ocean, *Nature, 333*, 651-655, 1988.

Khule, M., Subtropical mountain- and highland-glaciation as ice age triggers and the waning of the glacial periods in the Pleistocene, *GeoJournal, 14*, 393-421, 1987.

Krey, J., and Babenerd, B., *Phytoplankton Production Atlas of the International Indian Ocean Expedition*, Landesvermessungsamt Schleswig-Holstein, Keil, 1976.

Krissek, L. A., and Clemens, S. C., Mineralogic variations in a Pleistocene high-resolution eolian record from the Owen Ridge, western Arabian Sea (Site 722): implications for sediment source conditions and monsoon history, in *Proc. of the ODP, Scientific Results, 117*, edited by W. L. Prell, N. Niitsuma et al., pp. 197-213, Ocean Drilling Program, College Station, TX, 1991.

Kroon, D., Distribution of extant planktic foraminiferal assemblages in the Red Sea and northern Indian Ocean surface waters, in *Planktonic foraminifers as tracers of ocean-climate history*, edited by G. J. A. Brummer and D. Kroon, 229-267, VU Boekhandel/Uitgeverij, Amsterdam, 1988.

Kroon, D., Steens, T. N. F., and Troelstra, S. R., Onset of monsoonal related upwelling in the western Arabian Sea, in *Proc. ODP, Sci. Results, 117*, edited by W. L. Prell, N. Niitsuma et al., pp. 257-264, Ocean Drilling Program, College Station, TX, 1991.

Kutzbach, J. E., Monsoon climate of the early Holocene: Climate experiment with the Earth's orbital parameters for 9000 years ago, *Science, 214*, 59-61, 1981.

Kutzbach, J. E., and Guetter, P. J., The influence of changing orbital parameters and surface boundary conditions on climate simulations for the past 18,000 years, *J. Atmos. Science, 43*, 1726-1759, 1986.

Kutzbach, J. E., Guetter, P. J., Ruddiman, W. F., and Prell, W. L., The sensitivity of climate to Late Cenozoic uplift in southeast Asia and the American Southwest: Numerical experiments, *J. Geophys. Res., Atmospheres, 94*, 18,393-18,407, 1989.

Kutzbach, J. E., and Street-Perrott, F. A., Milankovitch forcing of fluctuations in the level of tropical lakes from 18 to 0 kyr BP, *Nature, 6033*, 130-134, 1985.

Kutzbach, J. E., Prell, W. L., and Ruddiman, W. F., Sensitivity of Eurasian climate to surface uplift of the Tibetan Plateau, *J. Geol.*, in press, 1992.

Luther, M. E., and O'Brien, J. J., A model of the seasonal circulation in the Arabian Sea forced by observed winds, *Prog. Oceanog., 14*, 353-385, 1985.

Luther, M. E., O'Brien, J. J., and Prell, W. L., Variability in upwelling fields in the northwestern Indian Ocean, 1, Model experiments for the past 18,000 years, *Paleoceanography, 5*, 433-445, 1990.

Mercier, J.-T., Armijo, R. Tapponnier, P. Carey-Gailhardis, E., and Lin, H. T., Change from late Tertiary compression to Quaternary extension in Southern Tibet during the India-Asia collision, *Tectonics, 6*, 275-304, 1987.

Mikkelsen, N., Cenozoic diatom biostratigraphy and paleoceanography of the western equatorial Indian Ocean, in *Proc. ODP, Sci. Results, 115*, edited by R. A. Duncan, J. Backman, L. C. Peterson et al., pp. 411-432, Ocean Drilling Program, College Station, TX, 1990.

Molnar, P., and England, P., Late Cenozoic uplift of mountain ranges and global climate change: chicken or egg? *Nature, 346*, 29-34, 1990.

Mountain, G., and Prell, W. L., Leg 117 ODP site survey: a revised history of Owen Basin, *EOS, Trans. Am. Geophys. Un., 68*, 424, 1987.

Murray, D. W., and Prell, W. L., Late Pliocene and Pleistocene climatic oscillations and monsoon upwelling recorded in sediments from the Owen Ridge, northern Arabian Sea, in *Evolution of Upwelling Systems since the Early Miocene, Geological Society Special Publication No. 62*, edited by C. Summerhayes, W. L. Prell, and K.-C. Emeis, Geological Society, London, 1992.

Murray, D. W., and Prell, W. L., Pliocene to Pleistocene variations in calcium carbonate, organic carbon, and opal on the Owen Ridge, Northern Arabian Sea, in *Proc. ODP, Sci. Results, 117*, edited by W. L. Prell, N. Niitsuma et al., p. 343-355, Ocean Drilling Program, College Station, TX, 1991.

Nair, R. R., Ittekkot, S.J. Manganini, S. J., Ramaswamy, V., Haake, B., Degens, E. T., Desai, B. N., and Honjo, S., Increased particle flux to the deep ocean related to monsoons, *Nature, 338*, 749-751, 1989.

Nigrini, C., Composition and biostratigraphy of radiolarian assemblages from an area of upwelling (northwestern Arabian Sea, Leg 117), in *Proc. ODP, Sci. Results, 117*, edited by W. L. Prell, N. Niitsuma et al., pp. 89-126, Ocean Drilling Program, College Station, TX, 1991.

Nigrini, C., and Caulet, J. P., Late Neogene radiolarian assemblages characteristic of Indo-Pacific areas of upwelling, *Micropaleontology, 38*, 139-164, 1992.

Oglesby, R. J., and Saltzman, B., Sensitivity of the equilibrium surface temperature of a GCM to systematic changes in atmospheric carbon dioxide, *Geophys. Res. Lett., 17*, 1089-1092, 1990.

Oglesby, R. J., and Saltzman, B., Equilibrium climate statistics of a general circulation model as a function of atmospheric carbon dioxide, Part I: Geographic distributions of primary variables, *J. Climate, 5*, 66-92, 1992.

Olson, D. B., Hitchcock, G. L., Fine, R. A., and Warren, B. A., Maintenance of the low-oxygen layer in the central Arabian Sea, *Deep-Sea Res.*, in press, 1992.

Prell, W. L., Variation of monsoonal upwelling: A response to changing solar radiation, in *Climatic Processes and Climate Sensitivity, 29*, edited by J. E. Hansen and T. Takahashi, pp. 48-57, American Geophysical Union Geophys. Monogr. Ser., Washington, D.C., 1984a.

Prell, W. L., Monsoonal climate of the Arabian Sea during the late Quaternary: A response to changing solar radiation, in *Milankovitch and Climate, 1*, edited by A. Berger, J. Imbrie, J. Hays, G. Kukla and B. Saltzman, pp. 349-366, D. Reidel, Hingham, 1984b.

Prell, W. L., and Curry, W. E., Faunal and isotopic indices of monsoonal upwelling: Western Arabian Sea, *Oceanologica Acta, 4*, 91-98, 1981.

Prell, W. L., and Kutzbach, J. E., Monsoon variability over the past 150,000 years, *J. Geophys. Res., 92*, 8411-8425, 1987.

Prell, W. L., and Kutzbach, J. E., Sensitivity of the Indian monsoon to changes in orbital parameters, glacial and tectonic boundary conditions and atmospheric CO_2 concentration, *Nature*, in press, 1992.

Prell, W. L., Marvil, R. E. Luther, M. E., and O'Brien, J. J., Variability in upwelling fields in the northwestern Indian Ocean, part 2: Data-model comparison at 9,000 years B.P., *Paleoceanography, 5*, 447-457, 1990.

Prell, W. L., and Niitsuma, N., editors, *Proc. ODP, Init. Repts., 117*,1236 pp.,Ocean Drilling Program,College Station,TX,1989.

Prell, W. L., Niitsuma N. et al., editors, *Proc. ODP, Scientific Results, 117*, 638 pp., Ocean Drilling Program, College Station, TX, 1991.

Prell, W. L., and Streeter, H. F., Temporal and spatial patterns of monsoonal upwelling along Arabia: A modern analogue for the interpretation of Quaternary SST anomalies, *J. Mar. Res., 40*, 143-155, 1982.

Quade, J., Cerling, T. E., and Bowman, J. R., Development of the Asian monsoon revealed by marked ecological shift during the latest Miocene in northern Pakistan, *Nature, 342*, 163-166, 1989.

Rao, R. R., Molinare, R. L., and Festa, J. F., Evolution of the climatological near-surface thermal structure of the tropical Indian Ocean, 1, Description of mean monthly mixed layer depth, and sea surface temperature, surface current, and surface meteorological fields, *J. Geophys. Res., 94*, 10,801-10,815, 1989.

Raymo, M. E., Geochemical evidence supporting T. C. Champerlin's theory of glaciation,*Geology,19*,344-347,1991.

Raymo, M., and Molnar, P., Himalayan and Tibetan Plateau uplift, II, in *AGU 1991 Fall Meeting Program and Abstracts*, supplement to EOS, American Geophysical Union, Washington, D.C., 1991.

Raymo, M. E., Ruddiman, W. F., and Froelich, P. N., Influence of the late Cenozoic mountain building on ocean geochemical cycles, *Geology, 16*, 649-653, 1988.

Rea, D. K., and Bloomstine, M. K., Neogene history of South Pacific tradewinds: Evidence for hemispheric asymmetry of atmospheric circulation, *Palaeogeog., Palaeoclimatol., Palaeoecol., 55*, 55-64, 1986.

Ruddiman, W. F., and Kutzbach, J. E., Late Cenozoic plateau uplift and climate change, *Transactions of the Royal Society of Edinburgh: Earth Sciences, 81*, 301-314, 1990.

Ruddiman, W. F., and Kutzbach, J. E., Forcing of Late Cenozoic northern hemisphere climate by plateau uplift in southern Asia and the American West, *J. Geophys. Res., 94*, 18,409-18,427, 1989.

Ruddiman, W. F., Prell, W. L., and Raymo, M. E., History of Late Cenozoic uplift in southeast Asia and the American Southwest: Rationale for general circulation experiments, *J. Geophys. Res., Atmospheres, 94*, 18,379-18,391, 1989.

Shimmield, G. B., Can sediment geochemistry record changes in coastal upwelling paleoproductivity? Evidence form northwest Africa and the Arabian Sea, in *Evolution of Upwelling Systems since the Early Miocene, Geological Society Special Publication No. 62*, edited by C. Summerhayes, W. L. Prell, and K.-C. Emeis, Blackwell Scientific Publications, London, 1992.

Shimmield, G. B., and Mowbray, S. R., The inorganic geochemical record of the northwestern Arabian Sea: A history of productivity variation over the last 400 k.y. from Sites 722 and 724, in *Proc. ODP, Sci. Results, 117*, edited by W.L. Prell, N. Niitsuma et al., pp. 409-429, Ocean Drilling Program, College Station, TX, 1991.

Shimmield, G. B., Mowbray, S. R., and Weedon, G. P., A 350 ka history of the Indian Southwest Monsoon-evidence from deep-sea cores, northwest Arabian Sea, *Transactions of the Royal Society of Edinburgh: Earth Sciences, 81*, 289-299, 1990.

Shukla, J., Interannual variability of monsoons, in *Monsoons*, edited by J. S. Fein and P. L. Stephens, pp. 399-463, John Wiley & Sons, New York, 1987.

Sirocko, F., and Lange, H., Clay-mineral accumulation rates in the Arabian Sea during the late Quaternary, *Mar. Geol., 97*, 105-119, 1991.

Sirocko, F., and Sarnthein, M., Wind-borne deposits in the northwestern Indian Ocean: A record of Holocene sediments versus modern satellite data, *Palaeoclimatology and Palaeometerology, NATO Advance Res. Workshop*, 401-434, 1989.

Sirocko, F., Sarnthein, M., Lange, H., and Erlenkeuser, H., Atmospheric summer circulation and coastal upwelling in the Arabian Sea during the Holocene and last glaciation, *Quat. Res., 36*, 72-93, 1991.

Slater, R. D., and Kroopnick, P., Controls of dissolved oxygen distribution and organic carbon deposition in the Arabian Sea, in *Marine Geology and Oceanography of Arabian Sea and Coastal Pakistan*, edited by B. U. Haq and J. D. Milliman, pp. 305-313, Van Nostrand Reinhold, New York, 1984.

Smith, R. L., and Bottero, J. S., On upwelling in the Arabian Sea, in *A Voyage of Discovery*, edited by M. Angel, Permagon Press, New York,1977.

Spencer, D., Broecker, W. S., Craig, H., and Weiss, R. F., *GEOSECS Indian Ocean Expedition, 6, Section and Profiles*, 140 pp., U.S. Government Printing Office, Washington, D. C., 1982.

Steens, T. N. F., Kroon, D., tenKate, W. G., and Sprenger, A., Late Pleistocene periodicities of oxygen isotope ratios, calcium carbonate contents, and magnetic susceptibilities of western Arabian Sea Margin Hole 728A, in *Proc. ODP, Sci. Results, 117*, edited by W. L. Prell, N. Niitsuma et al., pp. 309-320, Ocean Drilling Program, College Station, TX, 1991.

Steens, T. N. F., Ganssen, G., and Kroon, D., Carbon isotopes in planktonic foraminifera: A tool to discriminate between upwelling- and river-induced high productivity, in *Evolution of Upwelling Systems since the Early Miocene, Geological Society Special Publication No. 62*, edited by C. Summerhayes, W. L. Prell, and K.-C. Emeis, Blackwell Scientific Publications, London, 1992.

Street-Perrott, F. A., Mitchell, J. F. B., Marchand, D. S., and Brunner, J. S., Milankovitch and albedo forcing of the tropical monsoons: A comparison of geological evidence and numerical simulations for 9000 yBP, *Transactions of the Royal Society of Edinburgh: Earth Sciences, 81*, 407-427, 1990.

Suess, E., and Thiede, J., editors, *Coastal Upwelling, Its Sediment Record, Part A: Responses of the Sedimentary Regime to Present Coastal Upwelling*, 604 pp., NATO Scientific Affairs Division, Plenum Press, New York, 1983.

Summerhayes, C. P., Prell, W. L., and Emeis, K.-C., editors, *Evolution of Upwelling Systems since the Early Miocene, Geological Society Special Publication No. 62*, Blackwell Scientific Publications, London, 1992.

Thiede, J. and Suess, E., editors, *Coastal Upwelling, Its Sediment Record, Part B: Sedimentary Records of Ancient Coastal Upwelling*, 610 pp., NATO Scientific Affairs Division, Plenum Press, New York, 1983.

Warren, B. A., Medieval Arab references to the seasonally reversing currents of the North Indian Ocean, *Deep-Sea Research, 13*, 167-171, 1966.

Webster, P. J., The elementary monsoon, in *Monsoons*, edited by J. S. Fein, and P. L. Stephens, Wiley, New York, pp. 3-32, 1987.

Weedon, G. P., and Shimmield, G. B., Late Pleistocene upwelling and productivity variations in the northwest Indian Ocean deduced from spectral analyses of geochemical data from Sites 722 and 724, in *Proc. ODP, Sci, Results, 117*, edited by W. L. Prell, N. Niitsuma et al., pp. 431-443, Ocean Drilling Program, College Station, TX, 1991.

Whitmarsh, R. B., Weser, O. E., Ross, D. A. et al., *Init. Repts. DSDP, 23*, U.S. Government Printing Office, Washington, D.C., 1974.

Woodruff, F., and Savin, S. M., Miocene deepwater oceanography, *Paleoceanography, 4*, 87-140, 1989.

Wyrtki, K., *Oceanographic Atlas of the International Indian Ocean Expedition*, 531 pp., National Science Foundation, Washington, D.C., 1971.

Wyrtki, K., Physical oceanography of the Indian Ocean, in *The Biology of the Indian Ocean*, pp. 18-36, edited by B. Zeitschel, Springer-Verlag, New York, Heidelberg, Berlin, 1973.

Appendix 1

Indian Ocean Plate Reconstructions Since the Late Jurassic

the Alliance Exotique:

J. Y. ROYER[1], J. G. SCLATER[2], and D. T. SANDWELL[2]

Institute for Geophysics, The University of Texas at Austin, 8701 Mopac Blvd, Austin, TX, 78759-8345, USA

S. C. CANDE[2]

Lamont Doherty Geological Observatory, Palisades, NY, 10964, USA

R. SCHLICH, M. MUNSCHY, and J. DYMENT

Ecole et Observatoire de Physique du Globe de Strasbourg, 5, rue René Descartes, 67084 Strasbourg Cedex, France

R. L. FISHER, and R. D.MÜLLER

Geological Research Division, Scripps Institution of Oceanography, La Jolla, CA, 92093, USA

M. F. COFFIN[3]

Bureau of Mineral Resources, Geology and Geophysics, GPO Box 378, Canberra ACT 2601, Australia

P. PATRIAT

Institut de Physique du Globe de Paris, 4 place Jussieu, 75252 Paris Cedex 05, France

H. W. BERGH

Bernard Price Institute of Geophysical Research, University of Witwatersrand, Johannesburg 2001, South Africa

[1]*Now at: Laboratoire de Géodynamique sous-marine, Observatoire Océanologique de Villefranche, BP 48, 06230 Villefranche/mer, France*
[2]*Now at: Geological Research Division, Scripps Institution of Oceanography, La Jolla, CA, 92093, USA*
[3]*Now at: Institute for Geophysics, The University of Texas at Austin, 8701 Mopac Blvd, Austin, TX, 78759-8345, USA*

The evolution of the Indian Ocean (Figure 1) since the Late Jurassic can be summarized in three main periods of seafloor spreading separated by two major plate boundary reorganizations [Royer, 1992]. From breakup of Gondwana at about 160 Ma (Figure 2) to the Late Cretaceous (~90/95 Ma), the opening of the Indian Ocean coincided with divergence of two superblocks, Africa-

Synthesis of Results from Scientific
Drilling in the Indian Ocean
Geophysical Monograph 70
Copyright 1992 American Geophysical Union

South America and Madagascar-Seychelles-India-Australia-Antarctica, which further fragmented in the Early Cretaceous (~chron M0; Figure 3) into Africa-Madagascar-India and Australia-Antarctica. This two-plate system broke into a five-plate system during the Late Cretaceous with seafloor spreading between Madagascar and India-Seychelles and between Australia and Antarctica (Figures 4 to 8). The second major plate boundary reorganization occurred during the middle Eocene and coincided with the collision of India with Eurasia (Figures 8 and 9). The resulting plate boundary system has prevailed until present

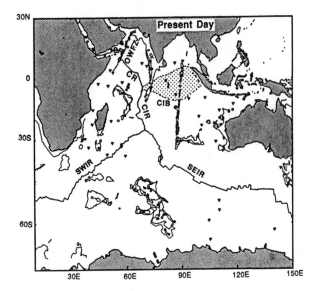

Fig. 1. Present-day chart of the Indian Ocean. The Indian Ocean is bordered by 5 plates: Africa, Arabia, India, Australia, and Antarctica, and comprises 5 plate boundaries: the Southwest Indian Ridge (SWIR), Sheba Ridge (SR), Owen FZ-Murray Ridge complex (OWFZ), Carlsberg Ridge (CR), Central Indian Ridge (CIR), diffuse plate boundary (stippled area) within the Central Indian Basin (CIB), and Southeast Indian Ridge (SEIR). In this figure and the following ones, drilling sites cored during the DSDP (1972-1974) and ODP (1987-1989) programs are indicated by triangles and circles, respectively. Major submarine ridges and plateaus are outlined by selected isobaths (x1000m).

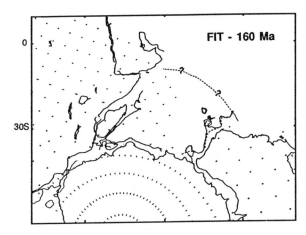

Fig. 2. Reconstruction of the Gondwana super-continent at 160 Ma after Lawver and Scotese [1987] in the African paleomagnetic reference frame [Besse and Courtillot, 1991].

(Figures 8 to 12, and 1). The final events of this last phase were opening of the Gulf of Aden at about Chron 5 (10 Ma) and onset of the intraplate deformation in the Central Indian Basin between India and Australia (~7 Ma). Evolution of the plate geometry is summarized in Figure 13.

The reconstructions (Figures 2 to 12) illustrating opening of the Indian Ocean since the Late Jurassic were presented

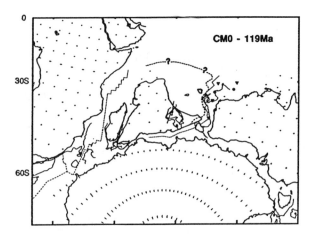

Fig. 3. Tentative reconstruction at chron M0 (118.7 Ma). Seafloor spreading has stopped in the Somali Basin between Africa and Madagascar while Greater India (outlined by a long-dashed line) is separating from the Australian-Antarctic block. Short-dashed lines show the location of the paleo-ridge axes.

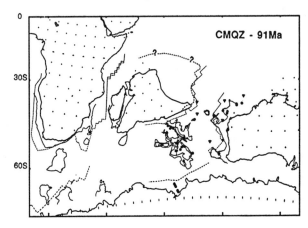

Fig. 4. Tentative reconstruction at 91 Ma during the Cretaceous magnetic quiet period (interpolated between the M0 and C34 reconstructions). This is the time of the first major plate boundary reorganization where the two major plates in Figure 3 broke into a five-plate system. Short-dashed lines show the M0 isochron.

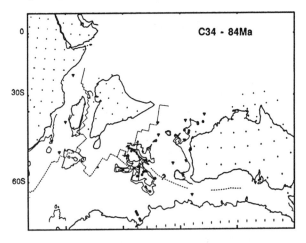

Fig. 5. Reconstruction at Chron 34 (84.0 Ma). Short-dashed lines show paleo-ridge axes.

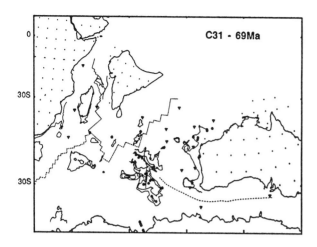

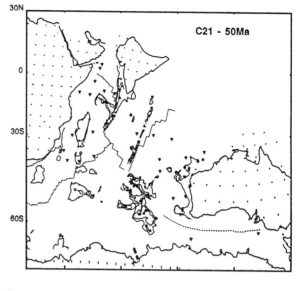

Fig. 6. Reconstruction at Chron 31 (68.5 Ma). The Seychelles micro-continent was progressively transferred from the Indian plate to the African plate. The ridge jumped from the Mascarene basin (south of Seychelles) to the Arabian Sea, coinciding with emplacement of the Deccan Traps (~66 Ma). At Chron 31, India started its rapid northward drift towards Eurasia.

Fig. 8. At Chron 21 (50.3 Ma) the second major plate boundary reorganization results from collision of the Indian plate with Eurasia.

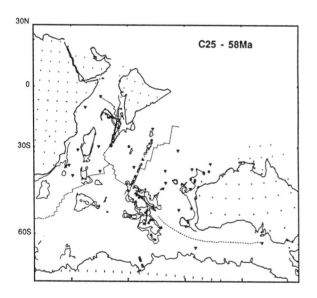

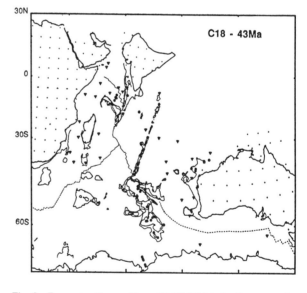

Fig. 7. Reconstruction at Chron 25 (58.2 Ma). The motion of the Indian plate relative to the other Gondwana fragments started to slow.

Fig. 9. Reconstruction at Chron 18 (42.7 Ma). Seafloor spreading stopped in the Wharton Basin (IND/AUS) and resumed between Broken Ridge and the Kerguelen Plateau while spreading rates increased between Australia and Antarctica. The plate boundaries are at their present-day configuration.

in July, 1989 at the 28th International Geological Congress in Washington DC [Royer et al., 1989a,b]. They are based on an early compilation of magnetic anomaly data and fracture zone traces from bathymetric and satellite altimeter data [Royer et al., 1989]. Unlike the Tertiary evolution of the Indian Ocean (0-84 Ma), reconstructions for the early opening are still poorly constrained [e.g., Lawver et al., 1991; Royer and Coffin, 1992]. Thus, only

two tentative reconstructions are shown for this period: at 91 Ma (Cretaceous Magnetic Quiet Period) and at chron M0 (119 Ma). Rotation parameters for the Cenozoic (0-84 Ma) are taken from Royer et al. [1988], Royer and Sandwell [1989] and Royer and Chang [1991], and for the Mesozoic (M0) from Royer and Coffin [1992]. Except for 160 Ma, this model for relative plate motion is combined with

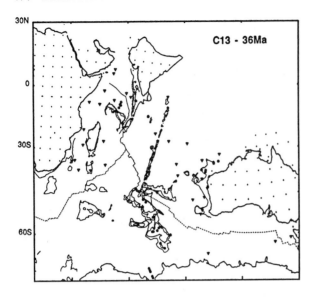

Fig. 10. Reconstruction at Chron 13 (35.5 Ma). The Chagos-Maldives and Mascarene Plateau breakup [Fisher et al., 1971].

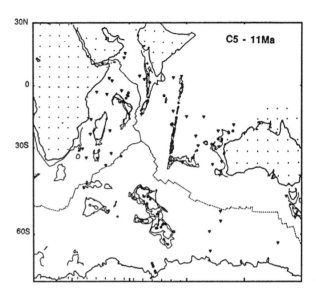

Fig. 12. Reconstruction at Chron 5 (10.4 Ma). Seafloor spreading has started in the Gulf of Aden [Laughton et al., 1970]. The next major event will be the onset of intraplate deformation within the Central Indian Basin at ~7 Ma.

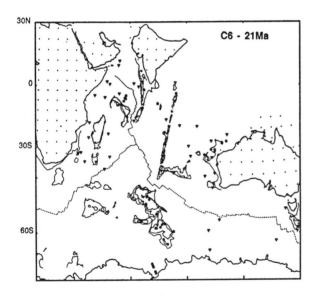

Fig. 11. Reconstruction at Chron 6 (20.5 Ma).

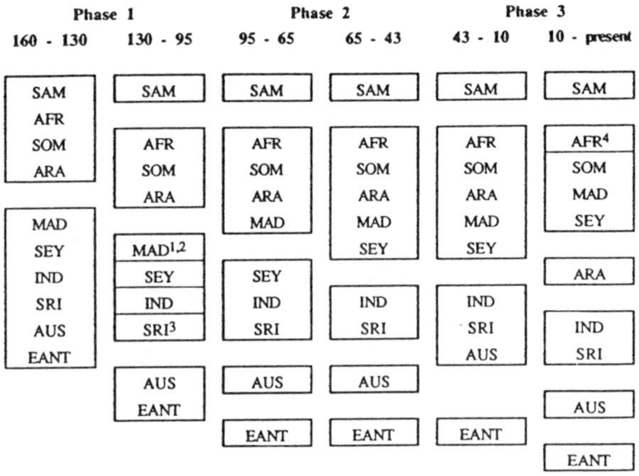

Fig. 13. Summary of the plate geometry in the Indian Ocean since the Late Jurassic [from Royer, 1992]. Ages are in Ma. AFR = African Plate, ARA = Arabian Plate, AUS = Australian Plate, IND = Indian Plate, EANT = East Antarctic Plate, MAD = Madagascar Plate, SAM = South American Plate, SEY = Seychelles Plate, SOM = Somalian Plate, SRI = Sri Lankan Plate. Notes:

(1) Strike-slip motions may have occurred between Madagascar and India from the break-up of India and Antarctica until the onset of seafloor spreading between India and Madagascar in the Late Cretaceous.

(2) Seafloor spreading between Africa and Madagascar stopped by 115 Ma [Ségoufin and Patriat, 1981; Cochran, 1988].

(3) A short rifting event between India and Sri Lanka occurred at about 130 Ma [see Lawver et al., 1991].

(4) The African and Somalian plates are separating along the East African Rift.

absolute motions of Africa relative to Atlantic and Indian hotspots from 0 to 120 Ma [Duncan and Richards, 1991].

A more thorough compilation of magnetic anomaly data and fracture zone traces is underway at Scripps Institution of Oceanography, Lamont-Doherty Geological Observatory, University of Texas at Austin Institute for Geophysics, Ecole et Observatoire de Physique du Globe de Strasbourg (France), and Bureau of Mineral Resources (Australia), and with collaboration with Institut de

Physique du Globe de Paris (France) and Bernard Price Institute for Geophysical Research (South Africa). This new compilation will produce an atlas presenting the compiled data set (bathymetry, magnetics, satellite-derived gravity), a tectonic chart and a more complete and accurate set of plate reconstructions (release scheduled for 1993/94).

Acknowledgments. The authors acknowledge support from the National Science Foundation (Grants OCE86-17193, PI John G. Sclater; OCE87-09137, PI Robert L. Fisher; OCE86-19862, PI Steven C. Cande), the Sponsors of the Paleoceanographic Mapping Project (POMP) at UTIG, and the Centre National de la Recherche Scientifique (CNRS). UTIG contribution 938.

REFERENCES

Besse, J., and Courtillot, V., Revised and synthetic apparent polar wander paths of the African Eurasian, North American and Indian plates, and true polar wander since 200 Ma, *J. Geophys. Res., 96*, 4029-4050, 1991.

Cochran, J. R., Somali Basin, Chain Ridge, and origin of the Northern Somali Basin gravity and geoid low, *J. Geophys. Res., 93*, 11985-12008, 1988.

Duncan, R. A. and Richards, M. A., Hotspots, mantle plumes, flood basalts, and true polar wander, *Rev. Geophys., 29*, 31-50, 1991.

Fisher, R. L., Sclater, J. G., and McKenzie, D. P., Evolution of the Central Indian Ridge, *Geol. Soc. Amer. Bull., 82*, 553-562, 1971.

Laughton, A. S., Whitmarsh, R. B., and Jones, M. T., The evolution of the Gulf of Aden, *Phil. Trans. Roy. Soc. London, A-267*, 227-266, 1970.

Lawver, L. A., Royer, J.-Y., Sandwell, D. T., and Scotese, C. R., Evolution of the Antarctic continental margins, in *Geological Evolution of Antarctica*, edited by M. R. A. Thomson, J. A. Crame, and J. W. Thomson,   533-539, Cambridge Univ. Press, U. K, 1991.

Lawver, L. A., and Scotese, C. R., A revised reconstruction of Gondwanaland, in *Gondwana Six: Structure, Tectonics, and Geophysics, AGU Geophysical Monogr., 40*, edited by G. D. McKenzie, 17-24, 1987.

Royer, J.-Y., and Chang, T., Evidence for relative motions between the Indian and Australian plates during the last 20 Myr from plate tectonic reconstructions, Implications for the deformation of the Indo-Australian plate, *J. Geophys. Res., 96*,   11,779-11,802, 1991.

Royer, J.-Y., and M. F. Coffin, Jurassic to Eocene plate reconstructions in the Kerguelen Plateau Region, in *Proc. ODP, Sci. Results 120*, edited by S. W. Wise ,Jr., R. Schlich et al., 917-928, Ocean Drilling Program, College Station, TX, 1992.

Royer J.-Y., Patriat P., Bergh H. and Scotese C.R., Evolution of the Southwest Indian Ridge from the Late Cretaceous (anomaly 34) to the Middle Eocene (anomaly 20), *Tectonophysics, 155*, 235-260, 1988.

Royer J.-Y., Cande S. C., Sclater J. G., Schlich R., Sandwell D. T., Patriat P., Fisher R. L., Coffin M. F. and Bergh H. W., A model for the evolution of the Indian Ocean from the Late Jurassic to present-day,   Abstract presented at the 20th

Internat. Geol. Congress, July 1989, vol. 2, p. 726, Washington D. C., 1989a.

Royer J.-Y., Cande S. C., Sclater J. G., Schlich R., Sandwell D. T., Patriat P., Fisher R. L., Coffin M. F. and Bergh H. W., An isochron chart of the Indian Ocean based on a tectonic fabric chart and plate reconstructions from the Late Jurassic to present-day, Abstract presented at the 20th Internat. Geol. Congress, July 1989, vol. 2, p. 726, Washington D. C., 1989b.

Royer, J.-Y., and Sandwell, D. T., Evolution of the eastern Indian Ocean since the Late Cretaceous: constraints from GEOSAT altimetry, *J. Geophys. Res., 94*, 13,755-13,782, 1989.

Royer, J.-Y., Sclater, J. G., and Sandwell, D. T., A preliminary tectonic fabric chart for the Indian Ocean, *Proceedings of the Indian Academy of Sciences (Earth and Planetary Sciences), 98*, 7-24, 1989.

Royer, J.-Y., The opening of the Indian Ocean since the Late Jurassic: an overview, in *Proc. of the Indian Ocean First Seminar on Petroleum Exploration*, edited by P. S. Plummer, 169-185, Seychelles, December 10-15, 1990, Published by United Nations Dpt. Techn. Coop. Development (UN/DTCD), New-York, NY, 1992.

Ségoufin, J., and P. Patriat, Reconstructions de l'océan Indien occidental pour les époques des anomalies M21, M2 et 34, Paléoposition de Madagascar, *Bull. Soc. Géol. France, 23*, 693-707, 1981.

Other references dealing with the breakup of Gondwana and opening of the Indian Ocean:

Fisher, R. L., and Sclater, J. G., Tectonic evolution of the Southwest Indian Ridge since the mid-Cretaceous: plate motions and stability of the pole of Antarctica/Africa motion for at least 80 Ma, *Geophys. J. R. Astr. Soc., 73*, 553-576, 1983.

Johnson, B. D., Powell, C. McA., and Veevers, J. J., Early spreading history of the Indian Ocean between India and Australia, *Earth Planet. Sci. Lett., 47*, 131-143, 1980.

König, M., Geophysical data from the continental margin off Wilkes Land, Antarctica, Implications for breakup and dispersal of Australia-Antarctica, in *The Antarctic Continental Margin: Geology and Geophysics of Offshore Wilkes Land*, edited by S. L. Eittreim, and M. A. Hampton, 117-146, CPCEMR Earth Science Series, 5A, Houston, TX, 1987.

Le Pichon, X., and Heirtzler, J. R., Magnetic anomalies in the Indian Ocean and sea-floor spreading, *J. Geophys. Res., 73*, 2101-2117, 1968.

McKenzie, D. P., and Sclater, J. G., The evolution of the Indian Ocean since the Late Cretaceous, *Geophys. J. R. Astron. Soc., 25*, 437-528, 1971.

Molnar, P., Pardo-Casas, F., and Stock, J., The Cenozoic and Late Cretaceous evolution of the Indian Ocean basin: uncertainties in the reconstructed positions of the Indian, African and Antarctic plates, *Basin Res., 1*, 23-40, 1988.

Norton, I. O., and Sclater, J. G., A model for the evolution of the Indian Ocean and the breakup of Gondwanaland, *J. Geophys. Res., 84*, 6803-6830, 1979.

Patriat, P., and Ségoufin, J., Reconstruction of the central Indian Ocean, *Tectonophysics, 155*, 211-234, 1988.

Powell, C. McA., Roots, S. R., and Veevers, J. J., Pre-breakup continental extension in East Gondwanaland and the early opening of the eastern Indian Ocean, *Tectonophysics, 155*, 261-283, 1988.

Schlich, R., The Indian Ocean: aseismic ridges, spreading centres and basins, in *The Ocean Basins and Margins:The Indian Ocean, 6*, edited by A. E. Nairn, and F. G. Stehli, 51-147, Plenum Press, New York, 1982.